ANNALS OF THE NEW YORK ACADEMY OF SCIENCES

Volume 361

EDITORIAL STAFF

Executive Editor
BILL BOLAND

Managing Editor
JOYCE HITCHCOCK

Associate Editor
MARY KATHERINE BRENNAN

The New York Academy of Sciences
2 East 63rd Street
New York, New York 10021

ORIGINS AND EVOLUTION OF EUKARYOTIC INTRACELLULAR ORGANELLES

ANNALS OF THE NEW YORK ACADEMY OF SCIENCES

VOLUME 361

ORIGINS AND EVOLUTION OF EUKARYOTIC INTRACELLULAR ORGANELLES

Edited by Jerome F. Fredrick

The New York Academy of Sciences
New York, New York
1981

Library of Congress Cataloging In Publication Data

Main entry under title:
Origins and evolution of eukaryotic intracellular organelles.
(Annals of the New York Academy of Sciences; v. 361)
Includes bibliographical references and index.
Papers presented at a conference held by the New York Academy of Sciences on Jan. 22–24, 1980.
1. Cell organelles—Evolution—Congresses.
I. Fredrick, Jerome F. II. Series.
Q11.N5 vol. 361 [QH573] 500s [574.87'34] 81-989
AACR2

Cover illustration—by Norman Tempia from an electron micrograph (×21,500) of the controversial alga, *Cyanidium caldarium*, by J. Seckbach, Hebrew University. The chloroplasts and mitochondria are emphasized in the illustration. The origin of these intracellular organelles is the subject of this *Annal*. The illustration will appear in *Cyanidium: A Comprehensive Treatise* by Joseph Seckbach and Jerome F. Fredrick. (Used with the authors' permission.)

PCP
Printed in the United States of America
ISBN 0-89766-111-7 (cloth)
ISBN 0-89766-112-5 (paper)

ANNALS OF THE NEW YORK ACADEMY OF SCIENCES

VOLUME 361

February 28, 1981

ORIGINS AND EVOLUTION OF EUKARYOTIC INTRACELLULAR ORGANELLES *

Editor and Conference Chairman

JEROME F. FREDRICK

CONTENTS

* This series of papers is the result of a conference held by The New York Academy of Sciences on January 22–24, 1980, which was entitled Origins and Evolution of Eukaryotic Intracellular Organelles.

Financial assistance was received from:

- E. I. DuPONT De NEMOURS & COMPANY
- HOFFMANN-LA ROCHE, INC.
- MONSANTO AGRICULTURAL PRODUCTS COMPANY
- NATIONAL SCIENCE FOUNDATION

EVOLUTIONARY THEORIES: DOGMAS OR SPECULATIONS?

Jerome F. Fredrick

The Research Laboratories
The Dodge Chemical Company
Bronx, New York 10469

We have come together for the purpose of indulging in some evolutionary "speculation." In doing so, it behooves us to pay careful attention to the fine line dividing "speculation" from "dogma." Too often, scientists, as do other human beings, fall into the trap of becoming extremely dogmatic in their speculations. Particularly so, today, the hazards inherent in the conviction of being in possession of *the* absolute truth, may lead to an all too easy abandonment of any further search for the truth.[1]

It may be well to keep in mind Roger Stanier's admonishment [2] as we examine the experimental data which will be presented during the next few days of this conference:

> Evolutionary speculation constitutes a kind of metascience, which has the same intellectual fascination for some biologists that metaphysical speculation possessed for medieval scholastics. It can be considered a relatively harmless habit, like eating peanuts, unless it assumes the form of an obsession; then it becomes a vice. The most appropriate response to such speculation, if it is plausible and logically constituted, is an Italian rejoinder, of which the amicable cynicism cannot be adequately translated:
>
> *Se non e vero, e ben trovato.*

The closest we can come to offering an idiomatic translation is: *"If it is not true, it is certainly fortuitous."* With this as our guide, it might be well to examine, very briefly, the history of the types of evolutionary speculations we are about to indulge in.

In 1938, Copeland [3] pointed out the existence of the large gap in evolutionary sequences between cells possessing a nucleus and those which were anucleate. The recent remarkable advances in ultrastructural and biochemical cytological knowledge, have demonstrated unequivocally that nucleated and anucleated forms constitute two sharply contrasted groups, the Eukaryota and the Prokaryota. In all probability, as suggested by Allsopp,[4] these two groups merit the status of "Superkingdoms," since the differences between them transcend those between the often-accepted, "Kingdoms."

Although there is still a great phylogenetic discontinuity between these two major groups, recent investigations have revealed striking similarities, not between *complete* cells of the Eukaryota and the Prokaryota, but rather between prokaryotic cells and *individual organelles* (plastids, mitochondria, etc.) of eukaryotic cells. These similarities have resulted in the wide acceptance of the interesting and plausible *speculation* that the eukaryotic cell represents a composite structure, the plastids and mitochondria of which, were originally derived from prokaryotes, which during the course of evolution became completely habituated to intracellular life in a host cell previously devoid of any such organelles.

In essence then, whether we admit it or not, we have two rival schools of biology based upon evolutionary speculations: the classical or "traditional" biologists who believe that the eukaryotic cell is the end-result of a gradual but continuous evolution from the prokaryotic cell via the accumulation of mutational events, and the endosymbiont proponents whose thesis is that the eukaryotic cell is the result of sudden (or "revolutionary") acquisition of whole prokaryotic cells, followed by the symbiotic evolution of guests and hosts.

The major goal of this conference is *not* to establish one school of evolutionary speculation over the other. Quite the contrary; we wish to start a dialogue, and a meaningful one, between the two. It is our sincere hope that the data to be presented will aid in establishing such a dialogue. However, we can only accomplish this *if* we admit to ourselves that neither of us is "in possession of *the* absolute truth."

References

1. Lemberg, R. & J. Barrett. 1973. The Cytochromes. Academic Press, Inc. New York, N.Y.
2. Stanier, R. Y. 1970. Symp. Soc. for Gen. Microbiol. **20:**1–38.
3. Copeland, H. F. 1938. Quart. Rev. Biol. **13:**383–420.
4. Allsopp, A. 1969. New Phytol. **68:**591–612.

UNIVERSALITY OF ENERGY-COUPLING PRINCIPLE: RELEVANCE TO EVOLUTION OF CELLULAR ORGANELLES

David E. Green

Institute for Enzyme Research
University of Wisconsin, Madison
Madison, Wisconsin 53706

All forms of present day life are based on the same three principles—the principle for replication of proteins and nucleic acid, the principle for energy mobilization and utilization, and the principle for programmed reproduction of identical copies of cells. The case for assuming that this set of three principles was operative throughout evolutionary history is sufficiently overwhelming that it requires no further elaboration.[1] By virtue of the invariance of the fundamental three principles it is reasonable to assume that the first cell, or cells that initiated the evolutionary process as soon as life became possible some billions of years ago,[2] already possessed the basic molecular structures and equipment required to translate these principles. Parenthetically, the capacity to undergo evolutionary change is in fact rooted in these very principles. My thesis is that in consequence of the invariance of the three principles a major sector of the total biological inventory and a considerable proportion of the systems that sustain life have been immune to the evolutionary process.

It is rather important to define as precisely as possible the nature of the three principles that we are invoking as universal and invariant in biological systems. Let us begin with the principle that underlies the replication of proteins and nucleic acid from templates thereof. The Watson-Crick principle of base-pairing[3] makes it possible to obtain an exact copy of a template molecule of DNA given the appropriate polymerizing system and the appropriate set of nucleotides (FIGURE 1).[4] The transcription of a protein molecule from messenger RNA is based on the genetic code and requires the ribosome-mediated interaction of the protein-synthesizing system with template RNA.[5] Basically the same principle with variations underlies the technique of making identical copies both of protein and nucleic acid. But tied in with this technique is an elaborate enzymic system and we must consider the enzymic machinery an integral part of the principle and as invariant as the principle itself. Among the essential components of the enzymic machinery for implementing the replication of proteins and nucleic acid are the ribosomes, the enzymes for generating the acylamino acids, the polymerases for synthesizing RNA and DNA, as well as the galaxy of enzymes that generate the amino acids or the nucleotides required for the syntheses.[4, 5]

The second principle underlies the technique by which energy is mobilized and utilized to carry out essential cellular functions. As far as present information is available, only one technique applies across the board to all biological systems present and past. The ultimate source of biological energy is the electronic energy intrinsic to the bonds that hold together the atoms in the molecules that provide the food for nutrition of the cells, in the specialized molecules that are intermediates between the foodstuff molecules and the systems that use

0077-8923/81/0361-0001 $01.75/0 © 1981, NYAS

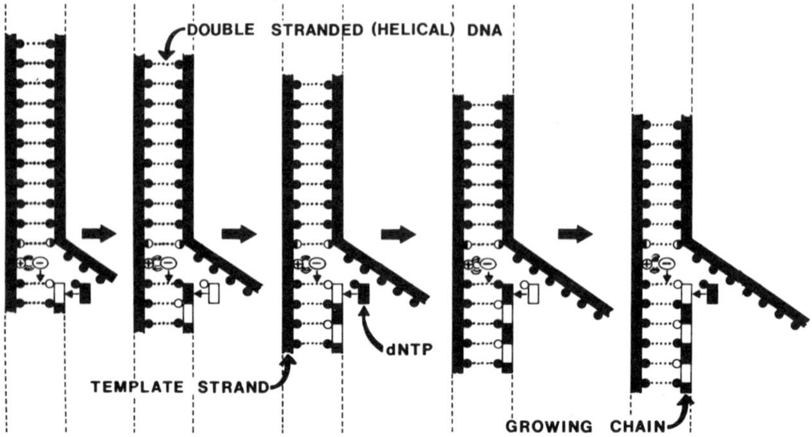

FIGURE 1. A model of the DNA replicating mechanism showing the stepwise growth of the new strand, the parallel stepwise separation of the old strand and the coupled maneuver whereby the template strand is moved a fixed distance to bring another nucleotide in place for bond formation.

the released free energy to do biological work.[6] The first step in the mobilization of energy is the enzymically mediated rupture of a bond and the short-range separation of a plus and negative charge. In enzymic catalysis, the charges are substituted so that the original atom pairs, which we shall refer to as A–B and C–D, are converted to B–C and A–D. In this transition, the drop in electronic energy allows the catalyzed reaction to proceed spontaneously in the desired direction. In plant cells, one unique chemical bond in chlorophyll can be ruptured by light so that the end result of the interaction of light and chlorophyll is identical with the end result of the interaction of enzyme and substrate. The bond in chlorophyll that is ruptured is a bond between an electron and a positively charged hydrogen ion.[7]

The short range enzyme-mediated separation of charge is always the first step in the long range separation of charge that is required for energy coupling—the technique by which free energy released in the downhill fall of an electron is conserved in the uphill flow of a positive charge (FIGURE 2). There is a transmembrane chain that permits the electron to transfer from one group to another in carefully programmed steps and an associated transmembrane chain * that permits the positively charged ionic species to move in parallel and simultaneously with the electron.[6] This parallel and simultaneous movement of the electron and positively charged species in two closely associated chains allows the efficient transfer of energy from the electron to the positively charged species. This technique of chain-mediated energy transfer applies to active transport, ATP synthesis, transhydrogenation, as well as to muscular contraction and template-dependent synthesis of proteins and amino acids. At the beginning of each energy-coupling cycle, charges are separated and at the end of each

* The notion of a chain for transmembrane movement of an ion is a new development that has been grafted onto the paired moving-charge model of energy coupling.[6]

cycle charges are united. The separation and uniting of charges is generally the prerogative of enzymes.

The machinery that implements the principle underlying the mobilization and utilization of the free energy intrinsic to bond energy in foodstuff molecules and their surrogate molecules would have to include not only a basic set of enzymes but also a basic set of energy coupling systems that would mediate active transport of cations across the cell membrane, ATP-energized synthesis of protein and nucleic acid, and synthesis of ATP energized by electron transfer. A large set of specialized molecules is required to consummate these coupled reactions: ATP, GTP, NADH, cytochrome *c*, Q (PQ), iron-sulfur proteins, etc. Since most energy coupling systems are housed in membranes, the apparatus for generating membranes would have to be included among the systems and molecules that are required to implement the energy principle.[8]

The third principle may be formulated as follows: in the same way that the base-pairing principle makes possible duplication of exact copies of particular proteins and nucleic acids, so an analogous principle makes possible duplication of exact copies of a particular cell. Just as an elaborate enzymic machinery is required to duplicate protein and nucleic acid molecules, so an even more elaborate machinery is required to duplicate a whole cell. You may well ask whether the base-pairing principle on a more grandiose level would be sufficient as a strategy for implementing cell duplication. This principle undoubtedly plays a major role in this strategy just as does the energy principle but there are new dimensions of this strategy and for that reason I am invoking a third

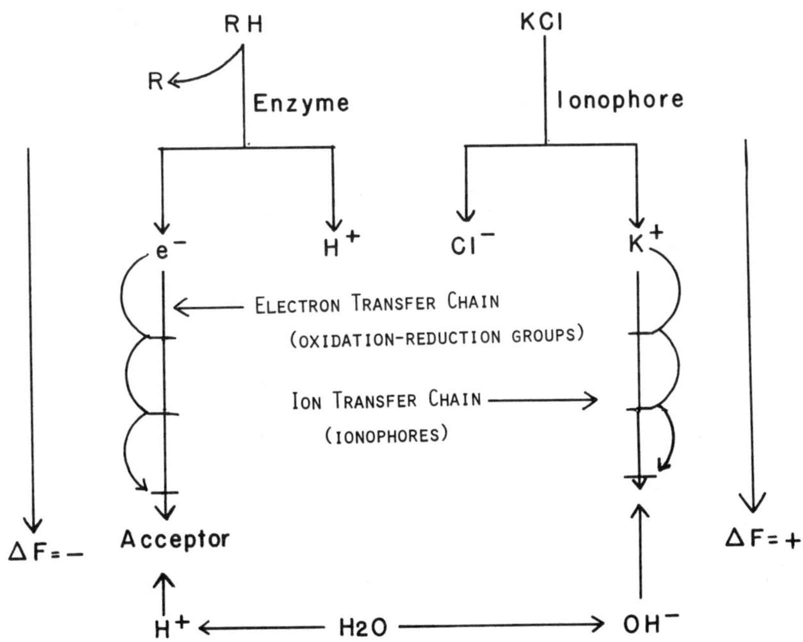

FIGURE 2. A model of the energy coupling mechanism for active transport of K^+ driven by electron transfer. Electron coupling via two chains: electron transfer (energetically down hill) and ion transfer (energetically uphill).

principle. What are the new dimensions? Replication of a cell requires a very precise time schedule, one event must follow another in an extended sequence. It also requires precise control of the amounts of each entity that is synthesized at the appointed time. Synthesis has to be turned on and off with great precision in respect to time and amount. Finally the necessity for intense activity at a given time and locale requires the availability of critical molecules in adequate supply. All of this has to be accurately programmed. It is difficult to believe that such an elaborate process does not have a master plan. For this reason I am invoking a new principle that can best be described as the orchestration principle.

The scope of the enzymic machinery that must translate programmed reproduction of cells is mind-boggling. It would certainly have to include the apparatus that determines cell division and some form of meiosis, the microtubular system, the system for generating membranes, as well as the programming system itself.

The overwhelming conclusion that must be drawn from this cataloging of the systems, structures, and molecules that translate the three universal principles of living systems is that these constitute a major part of the total inventory of living systems. This being the case, we may ask what is left in respect to the scope of the evolutionary process? I propose that evolution has produced innumerable variations on a theme but never variations of the theme itself. The conference is addressed to the consideration of these variations. I suggest that in no case is the theme itself, namely the three principles, compromised or altered.

Let us consider a few of these variations largely drawn from the domain of bioenergetics, which is my own specialty. The halobacteria can couple light-activated electron transport in membrane-bound rhodopsin to the synthesis of ATP mediated by the oligomycin-sensitive ATPase.[9] A cluster of rhodopsin molecules extending across the membrane of these bacteria can function as an electron-transfer chain. The transfer of electrons through the cluster is equivalent to the transfer of electrons through classical electron-transfer complexes.[10] This is not the only variation implicit in the rhodopsin molecule. Without any additions the rhodopsin system can couple electron flow to active transport of K^+ as shown by Montal *et al.*[11] This capability is referable to a transmembrane channel in the rhodopsin cluster that can transfer cations driven by electron transfer (FIGURE 1). Thus two substitutions have been successfully made. A transmembrane molecule for a transmembrane electron-transfer chain and a transmembrane channel for an ion-transfer chain.

The transition from energy coupling in the limiting cell membrane to energy coupling in an organelle, such as the chloroplast or the mitochondrion, is another example of a variation on a theme. The molecular actors in the cast are much the same in the two cases but the efficiency of the new arrangement is substantially greater as measured by the P/O ratio (ratio of molecules of ATP formed to oxygen atoms consumed). This is not to say that oxidative phosphorylation is less efficient in bacteria than in the mitochondrion. Coupling has multiple options in the bacterial cell membrane whereas ATP synthesis is the primary objective in the chloroplast and mitochondrion.

Cytochrome *c* has been varied every which way as far as amino acid changes have been made but the basic molecular strategy of the molecule, i.e., its role in electron transfer and energy coupling has been conserved throughout all these variations (TABLE 1).[12] The theme is constant but every variable detail has been varied in the course of evolutionary history.

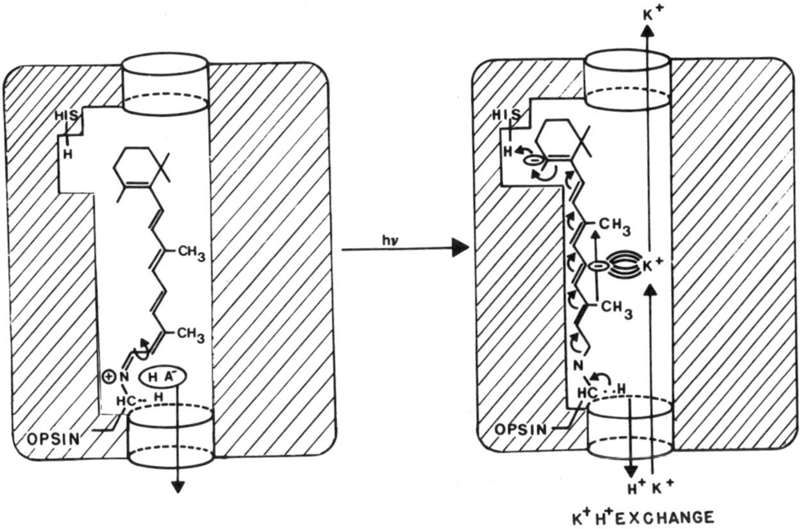

FIGURE 3. A model of the rhodopsin system for coupling electron flow to transport of a cation.

ATP is the preferred energy source for energy-coupling systems but in some bacteria phosphoenolpyruvate has replaced ATP as the energy source for the system that carries out active transport of sugars.[13] In some yeasts, pyrophosphate can replace ATP as the primary energy source.[14] In the ribosome-mediated protein system, GTP replaces ATP as the energy source.[15] In all cases, phosphoryl groups obtained by hydrolysis are the groups that drive the coupled reaction and the phosphoryl groups are the same whether generated by hydrolysis from ATP or from these alternative sources.

In cytochrome oxidase we find a surprising number of variations on a theme. In some bacteria only one subunit is found[16]; in others two subunits,[17] and in still others, as many as seven subunits as in mitochondria (TABLE 2).[18] But whether the cytochrome oxidase shows one, two, or seven subunits, the basic coupled reaction catalyzed by cytochrome oxidase is unaltered and the complement of prosthetic groups is basically unaltered. The difference between these

TABLE 1

INVARIANT FEATURES OF EUKARYOTIC CYTOCHROME *c*

1.	Redox potentials all close in value (+0.25 V)
2.	Absorption spectra indistinguishable
3.	Cytochrome oxidase of any eukaryotic cell is active with the cytochrome *c* of any eukaryotic cell.
4.	Invariance of 35 out of the 104 residues *

* Methionine 80, histidine 18, 2 heme-linked cysteine residues, residues 70–80, most of the glycine residues, lysine & arginine residues on surface.

different forms of the oxidase lies in the nature of the links between the different parts. If these are covalent, then the number of subunits is correspondingly reduced. This would explain why the number of subunits can vary between one and seven.

These above-mentioned variations on a theme are but a few samples of an extended list that is steadily growing. Others in the conference have documented and will be documenting equally cogent examples. I would like therefore to reserve a little of my remaining time to address the question of the origin of life on earth in the light of the invariance of the three fundamental principles throughout evolutionary history. For life to have started on earth within the relatively short span between the period when life was interdicted by virtue of excessive temperature and the period when the first recorded forms of life left fossil traces,[2] the transition had to be made from no life to a form of life that is essentially indistinguishable in broad principle from the form of life in the modern period. But it must also be borne in mind that the systems required to execute the three principles had to be evolved simultaneously. As

TABLE 2

MOLECULAR WEIGHTS OF SUBUNITS OF CYTOCHROME OXIDASE: PROKARYOTES VERSUS EUKARYOTES

	Prokaryote		Eukaryote		
	T. Novellus	*N. Agilis*	*S. cerevisae*	*N. crassa*	Cow
Subunit I	32,000	40,000	40,000	41,000	35,400
II	23,000	27,000	33,000	28,500	24,100
III			22,000	21,000	21,000
IV			14,500	16,000	16,800
V			12,700	14,000	12,400
VI			12,700	11,500	8,200
VII			4,600	10,000	4,400

(From Yamanaka *et al.*[17] By permission of the *Journal of Biochemistry.*)

we have pointed out, these systems account for a major part of the total inventory of biological systems and molecules.

In view of these considerations I consider the assumption that life began on earth as highly improbable. The physicist has learned to accept certain questions such as when did the universe begin as unanswerable, and the biologist may have to treat the corresponding question of when, where, and how life began in the universe as forever unanswerable.

REFERENCES

1. GREEN, D. E. & R. F. GOLDBERGER. 1967. Molecular Insights into the Living Process. Academic Press, Inc. New York, N.Y.
2. BRODA, E. 1975. The Evolution of the Bioenergetic Processes. Pergamon Press. Oxford, England.
3. WATSON, J. D. & F. H. C. CRICK. 1953. Molecular structure of nucleic acid. Nature **171:**737–738.

4. KORNBERG, A. 1974. DNA Synthesis. W. H. Freeman & Co. San Francisco, Calif.
5. HASELKORN, R. & L. B. ROTHMAN-DENES. 1973. Protein synthesis. Ann. Rev. Biochem. **42:**397–438.
6. GREEN, D. E. & G. A. BLONDIN. 1978. Molecular mechanism of mitochondrial energy coupling. BioScience **28:**18–24.
7. BOARDMAN, N. K. 1968. The photochemical system of photosynthesis. Adv. Enzymol. **30:**1–80.
8. GREEN, D. E. & H. BAUM. 1970. Energy and the Mitochondrion. Academic Press, Inc. New York, N.Y.
9. DANON, A. & W. STOECKENIUS. 1974. Proc. Natl. Acad. Sci. USA **71:**1234–1238.
10. GREEN, D. E. & G. A. BLONDIN 1975. A unifying model of bioenergetics. Chem. and Eng. News **53:**26–42.
11. MONTAL, M., A. BARSZON & H. W. TRISSL. 1977. Transmembrane channel formation in rhodopsin-containing bilayer membranes. Nature **267:**221–225.
12. DICKERSON, R. E. 1972. The structure and history of an ancient protein. Sci. Am. **226:**58–72.
13. ROSEMAN, S. 1972. Carbohydrate transport in bacterial cells. *In* Metabolic Transport. L. E. Hokin, Ed.: **4:**41–89. Academic Press, Inc. New York, N.Y.
14. BALTSCHEFFSKY, M. 1968. Inorganic pyrophosphate as energy donor, *In* BBA Library: **11:**277–286.
15. Cold Spring Harbor Symposium on Quantitative Biology. 1969. XXXIV Mechanism of Protein Biosynthesis. June 5–12. Cold Spring Harbor, New York.
16. SONE, N. & Y. KAGAWA. 1980. Thermostable Single-Band Cytochrome Oxidase Capable of Energy Transformation in Liposomes. International Symposium on Oxidases and Related Oxidation–Reduction Systems III.
17. YAMANAKA, T., K. FUJII & Y. KAMITA. 1979. Subunits of cytochrome *a*—type terminal oxidases derived from *Thiobacillus novellus* and *Nitrobacter agilis*. J. Biochem. **86:**821–824.
18. DOWNER, N. W., N. C. ROBINSON & R. A. CAPALDI. 1976. Characterization of a seventh different subunit of beef heart cytochrome *c* oxidase. Biochemistry **15:**2930–2936.

DISCUSSION OF THE PAPER

DR. J. M. OLSON (*Brookhaven National Laboratory, Upton, N.Y.*): I would like to challenge you on your third assumption about exact duplicates of cells. I think we have essentially similar duplicates of cells, but, I think "exact duplicates" is going a little bit far.

DR. D. E. GREEN: Yes. I don't insist on the use of the word exact.

EVOLUTION OF PHOTOSYNTHETIC AND RESPIRATORY PROKARYOTES AND ORGANELLES *†

John M. Olson

Biology Department
Brookhaven National Laboratory
Upton, New York 11973

INTRODUCTION

In the hypothesis presented in FIGURE 1, at least two common ancestors of all photosynthetic and respiratory prokaryotes and organelles existed more than three billion years ago. The older organism[1] contained protochlorophyll *a* (Pchl *a*), while its descendant[2] acquired the ability to synthesize chlorophyll *a* (Chl *a*). A subsequent series of mutations permitted the synthesis of bacteriochlorophyll *a* (Bchl *a*) by an organism[3] that was the common ancestor of purple and green bacteria.[4,5] Oxygen-respiring nonphotosynthetic bacteria arose from nonsulfur purple bacteria and mitochondria evolved from endosymbiotic respiring bacteria.

Most people have assumed that Bchl *a* came before Chl *a*, because the earliest photosynthetic organisms would have been anaerobic organisms without the ability to produce oxygen from water. I maintain that Chl *a* came first, because the biosynthetic pathway for Bchl *a* (FIGURE 2) contains three enzymic steps beyond the pathway for Chl *a*, and because it seems unlikely that this latter pathway was built backward.[4]

The evolution of organisms containing photosystems I and II and able to produce oxygen as a byproduct of photosynthetic carbon-dioxide fixation is thought to have required Chl *a* instead of Bchl *a* (FIGURE 3) because the maximum energy conversion by Chl *a*-containing reaction centers (donor to intermediate acceptor) is about 1.1 eV, whereas for Bchl *a*-containing reaction centers the maximum is about 0.8 eV. The energy required to transfer an electron from water to $NADP^+$ without losses is 1.1 eV, but this electron transfer is carried out with considerable energy losses by two reaction centers in series in Chl *a*-containing organisms. It is doubtful whether a photosystem II containing Bchl *a* could reduce the primary electron donor in a photosystem I containing Bchl *a* with a chain of several electron carriers between the two photosystems as shown in FIGURE 3. All chloroplasts and prokaryotes (*Prochloron* and cyanobacteria) able to evolve oxygen contain Chl *a* and are thought to be descended from a common ancestor[4] (FIGURE 1) containing Chl *a* and able to evolve oxygen.[5,6]

According to the hypothesis (FIGURE 1), the common ancestors[1,2] are

* Dedicated to the memory of Bessel Kok (1918–1979).

† This work was supported by the U.S. Department of Energy. By acceptance of this article the publisher and/or recipient acknowledges the U.S. government's right to retain a nonexclusive, royalty free license in and to any copyright covering this paper.

0077–8923/81/0361–0008 $01.75/0 © 1981, NYAS

thought to have contained Pchl *a*[1] or Chl *a*[2] in photochemical reaction centers that drove cyclic electron flow and phosphorylation through membrane-bound components including cytochromes *b* and *c* and quinone (FIGURE 4). The ATP thus formed was used for the assimilation of small organic molecules such as acetate for which the glyoxylate-cycle enzymes may have been present. In the Chl *a*-containing ancestral photosynthesizer,[2] a linear electron-transport chain

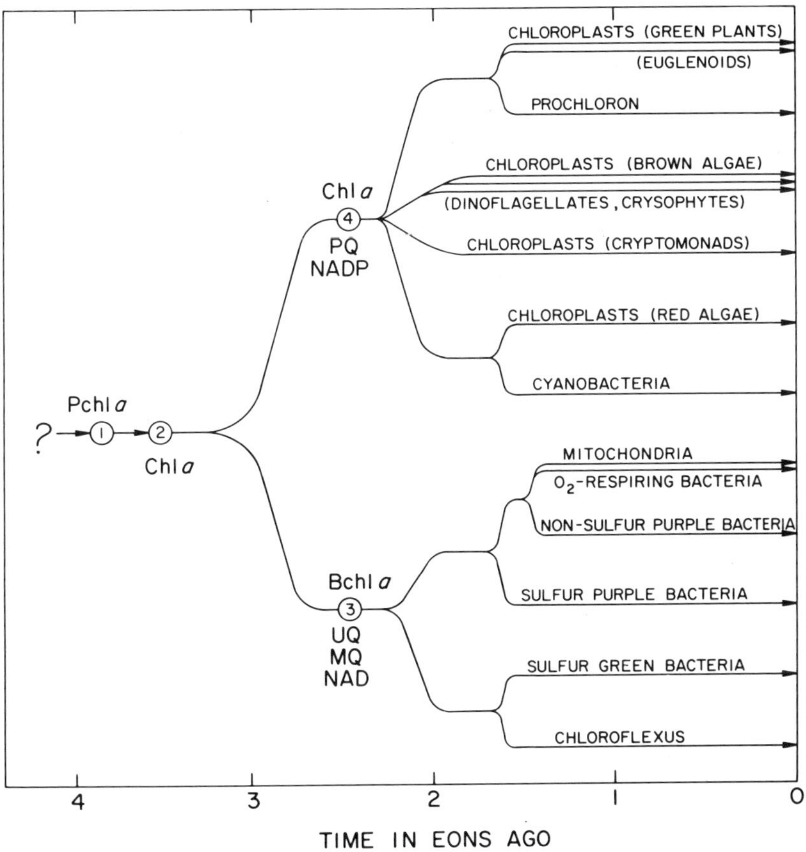

FIGURE 1. Evolutionary tree of photosynthesis and respiration. Bchl=bacteriochlorophyll; Chl=chlorophyll; Pchl=protochlorophyll; UQ=ubiquinone; MQ=menaquinone; and PQ=plastoquinone.

(in addition to the cyclic chain) had developed into photosystem I for the reduction of NAD^+ and/or $NADP^+$ and subsequent carbon-dioxide fixation via the Calvin cycle (FIGURE 5). However, oxygen was *not* evolved from water by these early photosynthetic carbon fixers; weak reductants such as hydrazine and hydroxylamine may have been the earliest hydrogen donors for photosynthesis.[4,5]

The main developments in my evolutionary theory of photosynthesis and respiration are shown in FIGURE 6.[6] Carbon-dioxide fixation should perhaps be moved up to 3.4 eons ago beside "Oldest Prokaryote Fossils" in view of recent

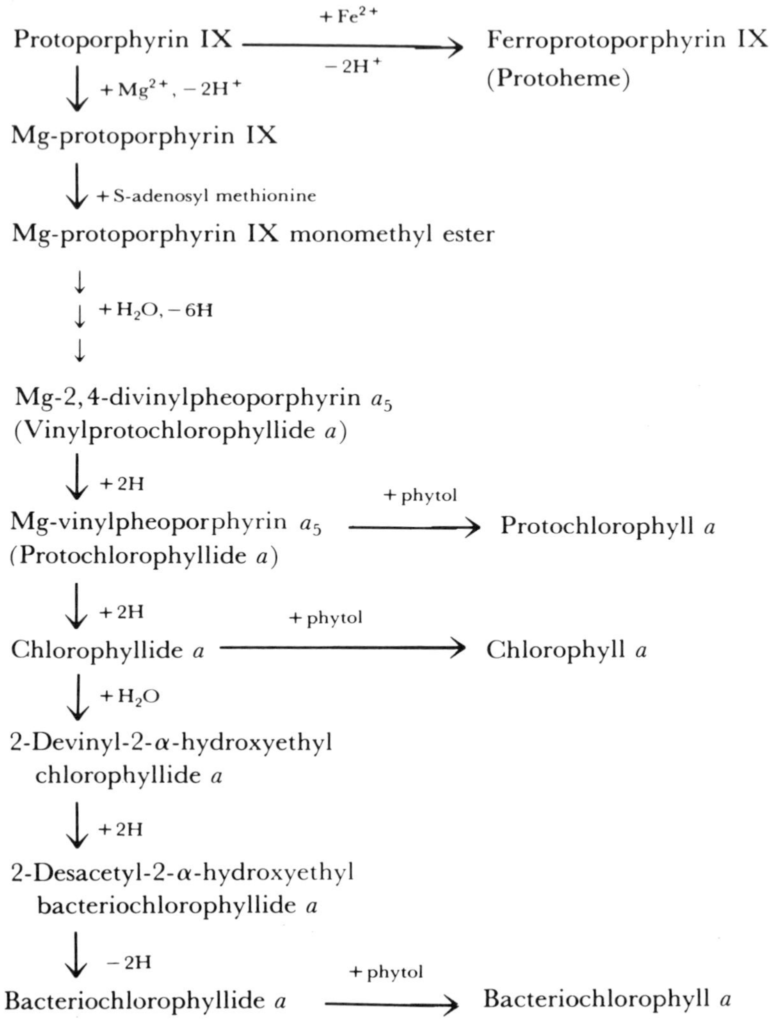

FIGURE 2. Biosynthetic pathway for bacteriochlorophyll *a* in *Rhodopseudomonas sphaeroides* (After Olson [5]).

estimates of the rate of organic carbon deposition in the sediments of the Sheba Formation of the Fig Tree Group.[7] The ancestral photosynthesizer [1] containing Pchl *a* should perhaps be moved up to 3.6 eons ago, since it presumably did not fix CO_2.

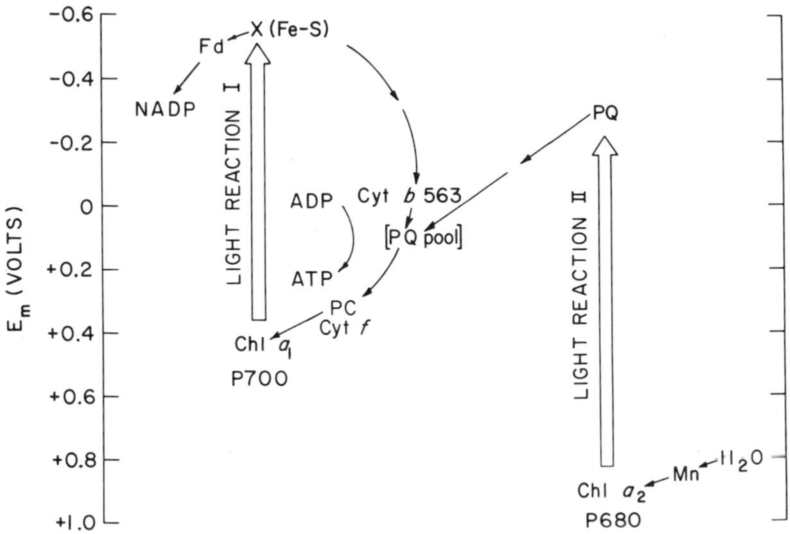

FIGURE 3. Electron transport scheme for oxygen-evolving photosynthesis. Cyt=cytochrome; Fd=ferredoxin; X=electron acceptor; and PC=plastocyanin. (From Olson.[6] By permission of Plenum Press.)

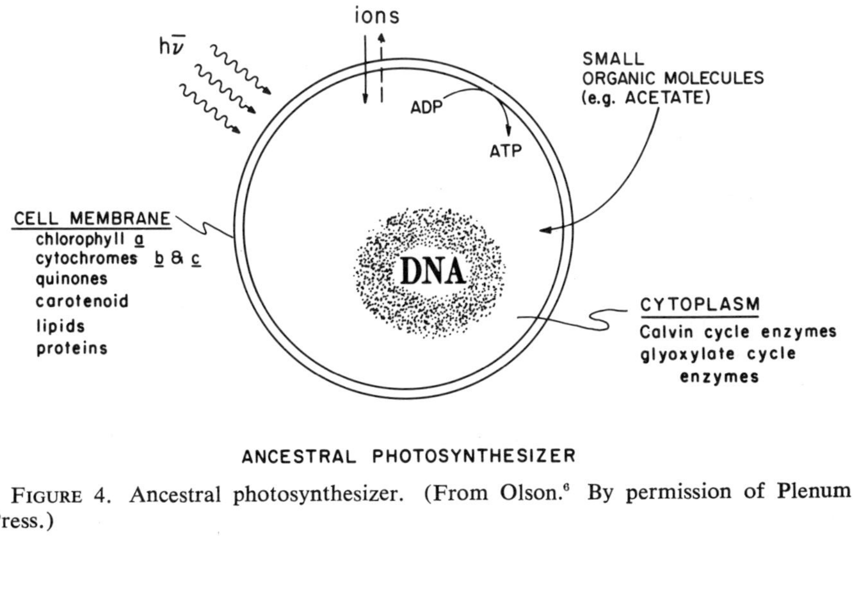

FIGURE 4. Ancestral photosynthesizer. (From Olson.[6] By permission of Plenum Press.)

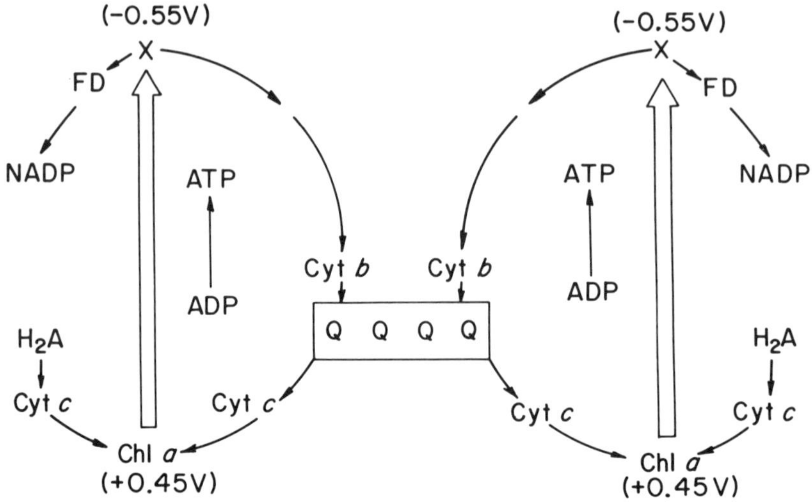

FIGURE 5. Electron transport scheme for both cyclic and linear electron flow in common ancestor. H_2A represents a class of fairly weak reductants such as hydrazine. (From Olson.[6] By permission of Plenum Press.)

OXYGEN EVOLUTION

It is uncertain when oxygen-evolving photosynthesis appeared. According to Reimer *et al.*[7] the presence of oxygenated minerals in Fig Tree and Moodies sediments indicates photosynthetic oxygen evolution 3.4 eons ago. There is no doubt that oxygen-evolving cyanobacteria were extant by the Middle Precambrian (2 eons ago). The finding of oxygenated minerals in sediments 3.4 eons old might be explained by postulating the direct oxidation of metal ions (e.g., Fe^{2+}) by photosynthetic autotrophs instead of assuming that molecular oxygen was produced. I would guess that the complex mechanism of water-splitting and oxygen evolution in photosynthesis evolved sometime between 3 and 2 eons ago.

Photosynthetic water-splitting requires a manganese-containing enzyme with four oxidation states corresponding to the four electrons that must be removed from two water molecules to release molecular oxygen. Photosystem II therefore must undergo four photochemical electron transfer reactions for each oxygen molecule. According to my view, photosystem II must have evolved from photosystem I in a series of small steps. In the earliest carbon-fixing autotrophs, photosystem II was not clearly differentiated from photosystem I. Weak hydrogen donors such as hydrazine and hydroxylamine may have reduced cytochrome *c* through appropriate reductases and thereby provided the electrons needed for the reduction of $NAD^+/NADP^+$ by a linear electron-transport chain *driven by a single photosystem* (FIGURE 5).

As the supply of weak hydrogen donors was depleted, the carbon-fixing autotrophs might have adapted to the use of even weaker hydrogen donors such as nitric oxide. Nitric oxide is a sufficiently poor electron donor that a photosystem II (FIGURE 7), clearly distinct from photosystem I, would have been required. The oxidation of nitric oxide to nitrite is a one-electron reaction requiring a redox potential of +0.37 V. The cytochrome *c* in photosystem I might have been replaced in photosystem II by manganese-cytochrome *c*, which can catalyze the oxidation of NO to NO_2^-.[2] Subsequent evolution of photosystem II might have involved the gradual replacement of NO by NO_2^- as the

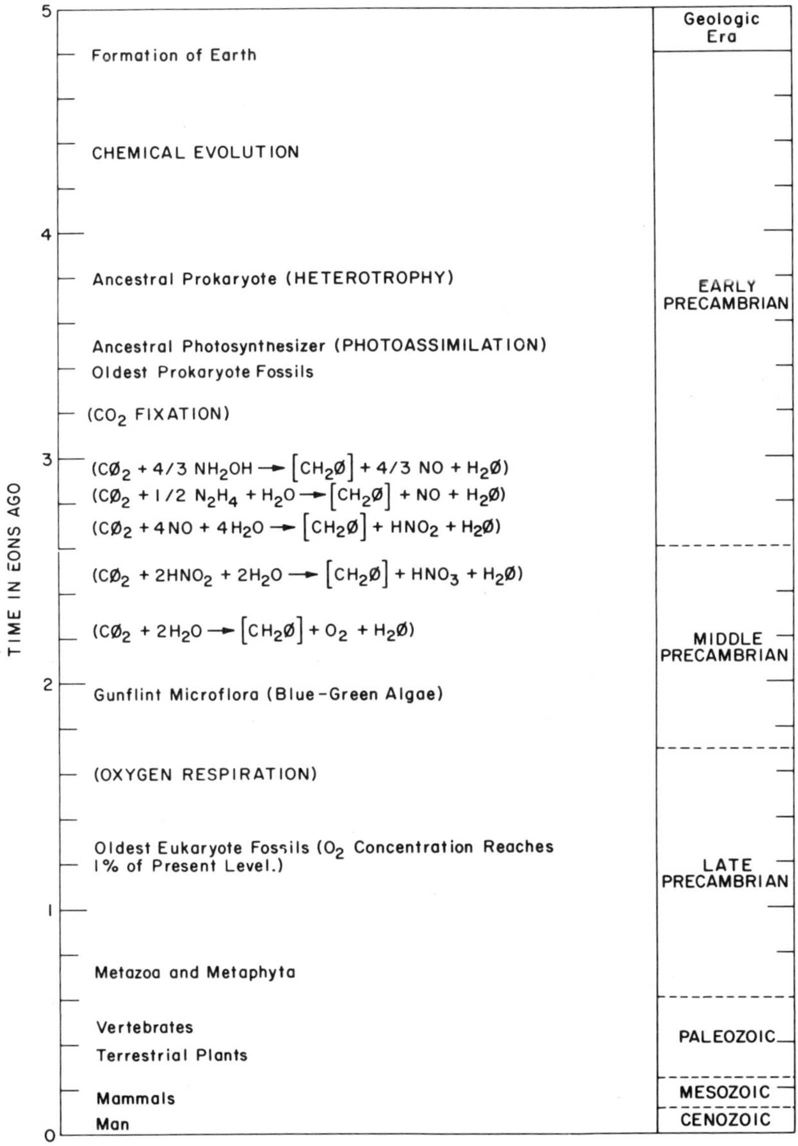

FIGURE 6. Major developments in the evolution of photosynthesis and respiration (From Olson.[6] By permission of Plenum Press.)

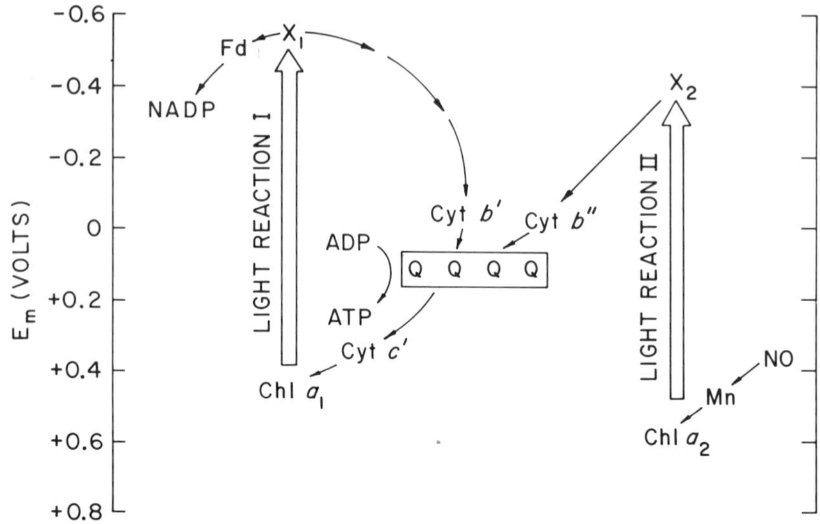

FIGURE 7. Electron transport scheme for nitric oxide-based photosynthesis. Q= quinone. (From Olson.[6] By permission of Plenum Press.)

electron donor and the development of an enzyme for handling a two-electron transfer at a redox potential of +0.42 V. Eventually nitrite and water may have replaced nitrite alone as electron donor at a redox potential of +0.69 V, and finally, water alone became the sole electron donor at a redox potential of +0.82 V. Some credibility is given to this hypothesis by the recent discovery that nitric oxide is formed photochemically in the nitrite-rich surface waters of the central equatorial Pacific.[13] Photosystem II may have evolved to its present state by a route different from that outlined here but still in accordance with the principle of gradual increases in complexity and in redox potential.

Oxygen Respiration

The view that photosynthetic oxygen evolution preceded oxygen respiration implies that respiration evolved *separately* in cyanobacteria, purple bacteria, and green bacteria.[1] Oxygen respiration by nonsulfur purple bacteria is assumed to have occurred about 1.6 eons ago in response to the increasing levels of photosynthetic oxygen in the atmosphere. The electron transport chain used for photosynthesis was adapted to respiration by the simple addition of cytochrome oxidase. Some bacteria dropped the synthesis of Bchl *a* and became nonphotosynthetic, oxygen-respiring organisms. Several extant "nonphotosynthetic" respiring bacteria have recently been found to synthesize Bchl *a* under appropriate conditions. The facultative methylotrophs *Protaminobacter ruber* and *Pseudomonas* AM1, which do not grow anaerobically in the light, do produce Bchl *a* under intermittent illumination in the presence of air with 1,2-propanediol as the sole carbon source.[8] Similarly, 16 strains of heterotrophic gram-negative marine bacteria (not methylotrophs) that do not grow under anaerobic conditions in the light or dark, will synthesize Bchl *a* when grown *aerobically* in the dark.[12]

Schwartz and Dayhoff[9] argue that a rudimentary form of aerobic respiration probably evolved *before* the development of oxygen-evolving photosynthesis. This respiratory capacity was necessary, they state, to protect the intracellular components of the oxygen-evolving organisms from oxidation. This argument is weak because there is little or no evolutionary pressure for aerobic respiration before the accumulation of significant oxygen in the atmosphere. Secondly, the evolution of photosystem II (required for oxygen production) was, in my view, a gradual process taking up to a billion years. This length of time afforded the organisms time to adjust the intracellular chemistry to cope with the oxidized products of photosystem II.

Phylogenetic Tree

Schwartz and Dayhoff[9-11] have put together a phylogenetic tree of prokaryotes and eukaryotes (Figure 8) derived from the sequences of ferredoxin, *c*-type cytochromes, and 5S ribosomal RNA. The three trees derived from each superfamily were scaled so that overlapping portions were equal, and a combined matrix of distances was calculated for the six species appearing on more than one tree. The heavy lines represent the tree calculated from the combined matrix of at least two trees, whereas the lighter lines represent branches scaled from a single tree and added to the combined tree. Therefore the data are incomplete for some organisms. The relative positions of *Chromatium* and *Chlorobium*, for example, are based on ferredoxin alone whereas the relative positions of *Chromatium* and *Rhodopseudomonas* are based on *c*-type cyto-

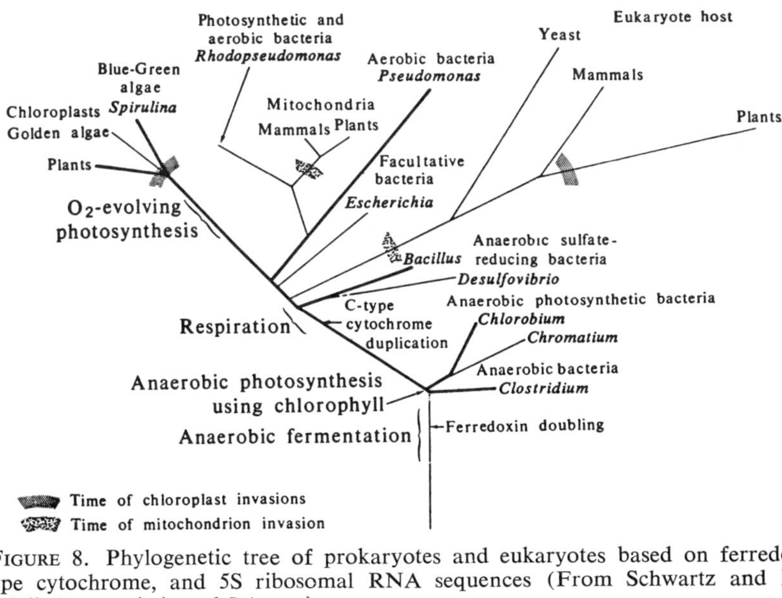

Figure 8. Phylogenetic tree of prokaryotes and eukaryotes based on ferredoxin, *c*-type cytochrome, and 5S ribosomal RNA sequences (From Schwartz and Dayhoff.[9, 10] By permission of *Science*.)

chromes alone. FIGURE 8 indicates that the sulfur purple bacterium *Chromatium* is more closely related to the sulfur green bacterium *Chlorobium* than to the nonsulfur purple bacterium *Rhodopseudomonas*. However this conclusion is inconsistent with the findings of Gibson *et al.*[3] that *Chromatium vinosum* 16S ribosomal RNA oligonucleotides resemble those of various *Rhodopseudomonas* (*Rps.*) species (binary relatedness = 0.35 to 0.41) more closely than those of *Chlorobium vibrioforme* (binary relatedness = 0.25). The purple bacteria (both sulfur and nonsulfur) so far investigated[3] fall into three subgroups. Group I contains *Rps. sphaeroides, Rps. capsulata, Rps. palustris, Rps. viridis, Rhodomicrobium vannielii,* and *Rhodospirillum rubrum;* group II contains *Rps. gelatinosa* and *Rps. tenue;* group III has *Chromatium vinosum* and *Thiocapsa pfennigii*. By the criterion of binary relatedness of 16S RNA oligonucleotides, *Chlorobium vibrioforme* is no more closely related to *Chromatium vinosum* than is the blue-green alga *Aphanocapsa* or the aerobic bacterium *Bacillus pumilus*.[3]

In spite of its serious deficiencies, the phylogenetic tree of Schwartz and Dayhoff is a beginning step toward an evolutionary tree derived solely from molecular sequences. At present such a tree is most easily understood in terms of the hypothesis that chloroplasts, mitochondria, and eukaryotic host cells each had an independent phylogeny up to the time of chloroplast and/or mitochondrion invasion. The main trunk appears to be a continuum of photosynthetic forms, and many nonphotosynthetic bacteria appear to be the descendants of a few ancestral forms that independently lost their photosynthetic capacity.[9] Gibson *et al.*[3] draw similar conclusions. They find that several gram-negative nonphotosynthetic bacteria are closely related to the purple bacteria as follows: Group I contains *Paracoccus dentrificans, Rizobium leguminosarum,* and *Aquaspirillum itersonii;* Group II contains *Alcaligenes faecalis* and *Sphaerotilus natans;* and Group III contains various species from the Enterobacteriaceae, Vibrionaceae, *Pseudomonas,* and *Acinetobacter*. All mitochondria appear to share a common ancestor with Group I purple bacteria and nonphotosynthetic bacteria (e.g., *Rps. sphaeroides* and *P. dentrificans*).[6]

SUMMARY

Common ancestors for mitochondria, chloroplasts, and photosynthetic bacteria (including cyanobacteria) probably existed more than three billion years ago. One ancestral prokaryote may have contained Pchl *a* in photochemical reaction centers that drove cyclic electron flow and phosphorylation through membrane-bound components including cytochromes and quinones. Substitution of Chl *a* for Pchl *a* and the development of linear electron-transport chains permitted the reduction of NAD^+ and/or $NADP^+$ for carbon-dioxide fixation. Evolution of photosystem II from photosystem I enabled one prokaryote to evolve oxygen as a byproduct of carbon-dioxide fixation. This organism was the common ancestor of cyanobacteria, *Prochloron,* and various chloroplasts. A photosynthetic bacterium containing Bchl *a* appears to have branched off from the Chl *a* line. This bacterium was the common ancestor of extant respiring bacteria, mitochondria, and purple and green photosynthetic bacteria.

ACKNOWLEDGMENT

I thank Dr. Sanford Lacks for helpful discussion.

References

1. Broda, E. & G. A. Peschek. 1979. Did respiration or photosynthesis come first? J. Theor. Biol. **81:**201–212.
2. Dickinson, L. C. & J. C. W. Chen. 1977. Manganese cytochrome *c*: structure and properties. J. Biol. Chem. **252:**6156–6162.
3. Gibson, J., E. Stackebrandt, L. B. Zablen, R. Gupta & C. R. Woese. 1979. A phylogenetic analysis of the purple photosynthetic bacteria. Curr. Microbiol. **3:**59–64.
4. Mauzerall, D. 1978. Bacteriochlorophyll and photosynthetic evolution. *In* The Photosynthetic Bacteria. R. K. Clayton & W. R. Sistrom, Eds.: 223–231. Plenum Press. New York, N.Y.
5. Olson, J. M. 1970. The evolution of photosynthesis. Science **168:**438–446.
6. Olson, J. M. 1978. Precambrian evolution of photosynthetic and respiratory organisms. *In* Evolutionary Biology. R. Dobzhansky, M. K. Hecht & W. C. Steere, Eds: **11:**1–37. Plenum Press. New York, N.Y.
7. Reimer, T. O., E. S. Barghoorn & L. Margulis. 1979. Primary productivity in an early Archean microbial ecosystem. Precambrian Res. **9:**93–104.
8. Sato, K. 1978. Bacteriochlorophyll formation by facultative methylotrophs, *Protaminobacter ruber* and *Pseudomonas* AM1. FEBS Lett. **85:**208–210.
9. Schwartz, R. M. & M. O. Dayhoff. 1978. Origins of prokaryotes, eukaryotes, mitochondria, and chloroplasts. Science **199:**395–403.
10. Schwartz, R. M. & M. O. Dayhoff. 1978. Cytochromes. *In* Atlas of Protein Sequence and Structure. M. O. Dayhoff, Ed: **5**(3):28–44. National Biomedical Research Foundation. Washington, D.C.
11. Schwartz, R. M. & M. O. Dayhoff. 1978. Other electron transport proteins. *In* Atlas of Protein Sequence and Structure. M. O. Dayhoff, Ed: **5**(3):45–57. National Biomedical Research Foundation. Washington, D.C.
12. Shiba, T., U. Simidu & N. Taga. 1979. Distribution of aerobic bacteria which contain bacteriochlorophyll *a*. Appl. Environ. Microbiol. **38:**43–45.
13. Zafiriou, O. C., M. McFarland & R. H. Bromund. 1980. Nitric oxide in seawater. Science **207:**637–638.

Discussion of the Paper

Dr. J. A. Schiff (*Brandeis University, Waltham, Mass.*): I have one or two questions about the coupling hypothesis of the two systems.

Dr. J. M. Olson: You mean, photosystem I and II?

Dr. Schiff: Yes. One of the difficulties I have is that in a large number of contemporary and oxygenic photosynthetic bacteria there is no cyclical electron flow, but rather a reverse electron flow driven by ATP from cyclic phosphorylation, which would indicate that the contribution of electrons was to reverse electron flow from external donors rather than to go through the photochemical chain itself.

Secondly, if you propose nitrous oxide as a proton donor to system II, there is a problem there biologically.

Dr. Olson: More than one.

Dr. Schiff: But, there is one interesting one and that is that the modern oxygenic photosynthetic bacteria don't use nitrogen compounds at all, even though their chemosynthetic counterparts do. For example, there are organisms that chemosynthetically fix CO_2 in the dark by oxidizing ammonium nitrate, but, there are no photosynthetic bacteria that do.

DR. OLSON: That fix nitrogen?

DR. SCHIFF: That use nitrogen as an electron donor.

DR. OLSON: As an electron donor?

DR. SCHIFF: They don't use ammonia, nitrite, or nitrate. Vishniac looked for years for these organisms in some very embarrassing places and never was able to find them because he was interested in them as a prototype for the Martian environment. But, I would rather turn the thing more or less around and say, isn't it possible that the prototype for the electron transport in oxygenic photosynthetic organisms from electron donors other than water really existed for both chemosynthesis and photosynthesis and really originated in reverse electron flow driven by ATP?

DR. OLSON: I am going to start with oxygenic photosynthesis in bacteria and I believe the point is made that electron flow, that is the formation of reduced NAD or NADP from external substrates is done by reverse electron flow that is driven by either ATP or an energized state that is provided by light-cycling electrons around a cycle. This is well established for the many, nonsulphur purple bacteria.

However, I think that if you are going to evolve a photosystem II from a photosystem I, you have to start with the configuration that I showed on the screen. I put it up there so that I could keep the systems apart and differentiate photosystem II from I, but, certainly, a reverse electron flow is well established for the oxygenic bacteria.

DR. SCHIFF: The other point is simply that no modern photosynthetic bacteria uses nitrogen as an electron donor.

DR. OLSON: That's right. I built my theory on nitrogen. There are thermodynamic reasons for using nitrogen to build a theory on—successively higher oxidation states of nitrogen. Dr. Schiff is perfectly correct that the experimental evidence for the utilization of these various intermediates which I postulate is sadly missing. The only thing I can say in my defense is to do a sort of end run about the compounds he mentioned; but I can mention that chloroplasts do oxidize hydroxylamine in the test tube, although I doubt that they do it in nature. That's about the best I can do on the use of nitrogen compounds as electron donors for photosystem II, but, the evidence for nitrite and nitric oxide as physiological donors doesn't exist that I know of and I regret that.

DR. D. E. GREEN: I am going to ask you an obvious question—that is the evolution of another type of chlorophyll. There is no good in making a chlorophyll if you haven't got a system to use it. So, you've got to simultaneously evolve a system that uses the chlorophyll.

DR. OLSON: Absolutely.

DR. GREEN: It means a rather major change and not just a matter of getting another chlorophyll. It is a matter of getting a whole system.

DR. OLSON: Yes. As I see it, the whole electron transport system has to evolve together and the chlorophyll is the photochemical part of that system, but, the whole system has to evolve.

DR. GREEN: Yes. I mean having a photochemical part without a system to translate it would be useless. So what it must mean is that each of these changes has to be a major change to be of any value. You don't have a holding operation where you produce a change in the chlorophyll.

Dr. Olson: You can modify one component at a time as long as you don't make too big a modification and as long as the modification doesn't destroy the function of the whole unit.

Dr. J. Locker (*The University of Chicago, Chicago, Ill.*): Could you comment on the evolution of cytochrome oxidase and, perhaps, elaborate a little bit more about the relationship of this to the evolution of the purple bacteria in general, like *Paracoccus,* and then finally, to mitochondria?

Dr. Olson: My view is just based on work of many others, especially the sequencers. I believe that the purple nonsulfur bacteria (especially the Rhodopseudomonas type), upon the advent of oxygen in the atmosphere, first probably developed cytochrome oxidase of the *o* type which is almost ubiquitous and seems to be found universally.

Dr. Green: What is the *o* type of setting for oxidase?

Dr. Olson: I don't know who named that cytochrome *o* anyway. Anyway, that is not satisfactory for *Paracoccus.* I think after cytochrome *o*–type evolved, then, the AA 111 type came subsequently in a nonsulfur purple bacteria, like Rhodopseudomonas and, then, one of its descendants lost its bacterial chlorophyll-synthesizing function, and became a straight respirer similar to *Paracoccus* and similar to the ancestor of mitochondria.

Dr. Locker: What about the evolution from two subunits in *Paracoccus* to the seven subunits in the mitochondria?

Dr. Green: It took me a little time to appreciate that's not a major change. From all we know, the cellular oxidase that has one subunit and the cellular oxidase that has several units are spitting images of one another. The only difference is that instead of having different subunits, you have covalent links between them. So, it is a trivial difference in a certain sense. It is nothing very fundamental because the molecular weights and the number of functional elements are the same and there are many cases where you can substitute for covalent bonds. You can simply have hydrophobic bonds, but, that's not a really major change.

It would be dramatic if instead of having heme, let us say you had ferridoxin, or instead of having copper you would have something utterly different, but, there isn't any variation of that kind that we know of. All cytochrome oxidases are based on the same ground plan chemically and there isn't any major change other than the happenstance of how many subunits you have.

Dr. Locker: Isn't it possible, for example in the case of ribulose triphosphate carboxylase, which has two subunits and is reputed in some prokaryotes to have only one, that perhaps the reason the extra subunits evolved was not to enhance function, but, to enhance regulation when the enzyme was in two separate compartments?

Dr. Green: That's an excellent point.

Dr. Locker: Because we know half the subunits of both enzymes are made in the nucleus, coded in the nucleus, and half coded in the organelle.

EVOLUTION OF ORGANELLE GENOMES AND PROTEIN-SYNTHESIZING SYSTEMS *

Nicholas W. Gillham and John E. Boynton

Departments of Zoology and Botany
Duke University
Durham, North Carolina 27706

Introduction

Today we know that neither chloroplasts or mitochondria are genetically autonomous. While the genomes of these organelles encode certain specific organelle components, most of the components of both organelles are specified by the nuclear genome.[1-3] Two general theories have been proposed to explain the origin of chloroplasts and mitochondria. According to the endosymbiotic theory, for whose revival in modern form we are largely indebted to Margulis,[4-7] mitochondria and chloroplasts arose by successive symbiotic events. The first step in the evolution of eukaryotic cells from prokaryotic cells was the invasion of a fermentative anaerobe by an aerobic eubacterial species. During the course of evolution the eubacterial member of this symbiotic protoeukaryote became the mitochondrion. At a later date certain populations of eukaryotic heterotrophs acquired the capacity to photosynthesize through the ingestion of blue-green algae which eventually became chloroplasts. The second, or nonsymbiotic, theory assumes that chloroplasts and mitochondria evolved by compartmentalization of existing genetic material within the ancestral protoeukaryote.[8-13] Assumptions vary concerning the mechanism by which compartmentalization of segments of the prokaryotic genome within membrane-limited vesicles occurred, but all nonsymbiotic theories agree that both chloroplasts and mitochondria were formed as a result of this process.

In both theories of organelle evolution the properties of organelle genomes and protein-synthesizing systems have assumed a central role. For example, the naked and usually circular genomes of chloroplasts and mitochondria have been likened to prokaryotic genomes. Organelle ribosomes have been described as prokaryotic in nature although we now know, particularly in the case of mitochondria, that organelle ribosomes have properties that distinguish them from ribosomes of both the eukaryote and prokaryote cytoplasms. In view of all of the speculation concerning the origin of chloroplasts and mitochondria, one would have thought that firmly grounded comparisons would have been made on phylogenetically important groups of existing organisms. Unfortunately, with certain special exceptions, the most thorough and detailed studies have been done with the most highly evolved eukaryotes instead of the most primitive eukaryotes. For example, as FIGURE 1 illustrates, although we have some information on the size of chloroplast and mitochondrial genomes in higher plants, there is a paucity of similar information among the progenitors of higher plants. At the same time, data have been accumulating that show that while the chloroplast protein-synthesizing system has many features common to that of

* This work supported by the National Institutes of Health (Grant GM-19427).

0077–8923/81/0361–0020 $01.75/0 © 1981, NYAS

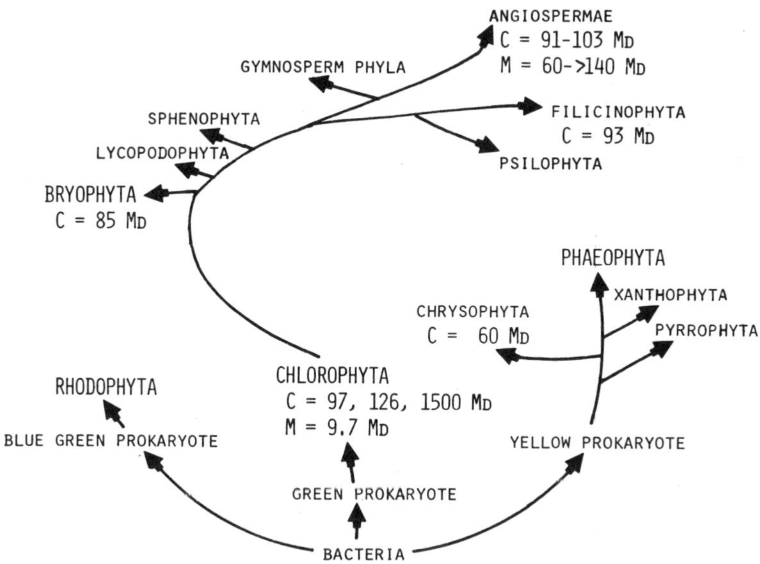

FIGURE 1. Size of chloroplast and mitochondrial genomes in different plant groups. Taxa in large capital letters correspond to the category Superphylum in the classification of Whittaker and Margulis.[17] Taxa in small capital letters correspond to phyla in the same classification scheme.

prokaryotes, the mitochondrial protein-synthesizing system is a melange of prokaryotic, eukaryotic, and unique features. Our purpose in this paper is to try to document these points and to focus attention on those groups of organisms where we think comparative phylogenetic investigation might be helpful.

Organelle Genomes and Protein-Synthesizing Systems from the Viewpoint of Phylogeny

Organelle genomes have been characterized to a reasonable extent (i.e., size and conformation as well as base composition) in over 100 species of eukaryotes.[1, 14–16] When these organisms are placed within a modern phylogenetic scheme, such as that proposed by Whittaker and Margulis,[17] it is immediately evident that representatives of only a fraction of the existing phyla have been studied. In the case of mitochondrial genomes, this is true of approximately 20 percent of all eukaryotic phyla (Table 1). Among chloroplast-containing phyla, chloroplast genomes from approximately 35 percent of the total have been examined (Table 2). A striking feature of the data is the paucity of information on organelle genome structure in the *Protista*. These are the organisms among which we might expect to find evolutionary relicts if they exist. Yet mitochondrial genomes have only been characterized in 21 percent of the phyla in this kingdom (Table 1) and in the thirteen protistan phyla whose species contain chloroplasts, the genomes of this organelle have been studied in only four phyla (Table 2).

TABLE 1

PHYLOGENETIC DISTRIBUTION OF ORGANISMS IN WHICH MITOCHONDRIAL GENOMES HAVE BEEN CHARACTERIZED *

Kingdom	Subkingdom or branch	Total number of phyla known	Percent of total phyla in kingdom	Phyla in which mitochondrial genomes have been characterized: Number	Phyla in which mitochondrial genomes have been characterized: Percent
Protista	Protophyta	13	18.6	2	15.3
	Protomycota	7	10.0	1	14.3
	Protozoa	8	11.4	3	37.5
	Total	28	40.0	6	21.4
Fungi	------	5	7.1	1	20.0
Animalia	Agnotozoa	2	2.85	0	0.0
	Parazoa	2	2.85	0	0.0
	Eumetazoa				
	Radiata	2	2.85	0	0.0
	Bilataria	27	38.6	6	22.2
	Total	33	47.1	6	18.2
Plantae	------	4	5.7	1	25.0
	Grand Total	70	—	14	20.0

* Phylogenetic distribution of organisms in which mitochondrial genomes have been characterized is based on the classification of Whittaker and Margulis.[17] Only those cases are included where the size of the mitochondrial genome is known from electron microscopic measurement, kinetic complexity, summation of restriction-fragment molecular weights, or a combination of these methods. Cases in which only the base composition of mitochondrial DNA is known are excluded. Lists of the organisms from which this tabulation was made are available in References 3, 15, and 16.

TABLE 2

PHYLOGENETIC DISTRIBUTION OF ORGANISMS IN WHICH CHLOROPLAST GENOMES HAVE BEEN CHARACTERIZED *

Kingdom	Subkingdom or branch	Total number of phyla known	Percent of total phyla in kingdom	Phyla in which chloroplast genomes have been characterized: Number	Phyla in which chloroplast genomes have been characterized: Percent
Protista	Protophyta	13	76.4	4	31.0
Plantae	------	4	23.6	2	50.0
	Grand Total	17	—	6	35.0

* Phylogenetic distribution of organisms in which chloroplast genomes have been characterized based on the classification of Whittaker and Margulis.[17] Cases to be included were determined by the same criteria described in TABLE 1. Lists of organisms from which this tabulation was made are available in References 3, 14, and 15.

Investigators have generally chosen to study organelle genomes in higher eukaryotes. Usually, organelle genomes have been studied in lower eukaryotes for one of two reasons. First, the organism in question has had a long history of being an experimental model for the study of either chloroplasts (e.g., *Acetabularia, Chlamydomonas, Euglena*) or mitochondria (e.g., *Aspergillus, Neurospora, Saccharomyces*). Second, the organelle genome has exhibited bizarre or unusual features (e.g., the minicircular and maxicircular DNAs of trypanosome kinetoplasts).

The strong phylogenetic biases that characterize the study of organelle genomes are evident in the data summarized in TABLE 3. Thus, at least 72 percent of the mitochondrial genomes characterized to date are from species representing only three phyla. One of these, the *Chordata,* is the most highly evolved animal phylum, the second, the *Ascomycota,* shows considerable variation in mitochondrial genome size and has been of interest for this reason

TABLE 3

NUMBER OF SPECIES PER PHYLUM IN WHICH ORGANELLE GENOMES HAVE BEEN CHARACTERIZED *

Genome	Number of species examined per phylum	Number of phyla	Phyla in which two or more species examined
Mitochondrial	0	55	------------------------------
	1	9	------------------------------
	2	2	Arthropoda, Ciliophora
	5	1	Tracheophyta
	10	2	Ascomycota, Zoomastigina
	11	1	Chordata
Chloroplast	0	10	------------------------------
	1	4	------------------------------
	2	2	------------------------------
	12	1	Angiospermae

* See TABLES 1 and 2 for references.

(TABLE 4). The third phylum, *Zoomastigina,* is the only phylum of protists in which the mitochondrial genomes of more than one or two species have been studied and all of the species characterized belong to the *Kinetoplastidae.* In these highly sophisticated Protozoa, the mitochondrial genomes are found in a specialized giant mitochondrion at the base of the flagellum called the kinetoplast. This organelle contains a complex array of DNA molecules including maxicircles, apparently the true mitochondrial DNA molecules, and thousands of tiny circular DNA molecules called minicircles, which are interlocked and catenated in a variety of esthetically pleasing ways (see Borst and Hoeijmakers[18] for a review). On the other hand, among the *Sarcodina,* which include very primitive protozoans, only the mitochondrial genome of *Acanthamoeba* has so far been studied in any detail. This same pattern of phylogenetic neglect is apparent wherever we look. Thus, mitochondrial genomes have not yet been studied in the most primitive eumetazoan phyla. Among the nine classes that compose the phylum *Tracheophyta* (vascular plants), chloroplast genomes have been characterized only in the *Angiospermae* and *Filicinophyta*

TABLE 4

VARIABILITY IN MITOCHONDRIAL GENOME SIZES IN THE *Ascomycota* *

Genus	Molecular weight of mitochondrial genome $\times 10^6$ daltons
Torulopsis	12.8
Hansenula	17.3
Schizosaccharomyces	17.8
Aspergillus	21.6
Candida	23.1
Kluyveromyces	24–31
Neurospora	40
Saccharomyces	50

* See TABLE 1 for references.

and mitochondrial genomes only in the former class (FIGURE 1). Since chlorophyte algae have small mitochondrial genomes while Angiosperms have the largest mitochondrial genomes known (TABLE 5), one would like to know whether there is a trend towards increasing mitochondrial genome size in the more primitive classes of *Tracheophyta.* As FIGURE 1 also shows, chloroplast genomes have only been characterized in the angiosperms, ferns, and bryophytes among the higher plants and in a few algae, with major groups such as the *Rhodophyta* and *Phaeophyta* being neglected entirely.

EVOLUTION OF MITOCHONDRIAL GENOMES AND PROTEIN-SYNTHESIZING SYSTEMS

The properties of mitochondrial genomes and protein-synthesizing systems have been extensively considered in recent symposia,[19-23] review articles,[15, 16, 24-28] and books.[3] Our purpose here is to make some general statements about these mitochondrial components as well as certain phylogenetic comparisons.

Mitochondrial genomes in all organisms studied to date are small, naked DNA molecules that are circular except in the case of the *Ciliophora,* which have linear mitochondrial genomes.[3] Using a combination of techniques (i.e., renaturation kinetics, measurement of contour length, and summation of restriction-fragment molecular weights) sufficient data are now available to suggest that in many organisms mitochondrial genomes are homogeneous collections of molecules. As shown originally by Quetier and Vedel,[29] higher plants are exceptional in this regard and appear to have heterogeneous mitochondrial DNAs. Recently, Bonen *et al.*[30] have pointed out that while maximum molecular weight estimates are in the range of 60–120 $\times$ 10^6 daltons based on electron microscopy or 70–140 $\times$ 10^6 daltons when determined by renaturation kinetics; restriction-enzyme analysis yields minimum molecular weight estimates which are usually appreciably higher than these values. In the *Kinetoplastidae,* the homogeneous maxicircles are, presumably, the true mitochondrial DNA molecules, while the minicircles, whose function is still obscure, are heterogeneous.[18] Even the largest mitochondrial genomes described to date (TABLE 5) are considerably smaller than the genome of a theoretical "minimal cell" such as a *Mycoplasma,* which has a genome size of around 5–8 $\times$ 10^8 daltons.[31]

No clear-cut trend towards increased mitochondrial genome size is evident in the *Protista* examined so far. However, the twofold or threefold difference in size of mitochondrial genomes in the *Protista* examined to date, with respect to the *Animalia* (TABLE 5), coupled with the structural diversity they exhibit, suggests that the study of mitochondrial genomes in other protists might yield useful phylogenetic insights. The unusual features that characterize the mitochondrial genomes of the *Ciliophora* and *Kinetoplastidae* have been mentioned already. In the case of *Acanthamoeba,* in the *Sarcodina,* and *Dictyostelium,* in the *Acrasiomycota,* mitochondrial genomes are circular and roughly two to three times the size of animal mitochondrial genomes. They are, however, similar in size to the mitochondrial genomes in several of the *Ascomycota.* On the other hand, the mitochondrial genome of *Chlamydomonas* in the *Chlorophyta* is the smallest known. The size of the *Euglena* mitochondrial genome although perhaps even smaller has been difficult to determine accurately. Not only have different molecular weights been reported by different workers, but

TABLE 5

COMPARATIVE MOLECULAR WEIGHTS OF MITOCHONDRIAL GENOMES FROM DIFFERENT EUKARYOTES *

Kingdom	Phylum	Number of species studied	Molecular weight of genome $\times$ 10^6 daltons
Protista	Chlorophyta	1	9.7
	Euglenophyta	1	1.6–4.5
	Sarcodina	1	26, 34
	Ciliophora	2	25–32
	Zoomastigina	5	13–23 †
	Acrasiomycota	1	35, 40
Fungi	Ascomycota	10	12.5–50
Animalia	Platyhelminthes	1	10
	Nematoda	1	10
	Echiurida	1	12
	Arthropoda	3	11–12
	Echinodermata	1	9.0
	Chordata	11	10–12
Plantae	Tracheophyta	5	60–140 ‡

* In most cases molecular weights have been determined from the contour length of isolated mitochondrial DNA molecules. In a number of cases molecular weights have also been determined by kinetic complexity measurements or by summation of molecular weights of restriction fragments. Lists of the organisms used for this tabulation together with original references are available in References 3, 15, and 16.

† All of these estimates come from the Kinetoplastidae, the kinetoplasts of which contain both maxicircles and minicircles. The molecular weights shown are for maxicircles and come from the review by Borst and Hoeijmakers.[18]

‡ Higher-plant mitochondrial DNA is heterogeneous and molecular weights obtained by restriction-enzyme analysis are higher than those estimated by measurement of contour length of individual molecules or by renaturation kinetics (see References 29 & 30).

a high level of A+T-rich spacer sequences has prevented satisfactory analysis of the kinetic complexity of this DNA.[16] As already mentioned, the *Angiospermae* representing the *Tracheophyta* are the only higher plants whose mitochondrial genomes have received attention. These are the largest mitochondrial genomes known (TABLE 5) and they are also heterogeneous.[29, 30] If, in fact, the kingdom *Plantae* was derived from a a protist ancestor in the *Chlorophyta,* one would have to postulate that higher-plant mitochondrial DNAs have been derived from a chlorophyte ancestor with mitochondrial DNA far larger than that of *Chlamydomonas* or that they have obtained additional new sequences from nuclear DNA. Thus, one would like to know whether large mitochondrial genomes are found among the *Bryophyta* and lower classes of *Tracheophyta* and whether these mitochondrial genomes are heterogeneous. Among the *Fungi* the mitochondrial genomes of the *Ascomycota* have been studied extensively and they cover nearly the whole gamut of sizes seen for mitochondrial genomes in other organisms (TABLE 4). Studies with baker's yeast, *Saccharomyces cerevisiae,* which has the largest ascomycete mitochondrial genome so far characterized (50×10^6 daltons), have shown that many of the genes contain intervening sequences.[1, 15] Mahler and Perlman[15] point out that one region of the mitochondrial genome can range from 6,400 base pairs in brewer's yeast, *S. carlsbergensis,* to 11,400 base pairs in different strains of *S. cerevisiae*—a difference of about 3.3×10^6 daltons. The importance of such sequences in explaining mitochondrial genome size variation in ascomycetes in particular, or in the overall sense, remains to be seen. In contrast, mitochondrial genomes in the *Animalia* have proved so far to be quite uniform in size and conformation (TABLE 5). Although most of these data come from the *Chordata,* the few mitochondrial genomes characterized from other eumetazoan phyla, including a tapeworm from the phylum *Platyhelminthes,* all are in the same size range as those of the *Chordata* (TABLE 5). Mitochondrial genomes from the most primitive *Eumetazoa,* including sponges and coelenterates, remain to be studied.

Mitochondrial DNA has been shown to code for a substantial number of mitochondrial tRNAs in HeLa cells, *Xenopus,* and *Saccharomyces.*[1, 3] In a recent review, Borst and Grivell[1] pointed out that the mitochondrial translation system may be able to function with an unusually low number of tRNAs. In *S. cerevisiae* only about 26–30 tRNA genes have been established with certainty. This number is fewer than the minimum of 32 predicted on the basis of Crick's wobble hypothesis. As pointed out by Van Ommen *et al.*[32] one solution may be that mitochondria use, wherever possible, tRNAs capable of recognizing all four bases in the third position of the codon. Recent studies of codon recognition by mitochondrial tRNAs and of the structure of these tRNAs do suggest that they have unique properties.[33–37] For example, in mitochondria the stop codon UGA is read by a mitochondrial tryptophan tRNA.[33–35] A number of yeast mitochondrial tRNAs have now been sequenced. Martin *et al.*[36] in reviewing these data point out that the tRNAs fall into two groups. The first is characterized by unique structural features and the second by conventional structure. For example, a threonine-tRNA anticodon recognizes CUA, the normal codon for leucine. Sequencing of this tRNA revealed that it differed from all previously sequenced tRNAs in having eight rather than seven nucleotides in the anticodon loop.[37] Three of these are complementary to CUA.[37] In discussing *Saccharomyces,* Borst and Grivell[1] point out that there is no evidence for import of cytoplasmic tRNAs, but the results of Suyama and Hamada[38] suggest that some cytoplasmic tRNAs may be imported into mitochondria in *Tetra-*

hymena. Unfortunately, in this ciliate no one knows whether mitochondrial DNA codes for the majority of mitochondrial tRNAs. We can summarize by saying that in two phyla, representing two kingdoms of eukaryotic organisms, careful studies with one or two model organisms indicate that the vast majority of mitochondrial tRNAs are products of the mitochondrial genome. Furthermore, certain mitochondrial tRNAs have unique structural features when compared to their prokaryotic and eukaryotic counterparts.

What little information there is on mitochondrial tRNA synthetases shows that at least some of these enzymes are immunologically distinct from their cytoplasmic counterparts and coded by the nuclear genome.[3] In other cases, one enzyme may serve to charge the cognate tRNAs in both the cytoplasm and the mitochondria.[3] While no mitochondrial tRNA synthetase has been shown to be a mitochondrial gene product, only a very limited number of synthetases have yet been looked at from this viewpoint. Mitochondrial polypeptide-chain initiation involves N-formyl methionine[3, 15] as is the case in bacteria, but the tRNA responsible exhibits significant differences from both the prokaryotic and eukaryotic species responsible for initiation.[39] For obvious reasons, detailed phylogenetic comparisons of mitochondrial tRNAs and tRNA synthetases have not been done. Thus any evolutionary conclusions we wish to reach must be drawn from the few model systems studied to date. With these limitations in mind, two conclusions seem appropriate at this time. First, at some point in mitochondrial evolution a division of labor occurred such that the genes for mitochondrial tRNAs became localized in the mitochondrial genome and their synthetases in the nuclear genome. This is likely to be a primitive characteristic of all mitochondria. Second, during the course of mitochondrial evolution, mitochondrial tRNAs have acquired certain unique features in terms of structure and codon usage that differentiate them from prokaryotic or eukaryotic tRNAs.

Borst and Grivell[1] have reviewed the reports on mitochondrial mRNA structure recently. In *Saccharomyces* the presence of split genes implies that mRNA processing must occur. Putative mRNAs in HeLa cells are uncapped, but do contain a poly(A) tail, which is added post-transcriptionally. Comparable data on capping seem not to exist for *S. cerevisiae* mitochondrial mRNA. The existence of poly(A) tails is, according to Borst and Grivell, disputable. Most mitochondrial messages in *S. cerevisiae* appear not to be polyadenylated. In summary, mitochondrial messages sometimes have features we associate with eukaryotes. This is not surprising in view of the presence of the enzymes necessary for processing and modification in eukaryotic cells.

From the viewpoint of phylogeny, mitochondrial ribosomes and rRNA have received the most satisfactory coverage of any of the components of the mitochondrial protein-synthesizing system. Since the properties of these ribosomes have been considered in detail in four recent reviews[3, 26–28] only the salient points need be considered here. With the exception of higher plants, mitochondrial ribosomes appear to be unique in lacking small (5S or 5.8S) rRNAs. The large rRNAs are also smaller than their counterparts in *E. coli* (1.1 and 0.6×10^6 daltons) except in the cases of the *Tracheophyta* and *Ascomycota* (TABLE 6).

As a group, mitochondrial ribosomes show great structural diversity and exhibit differences that may be useful phylogenetically (TABLE 6). So far the mitochondrial ribosomes of only two protists have been characterized in any detail. The mitochondrial ribosomes of the ciliate protozoans *Tetrahymena*

TABLE 6

SEDIMENTATION VALUES OF ORGANELLE RIBOSOMES AND MOLECULAR WEIGHTS OF THEIR rRNAs *

Source	Sedimentation value			Large RNAs molecular weight $\times 10^6$ daltons		Sedimentation value of small RNAs
	Monomer	Subunits: Large	Subunits: Small			
Mitochondrion						
Ciliophora	80	55	55	0.9	0.5	none
Zoomastigina	60	45	32	?	?	?
Euglenophyta	71	50	32	0.93	0.56	none
Ascomycota	67–75	50	30–40	1.2–1.3	0.7	none
Arthropoda	60	40	25	0.5	0.25	none
Chordata	55–60	40–45	30–35	0.5–0.8	0.3–0.5	none
Tracheophyta	77–80	60	40	1.1–1.3	0.65–0.8	5
Chloroplast						
Chlorophyta	68–70	52–54	37–41	1.1	0.56	3, 7
Euglenophyta	68	50	29	1.1	0.56	5
Tracheophyta	70	50	30	1.1	0.56	5, 4.5
E. coli	70	50	30	1.1	0.6	5
Rat liver	80	60	40	1.7	0.7	5, 5.5

* The data in this table are summarized from References 3, 16, 26, 27, and 28.

and *Paramecium* can be dissociated into subunits that are identical in sedimentation velocity (55S) and size, as measured by electron microscopy. The subunits can, however, be separated by buoyant density centrifugation in cesium chloride and the rRNAs shown to have different molecular weights. Unfortunately, since the *Ciliophora* are such highly evolved protozoans, results obtained with them probably are atypical for protists as a whole. This is suggested by the fact that *Crithidia* (*Kinetoplastidae*) and *Euglena* (*Euglenophyta*) have mitochondrial ribosome monomers with a lower sedimentation velocity than those of the ciliates and composed of subunits with different sedimentation velocities (TABLE 6). So far the only protist in which mitochondrial rRNA gene organization has been characterized is *Tetrahymena*. The genes for the large and small rRNAs are widely separated. While there is only one copy of the small rRNA gene, there are two or three copies of the large rRNA genes located in a terminal duplication-inversion in the mitochondrial DNA (FIGURE 2).[40, 41] This is the only exception to the rule that each rRNA gene is represented once per mitochondrial DNA molecule.

The mitochondrial ribosomes of two *Ascomycota* (*Neurospora* and *Saccharomyces*) have been characterized in detail. In *Saccharomyces* the genes coding for the large and small mitochondrial rRNAs are widely separated (approx. 30,000 base pairs) and there are many tRNA genes between them (FIGURE 2). Certain strains of *S. cerevisiae* designated ω^+ contain an intervening sequence of 1,050 base pairs in the large rRNA gene whereas other strains called ω^- lack this sequence. In *Neurospora* the large and small mitochondrial rRNAs are separated by 5,000 base pairs and the region between also contains many tRNA genes (FIGURE 2). In addition the large rRNA gene contains an intervening sequence of 2,300 base pairs and the unprocessed transcript of the entire large rRNA gene has been identified.[28] Thus, in contrast to the organization of rRNA genes in *E. coli* and eukaryote nuclei, these mitochondrial rRNA genes are widely separated from each other, contain many more tRNA genes in between, and the large rRNA gene is split in *Neurospora* and ω^+ strains of *S. cerevisieae*. The sites of synthesis of mitochondrial ribosomal proteins have been studied in detail in *Neurospora* and *Saccharomyces*[28] and in each case all mitochondrial ribosomal proteins, save one small subunit protein, have been found to be made on cytoplasmic ribosomes. These experiments suggest that, although mitochondrial rRNAs are mitochondrial gene products, virtually all mitochondrial ribosomal proteins are coded by nuclear genes.

Animal mitochondrial ribosomes have so far been characterized in chordates ranging from mammals to fish and in a couple of insects. One would like to know whether the mitochondrial ribosomes found in animals belonging to

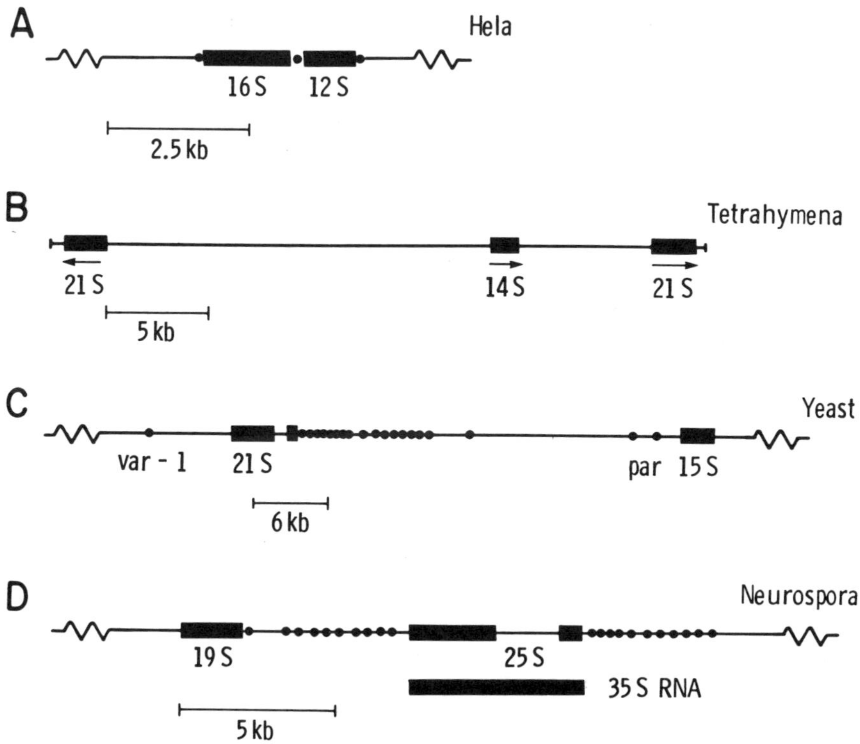

FIGURE 2. Organization of mitochondrial rRNA and mitochondrial tRNA (4S RNA) genes in several organisms. *var-1*=variant 1 gene and *par*=paromomycin-resistance gene in yeast. See text and Reference 28 for further discussion. (From Boynton *et al.*[28] By permission of the University Park Press.)

these two phyla are characteristic of the *Animalia* as a whole since they are totally unique (TABLE 6). Animal mitochondrial ribosomes have the lowest sedimentation velocities and the smallest rRNAs known.[28] Interestingly, the genes coding for animal mitochondrial rRNAs are also tightly linked in the mitochondrial genome, unlike the mitochondrial rRNA genes in the *Ascomycota* and the ciliate protozoan *Tetrahymena* (FIGURE 2). Mitochondrial ribosomes from animals are also the most protein-rich of any ribosomes known. Thus, O'Brien *et al.*[42] have shown that even under the most stringent washing conditions bovine mitochondrial ribosomes still contain 90 proteins as compared to 55 in the *E. coli* ribosome and 70 in rat liver or HeLa cell cytoplasmic ribosomes.

In contrast to animal mitochondrial ribosomes, the mitochondrial ribosomes of higher plants have sedimentation velocities in the range of eukaryotic cytoplasmic ribosomes (TABLE 6). These are also the only mitochondrial ribosomes that contain 5S rRNA and this rRNA has been shown to be distinct from cytoplasmic 5S rRNA by fingerprinting.[28] In addition, the large rRNAs of higher-plant mitochondrial ribosomes differ only slightly in molecular weight from their cytoplasmic counterparts. Again, all we know of mitochondrial ribosomes in plants comes from a couple of angiosperms. In a phylogenetic sense one would like to know whether large, 5S-containing mitochondrial ribosomes are characteristic of the *Tracheophyta,* perhaps even the kingdom *Plantae,* as a whole.

Finally, a few remarks on inhibitor sensitivity of mitochondrial ribosomes are appropriate. The supposed similarity of organelle ribosomes to bacterial ribosomes in sensitivity to inhibitors has played an important role in theories of mitochondrial evolution. While mitochondrial ribosomes are resistant to inhibitors of eukaryotic cytoplasmic ribosomes such as anisomycin and cycloheximide and are sensitive to many inhibitors of bacterial ribosomes, the degree of sensitivity is markedly different in some cases. Whether the inhibitors are acting by the classical mechanism is not clear. Certain pertinent studies have been reviewed recently by Boynton *et al.*[28] and several illustrations will suffice. Yeast mitochondrial ribosomes are as sensitive to lincomycin and carbomycin as bacterial ribosomes whereas mammalian mitochondrial ribosomes are 100- to 1000-fold less sensitive. There seems to be no simple phylogenetic pattern since mitochondrial ribosomes from *Neurospora* are similar to mammalian rather than yeast mitochondrial ribosomes with respect to carbomycin sensitivity. To date, no detailed study of the effect of aminoglycoside antibiotics on isolated mitochondrial ribosomes has been published. In yeast, neomycin and paromomycin have been found to be effective inhibitors of mitochondrial protein synthesis in whole cells and isolated mitochondria, but other aminoglycoside antibiotics (neamine, kanamycin, streptomycin, viomycin) are effective only at very high concentrations that also inhibit 80S cytoplasmic ribosomes. These examples merely serve to illustrate once again that mitochondrial ribosomes are a highly diverse group that cannot be labeled as either prokaryotic or eukaryotic. They are unique. Until mitochondrial ribosomes are studied in more phylogenetically important groups of organisms under standardized conditions we will have no idea of how these different classes of ribosomes evolved or whether they are characteristic of the majority of the phyla in their kingdom.

Evolution of Chloroplast Genomes and Protein-Synthesizing Systems

The properties of chloroplast genomes and protein-synthesizing systems have been thoroughly reviewed in recent symposia,[9, 43, 44] books,[3] and review articles,[14, 16, 28, 45–47] as was the case for mitochondria.

Chloroplast genomes like mitochondrial genomes are naked, circular DNA molecules, but they are invariably larger than all mitochondrial DNA molecules except those of higher plants (Table 7). So far all chloroplast genomes characterized appear to be homogeneous collections of molecules. Among the protists, chloroplast genomes have been studied in detail in three representatives of the superphylum *Chlorophyta* and one representative of the superphylum *Chromophyta*. The chloroplast genomes of the chlorophyte algae *Chlamydomonas* and *Euglena* are in the same general size range as those from the *Bryophyta* and *Tracheophyta* in the kingdom *Plantae* (Figure 3, Table 7). These genomes are about fivefold smaller than the genome of a *Mycoplasma*

Table 7

Comparative Molecular Weights of Chloroplast Genomes from Different Eukaryotes *

Kingdom	Phylum	Number of species studied	Molecular weight of genome $\times 10^6$ daltons
Protista	Chromophyta	1	60 †
	Chlorophyta	1	126
	Euglenophyta	1	97
	Siphonophyta	2	1,500
Plantae	Bryophyta	1	85 ‡
	Tracheophyta	7	91.1–103.2 ‡

* Molecular weights were determined as described in Table 5. Lists of organisms used for this tabulation together with original references are available in References 3, 14, and 16.

† For the *Olisthodiscus* (Chromophyta) estimate, see Reference 52.

‡ For the estimates for *Sphaerocarpos* (Bryophyta) and two ferns (Filicinophyta), see Reference 61.

and fifteenfold smaller than the genome of *E. coli*. In contrast, the chloroplast genome of *Acetabularia* has the same coding capacity as the *E. coli* genome (Table 7). Green *et al.*[48, 49] have done careful studies of renaturation kinetics of chloroplast DNA from *Acetabularia cliftonii* and *A. mediterranea* and have measured the renaturation rates of *Chlamydomonas* chloroplast DNA and *E. coli* DNA under the same conditions. The results they obtained with these standards are similar to those published by others so there is no reason to believe there is anything wrong with the measurements of Green *et al.* A molecule of 1.5×10^9 daltons would be 750 μm long and would not be expected to survive isolation intact. Hence, it is not surprising that only DNA fragments have been obtained in chloroplast lysates from *Acetabularia*, the longest being on the order of 200 μm, about 4.14×10^8 daltons, or approximately four times the size of other chloroplast DNA molecules. These experiments with *Acetabularia* underscore the importance of investigating organelle DNAs from additional protists. Theoretically, the chloroplast genome of *Acetabularia* is large enough to make the chloroplast of this alga independent of the nucleus although there is reason to believe that this may not be the case.[48, 49]

For some reason chloroplast DNA does not seem to have been characterized in algae from the superphylum *Rhodophyta*. In the superphylum *Chromophyta*, Gibbs *et al.* have shown that the chloroplast DNA is organized in a unique, peripheral ring-shaped nucleoid within the chloroplast.[50] Gibbs and Poole[51] estimated that a light-grown cell of *Ochromonas* contains a minimum of 10 and possibly 20 separate chloroplast DNA molecules. In another chromophyte alga, *Olisthodiscus*, Aldrich and Cattolico[52] reported a chloroplast genome size of

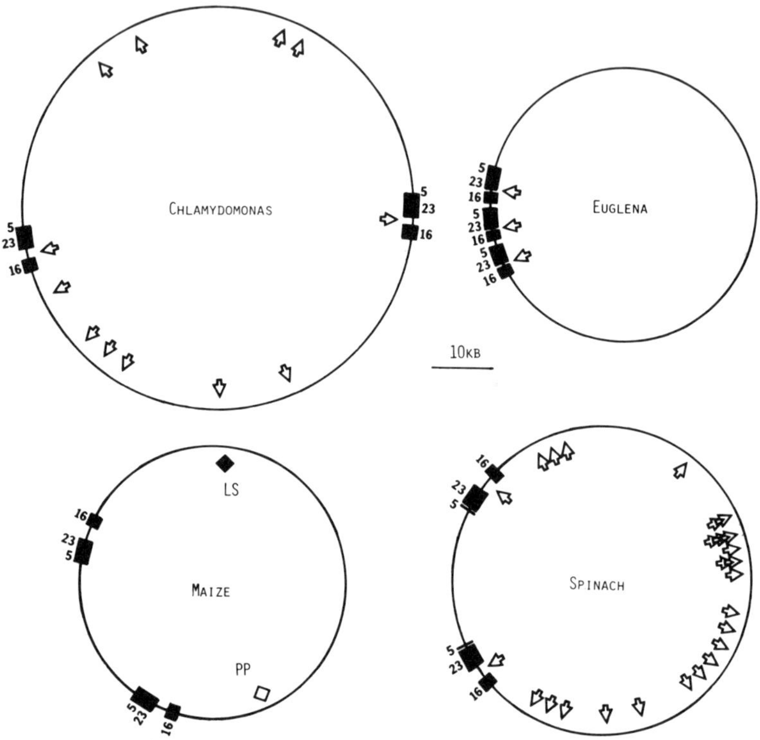

FIGURE 3. Chloroplast genome organization in four different plants. Location and orientation of rRNA genes are shown and arrows indicate the positions of 4S RNA genes in *Chlamydomonas* and *Euglena* and tRNA genes in spinach. In maize, LS=gene for large subunit of ribulose bisphosphate carboxylase and PP=gene for the maize photoprotein. See text and Reference 28 for further discussion. (From Boynton *et al.*[28] By permission of the the University Park Press.)

60×10^6 daltons based on preliminary restriction-enzyme analysis of the chloroplast DNA. Open circular molecules were also seen by electron microscopy corresponding to molecular weights of 22, 44, and 88×10^6 daltons. An interesting feature of certain chloroplast DNAs is the presence of a pair of large inverted repeats containing the rRNA cistrons. These inverted repeats range from 19,000 base pairs in *Chlamydomonas* to 24,500 base pairs in maize and spinach (FIGURE 3). These repeats are not an obligatory feature of chloroplast DNA since they are absent in chloroplast DNA from pea. In *Euglena* the three sets of rRNA genes are closely linked in straight tandem repeats (FIGURE 3). Whether these inverted repeats could be used in any sense as phylogenetic landmarks is not known.

Recent experimental results obtained with *Chlamydomonas, Euglena,* maize, and spinach all agree in showing that chloroplast DNA codes for a large number of tRNAs (FIGURE 3).[28] The experiments[53, 54] with spinach chloroplast tRNAs are particularly complete and most of the 35 tRNAs found within the spinach chloroplast have been mapped on the chloroplast genome. As in the case of mitochondria, distinct tRNA synthetases seem to be involved in charging chloroplast tRNAs and these appear to be made on cytoplasmic ribosomes.[3] Thus in both chloroplasts and mitochondria tRNA genes are localized in the organelle genome, but the corresponding synthetases are coded by nuclear genes. Polypeptide-chain initiation in chloroplasts involves N-formylmethionine. N-formylmethionyl tRNA and transformylases have been found in chloroplasts.[3] Ciferri *et al.* have purified elongation factor G ($EF\text{-}G_{chl}$) and elongation factor T($EF\text{-}T_{u\ chl}$) from spinach chloroplasts and shown that these soluble proteins are similar to prokaryotic elongation factors and are synthesized by isolated chloroplasts.[28] The two chloroplast mRNAs studied in detail so far appear to lack poly(A), but whether other chloroplast mRNAs are polyadenylated or not remains an open question.[16] No reports on whether the mRNAs are capped or whether they are ever transcribed from split genes and spliced have appeared to date.

The chloroplast ribosome is the most thoroughly characterized component of the chloroplast protein-synthesizing system. In contrast to mitochondrial ribosomes, chloroplast ribosomes are relatively uniform in structure and bear strong resemblances to prokaryotic ribosomes.[28] Chloroplast ribosomes sediment at 70S like bacterial ribosomes and consist of two subunits of approximately 50S (46–54S) and approximately 30S (26–41S). The large subunit contains 23S rRNA (1.1×10^6 daltons) and the small subunit 16S rRNA (0.56×10^6 daltons). In addition, the large subunit contains a 5S RNA that is distinct from plant cytoplasmic ribosomal 5S RNA. In this respect it is similar to the large subunit from bacterial but not mitochondrial ribosomes with the exception of those from higher plants. Higher-plant chloroplast ribosomes also contain 4.5S rRNA (2.5 or 3.3×10^4 daltons) associated with the large subunit and present in equimolar amounts with 5S RNA. In *Chlamydomonas,* Rochaix and Malnoe have identified three low molecular weight rRNAs associated with the large subunit: 7S, 5S, and 3S.[28] Neither the relationship of the 7S or 3S rRNAs to 4.5S rRNA nor the possible homology of these rRNAs to the 5.8S rRNA of eukaryotic cytoplasmic ribosomes is known. Buetow *et al.* have compared T_1 ribonuclease digests of 16S chloroplast rRNA from *Euglena* and the red alga *Porphyridium* with 18S cytoplasmic rRNA from yeast and L cells and with 16S rRNA from *Anacystis* and several bacteria.[28] The two chloroplast rRNAs had the most similar fingerprints and were typically prokaryotic in terms of their capacity for base modification. At least 12 of the oligomers referred to as prokaryotic "universals" (because they occur in prokaryotes, but not eukaryotes) are found in *Euglena* 16S rRNA.

As mentioned above, in *Euglena* there are three sets of rRNA genes per chloroplast genome in contrast to the two sets found in all other plants studied thus far (FIGURE 3). Organization of each set of chloroplast rRNA genes is reminiscent of what one sees in *E. coli.* Thus, the chloroplast rRNA genes are quite closely linked in the order 16S–tRNA–23S–5S (FIGURE 4). In *Chlamydomonas* the 7S and 3S genes are on the left hand side of the 23S gene and in spinach the 4.5S gene is very close to the 5S gene. The spacer between the 16S and 23S genes ranges from 2,100 base pairs in maize to about 300 base pairs in *Euglena.* In spinach, a large RNA precursor has been identified that includes all of the mature rRNAs. Interestingly, the gene for the large rRNA in *Chlamydomonas* contains an intervening sequence of 940 base pairs. In summary, rRNA gene organization in chloroplasts is reminiscent of prokaryotes in that the genes are close together and the spacer usually contains a tRNA gene. The presence of the intervening sequence in the large rRNA gene of *Chlamydomonas* and additional genes for low molecular weight rRNAs (not found in association with large ribosomal subunits from prokaryotes) emphasizes that chloroplast rRNA gene organization has certain unique features as well.

In tobacco Bourque *et al.* have found 20–24 proteins in the small subunit

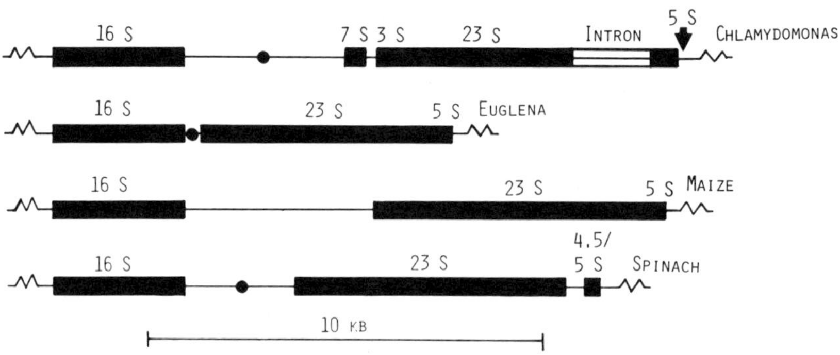

FIGURE 4. Organization of chloroplast rRNA genes in four plants. Small circles indicate chloroplast tRNA or 4S genes. See text and Reference 28 for discussion. (From Boynton *et al.*[28] By permission of the University Park Press.)

of the chloroplast ribosome and 34–38 in the large subunit.[28] This compares to 21 and 34 in the respective *E. coli* subunits. In *Chlamydomonas,* estimates for the large subunit range from 26–34.[28] Recent studies in our laboratory strongly suggest that the small subunit may contain as many as 34 proteins, which would make this subunit considerably richer in proteins than its counterpart in *E. coli.* The sites of synthesis of chloroplast ribosomal proteins have not been studied with the same rigor as was the case for mitochondrial ribosomal proteins. The best work is that of Freysinnet in *Euglena.*[28] His results indicate that at least 12 chloroplast ribosomal proteins are made on cytoplasmic ribosomes, but nine appear to be made on chloroplast ribosomes. If these results prove to be of general applicability, they will mean that the chloroplast, in comparison to the mitochondrion, plays a much more important role in the synthesis of organelle ribosomal proteins.

Chloroplast ribosomes show strong similarities to bacterial ribosomes in their spectra of antibiotic resistance.[28] Chloroplast ribosomes are resistant to inhibitors of eukaryotic cytoplasmic-protein synthesis, such as, cycloheximide and anisomycin, but sensitive to virtually all inhibitors of prokaryotic protein synthesis including chloramphenicol, the aminoglycosides (i.e. kanamycin, neamine, neomycin, spectinomycin, streptomycin), the macrolides (i.e. carbomycin, clindamycin, and erythromycin), and lincomycin. In *Chlamydomonas,* mutations resistant to these inhibitors have been isolated and studied in detail.[2, 3, 28] These mutations are scattered between both the chloroplast and nuclear genomes. Mapping and allele-testing experiments with chloroplast mutants have led to the identification of two recombinationally distinct loci for erythromycin resistance (*er-u*), one for neamine/spectinomycin resistance (*nr/spr-u*), and four for streptomycin resistance (*sr-u*), which map in the order: *er-u-1a—er-u-37—nr/spr-u—sr-u-2-23—sr-u-2-60—sr-u-sm3—sr-u-sm2*. Chloroplast mutations to streptomycin dependence (*sd-u*) have also been isolated and these map in three recombinationally distinct loci within this region. Nuclear mutations to streptomycin resistance (*sr-1*) map in a single gene, while nuclear mutations conferring resistance to erythromycin (*ery-M1, ery-M2, ery-M3*) map in three unlinked genes. In *E. coli,* mutations with similar phenotypes all confer resistance on either the small (e.g., streptomycin, spectinomycin) or large (e.g., erythromycin) subunit of the ribosome. The *Chlamydomonas* mutations show exactly the same localization of resistance with respect to the chloroplast ribosome. In *E. coli,* resistance mutations can be shown to alter specific ribosomal proteins. Similar alterations in chloroplast ribosomal proteins have also been reported in several cases for resistance mutations in *Chlamydomonas*. However, rigorous proof that a specific ribosomal protein is altered by resistance mutations has only been obtained for mutants at the nuclear *ery-M1* locus, which appears to be the structural gene for protein LC6 of the large subunit.[55] Nevertheless, the data obtained so far have revealed striking similarities between chloroplast and prokaryotic ribosomes with regard to antibiotic resistance mechanisms. The experiments with the resistance mutations illustrate clearly that chloroplast ribosome formation is under the dual control of chloroplast and nuclear genes. This conclusion is reinforced by the isolation of mutations that are assembly-defective in chloroplast ribosome formation which map in both genomes.[28]

Conclusions

There are three major conclusions we wish to draw. First, organelle genomes and protein-synthesizing systems have received insufficient attention from the phylogenetic viewpoint. This is true despite the enormous amount of speculation that has taken place on the evolutionary origins of chloroplasts and mitochondria, in which the properties of organelle genomes and protein-synthesizing systems have played a central role. Second, no organelle genome examined to date, with the exception of the chloroplast genome of *Acetabularia,* is sufficiently large to code for more than a fraction of the components of either a chloroplast or mitochondrion. Third, the genes coding for components of the organelle protein-synthesizing system are apportioned between the organelle and nuclear genomes in a highly specific manner. Thus, in both chloroplasts and mitochondria the stable RNAs of the organelle protein-synthesizing system are coded by the organelle genome, but most of the proteins are coded by the nuclear genome. Those proteins not coded by the nuclear genome include a few ribosomal proteins. This consistent division of labor is apparent for virtually all components coded by either the chloroplast or mitochondrial genome. For example, in all plants examined thus far, the large subunit of the CO_2-fixing enzyme ribulose bisphosphate carboxylase is coded by the chloroplast genome and the small subunit by the nuclear genome.[2] Similarly, certain polypeptides of cytochrome oxidase are coded by the mitochondrial genome whereas others are coded by the nuclear genome.[1] So far only one case is known where a given organelle component is coded by an organelle gene in one organism and the nuclear genome in another. This is the proteolipid of the mitochondrial F_1 ATPase, which is coded by a mitochondrial gene in yeast, but by a nuclear gene in *Neurospora*.[1]

Besides these major conclusions, we can reach a number of other more tentative conclusions always taking account of the phylogenetic biases inherent in the data. First, there is no evidence to indicate an increase in size and coding capacity of mitochondrial genomes as a function of more primitive phylogenetic position. Among the living organisms examined to date we face a major hiatus in genome size between the largest mitochondrial genome and the smallest cellular genome. With regard to the endosymbiont theory, these negative results may simply mean that the postulated intermediate cell types, with their protomitochondria containing much larger genomes, have long been extinct. On the basis of the nonsymbiotic theory these results could mean that the discontinuity is real because the DNA molecule that became the mitochondrial genome was about the size of the larger mitochondrial genomes we find today in nature. Presently, we cannot distinguish between these two possibilities.

Second, mitochondrial genomes show considerable structural diversity, particularly among protists. In terms of phylogeny, mitochondrial genome size might be a useful criterion in determining the protistan origins of the animal and plant kingdoms. In this regard we note that the available data suggest two opposing trends in animals and plants. Animal mitochondrial genomes have become reduced in size with respect to their putative protistan ancestors, whereas as higher-plant mitochondrial genomes have shown exactly the opposite trend. Furthermore, higher-plant mitochondrial genomes, unlike other mitochondrial genomes, are heterogeneous. One would like to know whether this trend towards increased size and heterogeneity is also evident in the *Bryophyta* and the lower classes of *Tracheophyta*. Mitochondrial ribosomes also show a great deal of structural diversity at the kingdom level that may be useful in formulating more firmly grounded conclusions concerning which protistan phyla were likely to have given rise to each of the higher kingdoms.

Third, mitochondrial protein-synthesizing systems have so many unique features that it is difficult to call them either prokaryotic or eukaryotic. The most useful guideline would seem to be to consider them as unique rather than trying to accommodate them within the prokaryotic framework. Although the prokaryotic model has proved useful in the past, this model has outlived its usefulness in our opinion because it tends to obscure the unique features of the mitochondrial protein-synthesizing system while emphasizing those seemingly prokaryotic features.

Fourth, chloroplast genomes, with the exception of *Acetabularia,* are all about the same size and roughly approximate the size of the bacteriophage T_4 genome. In the case of *Acetabularia,* the chloroplast genome actually approximates that of the *E. coli* genome. However, our knowledge of chloroplast genome sizes and properties among primitive protists is very limited so we do not know whether the large chloroplast genome of *Acetabularia* represents a derivative or primitive character. Examination of chloroplast genome structure among other *Protophyta* from the viewpoint of phylogeny might allow a decision between these possibilities. Since the chloroplast genome of *Acetabularia* is so large and because chloroplast ribosomes share so many properties with prokaryotic ribosomes, this organelle seems much more prokaryotic than the mitochondrion. That is, the prokaryote analogy is still useful for the chloroplast and of the two organelles the chloroplast seems the most likely to have arisen by endosymbiosis.

Fifth, if we assume that chloroplasts and mitochondria evolved independently either by endosymbiosis or by a nonsymbiotic mechanism, then we must note that the segregation of genes coding for the stable RNAs of the organelle protein-synthesizing system in the organelle genome and the genes coding for the essential proteins of this system principally in the nucleus represents a striking example of evolutionary convergence. Three explanations occur to us for this observation. First, the polyploid organelle genome provides an ideal system for gene amplification. Amplification of tRNA and rRNA genes and their products is provided automatically by localizing them in the organelle genome. Amplification of ribosomal proteins and tRNA synthetases, in contrast, is provided by the machinery of translation. Second, there is a general belief supported by some experimental data [3] that RNAs, particularly mRNAs, are not transported across organelle membranes. There is also abundant evidence that cytoplasmically synthesized proteins can be transferred across organelle membranes.[56] Third, in *E. coli* there are seven sets of rRNA "operons."[57] All of these transcription units contain tRNA genes in the spacer region between the 16S and 23S genes. Three contain the same isoaccepting form of $tRNA^{Glu}$ and two contain the same isoaccepting form of $tRNA^{Ala}$ as well as a $tRNA^{Ile}$ gene. In addition, other isoaccepting forms of tRNA are known in *E. coli* that have the same anticodon structure, such as $tRNA^{Met}$, $tRNA^{Gly}$, and $tRNA^{Tyr}$.[58, 59] Genes coding for ribosomal proteins in *E. coli* are not redundant [57, 60] and presumably the same is true of the genes coding for tRNA synthetases. On the nonsymbiotic theory of organelle evolution one might speculate that transfer of a subset of rRNA and tRNA genes might have occurred via a plasmid incorporated into a proto-organelle without lethality. Perhaps this first took place for an rRNA operon and the tRNAs associated with that operon and later occurred for other tRNAs. In this regard a remarkable observation is that in mitochondria of both yeast and *Neurospora* the two large rRNA genes are widely separated from each other and have many tRNA genes located between them. In contrast, transfer of genes for ribosomal proteins and tRNA synthetases would very likely not have occurred unless these genes were redundant. The first and second explanations predict evolutionary convergence whatever the origin of chloroplasts and mitochondria.

In conclusion, we note that the reason why organelles possess genomes and protein-synthesizing systems even in the most highly evolved living creatures is less clear now than it was several years ago. The general belief used to be that the most highly hydrophobic proteins of the mitochondrial inner membrane and the thylakoid membrane of the chloroplast ought to be synthesized inside the organelle close to their point of insertion.[3] While this may be part of the explanation, we now know that the proteolipid of the membrane factor of the F_1 ATPase of the mitochondrion, which is very insoluble, is a mitochondrial gene product in *Saccharomyces* and a nuclear gene product in *Neurospora*.[1]

REFERENCES

1. BORST, P. & L. A. GRIVELL. 1978. The mitochondrial genome of yeast. Cell **15:**705–723.
2. GILLHAM, N. W., BOYNTON, J. E. & N.-H. CHUA. 1978. Genetic control of chloroplast proteins. Curr. Top. Bioenerg. **8:**211–260.
3. GILLHAM, N. W. 1978. Organelle Heredity. Raven Press. New York, N.Y.
4. SAGAN, L. 1967. On the origin of mitosing cells. J. Theoret. Biol. **14:**225–274.
5. MARGULIS, L. 1970. Origin of Eukaryotic Cells. Yale University Press. New Haven, Conn.
6. MARGULIS, L. 1975. Symbiotic theory of the origin of eukaryotic organelles: criteria of proof. Symp. Soc. Exp. Biol. **29:**21–38.
7. MARGULIS, L. 1976. Genetic and evolutionary consequences of symbiosis. Exp. Parasitol. **39:**277–349.
8. ALLSOPP, A. 1969. Phylogenetic origins of the procaryota and the origin of the eucaryotic cell. New Phytol. **68:**591–612.
9. RAFF, R. A. & H. R. MAHLER. 1972. The nonsymbiotic origin of mitochondria. Science **177:**575–582.
10. RAFF, R. A. & H. R. MAHLER. 1972. The symbiont that never was: an inquiry into the evolutionary origin of the mitochondrion. Symp. Soc. Exp. Biol. **29:** 41–92.
11. UZZELL, T. & C. SPOLSKY. 1974. Mitochondria and plastids as symbionts: a revival of special creation? Am. Sci. **62:**334–343.
12. BOGORAD, L. 1975. Evolution of organelles and eukaryotic genomes. Science **188:**891–898.
13. BOGORAD, L. 1977. Genes for chloroplast ribosomal RNAs and ribosomal proteins: gene dispersal in eukaryotic genomes. *In* International Cell Biology 1976–1977. B. R. Brinkley & K. R. Porter, Eds.: pp. 175–182. The Rockefeller Press. New York, N.Y.
14. BEDBROOK, J. R. & R. KOLODNER. 1979. The structure of chloroplast DNA. Ann. Rev. Plant Physiol. **30:**593–620.
15. MAHLER, H. R. & P. S. PERLMAN. 1979. Mitochondrial biogenesis: evolution and regulation. *In* ICN-UCLA Symposium on Molecular and Cellular Biology: Extrachromosomal DNA. D. Cummings, P. Borst, I. Dawid & S. Weissman, Eds.: **15:**11–33. Academic Press, Inc. New York, N.Y.
16. EDELMAN, M. 1980. Nucleic acids of chloroplasts and mitochondria. *In* The Biochemistry of Plants: Proteins and Nucleic Acids. A. Marcus, Ed.: Vol. **6**. (In press) Academic Press, Inc. New York, N.Y.
17. WHITTAKER, R. H. & L. MARGULIS. 1978. Protist classification and the kingdoms of organisms. Bio Systems **10:**3–18.
18. BORST, P. & J. H. J. HOEIJMAKERS. 1979. Kinetoplast DNA. Plasmid **2:**20–40.
19. BÜCHER, TH., W. NEUPERT, W. SEBALD & S. WERNER, Eds. 1976. Genetics and Biogenesis of Chloroplasts and Mitochondria. Elsevier/North-Holland Biomedical Press. Amsterdam, The Netherlands.
20. BANDLOW, W., R. J. SCHWEYEN, D. Y. THOMAS, K. WOLF & F. KAUDEWITZ, Eds. 1976. Genetics, Biogenesis and Bioenergetics of Mitochondria. De Gruyter. Berlin, Federal Republic of Germany.
21. BANDLOW, W., R. J. SCHWEYEN, D. Y. THOMAS, K. WOLF & F. KAUDEWITZ, Eds. 1977. Mitochondria 1977. De Gruyter. Berlin, Federal Republic of Germany.
22. CUMMINGS, D., P. BORST, I. DAWID, S. WEISSMAN & C. F. FOX, Eds. 1979. Extrachromosomal DNA. ICN-UCLA Symposia on Molecular and Cellular Biology. Vol. 15. Academic Press, Inc. New York, N.Y.
23. BACILLA, M., B. L. HORECKER & A. O. M. STOPPANI, Eds. 1978. Biochemistry and Genetics of Yeast: Pure and Applied Aspects. Academic Press, Inc. New York, N.Y.
24. LINNANE, A. W. & P. NAGLEY. 1978. Structural mapping of mitochondrial DNA. Arch. Biochem. Biophys. **187:**277–289.

25. LINNANE, A. W. & P. NAGLEY. 1978. Mitochondrial genetics in perspective: The derivation of a genetic and physical map of the yeast mitochondrial genome. Plasmid **1:**324–345.

26. BUETOW, D. E. & W. M. WOOD. 1978. The mitochondrial translation system. Subcell. Biochem. **5:**1–85.

27. O'BRIEN, T. W. & D. E. MATTHEWS. 1976. Mitochondrial ribosomes. *In* Handbook of Genetics. R. C. King, Ed. **5:**535–580. Plenum Press. New York, N.Y.

28. BOYNTON, J. E., N. W. GILLHAM & A. L. LAMBOWITZ. 1980. Biogenesis of chloroplast and mitochondrial ribosomes. *In* Ribosomes: Structure, Function, and Genetics. G. Chamblis, G. R. Craven, J. Davies, K. Davis, L. Kahan & M. Nomura, Eds.: pp. 903–950. University Park Press. Baltimore, Md.

29. QUETIER, F. & F. VEDEL. 1977. Heterogeneous population of mitochondrial DNA molecules in higher plants. Nature **268:**365–368.

30. BONEN, L., T. Y. HUH & M. W. GRAY. 1980. Can partial methylation explain the complex fragment patterns observed when plant mitochondrial DNA is cleaved with restriction endonucleases? FEBS Lett. (In press).

31. MOROWITZ, H. J. 1967. Biological self replication systems. Prog. Theor. Biol. **1:**35–58.

32. VAN OMMEN, G. J. B., G. S. P. GROOT & P. BORST. 1977. Fine structure physical mapping of 4S RNA genes on mitochondrial DNA of *Saccharomyces cerevisiae*. Mol. Gen. Genet. **154:**255–262.

33. MACINO, G., G. CORUZZI, F. G. NOBREGA, M. LI & A. TZAGOLOFF. 1979. Use of the UGA terminator as a tryptophan codon in yeast mitochondria. Proc. Natl. Acad. Sci. USA **76:**3784–3785.

34. BARRELL, B. G., A. T. BANKIER & J. DROUIN. 1979. A different genetic code in human mitochondria. Nature **282:**189–194.

35. HALL, B. D. 1979. Mitochondria spring surprises. Nature **282:**129–130.

36. MARTIN, N. C., D. MILLER, J. HARTLEY, P. MOYNIHAN & J. E. DONELSON. 1980 The $tRNA_{AGY}^{ser}$ and $tRNA_{CGY}^{arg}$ genes form a gene cluster in yeast mitochondrial DNA. Cell **19:**339–343.

37. LI, M. & A. TZAGOLOFF. 1979. Assembly of the mitochondrial membrane system: sequences of yeast mitochondrial valine and an unusual threonine tRNA gene. Cell **18:**47–53.

38. SUYAMA, Y. & J. HAMADA 1976. Imported tRNA: its synthetase as a probable transport protein. *In* Genetics and Biogenesis of Chloroplasts and Mitochondria. Bücher, Th., W. Neupert, W. Sebald & S. Werner, Eds.: pp. 763–770. Elsevier/North-Holland Biomedical Press. Amsterdam, The Netherlands.

39. BIANCHETTI, R., G. LUCCHINI, P. CORSTI & P. TORTORA. 1977. Dependence of mitochondrial protein synthesis initiation on formylation of the initiator methionyl tRNA. J. Biol. Chem. **252:**2519–2523.

40. GOLDBACH, R. W., P. BORST, J. E. BOLLEN-DE BOER & E. F. J. VAN BRUGGEN. 1978. The organization of ribosomal RNA genes in the mitochondrial DNA of *Tetrahymena pyriformis* strain ST. Biochim. Biophys. Acta **521:**169–186.

41. GOLDBACH, R. W., J. E. BOLLEN-DE BOER, E. F. J. VAN BRUGGEN & P. BORST. 1978. Conservation of the sequence and position of the ribosomal RNA genes in *Tetrahymena pyriformis* mitochondrial DNA. Biochim. Biophys. Acta **521:** 187–197.

42. O'BRIEN, T. W., D. E. MATTHEWS & N. D. DENSLOW. 1976. Protein composition of the bovine mitochondrial ribosome. *In* Genetics and Biogenesis of Chloroplasts and Mitochondria. Bücher, Th., W. Neupert, W. Sebald & S. Werner, Eds.: pp. 741–748. Elsevier/North-Holland Biomedical Press. Amsterdam, The Netherlands.

43. BOGORAD, L. & J. H. WEIL, Eds. 1976. Nucleic Acids and Protein Synthesis in Plants. Plenum Press. New York, N.Y.

44. AKOYUNOGLOU, G. & J. H. ARGYROUDI-AKOYUNOGLOU, Eds. 1978. Chloroplast Development. Elsevier/North-Holland Biomedical Press. Amsterdam, The Netherlands.
45. ELLIS, R. J. 1977. Protein synthesis by isolated chloroplasts. Biochim. Biophys. Acta **463:**185–215.
46. KUNG, S.-D. 1977. Expression of chloroplast genomes in higher plants. Ann. Rev. Plant Physiol. **28:**401–437.
47. STUTZ, E. & A. BOSCHETTI. 1976. Chloroplast Ribosomes. *In* Handbook of Genetics. R. C. King, Ed.: **5:**425–450. Plenum Press. New York, N.Y.
48. GREEN, B. R., B. L. MUIR & U. PADMANABHAN. 1977. The Acetabularia chloroplast genome: small circles and large kinetic complexity. *In* Progress in Acetabularia Research. C. F. L. Woodcock, Ed.: pp. 107–122. Academic Press, Inc. New York, N.Y.
49. PADMANABHAN, U. & B. R. GREEN. 1978. The kinetic complexity of *Acetabularia* chloroplast DNA. Biochim. Biophys. Acta **521:**67–73.
50. GIBBS, S. P., D. CHENG & T. SLANKIS. 1974. The chloroplast nucleoid in *Ochromonas danica.* I. Three-dimensional morphology in light- and dark-grown cells. J. Cell Sci. **16:**557–577.
51. GIBBS, S. P. & R. J. POOLE. 1973. Autoradiographic evidence for many segregating DNA molecules in the chloroplasts of *Ochromonas danica.* J. Cell Biol. **59:**318–328.
52. ALDRICH, J. & R. CATTOLICO. 1979. Characterization of the chloroplast genome of the marine alga, *Olisthodiscus luteus.* J. Supramol. Struct. (Suppl. 3): 142.
53. STEINMETZ, A., M. MUBUMBILA, M. KELLER, G. BURKHARD & J. H. WEIL. 1978. Mapping of tRNA genes on the circular DNA molecule of *Spinacia oleracea* chloroplasts. *In* Chloroplast Development. G. Akoyunoglou & J. H. Argyroudi-Akoyunoglou, Eds.: pp. 573–580. Elsevier/North-Holland Biomedical Press. Amsterdam, The Netherlands.
54. DRIESEL, A. J., E. J. CROUSE, K. GORDON, H. J. BOHNERT, R. G. HERRMANN, A. STEINMETZ, M. MUBUMBILA, M. KELLER, G. BURKHARD & J. H. WEIL. 1979. Fractionation and identification of spinach chloroplast transfer RNAs and mapping of their genes on the restriction map of chloroplast DNA. Gene **6:** 285–306.
55. DAVIDSON, J. N., M. R. HANSON & L. BOGORAD. 1974. An altered chloroplast ribosomal protein in *ery*-M1 mutants of *Chlamydomonas reinhardi.* Molec. gen. Genet. **132:**119–129.
56. CHUA, N.-H. & G. W. SCHMIDT. 1979. Transport of proteins into mitochondria and chloroplasts. J. Cell Biol. **81:**461–483.
57. NOMURA, M. & L. E. POST. 1980. Organization of ribosomal genes and regulation of their expression in *Escherichia coli. In* Ribosomes: Structure, Function, and Genetics. G. Chamblis, G. R. Craven, J. Davies, K. Davis, L. Kahan & M. Nomura, Eds.: pp. 671–691. University Park Press. Baltimore, Md.
58. NISHIMURA, S. 1978. Modified nucleosides and isoaccepting tRNA. *In* Transfer RNA. S. Altman, Ed.: pp. 168–195. The Massachusetts Institute of Technology Press. Cambridge, Mass.
59. KÖRNER, A. M., S. I. FEINSTEIN & S. ALTMAN. 1978. Transfer RNA-mediated suppression. *In* Transfer RNA. S. Altman, Ed.: pp. 105–135. M.I.T. Press. Cambridge, Mass.
60. ISONO, K. 1980. Genetics of ribosomal proteins and their modifying and processing enzymes in *Escherichia coli. In* Ribosomes: Structure, Function, and Genetics. G. Chamblis, G. R. Craven, J. Davies, K. Davis, L. Kahan & M. Nomura, Eds.: pp. 641–669. University Park Press. Baltimore, Md.
61. HERRMANN, R. G., H. K. PALTA & K. V. KOWALLIK. 1980. Chloroplast DNA from three archegoniates. Planta **148:**319–327.

DISCUSSION OF THE PAPER

DR. L. MARGULIS (*Boston University, Boston, Mass.*): I wonder if there is any correlation between the fact that mitrochondria are so unique in their molecular biology and the fact that they are very highly standardized wherever we find them? Chloroplasts vary an enormous amount in morphology, but mitochondria seem to be fairly standardized.

DR. N. W. GILLHAM: In a functional sense, that's true, but it's almost as if evolution ran rampant at the level of genomes and protein synthesizing systems. It is not at all clear to me.

DR. M. V. SIMPSON: Just to add one piece of data: tRNA itself is pretty unique in mitochondria. It won't help a bit in any decision as to the origin of these organelles, and probably will turn out to be that way in chloroplasts too.

DR. GILLHAM: I am very glad you brought that up because I meant to mention that UGA is a tryptophan codon in mitochondria and not a "stop" codon.

DR. S. D. YOUNG (*St. John's University, Jamaica, N.Y.*): On the question of whether there are organisms that can survive without mitochondria, of course, the petite yeasts survive without the mitochondria.

DR. GILLHAM: They only survive without certain mitochondrial functions, namely, the respiratory chain functions, not the TCA cycle functions.

DR. YOUNG: The other point I wanted to bring up is I am a little bit bothered by sponges as lower organisms because ultimately, the evolutionary history of these organisms is just as long as that of a mammal and you would expect the evolutionary rate to be about equal. So, we just don't see morphologically the evolution for these organisms that we do see for vertebrates.

MR. M. LISANTI: I would like to know whether mitochondria or chloroplasts undergo any differentiation, like that of the nucleus during embryogenesis or tissue formation, and have transplantations been done of different mitochondria from different tissues to see if they can survive in tissues other than where they originated?

Is it possible to transplate chloroplasts into an animal egg cell and have the chloroplasts divide and, therefore, perform their functions in an animal cell?

DR. GILLHAM: With respect to your second question, which concerns organelle transplants, that's been done very nicely for mitochondria in ciliates. In fact, I think Dr. Eberhard was referring to those experiments earlier when he was talking about mitochondrial competition. I don't really know about higher cells.

As far as chloroplast differentiation goes, of course, one knows that proplasts must differentiate at some point into chloroplasts and one knows that etioplasts also differentiate into chloroplasts.

MR. LISANTI: Is it possible to transplant chloroplasts into an animal egg cell and have the chloroplasts divide and perform their functions just as if the chloroplasts had at one time been pro-cells; just as the mitochondria and chloroplasts could have evolved as just cells and might have been engulfed, might be cells within cells.

DR. GILLHAM: My guess would be no. On a long-term basis, I think it would be a very difficult thing. Chloroplasts have just lost so many essential functions that they share with the nucleus—I mean on a really long-term basis.

DR. J. L. CALAGAN (*State University at Stony Brook, Stony Brook, N.Y.*): Our lab at Stonybrook is now sequencing chloroplast tRNAs from spinach. We only have somewhat preliminary results, but it looks like the methionine initiator will be fairly prokaryotic in nature, but it also doesn't look like that's necessarily universal. They don't seem to be quite as odd as the mitochondrial tRNA sequence.

DR. J. A. SCHIFF: I just wanted to return to the question about the elimination of organelles. There is one case I know of in amoeba where mitochondria can be eliminated, but they are replaced by a symbiotic bacterium that apparently carries out the function of respiration and maybe one of the reasons mitochondria usually are not deleted is that most of the organisms that have them arose after the appearance of oxygen and don't have alternative means of making energy anaerobically. They are not fermentative organisms or they don't carry out anaerobic respiration.

In general, the only organism I know of in which you can get a very, very great reduction of organelle function and the elimination of the organelle DNA is *Euglena* and the reason for this seems to be that *Euglena* has been wise enough not to put any essential functions in its chloroplasts other than those for photosynthesis.

All of the chloroplasts of the higher plants require their chloroplasts for nitrite reduction, for starch synthesis, and for several other things and sulphate reduction. Well, *Euglena* makes its starch outside the chloroplasts. It does not carry out nitrate reduction, which requires ammonia, and it has its sulphate reduction in the mitochondria instead of in the chloroplasts.

DR. GILLHAM: I might just add that a number of us have tried for years without success to make *Chlamydomonas* colorless.

DR. D. E. GREEN: One of the things that intrigued me is this question of size. Size by itself would be meaningless without a lot of chemistry because you could have carbohydrate. You could have any number of things there unless one actually did precise chemical work and said look here, this is entirely one type of structure down the line and size by itself wouldn't be very meaningful. I presume that this has been looked into and that when they say the DNA is a particular size that they have actually done the appropriate chemical work to rule out the presence of anything else. The frequency with which they are doing these experiments makes me a little bit dubious whether they have actually gone to such pains.

DR. GILLHAM: Yes. I think that the sizes that I have been citing are based mostly on such things as renaturation kinetics, restriction-enzyme analysis, and electron microscopy. Most mitochondrial DNAs are overall fairly homogeneous and I think the same is true of chloroplast DNAs.

It is a much more difficult experiment to pin down function. Size is something that you can just measure a little faster than you can look at totality of function, which is a much more difficult sort of problem, and that's where model systems become so important.

Size, as you say, could be meaningless, but, it is a handle that we can at least begin to maybe get some ideas from. For example, I stressed that it seems to me that *Acetabularia* is an awfully interesting system to spend some more time on.

DR. MARGULIS: I just would like to clarify one of these points about the presence or absence of mitochondria. There is a striking absence of mitochondria in entire families of various flagellates. There are many hundreds of examples, I would say. Some of them have replacements, like *Pelomyxa,* of other bacteria that look as though they were serving mitochondrial functions in some way. Others have absolutely no evidence of mitochondria at all and these tend to not have Golgi and have other characteristics that are interesting. There is not a single organism known, at least to me, that has chloroplasts and lacks mitochondria. That's where the gap is which suggests that it was a mitochondrial-type organism that came first.

DR. L. BOGORAD (*Harvard University, Cambridge, Mass.*): I should like to recall some work of Woodcock's that showed enormous amounts of DNA in some chloroplasts of *Acetabularia* and virtually none within the limits of detection in other chloroplasts of *Acetabularia.* So, it is indeed an interesting organism and I think emphasizes how little we know about the biology of chloroplasts.

DR. R. SAGER (*Harvard University, Cambridge, Mass.*): There are two things that I think about as to the size of the genome. One is that we are probably going to find out that it is very misleading because the yeast mitochondrial DNA is a beautiful example of this, as Dr. Gillham pointed out. Maybe a half or less than a half of it has any coding function and it is certainly possible that a lot of these genomes, even though they are all genuine DNA, and even though repeating AT, AT, AT is still DNA, it doesn't convey very much information.

The other point that I think should be kept in mind is the possibility that there is more organelle DNA than you think and that what's being looked at as circles in, say mitochondrial circles or chloroplast DNA circles, in *Chlamydomonas* for example, may be amplified copies and there may be a template which is larger and which hasn't been found. I hesitate to say it because there is no good evidence that this is the case.

DR. GREEN: Where would it be found?

DR. SAGER: In the organelle, but you see it is very hard to rule this out because most of the separations are done on the basis of base composition, and purifying organelles is a very messy business. So, it is an open question, but since you raised the question of size, I think one should still think about it.

INTRAORGANISM COMPETITION INVOLVING EUKARYOTIC ORGANELLES *

William G. Eberhard

Smithsonian Tropical Research Institute
and Escuela de Biología
Universidad de Costa Rica
Ciudad Universitaria, Costa Rica

One of the most basic and interesting questions in studies of living organisms is "Why does the organism have the traits it has rather than some others?" Answers to such questions ultimately depend on the two basic concepts of evolution—gradual change from ancestral forms (the history of the evolutionary line) and the relative reproductive advantages of alternate traits (the natural selection that has acted during the evolutionary history). Recent work on eukaryotic organelle evolution has emphasized the historical aspect and, because of the lack of hard data, more or less neglected the functional aspects. Natural selection arguments concerning organelle traits are in general difficult to prove or disprove because of difficulties in quantifying selective advantages. Nevertheless, certain aspects of organelle biology directly affect their reproduction in unequivocal ways. The basic aim of this paper is to point out that several such characteristics are known, and to show how their effects on organelle DNA (oDNA) and nuclear DNA (nDNA) reproduction would affect the evolution of organelles and the cells in which they are found. The main conclusion is cautionary with respect to the commonly held belief that organelle relationships within cells are entirely symbiotic: there is good reason to believe that, in at least some organisms, the relationships between organelles and between organelles and nuclei are partially competitive in nature.

The arguments I will make depend on a clear understanding of the concept of natural selection, and it is worthwhile to begin with a reminder of the basic, deceptively simple, idea behind the phrase "natural selection." In any group of objects that are relatively stable, that are not all identical, and that reproduce themselves, it is mathematically inevitable that those varieties better able to reproduce themselves will eventually predominate. Thus, if A produces more copies than A′ each generation, it will inexorably replace A′ (FIGURE 1).

Traditionally biologists have thought about the effects of natural selection acting on molecules of nDNA. Those varieties best able to code for organisms capable of reproducing themselves (that is, their nDNA) eventually predominate (for a dramatic presentation of this point, see Reference 1). Since oDNA meets the same basic criteria for the action of natural selection, it seems inevitable, and at first unremarkable, that natural selection will act on it also. It has apparently not been realized, however, that the partial uncoupling which exists between the reproduction of oDNA and nDNA implies that natural selection on the two types will not be identical. By uncoupling I mean the following.

* Supported by the Comité de Investigaciones of the Universidad del Valle, Cali, Colombia.

0077–8923/81/0361–0044 $01.75/0 © 1981, NYAS

In the critical reproductive processes of mitosis, meiosis, and gamete fusion, nDNA is copied and passed on in rigid and very precise ways (FIGURE 2). On the other hand, the patterns of transmission of oDNA during these processes are generally much less rigid and variable, even within a single species. The process of organelle reproduction within a cell (organelle fission) also differs from nDNA reproduction in not usually being synchronized with that of other organelles or of nDNA. The important point here is that the process of nDNA reproduction can occur without that of oDNA occurring and *vice versa.* Since natural selection focuses on reproduction this uncoupling can have some important consequences.

Like nDNA, oDNA varieties can compete reproductively by furthering the reproduction of the individual organism in which they are present. Nevertheless, if genetic diversity exists in either mitochondrial or chloroplast DNA within an organism, it can lead to reproductive competition involving direct, intraorganism interactions. Selection in such cases could favor oDNA varieties that promoted any of the following types of behavior: selective destruction of genetically different organelles or their DNA; selective inclusion in newly formed embryos and/or the exclusion of genetically different organelles; selective inclusion in

FIGURE 1. The basic idea of natural selection: in any population of variable, self-replicating, and relatively stable entities, those that produce more copies of themselves will eventually predominate.

gametes and/or the exclusion of genetically different organelles; and differential reproduction within the organism, either by more rapid organelle multiplication or by more rapid DNA replication within organelles.

The necessary condition for these competitive interactions is obviously intraorganism genetic heterogeneity among organelles of a given type. The most likely way this could come about is for a given organism to inherit oDNA from each parent. How common is this in nature? Surprisingly, at least occasional, transient biparental inheritance seems to be very common. Data from both electron microscopy and genetic studies indicate that it occurs in algae,[2–6] *Paramecium*,[7, 8] mold,[9] yeast,[9] a fern,[10] gymnosperms,[11–14] monocot and dicot angiosperms,[10, 15–18] echinoderms,[19–21] an annelid,[22] a trematode,[23] a mollusc,[24] and mammals.[25–27]

Thus the conditions under which oDNA–oDNA competition would be expected to evolve seem to be very common.

Another, at first apparently paradoxical, possibility is that oDNA competes with nDNA of the same organism. Although the reproduction of oDNA is obviously linked to that of the nDNA of the cells in which it is present, the partial uncoupling of their reproduction implies that there are limits to the synonomy of their reproductive interests and that natural selection on oDNA could favor characteristics not favored by selection on nDNA in the following situations:

1. In species with strongly maternal inheritance of oDNA, being in a male rather than a female would imply reproductive death for oDNA, so selection on such DNA would favor biasing sex ratios in favor of females. In hermaphroditic species it would favor at least partial elimination of male

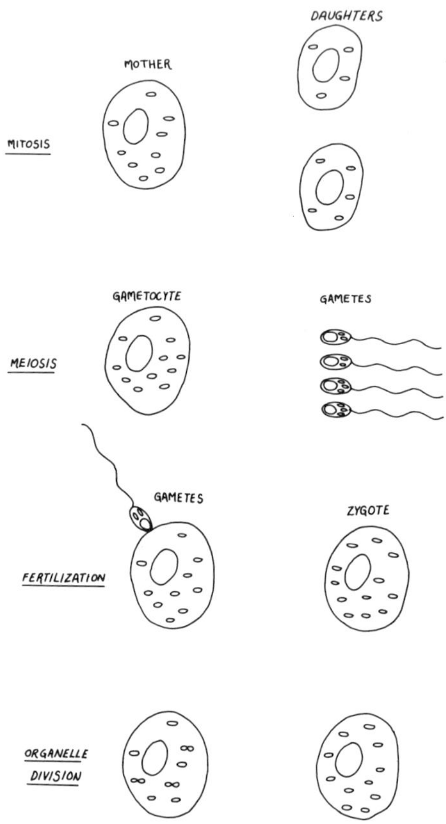

FIGURE 2. Typical patterns in the critical processes of propagation of nDNA and oDNA within eukaryotic organisms. The nuclear events are highly uniform, while those involving organelles are not.

reproduction. Selection on nDNA, on the other hand, generally favors equal investment in males and females.[28]

2. In species in which there is a premium on small size of male gametes, selection on nDNA would favor reduction in organelle numbers or even their complete elimination from male gametes. However, selection on oDNA would favor inclusion of organelles in gametes.

3. In species with strongly maternal inheritance of oDNA, selection on

oDNA in males would favor vegetative rather than sexual reproduction. The inverse would be true in species with strongly paternal inheritance. Selection on nDNA with respect to sexual versus asexual reproduction reflects other, independent factors.[29, 30]

4. In organisms with genetically heterogeneous mitochondria or chloroplasts, natural selection on all oDNA varieties would favor their own maintenance, whereas selection on nDNA might favor selective elimination of those oDNA varieties less efficient at the organism level.

5. In species in which heterotic vigor and/or recombinational advantage results from mixing oDNA varieties (e.g., from biparental inheritance of oDNA), selection on nDNA would favor incorporation of organelles from each gamete in the zygote. Selection on oDNA in each gamete would oppose the inclusion of other oDNA, at least when these advantages were outweighed by reduction in oDNA reproduction due to dilution.

6. In species in which mixing of oDNA reduces the fitness of the organism (due to incompatability between oDNA and nDNA, or competition between oDNA varieties), selection on oDNA in each gamete would favor its own inclusion in the zygote and the exclusion of the oDNA of the other, while selection on nDNA would favor inclusion of oDNA from one gamete only.

Conditions favorable to oDNA–nDNA competition are also common. Small gametes in one-sex, sexual reproduction, and strongly uniparental inheritance of oDNA are all well known.

I have now stipulated conditions under which competitive interactions might be expected to evolve, and have shown that these conditions are common in nature. Does such competition really occur in nature? If so, apparent examples of it should be found. I have found a number of such cases in the literature; there is not space to give a complete list, so a few examples will have to suffice (see Reference 31 for a more complete discussion of this and related points).

Selective elimination of chloroplasts derived from one of the parents has been seen in zygotes of the algae *Rhynchonema*,[6] *Spirogyra*, *Zygnema*,[2] *Hyalotheca*,[5] and *Ulva*.[3] The chloroplast DNA from one gamete, but not the rest of the chloroplast, is usually destroyed soon after fusion in the alga *Chlamydomonas*.[4]

The migration of egg mitochondria to cluster around the egg nucleus, thus favoring their inclusion in the "neocytoplasm" of the proembryon in the gymnosperms *Larix* and *Pseudotsuga*,[12] may result from competition for inclusion in the embryo. In at least *Larix* there is reason to believe that the parental organisms may be heteroplasmic for mitochondria.[12]

Differential reproduction of genetically different organelles seems to occur in *Paramecium*[7] and the angiosperm plant *Oenothera*.[32]

Cytoplasmic factors (mostly not yet associated with specific organelles) cause male sterility in many plant species (18 species in as many genera are listed by Grun[33]). Pollen sterility is the most commonly observed characteristic induced by cytoplasmic factors.[34] As predicted by the theory outlined here, species with cytoplasmic male sterility seem to have maternal inheritance of oDNA: 7 of the 18 species listed by Grun appear in the list of Tilney-Bassett[10] and all show strongly maternal inheritance; two congeneric species of an eighth are also listed and have maternal inheritance.

These possible examples are interesting, but they do not constitute proof that intraorganism competition occurs because each has other possible interpretations, and because the theory itself has, in common with many other evolutionary ideas concerning natural selection, the disadvantage of being able to explain any possible outcome—it seems unfalsifiable. However, it does make a number of specific predictions concerning mechanisms of inheritance of organelles. The last section of this paper describes these predictions and compares them with known data.

1. In apparent examples of oDNA–oDNA competition, products transcribed from oDNA should be involved in determining the outcomes much more frequently than the small fraction of oDNA genes in the organism's total genome would suggest.

2. Some DNA products are expected to function solely to suppress the effects of products of other kinds of DNA in the same organism (oDNA vs. oDNA, oDNA vs. nDNA) in situations in which the products' effects are in accord with the reproductive interests of their respective DNAs.

3. Mechanisms determining oDNA inheritance should tend to be complex rather than simple, and varied rather than uniform, due to the long history of coevolution of the genomes involved.

Although available data are scanty, they seem to be in accord with these predictions. oDNA and its products are directly involved in determining the outcomes of apparent oDNA–oDNA competitive interactions in at least four of the five best known organisms. In the alga *Chlamydomonas*, inhibition of oDNA transcription just prior to fertilization in the female gamete, which normally contributes most oDNA to the zygote, alters the pattern of oDNA inheritance in favor of the male.[35] In mitochondria of the yeast *Saccharomyces*, the mitochondrial DNA allele ω^+ causes the replacement of ω^- and sometimes adjacent mitochondrial genes when fusion occurs.[36] And in chloroplasts of the plants *Oenothera*[36] and *Pelargonium*,[10] crosses placing combinations of plastids in cells with different nuclei show that genetically different plastids have characteristic rates of reproduction independent (*Oenothera*) or partially independent (*Pelargonium*) of the nucleus. The relative ease of transfer of mitochondrial markers in *Paramecium* may also be influenced by oDNA, but the data are not conclusive.[37] Finally, indirect evidence from the electron microscopy study of fertilization in the sea urchin *Paracentrotus* implicates egg mitochondria in the elimination of sperm mitochondria (they surround it when it enters the interior of the egg, and it is then digested away while they remain intact).[21]

Antagonistic effects of oDNA products may occur in *Chlamydomonas*. Inhibition of female oDNA transcription results in predominantly male, rather than 50–50, transmission of chloroplast markers[4] suggesting that male gamete products, when not suppressed by female oDNA products, act to favor transmission of male chloroplast DNA. Antagonism between oDNA and nDNA products may occur in a number of normally hermaphroditic plant species in which nuclear genes ("restorers") inhibit the production of male sterility by cytoplasmic factors.[38, 34]

Finally, the mechanisms controlling organelle inheritance are complex in all of the five best studied organisms (*Chlamydomonas, Saccharomyces, Oenothera, Pelargonium,* and *Paramecium*), and vary substantially in detail. None is understood completely.

I will end by noting that although the ideas I have presented are obviously related to the endosymbiotic theory of organelle origins, they are not logically dependent on it. As soon as pieces of DNA begin reproducing regularly in the cytoplasm as entities separate from the nucleus, natural selection will begin to act on them in the ways I have described, whether they originally came from separate organisms, from pieces of nDNA, or from anywhere else.

The emphasis in this paper on possible reproductive competition between subcellular components is similar to the recent general reinterpretation of the roles of individuals in animal societies. Major advances in the understanding of a variety of kinds of animal behavior have resulted from the realization that natural selection does not always favor traits that benefit the good of the group, and that it also favors selfish behavior by group members which promote their own reproductive interests. This realization seems equally applicable to the "societies" of distinctive, reproductively semi-independent DNA molecules in eukaryotic cells, and may also be important in understanding why the cells are built and function as they do.

Acknowledgments

I thank the following for various kinds of help: E. C. Charnov, J. Coddington, M. J. W. Eberhard, F. García, C. Hahn, S. Hunt, E. G. Leigh, A. Levin, and H. K. Srivastava. Charnov had independently derived some of the ideas, especially those regarding sex ratio distortion.

References

1. Dawkins, R. 1976. The Selfish Gene. Oxford University Press. Oxford, England.
2. Granick, S. 1961. The chloroplasts: inheritance, structure, and function. *In* The Cell. J. Brachet & A. Mirsky, Eds. **1:**489–602. Academic Press, Inc. New York, N.Y.
3. Bråten, T. 1973. Autoradiographic evidence for the rapid disintegration of one chloroplast in the zygote of the green alga *Ulva mutabilis*. J. Cell Sci. **12:** 385–389.
4. Sager, R. 1977. Genetic analysis of chloroplast DNA. Adv. Genet. **19:**287–337.
5. Potthoff, H. 1927. Beitrage zur Kenntnis der Conjugaten. I. Untersuchungen uber die desmidiacee *Hyalotheca dissiliens* Breb. Forma Minor. Planta **4:**261–283.
6. Chmielevsky, O. 1890. Eine Notiz uber das Verhalten der Chlorophyllbander in den Zygoten der Spirogyraarten. Botan. Ztg. **48:**773.
7. Adoutte, A. & J. Beisson. 1972. Evolution of mixed populations of genetically different mitochondria in *Paramecium aurelia*. Nature **235:**393–396.
8. Beale, G. H., J. K. C. Knowles & A. Tait. 1972. Mitochondrial genetics in Paramecium. Nature **235:**396–397.
9. Birky, C. W. 1975. Mitochondrial genetics in fungi and ciliates. *In* Genetics and Biogenesis of Mitochondria and Chloroplasts. C. W. Birky, P. S. Perlman & T. J. Byers, Eds. : 182–224. Ohio State University Press. Columbus, Ohio.
10. Tilney-Bassett, R. A. E. 1975. Genetics of variegated plants. *In* Genetics and Biogenesis of Mitochondria and Chloroplasts. C. W. Birky, P. S. Perlman & T. J. Byers, Eds.: 268–308. Ohio State University Press. Columbus, Ohio.
11. Mangenot, G. 1938. Sur les oospheres, les tubes polliniques et la fécondation chez le Pin maritime. Compt. Rend. Acad. Sci. **206:**364–366.
12. Chesnoy, L. & M. J. Thomas. 1971. Electron microscopy studies of gametogenesis and fertilization in gymnosperms. Phytomorphology **21:**50–63.
13. Chesnoy, L. 1973. Sur l'origine paternelle des organites du proembryon du *Chamaecyparis lawsoniana* A. Murr. (Cupressacées). Carylogia **25:**223–232.

14. Ohba, K., M. Iwakawa, Y. Ohada & M. Murai. 1971. Paternal transmission of a plastid anomaly in some reciprocal crosses of Suzi, *Cryptomeria japonica* D. Don. Silvae Genet. **20:**101–107.
15. Semenuik, P. 1976. Transmission of plastids in *Browallia*. J. Hered. **67:**61–62.
16. Rao, M. K. & P. R. K. Koduru 1978. Biparental plastid inheritance in *Pennisetum americanum*. J. Hered. **69:**327–330.
17. Anton-Lamprecht, I. 1967. Elektronenmikroskopische Untersuchungen an Plasmonavanderungen von *Epilobium*-Bastarden IV. Mischzellen mit Normalen und atypischen Chondriosomen. Z. Naturf. **22**b:1340–1342.
18. Fleming, A. A. 1975. Effects of male cytoplasm on inheritance in hybrid maize. Crop. Sci. **15:**570–573.
19. Franklin, L. E. 1965. Morphology of gamete membrane fusion and of sperm entry into oocytes of the sea urchin. J. Cell Biol. **25:**81–100.
20. Longo, F. J. & E. Anderson. 1968. The fine structure of pronuclear development and fusion in the sea urchin, *Arbacia punctata*. J. Cell Biol. **39:**339–368.
21. Anderson, W. A. 1968. Structure and fate of paternal mitochondrion during early embryogenesis of *Paracentrotus lividus*. J. Ultrastruct. Res. **24:**311–321.
22. Colwin, A. L. & L. H. Colwin. 1961. Changes in the spermatozoan during fertilization in *Hydroides hexagonus* (Annelida). II. Incorporation with the egg. J. Biophys. Biochem. Cytol. **10:**255–274.
23. Burton, P. R. 1967. Fine structure of the reproductive system of a frog lung fluke. II. Penetration of the ovum by a spermatozoan. J. Parasitol. **53:**994–999.
24. Pasteels, J. J. 1965. Aspects structuraux de la fécondation vue au microscope électronique. Archs. Biol., Liége **76:**463–509.
25. Szollosi, D. G. 1965. The fate of sperm middle-piece mitochondria in the rat egg. J. Exp. Zool. **159:**367–378.
26. Presley, R. 1969. Some electron microscope observations on the fate of the sperm tail in rat and mouse ova. J. Anat. 104:584–585.
27. Swift, H. 1968. Addendum. In Biochemical Aspects of the Biogenesis of Mitochondria. E. C. Slater, J. M. Trager, S. Papa & E. Quagliariello, Eds.: 332. Adriatica Editrice, Bari, Italy.
28. Fisher, R. A. 1930. The Genetical Theory of Natural Selection. [Dover Publications, Inc. 1958 Reprint]. New York, N.Y.
29. Williams, G. C. 1975. Sex and Evolution. Princeton University Press, Princeton, N.J.
30. Maynard Smith, J. 1978. The Evolution of Sex. Cambridge University Press. London, England.
31. Eberhard, W. G. Evolutionary consequences of intracellular organelle competition. Quart. Rev. Biol. (In press.)
32. Schötz, F. 1968. Uber Plastidenkonkurrenz bei Oenothera. II. Biol. Zbl. **87:** 33–61.
33. Grun, P. 1976. Cytoplasmic Genetics and Evolution. Columbia University Press. New York, N.Y.
34. Watson, G. S. & E. Caspari. 1960. The behavior of cytoplasmic pollen sterility in populations. Evolution **14:**56–63.
35. Sager, R. & Z. Ramanis. 1973. The mechanism of maternal inheritance in *Chlamydomonas*: Biochemical and genetic studies. Theor. Appl. Genet. **43:** 101–108.
36. Dujon, B., P. P. Slonimski & L. Weill. 1974. Mitochondrial genetics. IX. A model for recombination and segregation of mitochondrial genomes in *Saccharomyces cerevisiae*. Genetics **78:**415–437.
37. Beisson, J., A. Sainsard, A. Adoutte, G. H. Beale, J. Knowles & A. Tait. 1974. Genetic control of mitochondria in *Paramecium*. Genetics **78:**403–413.
38. Lewis, D. 1941. Male sterility in natural populations of hermaphrodite plants. The equilibrium between females and hermaphrodites to be expected with different types of inheritance. New Phytol. **40:**56–63.

DISCUSSION OF THE PAPER

DR. D. E. GREEN (*Unversity of Wisconsin, Madison, Wisc.*): Were the genomes of the mitochondria from the male parent and the female parent demonstrated to be different?

DR. W. G. EBERHARD: Obviously, the answer is yes. In all of the cases the evidence that I am citing is genetic. There are a number of genetic studies, in *Paramecium* and in a number of species of plants.

DR. M. V. SIMPSON (*State University at Stony Brook, Stony Brook, N.Y.*): I would like to try to answer that question for mammals, but we only have a tentative answer. We discovered, some years ago, that in the laboratory rat population there were two kinds of DNA, although any individual seems to have just one type.

Now, that provides a system for showing maternal inheritance by reciprocal crosses and it also provides a marker for looking for male parental DNA in the offspring. We developed some rather sensitive methods, based on restriction-enzyme analysis and gel analysis, of looking for male parental DNA. We are not finished yet. I can give you the answer, in terms of one in five thousand, that there has to be less than one molecule of male parental DNA and five thousand of female in the offspring. In other words, that's the limit of our sensitivity. We think that in a short time we will be able to go up to fifty thousand. Tentatively, right now, we don't see it. We cannot find the male parental DNA.

DR. EBERHARD: What is the ratio of mitochondrial contribution?

DR. SIMPSON: That's an important question and we don't know with certainty yet. We are beginning studies on that. We think approximately one to eight hundred, but we'll bring our analysis up, if we can, to about one in fifty thousand. Dilution of course, is a factor.

I also wanted to ask you since you mentioned that there was evidence in mammals for two kinds of mitochondrial genes but didn't say mitochondrial DNA: What is this evidence?

DR. EBERHARD: What I am talking about is biparental inheritance, even transient biparental inheritance, and the critical moment in mammals is at the moment of fertilization. There are studies showing that the male mitochondria go into the egg and then disappear from sight.

DR. S. D. YOUNG (*St. John's University, Jamaica, N.Y.*): When you describe this, you are talking about competition. In some of the work with yeast there are indications that there is actual gene transfer between mitochondrial DNAs so that you may have more of an increase in the genome and a synthesis between the systems where this really would reduce the competition. You are just increasing the gene pool. Would you comment on this?

DR. EBERHARD: Agreed. That kind of phenomenon reduces the intensity of competition. But if there is, as has been shown, actual destruction of some alleles caused by other alleles, that seems to be out-and-out competition because the destroyed alleles will not be able to pass on to the next generation.

DR. YOUNG: But you don't know what the reasons for this destruction are.

DR. EBERHARD: I am proposing that the reasons for these varied phenomena may well be basic competition, reproductive competition, which derives from natural selection.

DR. YOUNG: But you don't have any way of saying what genes. Well, in some cases you said it may be from one of the parental strains that is preferentially selected. In other cases, it may not be. It may be something that happens after the fusion of the cytoplasm and so it may be a synthesis at that time.

DR. EBERHARD: So long as both genomes remain intact, then they've both won the competition or both survived, but if one is causing the destruction or elimination of the other, that is an apparent example of a competitive interaction and a reason to believe that natural selection favors that kind of competitive interaction.

MITOCHONDRIAL EVOLUTION: ORGANIZATION AND REGULATION OF MITOCHONDRIAL GENES *

Henry R. Mahler

Department of Chemistry *
and
Molecular, Cellular and Developmental Biology Program
Indiana University
Bloomington, Indiana 47405

INTRODUCTION

Mitochondria are essential organelles, found in all respiration-competent cells of contemporary eukaryotes. The structure and function of mitochondria appear to have remained virtually invariant throughout evolution. Investigations during the past decade have led to the unambiguous demonstration of a mitochondrial system for the maintenance, replication, and expression of a resident, separate, and distinctive genophore.[1-7] These considerations, together with some apparently prokaryotic features of this auxiliary genetic system (particularly in its translational apparatus), have provided the basis for the resuscitation of one of the two favored current hypotheses concerning origin—and initial evolution—of the organelle—the *endosymbiotic* hypothesis[8-14] (called the *exogenous* theory by F. J. R. Taylor).[15] The alternative is some version of a *developmental*[8, 16-18] (or *autogenous* in Taylor's terminology[15]) theory. One formulation of the latter, previously advanced and discussed at some length by Raff and myself[19, 20] might be described as a *plasmid-sequestration* hypothesis.[21, 22]

It is not my purpose in this presentation to compare these two theories and expound on their relative merits. Instead, I endeavor to describe a number of features disclosed by recent studies on mitochondria[1, 6, 7, 23-27] from a number of different species, mainly of fungi, protozoa, and animals, that suggest two rather surprising inferences: 1) In spite of a number of strong constraints resulting in the conservation of mitochondrial structure and function, the mitochondrial genetic system exhibits unmistakable signs of great inter- and intraspecies diversity. 2) These divergent aspects are placed within a context of the nature, organization, expression, and regulation of the mitochondrial genome that suggests that this system is unique and that its features are distinct from both its prokaryotic and eukaryotic counterparts.

SEQUENCE DIVERGENCE IN MITOCHONDRIAL DNA

Variability of Size and Base Composition

These parameters have been tabulated in a number of recent reviews and discussed in the present context by Mahler and Perlman.[24] The data presented

* Publication No. 3372. Supported by the Institute of General Medical Sciences (Research Grant GM 12228) of the National Institutes of Health. The author holds a Research Career Award (KO6 05060) from this Institute.

0077-8923/81/0361-0053 $01.75/0

there (TABLE 3) show that the size of mitochondrial (mt) DNA varies widely between kingdoms and families, with mtDNAs of metaphyta being the largest [74 megadaltons (Md) or 110,000 base pairs (110 kbp)] and that of metazoan animals among the smallest [10–12 Md (by hydrodynamic measurement) or 15–19 kbp (from restriction-endonuclease analysis)]. Their base composition spans the range between $\leq 20\%$ (G+C) (or a bouyant density of 1.683) in yeast, to $\geq 60\%$ (G+C) (bouyant density of 1.710) in several animal species. Even within some closely related species, such as, *Drosophila melanogaster, D. simulans,* and *D. virilis* the density of mtDNA has not only been reported to range all the way from 1.711 to 1.680, but also to exhibit considerable variability in size, with the first two $\geq 20\%$ greater in length than the third (≥ 12 Md or ≥ 18 kbp vs. 9.90 Md or 14.9 kbp).[33, 34] Similarly the mtDNA[24, 35, 36] for the closely related yeasts *Saccharomyces cerevisiae* and *S. carlsbergensis* vary in length from 68.0 to 78.3 kbp. The mtDNA of different strains of even *S. cerevisiae* range from 70.8 to 78.3 kbp, while the mtDNA of different strains of *Tetrahymena* vary from 28.2 and 32 kbp.[37]

Insertions and Deletions

Such gross variability might be expected to be caused by the insertion (or deletion) of sizable segments of DNA, and this has in fact turned out to be the case. In the case of *Saccharomyces,*[24, 35] Borst and his group have demonstrated the occurrence of both small (25–50 bp) and large (900–3,000 bp) inserts.[36] The latter have been localized precisely and shown to be positioned within known structural genes[24, 28, 29, 38] (FIGURE 1): insert VI (1,050 bp) closely linked to, and probably including, the genetic polarity locus ω, is part of the gene for the larger (21 S) ribosomal RNA; inserts III (3,000 bp)† and IX (25 bp), are within the gene for apocytochrome *b*; and inserts I (900 bp), II (2,600 bp), and IV (1,500 bp) are all within the *oxi3* gene.

In different strains of *Tetrahymena* there is evidence for the insertion of a subterminal inverted repeat as well as a terminal palindrome 12,700 bp in length.[37, 39] The profound differences in size and base composition of the mtDNA in different species of *Drosophila* are also due to site-specific insertions, consisting of a long (3 kbp), A+T-rich region inserted into a much shorter similar segment present in *D. virilis.*[33, 34]

Sequence Divergence from Restriction-Endonuclease Maps

Considerable additional information concerning sequence homologies (or divergences) is provided by restriction maps. Detailed studies of this type are, or are becoming, available for a large number of species,[23] varying from yeasts[35, 36] to primates[44] (including man[30, 45]).

Mutation frequencies for mtDNA in different species have been estimated by Shah and Langley[33] for *Drosophila,* by Upholt and Dawid[41] for individual sheep and goats, and by Brown *et al.*[44] for primates (including man) (16,000 bp). The latter two report rates (per nucleotide) of $\sim 10^{-2}$ and 2.0×10^{-2}

† Insert III is now known[29] to consist of two pieces, IIIa (1,900 bp) and IIIB (1,400 bp), separated by a conserved segment $\sim$100 bp in length.

substitutions per 10^6 years. The corresponding parameter for nuclear DNA is reported by Upholt and Dawid[41] (using protein data) to equal $1\text{–}3 \times 10^{-3}$, while Brown *et al.*,[44] using hybridization experiments, arrive at values of 2×10^{-3}. Sequence divergences in stable RNA transcripts, such as rRNA and tRNAs, have been examined in a number of species and compared with their nuclear equivalents.[46, 47] The general consensus appears to be that divergences have been more rapid in the mitochondrial cases.

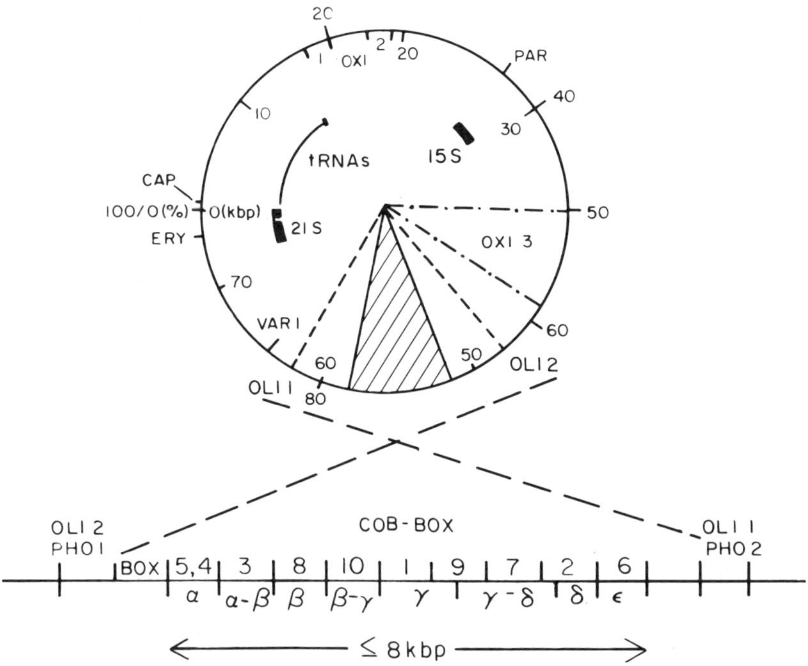

FIGURE 1. The genetic and physical map of *Saccharomyces cerevisiae* strains ID41-6/161 and KL 14-4A used in the author's laboratory. The fine structure map of the cytochrome *b* (*cob-box*) region is shown as well (not to scale). For references see text.

Conservative Features

However, in spite of these indications of heterogeneity—due probably to the maternal mode of inheritance of animal mtDNA[40, 43]—gene order and topography appear to have remained fixed for at least 350×10^6 years. This conclusion by Brown, George, and Wilson[44] is based on the demonstration of the invariance of a number of processes and structures in a variety of species, the evolutionary divergence of which can be dated over this time span. These species include humans,[30, 45] mice,[42] goats,[41] toads,[43] and probably even fruit flies,[40, 48] which would extend the period of mitochondrial constancy to at least 500×10^6 years. The invariant features include the position of the origin and the mode of replication of mtDNA; the topography, location (relative to the origin), and size of the genes for ribosomal RNA; the direction and selection of transcription of these and other genes which—except for a few small genes specifying tRNAs—always involves the same strand and a direction defined by smaller rRNA → larger rRNA. The transcribed strand is the pyrimidine-rich heavy strand and its transcripts can be hybidized only to the complementary light strand of the same, or different, species. In addition, it may be appropriate at this point to emphasize a similar invariance—with one or two notable exceptions involving ribosomal proteins and ATPase subunits (see below)—of the polypeptide products encoded in mitochondrial DNA.[19, 27, 45, 49, 50]

Unusual Features of the Genetic Code

Genes Examined

A completely unambiguous assignment of a nucleotide sequence to the specification of a polypeptide gene product requires the precise definition of the polynucleotide responsible by genetic and physical means, and the primary sequence information for both the DNA (or the stable messenger RNA furnishing the actual sequence translated) and the polypeptide. In the strictest sense, these criteria have been met in only one instance[52, 53]—the mitochondrial gene designated *oli1* (*pho2*) in *S. cerevisiae* responsible for the specification of subunit 9. Subunit 9 is a small, extremely hydrophobic, polypeptide (frequently referred to as a proteolipid) of the mitochondrial ATP synthetase[25] (oligomycin-sensitive ATPase). However, application of somewhat less stringent criteria has led to the identification of at least two additional DNA sequences for known polypeptides. The first is provided by the gene *oxi1*, the structural gene for the second largest subunit (subunit II, Cox II or CO II) of cytochrome oxidase in *S. cerevisiae*[54, 55] and the gene fulfilling the analogous role in human mitochondrial DNA.[30] In neither case has the protein been sequenced. However, in the case of the yeast protein, the physical and genetic localization of the segment of the mitochondrial genome involved in its specification is known with great precision,[25, 26, 29, 54, 55, 57] and the amino acid sequence deduced from it bears sufficient homology (81/226 residues) to the known sequence of this protein from bovine mitochondria,[56] to confer considerable confidence to the assignment. In spite of the complete absence of any genetic data, this degree of confidence is even higher in the correlation between the human DNA sequence and the bovine amino acid sequence because of the very extensive homology (164/226) between the two. For the second case, the protein specified by the mitochondrial *oli2* (*pho1*) gene of *S. cerevisiae*,[26] there is considerable genetic and DNA sequence information but no clues concerning amino acid sequence for this subunit of the mitochondrial ATPase. However, in spite of this relative paucity of correlative data, sufficient information is already in hand as to permit some rather startling inferences.

UGA Is Not A Terminator

All other systems examined so far, whether prokaryotic or eukaryotic, make use of three nonsense, or terminator, codons:[58] UAA (ochre), UAG (amber), and UGA (opal). In mitochondrial genes UGA is not a terminator, but instead provides a second codon for tryptophan.[26, 30, 54, 55] In fact among all the codons known to be utilized in mitochondrial genes for this amino acid (FIGURE 2) UGA is greatly favored over UGG (anticodon CCA on tRNA), its alternate used with high efficiency everyhere else.[58–60] The expected novel $tRNA^{Trp}$ with an UCA anticodon, capable of wobble-pairing with *both* tryptophan codons, has been identified (F. Sanger & A. R. Cookson, quoted in Barrell *et al.*[30]).

As pointed out by Barrell *et al.*,[30] this change in terminator patterns probably requires that mitochondria utilize a termination factor analogous to the *E. coli* factor R1 (which recognizes only UAA and UAG) rather than the promiscuous cytoplasmic termination factor.[61] Both terminators appear to be

	U				C				A				G				
		a	b	c		a	b	c		a	b	c		a	b	c	
U	Phe	24	8	3	Ser	8	4	1	Tyr	24	14	1	Cys	6	4	1	U
	Phe	16	4	7	Ser	0	0	4	Tyr	2	0	8	Cys	0	0	2	C
	Leu	83	27	4	Ser	23	7	4	UAA	2	1	0	Trp	?	5	3	A
	Leu	1	2	1	Ser	0	0	0	UAG		0	1	Trp	?	0	1	G
C	Leu	2	0	7	Pro	12	7	1	His	9	5	2	Arg	6	0	2	U
	Leu	0	0	6	Pro	0	0	9	His	0	0	4	Arg	0	0	1	C
	Leu	4 (△)	0	12	Pro	11	6	3	Gln	14	8	6	Arg	0	0	3	A
	Leu	0	0	3	Pro	0	0	2	Gln	1	1	1	Arg	0	0	0	G
A	Ile	61	24	7	Thr	11	6	5	Asn	23	8	2	Ser	6	2	1	U
	Ile	9	2	15	Thr	0	0	8	Asn	1	0	5	Ser	0	0	0	C
	Ile	6	3	8 (□)	Thr	20	10	8	Lys	14	7	4	Arg	10	5	0	A
	Met	18	5	2	Thr	0	0	0	Lys	2	1	0	Arg	0	0	0	G
G	Val	16	11	1	Ala	25	6	4	Asp	18	14	3	Gly	29	7	2	U
	Val	1	1	7	Ala	21	0	5	Asp	1	0	8	Gly	0	0	4	C
	Val	18	7	5	Ala	13	9	4	Glu	18	15	8	Gly	7	3	3	A
	Val	0	0	0	Ala	0	0	1	Glu	2	1	3	Gly	2	2	2	G

a = total, yeast △ = Thr
b = Co II, yeast
c = Co II, human □ = Met

FIGURE 2. The mitochondrial genetic code. For references see text. In addition to the symbols defined in the figure, the ovals indicate unutilized (open) and underutilized (shaded) triplets.

in use in mitochondria—UAA in the two yeast genes and UAG in the human gene.

Idiosyncratic Codons for Met in Humans and Thr in Yeast

In addition to these novel aspects of codon recognition shared by different mitochondrial genes, there also appear to be more individualistic substitutions in particular systems: AUA (anticodon UAU) has been suggested to code for methionine rather than isoleucine in human (but not in yeast) mitochondria; and CUA (anticodon UAG) is known to code for threonine, rather than leucine, in subunit 9 of yeast (but not in human) mitochondria. Yeast mitochondria contain two isoaccepting tRNAs for threonine.[62] The $tRNA_1^{Thr}$ responsible for this latter change has been sequenced and exhibits a most unusual structure[63] (FIGURE 3), containing *eight* rather than the customary seven nucleotides in its anticodon loop. This places a sequence capable of recognizing either the nonsense triplet UUA or UAG (a common leucine codon) into the central, mRNA-complementary position of the tRNA.

Biased Codon Assignments

In addition to these specific divergences from the universal code, the distribution of mitochondrial codons among the set of synonyms available for any given amino acid appears to exhibit considerable bias. This became first apparent to, and was commented on, by Tzagoloff and his collaborators,[26] who showed that among all the codons utilized in three yeast genes there was discrimination against G and C. The frequency of codons with G or C in the third position was only 10% and that of certain codons, such as the CGN triplet for arginine, appeared to be nil. It was recognized that yeast mitochondria, "limit organisms"[64] with 80% A+T in their mtDNA, might represent a special case, and the situation might not be as extreme in other species. However, examination of FIGURE 2 shows the persistence of at least some of these discriminatory aspects. G is consistently discriminated against in the third position in mitochondrial codons, as it is in viral codons.[65] These observations also suggest that the minimal number of isoaccepting tRNAs in mitochondria may be reduced by at least six below those required in other systems, to a number less than that fixed by the wobble rules (a total of 32).[30, 58] Twenty-five different tRNA molecules have already been claimed to be present in yeast, *Neurospora* and *Xenopus* mitochondria.[66-69] One of these, a RNA^{Lys} (anticodon CUU) in yeast may be of cytoplasmic origin.[70] This latter claim appears reasonable, since AAG is not a triplet utilized in any of the known yeast coding sequences (FIGURE 2).

These findings lead to some interesting evolutionary implications and genetic consequences. It is again evident that the mitochondrial genome has efficiently accommodated itself to a variety of constraints and functions in specifying the requisite number of essential, functionally related polypeptides; regardless of whether the organism in question is a relatively primitive unicellular fungus or the most highly evolved metazoan animal, whether the genome is large (75 kbp) or small (15 kbp), and whether it contains either 80% or 40% A+T.

The predicted genetic consequences—actually observed in the case of the *cob-box* region in yeast mitochondria, which is responsible for the specification of apocytochrome *b* (see below)—are the following: Mutational alterations producing a nontranslatable codon (FIGURE 2) should be relatively frequent—and provided ambiguous recognition by a noncognate tRNA (i.e., by wobble) is inefficient—ought to result in premature chain termination. In fact, point mutants accumulating *b*-fragment polypeptides are extremely prevalent in this region and many of them also produce, in reduced amounts, a polypeptide of equal mobility and closely homologous to the wild-type product, almost certainly due to ambiguous read-through. For instance, among the 15 chain-termination mutants studied by us, three exhibited this particular phenotype.

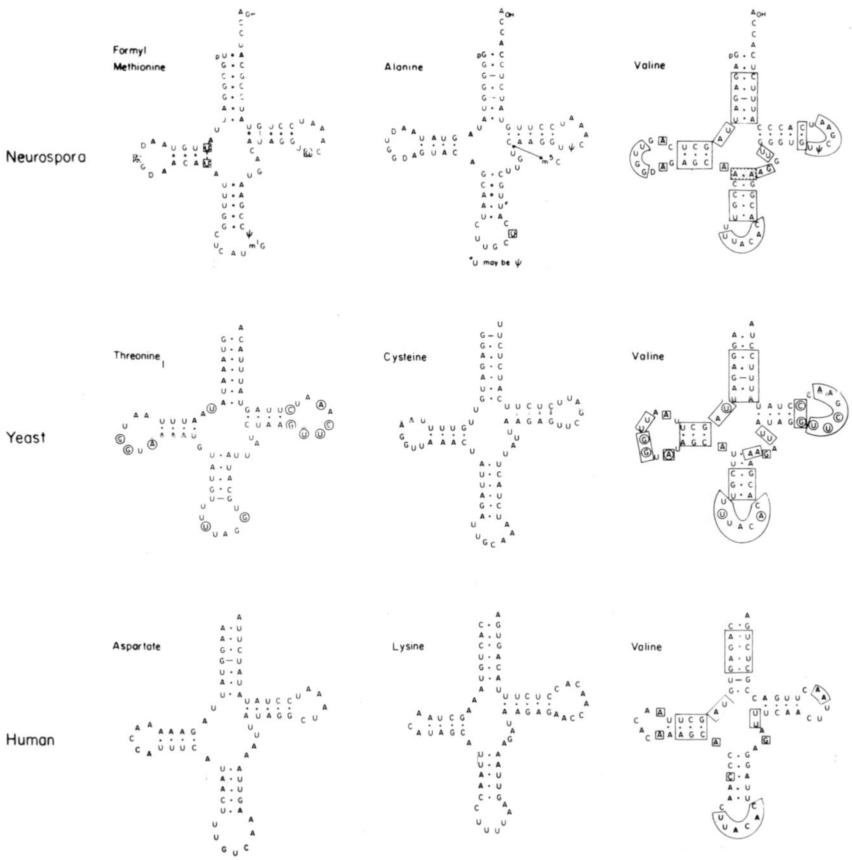

FIGURE 3. Mitochondrial transfer RNAs. *Top row: Neurospora crassa*–structures derived by Heckman, RajBhandary and collaborators (From Heckman *et al.*[73] By permission of *Cell.*). *Center row: S. cerevisiae*–threonine and valine (From Li and Tzagoloff.[63] By permission of *Cell.*) and cysteine (From Bos *et al.*[79] By permission of *Nucleic Acid Research.*). *Bottom* Human placenta–aspartate and lysine adapted (by substitution of U for T) from Barrell *et al.* (From Barrell *et al.*[30] By permission of *Nature.*), and valine (From Eperon *et al.* Private communication, to be submitted.). Valine tRNAs (third column) are compared with one another using the molecule from yeast as reference. Sequences also found in either of the other two are boxed. Circles indicate invariant residues.[63]

OTHER UNIQUE OR UNUSUAL PROPERTIES OF MITOCHONDRIAL GENES AND THEIR PRODUCTS

Initiator tRNA

Nascent mitochondrial polypeptide chains carry N-terminal formyl methionine [71] and are initiated [1, 19, 27, 31, 72] by fMet $tRNA_F^{Met}$. The structure of this initiator tRNA from *Neurospora* mitochondria has been investigated by Heckman *et al.*[73] and is shown in FIGURE 3. As pointed out by these investigators, this molecule exhibits several unusual features that set it apart from both its prokaryotic and eukaryotic counterparts. Unlike the prokaryotic type, its 5′-terminal nucleotide can form a base pair to the fifth nucleotide from the 3′-end (as shown in the figure), a property shared by initiators from eukaryotes and from the archaebacterium *Halobacterium cutirubrum*. The anticodon loop, however, contains all the unique features required for it to assume its functionally useful conformation.[74]

Other tRNAs–Isoaccepting Species

We have already directed attention to the problem of the number of tRNA species present in mitochondria. This question is actually composed of four separate items: 1) the total number of tRNAs; 2) the fraction of these actually encoded by mtDNA; 3) the number of mitochondrial genes responsible for the specification of the authentic mitochondrial tRNAs; and 4) the number of isoaccepting species for any given amino acid. Here we will expand on the third and fourth of these items in *Saccharomyces*.[62, 68] The existence of isoaccepting species in these mitochondria had already been suggested by separation and hybridization studies.[60, 67] In addition to the two tRNAs for methionine (the initiator $tRNA_F^{Met}$ (see above) and $tRNA_M^{Met}$), two separate isoaccepting species, mapping in different portions (FIGURE 1) of the mitochondrial genome, have recently been identified[75-77] for both $tRNA^{Ser}$ and $tRNA^{Thr}$. Sequence information for them is becoming available.[75, 77] These studies have raised the number of well-authenticated, discrete tRNA genes to 23. Of the isoaccepting species, $tRNA_2^{Ser}$ has been localized in the vicinity of *oli1*, while $tRNA_1^{Ser}$ maps within the cluster of the bulk of the mitochondrial tRNAs, where it is closely linked to $tRNA^{Arg}$. $tRNA_2^{Thr}$ is located at the beginning of the tRNA region, close to *cap*, while $tRNA_1^{Thr}$ maps between *oxi1* and *oxi2*, at a position linked to $tRNA^{Val}$. The unusual codon (AUA) recognized by $tRNA_1^{Thr}$ has already been discussed. Confirmation is provided by the lack of recognition of this tRNA for any of the four Thr codons *in vitro*.

Structures

The primary sequences of some typical mitochondrial tRNAs from *N. crassa, S. cerevisiae*, and humans are shown in FIGURE 3. Since only the first set is based on RNA rather than DNA sequencing studies, the modified bases are as yet unknown in the latter two instances. They also lack the posttranscriptionally added sequence CCA at the 3′-end. In yeast, a minimum of nine tRNA sequences, localized in different segments of the mitochondrial genome (FIGURE 1) have now been determined: two ($tRNA_1^{Thr}$ and $tRNA^{Val}$) by Lin and Tzagoloff,[63] two ($tRNA^{Cys}$ and $tRNA^{His}$) by Bos *et al.*,[79] and six (the same $tRNA^{Val}$, $tRNA_1^{Ser}$, $tRNA_2^{Ser}$, $tRNA^{Phe}$, $tRNA^{Arg}$, and $tRNA^{Gly}$) by Martin.[77, 78] Some of the homologous and conserved bases and sequences in the examples used are indicated in the figure.

Ribosomal RNAs–Size, Base Composition, and Location in the Genome

Another indication of unique properties and of evolutionary descent and development come from a consideration of mitochondrial ribosomes and especially their RNAs (rRNA).[1, 19, 50, 51, 80] The large rRNA has been reduced from 1.30×10^6 daltons (1.30 Md) or 4,060 nucleotides in certain fungi or higher plants, to 0.48 Md (1,500 nucleotides) in *Drosophila melanogaster,* coincident with a reduction in size of the small subunit from 0.70 Md (2,100 nucleotides) to 0.28 Md (860 nucleotides). The base composition of these RNA varies between a maximum of 54% (G+C) in higher plants to a minimum of 19.5% (G+C) in *Drosophila.*

Even more striking are alternative solutions to the disposition and fine structure of the coding sequences for these RNAs found in different mitochondrial DNAs.[24–31, 39–43] In all animals (vertebrate and invertebrate) examined so far, the sequences for the two subunits are adjacent and separated by only a short spacer region 100–200 nucleotides long. They are always transcribed from the same (heavy or H) strand in the same direction (small → large), coordinately, as part of the same transcriptional unit. In several instances the spacer segment codes for a 4S RNA, probably a tRNA, transcribed from the opposite strand in the opposite direction. In contrast, the two rRNA genes are separated by long stretches; 10 kbp, 5.5 kbp, and 30 kbp in *Tetrahymena, Neurospora,* and *Saccharomyces,* respectively.

In the two fungi, the large subunit is itself split by the presence of an intervening sequence; 1.1 kbp and 2.3 kbp long in *Saccharomyces* and *Neurospora,* respectively. While this sequence is present in all wild-type strains of *Neurospora* [31, 81, 82] and its removal (splicing) appears to be an essential part of large rRNA maturation in this organism,[83] the corresponding insert appears to be characteristic only of certain ω^+ strains of *S. cerevisiae.*[25, 28, 29, 84–86] It is absent in ω^- strains of this species, as well as in *S. carlsbergensis,* and thus does not appear to fulfill an *essential* role in rRNA biogenesis. The region between the two rRNAs contains genes for tRNAs: ≥9 in *Neurospora* and ~20 in yeast (FIGURE 1). In addition, this latter segment also contains at least two structural genes, *oxi1* and *oxi2,* which code for cytochrome subunits II (CO II) and III (CO III), respectively. These considerations, plus hybridization experiments of the isolated transcripts and the observation that *rho*⁻ mutants, which are deleted for large segments of mtDNA, can retain selectively one of the two rRNA genes and the capability to transcribe it, all lead to the inescapable conclusion that the two rRNA genes are transcribed independently of one another. Similar conclusions also apply to *Neurospora* where Lambowitz and his collaborators have studied transcription and processing in detail.[31, 81] Coincident with these profound evolutionary changes in ribosomal RNA, there appears to have been a compensating accommodation in the properties (but not the number, which may in fact have risen) of the ribosomal proteins interacting with them.[1, 19, 27, 31, 50, 51, 80] All of these entities—with the exception of a single polypeptide each in yeast and *Neurospora* [24, 31]—are encoded in nuclear DNA, synthesized on cytosolic ribosomes and must be imported into the mitochondria.

Divergence has been demonstrated in 56 out of 66 ribosomal proteins in a comparison between cow and rat; and in two species of toad, *Xenopus laevis* and *Xenopus borealis* (*mulleri*), the large subunit of the first contained three proteins absent from the second, while the converse was true for four proteins of the second.[88] No such profound differences have been seen in the proteins of eukaryotic cytosolic ribosomes, which, on the contrary, exhibit remarkably conservative features.[89]

Methylation of rRNA

Another distinguishing feature of mitochondrial rRNAs is their deficiency in methylated bases.[31, 50] In fact, early studies indicated that these residues were present in such small amounts in several species as to be below the level of detection. Dubin *et al.* have now demonstrated the presence of one each of m^4Cp (a novel species), m^5Cp, and m^5Up, and two residues of $m_2{}^6Ap$ in the small rRNA,[90] while the large subunit is reported to contain one equivalent each of the sequences GmpGp and UmpGmpUp.[91] In contrast, in yeast and *Neurospora* mitochondria, there is no methylation in the small subunit, and only two methylated bases per molecule in the large subunit.[31]

Absence of 5S RNA

A 5S RNA (120 residues) is a characteristic and essential component of the large ribosomal subunit from both bacteria and eukaryotic cytosol. Such a component has never been found in fungal or animal mitochondria.[19, 31] This observation defines yet another unique feature of the mitochondrial system. Three hypotheses have been advanced to reconcile this lack of a traditional 5S RNA with its apparent essentiality as part of the translational machinery[19, 31, 92]: 1) particularly easy detachment and resultant loss of this molecule in the course of mitochondrial isolation; 2) inseparability from the other small (4S, tRNA) species; or 3) covalent attachment to one end of the large ribosomal RNA. With the availability of sequencing techniques and the considerable conservation of primary structure characteristic of all 5S RNAs, such a "masked" version of this molecule, if it exists, should not remain hidden much longer.

Transcripts and Promoter Sequences

Examination of the major transcripts in a number of species has disclosed the existence, in addition to the tRNAs and rRNAs already discussed, of some 10 ± 3 species of RNAs probably engaged in functioning as messenger RNA.[27, 29, 42, 43, 45] It is very likely that in all these species most, if not all, of these mRNAs, as well as the two rRNAs and many of the tRNAs, are transcribed unidirectionally from the same strands. In many instances there is also evidence for the existence, in lower concentrations, of a smaller number of larger RNA molecules containing within them the sequences of one or more of the stable species. The actual number and kind of the true primary transcripts is not known with certainty, although in the case of the HeLa mtDNA, the studies of Attardi and his collaborators[45] have furnished strong inferential proof that initially both strands are transcribed in opposite directions over a considerable fraction of their total length, and perhaps their entirety. This is probably not the case in yeast where suggestive evidence exists for the separate transcription not only of the two rRNAs, but also of some structural genes.[24–26, 28, 29] In spite of a fairly diligent search, the nature of the promoter or other region(s) regulating the initiation and level of transcription remains as yet unknown.[26, 28–30, 52–55] This is due at least in part to the high proportion of T and A in this mtDNA, particularly in intergenic regions.‡ This makes it

‡ Yeast structural-gene sequences are immersed in a "sea of A and T" in various combinations,[24, 26, 52–55] while the known human sequence is closely adjacent to or abuts on one for tRNA.[30]

difficult to evaluate the significance of the presence of sequences of the type -TATA$^{AA}_{TT}$-, that have been suggested to be capable of functioning as eukaryotic promoter elements ("Pribnow box").

Caps and Tails

Most eukaryotic mRNAs are distinguished from their prokaryotic counterparts by at least three distinctive structural features[30, 93-95]: 1) The presence at the 5′-terminus of a distinctive "cap" structure containing as a minimum (cap-zero) the sequence mG(5′)ppp(5′)G(or A) followed by an untranslated leader sequence of varying length prior to the initiating AUG codon, which codes for the N-terminal methionine. 2) The absence within the leader not only of the heptanucleotide sequence (5′-AGGAGGU-3′) found in most prokaryotic mRNAs and complementary to the 3′-end of the small ribosomal RNA (16S) given by 3′-UCCUCCA-5′ ("Shine and Delgarno sequence"), but of *any* such common recognition elements responsible for interaction with the analogous site (sequence 5′-UAGGAAGGCGU-3′ on 18S rRNA). 3) The presence of a sizable sequence of adenylate residues linked to the 3′-OH (polyA tail) frequently adjacent to (within $<$20 nucleotides) of an -AAUAAA- sequence. Currently available information[25, 26, 28-30, 43, 45, 52-55, 77] suggests that mitochondrial mRNAs conform to *neither* of these two patterns. They contain neither caps, nor complemetary "Shine and Delgarno," nor other obvious rRNA recognition sites at their 5′-termini. Their 3′-termini, when occupied by polyadenylate,§ do not seem to be adjacent to the common eukaryotic recognition sequences. However, Barrell *et al.*[30] have drawn attention to the fact that sequences homologous to the prokaryotic rRNA recognition site may exist in the *interior* of structural genes, and they have identified a potential site for polyadenylation (i.e., the sequence -AAUAAA-) some 250 bases downstream from the *oxiI* structural gene. It should be mentioned that whenever ribosomal recognition sequences are actually found in an interior location in eukaryotic mRNAs, they are usually encountered within the first 20 nucleotides.[101]

Mosaic (Split) Genes in Mitochondria

Occurrence

As pointed out above, the rRNA genes in yeast and *Neurospora* exhibit an organizational pattern that requires the removal of the intervening sequence (splicing) in the course of the maturation or processing of the RNA specified by them. Analogous—but more complicated—organizational and processing features have recently been discovered to be the rule for a large number of structural genes in eukaryotic nuclear and viral genomes.[7, 93, 94, 102, 103] They have been referred to as split or mosaic genes, in which coding or expressed sequences (exons) are interspersed among intervening, noncoding sequences (introns).[104] These introns, in all cases examined so far, are transcribed initially

§ The presence of significant polyA tails, although on reasonably firm ground in the case of transcripts in animal mitochondria,[19, 42, 43, 45] is still a matter of dispute in yeast.[24]

but later removed by a splicing event or sequence, as part of the maturation of the mRNA. Surprisingly, this has also proven to be the case for at least one mitochondrial gene, the *cob-box* gene, responsible for the specification of the apocytochrome *b* of the respiratory chain in *Saccharomyces*.

The Cob-box Gene

The structure and organization of this gene has been investigated in a number of laboratories by a variety of techniques with essentially concordant results that have been summarized in a number of recent articles and reviews.[24, 25, 27–29, 32, 84] For our purpose, the presentation in FIGURE 1 will suffice with the following commentary: In some strains, such as KL14–4A (777–3A is similar) and ID41–6/161, this region of the genome is large and can occupy as much as $\geq 10\%$ of the total ($\geq 7{,}500$ base pairs). Cytochrome *b*-deficient mutants have been isolated and genetically and physically mapped in clusters of mutational sites located at various locations along this total stretch of mtDNA. Nine such clusters have been identified and designated in the manner shown on the diagram either as *cob-box* with a number, or by a Greek letter or pairs of such letters. Examination of the phenotypes of the mutants ordered into these clusters permits a reasonably unambiguous assignment of them into two classes. Those in the first class, consisting of *cob-box* clusters 5/4 (α), 8 (β), 1 (γ), 2 (δ), and 6 (ϵ), all contain as a prominent subset chain-termination mutants, producing fragment polypeptides homologous to cytochrome *b*, and increasing in size from left to right on the map (from ~14 to ≥ 30 kd). Not only does this observation define the direction of translation (and hence of transcription) of this region of the mtDNA, but it also permits the inference that in the aggregate the sequences in this class define the structural gene for this protein, and thus constitutes the class of expressed, or exon, sequences.

In contrast, the other clusters, *cob-box 3, 10,* and *7,* with an additional one between *2* and *6* (as yet to be defined in terms of actual mutants located within it) do not generate a simple, self-consistent set of polypeptide fragments. Quite on the contrary, mutants localized within them exhibit complex, pleiotropic phenotypes, producing one (or more) polypeptides larger than, and not obviously related to, apocytochrome *b,* as well as a profound inhibition of the expression of a second, unlinked mitochondrial gene.[105–109] This gene, *oxi3* (responsible for the specification of CO I, the largest subunit of cytochrome oxidase), is located (FIGURE 1) in a region of the genome that is not only physically separate from *cob-box* by at least one structural gene (*oli2* or *pho1*), but since it and *cob-box* are transcribed with the same polarity,[29] they can *not* form a single transcriptional unit.

These are the properties to be expected for a set of intervening sequences or introns. That mutants in them are in fact deficient in splicing, and accumulate RNA sequences normally removed during this processing act has been demonstrated in a variety of ways. Among them are hybridization to DNA sequences using restriction fragments and electron microscopy (R-loops),[28, 29] hybridization to a specific probe for *box 6,* 200 bp in length,[29a] and the demonstration that at least some of the polypeptides produced by them contain *both,* amino acid sequences related to apocytochrome *b,* as well as novel sequences not seen in this gene product, and thus probably due to readthrough from surviving intron RNA sequences.[108–111] This hypothesis has been verified by the demonstration[27, 84, 112] that the introduction of a second, chain terminating, exon mutation into such intron mutants, when located upstream (i.e., promoter or initiator-proximal) from it, extinguishes the intron phenotype and generates in its stead a phenotype indistinguishable from the single, exon mutant. The converse result, i.e., survival or dominance of the single mutant intron phenotype is obtained when the second mutation is introduced downstream from the intron.

How Prevalent Are Mosaic Patterns?

There are two parts to this question. The first asks whether and to what extent the pattern is essential and relevant to mitochondrial genes in general, the second repeats the question specifically for the cytochrome *b* gene. As concerns the first, a possible mosaic organization and complicated mRNA maturation pathways, perhaps involving splicing, appear to be operative also for *oxi3,* the structural gene for CO I.[27, 29, 84] In contrast, the sequences of at least two genes in *Saccharomyces,* i.e., *oxi1* and *oli1* (coterminus also with *oli3*[113] and *pho2*) have been shown to be completely devoid of inserts.[24–26, 28, 29, 36, 84] The presence of all large (>10 bp) intervening sequences has also been ruled out explicitly for *Xenopus* mitochondria by Rastl and Dawid,[43] and no intervening sequence has been detected in the human CO II gene.[30] So intervening sequences and mosaic organization are not *essential* prerequisites for mitochondrial genes in general.

How about the specific case of *cob-box*? It is known[24, 28, 29] that *S. carlsbergensis* and certain strains of *S. cerevisiae* (i.e., D273–10B) lack most of insert III, which overlaps, but is not necessarily identical with, intron *cob-box3* or α-β. Although the precise correspondence of these two entities remains to be established, it is already evident that the bulk of the sequences in this insert are dispensable. However, it remains to be established whether this statement holds for *all* the sequences in *S. carlsbergensis* or *S. cerevisiae* strain D273–10B, particularly those homologous to the existing intron–exon boundaries in strains KL14–4A and ID41–6/161, strains that retain the pair of large inserts IIIa and IIIb, and the intervening sequences α-β and β-γ. We have demonstrated[84] by appropriate genetic constructions that *absence* of these inserts has no effect on preexisting mutations in a different intervening sequence (*cob-box7* or γ-δ) or the regulatory effect on *oxi3,* but we have not yet demonstrated to what extent lack of it in the parent strain may affect splicing—and hence mutations—in the γ-δ intron.

One may therefore legitimately ask whether inserts in mtDNA can pop in or out of different sites in the genome more or less at random without any profound influence on its expression. Incidentally, the latter observation, together with their apparent site specificity rules out any *close* analogy between these mitochondrial inserts and the bacterial insertion sequence (IS) elements.[114, 115] There is one strong argument that militates against such an interpretation: Although these sequences may be permissive for certain strains, those others in which they *are* present, must possess a device, and a fairly intricate one at that, carefully designed to lead to their elimination (splicing) in the course of processing. Furthermore, there exists at least suggestive genetic evidence that these reaction sequences at different sites must differ in at least one component or reaction: for instance, mutations in or close to intron *cob-box 7* that interfere with splicing in this region still permit normal processing in intron *cox-box3*,[29a, 27, 84, 112] and mutations in these two introns can be shown to belong to two separate complementation groups.[29a, 32, 84]

Splicing Pathways

Ribosomal RNA

As already described briefly above, the large rRNA genes in some yeast (21S RNA), and all *Neurospora* strains (25S RNA) examined, contain an intervening sequence: in *Neurospora* it is 2.0–2.5 kbp long and separates the large (3 kbp) expressed region at the 5′-end from a small (0.5 kbp) one at the 3′-end. In yeast, the arrangement is similar, with a 1.05 kbp insert located some 2.3 kbp distal to the 5′-end separating it from a smaller 0.5 kbp segment at the 3′-end. Evidence for processing comes from studies in *Neurospora*, where several nuclear mutants have been shown to accumulate the complete primary transcript (3.5 kbp or 35S) of the gene,[83] which then functions as the precursor of the mature rRNA.

Cob-box RNA

The accumulation of large transcripts, 18.5S and 20S in length, and encompassing within them polynucleotide sequences known to form part of the *cob-box* region have been demonstrated by van Ommen et al.,[29] who also believe that the additional 600 bp sequence present in 20S RNA corresponds to the intervening sequence δ-ϵ (Figure 1) and is removed by splicing. Furthermore, R-loop techniques have provided evidence for the presence of sequences 600 (α), 100 (β), 200 (γ), 80 (δ), and 400 (ϵ) bp in length, interspersed among nonhybridized regions in the DNA (presumed intervening sequences of 1,900 (α-β), 1,400 bp (β-γ), 1,250 bp (γ-δ), and 650 bp (δ-ϵ); and a transcript 10S ($\sim$600 bp) in length has been identified as an early product in the processing of the α-β sequence. Furthermore, studies in several laboratories have suggested that intron mutants accumulate stable transcripts $\geq$25S ($\geq$5,000 bp) in length and can therefore be correlated with splicing defects. In particular Church, Slonimski, and Gilbert [29a] provided some evidence for accumulation of a 7.5 kbp precursor in a *box 3* mutant. While this observation and studies on polypeptide products in intron mutants provide strong evidence for the *occurrence* of splicing they say nothing about its nature or pathways.

From a consideration of the chemistry involved, it is clear that it consists of at least two separate reactions. 1) Two incisions or cleavages, at or near the two exon–intron boundaries, leading to the excision of the intron, followed by 2) the specific ligation of the two exon sequences. In addition to the specific features inherent in these presumably enzyme-catalyzed reactions, there must be some sort of provision to prevent premature diffusion of the two polynucleotides to be ligated (one at the 3′-terminus of the promoter-proximal exon, the other at the 5′-terminus of the promoter-distal exon) and for the preferential and efficient ligation between them. To accomplish this, one can conceive of two principal classes of models [27, 84, 102, 103]: A) unassisted or concerted and

B) assisted or guided. In models of type A, integrity is maintained by inherent structural features either 1) in the RNA (i.e., by a strongly hydrogen-bonded stem and flexible-loop structure) or 2) through the protein complex catalyzing the coupled cleavage-ligation. In models of type B, the structural integrity is preserved by a guide or bridge between the two exon segments, furnished either 1) by a complementary RNA or 2) some other specific bifunctional macromolecule, presumably a protein. Model A-2 is distinct from all others in being kinetically controlled, i.e., all the reactions must be completed prior to the dissociation of the complex between the polynucleotides and the enzymes in the reaction. What is the likelihood of one particular mechanism being applicable in the light of precedents in other systems and available evidence in the present one?

In the one case in which the reaction has been studied *in vitro,* that of the splicing of the cytosolic tRNAs of yeast, the mechanism adhered to appears to be A-1[102, 119] Specific structural integrity is maintained by the secondary structure in the (pre-tRNA) substrates and excision can be clearly separated from ligation in time and space—with excision proceeding in the absence of, and ligation requiring, the participation of ATP. The ends generated in the first step and required in the second step are the highly unusual set $\overset{5'}{\mathrm{P}}$———$\overset{3'}{\mathrm{P}}$ $\overset{5'}{\mathrm{HO}}$———$\overset{3'}{\mathrm{OH}}$. In another very interesting example, the splicing of the terminal "cap sequence" from the host (e.g., from its globin mRNA) to influenza virus mRNA, the viral transcriptase complex (which consists of a nucleoprotein plus three additional proteins) may be capable of performing this reaction *in vitro.*[120, 121]

There are some suggestive and revealing hints that in the case of *cob-box,* the mechanism involves an additional macromolecule. Among them are the following: 1) the demonstration by Slonimski and his collaborators (discussed in Reference 29a) that complementation of a defective intron by an exon (an event requiring a second diffusible species) invariably results in the expression of the phenotype of the *b* segment *cis* to the intron lesion; 2) some direct evidence for the participation of intron sequences in processing[29a]; 3) our finding[108] that mutants either in intron *box 7* (γ-δ), or in exon segments at or close to its boundary can generate identical or closely related phenotypes, involving production of read-through polypeptides, presumably by translation of unprocessed intron sequences together with the adjacent upstream exon; and 4) the plausible inference that similar proteins are formed also in the wild type where they participate in splicing.

Gene Transfer

The Case of the ATPase Proteolipid

One of the more surprising recent findings was the demonstration by Sebald and his collaborators[122–124] that a polypeptide in *Neurospora,* structurally and functionally homologous to the mitochondrially specified subunit 9 (proteolipid) of the F_0 part of yeast ATPase, is coded in the nucleus, synthesized on cytosolic ribosomes, and imported into the mitochondria.[25–27] The most parsimonious hypothesis to account for these observations is certainly that this represents gene transfer between mitochondria and nucleus. However, although most commentators[24] have postulated the transfer as having occurred from the mitochondrial to the nuclear genome in some common ancestor of these two ascomycetes, in the absence of compelling evidence to the contrary, the opposite directionality might be argued with equal plausibility. At the very least, one might wish to inquire into the prevalence of the two alternate sites in different species [25, 27, 125] in order to gain some tentative insight as to which represents the more "primitive" location.

The situation has already been complicated further by the discovery that the polypeptide in *Neurospora* is synthesized in the cytosol as a pre-protein ($M_r = 12{,}000$) and that the N-terminal leader sequence is removed in the course of the transfer and maturation of this species, which conforms to the pattern adhered to by other imported mitochondrial proteins.[127] Thus the nuclear gene must contain a minimum of 120 nucleotide pairs prior to the initiator, not present in its mitochondrial homolog. The sequence of and around the latter is known with good precision [52, 53] and there is no evidence, and indeed no room for such an arrangement. While it may be argued that there might have been acquisition of such a sequence (e.g., by gene fusion) in the course of the transfer of the gene from the mitochondria to the nucleus, its loss during the converse event appears somewhat more attractive.

Generality

As already pointed out by Fox,[55] the divergent utilization of the UGA codon renders generalized gene transfer between mitochondria and nucleus highly unlikely, and the proteolipid gene must represent a special case precisely because it contains no tryptophan in either yeast or *Neurospora*. Furthermore, such a shift would probably necessitate the acquisition of the additional sequences in the mRNA, and hence in the gene, required for eukaryotic recognition sequences for caps, tails, ribosomes, and transfer to the organelle.

Evolutionary Implications

Most of the points to be discussed here have already been mentioned in passing in the previous sections. In order to give some focus to their recapitulation and discussion one might pose the following four questions that form the subheadings of this part of the presentation.

Are Mitochondria Domesticated Prokaryotes?

This central theme of the endosymbiotic hypothesis [9–12, 14] has been critically examined on several recent occasions.[8, 15, 17–22, 27] To the points raised there, two additional ones elaborated here require attention. The presence in mitochondria of elements of a genetic code different from that found everywhere else in prokaryotes and eukaryotes alike. There is an essentially eukaryotic organizational pattern of certain mitochondrial genes in at least two organisms, and of the whole genome in yeast.[121] Together, these lines of evidence render the endosymbiotic hypothesis—in its usual formulation—unlikely, if not untenable.

Are Mitochondria Evolutionary Relicts?

Borst and Grivell,[24] and Barrell *et al.*[30] have resuscitated a hypothesis [129] that suggested that the mitochondrial genetic system is an evolutionary dead end, representing a frozen cross section of a point in evolution in which the hypothetical endosymbiont had transferred most, but not all, of its genes to the host chromosome. The observations, made in a large number of studies and summarized here in part, provide clear evidence of relatively rapid evolutionary changes in all classes of RNA specified by mitochondrial DNA, within a set of apparently powerful constraints: these are provided by the requirement for the RNA (but not most of the protein) components of a translational machinery designed to produce an invariant (but structurally diverse) set of polypeptides tightly integrated into the inner mitochondrial membrane.[24, 25, 27, 143] These minimally consist of the three largest polypeptides of cytochrome *c* oxidase, the apocytochrome *b* component of the reduced ubiquinone:cytochrome *c* reductase (Complex III, $b{:}c_1$ complex) and two (though not necessarily the same two) polypeptides of the F_0 part (membrane sector, protonophoric portion) of the ATP synthase (Complex V, oligomycin-sensitive ATPase).

Do Mitochondria Exhibit Archaic Features?

On the basis of the evidence available some ten months ago, we suggested that the most striking features of the organization and sequences of mitochondrial genes might represent a primitive pattern characteristic of an organism *preceding* rather than subsequent to the divergence of the contemporary eubacteria.[27] Several recent findings and conceptualizations have provided additional information that suggests that this hypothesis may deserve serious consideration. The mitochondrial initiator $tRNA_F^{Met}$ in *Neurospora* exhibits unique features, shared in part only by the corresponding species from the archaebacterium[130, 131] *Halobacterium cutirubrum*.[73] The genetic code in mitochondria exhibits features that may well be classified as more ancient than that operative in all other organisms. It utilizes only two (rather than three) stop codons and a minimal number (<32) of amino acid codons, which may be characteristic of a more primitive mechanism of codon recognition.[132–135] Mosaic organization of, and intervening sequences in, genes can plausibly be argued to represent the survival of a primitive character, rather than the acquisition of a more highly developed character in nuclear,[135–137] as well as mitochondrial chromosomes.[27] Reanney has now presented a cogent, tightly reasoned argument for why its corollary (the requirement for RNA splicing) may itself represent the presence of an even more ancient function, namely the most primitive mode of genetic recombination.[138] This mode perhaps exhibits its most striking aspect through the prerequisite for the guide sequence, inferred to be operative in the case of the *cob-box* gene. Furthermore, Reanney also proposes that the next still primitive step in the evolution of mechanisms for genetic recombination is the addition of "illegitimate" recombination in the DNA, which retains some of the sequence elements characteristic of RNA-dependent processes. Events formally or actually equivalent to "illegitimate" recombination have been proposed to be common occurrences for mitochondrial DNA in yeast.[100, 128, 139] Finally, recent speculations on the nature and timing of the prokaryotic–eukaryotic divergence have increasingly focused attention on the possibility that the eukaryotic cytoplasm may have diverged much earlier than had been commonly believed. The point of divergence may have involved a respiring (but not oxygen-dependent) organism reminiscent of a purple, nonphotosynthetic bacterium[19, 140] or an even more ancient common ancestor of the eubacteria, archaebacteria and the eukaryotic cytosol (urkaryote) called a progenote.[141] In the latter, the distinction between genotype and phenotype, so essential to contemporary organisms, had not yet become manifest, and more primitive mechanisms of biocatalysis and maintenance and reassortment of the genetic material (RNA?) must have been operative.[138, 142]

Why A Separate Genetic System?

Finally one might pose what may well be considered the central question concerning mitochondrial origin and evolution. What evolutionary advantage accrues to a cell, not in acquiring or developing this type of organelle, but in maintaining within it through all evolutionary vicissitudes a separate, semi-autonomous genetic system? After all, eukaryotic cells contain numerous other organelles, some of them of considerable complexity, that do not exhibit such a requirement. And perhaps most cogently, it can become entirely dispensable even in the case of mitochondria—*rho*^{0} cells lack all mtDNA.[1, 2, 19, 20]

A number of hypotheses have been advanced at various times to deal with these teleological problems (e.g., Reference 19). Slowly but surely all of them have been rendered untenable.[24, 27, 28] I now wish to propose a radically different point of view. Perhaps mitochondria contain and have retained their genetic system precisely because of the considerations presented in the earlier part of this section. Only by this device and its inherent enormous flexibility and adaptability could they have succeeded in maintaining functional invariance in a constantly evolving cellular background.

Outlook

Clearly, our perception concerning matters evolutionary will always be biased by a paucity of hard facts on which to base such speculations and make informed choices among competing hypotheses. But, clearly, some essential items of information could be obtained rather easily and will no doubt become available in short order. Among the most prominent is sequence information permitting identification of possible introns and intron–exon boundaries in mitochondrial DNA of various strains of *S. cerevisiae,* in other yeasts with a greatly reduced genomic size such as *Kluyveromyces lactis,* and in more primitive eukaryotes, such as the dinoflagellates. Similarly, additional information[144] concerning the molecular biology, including gene organization, in *Archaebacteria* would be most helpful.

Acknowledgments

The author is indebted to his colleagues P. S. Perlman (Columbus), R. Hageman (Jena), G. Bernardi (Paris), P. P. Slonimski (Gif), and, especially, R. Raff (Bloomington) for many helpful and stimulating discussions, and to U. L. RajBhandary and J. Heckman, as well as F. Sanger, I. C. Eperon, S. Anderson, and D. P. Nierlich for permission to use their as yet unpublished structures of tRNAs for *Neurospora* and human mitochondria, respectively. Experimental work in the author's laboratory, is principally by D. Hanson and J. Johnson.

REFERENCES

1. GILLHAM, N. W. 1978. Organelle Heredity. Raven Press. New York, N.Y.
2. BEALE, G. & J. KNOWLES. 1978. Extranuclear Genetics. Edward Arnold. London, England.
3. BIRKY, C. W., JR., P. S. PERLMAN & T. J. BYERS, Eds. 1975. Genetics and Biogenesis of Mitochondria and Chloroplasts. Ohio State University Press. Columbus, Ohio.
4. SACCONE, C. & A. M. KROON, Eds. 1978. The Genetic Function of Mitochondrial DNA. North Holland. Amsterdam, The Netherlands.
5. BÜCHER, T., W. NEUPERT, W. SEBALD & S. WERNER, Eds. 1976. Genetics and Biogenesis of Chloroplasts and Mitochondria. North Holland. Amsterdam, The Netherlands.
6. BANDLOW, W., R. J. SCHWEYEN, K. WOLF & F. KAUDEWITZ, Eds. 1977. Mitochondria 1977. Genetics and Biogenesis of Mitochondria. de Gruyter. Berlin.
7. CUMMINGS, D., I. DAWID, P. BORST & S. WEISSMAN, Eds. 1979. Extrachromosomal DNA. Academic Press, Inc. New York, N.Y.
8. DODSON, E. O. 1979. Can. J. Microbiol. **25:**651–674. (This article presents a concise yet comprehensive juxtaposition and synthesis of the various theories.)
9. MARGULIS, L. 1970. Origin of Eukaryotic Cells. Yale University Press. New Haven, Conn.
10. MARGULIS, L. 1975. Soc. Exp. Biol. Symp. **29:**21–38.
11. WHITTAKER, R. H. & L. MARGULIS. 1978. BioSystems **10:**3–18.
12. TAYLOR, F. J. R. 1974. Taxon **23:**229–258.
13. JOHN, P. & F. R. WHATLEY. 1975. Nature **254:**495–498.
14. SITTE, P. 1979. Nova Acta Leopold. (In press.)
15. TAYLOR, F. J. R. 1976. Taxon **25:**377–390.
16. ALLSOPP, A. 1969. New Phytol. **68:**591–612.
17. UZZELL, T. & C. SPOLSKY. 1974. Am. Sci. **62:**335–343.
18. CAVALIER-SMITH, T. 1975. Nature **256:**463–468.
19. MAHLER, H. R. & R. A. RAFF. 1975. Int. Rev. Cytol. **43:**1–124; *see also* RAFF, R. A. & H. R. MAHLER. 1972. Science **177:**575–582; 1975. Soc. Exp. Biol. Symp. **29:**41–92.
20. MAHLER, H. R. 1979. Nova Acta Leopold. (In press.)
21. MEYER, R. R. 1973. J. Theor. Biol. **38:**647–663.
22. BÖRNER, T. & R. HAGEMANN. 1976. Biol. Rundsch. **14:**249–268.
23. Abstracts for UCLA-ICN Symposium on Extrachromosomal DNA. 1979; J. Supramol. Struct. Suppl. 3:125–139; 153–160.
24. BORST, P. & L. A. GRIVELL. 1978. Cell **15:**705–723.
25. TZAGOLOFF, A., G. MACINO, & W. SEBALD. 1979. Ann. Rev. Biochem. **48:** 419–441.
26. TZAGOLOFF, A., C. MACINO, F. NOBREGA & M. LI. 1979. *In* Extrachromosomal DNA. D. Cummings, I. Dawid, P. Borst & S. Weissman, Eds. 339–355. Academic Press, Inc. New York, N.Y.
26a. MACINO, G., G. CORUZZI, F. G. NOBREGA, M. LI & A. TZAGOLOFF. 1979. Proc. Natl. Acad. Sci. USA **76:**3784–3785.
27. MAHLER, H. R. & P. S. PERLMAN. 1979. *In* Extrachromosomal DNA. D. Cummings, I. Dawid, P. Borst & S. Weissmann, Eds.: 11–33. Academic Press, Inc. New York, N.Y.
28. GRIVELL, L. A., A. C. ARNBERG, P. H. BOER, P. BORST, J. L. BOS, E. F. J. VAN BRUGGEN, G. S. P. GROOT, W. B. HECHT, L. A. M. HENSGENS, G. J. B. VAN OMMEN & H. F. TABAK. 1979. Ibid.: 305–324.
29. VAN OMMEN, G. J. R., G. S. P. GROOT & L. GRIVELL. 1979. Cell **18:**511–523.
29a. CHURCH, G. M., P. P. SLONIMSKI & W. GILBERT. 1979. Cell **18:**1209–1215.
30. BARRELL, B. G., A. T. BANKIER & J. BROUIN. 1979. Nature **282:**189–194.
31. BOYNTON, J. E., N. W. GILLHAM & A. M. LAMBOWITZ. 1979. *In* Proc. Steenbock Symp. Madison, Wisc. July, 1979. (In press.)

32. Slonimski, P. P., M. L. Claisse, Z. Kotylak, P. S. Perlman, R. J. Schweyen, M. Solioz & A. Spyrikakis. 1980. Cell. (Submitted for publication.)
33. Shah, D. M. & C. H. Langley. 1979. Nature **281:**696–699.
34. Zakour, R. A. & H. Bultmann. 1979. Biochim. Biophys. Acta **564:**342–351.
35. Prunell, A., H. Kopecka, F. Strauss & C. Bernardi. 1977. J. Mol. Biol. **110:** 17–52.
36. Sanders, J. P. M., C. Heyting, M. Ph. Verbeet, F. C. P. W. Meijink & P. Borst. 1977. Molec. Gen. Genet. **157:**239–261.
37. Goldbach, R. W., A. C. Arnberg, E. F. J. Van Bruggen, J. Defize & P. Borst. 1977. Biochim. Biophys. Acta **477:**37–50.
38. Dujon, B., A. M. Colson & P. P. Slonimski. 1977. *In* Mitochondria 1977. Genetics and Biogenesis of Mitochondria. W. Bandlow, R. J. Schweyen, K. Wolf & F. Kaudewitz, Eds.: pp. 579–669. de Gruyter. Berlin. Germany.
39. Goldbach, R. W., J. E. Bollen-De Boer, E. F. J. Van Bruggen & P. Borst. 1978. Biochim. Biophys. Acta **521:**187–197.
40. Klukas, C. K. & I. B. Dawid. 1976. Cell **9:**615–625.
41. Upholt, W. B. & I. B. Dawid. 1977. Cell **11:**571–583.
42. Battey, J. & D. A. Clayton. 1978. Cell **14:**143–156.
43. Rastl, E. & I. B. Dawid. 1979. Cell **18:**501–510.
44. Brown, W. M., M. J. George & A. C. Wilson. 1979. Proc. Natl. Acad. Sci. USA **76:**1967–1971.
45. Attardi, G., P. Cantatore, E. Ching, S. Crews, R. Gelfand, C. Merkel & D. Ojala. 1979. *In* Extrachromosomal DNA. D. Cummings, I. Dawid, P. Borst & S. Weissman, Eds.: 443–469. Academic Press, Inc. New York, N.Y.
46. Groot, G. S. P., R. A. Flavell & J. P. M. Sanders. 1975. Biochim. Biophys. Acta **378:**186–194.
47. Jakovcic, S., J. Casey & M. Rabinowitz. 1975. Biochemistry **14:**2037–2042.
48. Goddard, J. M. & D. R. Wolstenholme. 1978. Proc. Natl. Acad. Sci. USA **75:**3386–3390.
49. Koch, G. 1976. J. Biol. Chem. **251:**6097–6107.
50. Buetow, D. E. & W. M. Wood. 1978. Sub-Cell Biochem. **5:**1–86.
51. O'Brien, T. W. & D. E. Matthews. 1976. Handbook of Genetics. **5:**535–580. Plenum Publishing Corp. New York, N.Y.
52. Macino, G. & A. Tzagoloff. 1979. J. Biol. Chem. **254:**4617–4623.
53. Hensgens, L. A. M., L. A. Grivell, P. Borst & J. C. Bos. 1979. Proc. Natl. Acad. Sci. USA **76:**1663–1667.
54. Coruzzi, C. & A. Tzagoloff. 1979. J. Biol. Chem. **254:**9324–9330.
55. Fox, T. D. 1979. Proc. Natl. Acad. Sci. USA **74:**4537–4541.
56. Steffens, G. J. & G. Buse. 1979. Z. Physiol. Chem. **360:**613–619.
57. Fox, T. D. 1979. J. Mol. Biol. **130:**63–82.
58. Jukes, T. H. 1978. Adv. Enzymol. **47:**375–432.
59. Hirsch, D. 1971. J. Mol. Biol. **58:**439–458.
60. Harada, F., R. C. Sawyer & J. E. Dahlberg. 1975. J. Biol. Chem. **250:** 3487–3497.
61. Beaudet, A. L. & C. T. Caskey. 1971. Proc. Natl. Acad. Sci. USA **68:**619–624.
62. Wesolowski, M. & H. Fukuhara. 1979. Mol. Gen. Genet. **170:**261–275.
63. Li, M. & A. Tzagoloff. 1979. Cell **18:**47–53.
64. Woese, C. R. & M. A. Bleyman. 1972. J. Mol. Evol. **1:**227–229.
65. Grantham, R. 1978. FEBS Lett. **95:**1–11; 1972. Nature (London) New Biol. **237:**265–266.
66. Martin, R. P., J.-M. Schneller, A. J. C. Stahl & G. Dirheimer. 1977. Nucleic Acid Res. **4:**3497–3510.
67. Martin, N. C. & M. Rabinowitz. 1978. Biochemistry **17:**1628–1634.
68. De Vries, H., J. C. De Jonge, J.-M. Schneller, R. P. Martin, G. Dirheimer & J. C. Stahl. 1978. Biochim. Biophys. Acta **520:**419–427.

69. Ohi, S., J. L. Ramirez, W. B. Upholt & I. B. Dawid. 1978. J. Mol. Biol. **121:** 299–310.
70. Martin, R. P., J.-M. Schneller, A. J. C. Stahl & G. Dirheimer. 1979. Biochemistry **18:**4600–4605.
71. Feldman, F. & H. R. Mahler. 1974. J. Biol. Chem. **249:**3702–3709; *see also* Mahler, H. R. K. Dawidowicz & F. Feldman. 1972. J. Biol. Chem. **247:** 7439–7442.
72. Smith, A. E. & K. A. Marcker. 1968. J. Mol. Biol. **38:**241–248.
73. Heckman, J. E., L. I. Hecker, S. D. Schwartzbach, W. E. Barnett, B. Braustark, & V. L. Rajbhandary. 1978. Cell **13:**83–95.
74. Wrede, P., N. H. Woo & A. Rich. 1979. Proc. Natl. Acad. Sci. USA **76:** 3289–3293.
75. Macino, G. & A. Tzagoloff. 1979. Colec. Gen. Genet. **169:**183–188.
76. Colletti, E., L. Frontali, C. Palleschi, M. Wesoloski & H. Fukuhara. 1979. Molec. Gen. Genet. **175:**1–4.
77. Martin, N. C. 1979. *In* Extrachromosomal DNA. D. Cummings, I. Dawid, P. Borst & S. Weissman, Eds.: pp. 132. Academic Press, Inc. New York, N.Y.; *also* personal communication.
78. Miller, D. G., N. C. Martin, H. D. Pham & J. E. Donelson. 1979. J. Biol. Chem. **254:**11735–11740.
79. Bos, J. L., K. A. Osinga, G. Vanderhorst & P. Borst. 1979. Nucleic Acid Res. **6:**3255–3261.
80. Odintsova, M. S. & N. P. Yurina. 1977. Biokhimiya **42:**1347–1360.
81. Hahn, U., C. M. Lazarus, H. Lunsdorf & H. Küntzel. 1979. Cell **17:**191–200.
82. Heckman, J. & U. L. RajBhandary. 1979. Cell **17:**583–595.
83. Mannella, C., R. A. Collins, M. R. Green & A. M. Lambowitz. 1979. Proc. Natl. Acad. Sci. USA **76:**2635–2639.
83a. LaPolla, R. J. & A. M. Lambowitz. 1979. J. Biol. Chem. **254:**11746–11750.
84. Perlman, P. S., N. J. Alexander, D. K. Hanson & H. R. Mahler. 1980. *In* Gene Structure and Expression. D. H. Dean, L. F. Johnson, P. C. Kimbal & P. S. Perlman, Eds. (In press.) Ohio State University Press. Columbus, Ohio.
85. Jacq, C., C. Kugawa, C. Grandchamp & P. Netter. 1977. *In* Mitochondria 1977. Genetics and Biogenesis of Mitochondria. W. Bandlow, R. J. Schweyen, K. Wolf & F. Kaudewitz, Eds.: 255–270. de Gruyter. Berlin, Germany; Faye, G., N. Dennebouy, C. Kujawa & C. Jacq. 1979. Molec. Gen. Genet. **168:**101–109.
86. Heyting, C. & H. H. Menke. 1979. Molec. Gen. Genet. **168:**271–291.
87. Matthews, D. E., R. A. Hessler & T. W. O'Brien. 1978. FEBS Lett. **86:**76–80.
88. Leister, D. E. & I. B. Dawid. 1975. J. Mol. Biol. **96:**119–123.
89. Delaunay, J., F. Creusot & G. Schapira. 1973. Eur. J. Biochem. **39;** 305–312.
90. Dubin, D. T., R. H. Taylor & L. W. Davenport. 1978. Nucleic Acid Res. **5:** 4385–4397.
91. Dubin, T. D. & R. H. Taylor. 1978. J. Mol. Biol. **121:**523–540.
92. Lizardi, P. M. & D. J. C. Luck. 1971. Nature (London) New Biol. **73:**139–148.
93. Brownlee, G. G. 1979. Trans. Biochem. Soc. **7:**279–296.
94. Nishioka, Y. & P. Leder. 1979. Cell **18:**875–880; Homer, D. H. & P. Leder. 1979. Cell **18:**1299–1302.
95. Filipowicz, W. 1978. FEBS Lett. **96:**1–11.
95a. Kozak, M. 1978. Cell **15:**1109–1123.
96. Carlin, R. K. 1978. J. Theor. Biol. **71:**323–338.
97. Hagenbüchle, O., M. Santer, J. A. Steitz & R. J. Mans. 1978. Cell **13:** 551–563.
98. Grohmann, K., F. Amalric, S. Crews & G. Attardi. 1978. Nucleic Acid Res. **5:**637–651.

99. QUAGLIARELLO, C., R. GALLERANI, C. GADALETA & C. SACCONE. 1978. Biochem. Biophys. Res. Comm. **84:**45–81.
100. GAILLARD, C., F. STRAUSS & G. BERNARDI. 1980. Nature (In press.); *see also* BERNARDI, G., A. PRUNELL, C. FONTY, H. KOPECKA & F. STRAUSS. 1976. *In* The Genetic Function of Mitochondrial DNA. C. Saccone & A. M. Kroon, Eds.: pp. 185–198. North-Holland. Amsterdam, The Netherlands.
101. BOTH, G. W. 1979. FEBS Lett. **101:**220–224.
102. ABELSON, J. 1979. Ann. Rev. Biochem. **48:**1035–1069; AXEL, R., T. MANIATIS & C. F. FOX, Eds. 1979. Eucaryotic Gene Regulation. ICN-UCLA Symposium on Molecular & Cellular Biology, Vol. 14. Academic Press, Inc. New York, N.Y.
103. CRICK, F. H. C. 1979. Science **204:**264–271.
104. GILBERT. W. 1978. Nature **271:**501.
105. CLAISSE, M. L., A. SPYRIDAKIS, M. L. WAMBIER-KLUPPEL, P. PAJOT & P. P. SLONIMSKI. 1978. *In* Biochemistry and Genetics of Yeast. M. Bacila, B. L. Horecker & A. O. M. Stoppani, Eds.: pp. 369–390. Academic Press, Inc. New York, N.Y.
106. ALEXANDER, N. J., R. D. VINCENT, P. S. PERLMAN, D. H. MILLER, D. K. HANSON & H. R. MAHLER. 1979. J. Biol. Chem. **254:**2471–2479.
107. HAID, A., R. J. SCHWEYEN, H. BECHMANN, F. KAUDEWITZ, M. SOLIOZ & G. SCHATZ. 1979. Eur. J. Biochem. **94:**451–464.
108. CLAISSE, M., P. P. SLONIMSKI, J. JOHNSON & H. R. MAHLER. 1980. Mol. Gen. Genet. **177:**375–387.
109. MAHLER, H. R., D. HANSON, D. MILLER, C. C. LIN, N. J. ALEXANDER, R. D. VINCENT & P. S. PERLMAN. 1978. *In* Biochemistry and Genetics of Yeasts. Pure and Applied Aspects. M. Bacilla, B. L. Horecker & A. O. M. Stoppani, Eds.: pp. 513–547. Academic Press, Inc. New York, N.Y.
110. MAHLER, H. R. & J. JOHNSON. 1979. Molec. Gen. Genet. **176:**25–31.
111. SOLIOZ, M. & G. SCHATZ. 1979. J. Biol. Chem. **254:**9331–9334.
112. ALEXANDER, N. J., P. S. PERLMAN, D. K. HANSON & H. R. MAHLER. 1980. Cell **20:**199–206.
113. SEBALD, W., E. WACHTER & A. TZAGOLOFF. 1979. Eur. J. Biochem. **100:**599–607.
114. REANNEY, D. 1976. Bact. Rev. **40:**556–590.
115. NISEN, P., M. PURUCKER & L. SHAPIRO. 1979. J. Bact. **140:**588–596.
116. KONKEL, D. A., S. M. TILGHMAN & P. LEDER. 1978. Cell **15:**1125–1133.
117. MURRAY, V. & R. HOLLIDAY. 1979. FEBS Lett. **106:**5–7.
118. DAVIDSON, E. H. & R. J. BRITTEN. 1979. Science **204:**1052–1059.
119. KNAPP, G., R. C. OGDEN, G. L. PEEBLES & J. ABELSON. 1979. Cell **18:**37–45; PEEBLES, C. L., R. C. OGDEN, G. KNAPP & J. ABELSON. 1979. Cell **18:**27–35.
120. PLOTCH, S. J., M. BOULOY & R. M. KRUG. 1979. Proc. Natl. Acad. Sci. USA **76:**1618–1622.
121. KRUG, R. M., B. A. BRONI & M. BOULOY. 1979. Cell **18:**329–334.
122. SEBALD, W. 1977. Biochim. Biophys. Acta **463:**1–27.
123. SEBALD, W., M. SEBALD-ALTHAUS & E. WACHTER. 1977. *In* Mitochondria 1977. Genetics and Biogenesis of Mitochondria. W. Bandlow, R. J. Schweyen, K. Wolf & F. Kaudewitz, Eds.: pp. 433–440. de Gruyter. Berlin, Germany.
124. SEBALD, W., TH. GRAF & H. B. LUKINS. 1979. Eur. J. Biochem. **93:**587–599.
125. DE JONG, L., M. HOLTROP & A. M. KROON. 1979. Biochim. Biophys. Acta **548:** 48–62.
126. MICHEL, R., E. WACHTER & W. SEBALD. 1979. FEBS Lett. **101:**373–376.
127. SCHATZ, G. 1979. FEBS Lett. **103:**203–211.
128. BERNARDI, G. 1979. TIBS **4:**197–201.
129. BORST, P. 1972. Ann. Rev. Biochem. **41:**333–376.
130. WOESE, C. R., L. J. MARGRUM & G. E. FOX. 1978. J. Mol. Evol. **11:**245–252; WOESE, C. R. 1977. Proc. Natl. Acad. Sci. USA **74:**5088–5090.
131. BAYLEY, S. T. 1979. TIBS **4:**223–225.

132. LAGERKVIST, U. 1978. Proc. Natl. Acad. Sci. USA **75:**1759–1762.
133. HOPFIELD, J. J. 1978. Proc. Natl. Acad. Sci. USA **75:**4334–4338.
134. GROSJEAN, H. J., S. DE HEUAM & D. M. CROTHERS. 1978. Proc. Natl. Acad. Sci. USA **75:**610–614.
135. DARNELL, J. E. J. 1978. Science **202:**1257–1259.
136. DOOLITTLE, W. F. 1978. Nature **272:**581–582.
137. REANNEY, D. 1974. J. Theor. Biol. **48:**243–251.
138. REANNEY, D. 1979. Nature **277:**596–600.
139. SLONIMSKI, P. P. & J. LAZOWSKI. 1977. *In* Mitochondria 1977. Genetics and Biogenesis of Mitochondria. W. Bandlow, R. J. Schweyen, K. Wolf & F. Kaudewitz, Eds.: pp. 39–52. de Gruyter. Berlin, Germany.
140. BRODA, E. & C. A. PESELAH. 1979. J. Theor. Biol. **81:**201–212.
141. WOESE, C. R. & G. E. FOX. 1977. J. Mol. Evol. **10:**1–6.
142. KUPPERS, B. 1979. Naturwissenschaften **66:**228–243.
143. SCHATZ, G. & T. MASON. 1974. Ann. Rev. Biochem. **43:**51–87.
144. MITCHELL, R. M., L. A. LOEBLICH, L. C. KLOTZ & A. R. LOEBLICH, III. 1979. Science **204:**1082–1084.

CO-EVOLUTION OF LUMINOUS BACTERIA AND THEIR EUKARYOTIC HOSTS

Kenneth Nealson, Daniel Cohn, Gary Leisman, and Bradley Tebo

Scripps Institution of Oceanography
La Jolla, California 92093

INTRODUCTION

Bioluminescence, the capacity to emit visible light, is a characteristic of certain groups of bacteria, fungi, protists, invertebrates, and vertebrates.[1] Bacteria, fungi, protists, and many animals synthesize their own luciferin and luciferase and thus produce their own chemical light, while other animals are luminous as a result of symbiotic associations with luminous bacteria. Large light organs, specialized tissue compartments that serve to culture luminous bacteria, were first described in the early 1900s in the flashlight fish, *Anomalops katoptron*,[2] and in the Japanese pinecone fish, *Monocentris japonicus*.[3] There are now examples in many widely distributed families of marine animals.[4] Symbiotically luminous marine animals of shallow waters are restricted mainly to warm tropical waters, whereas in the deep sea luminous animals are ubiquitous.

Despite their prevalence, especially in marine environments, luminous bacterial symbioses have not been extensively studied. Luminous fish are difficult to maintain, and the complete life cycle of a luminous fish in captivity has never been observed. With few exceptions, the physiology of the luminous symbionts, especially studies of those properties that might be relevant to symbiosis, has not been documented. Many luminous bacteria have been cultured outside their hosts. Oxygen tension is important in regulating bacterial bioluminescence and significant differences exist between different symbionts.[5] Some species of symbiotic bacteria excrete pyruvate during growth on glucose.[6] These observations have been put together in a model of symbiosis.[7, 8] The ultrastructure of several luminous bacteria–fish symbioses has been carefully studied.[9–14] Apparently, many of the general properties usually attributed to other symbiotic systems, including host specificity, structural adaptations in both partners, and nutrient exchanges also occur in the luminous symbioses.

Three closely related genera of luminous bacteria have been described; these gram-negative rods, members of the family *Enterobacteriaceae*, all use an identical mechanism of light emission.[15–17] Light emission is catalyzed by bacterial luciferase, an enzyme that is different from all eukaryotic luciferases in its structure, substrates, and decay kinetics [18] (see below). Of the two marine genera, *Photobacterium* and *Beneckea*, only *Photobacterium* species have been isolated as light organ symbionts. Luminous bacteria of the genus *Xenorhabdus* [19] are found as gut symbionts of terrestrial nematodes. Marine luminous bacteria in general occupy several niches. Some are found free-living in seawater, some are saprophytes growing on animal carcasses, some are gut symbionts in many different marine animals, and others are highly specific symbionts of fish or other animals.[16] TABLE 1 summarizes the associations of luminous

0077–8923/81/0361–0076 $01.75/0 © 1981, NYAS

TABLE 1

ASSOCIATIONS OF LUMINOUS BACTERIA

Type of Association	Hosts	Bacteria	Kinetics of Bacterial Luciferase *	References
Saprophytic	Decaying organic matter	*Beneckea* spp.	S	1,16,32
		Photobacterium spp.	F	
Gut symbionts	Marine fish, squids, crustaceans †	*Beneckea* spp.	S	1,16,32
		Photobacterium spp.	F	
	Nematodes ‡	*Xenorhabdus* spp.	S	19,34,40
Light organ symbionts				
Extracellular facultative	Fish: *Monocentridae*	*P. fischeri*	F	7,21
	Leiognathidae §	*P. leiognathii*	F	24
	Apogonidae	*P. leiognathii*	F	4,16
	Macrouridae	*P. phosphoreum*	F	4,16,25
	Merluccidae	*P. phosphoreum*	F	4,16
	Opisthoproctidae	*P. phosphoreum*	F	4,16,25
	Trachichthydae	*P. phosphoreum*	F	4,16
	Moridae	*P. phosphoreum*	F	4,16
	Squid: *Sepiolidae* ¶	*P. fischeri*	F	Leisman, unpublished
Extracellular non-culturable	Fish: *Ceratioidea* **	NC ††	F	47
	Anomalopidae	NC	S	28,47
	Squid: *Sepiolidae* ¶	NC	F	47
Intracellular	Pyrosomes	NC	F	47

* F = Fast; S = Slow

† This association is nonspecific with a wide variety of marine animals

‡ This association is confined to nematodes, but its species specificity is not documented.

§ One member of this fish group, *Leiognathus elongatus* has symbiotic luminous bacteria that have been reported to be nonculturable.[22]

¶ Leisman (unpublished) has determined that two genera (*Sepiola* and *Euprymna*) have *P. fischeri* as their symbiont. A member of another genus *Heteroteuthis hawaiiensis* has nonculturable symbionts.

** This suborder consists of 13 families.[4]

†† NC=Nonculturable

bacteria, which range from loose associations of culturable bacterial symbionts that closely resemble free-living isolates, to nonculturable intracellular symbionts that can be detected only by structural or enzymatic analyses.

CRITERIA OF SYMBIOSIS

Three criteria can be used to establish that luminescence in animals is due to symbiotic luminous bacteria: (1) the culturing of high numbers (10^7/ml or more) of luminous bacteria from the host's light organ; (2) the microscopic visualization of high numbers of bacteria in localized areas and/or in specialized light-emitting structures of the host; and (3) the detection of bacterial luciferase in host tissue.

The isolation and growth of luminous bacteria from light organs has been the most widely accepted proof that the animal's luminescence results from bacterial symbiosis. Host tissue samples are excised, washed with sterile solutions, and homogenized. The samples are diluted and plated onto agar media upon which colonies of bacteria grow. The original numbers of bacteria in the host tissue are inferred from the number of luminous colonies.[20] Since the bacteria grow either in light organs or alone on standard bacteriological media, they are called facultative symbionts. High numbers of luminous colonies have been obtained from the light organs of many fish, indicating that the light is produced by symbiotic luminous bacteria.[7, 11, 21–25] However, since not all luminous bacterial symbionts are easily cultured outside their hosts, one must rely on the other two criteria listed above to reveal the presence of the symbionts.

Bacteria associated with light organs have been studied by light, transmission electron (TEM), and a scanning electron (SEM) microscopy.[9–14, 22, 26–30] While these studies provide a basic knowledge of the structure of bacteria-associated luminous systems, structural observations alone cannot be regarded as proof that the bioluminescence is due to bacterial symbiosis. Luminous bacteria are not structurally distinguishable from other gram-negative rods, so direct observation of host-associated bacteria does not by itself prove that those seen are actually luminous bacteria. Furthermore, in many cases, nonculturable bacteria-like particles, called bacteroids, are seen in light organs. These may appear structurally identical to bacteria or they may be swollen and irregular in shape, sometimes with altered membranes. As a sole criterion, the presence of bacteroids in light organ tissue is inadequate proof that luminescence is of symbiotic origin. In these cases, the third criterion, the biochemical assay for bacterial luciferase, correlated with the structural presence of bacteroids, can prove that luminescence is due to associated bacterial symbionts.

Bacterial luciferase is unique to prokaryotes[18] and can thus be detected in eukaryotic tissue homogenates without purification. It catalyzes the following reaction:

$$FMNH_2 + RCHO + O_2 \rightarrow FMN + RCOOH + H_2O + \text{light}$$

Two of the substrates, reduced flavin mononucleotide ($FMNH_2$) and a long chain aliphatic aldehyde (RCHO), are oxidized by oxygen. An oxidized flavin (FMN), an aliphatic acid, water, and light are produced.[31, 32] From the aldehyde substrate specificity, the FMN requirement, and the kinetic decay profile (i.e., the rate at which luminescence decreases after substrate addition), bacterial luminescence can be easily distinguished from intrinsic animal bioluminescence. The characteristic kinetic decay profile[16, 20, 23, 31] is a result of a long-lived enzyme intermediate with a half-time of decay on the order of seconds. The fact that excess $FMNH_2$ auto-oxidizes to FMN in milliseconds allows each enzyme molecule to turn over only once during the assay. The decay profile is a function of the chain length of aldehyde used,[33] but with a given aldehyde, all bacterial luciferases do not behave identically; on the basis of the decay kinetics, luciferases from *Photobacterium* can be distinguished from both *Beneckea* and *Xenorhabdus,* which are similar to each other. When dodecanal is used, luciferases isolated from *Photobacterium* species display "fast kinetics" while those from the genera *Beneckea* and *Xenorhabdus* display "slow kinetics."[16, 20, 23, 34] Thus, the bacterial luciferase assay permits identification of symbiotic bacteria to genus even when they cannot be cultured. This is of special value in distinguishing adventitious luminous bacteria that are not removed simply by sterile washing. The adventitious bacteria are cultured and their luciferases kinetically characterized. When their luciferase kinetics differ from those of the animal organ, it follows that they are not symbionts specifically associated with the host. This technique has been used to prove that the bacteria occasionally cultured from pyrosomes (see below) are adventitious forms rather than specific light organ symbionts.

Taken together these criteria establish that animal luminescence is due to symbiotic bacteria. Such studies also lend clues as to the physiological states of both symbiotic partners, and to the type of bacteria present. We have placed those systems that qualify as bacterial luminous symbioses into five divisions based on the apparent degree of physiological and morphological integration of the luminous bacteria with their hosts (TABLE 1). They range from extracellular facultative associations to nonculturable intracellular ones.

NONSPECIFIC ASSOCIATIONS OF LUMINOUS BACTERIA

Saprophytes and Gut Symbionts

Luminous bacteria are widespread in the ocean as free-living or nonspecific associates of dead and decaying organic matter or as gut symbionts in the digestive tracts of a wide variety of fish, squid, and other animals. The distribution of these bacteria is a function of several environmental factors including temperature, nutrients, and visible light.[6, 16, 35–38] The digestive tracts of marine animals are probably the most common niche of the marine luminous bacteria. Free-living luminous isolates may simply be excreted symbiotic bacteria.[39] Most taxonomic and physiological studies have been done on free-living isolates, and these studies have served as the basis on which the bacterial symbionts are evaluated.[15, 17]

Nematode Gut Symbionts

Luminous bacteria are not at all common in terrestrial environments.[1] One genus of terrestrial bacteria, *Xenorhabdus,* has been identified; the species are gut symbionts of certain nematodes.[19, 34, 40] It is not known whether they are species specific. The bacteria are carried as nongrowing symbionts within the lumen of the pharynx and intestine of nonfeeding third-stage juvenile nematodes. The nematodes parasitize a variety of insect larvae, spanning many taxonomic orders. Nematodes locate suitable hosts, enter through the mouth, and penetrate the coelomic wall. The luminous bacteria are released into the hemocoel, where they grow rapidly, allowing the nematode to complete its life cycle in the hemocoel. Nematodes that do not harbor bacteria are not pathogenic. While the bacteria proliferate in the hemocoel the insect is visibly luminous. It has been postulated that the luminescence may aid in the distribution of the nematodes by attracting either scavengers or other hosts.[16, 34]

Light Organ Symbioses

The remaining symbioses to be discussed are those of fish, squid, and certain urochordates (pyrosomes), in which luminous bacteria are found in specialized light organs. Since the luminous animals use the light for a variety of purposes, including intraspecific communication, attraction of prey, escape from predators, and visual location,[2, 4, 41, 42] it is not surprising that the light organs are highly variable in structure, size, and location on the host. Yet several features are common to most light organs of fish and squid. Light organs are complex structures, made up of canals or tubules in which luminous bacteria are maintained as extracellular symbionts. The organs, which are usually highly vascularized, have apparently evolved to contain the bacteria in a luminous state, and to transmit and control the emitted bacterial light. In some light organs, the cells lining the tubules are highly specialized for the physiology of symbiosis; involvement in nutrient exchange, regulation of oxygen tension, and maintenance of the proper culture environment for the bacteria have been postulated.[7, 8, 12, 14] Bacterial luminescence is continuous and nondirectional. With time, special structures have evolved in light organs for reflecting and/or occluding light.[2, 4, 9–12, 14, 43, 44] Pores or openings are generally present; they lead either directly to the exterior environment (seawater) or empty into the digestive tract. These pores may serve for the exit of excess bacteria.[2, 9, 11, 12, 14]

The light organs of pyrosomes differ from those of fish and squid.[1, 26, 30] The symbionts are intracellular and the light organs lack tubules, vascularization, and pores. Since no structures for regulating the bacterial light are present, it has been suggested that the animals control the bacterial luminescence at the biochemical level.[30] Intracellular bacteria have been clearly seen in only one pyrosome, *Pyrosoma atlanticum*,[30] although the presence of intracellular symbionts in other species has been inferred from light microscopic studies.[26]

Facultative Light-Organ Symbionts

By definition, bacterial symbionts called facultative can grow outside their hosts on artificial media. All are extracellular in light organs of several fish families, as well as one family of squids (Table 1). The symbiotic bacteria of a given fish family are always the same species. Squid light-organ symbionts are probably family specific as well, although only two examples have been studied (G. Leisman, unpublished). Each fish appears to maintain a pure culture of luminous bacteria in an individual light organ. By the criteria of numerical taxonomy, the bacteria within a light organ are indistinguishable from one another, suggesting that each fish may establish its light organ population from a single bacterium of the appropriate species. However, strain variations are seen between the light organ populations of individuals of a given fish species. How species specificity is established and maintained is unknown.

The association between the Japanese pinecone fish, *Monocentris japonicus*, and its facultative extracellular symbiont, *Photobacterium fischeri*, has been studied the most. The ease with which the bacteria are cultured has permitted extensive taxonomic, nutritional, and physiological studies on pure cultures in the laboratory.[7] The bacteria produce large amounts of pyruvate when grown on glucose,[6] which is the major carbohydrate of fish blood[45] and probably serves as the substrate for the luminous bacterial growth. It has been postulated that the bacteria grow on blood-derived glucose and excrete pyruvate into the tubule lumina [6-8] establishing a nutrient exchange-system reminiscent of other symbioses. Since *P. fischeri* also has a characteristic low oxygen optimum for luciferase synthesis,[5] it has been proposed further [7, 8] that the fish regulates bacterial growth and light emission by controlling the oxygen tension and/or glucose concentration in the light organ. The oxidation of excreted pyruvate by the cells lining the tubules may serve as the mechanism whereby the oxygen tension is lowered.[7, 8]

The concept that the luminescence is regulated by the fish via the fish control of oxygen and glucose supplied to the bacterial population of the light organ is consistent with ultrastructural studies of the light organs of *M. japonicus* [14] (FIGURE 1). The bacteria are contained in paired light organs on the lower jaw; within the organs, the bacteria are packed tightly into tubules (FIGURE 1D) that are surrounded by an extensive network of capillaries that probably supply oxygen and nutrients to the bacteria. The tubules are lined with mitochondria-rich cells that are thought to oxidize pyruvate and thus regulate oxygen tension.[7, 8, 14]

The symbioses in the leiognathid fish differ from those of the monocentrids.[9, 11, 22] Their symbiotic bacteria, *Photobacterium leiognathi,* do not excrete pyruvate in laboratory culture and do not have a low oxygen optimum for bioluminescence.[5, 6] The cells lining the light organs appear to lack the mitochondria-rich cells characteristic of *M. japonicus,* although in other ways, the light organs are structurally similar to those of the other luminous fish.[9, 11] The leiognathid light organs all have openings that empty into the digestive tract of the fish, suggesting that the fish may digest the excess bacteria from the light organ.[8] Although the contacts between the fish and its symbionts are quite close [9, 11] (FIGURE 2), no intracellular symbionts of the *Leiognathidae* have been seen (J.-M. Bassot, personal communication). The light organ structure of one leiognathid species, *L. elongatus,* while structurally similar to that of the others, possesses luminous symbionts that cannot be cultured.[22]

Whereas, structural studies of other facultative symbioses are rare,[1] some cultural data are available. *Photobacterium phosphoreum* is a psychrophilic (cold-loving) luminous bacterial species that is found as a specific light-organ symbiont of many midwater and deep water fishes.[4, 16, 25] Since *P. phosphoreum* has a low oxygen optimum and excretes pyruvate, studies of the light organs of these symbioses will be of great interest from a comparative point of view. Similarly, two genera in the family *Sepiolidae* are known to harbor *P. fischeri* (TABLE 1).

Light Organ Symbionts That Have Not Been Grown in the Laboratory

Several groups of luminous animals contain luminous symbionts (bacteroids) that have not yet been cultured (TABLE 1). In the fishes of the *Anomalopidae* and Ceratioidea, the bacteroids are morphologically similar to bacteria, while in the heteroteuthid squids they are not. Debate still exists as to whether such squids are intrinsically or symbiotically luminous. Within each fish family, the light organs serve similar functions, are similar in location and size, and the bacteroids are structurally similar. However, between families great differences exist.

The Ceratioidea (angler fish) is a suborder of teleosts consisting of 13

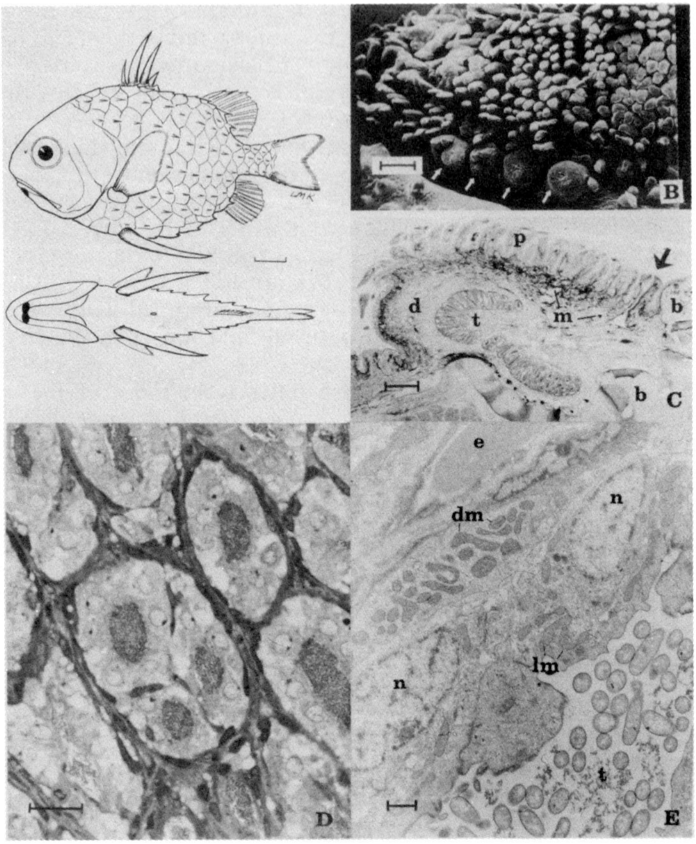

FIGURE 1. Structural features of a light organ of the luminous fish, *Monocentris japonicus* (A) Line drawing of fish and ventral view of the lower jaw, showing location of the light organs (solid black area). Bar = 1.0 cm. (B) Scanning electron micrograph of the dorsal surface of the light organ. Numerous dermal papillae can be seen. The emissary ducts from the light organ emerge at the tips of the four large papillae (arrows). Bar = 0.2 cm. (C) Light micrograph of a sagittal section of the lower jaw. m, melanocytes; t, tubules with bacteria; b, mandibular bone; d, dermal layer; p, dermal papillae. The arrow points to an emissary duct. Bar = 50 μm. (D) Light micrograph showing the tubules of the light organ filled with bacteria. Tubules are aligned with a single layer of cuboidal epithelial cells displaying loose nuclear chromatin and prominent nucleoli supported by connective tissue cells. Blood capillaries are sparse and not readily visible. Bar = 15 μm. (E) Electron micrograph showing the major features of the tubule epithelium. Epithelial cells that make up the lining of the tubules have light-staining mitochondria (lm) with fine cristae. Epithelial cells that are farther away from the tubule lumen and next to the blood capillaries have dark-staining mitochondria (dm) with thick cristae. t, tubule with luminous bacteria; e, erythrocyte visible in capillary; n, nuclei of tubule epithelial cells. Bar = 1.0 μm.

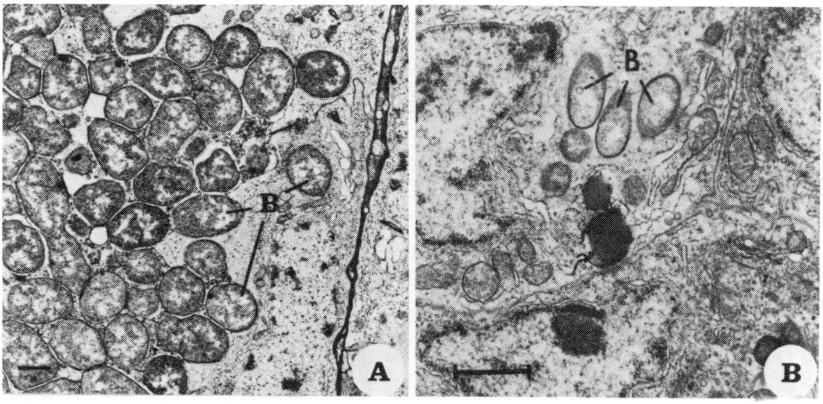

FIGURE 2. Association of luminous bacteria with eukaryotic hosts. These two electron micrographs of the light organs of luminous fishes demonstrate the close association of the culturable light organ symbionts with their eukaryotic hosts. (A) *Photobacterium leiognathi* in the light organ of a pony fish. It is reported that although extensive invagination of the light organ occurs, the bacteria are never truly intracellular (micrograph kindly supplied by J-M Massot). (B) *Photobacterium fischeri* in the light organ of the pinecone fish, *Monocentris japonicus*. The bacteria appear to be intracellular, although only extensive serial sections would prove this. Bar = 1.0 μm.

families, most of which are luminous. In the females of all luminous species, the first dorsal fin ray is elongated and bears a terminal luminous bulb, the esca, which serves to lure prey within reach of the jaws. It may also serve to attract nonluminous males.[44] In one family, Ceratiidae, there are also caudal luminous organs called caruncles, which are modified posterior dorsal fin rays.[4]

Using the light microscope, several workers identified bacteroids in the light organs of ceratioids.[4, 26, 46] Recently, TEM analysis has revealed that they are gram-negative rods, 4–10 μm long and 1 μm in diameter[9, 13, 46] (FIGURE 3). While these extracellular bacteroids have not been cultured, bacterial luciferase activity has been detected in extracts of escas from four different species in three families.[47] Luciferases from escas and caruncles display the fast kinetics characteristic of the genus *Photobacterium*. Furthermore, the caruncles contain bacteroids that are morphologically similar to those in the escal light organs.[4]

The anomalopid family (flashlight fish) contains three genera, all of which are luminous (TABLE 1). These fish all have similar paired light organs located in the orbital sockets beneath the eyes; the luminescence is used for attraction of prey, escape from predators, intraspecific communication, and vision.[42] Harvey[2] postulated that the light organs of *Photoblepharon* contain bacteriods. Bacteroids and light organ ultrastructure have been described for *Anomalops* by Bassot[10] and for *Photoblepharon* by Kessel.[12] Although the bacteria have never been grown in culture, bacterial luciferase activity[28] with slow kinetics[47] has been detected. This supports the view that the bacteroids are derived from luminous bacteria of the genus *Beneckea*; however, on morphological and other grounds, assignment of the bacteroids to any known bacterial genus is not yet possible. Many attempts to culture the bacteria have been made using a large variety of liquid and agar media in which the carbon source, exposure to

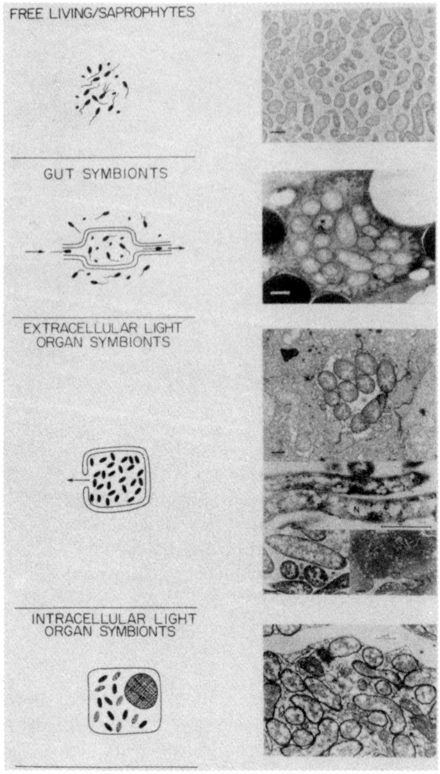

FIGURE 3. Symbiosis of the luminous bacteria. The diagrams on the left schematically depict the various niches that luminous bacteria occupy. These begin at the top with the free living or saprophytic forms. Like the gut symbionts, the saprophytic associations are neither species specific nor permanent. The associated luminous bacteria are usually not found as pure cultures, and grow equally well in the unassociated state. The extracellular light organ symbionts can be either nonculturable or culturable. For the latter group, the symbionts exhibit species specificity and are usually found as pure cultures. Morphological studies indicate that the nonculturable symbioses are also highly specific. The intracellular symbionts represent the most highly adapted of the luminous bacterial symbionts found lying near the mitochondria and inside of pyrosome cells.

On the right, several electron micrographs representative of the diagrams on the left are shown. The free-living bacteria are *Photobacterium fischeri,* while the gut symbionts are *Xenorhabdus luminescens* seen in the pharyngeal cavity of a nematode (courtesy of Dr. G. Poinar, Jr.). Several examples of extracellular light organ symbionts are shown: top, *Photobacterium leiognathi* from the light organ of a pony fish (courtesy of Jean-Marie Bassot); middle, nonculturable symbionts of *Photoblepharon palpebratus* (courtesy of Dr. M. Kessel); lower left, nonculturable symbionts of the angler fish *Oneirodes acanthias* (courtesy of Dr. W. T. O'Day); lower right, nonculturable symbionts of the squid *Heteroteuthis hawaiiensis* (courtesy of Dr. R. E. Young). The intracellular nonculturable symbionts are from the genus *Pyrosoma* (courtesy of Dr. G. E. Mackie). Note the cellular mitochondria lying next to the intracellular bacteria. All bars = 1.0 μm.

oxygen, temperature, and other factors were independently varied.[1, 12, 20, 28, 47] Even extracts of squid[28] or of the host fish (Nealson, unpublished) could not support growth of the symbionts.

The internal features of the light organs are similar in the anomalopid genera, although the mechanisms for occluding the light differ.[43] The light organs are highly vascularized and backed with darkly pigmented tissue overlaid with a layer of guanine crystals that serve to reflect and direct the light. A series of tubules with mitochondria-rich cells at their bases funnel into a number of pores at the surface of the organ. The tubules contain densely packed luminous bacteroids that are typically rod-shaped, about 0.5 μm in diameter and from 1–5 μm in length (FIGURE 3). They have tufts (triads) of sheathed polar flagella, and contain inclusions that are assumed to be polybetahydroxybutyric acid (PHB) granules. On morphological grounds, the bacteroids in the anomalopids are not assignable to either *Beneckea* or *Photobacterium*. Luminous *Beneckea* species have only a single sheathed polar flagellum and lack PHB granules. The only *Photobacterium* species that forms PHB is *P. angustum,* a nonluminous species with an unsheathed flagellum. The only *Photobacterium* species with sheathed flagella, *P. fischeri,* is unable to form PHB under any conditions tested.[17]

A final group of nonculturable light-organ symbionts are found associated with squids of the genus *Heteroteuthis.* This genus is in the pelagic subfamily Heteroteuthinae of the family Sepiolidae. Members of a related subfamily, the Sepiolinae, contain culturable luminous symbionts (TABLE 1). Although attempts to culture bacteria from the Heteroteuthinae have failed,[27, 47] bacterial luciferase with fast kinetics has been detected in homogenates of the light organs of several specimens of *Heteroteuthis hawaiiensis.*[47] Electron microscopy of the light organ has revealed bacteria-like particles with two membranes, a plasma membrane of the eukaryotic host cell and an outer bacterial membrane (R. E. Young, personal communication). The size of these bacteroids in the light organ of *H. hawaiiensis* is 1.5 × 3 μm (FIGURE 3). *Heteroteuthis dispar,* an Atlantic squid similar to *H. hawaiiensis,* has also been examined.[27] Bacteria-like particles were described, but were interpreted as spherical elongated vesicles, not bacteroids. Light-organ extracts of *H. dispar* have not yet been analyzed for bacterial luciferase.

Intracellular Light-Organ Symbionts

Harvey[1] thought all symbiotic luminous bacteria were extracellular. Since the light organs of the urochordates lack the tubules, vascularization, and pores characteristic of fish and squid light organs, which harbor extracellular symbionts, it was concluded that the urochordates were not symbiotically luminous. Other workers reported intracellular symbionts in these systems (reviewed by Büchner[26]). According to Büchner, most, and perhaps all, of the luminous marine urochordates contain intracellular luminous symbionts. Harvey, however, discounted the possibility of intracellular symbiosis in urochordates because: (1) the morphology of the intracellular inclusions was unlike free-living luminous bacteria; (2) no biochemical evidence supported the contention that luminous bacteria were involved; and (3) the luminescence of the pyrosomes is not continuous as are luminous bacteria in culture or fish light organs.

New evidence has emerged that suggests that intracellular luminous symbionts are indeed present in pyrosomes. Ultrastructural studies of the light organs of *Pyrosoma atlanticum*[30] (FIGURE 3) show intracellular bacteroids similar to the bacteria described by Büchner.[26] These bacteroids are large (2–3 μm × 20 μm) and quite distinct from other symbiotic luminous bacteria. Although no mechanism is suggested, it is also proposed that the control of the luminous flashes, characteristic of the pyrosomes, is at the biochemical (substrate) level, and is mediated by the pyrosome.[30] That luminous bacteria are involved is consistent with our recent discovery that homogenates of several different pyrosomes contain bacterial luciferase.[47] Although luminous bacteria were occasionally cultured from the pyrosomes, which have such small light organs that pyrosome tissue and fluid were always contained in the homogenates, their numbers were never in proportion to the luminescence obtained. Furthermore, the luciferase assayed from the disrupted pyrosomes displayed fast kinetics, while that of the cultured bacteria showed slow kinetics, proving that those bacteria growing in culture were not those involved with the luminescence measured from the disrupted pyrosomes.

If the pyrosome particles are symbiotic luminous bacteria, they are unique. When the pyrosomes were gently homogenized in either sea water or bacterial-lysis buffer,[20] a continuous glow characteristic of luminous bacteria was seen. All other symbiotic bacteria suspended in lysis buffer stop glowing abruptly because they lyse and release their luciferase into solution. The intensity of the glow in the homogenate was not affected by the addition of any of the substrates of the bacterial luminescent reaction. If the pyrosome homogenate was gently sonicated for 2–5 seconds, the glow ceased, and as with usual bacterial lysates, bacterial luciferase was detectable upon substrate addition. The release of luciferase from pyrosome homogenates on gentle sonication occurred in either seawater or in lysis buffer. Normal luminous bacteria suspended in seawater do not release their luciferase on gentle sonication. These results suggest that pyrosome luminescence is due to a unique bacteria-derived system. The bacteroids are intracellular and the bacterial luciferase is apparently both packaged and controlled in a way different from all other prokaryotes.

FROM EXTRACELLULAR TO INTRACELLULAR: THE ESTABLISHMENT OF ORGANELLES

In considering the establishment of intracellular organelles via symbiotic processes, symbioses between luminous bacteria and marine animals may provide several insights. Two questions arise: (1) What kinds of structural and physiological modifications occur in both partners as a result of symbiosis? (2) By what mechanisms are intracellular symbioses established? To elucidate the discussion, we have arranged some exemplary luminous bacteria–animal symbioses in a series from extracellular to intracellular (FIGURE 3).

With regard to host modifications, the animals with light organs are greatly modified, but there seems to be little direct relationship between the extent of the symbiotic integration as determined by the metabolic independence of the bacteria and the complexity of the light organs. Light organs of fish in which the symbionts can be cultured are as complex as those from which the symbionts have not yet been cultured. This generalization may become obsolete when the physiological and biochemical details of the light organ tissue and the nutritional and environmental requirements of the bacteria are better understood.

However, there does appear to be a relationship between the extent to which the bacteria are integrated into the host tissue and the degree of bacterial modification. With the exception of a few methionine auxotrophs,[17] the free-living forms are prototrophic (they have no special nutritional requirements). In this respect, the facultative symbionts are virtually identical to the free-living bacteria. Thus, these forms, which may be ecologically closely related, show little nutritional adaptation to their symbioses. In addition, little structural modification is seen in the facultative symbionts, which are morphologically indistinguishable from their free-living counterparts. Nevertheless, there are some metabolic properties of the facultative symbionts that may make them physiologically suited for a symbiotic existence.[8] Both *P. fischeri* and *P. phosphoreum* exhibit luminescence maxima at low oxygen concentrations,[5] and both excrete pyruvate when grown on a glucose medium.[6] The quantitative aspects of these features are probably adaptations to life in the light organ, specifically involvement in nutrient exchange and the control of luminescence.[6–8] Since free-living *P. fischeri* and *P. phosphoreum* isolates, as well as some *Beneckea* species (which have not yet been proved to be light organ symbionts), also excrete pyruvate, this argument must be viewed with caution. With additional information, it may become clear how metabolic properties (oxygen sensitivity, pyruvate production, etc.) of these gram-negative rods may have preadapted these two genera of marine bacteria to form associations with animals.

The various families of luminous animals tend to be uniform; if one member possesses either facultative or nonculturable symbionts, they all do. As one proceeds to the nonculturable symbionts, significant modifications that preclude growth in culture must be occurring, although these have not been identified. However, one fish and one squid family have some genera with culturable and others with nonculturable symbionts. In the leiognathids, *Leiognathus elongatus* contains nonculturable bacteria while the other species have facultative ones.[22] In sepiolid squids, *Heteroteuthis hawaiiensis* harbors nonculturable symbionts, while other genera contain culturable ones (TABLE 1). Such observations suggest that the growth requirements of these nonculturable symbionts may be among the easiest to determine. Bacterial morphological variation is also seen between the groups of extracellular light organ symbionts. While all facultative symbionts are structurally similar to bacteria, the nonculturable ones vary from obvious bacteria in the anomalopids and leiognathids to highly modified bacteroids in the heteroteuthids. In the case where members of a given taxon differ (for example, in the squids), the various related species provide material for the study of the changes that occur in partners as a result of the symbiotic life style. New symbioses may be found that represent states between those outlined here.

The luminous symbionts of pyrosomes exhibit the most modification relative to free-living luminous bacteria. The most obvious change is their location within the host cells. Concomitant with this movement to an intracellular existence, other changes are seen; the bacteroids are nonculturable, structurally highly modified, and altered with respect to the control of bioluminescence. This latter change may imply the loss by the symbiont, and acquisition by the host, of some component of the bacterial luminescent system. Whether the integration is at the gene, gene product, and/or metabolic level,[48] is yet to be determined. This symbiosis is potentially of major importance for understanding the details of biochemical and structural host–symbiont integration.

How do intracellular symbioses become established? Both the studies of Bassot[9-11] and Tebo *et al.*[14] with facultative symbionts are relevant. Extensive close contacts between the extracellular bacteria and animal tissue were seen in both cases. In the leiognathid fish, Bassot observed infolding of light organ membranes such that the bacteria were completely surrounded by animal membranes yet were still lying topologically exterior to the animal cell (FIGURE 2A). In *M. japonicus*, bacteria were observed within animal cells[14] (FIGURE 2B). Extensive anastomoses of the light organ membranes, with subsequent pinching off, might be responsible for these occurrences; phagocytosis by the fish light-organ epithelial cells might also be a mechanism by which the bacteria become intracellular (FIGURE 2). With these mechanisms in mind, it is easy to visualize the transitions from extracellular to intracellular symbiosis (FIGURE 3). Comparable bacterial and algal symbionts with both extracellular and intracellular phases are known in both marine[49] and fresh water[50] systems and suggest questions similar to those raised by the systems discussed here. Are bacteria always present? Do they have an extracellular phase? How are they transmitted to pyrosome progeny? Such questions form the basis for further research.

SUMMARY

Several species of three genera of luminous bacteria from marine and soil environments are known to form specific symbioses with fish, squid, urochordates, and nematodes. These bacteria contain a unique and easily detectable enzyme, bacterial luciferase, which allows the detection of the bacteria even when they cannot be cultured from animal tissue. The bacteria–animal associations vary; they range from transient, nonspecific gut symbionts in many fishes to highly specific, nonculturable, intracellular symbionts in pyrosomes. The study of these microbe–animal symbioses may allow understanding of the alterations that occur in the partners during the establishment of intracellular organelles. These studies are incomplete, in particular the level of partner integration, the nature of the metabolic exchange, the mechanism of transmission of symbionts to offspring, and the identification of the cellular inclusions as modified bacteria are not always known. However, the outline of what is likely to be a continuous evolutionary sequence of luminous bacterial symbionts with their various animal hosts is becoming clear. As more of these symbioses are studied, we anticipate that new systems will be found that represent states between those described here, and that the luminous bacteria will provide a living model for the gradual evolution from free-living microbes to intracellular organelles.

ACKNOWLEDGMENTS

We gratefully acknowledge the generous cooperation of our colleagues, without whose help we would not have been able to assemble this paper. This help included statements, ideas, and unpublished micrographs from Drs. Jean-Marie Bassot, Q. Bone, P. N. Dilly, P. J. Herring, M. Kessel, G. O. Mackie, G. O. W. T., O'Day Poinar, Jr., G. Thomas, and R. E. Young. Thanks also go to Dr. L. Margulis for help with the manuscript, and to the U.S. National Science Foundation for financial support to K.H.N. (PCM 74–14788).

REFERENCES

1. HARVEY, E. N. 1952. Bioluminescence. Academic Press, Inc. New York, N.Y.
2. HARVEY, E. N. 1921. A fish with a luminous organ designed for the growth of luminous bacteria. Science **53:**314–315.
3. YASAKI, Y. 1928. On the nature of the luminescence of the knight fish, *Monocentris japonicus.* J. Exp. Zool. **50:**495–505.
4. HERRING, P. J. & J. G. MORIN. 1978. Bioluminescence in fishes. *In* Bioluminescence in Action. P. J. Herring & J. G. Morin, Eds.: 272–329. Academic Press, Inc. New York, N.Y.
5. NEALSON, K. H. & J. W. HASTINGS. 1977. Los oxygen is optimal for luciferase synthesis in some bacteria: ecological implications. Arch. Microbiol. **112:** 9–16.
6. RUBY, E. G. & K. H. NEALSON. 1977. Pyruvate production and excretion by the marine luminous bacteria. App. Env. Microbiol. **34:**164–169.
7. RUBY, E. G. & K. H. NEALSON. 1976. Symbiotic association of *Photobacterium fischeri* with the marine luminous fish *Monocentris japonicus:* a model of symbiosis based on bacterial studies. Biol. Bull. **151:**574–586.
8. NEALSON, K. H. 1979. Alternative strategies of symbiosis of marine luminous fishes harboring light-emitting bacteria. Trends in Biochem. Sci. **4:**105–110.
9. BASSOT, J.-M. 1966. On the comparative morphology of some luminous organs. *In* Bioluminescence in Progress. F. H. Johnson & Y. Haneda, Eds.: 557–610. Princeton University Press. Princeton, N.J.
10. BASSOT, J.-M. 1968. Les organes lumineux a bacteries symbiotiques du Teleosteen *Anomalops,* Donnees histiologiques. Bull. Soc. Zool. France **93:**569–579.
11. BASSOT, J.-M. 1975. Les organes lumineux a bacteries symbiotiques de quelques teleosteens Leiognathides. Arch. Zool. Exp. Gen. **116:**359–373.
12. KESSEL, M. 1977. The ultrastructure of the relationship between the luminous organ of the teleost fish *Photoblepharon palpebratus* and its symbiotic bacteria. Cytobiologie **15:**145–158.
13. O'DAY, W. T. 1974. Bacterial luminescence in the deep-sea angler anglerfish *Oneirodes acathias* (Gilbert, 1915). Contr. Sci. 255, Nat. Hist. Mus., Los Angeles, Calif.
14. TEBO, B. M., D. S. LINTHICUM & K. H. NEALSON. 1979. Luminous bacteria and light emitting fish: ultrastructure of the symbiosis. Biosystems **11:**269–280.
15. BAUMAN, P. & L. BAUMANN. 1977. Biology of the marine enterobacteria: Genera *Beneckea* and *Photobacterium.* Ann. Rev. Microbiol. **31:**39–61.
16. NEALSON, K. H. & J. W. HASTINGS. 1979. Bacterial bioluminescence: its control and ecological significance. Microbiol. Rev. (In press.)
17. REICHELT, F. L. & P. BAUMANN. 1973. Taxonomy of the marine luminous bacteria. Arch. Microbiol. **94:**283–330.
18. HASTINGS, J. W. 1968. Bioluminescence. Ann. Rev. Biochem. **37:**597–630.
19. THOMAS, G. M. & G. O. POINAR, JR. 1979. A new entomopathogenic, nematophilic, bioluminescent bacterium, *Xenorhabdus luminescens* sp. nov. (Enterobacteriaceae: Eubacteriales) associated with the terrestrial nematode *Heterorhabditis bacteriophora.* Intern. J. Syst. Bacteriol. (In press.)
20. NEALSON, K. H. 1978. Isolation, identification, and manipulation of luminous bacteria. *In* Methods in Enzymology. M. DeLuca, Ed. **57:**153–166. Academic Press, Inc. New York, N.Y.
21. FITZGERALD, J. M. 1977. Classification of luminous bacteria from the light organ of the Australian pinecone fish, *Cleidopus gloriamaris.* Arch. Microbiol. **112:**153–156.
22. HANEDA, Y. & F. I. TSUJI. 1976. The luminescent systems of pony fishes. J. Morphol. **150:**539–552.
23. HASTINGS, J. W. & G. MITCHELL. 1971. Endosymbiotic bioluminescent bacteria from the light organ of pony fish. Biol. Bull. **141:**261–268.

24. Reichelt, J. L., K. H. Nealson & J. W. Hastings. 1977. The specificity of symbiosis: Pony fish and luminescent bacteria. Arch. Microbiol. **112:**157–161.
25. Ruby, E. G. & J. G. Morin. 1978. Specificity of symbiosis between deep-sea fishes and psychrotrophic luminous bacteria. Deep-Sea Res. **25:**161–167.
26. Büchner, P. 1965. Endosymbiosis of Animals with Plant Microorganisms. John Wiley & Sons, Inc. New York, N.Y.
27. Dilly, P. N. & P. J. Herring. 1978. The light organ and ink sac of *Heteroteuthis dispar* (Mollusca:Cephalopoda). J. Zool. **186:**47–59.
28. Haneda, Y. & F. I. Tsuji. 1971. Light production in the luminous fishes *Photoblepharon* and *Anomalops* from the Banda Islands. Science **173:**143–145.
29. Herring, P. J. 1978. Bioluminescence of invertebrates other than insects. *In* Bioluminescence in Action. P. J. Herring, Ed.: 199–240. Academic Press. London.
30. Mackie, G. O. & Q. Bone. 1978. Luminescence and associated effector activity in *Pyrosoma* (Tunicata:Pyrosomida). Proc. R. Soc. London Ser. B **202:** 483–495.
31. Hastings, J. W. & Q. Gibson. 1963. Intermediates in the bioluminescent oxidation of flavin mononucleotide. J. Biol. Chem. **238:**2537–2554.
32. Hastings, J. W. & K. H. Nealson. 1977. Bacterial bioluminescence. Ann. Rev. Microbiol. **31:**549–595.
33. Hastings, J. W., J. Spudich & G. Malnic. 1963. The influence of aldehyde chain length upon the relative quantum yield of the bioluminescent reaction of *Achromobacter fischeri*. J. Biol. Chem. **238:**3100–3105.
34. Poinar, G. O., G. Thomas, M. Haygood & K. H. Nealson. 1979. Properties of the luminescent bacterium associated with the terrestrial nematode, *Heterorhabditis bacteriophora*. Soil Biol. Biochem. (In press.)
35. Ruby, E. G., E. P. Greenberg & J. W. Hastings. 1979. Planktonic luminous bacteria: patterns of species distribution in the water column (In press.)
36. Ruby, E. G. & K. H. Nealson. 1978. Seasonal changes in the species composition of luminous bacteria in nearshore seawater. Limnol. Oceanogr. **23:**530–533.
37. Shilo, M. & T. Yetinson. 1979. Physiological characteristics underlying the distribution patterns of luminous bacteria in the Mediterranean Sea and the Gulf of Elat. Appl. Env. Microbiol. (In press.)
38. Yetinson, T. & M. Shilo. 1979. Seasonal and geographic distribution of luminous bacteria in the eastern Mediterranean Sea and the Gulf of Elat. Appl. Env. Microbiol. (In press.)
39. Ruby, E. G & J. G. Morin. 1979. Luminous enteric bacteria of marine fishes: a study of their distribution, densities and dispersion. Appl. Env. Microbiol. **38:**406–411.
40. Khan, A. & W. M. Brooks. 1976. A chromogenic bioluminescent bacterium associated with the entomophilic nematode *Chromonema heliothidis*
41. Buck, J. 1978. Functions and evolutions of bioluminescence. *In* Bioluminescence in Action. P. J. Herring, Ed.: 419–460. Academic Press, Inc. London.
42. Morin, J. G., A. Harrington, N. Krieger, K. H. Nealson, T. O. Baldwin & J. W. Hastings. 1975. Light for all reasons. Science **190:**74–76.
43. McCosker, J. E. 1977. Flashlight fishes. Sci. Amer. 236:106–114.
44. Pietsch, T. W. 1974. Osteology and relationships of ceratioid anglerfishes of the family Oneirodidae, with a review of the genus *Oneirodes* Lutken. Bull. Los Angeles County Mus. Nat. Sci. No. 28, 113 pp.
45. Dittmer, D., Ed. 1961. Blood and Other Body Fluids. Federation of American Societies for Experimental Biology. Washington, D.C.
46. Hanson, K. & P. J. Herring. 1977. Dual bioluminescent systems in the anglerfish genus *Linophryne* (Pisces:Ceratioidea). J. Zool. **182:**103–124.
47. Leisman, G., D. H. Cohn & K. H. Nealson. 1980. Bacterial origin of luminescence in marine animals. Science **208:**1271–1273.

48. MARGULIS, L. 1976. Genetic and evolutionary consequences of symbiosis. Exp. Parasit. **39**:277–349.
49. THORINGTON, G. 1980. Transmission of the algal and bacterial symbionts through the sexual cycle of *Hydra viridias*. Ph.D. thesis. Boston University. Boston, Massachusetts.
50. SMITH, D. C. 1979. From extracellular to intracellular: The establishment of a symbiosis. Proc. R. Soc. London Ser. B **204**:115–130.

EVIDENCE ON THE ORIGIN OF EUKARYOTIC MITOCHONDRIA FROM PROTEIN AND NUCLEIC ACID SEQUENCES *

M. O. Dayhoff and R. M. Schwartz

National Biomedical Research Foundation
and the
Department of Physiology and Biophysics
Georgetown University Medical School
Washington, D.C. 20007

Many proteins and nucleic acids are "living fossils" in the sense that their structures have been dynamically conserved by evolution over billions of years—recognizably related forms are found in eukaryotes and prokaryotes. These eukaryotes and prokaryotes are believed to be of common evolutionary origin, having evolved by a great number of small changes in their genetic sequences.[1, 2] Recently, enough sequence information has become available from prokaryotes to permit the construction of a biologically comprehensive phylogenetic tree.[3–5] This tree was derived by objective mathematical methods from the chemical structures of the sequences; a line of reasoning independent of information from biological traits, metabolic capacities, or fossil evidence.

Tree Representation of Evolutionary History

Evolutionary history is conveniently represented by a tree. Each point on the tree corresponds to a time, a macromolecular sequence, and a species within which the sequence occurred. There is one point representing the first divergence that corresponds to the earliest time and to the ancestral organism and sequence. Time advances on all branches of the tree emanating from this point. The topology of the branches gives the relative order of events. During evolution, sequences in different species have gradually and independently accumulated changes yielding the sequences found today in the various organisms represented at the ends of the branches. On the trees in this paper we have drawn the lengths of the branches proportional to the amount of evolutionary change in the sequences. The topological connections between the sequences are determined by the methods described below. It is not usually possible to infer the position of the point of earliest time from the sequence data. An exception occurs when there has been a duplication of a sequence in an ancestor common to all species on the tree and this duplication has been well preserved. The chronological dates of points on the tree cannot be inferred from sequences, but a number of divergence times have been deduced from fossil evidence.

Testing Validity of Methods

The mechanisms of evolution acting on the amino acid sequences are sufficiently well understood that the usefulness of any proposed method for recon-

* Supported by NASA (Contract NASW-3317) and the National Institutes of Health (Grants GM-08710 and RR-05681).

0077–8923/81/0361–0092 $01.75/0 © 1981, NYAS

structucting phylogeny can be tested by simulating sequences connected by a known history of divergence and amount of evolutionary change. The computer model that we have used is simple, yet adequate for this purpose.[6] Starting with a sequence of average amino acid composition, a residue is selected for change by use of a random number. The probability of selecting a given residue is proportional to the average mutability of this amino acid in protcins. The replacement amino acid is likewise selected by a random number, the probability of selection being proportional to its probability of replacing the original amino acid. The mutated sequence is now subjected to a subsequent mutation and so on until sequences corresponding to all evolutionary successions on the tree have been formed. A matrix of differences between terminal or "observed" sequences can be calculated and the accuracy of the reconstructed history using matrix information can be assessed. The fidelity of the reconstruction of the ancestral nodal sequences by various methods can also be tested. By considering many groups of simulated sequences, with each group related by the same evolutionary history but generated with different random numbers, a measure of the accuracy of a given method can be assessed and different methods compared.

Methods for Reconstructing Phylogeny

There are two major approaches to reconstructing evolutionary history; one is based on the matrix of differences between sequences[7,8] and the other is based on inferred ancestral sequences at the nodes of the tree.[9,10] Many variations of each method have been suggested.

We have found the least-squares matrix method most reliable for distantly related sequences: it was used for constructing the trees in this paper. In this method, sequences are aligned to reflect the genetic changes in evolution and a matrix of percentage difference between them is calculated. These percentages are corrected for inferred superimposed mutations to yield the "observed" matrix of accepted point mutations per 100 residues (PAMs). The computer program determines branch lengths on a tree of given topology that make a weighted least-squares fit of the reconstructed matrix to the observed matrix. (The weights are introduced to compensate for the unequal precision with which the observed matrix elements reflect the actual number of mutations. These weights are inversely proportional to the size of the matrix element.) We have evaluated many variations of criteria to infer the true evolutionary history. Of these, that topology for which the sum of the absolute values of the branch lengths is smallest gives the best estimate.

The least-squares matrix method makes no assumptions about rates of change on the various branches in a tree and says nothing about the point of earliest time. By contrast, if the classification based on the unweighted pair-group matrix method[11] is identified as the evolutionary tree, equality of rates is assumed, and the point of earliest time in the tree is located between the most distantly related groups. Unfortunately, this procedure is also sometimes referred to as the matrix method.[12] Clearly, assuming equality of rates is not valid, especially for major prokaryote groups that diverged between one and one-half and three billion years ago and for which there has frequently been twice as much, and sometimes ten times as much, change in one line as in another during the same time interval.

In the ancestral sequence method, for a given topology, ancestral sequences, one for each node, are constructed in such a way that the number of mutations on the whole tree is minimized. All likely topologies are considered, and that one requiring the smallest total number of mutations is chosen as best. Our simulations have shown that the ancestral sequence method is not nearly as reliable for sparse trees of distantly related sequences as the matrix method (FIGURE 1).

THE EVOLUTION OF ORGANISMS AND THEIR SEQUENCES

The mathematical methods yield the best topological relationship between sequences. The sequence divergences may have arisen from gene duplications within a species as well as from species divergences, and the gene duplications (except for those confined to a single branch) must be identified before inferences about the phylogenetic history of the organisms can be made on the basis of the sequences. Whenever two related genes are found in one species, a gene duplication must be inferred at the earliest node joining the sequences. The same evolutionary history subsequent to the duplication will be reflected by each gene. Confusion of interpretation arises only when a duplication has occurred but cannot be identified from the data. In this case an earlier gene duplication may be misinterpreted as a species divergence. As data accumulate, the significant gene duplications will be identified because both forms will persist in some species.

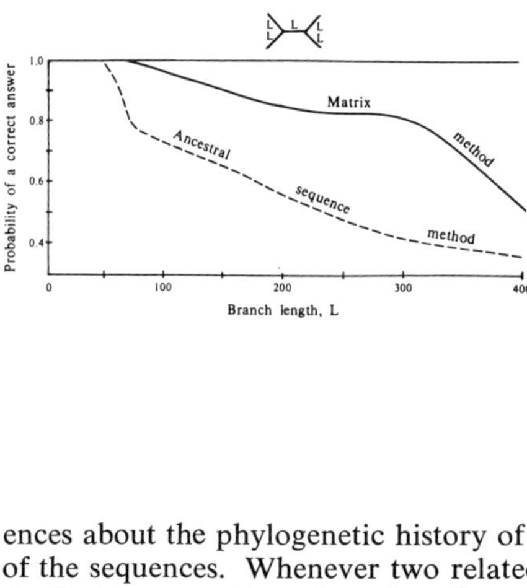

FIGURE 1. Probability of inferring the correct topology from sequences of simulated evolutionary connection. A history of five equal intervals of mutational distance was used, as shown at the top of the figure. Ten sets of sequences of 100 residues were generated for each mutational distance (L=25, 50, 75, 100, 125, 150, 175, 200, 225, 250, 300, 400). Random events were assigned according to the average mutability and mutation pattern of each amino acid. In totaling the number of correct topologies inferred, a single correct answer counted 1, a wrong answer 0, a two-way tie 0.5, and a three-way tie 0.33. Smoothed curves through the data points are shown. The matrix method is clearly superior. This figure is adapted from Dayhoff, M. O.[13]

PHYLOGENY OF MAJOR PROKARYOTE GROUPS

Three and one-half years ago there was sufficient sequence data for us to make a comprehensive evolutionary tree including the position of the root, derived objectively from sequence data.[3-5] This tree, based on sequences of the *c*-type cytochromes, ferredoxins, and 5S ribosomal RNA molecules (FIGURE 2), is consistent with newer sequence data. Our working hypothesis was that, for these sequences that perform basic metabolic functions, genetic transfer between major types of bacteria, followed by permanent acceptance in evolution, is very infrequent. All such basic sequences within an organism share the same phylogenetic history, and any one of them can be used to infer its course. There are, then, hundreds of kinds of fundamentally important protein and nucleic acid sequences in a bacterium, each reflecting the same evolutionary history. If accepted genetic transfer between species has been infrequent, each event would be recognized as more data accumulate. If, on the other hand,

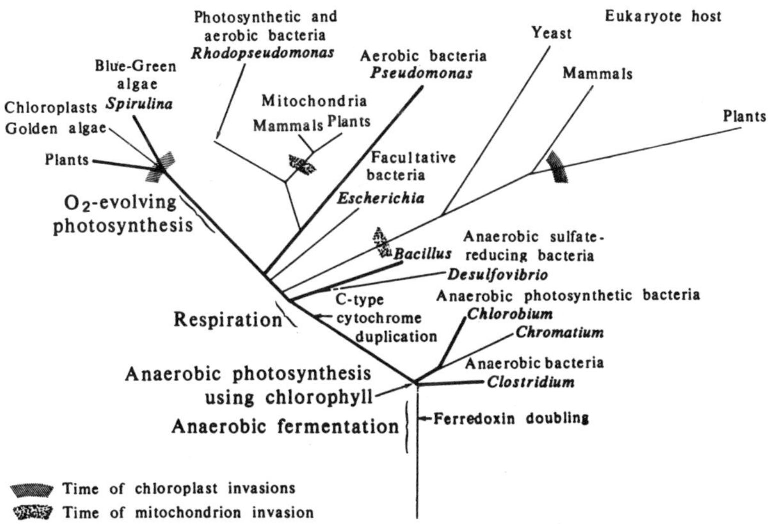

FIGURE 2. Phylogenetic tree of prokaryotes and eukaryotes. This tree presents an overview of evolution based on ferredoxin, *c*-type cytochrome, and 5S ribosomal RNA sequences. The trees derived from each superfamily were scaled so that overlapping portions were comparable. A combined matrix of distances was calculated for the six species appearing on more than one tree. The heavy lines represent the tree calculated from this combined matrix. The lighter lines represent branches scaled from a single tree and added to the combined tree. (From Schwartz, R. M. & M. O. Dayhoff.[5] By permission of *Science*. Copyright 1978 by the American Association for the Advancement of Science.)

accepted genetic transfer has been frequent, phylogenies based on different molecules would conflict. However, large numbers of conflicts are not evident in the known data and therefore this last course of events seems unlikely.

The genetic doubling, most obvious in the clostridial-type ferredoxins but observed in the sequences from all of the species, allows us to locate the base on the tree of FIGURE 2. The hypothetical ancestral sequence that was formed just after the duplication was included with the present day sequences in the calculations. Moreover, because the species whose sequences have changed least since this doubling event were all anaerobic, heterotrophic bacteria, it is likely that the ability to live fermentatively is primitive.

The tree describes the evolution of photosynthesis using chlorophyll, starting with the development of the ability to synthesize this class of compounds. Three families of photosynthetic bacteria are represented: Chromatiaceae, Chlorobiaceae, and Rhodospirillaceae. The divergence of these from the anaerobic nonphotosynthetic bacteria was quite early in evolution. Sequence data from the recently described aerobic, photosynthetic bacterial genus *Chloroflexus* would make an interesting addition.

The next major event shown on the trunk of the tree leading to the blue-green algae is the development of respiration. The time of divergence of *Bacillus* and *Desulfovibrio* probably marks the presence of major components of this adaptation. The final elements were evolved separately because these groups differ in their terminal electron acceptor in respiration.

The branch that we identify as the eukaryote host diverged at about the same time as *Bacillus* and *Escherichia.* The recently elucidated 5S ribosomal RNA sequence from *Halobacterium salinarium*[14] diverges from the eukaryote host branch near its base (FIGURE 1 in Reference 30). That *Halobacterium* should be our close relative is further suggested by its use of two biochemical systems that are found in eukaryote cytoplasmic systems but are very rare in bacteria. Methionine transfer RNA, rather than *N*-formylmethionine transfer RNA, is used for the initiation of protein synthesis and a light-sensitive system using retinal is found.

The mitochondria are most closely related to the third family of photosynthetic bacteria, the Rhodospirillaceae. The topology of the tree invites the speculation that this ancestral bacterium was photosynthetic until shortly before or just after it invaded the host.

The blue-green algae and chloroplasts of the eukaryotes are capable of oxygen-releasing photosynthesis, and the available sequence data point to this capacity as having evolved only once. The chloroplasts are grouped together with free-living prokaryotes. Humans would never have had an oxygen-producing algal ancestor.

The main stream of evolution was very probably carried by species that performed bacterial photosynthesis, contrary to most classifications of bacteria. Bacterial photosynthesis was independently lost in a number of lines: *Desulfovibrio–Bacillus, Escherichia coli,* the eukaryote host, *Paracoccus denitrificans,* and *Pseudomonas–Azotobacter.*

SYMBIOTIC ORIGIN OF MITOCHONDRIA AND CHLOROPLASTS

There are two schools of thought concerning the origin of eukaryote organelles: one is that they arose by the compartmentalization of the DNA within the cytoplasm of an evolving protoeukaryote;[15,16] the other is that they arose from free-living forms that invaded a host and established symbiotic relationships with it.[17,18] In the nonsymbiotic theory, all genes arose within the organism; homologues found in both the nucleus and the organelles arose by gene duplication. Organelles formed in this way would probably not be capable of independent existence outside of the host cell, and their continued competitive extracellular existence in evolution would be most unlikely. The host also would seldom survive competitively in evolution after the loss of well-established mitochondria and chloroplasts because of uniquely useful genes found only in these organelles. Therefore, this theory would place the mitochondria and chloroplasts together close to the host on the tree, with no intermixed prokaryote branches. The symbiotic theory, on the other hand, proposes that the chloroplast was originally a free-living blue-green alga; other symbionts include the mitochondrion, which was originally a free-living aerobic bacterium. Their current status as organelles would have gradually evolved from these symbioses. In this theory, mitochondrial, chloroplast, and host genes are expected to show evidence of recent common ancestry with separate types of contemporary free-living prokaryote forms; the host and organelles would occur on different branches that could also contain free-living forms. The tree of FIGURE 2 clearly supports the symbiotic theory.

EVOLUTION OF MITOCHONDRIAL SEQUENCES

Of the several thousand genes that may have been present in the free-living form that has become the mitochondrion, only a few remain in the DNA of the mitochondrion today. These are transcribed and the messenger translated by the mitochondrial apparatus. An excellent review of these proteins is given by Tzagoloff *et al.*[19] Many other genes are thought to have originated in the mitochondrion but been transferred to the nucleus. Their messengers are translated by the cytoplasmic apparatus and the mature proteins are transported to the inside of the mitochondrion or to its inner membrane where they form functional complexes with the gene products produced in the mitochondrion. Enough sequence information to be informative about mitochondrial evolution is presently known from the ATP-synthetase proteolipid subunit, from phenylalanine and initiator methionine transfer RNAs, and from *c*-type cytochromes.

ATP-SYNTHETASE PROTEOLIPID SUBUNIT

The evolutionary tree constructed from the sequences of the ATP-synthetase proteolipid subunit[20] (FIGURE 3) is consistent with that expected from our general scheme of FIGURE 2. The point of earliest time and the connection of the eukaryote host should be near the nodal ancestor of *E. coli* and thermophilic bacterium PS3. The proteins that function in the mitochondria appear together on a separate branch, showing the same recent evolutionary history as cytochrome *c*, even though the sequence from baker's yeast is coded in the mitochondrial genome and translated in the mitochondrion, whereas the sequence in *Neurospora crassa* is coded in the nucleus and translated in the cytoplasm. The close similarity of the two fungal sequences supports the view that the *N. crassa* gene was originally from the mitochondrion. In this case, the place where the protein functions rather than the location of its gene is indicative of evolutionary history. Clearly, the amount of change in the mitochondrial line is greater than that in the chloroplast or the bacterial lines.

TRANSFER RNA SEQUENCES

We have reviewed the literature and constructed an evolutionary tree from phenylalanine transfer RNA sequences.[21] The chloroplast sequences, the bacterial sequences, and the eukaryote cytoplasmic sequences from plant, animal,

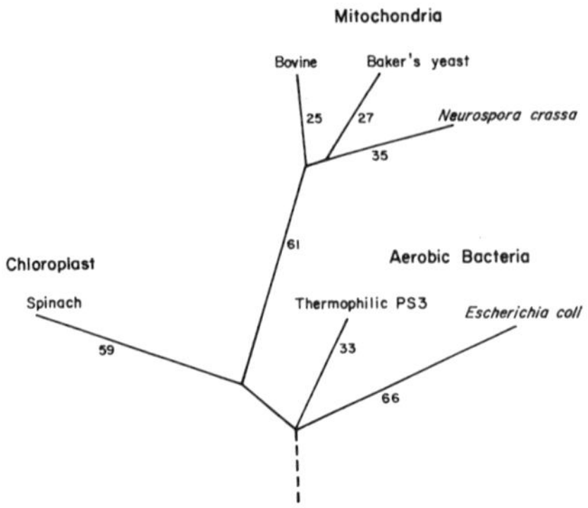

FIGURE 3. Evolutionary tree from ATP-synthetase, proteolipid subunit of membrane factor, F_0. The root of the tree has been placed in accord with that in FIGURE 2. The mitochondrial, chloroplast, and bacterial sequences form three distinct groups, as expected.

and fungus form three distinct groups. The recently determined yeast mitochondrial sequence is about 50% different from all of the others. It would form a very long, separate branch on the tree for which the position of connection cannot be deduced accurately.

We have also reviewed the literature and constructed an evolutionary tree for the initiator methionine tRNA sequences of bacteria, blue-greens, and the eukaryote cytoplasmic constituents.[21] The initiator methionine tRNA sequence from the mitochondrion has also changed more than the other sequences and it is very different from any of them. Its connection to the tree also cannot be placed precisely.

C-TYPE CYTOCHROME SEQUENCES

The only molecules whose sequences give us great insight into the origin of mitochondria are the *c*-type cytochromes. In this case, at least for multicellular forms, the cytochrome *c* gene is located in the nucleus, and the protein is translated in the cytoplasm and is transported into the mitochondrion where it functions. The rate of change of this gene is comparable with that of the free-living bacterial forms to which it is closely related. We have reviewed the literature on *c*-type cytochrome sequences.[22] The iso-2 sequence from *Rhodospirillum* (*Rs.*) *fulvum* has recently been determined.[23]

The details of the region of the tree of FIGURE 2 involving the Rhodospirillaceae, *Pseudomonas,* and *Paracoccus* are shown in FIGURE 4. All 12 species of purple, nonsulfur photosynthetic bacteria, the Rhodospirillaceae, that are described in *Bergey's Manual of Determinative Bacteriology*[24] are represented. Apparently, morphology is not an accurate guide to the evolution of these bacteria. Sequences from all three genera of Rhodospirillaceae are found intermixed on the tree. There was a gene duplication in the line leading to *Rs. fulvum* and *Rs. molischianum;* iso-1 and iso-2 cytochromes c_2 are found in each of these species. The gene duplication is marked by a diamond in the figure. Both genes show the species divergence to be rather recent.

It has been suggested that the c_2 and c_{551} sequences in the Rhodospirillaceae have resulted from a gene duplication.[25] Sequences from 13 species are now known, and in no case is there sequence evidence for both types of protein in any one species. This lack of evidence, in spite of extensive investigation, speaks against a gene duplication. Furthermore, evidence from 16S ribosomal

RNA catalogues[26] confirms that the sequences from *Rhodopseudomonas* (*R.*) *gelatinosa* and *Rs. tenue* form one group of similar sequences, whereas those from *R. sphaeroides, R. capsulata, Rs. rubrum, R. viridis, Rhodomicrobium* (*Rm.*) *vannielii,* and *R. palustris* form a second group, just as is found for the cytochromes. This confirms our suggestion that the branching of these Rhodospirillaceae is a phylogenetic one, not caused by a gene duplication, because it is very unlikely that, for two different macromolecules, the same two species out of eight would be found to form a distinct group by chance. Furthermore, it supports our hypothesis that the integrity of the genome coding for structures of basic importance has been conserved.

The details of the region of the tree of FIGURE 2 showing the eukaryote mitochondria are shown in FIGURE 5. The divergence of the mitochondria of animals, plants, and fungi from one another must have been very close in time. The invertebrate phyla shown probably share a single most recent common ancestor with the vertebrates. The cytochrome *c* sequences place the divergence of the prawn (a crustacean) early on the insect line, as expected, and the divergence from one another of the garden snail (a mollusc), the starfish (an echinoderm), and the brandling worm (an annelid) as indistinguishably close in time. The cytochrome *c* sequence clearly places *Enteromorpha intestinalis,* a multicellular chlorophyte alga, on the plant branch.

The phylogenetic tree derived from selected *c*-type cytochrome sequences is shown in FIGURE 6 with the very interesting sequences from *R. globiformis*[27] and *Ectothiorhodospira halophila*[28] recently determined by Richard Ambler.

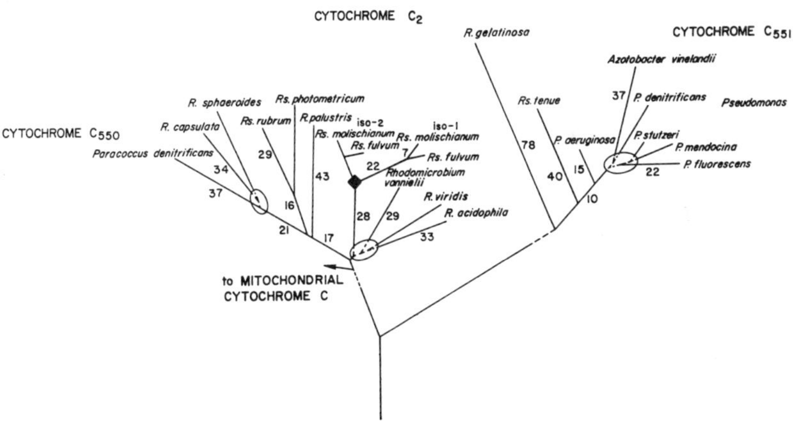

FIGURE 4. Evolutionary tree of cytochromes c_2 and c_{551}. This tree shows the corresponding branches of FIGURE 2 in more detail. The two subtrees were constructed separately. The matrix elements for the mitochondrial *c*-type cytochromes were averaged for this calculation. The base of the c_2–c_{550} subtree was placed between these sequences and those of mitochondrial cytochrome *c* because these c_2 and c_{550} sequences share a unique deletion close to one of their heme-binding cysteines. The base of the c_2–c_{551} subtree was centered because these sequences are distant from the rest of the tree and the point of their earliest divergence could not be precisely estimated. The distance between the two subtrees was estimated from an alignment omitting multiple-residue insertions and deletions. Branch lengths are given in PAMs. The genera *Rhodopseudomonas, Rhodospirillum,* and *Rhodomicrobium* are abbreviated *R., Rs.,* and *Rm.*, respectively. Adapted from Schwartz, R. M. & M. O. Dayhoff.[22]

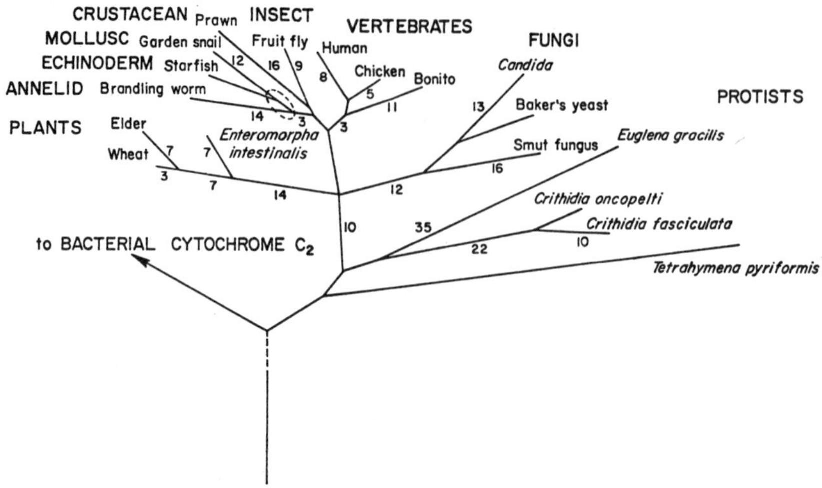

FIGURE 5. The mitochondrial cytochrome *c* evolutionary tree, showing the corresponding branch of FIGURE 2 in more detail. The single ciliate sequence, from *Tetrahymena,* appears to diverge first. The three flagellate sequences share a unique substitution of an alanine for one of the two cysteines that bind the heme group, substantiating their appearance as a separate group. Adapted from Schwartz, R. M. & M. O. Dayhoff.[22]

The *Ectothiorhodospira* sequence falls in the anaerobic portion of the tree as expected for this anaerobic, photosynthetic organism. There is not yet sufficient information to fix the base of the tree exactly, but it is close to the position shown. The blue-green and chloroplast sequences are very similar and their matrix elements have been averaged in making this tree. A gene duplication, which yielded cytochromes c_5 and c_{551} in *Pseudomonas,* occurred very early on the cytochrome tree. The blue-greens, Rhodospirillaceae, and mitochondria all arose after the duplication and their evolution is deduced from one of the genes. We would expect that a second *c*-type cytochrome would still be found in many of the species on the tree, and a second main branch repeating the upper part of the tree shown would then be inferred from this other gene. This duplication causes no errors of phylogenetic interpretation in the tree shown here.

The group from which mitochondria originated, the Rhodospirillaceae, are all shown at the top of the tree. The *R. globiformis* sequence clearly diverges from the *Euglena* and *Crithidia* branches close to the point of their divergence from one another. The best solution would have *R. globiformis* coming from the *Crithidia* branch. However, the cytochrome *c* sequences from *Euglena* and *Crithidia* share a rare substitution of an alanine for one of the heme-binding cysteines, an event that has not been observed in any other cytochrome *c* sequence examined. We presume that this change occurred once in an ancestor to *Crithidia* and *Euglena* just subsequent to the separation from *R. globiformis.* At the *R. globiformis* divergence, the ancestral cytochrome occurred in a free-living form. Very soon thereafter, an endosymbiosis occurred to form the mitochondrion. The occurrence of a separate symbiosis among flagellates is consistent with the lack of mitochondria in the Trichomonadida, an anaerobic flagellate group. This group either diverged before the endosymbiosis or else the symbiosis was so newly established that the mitochondrion could be easily discarded in an anaerobic niche.

It seems likely that in the species shown (FIGURE 6) there were at least two other symbioses established, to account for the *Tetrahymena* mitochondrion and that of the multicellular forms. These could both have been established by direct ancestors of *R. globiformis*. More likely they were produced by related forms some time after their divergences from the *R. globiformis* line, as suggested in the figure. It is possible that these related forms have also given rise to other extant bacteria that are yet to be described. In each line photosynthetic ability was lost, either before or soon after the symbiosis. The connections on the tree drawn in a single line show the divergences among the free-living Rhodospirillaceae. The parallel lines indicate that the biological form is now the mitochondrion within the specified organisms. The solid lines need not reflect the evolutionary connections of the cytoplasmic constituents of the species shown.

Possibly, the mitochondria of animals, plants, and fungi originated from a single symbiosis preceding the divergences of these kingdoms. The mitochondria of these lines show a great similarity in the genetic material that has been

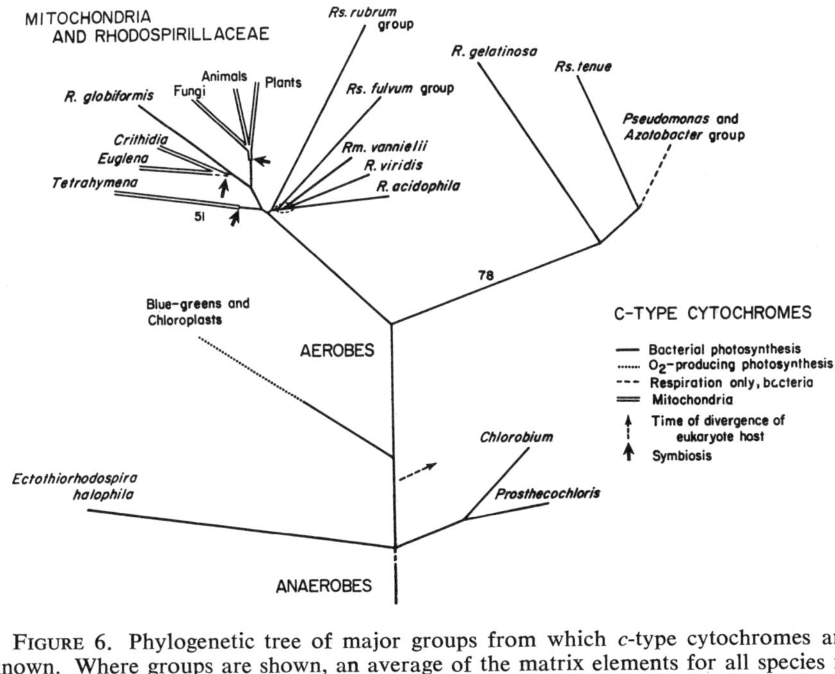

FIGURE 6. Phylogenetic tree of major groups from which *c*-type cytochromes are known. Where groups are shown, an average of the matrix elements for all species in the group was used. The *Rs. rubrum* group includes five other sequences most closely related to *Rs. rubrum* as shown in FIGURE 4. The *Rs. fulvum* group includes the four related sequences from *Rs. fulvum* and *Rs. molischianum*. The *Pseudomonas* and *Azotobacter* group includes the six sequences shown in FIGURE 4. Animals include human, chicken, and screw-worm fly sequences. Fungi include the three sequences of FIGURE 5 and the *Neurospora* sequence. Plants include elder, wheat, and *Enteromorpha intestinalis*.

retained and in that transferred to the nucleus. There have been considerable further independent changes in the genetic information in the mitochondria since the divergences of the kingdoms.

There are other possible sequences of events, one involving the gain of the mitochondrion at the bottom of the eukaryote tree, followed by the divergence of the lines leading to *R. globiformis* and the flagellates. This would mean that *R. globiformis* had a eukaryote ancestor, a most unlikely explanation.

In conclusion, the sequence evidence overwhelmingly favors the origin of the eukaryote mitochondria by endosymbiosis. The eukaryote mitochondria shown were all derived from ancestors of *R. globiformis* and their divergence times from these ancestors were more recent than their divergence times from any of the other 12 species of Rhodospirillaceae. Therefore, we predict that *R. globiformis* is, on the whole, more similar to the eukaryote mitochondria, particularly those of the flagellates and higher organisms, than are any of the other Rhodospirillaceae shown.

Acknowledgments

We would like to thank Dr. Richard Ambler for supplying us with the cytochrome c_{551} sequence from *Ectothiorhodospira halophila* in advance of its publication, and K. Lawson for drafting the figures.

References

1. Bryson, V. & H. J. Vogel, Eds. 1965. Evolving Genes and Proteins. Academic Press, Inc. New York, N.Y.
2. Dayhoff, M. O., W. C. Barker & L. T. Hunt. 1976. Atlas of Protein Sequence and Structure. M. O. Dayhoff, Ed. Vol. **5** (Suppl. 2): 9–19. National Biomedical Research Foundation. Washington, D.C.
3. Schwartz, R. M. & M. O. Dayhoff. 1978. Comparative Planetology. C. Ponnamperuma, Ed.: 225–242. Proceedings of the Third College Park Colloquium on Chemical Evolution. College Park, Md. Sept. 1976. Academic Press, Inc. New York, N.Y.
4. Dayhoff, M. O. & R. M. Schwartz. 1978. Origin of Life. H. Noda, Ed.: 547–560. Proceedings of the Second ISSOL Meeting and the Fifth International Conference on the Origin of Life. Kyoto, Japan. April 1977. Center for Academic Publications Japan/Japan Scientific Societies Press. Tokyo, Japan.
5. Schwartz, R. M. & M. O. Dayhoff. 1978. Science **199**:395–403.
6. Dayhoff, M. O. & R. M. Schwartz. 1979. Atlas of Protein Sequence and Structure. M. O. Dayhoff, Ed. Vol. **5** (Suppl. 3): 345–352. National Biomedical Research Foundation. Washington, D.C.
7. Fitch, W. M. & E. Margoliash. 1967. Science **155**:279–284.
8. Dayhoff, M. O. 1979. Atlas of Protein Sequence and Structure. M. O. Dayhoff, Ed. Vol. **5** (Suppl. 3): 1–10. National Biomedical Research Foundation. Washington, D.C.
9. Eck, R. V. & M. O. Dayhoff. 1966. Atlas of Protein Sequence and Structure. Vol. **2**:164–171. National Biomedical Research Foundation. Washington, D.C.
10. Fitch, W. M. & J. S. Farris. 1974. J. Mol. Evol. **3**:263–278.
11. Sneath, P. H. A. & R. S. Sokal. 1973. Numerical Taxonomy. 230–235. W. H. Freeman & Co. San Francisco, Calif.
12. Hori, H. & S. Osawa. 1979. Proc. Natl. Acad. Sci. USA **76**:381–385.
13. Dayhoff, M. O. 1976. Fed. Proc. **35**:2132–2138.

14. Nazar, R. N., A. T. Matheson & G. Bellemare. 1978. J. Biol. Chem. **253:** 5464–5469.
15. Raff, R. A. & H. R. Mahler. 1972. Science **177:**575–582.
16. Uzzell, T. & C. Spolsky. 1974. Am. Sci. **62:**334–343.
17. Margulis, L. 1970. Origin of Eukaryote Cells. Yale University Press. New Haven, Conn.
18. Margulis, L. 1974. *In* Handbook of Genetics. R. C. King, Ed. Vol. **1:**1–41. Plenum Press, New York, N.Y.
19. Tzagoloff, A., A. Macino & W. Sebald. 1979. Annu. Rev. Biochem. **48:**419–441.
20. Sebald, W., J. Hoppe & E. Wachter. 1979. *In* Function and Molecular Aspects of Biomembrane Transport. E. Quagliariello *et al.*, Eds. 63–74. Elsevier North-Holland. Amsterdam. (The sequences are reported in this paper.)
21. Schwartz, R. M. & M. O. Dayhoff. 1979. Atlas of Protein Sequence and Structure. M. O. Dayhoff, Ed. Vol. **5** (Suppl. 3): 313–325. National Biomedical Research Foundation. Washington, D.C. (The tRNA sequence literature is reviewed here and our alignments and trees are shown.)
22. Schwartz, R. M. & M. O. Dayhoff. 1979. Atlas of Protein Sequence and Structure. M. O. Dayhoff, Ed. Vol. **5** (Suppl. 3): 28–43. National Biomedical Research Foundation. Washington, D.C. (The c-type cytochrome sequence literature is reviewed here and our alignment and phylogenetic trees are shown.)
23. Ambler, R. P., M. Daniel, J. Hermoso, T. E. Meyer, R. G. Bartsch & M. D. Kamen. 1979. Nature. **278:**659–660.
24. Buchanan, R. E. & N. E. Gibbons, Eds. 1974. Bergey's Manual of Determinative Bacteriology. 8th Ed. Williams and Wilkins Co. Baltimore, Md.
25. Uzzell, T. & C. Spolsky. 1980. Ann. N.Y. Acad. Sci.
26. Woese, C. R., J. Gibson & G. E. Fox. 1980. Nature **283:**212–214.
27. Ambler, R. P. 1979. Abstr. Third Int. Symp. Photosynthetic Prokaryotes (Oxford). J. M. Nichols, Ed.: Abstr. E17. University of Liverpool. Liverpool, England.
28. Ambler, R. P. Sequence of cytochrome c_{551} from *Ectothiorhodospira halophila,* strain BN 9626. Submitted to the Atlas of Protein Sequence and Structure Reference Data Collection. August 1979.
29. Schwartz, R. M. & M. O. Dayhoff. 1980. Ann. N.Y. Acad. Sci.

Discussion of the Paper

Dr. A. Cronquist (*New York Botanical Garden, Bronx, N.Y.*): I am in a bit of a quandary on the interpretation of your first major phylogenetic tree. I want to concentrate attention on one particular anomaly and ask about it.

I see the word "plants" in three places on that tree without further identification. In one of those places it is associated closely with golden algae and blue-green algae which is all right from the botanical standpoint, but in the other two places, quite different places, it is associated with the word mammals.

Dr. M. O. Dayhoff: There was a common ancestor between plants and animals and fungi that was rather recent. The 5S RNA says that and that's basic to the definition of a species.

Dr. Cronquist: Do you mean that the plants that you are referring to come from three quite different origins?

Dr. Dayhoff: Let me point out that there are three phylogenetic histories in plants. There is the phylogenetic history of the chloroplast. There is the phylogenetic history of the mitochondrion, and there is the phylogenetic history of the nuclear material, and these are different. The recent plants embody all these three, the organelles and the nucleus. A billion years ago however, these were free-living forms and they had different histories from the nuclear material. Therefore, we need really to draw three trees.

DR. L. MARGULIS: That's exactly the same question I was going to ask. You have this tree with plants three times and you have three invasion steps on it. Could you explain all of that?

DR. DAYHOFF: This is the eukaryote host. That's the cytoplasmic constituents the 5S ribosomal RNA, and so on. Plants have a nucleus, and so on, and this is that part of plants, the host part of plants. If you are going to have a symbiosis, there is one organism that engulfed the chloroplast and the mitochondrion. So, this is the host. That's the evolutionary history of the host. This is the evolutionary history of the chloroplast part of the plant, how it connects to the bacteria and the mitochondrion connects in yet another way to the bacteria.

DR. CRONQUIST: You are suggesting a very close relationship between mammals and plants with regard to the mitochondria?

DR. DAYHOFF: Yes.

DR. F. CASTORA (*State University at Stony Brook, Stony Brook, N.Y.*): I want to ask about the evidence that's accumulating of the rapid evolution of organelles with respect to the evolution in the nucleus. I am intrigued by the interaction of the nuclear components with the organelle components and how one could possibly have a five to ten times greater evolutionary rate in the mitochondrion, and still be able to get this intimate interaction among the various subunits of certain proteins like the cytochrome oxidase subunits. I can think of a possible theory that would say something like there is rapid evolution in the organelle and local rapid evolution of those genes in the nucleus which are involved with subunits which will assemble in the organelle and whether the rapid evolution in the mitochondrion just gives rise to silent kinds of mutations.

DR. DAYHOFF: We don't know why. We just count them in the sequences we have and as we put them on the tree, certainly, the two transfer RNAs were changing very rapidly. The cytochrome has been taken into the nucleus that is not, you know, changing particularly rapidly.

EUKARYOTIC FEATURES OF MITOCHONDRIAL TRANSCRIPTION AND GENE STRUCTURE IN YEAST *

Joseph Locker, Richard M. Synenki, Sylvie Merten, and Murray Rabinowitz

Departments of Pathology, Medicine, Biochemistry, and Biology
The University of Chicago
Chicago, Illinois 60637

INTRODUCTION

Mitochondrial biogenesis results from a dialogue between two cellular genomes, one in the nucleus and one in the mitochondrion itself (see References 1–3 for recent reviews). Several differences in the protein synthesizing machinery of these two cellular systems suggest a different evolutionary origin for each (see Reference 4 for an extensive discussion). For example, mitochondrial ribosomes are sensitive to chloramphenicol and erythromycin and are insensitive to cycloheximide, like bacterial ribosomes, and unlike cytoplasmic ribosomes. Similarly, mitochondrial protein synthesis is initiated by formylmethionyl tRNA, as in bacterial protein synthesis, while cytoplasmic protein synthesis is initiated by methionyl tRNA. The existence of two genomes and the bacteria-like qualities of the mitochondrial system have led to extensive speculation about a distinct origin for the mitochondrial system, including the possibility that mitochondria arose as an endosymbiotic prokaryote in the cytoplasm of a primitive eukaryotic ancestor. A number of laboratories, including our own, have been studying the biogenesis and molecular biology of mitochondria in the yeast *Saccharomyces cerevisiae* not from a desire to understand the evolutionary relationships in a primitive eukaryote, but because of the ease with which yeast mitochondria may be manipulated both biochemically and genetically. The detailed analysis of the yeast mitochondrial genome has shown a number of surprising features. Mitochondrial genes are organized like nuclear genes in higher eukaryotes and not like bacterial genes. They contain intervening sequences, are surrounded by relatively nontranscribed spacers, and maintain the enzymatic machinery for processing the transcripts from such genes. Theoretical analysis of the origin of mitochondria must now consider the eukaryotic as well as the prokaryotic features of the mitochondrial genome. In this presentation, after reviewing the mitochondrial genome and its gene products, the molecular biological analysis from our laboratory that demonstrates the eukaryotic features of the yeast mitochondrial genome is reviewed. The implications of these findings on the origin of mitochondria and the possible position of the yeast mitochondrion in mitochondrial evolution are discussed.

* Supported by the National Institutes of Health (Grants HL-04442 and HL-09172), the American Cancer Society (Grant NP-281), and the Louis Block Fund of the University of Chicago.

0077–8923/81/0361–0105 $01.75/0 © 1981, NYAS

Organization of the Mitochondrial Genome

The contribution of the mitochondrial genome to mitochondrial biogenesis is rather limited; it consists of only a few protein subunits of the major respiratory complexes and some structural RNAs required for mitochondrial protein synthesis (TABLE 1).[1-4] Yeast mitochondrial gene products are essentially the same as those of the animal mitochondrial genome: a large and a small ribosomal RNA, about 25 tRNAs, 3 subunits of the cytochrome oxidase complex, one cytochrome *b* subunit, and 4 ATPase subunits. These gene products account for only 20% of the 75,000 base-pair yeast mitochondrial genome and about 60% of the 15,000 base-pair animal genome. A few years ago, this observation was often interpreted to mean that some mitochondrial genes were yet to be discovered, but several years of additional research have provided little evidence of such missing mitochondrial genes. In yeast, part of the extra DNA seems to be accounted for by the presence of numerous intervening sequences in mitochondrial genes. A single intervening sequence is present in the gene for the large ribosomal RNA,[5,6] and the cytochrome *b* gene is a mosaic of multiple exons and introns.[2,4] Our preliminary experiments indicate the same kind of structure for the gene encoding the largest cytochrome oxidase peptide. There appear to be few, if any, genes left to be found in the spacious yeast mitochondrial genome, but there is a great deal to be learned about the organization of the genes already discovered.

Mitochondrial Transcription in Grande and Petite Yeast[7-9]

Electrophoretic analysis of RNA purified from grande mitochondria[7,9-11] demonstrated over 40 mitochondrial RNA species other than tRNA, ranging in size from 200 to 9,500 nucleotides (FIGURE 1). The large number and the

TABLE 1

YEAST GENOME: MITOCHONDRIAL GENE PRODUCTS[1-4]

	Animal		Yeast	
		15,000 base pairs		75,000 base pairs
DNA	amino acids	nucleotides	amino acids	nucleotides
rRNA		1700		3600
rRNA		1000		2000
tRNA (25)		2000		2000
cytochrome oxidase peptides				
I	320	960		960
II	240	720		720
III	160	480		480
cytochrome *b*	250	720		720
ATPase subunits				
	240	720		720
	160	480		480
	160	480		480
	60	170		170
ribosomal protein	–	–	320	960
TOTAL (% of genome)		9,400 (63%)		14,300 (19%)

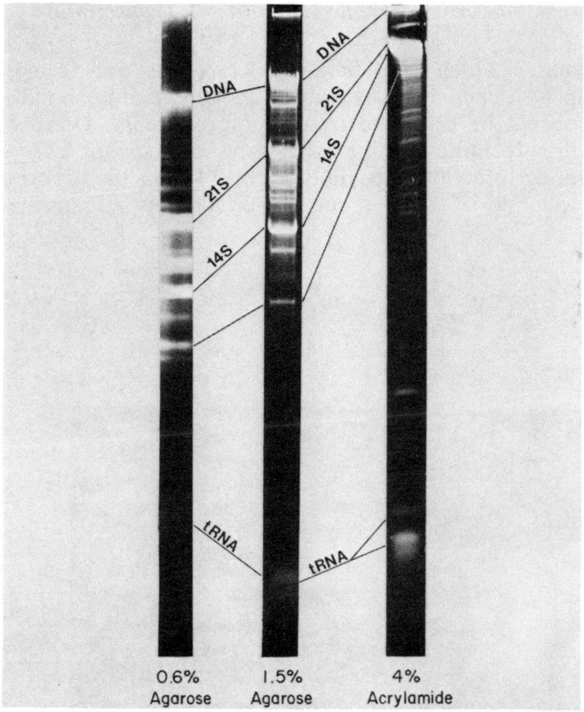

FIGURE 1. Electrophoresis of mitochondrial RNA in 6 M urea.[11] More than 40 mitochondrial RNA species other than tRNA have been resolved by using different gel formulations.

large size of the transcripts suggested large transcription units and extensive posttranscriptional processing of RNA.

We then analyzed mitochondrial RNA from cytoplasmic petite mutants.[8] Petites are deletion mutants in which a segment (usually small) of the mitochondrial genome is retained and amplified (see Reference 12 for review). Because of the extensive loss of genetic information, petites do not have mitochondrial protein synthesis. The petite mitochondrial genome may range from as little as 0.1% to as much as 80% of the grande mitochondrial genome, but petites with such large genomes as the latter are unusual. Our study was confined to a series of petites with large, unrearranged genomes that had been genetically and physically mapped. We found that these petites, despite the lack of mitochondrial protein synthesis, actively transcribed mitochondrial RNA and processed it in a fashion similar to the grande (FIGURE 2). Using the defined petite genomes, we were able to construct a deletion map of about 20 mitochondrial RNA species (FIGURE 3). Although the map is qualitative in many ways, it shows groupings of different sized transcripts into the same regions of the genome in a fashion that seems to indicate extensive RNA processing from large primary transcripts.

Cytochrome Oxidase Peptide 1 Transcription

Experiments in which mitochondrial RNA was transferred from gels to diazotized paper [8, 13] and then hybridized to labeled mitochondrial DNA confirmed this concept of extensive transcript processing. Genetic and physical studies had already indicated that this gene was spread over a fairly large region, amounting to about 10.5 kilobases or 14% of the genome, even though it encodes a peptide of only about 40,000 daltons. Transferred RNA was

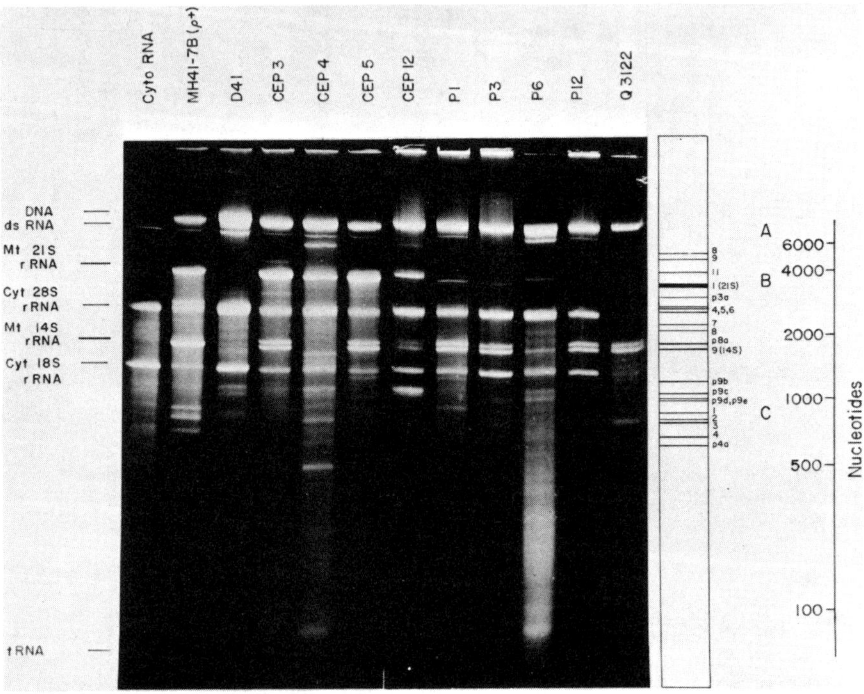

FIGURE 2. Electrophoresis of mitochondrial RNA from 10 petite strains and grande MH41–7B.[8] Petites show three types of mitochondrial RNA bands: 1) RNA species that comigrate with grande species, such as 21S and 14S rRNA; 2) "group-specific" new bands that are found in multiple petite strains (designated with a lower case p in the schematic diagram on the right); and 3) "strain-specific" new bands found only in individual petite strains.

hybridized to specific DNA segments from this region (FIGURE 4) and showed a group of 12–14 transcripts ranging from 800 to 6,600 nucleotides. DNA hybridizations from across the entire 10 kilobase region showed a consistent pattern of bands distinct from that of other regions of the genome. Further, the most abundant species, a 2,600 base transcript, hybridized across the whole 10 kilobase region, requiring the presence of several large introns. More detailed analysis of this gene and its transcripts is now being carried out.

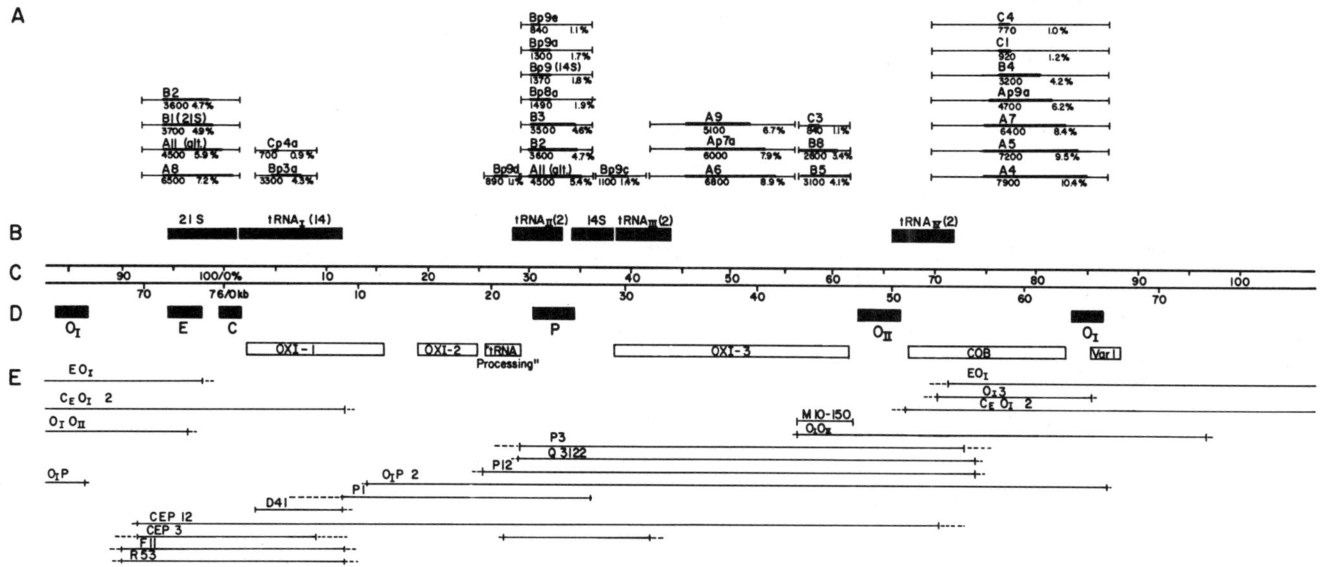

FIGURE 3. Deletion map of mitochondrial transcripts.[8, 11] The upper region of the figure shows the size and approximate location of about 20 mitochondrial transcripts. Transcript size is indicated by the heavy line, and the deletion interval to which it has been mapped, by a lighter line. The middle region of the figure shows a current map of mitochondrial genes, and the bottom part of the figure shows the extent of the petite genomes used to construct the deletion intervals. The size, grouping, and number of transcripts seem to indicate extensive RNA processing and large transcript units.

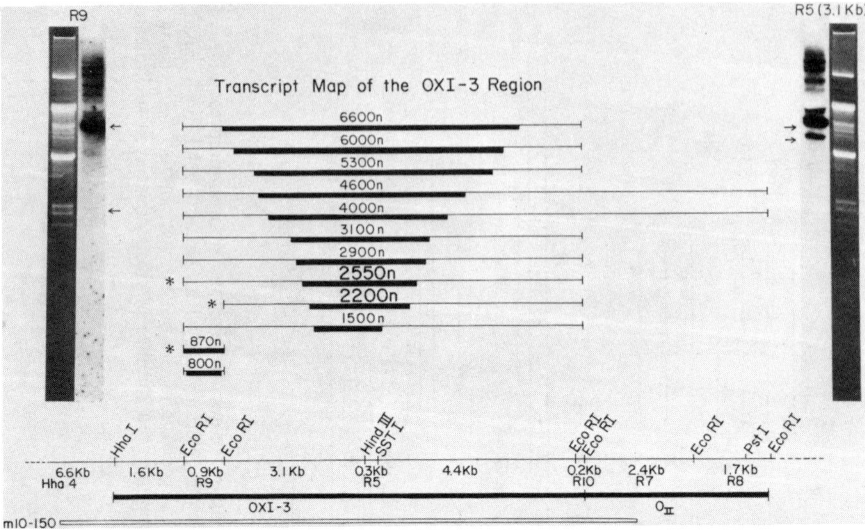

FIGURE 4. Transcript map of the OXI-3 region. Mitochondrial RNA was transferred to diazobenzyloxymethyl paper and hybridized to labeled DNA fragments. Two such hybridizations are illustrated, to *Eco*R1 fragment R9, and a 3.1 kb segment of *Eco*R1 fragment R5. Fragments all across the 10 kb OXI-3 region give a characteristic hybridization pattern that is distinct from that of other regions of the genome. The most prominent transcript (calculated to be 2,550 nucleotides) hybridizes all across the entire 10 kb region, indicating the presence of multiple introns in the gene that has been removed from this transcript.

21S rRNA Transcription and Processing

For analysis of the 21S rRNA gene, we relied extensively on petite F11, which has a 12 kilobase genome surrounding the 21S gene. This gene contains a single large intron. For this study, we constructed a high resolution restriction-enzyme map of the F11 genome. FIGURES 5 and 6 show hybridization of purified 21S rRNA and total RNA to F11 DNA. A map of the gene was constructed using this information. Transfer hybridization of mitochondrial RNA was then carried out, using a DNA fragment located within the intron and another fragment from within the large exon at the 5′-end of the molecule (FIGURE 7). These hybridizations demonstrated several RNA species larger than 21S rRNA containing an intron and/or an extension. We also carried out electron microscopic analysis of R-loop hybrids using purified high molecular-weight RNA, and demonstrated several RNA species containing the unexcised intron segment and/or an extension of the 3′-end (FIGURES 8 & 9). This extension may be removed by a single cleavage or in segments. A processing scheme is summarized in FIGURE 10. The same RNA species were identified in both F11 and the grande, although the relative amounts are slightly different. A map summarizing the restriction-enzyme data and several different kinds of hybridization experiments is shown in FIGURE 11. The ribosomal and tRNA transcription units are adjacent, and we are currently trying to identify even larger precursors that may contain both transcription units. A striking feature of these transcription units is that they are surrounded by extensive nontranscribed regions. Genetic analysis, incidentally, has not indicated any additional gene products from these regions.

Origin of the Mitochondrial Genome

The data presented in this paper demonstrates several features of gene structure and RNA processing common to both the yeast mitochondrion and the nucleus of higher eukaryotes, features not found in the prokaryotes so far analyzed. A simple conclusion might be that mitochondria did not have a prokaryotic origin. However, this is really an unwarranted conclusion. The identification of introns and RNA splicing in mitochondria presents the same problem as finding these features in the nucleus. How did these special features of eukaryotic gene structure arise in the first place? Certainly the prokaryotic genome was the evolutionary precursor to the eukaryotic nuclear genome. The fact that introns and RNA splicing have not been found in *E. coli* or

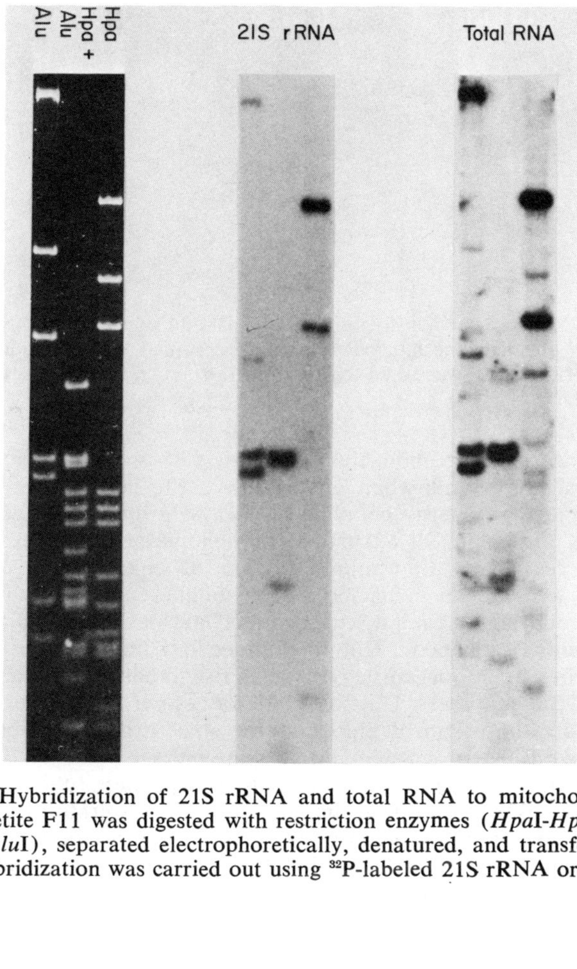

FIGURE 5. Hybridization of 21S rRNA and total RNA to mitochondrial DNA.[14] DNA from petite F11 was digested with restriction enzymes (*Hpa*I-*Hpa*II, *Alu*I, and *Hpa*I-*Hpa*II-*Alu*I), separated electrophoretically, denatured, and transferred to nitrocellulose. Hybridization was carried out using ^{32}P-labeled 21S rRNA or total RNA.

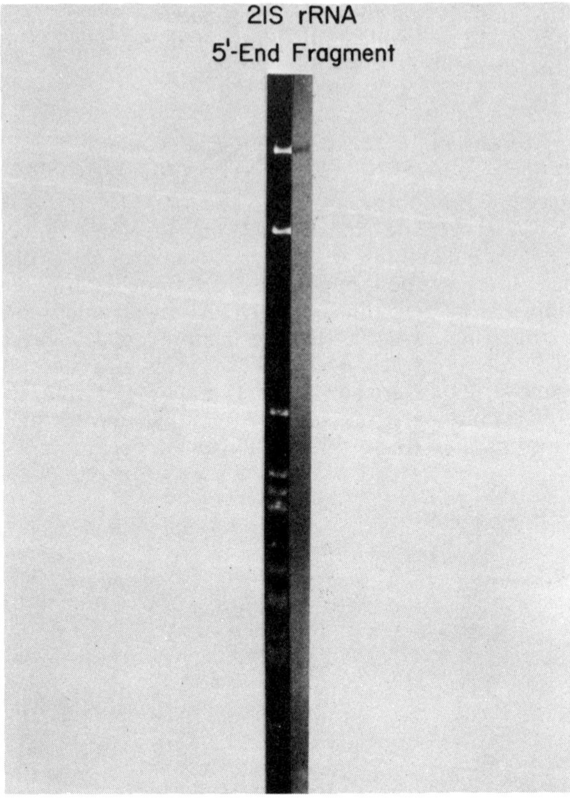

FIGURE 6. DNA transfer-hybridization of the 5′-end of the 21S rRNA molecule.[14] 21S rRNA was purified, end-labeled with ^{32}-P, repurified, and degraded to 200 nucleotides. Hybridization was then carried out to an *Hpa*I-*Hinc*II digest of F11 mitochondrial DNA.

blue-green algae says little about the huge variety of unstudied bacterial genera. This is especially apparent when one contrasts the large variety of different electron transport and energy coupling pathways in different bacteria with the single pathway present in all eukaryotic mitochondria. The presence of intervening sequences is not the unique feature of eukaryotic genes; insertion sequences are well known in bacteria. The unique eukaryotic ability, present in both mitochondria and the nucleus, is a mechanism for specific RNA splicing around intervening sequences. One might speculate that specific RNA splicing first appeared in a line of organisms that gave rise to both the eukaryotic nuclear and mitochondrial genomes. This speculation does not favor any of the current hypotheses about the origin of mitochondria, since it is compatible with all of them. Extensive RNA processing and splicing and the presence of introns and intergenic spacers both seem rather wasteful; perhaps these functions arose in an energy-rich organism that utilized mitochondria-like respiration. And if a relative of such a eukaryotic progenitor still exists, then this organism is a likely place to find intronic genes and specific RNA splicing. Thus, finding eukaryotic structural features in mitochondrial genes presents the same problem as finding these features in nuclear genes. The observation does little to explain the origin of introns and RNA splicing in either genome, nor does it explain the origin of mitochondria.

The Relationship of the Yeast Mitochondrial Genome to Mitochondrial Evolution

This concept of organization of the mitochondrial genome into genes and spacers was first proposed by Bernardi [15] and his co-workers, on the basis of fractionation of this unusually high A+T DNA. The yeast mitochondrial DNA is only 18% G+C, making it probably the most A+T-rich genome in nature. Fractionation of the DNA indicated that it consisted of large blocks of almost pure A+T interspersed with blocks of a similar size of "normal" G+C content, averaging more than 30% G+C.[15] Since it was difficult to imagine pure A+T DNA encoding genes, Bernardi deduced that these regions were spacers between genes of normal G+C content. DNA sequence analysis has indeed shown that genes are surrounded by extensive tracts of virtually pure A+T

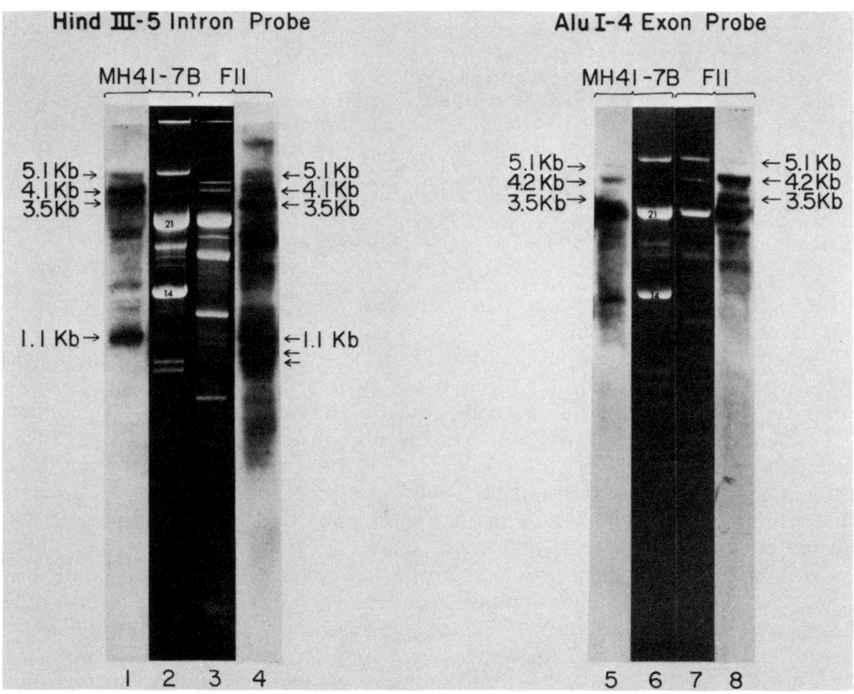

FIGURE 7. Transfer-hybridization of the 21S rRNA region.[6] RNA from the grande and from petite F11 was transferred to diazobenzyloxymethyl paper and hybridized to DNA fragments from the 5′-exon (*Alu*I-4) and the intron (*Hin*dIII-6) of the 21S rRNA gene. The hybridizations demonstrate multiple intron and extension containing precursors.

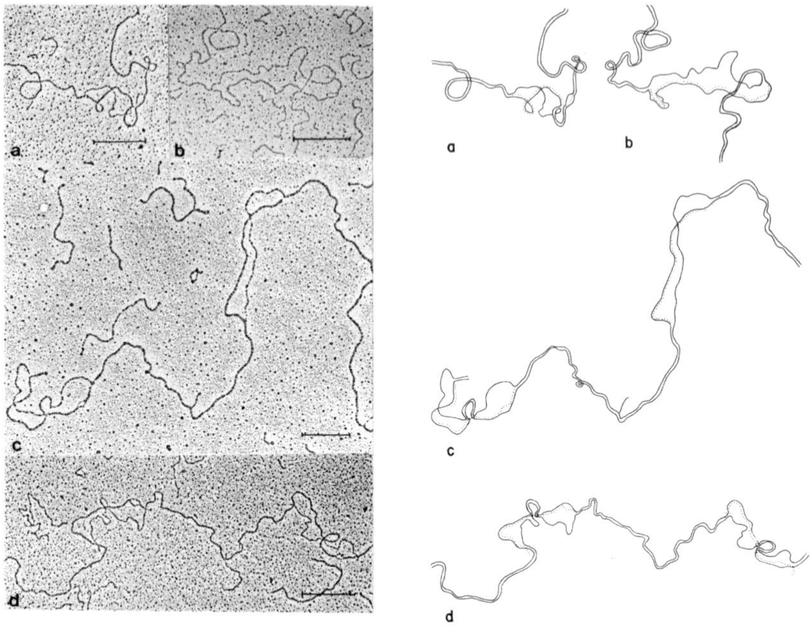

FIGURE 8. "R-loop" hybrids from the 21S rRNA gene.[6] a) 21S rRNA R-loop hybrid contains a characteristic double-stranded intron loop flanked by two single-stranded exon loops. b) A single loop structure, indicating an RNA species from which the intron has not been excised. c) An oligomeric molecule containing two R-loop structures; one is a single loop form and the other has an intron loop and an enlarged small exon relative to 21S rRNA. d) An oligomeric molecule containing two R-loop structures, both of which have an enlarged small exon.

DNA.[16-18] However, the genes themselves are not of normal G+C content, but show a striking A+T bias.

Analysis of mitochondrial codon usage in an ATPase subunit[16] gene and a cytochrome oxidase subunit[18] gene has shown a marked tendency towards A+T-rich codons (93% of the codons are of highest possible A+T content, 7% of higher G+C content). Further, the ribosomal RNA is 24–26% G+C,[19] and tRNAs are as low as 18% G+C.[20-22] If we assume that conservation of protein sequence is a constraint on nucleic-acid sequence evolution, it appears that protein coding genes must contain a minimum G+C content of about 30%. Structural genes for rRNA or tRNA would also have some minimal G+C content below which their products could not assume appropriate configurations, but this may be considerably lower than 30%. Finally, spacers would have no constraints and could eventually become pure A+T. Thus the mitochondrial genome in yeast does not consist of high A+T spacers interspersed with normal A+T genes, but rather the entire genome has evolved to approximately the maximum A+T content that is compatible with functional genes.

These conclusions suggest either that G to A or C to T mutations are more frequent in yeast mitochondria, or that there is some selective pressure for high A+T DNA. One possibility to be considered is that yeast evolved to grow in a low-temperature environment, by analogy to the very high G+C content of thermophilic bacteria. The hypothesis is testable, since the nuclear genome of yeast would be exposed to the same environmental conditions as the mitochondrial genome. In fact, however, the G+C component of the nuclear genome is 40%, and examination of the DNA sequence for iso-1-cytochrome *c* [23] shows normal codon utilization without an unusual A+T bias (52% of the codons are of lowest G+C content, 48% of higher G+C content). It thus appears that the high A+T content of yeast mitochondrial DNA is due to selective A+T mutagenesis resulting from some undefined chemical property of the environment within the yeast mitochondrion. If this condition were reversed, then G+C content would gradually drift upward, but unusual codon usage would persist. Once a codon had been eliminated, its tRNA would become useless, and over a period of time the gene would be converted into spacer. Both the yeast and animal mitochondrial genomes have a reduced number of tRNAs with a limited range of isoacceptors. Further, recent sequence analysis of human mitochondrial DNA [24] has shown a novel UGA tryptophan codon (normally UGG) and other types of anomalous codon usage also present in yeast mitochondria.[18] The difference between the high

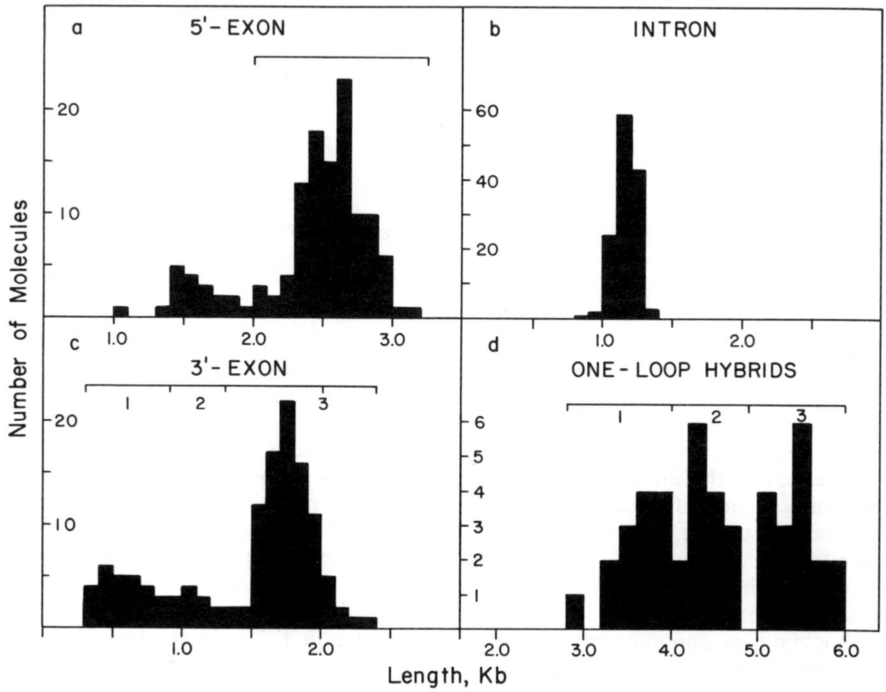

FIGURE 9. Length analysis of R-loop regions.[6] a) The large (5′) exon loop measured 2.5±.22 kb. The distribution was arbitrarily defined and smaller molecules were presumed to represent degraded RNA. b) The intron loop was measured at 1.18±0.05 kb. c) The length distribution for the small (3′) exon was divided into three regions: 1, L=0.58±0.16 kb; 2, L=1.12±0.13 kb; and 3, L=1.77±0.18 kb. d) The length distribution for single R-loop molecules was divided into three regions: 1, L=3.70±0.30 kb; 2, L=4.43±0.20 kb; and 3, L=5.43±0.26 kb.

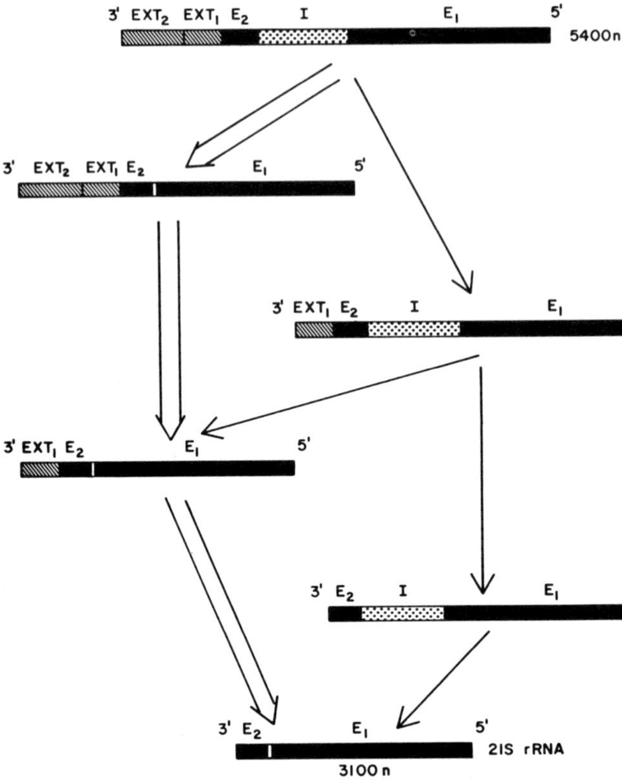

FIGURE 10. Scheme for the processing of 21S rRNA. The large arrows show the predominant pathway in petite F11.

A+T 75,000 base-pair yeast mitochondrial genome and the normal A+T 15,000 base-pair animal mitochondrial genome are probably much less important than their similarities. The type of sequence drift that seems to have taken place in the yeast mitochondrion would readily account for the gene sequences of the animal mitochondrian, and both genomes contain essentially the same genes. The most important difference appears to be deletion of most of the spacer and intron sequences in the animal mitochondrial genome. The similarities in genes and tRNAs suggest that a yeast-like mitochondrion was probably the ancestor of the animal mitochondrion, and that a process of sequence drift, inferred by the extremely high A+T content of the yeast mitochondrial genome, but no longer occurring in the animal mitochondrial genome, has caused the simplification of the genetic code found in both genomes.

In conclusion, the yeast mitochondrial genome, despite its larger size and high A+T content, is actually quite similar to the animal mitochondrial genome. The additional space in the yeast mitochondrial genome is taken up by features that emphasize the eukaryotic nature of the mitochondrion, introns and intergenic spacers, but the presence of these features does little to clarify the origin of the mitochondrial genome in eukaryotes.

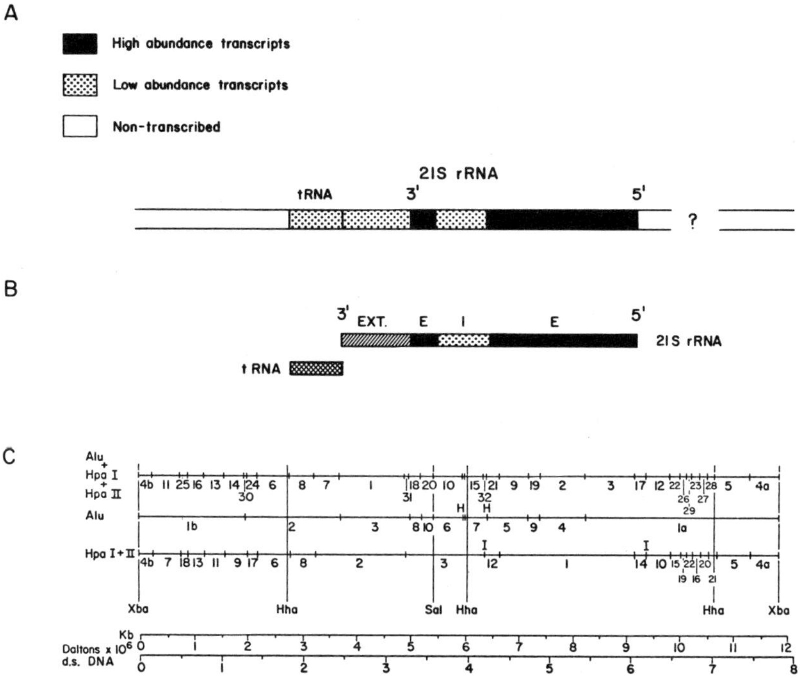

FIGURE 11. Map of the 21S rRNA region of the mitochondrial genome. a) The hybridization map of total RNA. The "nontranscribed" regions show an almost undetectable level of hybridization. The area indicated by a question mark has not been evaluated, because the small DNA fragments in this region do not stick to nitrocellulose. b) The localization of the 21S rRNA gene, its intron and 3′-extension, and nearby tRNA genes. c) The restriction-enzyme map of the 12 kb genome of petite F11.

REFERENCES

1. LOCKER, J. & M. RABINOWITZ. 1979. Methods in Enzymology **56:**3–16.
2. BORST, P. & L. A. GRIVELL. 1978. Cell **15:**705–723.
3. TZAGOLOFF, A., G. MACINO & W. SEBALD. 1979. Ann. Rev. Biochem. **48:**419–443.
4. MAHLER, H. 1981. Ann. N.Y. Acad. Sci. This volume.
5. BOS, J. L., C. HEYTING, P. BORST, A. C. ARNBERG & E. F. J. VAN BRUGGEN. 1978. Nature **275:**336–338.
6. MERTEN, S., R. M. SYNENKI, J. LOCKER, T. CHRISTIANSON & M. RABINOWITZ. 1980. Proc. Natl. Acad. USA **77:**1417–1421.
7. LOCKER, J., R. MORIMOTO, R. M. SYNENKI & M. RABINOWITZ. 1980. Curr. Gen. **7:**163–172.
8. MORIMOTO, R., J. LOCKER, R. M. SYNENKI & M. RABINOWITZ. 1979. J. Biol. Chem. **254:**12461–12470.
9. LEVENS, D., J. EDWARDS, J. LOCKER, A. LUSTIG, S. MERTEN, R. MORIMOTO, R. SYNENKI & M. RABINOWITZ. 1979. *In* Extrachromosomal DNA. D. Cummings, P. Borst, I. Dawid, S. Weissman & C. F. Fox, Eds. Vol. **15:**207–304. Academic Press, Inc. New York, N.Y.
10. LOCKER, J. 1979. Anal. Biochem. **98:**358–367.

11. Bailey, J. M. & N. Davidson. 1976. Anal. Biochem. **90:**75–85.
12. Locker, J., A. Lewin & M. Rabinowitz. 1979. Plasmid **2:**155–181.
13. Alwine, J. C., D. J. Kemp & G. R. Stark. 1977. Proc. Natl. Acad. Sci. USA **74:**5350–5354.
14. Locker, J. & M. Rabinowitz. 1980. (Manuscript in preparation.)
15. Prunell, A. & G. Bernardi. 1974. J. Mol. Biol. **86:**825–841.
16. Hensgens, L. A., L. A. Grivell, P. Borst & J. L. Bos. 1979. Proc. Natl. Acad. Sci. USA **76:**1663–1667.
17. Macino, G. & A. Tzagoloff. 1979. Proc. Natl. Acad. Sci. USA **76:**131–135.
18. Coruzzi, G. & A. Tzagoloff. 1979. J. Biol. Chem. **254:**9324–9330.
19. Reijnders, L., C. M. Kleisen, L. A. Grivell & P. Borst. 1972. Biochim. Biophys. Acta **272:**396–407.
20. Bos, J. L., K. A. Osinga, G. Vanderhorst & P. Borst. 1979. Nucleic Acid Res. **6:**3255–2361.
21. Li, M. & A. Tzagoloff. 1979. Cell **18:**47–53.
22. Martin, N. C., D. L. Miller, J. E. Donelson, C. Sigurdson, J. L. Hartley, P. S. Moynihan & H. D. Pham. 1979. *In* Extrachromosomal DNA. D. Cummings, I. Dawid, S. Weissman & C. F. Fox, Eds. Vol. **15:**357–375. Academic Press, Inc. New York, N.Y.
23. Smith, M., D. W. Leung, S. Gillam, C. R. Astell, D. L. Montgomery & B. D. Hall. 1979. Cell **16:**753–761.
24. Barell, B. G., A. T. Bankier & J. Drouin. 1979. Nature **282:**189–194.

Discussion of the Paper

Dr. D. Beattie (*Mt. Sinai School of Medicine, New York, N.Y.*): You show in your map that there is a transfer RNA gene which maps just adjacent to the 21S ribosomal RNA. Do you have any evidence that this is present in the initial nucleotide transcript?

Dr. J. Locker: Actually, there is a cluster of tRNA genes. There are about six or eight. The regions are fairly large. We have not identified them in precursors and, in fact, we have considerable evidence for separate promoters for these two groups of genes which are right next to each other. But we are looking quite extensively right now for some bigger transcript that might go across the whole region.

MECHANISMS OF EVOLUTION IN ANIMAL MITOCHONDRIAL DNA *

Wesley M. Brown

Division of Biological Sciences
University of Michigan
Ann Arbor, Michigan 48109

The mitochondrial genome of multicellular animals is a covalently closed-circular duplex DNA that is, in many respects, ideally suited for evolutionary studies. It is relatively easy to prepare in highly purified form and in amounts sufficiently large to allow its analysis by a number of different methods. It is small and simple enough to be easily characterizable by these methods. It appears to lack many of the complicating features—e.g., large size, intervening sequences, and the presence of different repetition classes—that are found in nuclear DNA, chloroplast DNA, and the mitochondrial DNA of lower eukaryotes and plants. Also, it is possible to correlate the changes observed in animal mitochondrial DNA with time, because the evolutionary history of the animals themselves is known from the fossil record.

There is evidence that, at least in vertebrates, the mitochondrial gene content and relative gene order is very similar, if not identical, among different species.[1-7] Despite this apparently conservative feature, the sequence of animal mitochondrial DNA (mtDNA) appears to evolve very rapidly, at least among vertebrates.[1-3, 8-11] Comparison of the rates of DNA base substitution among mammalian species indicates that mtDNA evolves 5 to 10 times faster than single-copy nuclear DNA.[3]

Three central questions regarding the evolution of animal mtDNA are: What kind of evolutionary change does animal mtDNA undergo? What mechanisms are responsible for the generation of change? How can the high rate of change observed be tolerated in the genome of an intracellular organelle whose function is so vital for the cell? Fairly substantial answers to the first two questions can be provided. At present, only a speculative answer to the third question is possible, although a more substantial answer may be available in a few years when more is known about the exact functions carried out by the mtDNA gene products.

Although this report contains previously unpublished data that bear on the above questions, I have also relied on past studies (my own and those of others) to provide support for the conclusions presented. In this respect I have tried to select papers that are representative of the studies that bear on a given point, and have made no attempt to provide a comprehensive survey of the literature. As a consequence, many contributions to this area of investigation have not been cited.

* Supported by the National Science Foundation (Grants DEB78–02841 and DEB76–20599).

0077–8923/81/0361–0119 $01.75/0 © 1981, NYAS

Materials and Methods

Sources of mtDNA

The tissues and cell lines used for mtDNA preparation for many of the species examined have been described previously.[3, 11-14] *Xenopus laevis* and *X. borealis* mtDNAs were gifts from Richard Hallberg. Primary cultures of chimpanzee and Talapoin monkey cells were obtained from Adeline Hackett (Naval Biomedical Research Laboratory, Oakland, Calif.), Bush Baby liver was obtained from John Allman (Biology Division, California Institute of Technology, Pasadena, Calif.). The remaining sources were tissues (usually liver) that were taken from animals obtained commercially or collected in their native habitats. A complete listing of the species examined and the sources of mtDNA is given in Table 1 (see Results).

Preparation, Labeling and Strand Separation of mtDNA

The preparation of mtDNA from cells and tissues was as described.[3, 11-14] For base compositional analysis, mtDNA was prepared from cultured cells grown for 96 hr in medium containing 5×10^{-4} M sodium phosphate, pH 7.4 and 5 μCi/ml of carrier-free ^{32}P to ensure uniform labeling of all bases. The CsCl, propidium diiodide gradient profiles obtained during mtDNA preparation are shown in Figure 1. The purity of the mtDNA obtained is indicated by the

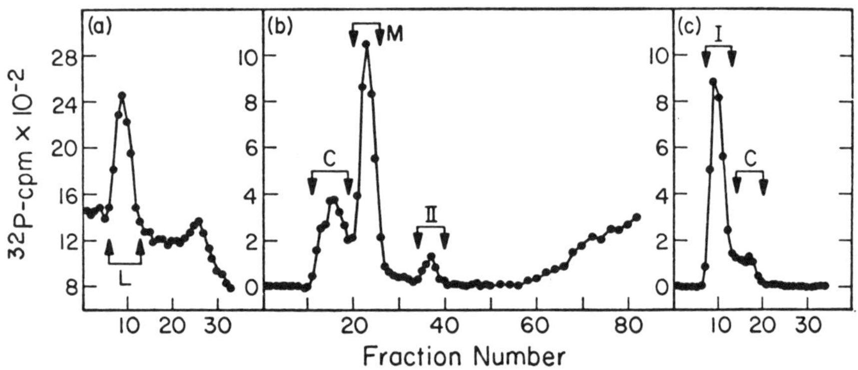

Figure 1. Purification of mtDNA from a crude mitochondrial fraction of ^{32}P-labeled human (HeLa) cells by centrifugation in CsCl, propidium diiodide gradients. (a) Buoyant density profile of the crude lysate. (b) Sedimentation profile of the material from the lower band, L, from (a). (c) Buoyant density profile of the material contained in fractions 1–32 of (b). Electron microscopy revealed that the major constituents of the fractions pooled as C, M, and II in (b) were catenated mtDNA, monomeric closed-circular mtDNA and monomeric open-circular mtDNA, respectively. Similar analyses in (c) revealed closed-circular mtDNA (both catenanes and monomers) in I and replicating forms of mtDNA plus catenated forms containing closed joined to open circles in C. The lack of detectable radioactivity in the final gradient, (c), except where expected for closed-circular, partially nicked catenated and replicating mtDNA molecules, indicates the efficiency of the purification. Each fraction represents 10 drops (~100 μl) in (a) and (c) and 5 drops (~50 μl) in (b). 1-μl aliquots were spotted onto Whatman GF/A filters, dried, and counted.

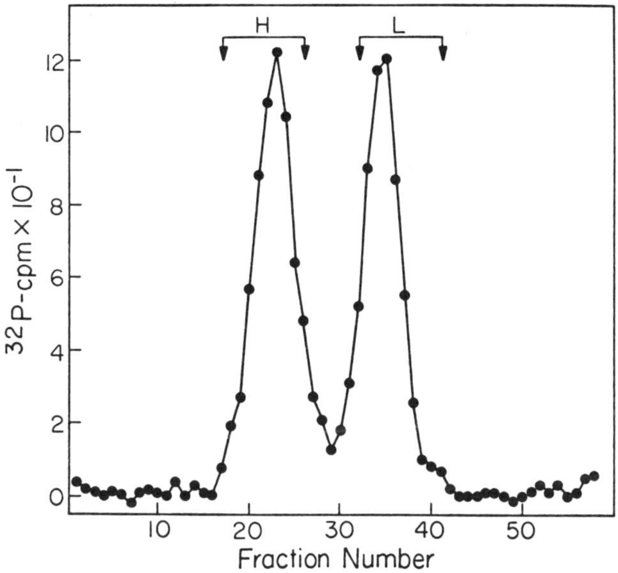

FIGURE 2. Separation of the complementary strands of ^{32}P-labeled human (HeLa) mtDNA in buoyant alkaline CsCl. The heavy (H) and light (L) strands were pooled as indicated. Each fraction represents 5 drops (~50 μl). 1-μl aliquots were taken for counting.

final sedimentation equilibrium profile (FIGURE 1c) and by the characterization with electron microscopy.[15]

The complementary single strands of purified mtDNA were separated in buoyant preparative alkaline (pH12.8) CsCl gradients and isolated by gradient fractionation as described.[15] FIGURE 2 is a profile of a typical preparation, showing the fractions pooled for base compositional analysis.

Base Compositional Analysis of mtDNA

^{32}P-labeled mtDNA (in 10 mM Tris, pH 7.5, 8 mM $MgCl_2$) was digested with DNase I (Worthington Biochemical, Inc.) at 250 μg/ml for 75 min at 37° C, then made 0.1 M in glycine, pH 9.2, and digested for 60 min at 37° C with 125 μg/ml of venom phosphodiesterase (Worthington Biochemicals, Inc.) that had been freed of detectable 5′-nucleotidase activity by the method of Sulkowski and Laskowsi.[16] Samples were made 20 mM in EDTA, the volumes were reduced to 50 μl by evaporation under N_2, nonradioactive marker 5′-deoxymononucleotides were added, and the mixture was spotted onto 7.5 cm-wide strips of Whatman 3MM chromatography paper. The strips were placed in a Gilson high voltage electrophorator containing an aqueous buffer composed of 5% acetic acid, 0.5% pyridine, 10 mM EDTA, pH 3.5, under Varsol, and electrophoresed at 3,000 volts for 90–100 min. The strips were steamed in an autoclave to remove pyridine, dried, cut into 1 cm slices, and the ^{32}P counted in a Beckman scintillation counter, using a toluene-based scintillation mixture. Radioactivity was always coincident with marker nucleotide fluorescence and none was detected at the migration position for free ^{32}P.

TABLE 1

SPECIES FROM WHICH MTDNAS WERE OBTAINED, WITH THE GENOME SIZES AND BUOYANT DENSITIES OF THE MTDNAS

	Common Name	Species Name	mtDNA Source *	Genome Size †	Buoyant Density, g/ml Neutral CsCl	Alkaline CsCl Heavy (H)	Alkaline CsCl Light (L)	H–L
1.	Human	*Homo sapiens*	HeLa cells	16.5 ± .3 (61)‡	1.6996 (12)§	1.779 (4)§	1.738 (4)	.041 (4)
	Human	*Homo sapiens*	placenta	–¶	1.6999			
	Human	*Homo sapiens*	0–717 cells	16.5 ± .5 (12)	1.6993 (2)	1.779	1.738	.041
2.	Chimpanzee	*Pan troglodytes*	primary cells	16.4 ± .3 (19)	1.6989	1.780	1.738	.042
3.	Guenon	*Cercopithecus aethiops*	BSC-1 cells	16.4 ± .5 (36)	1.6971 (3)	1.779	1.738	.041 (2)
4.	Talapoin monkey	*Miopithecus talapoin*	primary cells	–	1.6973	1.779	1.738	.041
5.	Woolly monkey	*Lagothrix cana*	primary cells	16.3 ± .4 (47)	1.6942	1.772	1.739	.033
6.	Bush baby	*Galago senegalensis*	liver	16.5 ± .3 (19)	1.6962	–	–	–
7.	Rabbit	*Oryctolagus cuniculus*	liver, kidney	17.3 ± .4 (21)	–	–	–	–
8.	Mouse	*Mus musculus*	LA9 cells	16.3 ± .4 (67)	1.6920 (3)	1.767 (2)	1.736 (2)	.031 (2)
9.	Rat	*Rattus norvegicus*	liver	16.4 ± .3 (74)	–	–	–	–
10.	Hamster	*Mesocricetus auratus*	liver	16.3 ± .5 (21)	–	–	–	–
11.	Pekin duck	*Anas platyrhynchos*	liver, spleen	16.4 ± .3 (19)	–	–	–	–
12.	Chicken	*Gallus gallus*	liver, spleen	–	1.7009 (2)	1.783	1.740	.043
13.	Lizard	*Uma notata*	liver, heart, spleen	–	1.6831	–	–	–
14.	Clawed frog 1	*Xenopus laevis*	ovaries, liver	17.7 ± .2 (20)	1.6940 (3)	1.762	1.739	.023
15.	Clawed frog 2	*Xenopus borealis*	ovaries, liver	17.6 ± .2 (15)	–	–	–	–
16.	Trout	*Salmo gairdneri*	liver, heart, ovaries	–	1.6954	–	–	–
17.	Mussel	*Mytilus californianus*	eggs	–	1.6986	–	–	≤.010
18.	White sea urchin	*Lytechinus pictus*	eggs	15.7 ± .5 (28)	–	–	–	≤.005
19.	Purple sea urchin	*Strongylocentrotus purpuratus*	eggs	15.7 ± .5 (54)	1.6940 (2)	–	–	≤.005

* Established cell lines are named or numbered. Primary cells were all fibroblast in morphology.

† Expressed as base pairs $\times 10^{-3} \pm 1$ SD, estimated from contour length measurements versus ΦX174 RF DNA, taken to be 5,375 base pairs.

‡ Numbers in parentheses indicate the number of molecules measured.

§ Numbers in parentheses indicate the number of independent determinations. *Micrococcus luteus* DNA, at 1.726 g/ml (neutral) and 1.788 g/ml (alkaline) used as a density standard.

¶ A dash (–) indicates that the mtDNA was not investigated with respect to the indicated characteristic.

Size and Buoyant Density Analysis of mtDNA

The sizes of the mtDNAs were determined by contour-length measurements from electron micrographs using bacteriophage ØX174 RF DNA as an internal size standard as described.[3, 11, 15] Buoyant density analysis in neutral and alkaline CsCl was performed as previously described.[15] Using these conditions, *E. coli* DNA has a neutral buoyant density of 1.704 g/ml.

Results

Sizes of the mtDNAs

The range in the size of the mitochondrial genome among the species examined in this study was from 15.7 to 17.7 kilobases (kb) (Table 1). Primates, rodents, and one bird species had mitochondrial genome sizes that did not differ significantly from 16.4 ± 0.1 kb. Among the mammals surveyed only rabbit mtDNA differed from this value, being ~1 kb larger. The two *Xenopus* species surveyed had mtDNAs that were ~17.6 kb. Two distantly related sea urchin species (Table 1) have rather small mtDNAs (~15.7 kb).

Buoyant Densities of the mtDNAs

The neutral buoyant densities of mtDNAs from 13 species were examined. The results (Table 1) indicate that the range of the neutral buoyant densities among the six primate mtDNAs (1.694 to 1.700 g/ml) is nearly the same as the range for the entire sample (1.692 to 1.700 g/ml). This corresponds to a 9% range in the guanine-plus-cytosine (G+C) content of the mtDNAs, as determined by this method.[17] The alkaline buoyant densities of mtDNAs from eight species were examined (Table 1). The separations of the complementary strands ranged from 23 mg/ml (*Xenopus laevis*) to 43 mg/ml (chicken). The separations and the respective strand densities were the same among the Old World monkeys, chimpanzee, and man (entries 1–4, Table 1), but both the relative strand separation and buoyant density of the heavy strand are less in woolly monkey, a New World species. In the mollusk and sea urchins examined, the mtDNAs formed a broad, flat-topped peak in buoyant alkaline CsCl, as previously reported for one of the species[18, 19] with estimated strand-density differences of less than 10 mg/ml.

Double- and Single-Stranded mtDNA Base Compositions

Uniformly ^{32}P-labeled mtDNAs and complementary strands from human (HeLa), guenon (BSC-1), and house mouse (LA9) cells were prepared. The DNAs were digested to 5′-mononucleotides with the enzymes DNase I and venom phosphodiesterase. After separation of the four mononucleotides by high voltage electrophoresis (Figure 3) the amount of radioactivity in each mononucleotide was counted. Results of all determinations are summarized in Figure 4 and Tables 2 and 3. The base compositions of the primate mtDNAs were very similar, if not identical. In these DNAs, the heavy strand was more

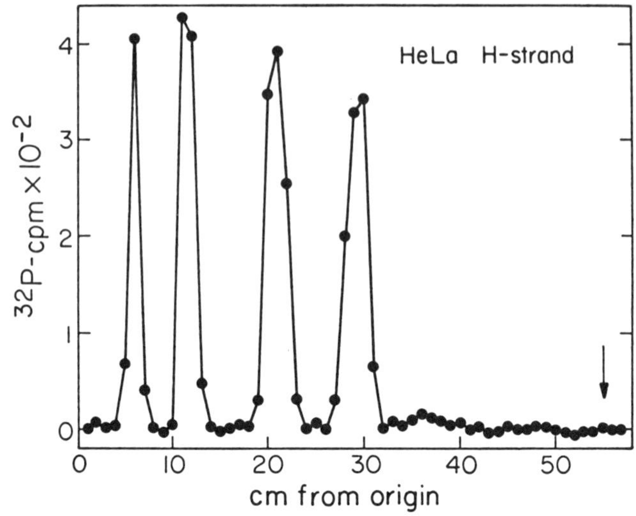

FIGURE 3. Separation of ^{32}P-labeled 5′-deoxymononucleotides by high-voltage electrophoresis. The nucleotides were produced by the enzymatic digestion of the heavy strand of human (HeLa) mtDNA as described in MATERIALS AND METHODS. The four peaks correspond to the 5′-monophosphates of deoxycytidine (dC), deoxyadenosine (dA), deoxyguanosine (dG), and thymidine (dT) in order of increasing mobility. The arrow indicates the peak position of $^{32}P_i$ run under identical conditions.

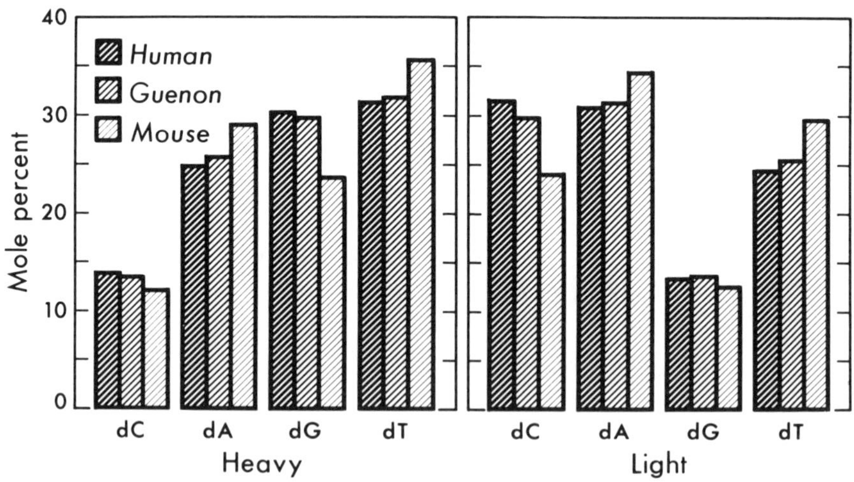

FIGURE 4. Base compositions of the complementary mtDNA strands of human, guenon, and house mouse. The complementary strands were prepared, digested to 5′-deoxymononucleotides and analyzed as described in MATERIALS AND METHODS. The values shown are averages for multiple determinations, as detailed in TABLE 3. The abbreviations for the four bases are given in the legend to FIGURE 3.

than 2.2 times richer in guanine and more than 1.2 times richer in thymine than the light strand. Mouse mtDNA differed from these not only in the overall base composition, but also in the degree of guanine bias between the complementary strands.

Discussion

DNA undergoes two kinds of evolutionary change: structural rearrangements and base substitutions. The former includes additions (duplications), deletions, inversions, and transpositions (including translocations). Structural rearrangements occur in animal mtDNA, but such rearrangements appear to be restricted both in kind and in place of occurrence.

Table 2
Base Compositions of Three Native Duplex mtDNAs

	Base Compositions *				Molar Ratios †				
	C	A	G	T	Py:Pu	T:A	C:G	(G+C) %	N
Human	21.9±1.4	28.1±1.6	22.1±1.6	28.0±1.9	0.99	1.00	0.99	44.0	12
Guenon	21.5±0.5	28.2±0.5	21.7±0.5	28.6±0.6	1.00	0.99	0.99	43.2	5
Mouse	17.8±0.5	30.5±0.8	19.0±0.8	32.8±0.9	1.03	0.93	0.92	36.8	3

* Base compositions are expressed as mole % ±1 SD. C, A, G, and T refer to the 5′-deoxymononucleotides of cytosine, adenine, guanine, and thymine, respectively.

† Py:Pu is the ratio of pyrimidines (C+T) to purines (A+G) and N is the number of independent determinations.

Additions and Deletions Occur Predominantly in the Region of the Origin of Replication of Animal mtDNA

The marked variation in mitochondrial genome size that exists among species (Table 1) is *prima facie* evidence for the occurrence of additions and deletions. Although apparent size homogeneity of mtDNA exists among species in some taxonomic groups, e.g., Primates, others show marked heterogeneity in size. The range of sizes reported for mtDNA among 39 species of *Drosophila* was 15.6 to 19.4 kb,[20] a range greater than that for all other animal species examined. Although the available data suggest that, in general, mtDNA is homogeneous in size within species, at least one exception to this has been found. Comparison of mtDNA size among five geographically different populations of a lizard, *Cnemidophorus sexlineatus* indicated that individuals from one population had mtDNA that was ~1200 base pairs larger than mtDNA from individuals in the remaining populations (Brown & Wright, unpublished data).

In a comparison of the mtDNAs of gorilla, chimpanzee, human, orangutan, and lar gibbon, gorilla mtDNA has been shown to differ in size from the remaining mtDNAs by a deletion of ~100 base pairs (Ferris, Wilson & Brown,

TABLE 3

BASE COMPOSITIONS OF THE COMPLEMENTARY STRANDS OF THREE MTDNAS *

	Heavy Strand				Light Strand			
	C	A	G	T	C	A	G	T
Human	13.9±0.5	24.7±0.6	30.2±0.7	31.3±1.0	31.5±0.5	30.5±1.2	13.3±1.2	24.7±1.4
Guenon	13.2±0.4	25.9±0.2	29.3±0.5	31.6±0.3	29.6±0.6	31.4±0.3	13.6±0.2	25.4±0.3
Mouse	12.5±0.4	28.5±0.8	23.5±0.3	35.5±0.3	24.1±1.2	34.2±1.0	12.3±1.1	29.4±1.2

	Complementarity †				Strand Bias †					
	C_H:G_L	G_H:C_L	T_H:A_L	A_H:T_L	C_L:C_H	G_H:G_L	A_L:A_H	T_H:T_L	% (G+C) ‡	N §
Human	1.05	0.96	1.03	1.00	2.27	2.27	1.23	1.27	44.5	14
Guenon	0.97	0.99	1.01	1.02	2.24	2.15	1.21	1.24	42.9	4
Mouse	1.02	0.98	1.04	0.97	1.93	1.91	1.20	1.21	36.2	3

* Base compositions are expressed and bases abbreviated as in TABLE 2.
† The subscripts H and L refer to the origin of the base from the heavy and light strands, respectively.
‡ % (G+C) is the average of the mole % (G+C) of the complementary strands.
§ N is the number of independent determinations.

manuscript submitted). This deletion occurs in the region near the heavy strand origin of replication, or D-loop.[22-25] The size difference between frog (*Xenopus*) and human mtDNA, ~1,150 base pairs, (TABLE 1) can be accounted for by a size difference in the regions directly adjacent to and including the D-loop. The D-loop of *X. laevis* mtDNA is ~900 base pairs longer than that of human mtDNA[4, 12, 26, 27] and the region from the proximal end of the smaller ribosomal gene to the origin of the D-loop is ~300 base pairs longer.[4, 27] A heteroduplex analysis of the region near the D-loop of sheep and goat mtDNA[10] indicated that addition-deletion types of rearrangements were present, and possibly involved the duplication or deletion of an iterated sequence.

Contour length and restriction-endonuclease fragment size analyses indicate that most, perhaps all, of the variation observed among *Drosophila* species is due to differences in the size of the A+T-rich region of the mtDNA and that the remainder of the genome is not distinguishable in size among the species studied.[20, 21, 28-30] The A+T-rich region of *Drosophila* mtDNA contains the origin of replication.[21, 31] It was suggested, based on contour-length measurements of this region, that the size differences might have arisen as duplications or deletions of an iterated unit sequence.[20] However, heteroduplex studies of the A+T-rich regions indicate that there is almost no homology between species and also provide no evidence for the existence of iterated sequences within species.[21, 29]

Taken together, these data suggest strongly that addition-deletion events in animal mtDNA occur predominantly in the relatively small portion of the genome that includes the origin of heavy strand replication.

Inversions and Transpositions May Not Occur in Animal mtDNA

There is substantial evidence that, at least in vertebrates, the mitochondrial gene content and relative gene order is similar, perhaps identical, among different species. The relative positions of the origin of heavy strand replication and the genes coding for the small and large mitochondrial ribosomal RNAs are the same in primates, rodents, and frogs.[2, 5, 6] The similarity in the genomic positions of the mitochondrial transfer RNA genes among representatives of these groups is also striking.[1, 4, 6] Further evidence for the conservation of gene order in mtDNA is provided by comparisons of restriction-endonuclease cleavage maps. Although the evolutionary rate of base substitution in animal mtDNA is too high to permit long-range comparisons of cleavage sites (e.g., between primates and amphibians), such comparisons can and have been made among more closely related taxa. The first published study of mtDNA cleavage maps indicated that human, guenon, and mouse mtDNAs each contained three *Hin*dIII sites, two of which appeared to be at identical positions in the three genomes.[11] Further studies have shown that *Hin*dIII sites occur at these same two positions in 10 of 11 primate species and in 3 of 5 rodent species investigated[3, 32] (Ferris, Wilson & Brown, manuscript submitted; S. Ferris and M. George, Jr., unpublished data). Other highly conserved sites have also been identified. These sites occur infrequently and are distributed at positions throughout the genome, outside of as well as within the mitochondrial ribosomal RNA genes. These data suggest that the order of sequences in the mitochondrial genome is conserved and that transpositions and inversions have not occurred in these taxa.

Finally, a hybridization study[33] in which each of the labeled, separated strands of human mtDNA annealed exclusively with the corresponding complementary strand of frog (*Xenopus laevis*) mtDNA indicates that the hybridizable sequence information contained in each of the complementary strands has remained in that strand since the divergence of the human and frog lines from a common ancestor, an event that occurred $\sim 3.5 \times 10^8$ years ago.[34]

Studies of mtDNA transcription in both house mouse[5] and *Xenopus*[4, 27] indicate that intervening sequences are probably not present, at least not in the genes for ribosomal or messenger RNAs. No intervening sequences occur in the COII gene of human mtDNA or in the transfer RNA genes flanking it.[35] These studies, and others dealing with the location of transfer RNA genes,[1, 4, 6] also indicate that intergenic spacer regions are extremely small or absent. This has been directly confirmed by partial sequence data for human[35] and house mouse (D. Clayton, personal communication) mtDNAs. This contiguous arrangement of the genes may be the reason for the lack of occurrence of structural rearrangements outside the region of the heavy strand origin of replication in animal mtDNA, since any rearrangement outside this region would almost certainly inactivate one or more genes.

Sequence Evolution in Animal mtDNA Is Rapid

Despite the apparent conservation of gene content and order in animal mtDNA, the rate of base substitution appears to be rapid. The substitution rate, estimated to be 0.5 to 1% per genome per 10^6 years, is 5 to 10-fold faster than the rate for single-copy nuclear DNA, based on a comparison among mammals.[3] A similar rate estimate is obtained using data from a comparison of the mtDNAs from two amphibian species (*Xenopus laevis* and *X. borealis*)[8] that diverged $\sim 10 \times 10^6$ years ago.[36] Thermal dissociation of mtDNAs hybridized *in vitro*[3, 8, 33] and comparisons of cleavage maps[3] (Ferris, Wilson & Brown, submitted) indicate that all portions of the mitochondrial genome accumulate base substitutions relatively rapidly, at least among vertebrate species. No evidence for clustering of changed or conserved sequences is provided by either method of analysis, although both indicate that some rate heterogeneity exists for different regions of the mtDNA. The available evidence indicates that the genes for mitochondrial ribosomal RNA[8, 30] (Ferris, Wilson & Brown, submitted) and transfer RNA[37] are relatively more conserved than other mtDNA sequences, but that even these genes evolve much more rapidly than their nuclear counterparts.

Among species of invertebrates, only the mtDNA of *Drosophila* has been well studied. In this genus, the A+T-rich region has been shown by heteroduplex analysis to change much more rapidly than the remainder of the mtDNA.[21, 29] The rate of base substitution for *Drosophila* mtDNA is unknown. However, a comparison of cleavage maps for four restriction endonucleases in mtDNAs of three *Drosophila* species has been made.[30] These data were compared with comparative mobility data from starch-gel electrophoresis of proteins, and it was concluded that the rate of base substitution in *Drosophila* mtDNA (exclusive of the A+T-rich region) was the same as the rate for nuclear genes.[30] This is a surprising result, since it is very different than the result obtained from comparisons of vertebrate DNAs. It is especially surprising, however, because the coding regions of nuclear genes (as represented by mature, cytoplasmic messenger RNAs) have been shown to accumulate base substitutions more slowly than does single-copy nuclear DNA,[38] which contains both coding and noncoding regions (e.g., intervening and flanking "spacer" sequences). While it is possible that the substitution rate varies among different taxonomic groups, to draw such a conclusion now would be premature. The *Drosophila* study included relatively few cleavage sites for comparison; this can lead to a large error.[39] Also, the comparison was indirect, with nuclear proteins, rather than direct, with nuclear DNA. This result is, nevertheless, highly interesting and a more rigorous comparison ought to be undertaken.

Changes in the Distribution of Bases between Strands

All animal mtDNAs so far examined show an unequal distribution of guanine plus thymine residues between the complementary strands. This has been demonstrated both directly, by base compositional analysis (TABLE 3) and indirectly, by equilibrium sedimentation in alkaline CsCl gradients (TABLE 1). The difference in the buoyant densities of the complementary strands in alkaline CsCl can vary among animal species from ≤ 5 to 43 mg/ml (TABLE 1). The invertebrate species show much smaller interstrand differences than do the vertebrate species. Among vertebrates, the "warm blooded" species (birds and mammals) exhibit greater interstrand differences than does the one "cold-blooded" species examined (*Xenopus*). Although the amount of data available is limited, one may speculate that there is an evolutionary trend toward an increasing strand bias for guanine plus thymine content in going from invertebrates to vertebrates and, within the vertebrates, in going from less metabolically active, "cold-blooded" forms to more metabolically active, "warm blooded" forms.

How Are Rapid Changes in mtDNA Generated?

It seems unlikely that the rates of mitochondrial and nuclear mutations caused by external physical events (e.g., radiation) or by externally applied chemical mutagens would differ (although one could argue that mtDNA, being shielded by an additional membrane barrier, might be less exposed to chemical mutagens). It has been proposed, however, that chemical events within the mitochondrion result in the production of mutagenic compounds, such as free radicals.[40] If this proposal is correct, these would increase the rate of substitutions in mtDNA without affecting nuclear DNA, since the mutagenic compounds generated would be confined to the mitochondrion.

DNA base substitutions can arise as a result of errors generated during replication. There is evidence that in vertebrates the occurrence of replication-generated errors is higher in mtDNA than in nuclear DNA. The incorporation of ribonucleotides into replicating mtDNA and the presence of these in mature (i.e., nonreplicating) mtDNA have been well documented.[41] Although the frequency of ribonucleotide incorporation may be greatest near the respective origins of replication of the two strands,[42, 43] ribonucleotides also occur more or less randomly throughout the remainder of the genome.[33, 44] The presence of ribonucleotides in mature mtDNA suggests strongly that the enzymatic replication complex for mtDNA, unlike its nuclear counterpart, is either inefficient or lacking in the ability to edit newly-replicated DNA. Unlike vertebrate mtDNA, mature *Drosophila* mtDNA is not susceptible to strand scission at high pH, and therefore does not appear to contain ribonucleotides (J. Battey, personal communication).

The turnover rate of mtDNA is higher than that of nuclear DNA in the somatic cells of vertebrates.[45] If this is also true for germ-line cells it would contribute to a higher substitution rate, since mtDNA would undergo more rounds of replication than nuclear DNA, thus providing more opportunities for the occurrence of replication-associated substitutions.

The apparent lack of post-replicative repair mechanisms in vertebrate mitochondria may also contribute to the elevated rate of base substitution in mtDNA. No mechanism for the excision and repair of ultraviolet radiation-induced thymine dimers is demonstrable.[46] Studies of the ability of thymine kinase-deficient cells to recover from photoinduced lesions in bromouracil-substituted mtDNA indicate indirectly that mechanisms for the efficient repair of these lesions are lacking.[47] The persistence of ribonucleotide bases in mtDNA also indicates the lack of efficient post-replicative repair mechanisms in vertebrate mitochondria.

In cultured mammalian cells, both replicative and post-replicative repair syntheses of DNA have been shown to depend on the presence of the enzyme α-DNA polymerase.[48] No DNA repair was observed in its absence, even though β- and γ-polymerase were present.[48] In vertebrates, only the γ-polymerase has been isolated from mitochondria and this enzyme seems to be the only polymerase involved in the replication of mtDNA.[49] The lack of both replicative and post-replicative repair of vertebrate mtDNA may thus be due to the lack of an α-polymerase in the mitochondria.

Three forms of DNA polymerase have been obtained from *Drosophila*.[50,51] Unlike the separately coded polymerases in vertebrates, in *Drosophila* the three activities appear to result from the specific proteolysis of the product of a single gene product.[51,52] It is not known which form of polymerase is associated with mitochondria. If the rate of nucleotide substitution in *Drosophila* mtDNA differs from that in vertebrate mtDNA, characterization of the *Drosophila* polymerase and comparison of it with vertebrate γ-polymerase can provide not only a mechanistic interpretation for any rate difference, but also a quantitative understanding of the role that DNA polymerases play in the generation of mutations.

The Rate of Evolution of Vertebrate mtDNA May Be Mutationally Driven

The rate of evolution is the product of the separate rates for mutation and fixation. In comparisons of nuclear genes, a rate difference is usually attributed to a difference in the rate of fixation, due to functional constraints imposed by selection on the gene products. Mitochondrial gene products function as components in complex, multisubunit structures (e.g., ribosomes, cytochrome oxidase complex, etc.), most of the components of which are products of nuclear genes. It is thus difficult to accept an argument for lower functional constraints on mitochondrial gene products as a valid explanation for the high rate of mtDNA evolution. It seems at least as likely that the rapid evolution of vertebrate mtDNA is due exclusively to its higher mutation rate and that the rate of fixation is the same as that for an analogous set of nuclear DNA sequences.

ACKNOWLEDGMENTS

I am deeply grateful to the late Jerome Vinograd, Divisions of Biology and Chemistry, California Institute of Technology, Pasadena; to Howard M. Goodman, Department of Biochemistry and Biophysics, University of California, San Francisco; and to Allan C. Wilson, Department of Biochemistry, University of California, Berkeley, for their interest, support and encouragement of my research and for the use of their facilities; to Jim Battey, David Clayton, Steve Ferris, and Matt George for sharing their unpublished results; to Allan Wilson, David Clayton, C. Allen Smith, Harold Varmus, John Wright, and Susan Brown, and to the members of the Vinograd, Goodman, and Wilson labs for many helpful and enlightening discussions; and to Lori Erdley for assistance in the preparation of this paper.

REFERENCES

1. DAWID, I. B., C. K. KLUKAS, S. OHI, J. L. RAMIREZ & W. B. UPHOLT. 1976. *In* The Genetic Function of Mitochondrial DNA. C. Saccone & A. M. Kroon, Eds. Elsevier, Amsterdam.
2. RAMIREZ, J. L. & I. B. DAWID. 1978. J. Mol. Biol. **119:**133–146.
3. BROWN, W. M., M. GEORGE, JR. & A. C. WILSON. 1979. Proc. Natl. Acad. Sci. USA **76:**1967–1971.
4. RASTL, E. & I. B. DAWID. 1979. Cell **18:**501–510.
5. BATTEY, J. & D. A. CLAYTON. 1978. Cell **14:**143–156.
6. ATTARDI, G., M. ALBRING, F. AMALRIC, R. GELFAND, J. GRIFFITH, D. LYNCH, C. MERKEL, W. MURPHY & D. OJALA. 1976. *In* Genetics and Biogenesis of Chloroplasts and Mitochondria. T. Bucher, W. Neupert, W. Sebald & S. Werner, Eds. Elsevier, Amsterdam.
7. BORST, P. 1977. Trends in Biochem. Sci. **2:**31–34.
8. DAWID, I. B. 1972. Devel. Biol. **29:**139–151.
9. JAKOVCIC, S., J. CASEY & M. RABINOWITZ. 1975. Biochemistry **14:**2037–2042.
10. UPHOLT, W. B. & I. B. DAWID. 1977. Cell **11:**571–583.
11. BROWN, W. M. & J. VINOGRAD. 1974. Proc. Natl. Acad. Sci. USA **71:**4617–4621.
12. BROWN, W. M., J. SHINE & H. M. GOODMAN. 1978. Proc. Natl. Acad. Sci. USA **75:**735–739.
13. BROWN, W. M. & H. M. GOODMAN. 1979. *In* Extrachromosomal DNA. D. Cummings, P. Borst, I. Dawid, S. Weissman & C. F. Fox, Eds. Vol. **15:**485–500. Academic Press, Inc. New York, N.Y.
14. BROWN, W. M. & J. W. WRIGHT. 1979. Science **203:**1247–1249.
15. BROWN, W. M., R. M. WATSON, J. VINOGRAD, K. M. TAIT, H. W. BOYER & H. M. GOODMAN. 1976. Cell **7:**517–530.
16. SULKOWSKI, E. & M. LASKOWSKI, SR. 1971. Biochim. Biophys. Acta **240:**443–447.
17. SCHILDKRAUT, C. L., J. MARMUR & P. DOTY. 1962. J. Mol. Biol. **4:**430–443.
18. PIKO, L., A. TYLER & J. VINOGRAD. 1967. Biol. Bull. **132:**68–90.
19. PIKO, L., D. G. BLAIR, A. TYLER & J. VINOGRAD. 1968. Proc. Natl. Acad. Sci. USA **59:**838–845.
20. FAURON, C. M-R. & D. R. WOLSTENHOLME. 1976. Proc. Natl. Acad. Sci. USA **73:**3623–3627.

21. Wolstenholme, D. R., J. M. Goddard and C. M-R. Fauron. 1979. *In* Extrachromosomal DNA. D. Cummings, P. Borst, I. Dawid, S. Weissman & C. F. Fox, Eds. Vol. **15:**409–425. Academic Press, Inc. New York, N.Y.
22. Kasamatsu, H., D. L. Robberson & J. Vinograd. 1971. Proc. Natl. Acad. Sci. USA **68:**2252–2257.
23. Robberson, D. L., H. Kasamatsu & J. Vinograd. 1972. Proc. Natl. Acad. Sci. USA **69:**737–741.
24. Kasamatsu, H. & J. Vinograd. 1973. Nature New Biol. **241:**103–105.
25. Robberson, D. L. & D. A. Clayton. 1972. Proc. Natl. Acad. Sci. USA **69:** 3810–3814.
26. Gillum, A. M. & D. A. Clayton. 1978. Proc. Natl. Acad. Sci. USA **75:**677–681.
27. Rastl, E. & I. B. Dawid. 1979. *In* Extrachromosomal DNA. D. Cummings, P. Borst, I. Dawid, S. Weissman & C. F. Fox, Eds. Vol. **15:**395–407. Academic Press, Inc. New York, N.Y.
28. Goddard, J. M. & D. R. Wolstenholme. 1979. J. Cell Biol. **83**(2): Abstr. MC1936.
29. Fauron, C. M-R. & D. R. Wolstenholme. 1979. J. Cell Biol. **83**(2): Abstr. MC1937.
30. Shah, D. M. & C. H. Langley. 1979. Nature **281:**696–699.
31. Goddard, J. M. & D. R. Wolstenholme. 1978. Proc. Natl. Acad. Sci. USA **75:** 3886–3890.
32. Parker, R. C. & R. M. Watson. 1977. Nucleic Acids Res. **4**(5):1291–1299.
33. Brown, W. M. 1976. Ph.D. Thesis. California Institute of Technology. Pasadena, Calif.
34. Romer, A. S. 1966. Vertebrate Paleontology. University of Chicago Press. Chicago, Ill.
35. Barrell, B. G., A. T. Bankier & J. Drouin. 1979. Nature **282:**189–194.
36. Bisbee, C. A., M. A. Baker, A. C. Wilson, I. Hadji-Azimi & M. Fischberg. 1977. Science **195:**785–787.
37. Jakovcic, S., J. Casey & M. Rabinowitz. 1975. Biochemistry **14:**2043–2050.
38. Rosbash, M., M. S. Campo & K. S. Gummerson. 1975. Nature **258:**682–686.
39. Nei, M. & W-H. Li. 1979. Proc. Natl. Acad. Sci. USA **76:**5269–5273.
40. Borst, P. & A. M. Kroon. 1969. Int. Rev. Cytol. **26:**107–190.
41. Grossman, L. I., R. M. Watson & J. Vinograd. 1973. Proc. Natl. Acad. Sci. USA **70:**3339–3343.
42. Martens, P. A. & D. A. Clayton. 1979. J. Mol. Biol. **135**. (In press.)
43. Gillum, A. M. & D. A. Clayton. 1979. J. Mol. Biol. **135**. (In press.)
44. Flory, P. J., Jr. 1974. Ph.D. Thesis. California Institute of Technology. Pasadena, Calif.
45. Rabinowitz, M. & H. Swift. 1970. Physiol. Rev. **50:**376–427.
46. Clayton, D. A., J. N. Doda & E. C. Friedberg. 1974. Proc. Natl. Acad. Sci. USA **71:**2777–2781.
47. Lansman, R. A. & D. A. Clayton. 1975. J. Mol. Biol. **99:**761–776.
48. Ciarrocchi, G., J. G. Jose & S. Linn. 1979. Nucleic Acids Res. **7**(5):1205–1219.
49. Weissback, A. 1977. Annu. Rev. Biochem. **46:**25–47. E. Snell, P. Boyer, A. Meister and C. Richardson, Eds. Annual Reviews, Inc. Palo Alto, CA.
50. Brakel, C. L. & A. B. Blumenthal. 1977. Biochemistry **16:**3137–3143.
51. Brakel, C. L. & A. B. Blumenthal. 1978. Eur. J. Biochem. **88:**351–362.
52. Brakel, C. L. & A. B. Blumenthal. 1978. Nucleic Acids Res. **7**(5):2565–2575.

Discussion of the Paper

Dr. D. Ruiz: Continuing on your speculation about the higher rate of fixation of mutations in the mitochondria, it is intriguing to me to postulate that there could be a much higher rate of fixation of mutations based on the fact that selective pressure against an unfavorable mutation would not be nearly so high as against a mutation in the nuclear DNA which only exists in one copy.

Dr. W. M. Brown: I forgot to mention that there are several hundred copies of mitochondrial DNA in the several hundred mitochondria per cell.

Dr. Ruiz: The same kind of theory that has been applied to greater rates of evolution in duplicated genes in the nuclear DNA could be applied to this sort of situation.

Dr. Brown: I think that in nuclear DNA the situation is that duplicated genes evolve much more slowly. That's certainly the case with ribosomal genes. In fact, the mechanism that has been postulated is unequal crossing over for keeping things the same. You see these, for instance, in duplicate genes like duplicate alphaglobin genes.

Dr. R. Sager: I wanted to return to this question of heterogeneity because I think it is really a fascinating and very important question. How, in the absence of good repair system, is the amount of homogeneity retained that is retained? We've been looking at mouse cell lines with a somewhat more sensitive method. What we found with the cell lines that we've been working with is that there is some micro-heterogeneity. The reason I am bringing this up now is that there is a very perplexing genetic question. We've been selecting mutations to antibiotic resistance which we think are mitochondrial such as chloramphenicol resistance, oligomycin, and a few others; and they can be selected without mutagenizing the cells. This has suggested to us that there is some genetic heterogeneity there that we are selecting. Chloramphenicol resistance is very dominant in the sense that you don't need many copies to confer resistance to the cell. Oligo is different. You need a lot of copies. So we can watch segregation and where did it come from? I think it must be coming from some heterogeneity that hasn't been detected.

Dr. Brown: How many cell generations are there between the time you start the selection and the time you actually see colonies?

Dr. Sager: I can't give you an easy answer to that.

Dr. Brown: I know that the mutations to antibiotic resistance occur at a very low frequency so it is possible that they are occurring during the selection procedure itself. You know, that's an alternative.

I am really starting to work on the problem within individual heterogeneity in a serious way and I hope to have some data soon that will sort of indicate what's going on.

Dr. M. V. Simpson: I thought in this connection I would remind everyone of a fascinating experiment done many years ago by David Luck in which he introduced into a healthy *Neurospora* cell a sick mitochondrion. This mitochondrion was not a *pokey*, but, it was a *pokey*-like mutant and in three generations the *Neurospora* caught the disease. All the mitochondria in the *Neurospora* had mutated. It was a rather fascinating experiment.

Dr. Brown: We really need to know something about the transmission genetics of animal mitochondrial DNA.

Dr. K. S. Chiang (*University of Chicago, Chicago, Ill.*): I am also interested in the micro-heterogeneity. I wonder whether you would like to comment on whether micro-heterogeneity is necessarily coming from the mechanism you suggested. I think the question of whether or not the organelle unit has a repair system has to be scrutinized more carefully. Whether mitochondria have a similar trend with lower UV doses, I think is an open question. Micro-heterogeneity does exist. Perhaps the mechanism for explaining such existence should be broadened rather than saying only that repair is absent.

DR. BROWN: You still have to be able to explain the retention of ribonucleotides in mature mitochondrial DNA. I think that's pretty strong evidence that the repair system is either not there at all or very inefficient.

MITOCHONDRIAL DNA POLYMORPHISM: EVOLUTIONARY STUDIES ON THE GENUS *RATTUS* *

Gregory G. Brown and Frank J. Castora

Department of Biochemistry
State University of New York at Stony Brook
Stony Brook, New York 11794

Stephen C. Frantz

Rodent Control Evaluation Laboratory
State of New York Department of Health
Troy, New York 12180

Melvin V. Simpson

Department of Biochemistry
State University of New York at Stony Brook
Stony Brook, New York 11794

Several years ago, a number of groups independently discovered that more than one type of mtDNA occurs within the population of a mammalian species.[1-8] The technique employed was restriction-endonuclease analysis and the phenomenon was first observed in horse and man by Potter *et al.*,[1] in sheep and goat by Upholt and Dawid,[2] and in the laboratory rat by Francisco and Simpson.[3] This discovery of mtDNA polymorphism combined with the speed and convenience of restriction-enzyme analysis, used to detect low degrees of sequence divergences, has led our and other laboratories to studies that indicate that this approach can serve as a powerful tool in evolutionary studies, particularly those concerned with the short-term evolution of closely related species, and with the evolution of mtDNA within a species, a phenomenon that bears on population genetics.

Our evolutionary studies are reported here. In addition, however, the mtDNA polymorphism provides the opportunity to study certain aspects of mtDNA gene products as well as structure-function relationships between mtDNA-encoded gene products and certain nDNA-encoded products that function in the mitochondrion. This work is in progress.

Results and Discussion

Our early work [3] on mtDNA polymorphism showed that there were two types of mtDNA, which we colorfully termed A and B, distributed in the laboratory *Rattus norvegicus* population; an individual animal appeared to possess only one of these types, irrespective of the tissue examined. The two types could be distinguished by the restriction enzyme *Eco*RI, type A mtDNA

* Supported by the National Institutes of Health (Grant GM22333) and the American Cancer Society (Grant NP87K). G. G. B. is the recipient of a National Research Service Award (GM-06972).

0077–8923/81/0361–0135 $01.75/0 © 1981, NYAS

showing two or more recognition sites than the B type. We examined[3] about sixty animals at that stage and reached the conclusion that the DNA types are not strain specific; that is, we found both types but no others in any of the outbred strains we tested, namely, Sprague Dawley, Wistar, and Long Evans hooded, and this was true irrespective of the commercial supplier (TABLE 1). Since that time, we have analyzed several hundred additional commercially bred laboratory rats and have found no evidence that challenges this conclusion. The situation seems to be similar for other outbred stocks as well, namely the Donryu rat,[7] a Japanese strain, and the SASCO rat,[8] another American strain. Recently, however, several inbred strains have been found to have different mtDNA types.[20]

It was not evident why the mtDNA polymorphism in the standard outbred laboratory strains should consist of two and only two mtDNAs. One hypothesis which we considered, but which merely pushes the problem back one step, is that only two mtDNA types exist in the entire *R. norvegicus* population, laboratory or wild. However, we felt that a more likely hypothesis is that the wild population consists of many mtDNA types and that, simply by chance, female breeders of types A and B were selected from this population and were bred with any wild males to initiate the first colony. A second colony, started at a later date, could then have been partially derived from the first, say by matings of wild males with laboratory females from the first colony, and a third stock similarly derived from either the first or the second colony, always using females from a previously existing colony as breeders.

TABLE 1

THE OCCURRENCE OF TYPES A AND B DNA IN THE LIVER MITOCHONDRIA OF DIFFERENT RAT POPULATIONS

Rat strain	Commercial source *	Number of Rats Possessing Type A DNA	Type B DNA
Sprague Dawley	CR	26	4
	TF	6	2
Wistar	CR	0	9
	MA	3	3
Long Evans Hooded	CR	0	1
	MBF	4	2

* Consult Reference 3 for abbreviations.

The latter hypothesis makes use of the phenomenon of the maternal inheritance of mtDNA to explain the retention in the entire laboratory rat population of the two mtDNAs that remain unaltered through many generations. Evidence for the occurrence of this phenomenon in mammals will be presented later.

While detailed information about the origins of outbred rat colonies is sparse, what information there is (see Reference 9 for summary review) indicates that a close relationship exists at least between some colonies. The first American rat colony had its beginnings in England prior to 1900. The rats used to start the colony were derived from albino rats that were kept for show.

These rats were originally obtained in connection with the sport of rat baiting, popular between 1800 and 1870, in which the time taken for a terrier to kill a number of rats was recorded. H. H. Donaldson brought breeders from the English colony to America in 1900, established the stock at the University of Chicago, and transferred it to the Wistar Institute in 1907. The Wistar colony seems to have been the progenitor of a number of new stocks. It seems certain that the Long Evans hooded strain, established in 1915, originated from a cross between Wistar females and wild males. The Sprague Dawley stock established nine years later, also was derived from the Wistar strain.

If the second hypothesis is correct, then a laboratory rat colony that has been started independently of Donaldson's, would probably show a new type of mtDNA. Such a colony is the McCollum strain that was started in America by McCollum in 1907 and is still being maintained.

We tested 12 of these animals. Surprisingly, we found no new mtDNA types; all the rats were of the B type. However, it is known that the albino females which were used to start the colony were obtained in a pet shop, and the pet shop animals may in turn have been obtained either from Donaldson's colony or from a European dealer who stocked an albino stock from which Donaldson's rats originated.

A third hypothesis, that albino rats have special types of mtDNA, was considered. However, albinism results from a single recessive nuclear gene and any connection to the mitochondrial genome, even indirect, is unlikely.

Realizing that the key to the mystery of the two mtDNA types lay in the mtDNA of the wild rat population, we initiated a survey of this population, the results of which will be discussed later.

Maternal Inheritance in Mammals

While the occurrence of uniparental inheritance has for some time been established in eukaryotic microorganisms, maternal inheritance in the animal kingdom has been demonstrated only recently (in the amphibian, *Xenopus*[10]). While the phenomenon has also been demonstrated to occur in mammals even more recently,[11] it had to be done by an interspecific cross, donkey × horse. The experiment, while very cleverly designed and controlled, nevertheless could be subject to ambiguous interpretation because of the interspecific nature of the cross. The occurrence of the two types of mtDNA within a mammalian species, especially a laboratory animal, provided the opportunity for a definitive demonstration of mammalian maternal inheritance and a number of laboratories, including ours, rapidly took advantage of this opportunity.[5, 8, 12, 13] Table 2 shows the results of type A and B double reciprocal crosses. The mtDNA type of the 58 resulting progeny is exclusively that of the female parent. All the laboratories that have done this experiment reported similar findings—the total number of progeny being well over 100.

Types A and B mtDNA Differ in Nucleotide Sequence

We have tacitly assumed thus far that the differences between the two types of mtDNA result from nucleotide sequence differences rather than from differential methylation or from one or another possible artifacts. The question is

TABLE 2

INHERITANCE OF TYPES A AND B RAT LIVER MTDNA

Cross	Parental DNA type ♀	♂	Number of offspring of each DNA type ♀A	♂A	♀B	♂B
1	A	B	5	4	0	0
2	A	B	5	5	0	0
3	A	B	5	7	0	0
4	B	A	0	0	2	9
5	B	A	0	0	2	5
6	B	A	0	0	6	3
7	A	A	3	5	0	0
8	A	A	5	7	0	0
9	B	B	0	0	6	3
10	B	B	0	0	3	5

an important one because it also bears on evolutionary studies of various wild species in this and other laboratories. To obtain evidence in support of this assumption, we decided to sequence[14, 15] appropriate corresponding segments of the two DNAs. Our restriction maps[5] of types A and B DNAs are shown in FIGURE 4. At 30 map units, there is a variant *Eco*RI site closely flanked by two conserved *Hin*dIII sites. To obtain sufficient amounts of this small *Hin*dIII fragment, the mtDNAs were cloned (in collaboration with Norman Arnheim's group at Stony Brook). A mixture of the two DNAs was first cleaved with *Bam*HI, which cuts the genome into a small and a large fragment. These fragments were then cloned in the usual manner using pBR322 as the vector and *E. coli* K12 as the host. The amp^r tet^S colonies were assayed for the presence of the small type A and small type B fragments by *Bam*HI–*Eco*RI double digestion of the plasmid preparations.

This assay procedure provided some evidence as a side product. The very fact that the cloned A and B fragments still could be differentiated by *Eco*RI, yielding products of the expected size, strongly suggested that methylation does not play a role; it would not be expected that *E.coli* methylases would methylate at the same sites as would (putative) mitochondrial methylases. Nevertheless, we went on to obtain conclusive proof by sequencing the *Hin*dIII fragments.

The plasmids pFC111A and pFC101B, which contained the type A and the type B small *Bam*HI fragments respectively, were used to prepare the *Hin*dIII fragments and these were then sequenced (FIGURE 1).[14, 15] Three differences between the 169 base pair (bp) type A and B fragments are revealed: A/T ↔ G/C at position 56, A/T ↔ G/C at position 80 and T/A ↔ G/C at 122. The base change at position 80 accounts for the absence of the *Eco*RI site in the type B fragment since it alters the recognition sequence from GAATTC to the nonrecognition sequence GGATTC. We thus conclude that differences between the two mtDNAs are, at least in part but probably completely, the result of nucleotide substitutions.

Other Aspects of the Sequences

Two other aspects of the sequences that are relevant here deserve discussion. First, the three nucleotide substitutions found represent a sequence divergence of 1.8% for these 169 bp fragments, a figure that is in good agreement with the value of 1.6% [5] (TABLE 5) obtained by restriction site analysis (scc later section). Second, we found that if we were to make two assumptions, one with respect to which strand is the transcriptional template, and the other, as to which is the correct reading frame, the three base substitutions we have observed would have an interesting consequence with respect to possible translation products. We assume first that the upper strand (FIGURE 1) is the transcriptional template. This is not an unreasonable hypothesis since the heavy strand of mtDNA is known to be the one on which nearly all transcription occurs [16] and the high G+T content of the upper strand of the fragment, in comparison with that of the lower (60% vs. 40%), suggests that the upper strand is derived from the heavy strand of mtDNA. Secondly, we assume that the correct DNA reading frame starts with the first nucleotide at the 3′-end of the heavy (upper) strand (the complementary lower strand thus reflecting the structure of the putative mRNA). This assumption is supported by the fact that one of the other reading frames (starting with position 167) gives rise to multiple termination codons, and the remaining frame (starting with position 168) specifies about 50% fewer hydrophobic amino acids than does the pre-

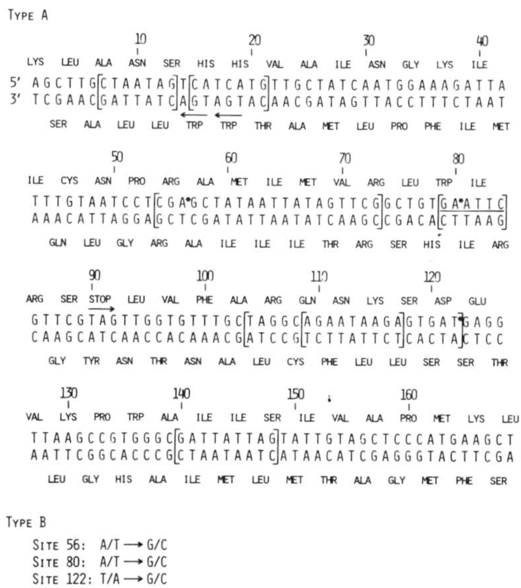

FIGURE 1. The nucleotide sequences of the corresponding small *Hin*dIII fragments from types A and B mtDNA. *, sites of nucleotide replacements; *GAATTC*, *Eco*RI recognition site; ←, opal termination codons shown to be read as tryptophan in mitochondria [17, 18]; →, termination codon; [ ], palindromic or near-palindromic sequences.

ferred frame, whereas hydrophobicity is a characteristic of mitochondrially synthesized proteins. The third frame also leads to high G or C in the codon third positions, not in accord with either yeast[17] or mammalian[18] mtDNA codon usage.

Thus, if our preferred frame is used to assess the phenotypic effects of the three variant codons, UCA ↔ UCC, AUU ↔ AUC, GCU ↔ GCC, we see that all the changes are in the third position and that in no case is there a change in the amino acid encoded.

The same conclusion is reached if the lower strand is assumed to be the transcriptional template. Two reading frames can be eliminated on the grounds of multiple termination codons. The remaining frame, starting at position 3, codes for 30 amino acids until a stop signal at 90–92, with no start signal visible thereafter. Two variant codons fall within this region, CGA ↔ CGG and UGA ↔ UGG. Again, the changes are in the third position with no amino acid shifts. In the unlikely event that GUG acts as a start signal (position 96–98), the third variant codon produces an aspartic–glutamic change, a relatively harmless replacement. Thus, these arguments, albeit thus far based on limited data, support the hypothesis that the variant nucleotides in the two mtDNA variants occur at positions that result in few important changes in the amino acid sequence of gene products. We plan to test this hypothesis.

mtDNA Polymorphism in Wild R. norvegicus Populations

To answer the questions raised earlier about the origin of types A and B mtDNA, we undertook a study of the wild *R. norvegicus* population. However, we felt that there was an even more important reason for doing so. Were such studies to show that a great many types of mtDNA exist in the wild population, perhaps even a continuum, then mtDNA restriction analysis might prove to be a tool of high resolving power and of great convenience for the study of the evolution of not only closely related species but evolution within a species as well.

At that point, there were as yet no evolutionary studies of this type published, but since that time several publications have appeared from our and other laboratories reporting evolutionary studies on a number of species, namely rats,[19-21] cricetid rodents (*Peromyscus*),[22] primates,[23] and man.[24] Our results on *R. norvegicus,* and then on *R. rattus,* follow.

Wild Norway rats were obtained by a two-step procedure. The first step was an attempt by some of us (G.G.B. and M.V.S.) to trap the rats ourselves. After a number of unsuccessful excursions, we proceeded immediately to the second step, which was to obtain them from laboratories that are involved in studies on wild rats. Eighteen rats became available to us, trapped in various parts of the U.S.A. and Puerto Rico, along with one Japanese rat. A number of different restriction-enzyme fragment patterns, for which we employ the term restriction morph, of the mtDNAs of these and of types A and B animals are shown in FIGURE 2A. While *Hin*dIII or the *Bam*HI+*Sst*I pair each produces a single morph, each of the other three enzymes elicits three or four different morphs (including those from the laboratory mtDNA types, as will be seen in TABLE 3). We have given each characteristic series of morphs a particular mtDNA designation, thus far A and B for the laboratory types and C and D for the new wild types, and these are shown in column 3. Subscripts

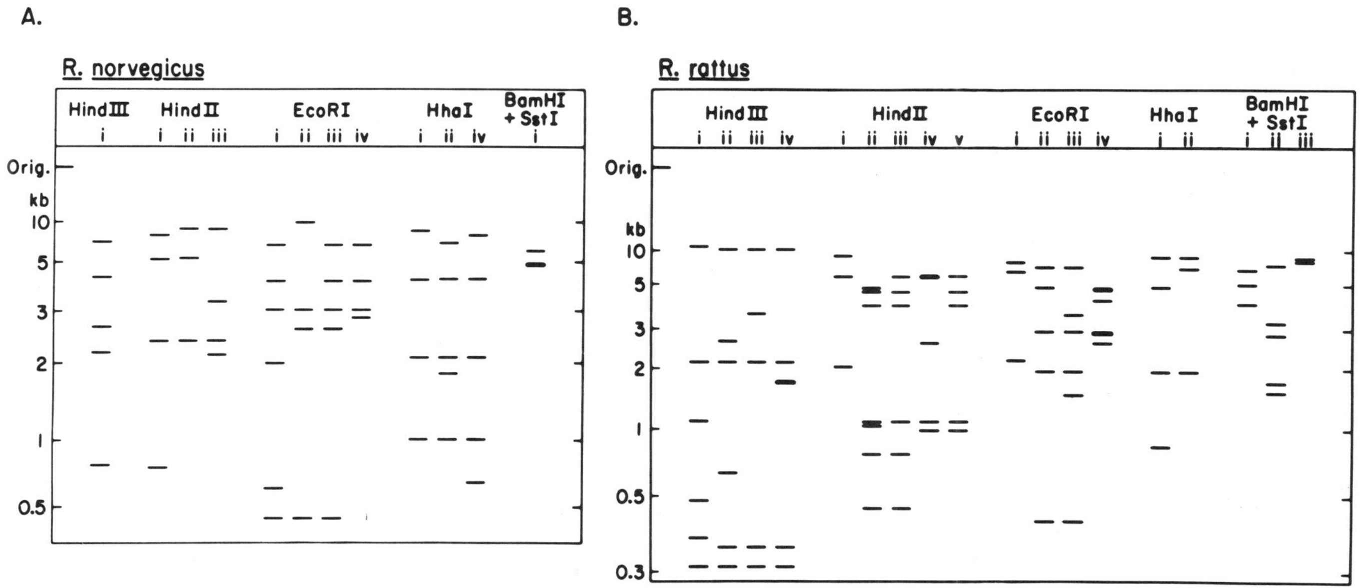

FIGURE 2. Restriction morphs of *R. norvegicus* and *R. rattus* mtDNAs. The lower case numerals heading the columns designate particular restriction morphs found in one or more individuals. These designations are arbitrary; restriction morphs bearing identical numerals but produced by different enzymes (e.g., *Hin*dII-i and *Eco*RI-i) are not necessarily derived from the same DNA. *Sst*I and *Bam*HI were used in combination because many of the mtDNAs were cut only once by either enzyme alone and thus a possible change in position of the cleavage site would fail to be revealed. Heavy lines indicate two or more bands that are poorly or incompletely resolved. Fragments smaller than about 0.5 kb in diagram A and 0.3 kb in diagram B have not been included.

TABLE 3
RESTRICTION MORPH TYPES OBTAINED
FROM INDIVIDUAL *R. norvegicus* ANIMALS

Sample Origin	Number of Animals	mtDNA type	Restriction Morph Types					
			*Eco*RI	*Hha*I	*Hind*II	*Hind*III	*Bam*HI +*Sst*I	*Hinf*I
Troy N.Y.	1	Cx *	iii	i	ii	i	i	
Boston, Mass.	1	Cx *	iii	i	ii	i	i	
San Juan, P.R.	1	Cx *	iii	i	ii	i	i	
New York, N.Y.	1	Co	iii	i	ii	i	i	iii
Chesapeake, Va.	1	Co	iii	i	ii	i	i	iii
Omaha, Neb.	1	Co	iii	i	ii	i	i	iii
Houston, Tex	2	Co	iii	i	ii	i	i	iii
Ponce, P.R.	1	Co	iii	i	ii	i	i	iii
Brooklyn, N.Y.	4	C_1	iii	i	iii	i	i	iii
Brooklyn, N.Y.	1	C_2	iii	i	ii	i	i	iv
Ponce, P.R.	1	C_2	iii	i	ii	i	i	iv
Camden, N.J.	1	C_3	iii	i	ii	i	i	v
Cohoes, N.Y.	1	C_4	iii	i	ii	i	i	vi
Harrisburg, Pa.	1	C_4	iii	i	ii	i	i	vi
Japan	1	D	iv	iii	ii	i	i	
Lab A		A	i	i	i	i	i	i
Lab B		B	ii	ii	ii	i	i	ii

* Subtype not assigned since *Hinf* I analysis was not performed.

designate mtDNA *subtypes* that differ from *types* in that their sequences diverge much less from one another than do the sequences between *types*. These subtypes, along with the *Hin*fI restriction morphs that brought three of these to light, will be discussed in a succeeding section, as will the type–subtype system of classification.

A striking feature of the results is the frequency of appearance of type C DNA throughout the U.S.A. and Puerto Rico. The sampling distribution is probably sufficient to designate this type as the common U.S. type. The Japanese rat mtDNA (type D) differs from type C by the gain of one *Hha*I site and the loss of one *Eco*RI site (FIGURE 2 and TABLE 3). Both types A and B also differ from C by two restriction sites. Interestingly, one additional morph, *Hin*dII-iii, was found only within a small neighborhood in New York City, and this type (C_1) differs from C by one restriction site. As will be seen later, its sequence diverges from that of C by only a small extent.

mtDNA Polymorphism in the Wild R. rattus Population

Because of their close relationship, it was of interest to assess the degree of sequence heterogeneity between the mtDNAs of the two species, *R. norvegicus* and *R. rattus*. In addition, although the number of *R. norvegicus* animals was limited, the results do leave the impression that mtDNA sequences between animals whose habitats were geographically close diverged much less than when two populations were separated by a geographical barrier such as an ocean. This view could be tested with *R. rattus* since the animals that became available to us were from geographically widely distributed habitats.

It might be mentioned at this point that most of the *R. norvegicus* restriction morphs were obtained by isolating lower band material after CsCl-ethidium bromide centrifugation of the crude mtDNA, restricting the DNA, and visualizing the gel pattern by the standard procedure of ethidium bromide staining and fluorescence photography. However, when greater sensitivity was required, as in the case of the *Hin*fI digests, [^{32}P]-end labeling and autoradiography was used. With *R. rattus*, however, in many cases only frozen tissues or frozen homogenates in which the mitochondria were disrupted were available to us. In these cases, not only did we find that most or all of the mtDNA molecules were nicked, but that this material was overwhelmingly contaminated with nDNA. This problem was overcome by using the Southern blot technique in which the restriction fragments produced by crude bulk DNA were detected with ^{32}P-labeled (by nick translation) purified *R. rattus* mtDNA as the probe, and subsequent autoradiography.

The analyses of *R. rattus* mtDNA (FIGURE 2B) elicited a greater number of restriction morphs than was evident with the Norway rats. For example, *Hin*dIII exposes four and the *Bam*HI-*Sst*I pair exposes three different morphs. It is also evident that, except for *Hha*I, the restriction morphs produced by a given enzyme differ from one another to a greater extent in the number of nonconserved fragments than do the *R. norvegicus* morphs.

TABLE 4 shows the geographical origin of each *R. rattus* animal along with its restriction morph spectrum. Five types of mtDNA are revealed, A to E. (These designations bear no relation to the *R. norvegicus* designations.) D appears to be the type most commonly found in the U.S. and possibly Puerto Rico. The wide spectrum of morphs found in the Sri Lanka animals is unusual;

TABLE 4
RESTRICTION MORPH TYPES OBTAINED FROM INDIVIDUAL *R. rattus* ANIMALS

			Restriction Morph Types					
Sample Origin	Number of Animals	mtDNA type	*Eco*RI	*Hha*I	*Hind*II	*Hind*III	*Bam*HI +*Sst*I	*Hinf*I
Sri Lanka	5	A	i	i	i	i	i	
Hong Kong	2	B	ii	ii	ii	ii	ii	
Japan	3	C	iii	ii	iii	ii	ii	
Tampa, Fla.	1	Dx *	iv	ii	iv	iv	iii	
Alameda Co. Cal.	4	Dx *	iv	ii	iv	iv	iii	
Alameda Co. Cal.	5	D_1	iv	ii	iv	iv	iii	ii
Alameda Co. Cal.	2	D_2	iv	ii	iv	iv	iii	iii
San Joaquin Co. Cal.	1	D_3	iv	ii	iv	iii	iii	i
Chesapeake, Va.	1	D_4	iv	ii	iv	iv	iii	iv
San Juan, P.R.	1	D_5	iv	ii	iv	iv	iii	v
Ft. Lauderdale Fla.	1	E	iv	ii	v	iv	iii	

* See footnote to TABLE 3

each enzyme yielded a unique morph. While it is tempting to speculate from these results that more diversity exists in the *R. rattus* than in the *R. norvegicus* population, much more data is needed from *R. rattus* individuals that inhabit a more limited geographical area than is reflected in our present sampling.

mtDNA Variants Revealed by HinfI

The enzymes discussed thus far make few cuts in mtDNA and are valuable for the construction of cleavage maps. However, an enzyme that makes many cuts would be expected to possess more resolving power and would therefore be more likely to expose mtDNA subtypes. *Hin*fI, a four-base enzyme, yields about 30 fragments. Although containing too many fragments to map easily, the digest nevertheless proved valuable in providing information about additional variants.

Type C DNA is the general type of mtDNA found in the U.S.A. and this type along with types A and B were subjected to *Hin*fI analysis. Six different morphs were revealed (FIGURES 3A & 3C). (Morph iii appears twice for comparison). Two of these (morphs i and ii) were derived from types A and B mtDNA (TABLE 3). The remaining four were products of general type C animals. Morphs iii, v, and iv, corresponding to subtypes C_0, C_3, and C_2 (TABLE 3) are unusual and, therefore, we have included a photograph of the gel (FIGURE 3C). The band differentiating the three morphs suggests the occurrence of deletions/insertions, since it differs by increments of only about 3 bp in the three morphs. All the other bands seem to be identical in size, so it is highly probable that these bands correspond to one another. It is unlikely that these fragments differ in cleavage sites close to the fragment terminus rather than by deletions/insertions. In *R. rattus,* four previously indistinguishable mtDNAs were shown by *Hin*fI analysis to be variants of the general D type found in the U.S.A. (FIGURE 3B). Morphs ii, iii, iv, and v characterize subtypes D_1, D_2, D_4 and D_5, respectively (TABLE 2). Morph i, closely related to morphs ii and iii, was obtained from the subtype D_3, which is also variant in its *Hin*dIII morph.

Restriction-Enzyme Cleavage Maps

Cleavage maps were constructed for all the variant mtDNAs of both species discerned thus far by enzymes other than *Hin*fI, and these maps are presented in linearized fashion for comparative purposes in FIGURE 4. Three salient features of mitochondrial DNA variation emerge from the maps. The first is that there is a relative paucity of variant sites occurring between 82.5 and 100 map units—the region of the molecule that codes for ribosomal RNA.[25] This can be taken as evidence that the ribosomal RNA genes constitute a more slowly evolving region of the genome, and evidence from interspecific hybridization studies [26] also indicates that this is the case. The second is that variant sites appear to be more or less randomly located throughout the remainder of the genome—that evolutionarily "hot" regions do not seem to exist. The third feature is that intraspecific variation differs from interspecific only in extent and not in kind—the distribution of variant sites within the *R. rattus* group does not differ significantly from that distribution found between the two species.

Some specific sites deserve comment. Sites at 66.5, 73, and 83.5 map units are present in all the rat mtDNAs we have examined, as well as in mouse mtDNA.[27] With respect to the two *Rattus* species alone, ten sites are conserved in all the mtDNAs analyzed. Other sites are confined to the mtDNAs of one or the other of the two species and some sites are distributed among certain mtDNAs of both species.

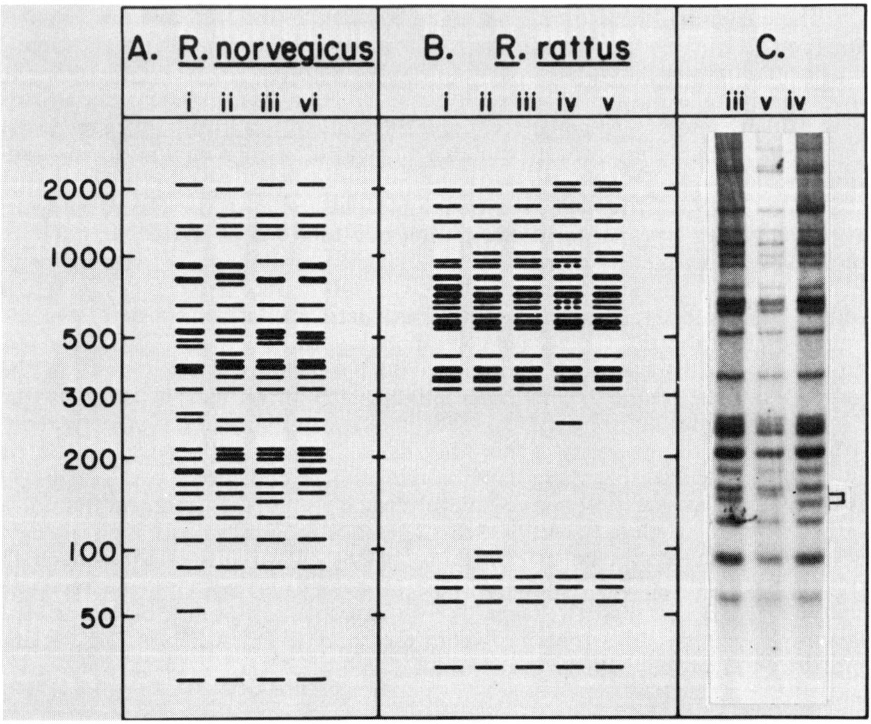

FIGURE 3. Restriction morphs obtained by *Hin*fI digestion of certain types of *R. norvegicus* and *R. rattus* mtDNAs. The mtDNAs from various type C (as well as laboratory type A & B) *R. norvegicus* individuals and from various *R. rattus* type D individuals were subjected to *Hin*fI analyses. The lower case numerals heading the columns designate particular restriction morphs found in one or more individuals. Heavy lines indicate poorly or incompletely resolved bands. The dotted lines in *R. rattus* morph iv reflect bands about which some uncertainty still exists. Fragments smaller than about 0.03 kb would not have been detected. Frame C, radioautograph; morphs iii, iv, and v, correspond to *Rattus norvegicus* subtypes C_0, C_2, and C_3, respectively. Morph iii appears in both frames A and C for purposes of comparison.

Extent of Sequence Divergence

Shortly after our discovery of the two mtDNA types in the laboratory rat, two methods for computing the extent of sequence data from the results of restriction-enzyme analysis were reported by Upholt.[28] One of these is based

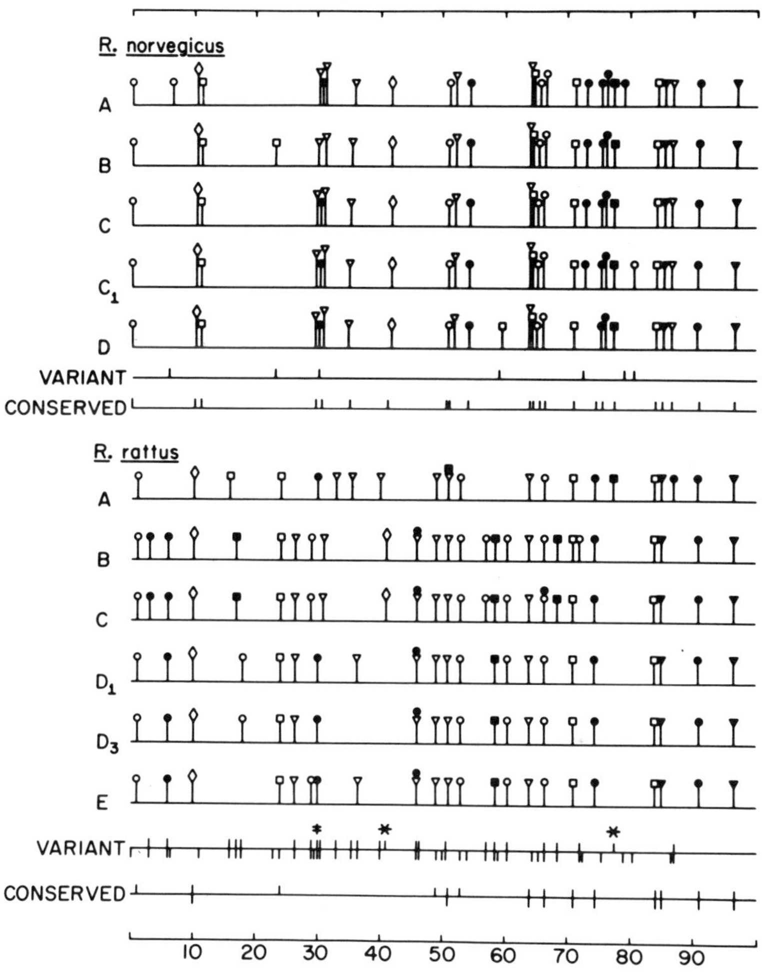

FIGURE 4. Restriction maps of various *R. norvegicus* and *R. rattus* mtDNA types and subtypes. The circular maps have been linearized for more convenient comparison. The scale begins at the origin of H-strand replication and proceeds from left to right in the direction of replication. Positions of variant and conserved sites for each species are indicated on the horizontal axes beneath each set of maps. The indicators extending downward from the *R. rattus* axes represent a composite of the variant and conserved sites in both species. *, sites present in all *R. norvegicus* mtDNAs but only in some *R. rattus* mtDNAs; ‡, sites present in certain *R. norvegicus* and certain *R. rattus* mtDNAs. (●), *Eco*RI; (▼), *Sst*II; (■), *Sst*I; (○), *Hind*II; (▽), *Hind*III; (□), *Hha*I; (◇), *Bam*HI.

on the conservation of restriction-enzyme fragments, the other on the conservation of sites. The latter method requires a restriction map for each enzyme used. The amount of data that can be harvested by the site method is therefore subject to practical limitations, and the statistical variance of a result obtained by this method will therefore usually be less than by the fragment method, a large amount of data for which is easy to obtain. On the other hand, the accuracy of the fragment method can be seriously flawed by errors in interpretation of gel patterns. For this reason, we have chosen to employ the restriction-site method in those instances where we were able to construct a map, namely, for all enzymes except *Hin*fI. The results obtained after pairwise comparison of the mtDNAs using the site method were used to construct a resemblance matrix [29] (TABLE 5). A rather startling spectrum of divergence can be seen

TABLE 5

RESEMBLANCE MATRIX OF MTDNAS FROM POPULATIONS OF LABORATORY AND WILD *R. norvegicus* AND WILD *R. rattus*

	R. rattus						*R. norvegicus*				
	A	B	C	D_1	D_3	E	A	B	C	C_1	D
A	—										
B	10.5	—									
C	10.5	0.7	—								
D_1	7.3	4.5	4.4	—							
D_3	6.9	4.2	4.0	0.4	—						
E	7.3	3.4	3.4	1.0	1.4	—					
A	14.7	18.4	18.0	16.4	16.1	16.4	—				
B	15.4	17.8	17.3	17.1	16.8	17.1	1.6	—			
C	13.7	17.5	17.1	15.4	15.1	15.4	0.7	0.9	—		
C_1	14.5	18.2	17.8	16.1	15.9	16.1	1.2	1.4	0.4	—	
D	14.1	17.8	16.7	15.7	15.2	15.7	1.6	1.8	0.9	1.3	—

ranging from 0.4% to 18.4%. Based on the additional data from *Hin*fI (data not shown in matrix) the range of divergences is 0.1%–0.4% for *R. norvegicus* subtypes and 0.17%–0.99% for *R. rattus* subtypes.

For convenience, we have used these divergence values to tentatively classify mtDNA into types and subtypes. The validity of this classification will be discussed later. We have arbitrarily assigned the term *type* to a *R. norvegicus* mtDNA whose sequence diverges 0.5% or more from any other type in this species, and the term *subtype* to divergence less than 0.5%. Because of the generally higher level of sequence divergence thus far found in *R. rattus*, we have used a higher value, 1%, for this species to divide type from subtype.

Several remarkable results are evident from the matrix.[19, 21] First, the maximum sequence divergence between any of the *R. norvegicus* and any of the *R. rattus* mtDNAs is 18.4%, with the divergence between all mtDNA types of one species and all mtDNA types of the other species averaging 16.5%. These values are clearly extremely high for closely related species *of the same genus* and reflect a surprisingly high evolutionary rate for rat mtDNA. Values of this order (ranging from 21% to 29%) have been found by Brown *et al.*[23] for four primate species, human, rhesus, baboon, and guenon, but these are members of four different genera. However, both sets of data point to the rapid evolution of mtDNA and both our laboratories are in accord on this point.[23] Moreover, these conclusions are in accord with those of Dawid based on mtDNA hybridization results on *Xenopus*.[26]

Even more striking are the sequence divergence values within the *R. rattus* species. Our experience with *R. norvegicus* had prepared us for values of the order of 2%; here we see *R. rattus* individuals from different geographical areas with mtDNAs that diverge in sequence by more than 10%. This was completely unexpected. Such an extensive divergence within a species poses a special type of mating problem. While maternal inheritance avoids the problem of the pairing of male and female parental mtDNA strands of rather high heterogeneity, the difficulty of mismatched mitochondrial and nuclear gene products remains. An example of this is the mutual recognition required between mitochondrial rRNA (encoded by mtDNA) and mitochondrial ribosomal proteins (encoded almost entirely by nDNA). A 10% change in the nucleotide sequence of the rRNA could destroy this recognition. True, rRNA genes tend to be conserved, but an entirely analogous recognition problem exists between mtDNA-encoded tRNAs and nDNA-encoded mitochondrial aminoacyl tRNA synthetases, and between the mtDNA-encoded and nDNA-encoded subunits of any of a number of mitochondrial membrane proteins, e.g., cytochrome oxidase and oligomycin-sensitive ATPase.

There are escapes from this difficulty. One is the consideration that half the nuclear DNA of the offspring will also be derived from the female parent so that recognition of 50% of the gene products in question would occur and this may be sufficient for viability. Alternatively, it is possible that all, or a good portion, of the nucleotide replacements are silent, as was discussed elsewhere.

One might ask at this stage of our work, whether we are really dealing with types and subtypes, or whether a virtual continuum of sequence divergences exists of which we have more or less randomly sampled only a few. Our *R. rattus* data, obtained from a few animals derived from a huge geographical area, is clearly insufficient to answer this question. The *R. norvegicus* results tentatively suggest a limited answer, that at least in the U.S.A., the number of mtDNA types is few in number. However, exploration with a much larger array of restriction enzymes may expose a great many more variants.

Construction of Dendrograms

The inter- and intraspecific relationships among the various DNAs we have been discussing can best be seen by the construction of a relationship tree. In addition, such a tree would provide a testing ground for our data; calculation of the cophenetic correlation[29] for the tree would tell us how closely all of our data fits the generated tree. The resemblance matrix (TABLE 5) can be used to construct such a tree and this was done using the similarity clustering program of James Rolhf,[30] which is based on the unweighted pair method (i.e., UPGMA).

This dendrogram along with two others which show the mtDNA subtypes generated by *Hin*fI, are shown in FIGURE 5. We chose not to merge the three trees because the dendrograms shown in both frames B and C were constructed using data from just one additional enzyme (*Hin*fI), and since the sequence divergences are very small among these subtypes, the statistical error may be larger than that in frame A.

The three calculated cophenetic correlations are: frame A, 0.99; frame B, 0.96; frame C, 0.67. Clearly the data fits in the first two trees are excellent. However, more restriction data is required in order to increase the reliability of the *R. rattus* subtype tree.

Inspection of the three dendrograms reveals several interesting features. First, the various mtDNA types within each species are more closely related

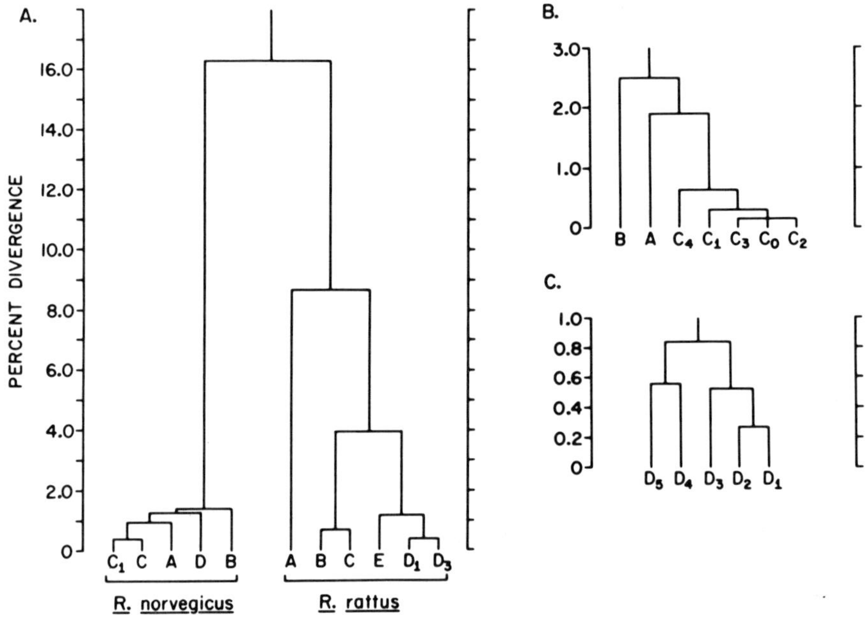

FIGURE 5. Phenograms of the various *R. norvegicus* and *R. rattus* mtDNA types and subtypes. A, constructed from data obtained with all enzymes in TABLES 3 and 4 except *Hin*fI; B, constructed from data of all enzymes including *Hin*fI.

to one another than to any type in the sister species. Second, the *R. norvegicus* types are more closely related to one another than are the *R. rattus* types, although this may be related to a wider geographical distribution of the latter. Nevertheless, the Japanese *R. norvegicus* type (type D) diverges in sequence from the major U.S. type (type C) by only 1.6%. Finally, the Sri Lanka type, while clearly more closely related to its own species than to *R. norvegicus*, shows an unusually large divergence from the remaining *R. rattus* types.

With respect to the significance of the dendrograms, the information used to construct them is phenetic in character, a phenetic relationship being defined as one that records the similarity between present properties of organisms, without reference as to how they came to possess them.[29, 31] Thus, the dendrograms we have constructed, termed phenograms, show phenotypic relationships between the different rat types but without further assumptions do not imply phylogenetic relationships, that is, the evolutionary pathways leading to these populations.

Conversion of this phenogram to a phylogenetic tree would require two basic assumptions, one, the absence of convergence in the evolution of these mtDNA types and the other, the constancy of evolutionary rates. However, data on mtDNA evolution is still scarce and it is difficult to find experimental support for these assumptions at this time. In particular, while a good deal of information is available indicating that the evolution of a number of nuclear genes is constant for each within a factor of 2,[32] no such firm conclusion can be drawn for the rapidly evolving mtDNA data at this time, although an initial attempt to study this problem has been made.[23]

However, the very high cophenetic correlation obtained for some of our data, e.g., for the tree in frame A (TABLE 5), does suggest, at least tentatively, that the aforesaid assumptions may be valid within limits. If this is the case, the tree would supply some additional information. For example, in the *R. rattus* genus (frame A), the Sri Lanka (A) lineage branched off prior to the separation of the Japan–Hong Kong types (B and C) from the American types (E, D_1, and D_3). In addition, it is clear that the separation of the types within each species took place after the separation of the two species.

As an alternative to making these assumptions, we considered the construction of a tree by cladistic analysis, in this case by following the gain or loss of restriction sites, using procedures analogous to those of Fitch and Margoliash.[33] While the actual sequencing of large stretches of mtDNA would produce the type of data needed for such analysis, methods are not available at the present time for the treatment of restriction enzyme data for this purpose.

GENERAL DISCUSSION

The power of the method of restriction enzyme analysis for studying various aspects of evolution, particularly evolution of the mtDNA molecule itself, as well as evolution of closely related species and evolution of populations within a species, lies in two factors. The first of these is the convenience and rapidity of restriction enzyme analysis. The second is the rapid rate of evolution of mtDNA. The possible reasons for the latter deserve some discussion.

Formulating a reasonable hypothesis as to why mtDNA might undergo a higher rate of mutation than nDNA is not difficult. In the first place, the half life of mtDNA is relatively short; it turns over many times during the lifetime of those cells in which it has been studied,[34] whereas nDNA is replicated only during cell division. It is probably a safe assumption that this situation also holds true for all the cells involved in the various stages of oocyte development, which are presumably the only cells relevant to mtDNA evolution. Secondly, there does not seem to be a DNA repair system in mitochondria.[35]

The question of why so many mutations are fixed is a more difficult one to answer. The mitochondrion is a rather tightly knit biochemical and structural network and one would not predict that such a highly integrated machine would permit rapid evolutionary changes. Moreover, mitochondria are, from the molecular point of view, very closely structurally integrated into the rest of the cell, as evidenced by a number of cases (discussed earlier) best exemplified by mitochondrial proteins whose encoding is shared by both nuclear and mitochondrial DNA, and this would speak against "runaway" evolution of mtDNA. While an easy explanation is that most or all of the mutational changes are silent—indeed, our somewhat speculative arguments, based on only three nucleotide replacements, supports this conclusion—we feel that this view, with further work, will not be substantiated.

With respect to the seeming absence of types A and B in the wild population, the absence of European rats in our survey will have been noted. Inasmuch as our laboratory strains were derived, in whole or in part, from European rats, efforts are being made to obtain a representative sample from this continent.

Acknowledgments

We are indebted to Dr. Allan Wilson for stimulating discussions, to Dr. John Patton for sending us wild rat tissue specimens and for suggesting other sources, to Dr. T. H. Yosida for wild rat specimens and to Dr. Larry Grossman for sending us rats from the Johns Hopkins University McCollum Colony. We are particularly grateful to Dr. James Rohlf of our Department of Ecology and Evolution for making available to us his unpublished computer program and for spending a considerable amount of time with us discussing the numerical taxonomic implications of our results. We also thank Dr. Walter Fitch for valuable advice.

References

1. Potter, S. S., J. E. Newbold, C. A. Hutchison, III & M. H. Edgell. 1975. Proc. Natl. Acad. Sci. USA **71**:4447–4451.
2. Upholt, W. B. & I. B. Dawid. 1977. Cell **4**:571–583.
3. Francisco, J. F. & M. V. Simpson. 1977. FEBS Lett. **79**:291–294.
4. Francisco, J. F., F. F. Vissering & M. V. Simpson. 1977. *In* Mitochondria 1977: Genetics and Biogenesis of Mitochondria. W. Bandlow, R. J. Schweyen, K. Wolf & F. Kaudewitz, Eds.: 25–37. de Gruyter, Berlin.
5. Francisco, J. F., G. G. Brown & M. V. Simpson. 1979. Plasmid **2**:426–436.
6. Saccone, C., G. Pepe, H. Bakker & A. Kroon. 1977. *In* Mitochondria 1977: Genetics and Biogenesis of Mitochondria. W. Bandlow, R. J. Schweyen, K. Wolf & F. Kaudewitz, Eds.: 303–315. de Gruyter, Berlin.
7. Hayashi, J., H. Yonekawa, O. Gotoh, J. Motohashi & Y. Tagashira. 1978. Biochem. Biophys. Res. Commun. **81**:871–877.
8. Buzzo, K., D. L. Fouts & D. R. Wolstenholme. 1978. Proc. Natl. Acad. Sci. USA **75**:909–913.
9. Inbred and Genetically Defined Strains of Laboratory Animals: Part I Mouse and Rat 1979. P. L. Altman & D. D. Katz, Eds. Federation of American Societies for Experimental Biology. Bethesda, Maryland.
10. Dawid, I. B. & A. W. Blackler. 1972. Devel. Biol. **29**:152–161.
11. Hutchison, C. A. III, J. E. Newbold, S. S. Potter & M. H. Edgell. 1974. Nature (London) **251**:536–538.
12. Hayashi, J., H. Yonekawa, O. Gotoh, J. Watanabe & Y. Tagashira. 1978. Biochem. Biophys. Res. Commun. **83**:1032–1038.
13. Kroon, A. M., W. M. deVos & H. Bakker. 1978. Biochim. Biophys. Acta **519**: 269–273.
14. Castora, F. J., N. Arnheim & M. V. Simpson. 1980. Fed. Proc. **39**:1914.
15. Castora, F. J., N. Arnheim & M. V. Simpson. Proc. Natl. Acad. Sci. USA (In press.)
16. Ojala, D. & G. Attardi. 1974. J. Mol. Biol. **88**:205–219.

17. MACINO, G., G. CORUZZI, F. NOBREGA, M. LI & A. TZAGALOFF. 1979. Proc. Natl. Acad. Sci. USA **76:**3784–3785.
18. BARRELL, B. G., A. T. BANKIER & J. DROUIN. 1979. Nature (London) **282:** 189–194.
19. BROWN, G. G. & M. V. SIMPSON. 1979. J. Cell Biol. **83:**357a.
20. HAYASHI, J., H. YONEKAWA, O. GOTOH, Y. TAGASHIRA, K. MORIWAKI, T. H. YOSIDA. 1979. Biochim. Biophys. Acta **564:**202–211.
21. BROWN, G. G. & M. V. SIMPSON. 1980. (Submitted for publication.)
22. AVISE, J. C., R. A. LANSMAN & R. O. SHADE. 1979. Genetics **92:**279–295.
23. BROWN, W. M., M. GEORGE & A. C. WILSON. 1979. Proc. Natl. Acad. Sci. USA **76:**1967–1971.
24. BROWN, W. M. & H. M. GOODMAN. 1979. *In* Extrachromosomal DNA. D. J. Cummings, P. Borst, I. B. Dawid, S. M. Weissman & C. F. Fox, Eds.: 485–500. Academic Press, Inc. New York, N.Y.
25. KROON, A. M., G. PEPE, H. BAKKER, M. HOLTROP, J. E. BOLLEN, E. F. J. VAN BRUGGEN, P. CANTORE, P. TERPSTRA & C. SACCONE. 1977. Biochim. Biophys. Acta **478:**128–145.
26. DAWID, I. B. 1972. Dev. Biol. **29:**139–151.
27. PARKER, R. C. & R. M. WATSON. 1977. Nucl. Acids Res. **4:**1291–1299.
28. UPHOLT, W. B. 1977. Nucl. Acids Res. **4:**1257–1265.
29. SNEATH, P. H. A. & R. R. SOKAL. 1973. *In* Numerical Taxonomy: 190. W. H. Freeman and Co. San Francisco, Calif.
30. ROLHF, F. J. Personal communication.
31. SNEATH, P. H. A. 1974. *In* Evolution in the Microbial World: Twenty-fourth Symposium of the Society for General Microbiology: 1–39. Cambridge University Press. Cambridge, England.
32. WILSON, A. C., S. S. CARLSON & T. J. WHITE. 1977. Ann. Rev. Biochem. **46:** 573–637.
33. FITCH, W. M. & E. MARGOLIASH. 1969. *In* Structure, Function and Evolution in Proteins: Brookhaven Symposia in Biology. **21:**217–242. Brookhaven National Laboratory. Upton, N.Y.
34. RABINOWITZ, M. & H. SWIFT. 1970. Physiol. Rev. **50:**376–427.
35. LANSMAN, R. A. & D. A. CLAYTON. 1975. J. Mol. Biol. **99:**761–776.

DISCUSSION OF THE PAPER

DR. R. E. CALHOON (*Queens College, CUNY, Flushing, N.Y.*): I was wondering, in the construction of the dendrogram, are you using a similarity index? Is this an unweighted pair-groups analysis or was there some kind of weighting involved?

DR. M. V. SIMPSON: No, As far as I know it is unweighted. I must say that I don't want to pose yet as an expert on this program. It is a very complex program. It took a very long time to work out, but, as I understand it, it is not weighted.

DR. F. CASTORA: Could it be that even in the nucleus there might be local pockets of rapidly evolving DNA? Being interested in mitochondria that are evolving rapidly also, I wonder if anybody knows anything about mitochondrial proteins that might be coded in the nucleus and that might be evolving locally more rapidly than nuclear DNA.

DR. SIMPSON: Well, the only thing I could say about that is that when one talks about the rate of evolution of single-copy genes, it is an average figure. There are some genes that evolve fast, and some nuclear genes that evolve much faster than others.

CHLOROPLAST EVOLUTION—ANCIENT AND MODERN

Jean M. Whatley

Botany School
South Parks Road
Oxford OX1 3RA
England

If the theory of endosymbiosis is valid, then the general similarity of all mitochondria suggests that they evolved from bacteria following a single act of endosymbiosis, whereas the diversity of chloroplasts suggests for them a multiple origin from a variety of algal symbionts. Traditionally the theory for chloroplasts considers that the direct precursors of all plastids were prokaryotic algae. This hypothesis may reasonably hold for chloroplasts of the red and most of the green algae, i.e., those algae with plastids surrounded by two membranes, which are considered to be homologous with the plasma membrane of the prokaryotic symbiont and the vacuolar membrane provided by the host cell. However, the theory as it stands is less satisfactory for other algal groups in which the plastids are surrounded by more than two membranes.

Within the last decade information has accumulated from a variety of sources that has prompted several authors to propose that the more complex chloroplasts of these other algae may have arisen following a second act of endosymbiosis in which the symbionts were eukaryotes rather than prokaryotes.[1-4]

Growing interest in the mode of action of symbiotic associations between plants and animals has led to the realization of how frequent these associations are. As a result, the hypotheses that plastids had a multiphyletic origin and that this included repeated acts of symbiosis seem more reasonable than was formerly thought. Electron microscope studies of different algae and of different symbiotic associations have shown how diverse are the organisms involved and how varied in number are the membranes that separate the photosynthetic apparatus from the host cytoplasm.[5-8] It has become clear that the postulated homologies between the membranes surrounding symbionts and organelles that form a basis for the endosymbiotic theory must be extended and reinterpreted to include these newly discovered membrane relationships.

An important factor in this reinterpretation for one group of algae was the description by Greenwood[9] of the nucleomorph (FIGURE 1), a small structure enclosed by two membranes with pores, which is situated in the perichloroplastic space of cryptomonads. If this structure indeed represents a reduced or degenerate nucleus, then the perichloroplastic space may be equated with the reduced cytoplasm of a eukaryotic symbiont.[1] The complex plastids of cryptomonads (as well as those of most other algae that contain chlorophyll *c*) are surrounded by four membranes. The innermost pair represent the true chloroplast double envelope; the outer pair are usually referred to as chloroplast endoplasmic reticulum.[10, 11] However, if the perichloroplastic space is indeed the reduced cytoplasm of a eukaryotic symbiont, then the inner membrane of this outer pair should be reinterpreted as homologous with the plasma membrane of that symbiont. The fourth or outermost membrane is usually con-

0077–8923/81/0361–0154 $01.75/0 © 1981, NYAS

tinuous with the outer nuclear envelope and would thus on its own be homologous with endoplasmic reticulum. According to this interpretation the eukaryotic symbiont is sited within a channel of that endoplasmic reticulum (FIGURE 1, ③). Some of the algae with four encircling membranes around the plastids have more than one chloroplast in each cell. In these, the outermost membrane is usually not continuous with the nuclear envelope and the complex plastids (or symbionts) apparently lie inside isolated vacuoles within the host cytoplasm (FIGURE 1, ④).[12] The giant multinucleate amoeba, *Pelomyxa palustris,* contains endosymbiotic bacteria. During most of the life cycle of the amoeba the bacteria lie within vacuoles dispersed throughout the cytoplasm (FIGURE 1, ②), but when the amoeba is about to divide the bacteria in their vacuoles migrate towards the nuclei. At this time the vacuolar membranes become attached by

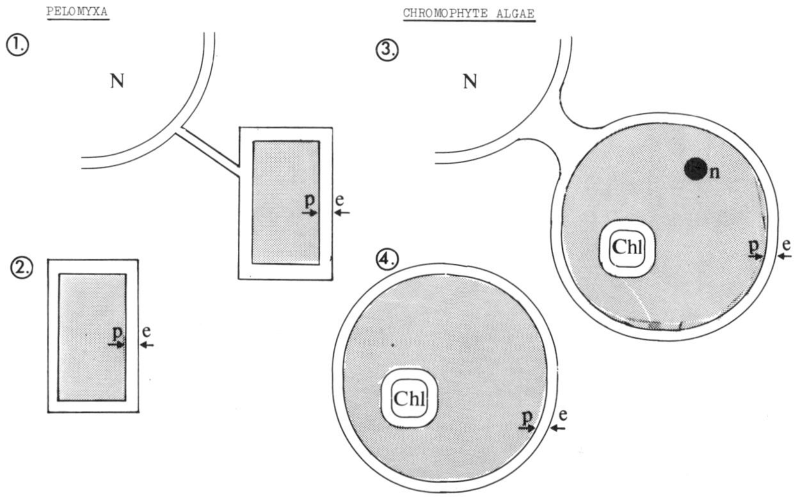

FIGURE 1. The location of symbionts in Pelomyxa and Chromophyte Algae. Chl=Chloroplast, e=endomembrane (endoplasmic reticulum or vacuolar), n=nucleomorph, N=nucleus, p=plasma membrane of symbiont.

membranous tubules to the outer nuclear membrane and the bacterial symbionts then lie within channels of the endoplasmic reticulum (FIGURE 1, ①).[13] Thus at different stages during the life cycle of *Pelomyxa* the symbionts occupy two sites within the endomembrane system of the cell, equivalent to those occupied by the chloroplasts in the different algal species.

By interpreting the homologous relationships between membranes that surround plastids and those that surround present day symbionts, one may obtain a guide (but only a guide) to the possible origin of plastids from prokaryotic or eukaryotic algae.[4] The membrane arrangements of three different types of chloroplast have their counterparts among present day symbionts (TABLE 1): a) plastids of red and most green algae, like modern prokaryotic symbionts, have two encircling membranes (plasma membrane of symbiont plus vacuolar membrane of first host); b) plastids of euglenoids and dinoflagellates, like isolated chloroplast symbionts from siphonaceous algae found in

TABLE 1

A COMPARISON BETWEEN MODERN AND ANCIENT SYMBIOSES

Number of Successive Symbioses	Host and Endosymbiont combinations	Number of Membranes around the Thylakoids	Organisms	
			Examples from Ancient Endosymbioses	Examples from Modern Endosymbioses
0	Free-living prokaryotes	1	Blue-green algae Prochlorons *	Blue-green algae
1	Prokaryotic algal symbionts or the chloroplasts, sensu stricto, derived from this combination inside eukaryotic host no. 1	2	Chloroplasts of Red algae and most Green algae	Blue-green algae inside sponges, etc. Cyanelles inside *Cyanophora,* etc.
2	Eukaryotic algal symbionts (containing true chloroplasts) or the compound chloroplast units derived from this combination inside eukaryotic host no. 2	3	Chloroplast units of Dinoflagellates and Euglenoids	Red algal or siphonaceous algal chloroplasts inside Opisthobranchs, e.g., *Elysia.*
		4	Chloroplast units of Cryptomonads and other Chromophyte algae	Chlorellae inside *Hydra,* etc. and *Platymonas* inside *Convoluta.*
3	Eukaryotic algal symbionts (containing compound chloroplast units) inside eukaryotic host no. 3	5 (+)	None known	Cryptomonad inside *Mesodinium*: Dinoflagellates inside sea anemones or giant clams etc: Diatom inside *Convoluta*?
		5+3 (Two forms of plastid present)		Chrysophyte inside the dinoflagellate, *Peridinium*

* Epizooic.

opisthobranchs, have three encircling membranes (plasma membrane of the initial prokaryote, and vacuolar membranes of both first and second host); c) plastids of the remaining algae, including cryptomonads, like eukaryotic symbionts such as zoochlorellae, have four encircling membranes (plasma membranes of both the initial prokaryotic and the subsequent eukaryotic symbionts and the vacuolar membranes of the two successive hosts). There are, in addition, more complex modern associations involving three successive symbioses. For these there are apparently no chloroplast counterparts.

It is tempting to suggest that the three groups of present-day chloroplasts listed above arose from symbionts that, like present-day symbionts, were prokaryotes, isolated chloroplasts, and complete eukaryotes respectively. However it is much more important to recognize that chloroplasts probably evolved from eukaryotic as well as from prokaryotic symbionts than it is to try to subdivide eukaryotic symbioses into different types. Indeed there may be no sharp distinction between the two proposed forms of eukaryotic symbiosis since present-day isolated chloroplast symbionts may, when they are incorporated in host cells such as *Elysia,* have associated with them traces of a fourth membrane and varying amounts of cytoplasm, including mitochondria, belonging to the siphonaceous algae from which they were derived.[6]

In this paper I am concerned not with algal phylogeny or with characteristics of the host cells but only with characteristics associated either directly or indirectly with the chloroplasts. For this reason I have used for chloroplast nomenclature the algal classification of Christensen,[14] which emphasizes chloroplast features. Christensen distinguishes three main phylogenetic lines, the chloroplasts of each having the same primary photosynthetic pigment, chlorophyll *a*, but a different accessory pigment. Plastids of the red algae have accessory phycobilins, those of the green algae have chlorophyll *b*, and those of the remainder, the chromophyte algae, have chlorophyll *c* (FIGURE 2 & TABLE 2). Descendants of possible prokaryotic ancestors exist today for the double membrane-bounded chloroplasts of the red and green algal lines; blue-green algae have chlorophyll *a* and phycobilins, and prochlorons have chlorophylls *a* and *b*.[15, 16] However no prokaryotic alga is known that has accessory chlorophyll *c*. Nor is any eukaryotic alga known that contains chlorophyll *c* and whose chloroplasts have two surrounding membranes, suggesting a prokaryotic ancestor. We therefore have at present no means of knowing at what stage chlorophyll *c* evolved. It might have risen either in a free-living prokaryote or, later, within the chloroplast of a postendosymbiotic eukaryote, perhaps a red alga, and both possibilities are shown in FIGURE 2. Such a modified red or "pre-chromophyte" alga with both phycobilins and accessary chlorophyll *c* (specifically c_2) when taken up by a second host would be, in effect, a cryptomonad. The subsequent loss of phycobilins from and, later, the evolution of chlorophyll c_1 in this pre-chromophyte alga, followed by the uptake of its chloroplasts or of the whole pre-chromophyte alga by other secondary hosts would lead respectively to dinoflagellates and to the other forms of chromophyte algae. The secondary uptake of the chloroplasts from a green alga would give rise to the euglenoids.

If chloroplasts evolved from photosynthetic symbionts, then algal phylogeny must take account not only of the individual characteristics of both the host cell, including any earlier symbionts (e.g., cell covering, nucleus, or type of flagellum), and of the photosynthetic symbionts, but must also take account of characteristics resulting from their interaction (TABLE 2). It is therefore con-

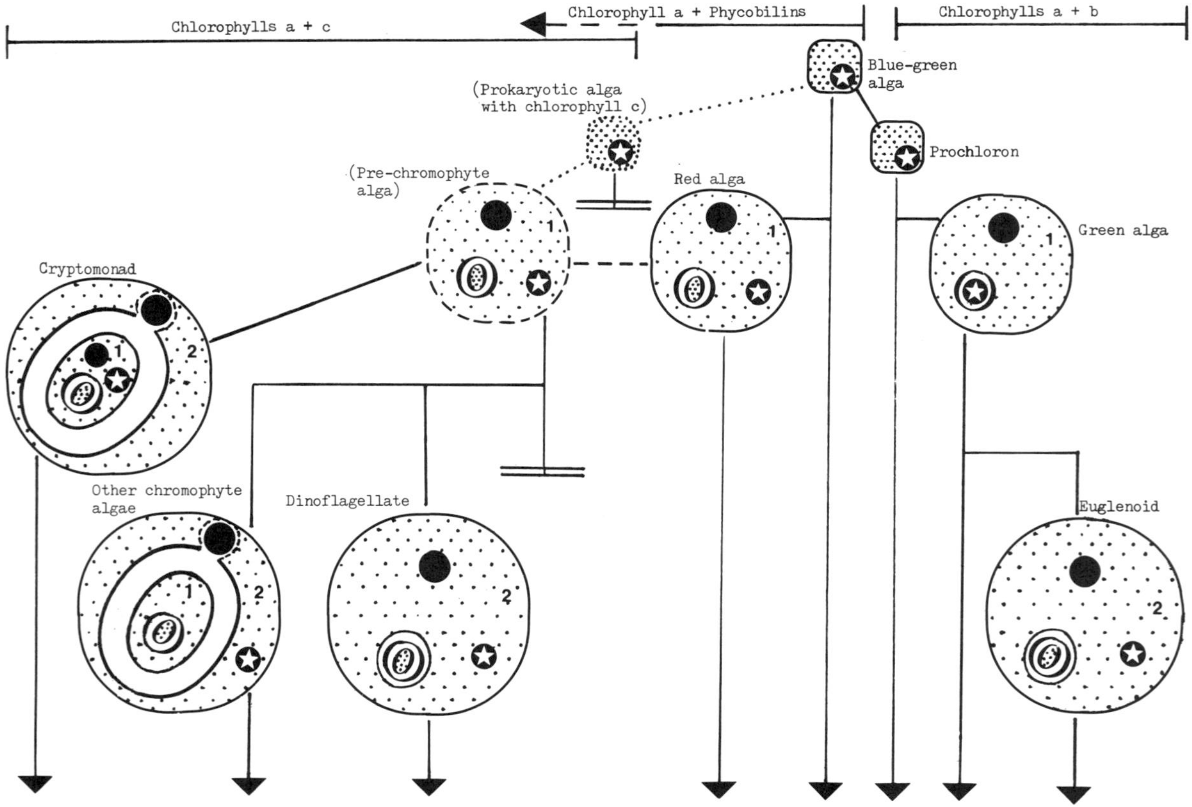

FIGURE 2. A scheme for plastid evolution. 1=First host, 2=Second host, ● =Nucleus of host, ●=Nucleomorph, ☆—Site of storage product, [stroma symbol] =Stroma of chloroplast, [cytoplasm symbol] =Cytoplasm of eukaryote, ······=Prokaryotic alga unknown today, ---=Eukaryotic alga unknown today. The spacing between groups does *not* reflect their evolutionary relationship but is merely a convenient diagrammatic arrangement.

TABLE 2

CHLOROPLAST-ASSOCIATED CHARACTERISTICS

Algal Classification	Plastid-Associated Characteristics: ← Ancient →						← Modern →				Plastid—Host Combined Characteristics		
	Primary pigment	Acces-sory pigment	Number of thyla-koids per band	Thylakoid appression	Inter change between bands	Evagina-tion etc. of thyla-koids	Chloro-phyll c_1 & c_2	Inner limit-ing disc	Grana	Girdle lamella	Number of sur-rounding mem-branes	Nature of storage product	Site of storage product
* Blue-green algae	Chl *a*	Phycob.	1	—	—	+	—	—	—	—	1	α-1,4 glucan	Cytoplasm of Prokaryote
Red algae	Chl *a*	Phycob.	1	—	—	+	—	+—	—	—	2	α-1,4 glucan	Cytoplasm of Host 1
† Prochlorons	Chl *a*	Chl *b*	$2 \rightarrow \infty$	Tight	Common but irregular	+	—	—	—	—	1	α-1,4 glucan	Cytoplasm of Prokaryote
Green algae and land plants	Chl *a*	Chl *b*	$2 \rightarrow \infty$	Tight	Common but irregular	+	—	—	+—	—	2	α-1,4 glucan	Stroma (Cyto-plasm of Pro-karyote)
Euglenoids	Chl *a*	Chl *b*	$2 \rightarrow \infty$ (3)	Tight	Common but irregular	+	—	—	—	—	3	β-1,3 glucan	Cytoplasm of Host 2
Cryptomonads	Chl *a*	Phycob. + Chl *c*	2	Loose	Uncommon	—	C_2	—	—	—	4	α-1,4 glucan	Perichloroplastic space (Cyto-plasm of Host 1)
Dinoflagellates	Chl *a*	Chl *c*	3	Variable	Uncommon	—	C_2	—	—	—	3	α-1,4 glucan	Cytoplasm of Host 2
Other chromophyte algae	Chl *a*	Chl *c*	3	Loose-Variable	Common (and regular?)	—	$C_1 + C_2$	—	—	+—	4	β-1,3 glucan	Cytoplasm of Host 2

* Free-living prokaryotic algae.
† Epizoic prokaryotic algae.
After S. P. Gibbs.[17]
Note: Chl *a* = Chlorophyll *a*; Chl *b* = Chlorophyll *b*, Chl *c*, c_1, and c_2 = Chlorophylls *c*, c_1, and c_2, respectively; and Phycob. = Phycobilins.

venient to consider chloroplast evolution under the two headings: 1) evolutionary changes within the plastid itself (e.g., pigmentation and thylakoid structure) and 2) evolutionary changes resulting from the interaction of host and photosynthetic symbiont (e.g., the number of membranes surrounding the chloroplasts). However it is recognized that this distinction is an arbitrary one as the plastid is in part under the control of the (host) nucleus, if only by reason of transfer of genetic information from the organelle.

The three main phylogenetic algal lines drawn by Christensen on the basis of differences in the accessory pigments (phycobilins, chlorophyll *b*, and chlorophyll *c*) are, as described by Gibbs,[3, 17] paralleled by important differences in thylakoid structure—e.g., the extent of thylakoid appression or pattern of

TABLE 3

METABOLIC RESULTS OF SYMBIOTIC ASSOCIATION *

Donors=Symbiotic Algae			Recipients=Animals
Effect of symbiotic association	Carbohydrate released		Immediate fate of carbohydrate
1. Growth rate restricted more than capacity for photosynthesis	*Zoochlorellae* maltose) glucose)	→	glycogen & pentoses
2. Fixed carbon diverted from insoluble to soluble.	*Zooxanthellae* glycerol	→	lipids & proteins
3. "Special" carbohydrates produced which are transferred to recipients.			
Carotenoids synthesized by plastids		→	Carotenoids modified by recipients.

* After Smith *et al.*[18]

thylakoid interconnection (TABLE 2). These major differences in pigmentation and thylakoid structure appear to represent ancient divergences in evolution that took place prior to the final symbiosis, some much earlier. Other variations (e.g., chlorophyll c_1 or grana) occur in only occasional subgroups within the major classes and presumably reflect more recent changes in plastid structure, most of which have taken place after the various symbiotic acts had been completed. These pigment and thylakoid modifications represent evolutionary changes within the plastids themselves.

Characteristics representing the interaction between host and plastid are those about which we know least. Several years ago Smith, Muscatine, and Lewis[18] published a table (partly incorporated in TABLE 3) that illustrates the fate of some carbohydrates produced by various photosynthetic organisms when incorporated as symbionts in a variety of hosts. It shows how the carbohydrate of the dual organism differs from that of each organism on its own.

TABLE 2 suggests that the early forms of storage carbohydrate were all α-1,4 glucans. Presumably host and symbiont were both initially capable of storing these carbohydrates. In green algae, starch is accumulated only within the chloroplast, suggesting that the accumulation site of the symbiont survived at the expense of that of the host. In red algae, on the other hand, starch accumulates only in the cytoplasm, i.e., the host site survived. The cryptomonads (modified via the pre-chromophyte algae from red algae) accumulate starch in the perichloroplastic space—i.e., in the area homologous with the red algal cytoplasm. In all other algae, carbohydrate, whether in the earlier α-1,4 or in the later β-1,3 form, appears to accumulate in the cytoplasm of the second host. The precise nature of each storage product presumably reflects an interaction between host and symbiont similar to that suggested in TABLE 3.

The carotenoids, like the chlorophylls, have been used in algal systematics.[19, 20] Within the algae the carotenoids are principally manufactured *de novo* inside the photosynthetic prokaryotes or in the chloroplasts. However some nonphotosynthetic organisms also synthesize carotenoids, e.g., nonphotosynthetic bacteria and fungi, while others modify the carotenoids that they obtain from photosynthetic organisms in their food, e.g., crustaceans and man. It is apparently not known if any carotenoids are manufactured in primitive, nonphotosynthetic, unicellular organisms such as might be considered the modern equivalent of the host cells of early algal symbionts. Nevertheless it is logical to suggest that the carotenoids present today in chloroplast-containing algae might at least reflect both the primary carotenoids, synthesized by the chloroplasts, and the secondary carotenoids, as modified by the host cells. Indeed the possible dual nature of carotenoid content following the uptake of one algal symbiont (a chrysophyte) by another photosynthetic alga (the dinoflagellate, *Peridinium foliaceum*) has already been recognized[21] though here the carotenoid components are apparently derived from the two different types of chloroplast present. Analysis of the biosynthetic pathways of carotenoid formation might permit a distinction to be made between primary and secondary carotenoids and between those carotenoids that evolved early and are thus associated with the classification of the major algal groups (e.g., simple non-epoxide xanthophylls found in red algae) and those that evolved later and so are associated with subgroupings (e.g., peridinin within the dinoflagellates).

It is likely that plastid evolution still continues to take place by the two different methods outlined above: by the establishment of new symbiotic associations (TABLE 1) and by modifications of the structure of existing plastids (TABLE 2). The cyanelles of anomalous algae, such as, *Cyanophora* and *Glaucocystis*, may be taken to represent intermediate stages in the evolution of chloroplasts from prokaryotic algal symbionts; the complex symbiotic associations, such as *Mesodinium* (TABLE 1), which appear to have arisen from *three* successive symbioses, may eventually evolve into even more complex chloroplast units than those we know today.

One of the major modifications in plastid structure has been the evolution of plastids that lack thylakoids and photosynthetic pigments. The prokaryotic algae and the plastids of most eukaryotic algae usually retain their thylakoids together with their pigments at all times. However individual species from most algal classes have plastids in which the thylakoid system can be reduced or increased in response to changes in environmental factors, such as, nutrient supply, light quality, or light intensity.[22-24] Other taxonomically diverse species have plastids that appear permanently to lack thylakoids.[25] Although the capacity of plastids to dedifferentiate or to redifferentiate is thus widespread, it is unusual for plastids to undergo these two processes as a regular event during the normal algal life cycle. In specialized cells, such as gametes (particularly male gametes), the chloroplasts often lose their thylakoids but no subsequent redifferentiation takes place. It is of particular interest that redifferentiation of nonpigmented plastids (proplastids) regularly occurs only in those multicellular algae and their successors that have evolved meristems (i.e., brown algae, the more complex red and green algae, and the land plants). However the enormous proliferation of different types of plastids that lack photosynthetic pigments and are found within a wide variety of different types of cell, seems only to be associated with the evolution of land plants, particularly angiosperms.[26] Among the angiosperms plastid specialization within some types of cell is sufficiently well established that differences in plastid ultrastructure can be used to assist in classification, as in the use of sieve-element plastids by Behnke.[27]

The controlled formations of nonpigmented plastids at regular intervals during the plant life cycle or their production in particular types of cell (like sieve elements) or organs (like roots) perhaps represents an adaptation to colonization of the "adverse" terrestrial environment by maintaining many of the plastids in the energetically "less expensive" non-green form. Such plastids are common in parts of plants where photosynthesis would be ineffective and there they may be restricted to other metabolic functions. The evolution of complete organs containing non-green plastids is of course dependent on the evolution of an efficient vascular system to permit the movement of photosynthate from the green to the non-green parts of the plant.

Summary

The traditional theory of serial endosymbiosis envisages the origin of all chloroplasts from prokaryotic algal symbionts. Of the two membranes that immediately surround plastids, the inner is considered homologous with the plasma membrane of the symbiont and the outer homologous with the vacuolar membrane provided by the host. This theory has been modified to suggest that those chloroplasts that are surrounded by more than two membranes were derived, following a second act of symbiosis, from eukaryotic rather than prokaryotic symbionts; these may have been whole eukaryotic algae or chloroplasts isolated from them. This suggestion is partly based on homologies between the various membranes that surround chloroplasts of different taxonomic groups and those that surround photosynthetic symbionts known today. If chloroplasts did indeed evolve from photosynthetic symbionts, then algal phylogeny must take account not only of the individual characteristics of the host cells and their symbionts but also of modifications of these characteristics arising from interactions between host and symbiont.

References

1. Greenwood, A. D., H. B. Griffiths & U. J. Santore. 1977. Chloroplasts and cell compartments in Cryptophyceae. Br. Phycol. J. **12:**119.
2. Lee, R. E. 1977. Evolution of algal flagellates with chloroplast endoplasmic reticulum from the ciliates. S. Afr. J. Sci. **73:**179–182.

3. Gibbs, S. P. 1978. The chloroplasts of Euglena may have evolved from symbiotic green algae. Can. J. Bot. **56:**2883–2889.
4. Whatley, Jean M., P. John & F. R. Whatley. 1979. From extracellular to intracellular: the establishment of mitochondria and chloroplasts. Proc. R. Soc. London Ser. B. **204:**165–187.
5. Dodge, J. D. 1973. The Fine Structure of Algal Cells. Academic Press, Inc. London & New York.
6. Hawes, C. R. 1979. Ultrastructural aspects of the symbiosis between algal chloroplasts and *Elysia viridis.* New Phytol. **83:**445–450.
7. Hibberd, D. J. 1977. Observations on the ultrastructure of the cryptomonad endosymbiont of the red-water ciliate, *Mesodinium rubrum.* J. Mar. Biol. Assoc. U.K. **57:**45–61.
8. Jeffrey, S. W. & M. Vesk. 1976. Further evidence for a membrane-bound endosymbiont within the dinoflagellate *Peridinium foliaceum.* J. Phycol. **12:** 450–455.
9. Greenwood, A. D. 1974. The cryptophyta in relation to phylogeny and photosynthesis. *In* 8th Int. Congress of Electron Microscopy, Canberra II, pp. 566-567.
10. Gibbs, S. P. 1962. Nuclear envelope-chloroplast relationships in algae. J. Cell Biol. **14:**433–444.
11. Bouck, G. B. 1965. Fine structure and organelle associations in brown algae. J. Cell Biol. **26:**523–527.
12. Bisalputra, T. 1974. Plastids. *In* Algal physiology and biochemistry; Botanical monographs. W. D. P. Stewart, Ed. **10:**124–160.
13. Whatley, Jean M. 1976. Bacteria and nuclei in *Pelomyxa palustris*: comments on the theory of serial endosymbiosis. New Phytol. **76:**111–120.
14. Christensen, T. 1964. The gross classification of algae. *In* Algae and man. D. F. Jackson, Ed.: 59–64. Plenum Publishing Corp. New York, N.Y.
15. Lewin, R. A. & N. W. Withers. 1975. Extraordinary pigment composition of a prokaryotic alga. Nature (London) **256:**735–737.
16. Newcomb, E. H. & T. D. Pugh. 1975. Blue-green algae associated with ascidians of the Great Barrier Reef. Nature (London) **253:**533–534.
17. Gibbs, S. P. 1970. The comparative ultrastructure of the algal chloroplast. Ann. N.Y. Acad. Sci. **175:**454–473.
18. Smith, D. C., L. Muscatine & D. Lewis. 1969. Carbohydrate movement from autotrophs to heterotrophs in parasitic and mutualistic symbiosis. Biol. Rev. **44:**17–90.
19. Liaaen-Jensen, S. 1977. Algal carotenoids and chemosystematics. *In* Marine Natural Products Chemistry. D. J. Faulkner & W. H. Fenical, Eds.: 239–259. Plenum Pub. Co. London.
20. Liaaen-Jensen, S. 1979. Carotenoids—a chemosystematic approach. Pure Appl. Chem. **51:**661–675.
21. Withers, N. & F. T. Haxo. 1975. Chlorophyll c_1 and c_2 and extraplastidic carotenoids in the dinoflagellate, *Peridinium foliaceum* Stein. Plant Sci. Lett. **5:**7–15.
22. Gibbs, S. P. 1962. Chloroplast development in *Ochromonas danica.* J. Cell Biol. **15:**343–361.
23. Vesk, M. & S. W. Jeffrey. 1977. Effect of blue-green light on photosynthetic pigments and chloroplast structure in unicellular marine algae from six classes. J. Phycol. **13:**280–288.
24. Callow, M. E. & L. V. Evans. 1979. Polyphosphate accumulation in sulphur-starved cells of *Rhodella maculata.* Br. Phycol. J. **14:**327–337.
25. Pringsheim, E. G. 1963. Farblosen Algen—ein Beitrag zur Evolutionsforschung. Gustav Fischer Verlag. Stuttgart, West Germany.
26. Thomson, W. W. & Jean M. Whatley. 1980. Development of non-green plastids. Ann. Rev. Plant Physiol. **31:**375–394.

27. BEHNKE, H. D. 1972. Sieve-tube plastids in relation to angiosperm systematics—An attempt towards a classification by ultrastructural analysis. Bot. Rev. **38:** 155–197.

DISCUSSION OF THE PAPER

DR. N. WEEDEN (*University of California at Davis*): I was wondering if we have a green algae with the ribulose diphosphate carboxylase genes in two places and it was used as a secondary endosymbiont what has happened to the small subunit that was in the nuclear DNA? Is there any DNA that is outside the chloroplast say in the peri-chloroplastic space?

DR. J. M. WHATLEY: I don't know, Dr. Gibbs might; she was looking at something of this sort but I don't know what has turned up there.

DR. S. P. GIBBS (*McGill University, Montreal, Canada*): Jean Whatley's theory and mine on the origin of the chloroplast in *Euglena* differ. I think *Euglena* took up a whole green algae and she thinks that *Euglena* just took up a chloroplast. If *Euglena* took up a whole green algae and the information for the small subunit was in the nucleus then with many eons of time that symbiosis persisted; the information was in the nucleus of the *Euglena* and so when you lost the nucleus of the green alga, the *Euglena* already had the information.

DR. D. E. GREEN: I would like to raise a point about endosymbiosis, which I see from the standpoint of function. There are many, many ways available by which you can take a mitochondrion or a nucleus, or some organelle from one cell and fuse it with another cell; we would call that trivial. In other words, this is a little trick you can play. It's called fusion and you can play it with all kinds of membranes and it really doesn't tell you very much other than that fusion is a very nice process. The more fundamental problem which few people seem to address is, how would you perpetuate such a relationship? How would you do it one generation after another? Now, in the mitochondrion we are aware that there has to be genetic information.

In other words there's a certain precise relationship between proteins made by the mitochondrion and the proteins made by the cytoplasm and unless that relationship is established it's a very complicated thing and the trivial entry or exit of an organelle into another cell is a completely impermanent thing. The real problem is not this getting in and out, but perpetuating that and providing the information in the DNA of the host cell so that this can be a permanent relationship.

DR. J. M. WHATLEY: I think there are two points here, the first is the work done on the amoeba. I think where endosymbiotic bacteria infected the amoeba, within a very short space of time the amoeba became nonviable unless the bacteria were present. So a mutualistic relationship seemed to become established very quickly.

The second point I think comes from the other amoeba *Pelomyxa pelustris* which also has symbiotic bacteria. There one sees an extraordinarily close relationship structurally between the bacteria, the host membrane, and the cell nucleus. The bacteria in their vacuoles in fact, get tied on to the nucleus and carried through to the next generation. They only become tied on when the amoeba is about to divide and the vacuolar membrane, which is part of the host cell, actually participates in division of the bacterium.

I think the actual act of taking up an organism is extraordinarily common but I think it's also pretty common that permanent relationships do become established.

DR. A. W. COLEMAN (*Brown University, Providence, R.I.*): I wanted to emphasize one thing that we already know about these chloroplasts, and that is there are two major groups of them in terms of the cytology of their genephore. In the chlorophyll *a* and *b* forms the genephore appears as concentrated dots of DNA within the chloroplast. The other form is the ring form of genephore which is found in the chlorophyll *a* and *c* types but it is not true in the dinoflagellate chloroplasts.

DR. K.-S. CHIANG: Dr. Whatley just referred to one case where the host is *amoeba proteus* and there are two things about Dr. Green's question. One is the relationship between the host and the symbionts; somehow in the beginning the symbionts were virulent so that when a new amoeba was infected either by injection or by phagocytosis, the newly infected hosts would die. But after some time they lost the virulence so that they no longer killed the host. So there is some kind of mutual accommodating change that has to occur for the symbiotic relationship to be permanent. In terms of the multiplication, there seems to be some kind of mutual control so that when an amoeba is first infected with very small symbionts they will multiply at a much faster rate until the number of symbionts per host cell reaches a maximum carrying capacity, and once that is reached, then the replication will sort of parallel that of the host. These two aspects have to be established before they can be permanent.

DR. R. A. LEWIN (*Scripps Institute, La Jolla, Calif.*): There was a report of dinoflagellates with bilin pigment, is this wrong?

DR. F. J. R. TAYLOR (*University of British Columbia, Vancouver, B.C.*): There are blue dinoflagellates known, but nobody has analyzed their pigment.

ORIGIN AND EVOLUTION OF THE PLASTID AND ITS FUNCTION

Jerome A. Schiff

Institute for Photobiology of Cells and Organelles
Brandeis University
Waltham, Massachusetts 02154

INTRODUCTION: PLASTIDS AND MITOCHONDRIA AS PROKARYOTIC RESIDENTS IN EUKARYOTIC CELLS

Any modern discussion of the evolution of cellular organelles such as the mitochondrion and chloroplast must begin with a consideration of their prokaryotic properties.[1] The more primitive free-living cells or prokaryotes, lack microscopically visible internal organelles. Some of their properties are summarized in TABLE 1. Eukaryotic cells contain conspicuous organelles when viewed in the light microscope, and electron microscopy shows that these organelles (among them the mitochondrion and the chloroplast) are surrounded by limiting membranes. As information has accumulated, biologists have realized that the mitochondria and chloroplasts resemble prokaryotic cells within eukaryotic cells (TABLE 1).

These organelles are semiautonomous. They perpetuate themselves by division but are subject to regulation by the eukaryotic cell and, in turn, regulate the cell in which they are residents. Genetic studies with higher plants, algae and fungi, some of them quite early in this century, have shown that the genes determining organelle phenotypes are situated in both the nuclear genome and an organelle's genome.[1, 2]

The fact that plastids and mitochondria resemble prokaryotic cells within eukaryotic cells suggests that they may have originated from the endosymbiotic invasion of primitive nonphotosynthetic, nonrespiratory eukaryotic cells by free-living prokaryotes.[3, 4] In this way, cyanobacteria, *Prochloron,*[5] and as yet undiscovered pigmented prokaryotes may have established the evolutionary lines leading, respectively, to the chloroplasts of: the red algae; green algae, higher plants and *Euglenas;* and other modern algae such as the brown and golden-brown organisms. Similarly, mitochondria may have begun as the endosymbiotic establishment of primitive aerobic bacteria in such primitive eukaryotic cells.[3, 4]

ARRESTED DEVELOPMENT OF BIOENERGETIC STRUCTURES AND ORGANELLES

The development of bioenergetic processes is under substrate control in many systems.[1] Certain environmental conditions arrest their formation in prokaryotes (FIGURE 1). Thus the presence or absence of oxygen determines whether structures carrying out anoxygenic photosynthesis will form in photosynthetic bacteria; in these organisms photosynthesis only takes place under anaerobic conditions.[6] Facultatively aerobic/anaerobic photosynthetic bacteria carry out respiration, but not photosynthesis, in the presence of oxygen. Al-

0077–8923/81/0361–0166 $01.75/0 © 1981, NYAS

TABLE 1

SOME PROPERTIES OF PROKARYOTES COMPARED WITH MITOCHONDRIA, PLASTIDS, AND THE EUKARYOTIC CELLS CONTAINING THEM [1]

Property	Prokaryotes (Bacteria, Blue-Green Algae or Cyanobacteria)	Mitochondria	Plastids	Rest of Eukaryotic Cell
Size	~ 1–10 μm	~ 1–2 μm	~ 1–10 μm	~ 1–100 μm
Membrane-bounded organelles within	None	None	None	Yes
Cell Wall	Usually present	Absent	Absent	Present or Absent
Endoplasmic Reticulum and Golgi	Absent	Absent	Absent	Present
Ribosomes	70S type	70S type (but variable in size)	70S type	80S type
Ribosomal Proteins	~ 55	18–107	~ 55	~ 75
tRNAs	Complete set	Unique complete set	Unique complete set	Unique complete set
Inhibition of Protein Synthesis by Chloramphenicol, Streptomycin, etc.	Yes	Yes	Yes	No
Inhibition of Protein Synthesis by Cycloheximide, etc.	No	No	No	Yes
DNA Genome	Single molecule, circular	Single molecule, often circular	Single molecule, often circular	Many molecules in chromosomes
Spindle or Comparable Mitotic Mechanism	No	No	No	Present
Reproduction	Fission, fragmentation, directional parasexual recombination	Fission, fragmentation	Fission	Mitosis, meiosis, sexual reproduction
Phosphorylative Cell Repiration (when present)	In cell membrane	In inner membrane	—	No, in mitochondria
Photosynthesis (when present)	In cell membrane, extensions, or free thylakoids	—	In thylakoids	No, in chloroplasts

I. Anoxygenic Photosynthetic Bacteria
Photosynthetic Machinery:
Repressed by O_2 in Light

II. Facultative Oxygenic-Anoxygenic Cyanobacteria
(Blue-Green Algae)
Photosynthetic Machinery Largely Constitutive, but in some Cases System II Activity:
Repressed by anaerobic conditions and H_2S
Induced by aerobic conditions

FIGURE 1. Patterns of arrested development of bioenergetic structures in prokaryotes.

though all cyanobacteria (blue-green algae) carry out oxygenic photosynthesis and aerobic respiration, certain of them can carry out anoxygenic photosynthesis when presented with hydrogen sulfide under anaerobic conditions; system II (the water-splitting oxygen-evolving photosystem of oxygenic photosynthesis) becomes nonfunctional and hydrogen sulfide is used as the electron donor for a bacterial type of photosynthesis.[7]

In eukaryotes, arrested development of organelles is encountered.[1] In order to test for the presence of this type of regulation, the organism in question must be capable of growth on an alternative substrate (FIGURE 2). Thus plastid-

I. Must be able to grow organism under conditions gratuitous for photosynthesis or respiration.

II. Mitochondria
Most: constitutive
Some [e.g. yeast]:
repressed by anaerobic conditions (promitochondrion)
promitochondrion ⟶ mitochondrion (substrate induction by O_2)

III. Chloroplasts
Many (most?) species: constitutive [e.g. green algae (Chlorella, Chlamydomonas, Scenedesmus); Diatoms (two species); Gymnosperms (cotyledons)]
Some inducible: [e.g. Euglena (2 or 3 species), Cyanidium, Ochromonas, all angiosperms]:
repressed by darkness and utilizable substrates (proplastid or etioplast)
proplastid or etioplast ⟶ chloroplast (substrate induction by light)
Mutation of constitutive to inducible: [e.g. Chlamydomonas Y_1, Scenedesmus C-2A', etc.]

FIGURE 2. Patterns of arrested development of chloroplasts and mitochondria in eukaryotes.

containing organisms must be capable of growth on a reduced carbon source as an alternative to photosynthesis, and mitochondrion-containing organisms must be able to use fermentation or some other anaerobic process to replace oxygen-utilizing respiration. Judging from those favorable systems in which studies of this sort can be done (FIGURE 2) most mitochondria appear to be constitutive, i.e., they are always present and do not require induction. This is to be expected of an organelle that apparently became established in the eukaryotic cell after oxygen entered the primitive anaerobic atmosphere through the activities of primitive oxygenic photosynthetic prokaryotes. In certain cases, however, mitochondrial development has become subject to substrate control. Yeast, for example, when grown under anaerobic conditions, has a highly reduced promitochondrion and is only induced to form a mitochondrion when oxygen is present as an inducing substrate (oxygen being the substrate for mitochondrial respiration).[8, 9]

In the case of chloroplasts, most are probably constitutive (FIGURE 2) and are present in darkness or in light. In certain cases, plastid development has become subject to regulation by light or other substrates.[1] In these organisms a proplastid or etioplast is found as the arrested state in darkness and induction to form a chloroplast ensues on illumination (light being a substrate for photosynthesis by the chloroplast). Mutation in the laboratory can yield inducible strains from constitutive ones.[4, 10–12]

The presence or absence of inducible organelles does not appear to follow a strictly phylogenetic pattern. Rather, arrested development and substrate induction appear to have arisen separately many times in response to secular ecological selection. Thus the chloroplasts are inducible by light in certain *Euglena* strains which are facultative organotrophs and live in environments rich in reduced-carbon sources.[4] Arrested development of the chloroplasts and induction by light are widespread among Angiosperms probably because the seed habit introduces a dark period into the life cycle and until the seedling emerges into the light there is no strong selection for the presence of a mature photosynthetic chloroplast.[13] In the case of yeast, growth under anaerobic conditions is highly adaptive under certain circumstances where the organism uses fermentation to supply its needs and turns to respiration by fully developed mitochondria only when oxygen is available.

Not all arrested states are the same, particularly in the case of chloroplasts (FIGURE 3). The most primitive occur in higher plants in the embryonic and apical cells.[14] As leaves are formed in darkness the proplastids enlarge somewhat and form a few internal membranes. The structure of the proplastid in dark-grown *Euglena* and *Ochromonas* is not very different from these leaf proplastids at about two to three days after germination (TABLE 2). In the case of *Euglena,* the protochlorophyll(ide) system for forming chlorophyll is also very similar to that in the two- to three-day-old etiolated leaf.[15–20] On prolonged etiolation, however, the leaf proplastid enlarges and the internal membrane structures become more complex, including the formation of a prominent crystalline prolamellar body. These etioplasts are not obligatory stages in normal plastid development, since the proplastids of the algae and younger leaves develop perfectly well when exposed to light. Rather, the etioplast appears to be an adaptation to permit more rapid development of chloroplasts and photosynthesis when the plant, nearing the limit of its stored reserves after prolonged dark growth, finally finds the light.[18]

The formation of plastids can be repressed in light if utilizable organic substrates are present.[21, 22] In some cases, particularly in *Euglena,* plastid DNA and most plastid structure can be eliminated from cells growing on an organic substrate, and other mutants blocked in chloroplast development can be obtained.[1, 4, 23] In higher plants, proplastids and chloroplasts can differentiate into other plastid types such as chromoplasts, amyloplasts, and elaioplasts.[1, 13]

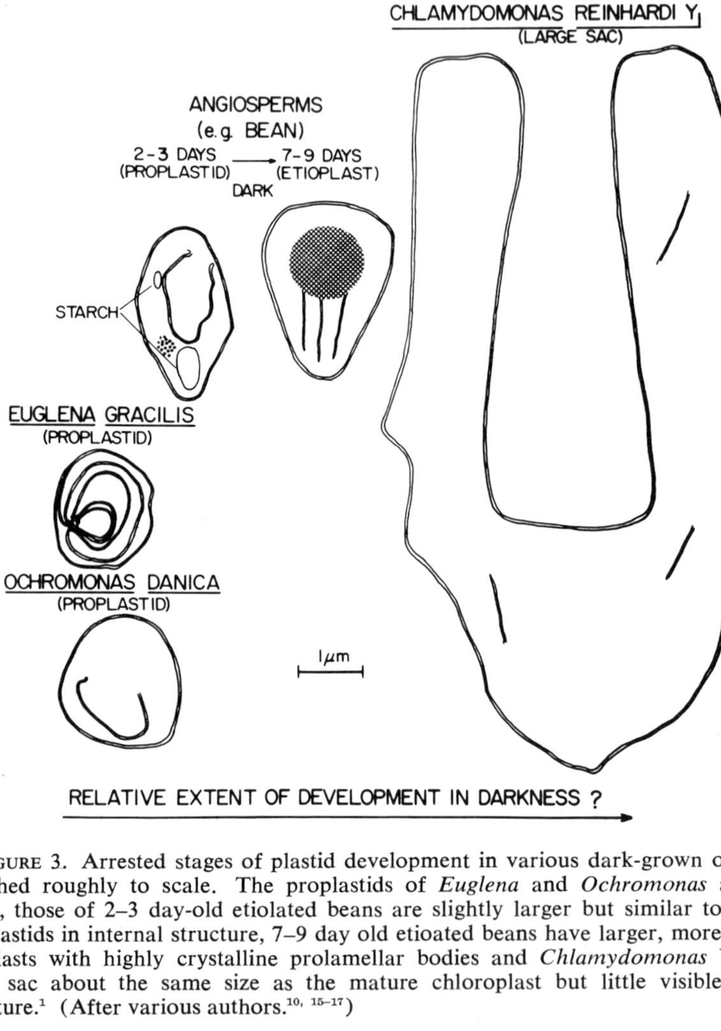

FIGURE 3. Arrested stages of plastid development in various dark-grown organisms sketched roughly to scale. The proplastids of *Euglena* and *Ochromonas* are quite small, those of 2–3 day-old etiolated beans are slightly larger but similar to *Euglena* proplastids in internal structure, 7–9 day old etioated beans have larger, more complex etioplasts with highly crystalline prolamellar bodies and *Chlamydomonas* Y_1 has a large sac about the same size as the mature chloroplast but little visible internal structure.[1] (After various authors.[10, 15–17])

The mature plastids of various groups of organisms show a trend from plastids resembling free-living prokaryotes, such as cyanobacteria in the case of the red algae, to more elaborate forms in other groups.[1] The thylakoids of cyanobacteria and red algae have phycobilisomes and are unstacked. In the brown algae, the plastids are still fairly simple but stacking of thylakoids is now

TABLE 2

COMPARISON OF PROPERTIES OF DARK-GROWN EUGLENA, 2–3 DAY-OLD AND 7–9 DAY-OLD BEAN LEAVES [18]

Property	Dark-grown *Euglena*	2–3 day etiolated bean	7–9 day etiolated bean
Structure	Proplastid	Proplastid	Etioplast
Size (μm)	1–2	2–3	~ 4
Prolamellar Body	Small, noncrystalline	Absent, or small noncrystalline	Large, crystalline
Predominant absorption *in vivo*	$Pchl(ide)_{635}$	$Pchl(ide)_{635}$	$Pchl(ide)_{650}$
Total Protopigment (pg/plastid)	$\sim 1 \times 10^{-4}$	$\sim 0.5 \times 10^{-4}$	$\sim 7\text{–}10 \times 10^{-4}$
Protopigments present	Pchlide and Pchl	Pchlide and Pchl	Predominantly Pchlide
Ratio: moles -ide/moles -yll	~ 3	~ 1	~ 6
First stable photoproduct *in vivo*	$Chl(ide)_{675}$ directly	$Chl(ide)_{675}$ directly	$Chlide_{685}$ Shibata shifting to 675 nm
Pigments produced directly on illumination	Chlide and Chl	Chlide and Chl	Chlide (predominantly)
% Conversion Pchl(ide) to Chl(ide)	~ 10–50%	~ 40%	~ 80–90%
Rate of Chlide esterification	Fast (no lag)	Fast (no lag)	Slow (lag)
Rate of protopigment regeneration	Fast (short lag)	Fast (no lag)	Slow (lag)

present. In the dinoflagellates, *Euglenas,* green algae, and higher plants, large numbers of thylakoids showing elaborate stacking is the usual situation. We have already suggested that this phylogenetic pattern may be recapitulated by the normal development of a proplastid into a chloroplast.[15] If the original invading plastid resembled a primitive photosynthetic prokaryote, and subsequent evolution has led to the more complex plastids we see today, we might expect to see the primitive condition when plastid development is arrested and proplastids are formed.

Evolution of Photocontrol Systems

Several evolutionary patterns of photocontrol can be distinguished in contemporary organisms (Table 3). Among the prokaryotes, two major patterns in phototactic and phototropic responses can be discerned. One begins with the use of the photosynthetic pigments of the anoxygenic photosynthetic bacteria and the oxygenic cyanobacteria as a means of finding the light for photosynthesis; the action spectrum for phototaxis in these organisms is generally the same as the action spectrum for photosynthesis.[24, 25] The ATP produced from photophosphorylation is undoubtedly made available for cell motility. It would have been very adaptive during evolution for cells to use the same pigments for phototaxis and photosynthesis since this would insure that the light, once found, would be of a quality suitable for photosynthesis. Later in the evolution of algal eukaryotes, a separation of perception and photosynthesis took place leading to the stigma/photoreceptor system for control of flagellar movement, as in *Euglena.*[26-29] With this anatomic specialization came a specialization in pigments as well since blue light is effective in orienting these organisms, but red light is not. The unknown blue-light absorbing pigments variously suggested to be flavoproteins carotenoids, and other molecules are collectively called "cryptochrome" as a short-hand notation for our ignorance.[30] This system was apparently preserved in the evolution of multicellular plants as a cryptochrome system in the ferns and the tip of the stem, or coleoptile, of Angiosperms where it controls the transport of auxin, leading to differential growth below the tip and consequent growth towards or away from the light.[31, 32]

Also found among the prokaryotes is a bacteriorhodopsin system in *Halobacterium* that mediates phototaxis, ion transport, and ATP formation (Table 3). This may well have become the perceptual system of invertebrates and vertebrates culminating in the rhodopsin of the rods and cones of vertebrates and the squid and octopus.[34] Although a nervous system serves as a mediator between the absorption of light in the thylakoids of the rods and cones and the muscles of these organisms, the system is analogous to the more primitive association of perceptor and flagellum or cilium in the more primitive cells. It's worth remembering that in the rods and cones, there are immobile cilia that may be vestigial remnants of the motile structures in more primitive organisms.

The developmental cycle of *Nostoc* and chromatic adaptation in other cyanobacteria (Table 3) are controlled by primitive phytochromes or "prophytochromes," which are based on red-green reversible chromoproteins.[35-39] Red light induces the formation of seriate filaments in *Nostoc;* green light reverses the effect of red.[37] Similarly, phycocyanin synthesis is induced by orange-red light and phycoerythrin synthesis by green light.[38, 39] It seems likely

TABLE 3

PHOTOCONTROL AND PHYLOGENY

	Phototaxis, Phototropism, Vision	Chloroplast Development: Non Chloroplast	Chloroplast Development: Chloroplast	Photomorphogenesis	Chromatic Adaptation	Plastid Rotation
Eukaryotes						
Higher Animals	Rhodopsin	—	—	—	—	—
Higher Plants Angiosperms	Cryptochrome	Phytochrome	Protochlorophyll(ide)	Phytochrome- (± Cryptochrome?)	—	—
Ferns	Phytochrome? Cryptochrome?	?	?	Phytochrome-Cryptochrome	—	—
Algae	Cryptochrome	Cryptochrome	Protochlorophyll(ide)	Phytochrome-Cryptochrome	—	Phytochrome-Cryptochrome
Fungi	Cryptochrome	—	—	Cryptochrome-Phytochrome	—	—
Prokaryotes						
Cyanobacteria (Blue-Green Algae)	Photosynthetic Pigments	—	—	Prophytochrome (red-green reversible)	Prophytochrome (red-green reversible)	—
Photosynthetic Bacteria	Photosynthetic Pigments	—	—	—	—	—
Halobacteria	Bacteriorhodopsin	—	—	—	—	—

[Cryptochrome: Near UV-Blue; Phytochrome: Red Far-Red Reversible; Protochlorophyll(ide): Blue-Red]. ("Cryptochrome" designates pigments unidentified or tentatively identified as flavoproteins or carotenoids).

that these prophytochromes evolved after open-chain tetrapyrrole pigments were available from the evolution of phycoerythrin and phycocyanin as photosynthetic accessory pigments. Linkage of modified open-chain tetrapyrroles to other proteins would have led to the formation of the prophytochromes for cellular control functions. Later, these prophytochromes could have become modified during evolution to become the familiar red/far-red reversible phytochrome chromoprotein of *Mougeotia* and multicellular plants.[40–42] Phytochrome and cryptochrome frequently act in concert to control cellular morphogenesis as in prothalial development in ferns, spore germination in fungi, and various responses in higher plants.[31, 40, 41]

This brings us, finally, to the control of chloroplast development by light (TABLE 3). In algae such as *Euglena* and *Scenedesmus*, many early functions connected with chloroplast development occur external to the chloroplast during the lag period and are under control of blue-light absorbing cryptochrome systems.[30, 43] In higher plants this function appears to have been largely assumed by the red/far-red phytochrome system.[44] The later, plastid-localized functions of the linear period of plastid development in both *Euglena* and higher plants appear to be under the control of blue and red light absorbed by protochlorophyll(ide) or some pigment with a similar absorption spectrum. Since the photoconversion of protochlorophyll(ide) to chlorophyll(ide) is often an obligatory step in chlorophyll formation, this control point is very strategically located.

Cellular Coordination by Light During Organelle Development

Within the cell, photocontrol influences many processes related to organelle development (FIGURE 4).[43] The developing chloroplast obtains energy, reducing power, and small molecules from the rest of the cell during the lag period early in light-induced plastid development. This involves the activation of the central metabolism of the cell in the cytoplasm and mitochondrion, including the activation of the breakdown of storage polysaccharides and an increase in cellular respiration. This process appears to be under the control of nonplastid photoreceptors, cryptochrome in *Euglena* and *Scenedesmus*, and phytochrome in higher plants. At the same time photocontrol of the transcription of nuclear DNA by the same pigment systems makes available those organelle proteins that are synthesized on cytoplasmic ribosomes.[45] These proteins appear to have a signal polypeptide appended that is thought to allow attachment to, and transport through, the organelle membrane at which point the signal is removed.[46–48] Subsequently, probably after enough materials have entered the developing plastid from outside, synthetic reactions within the organelles increase and plastid development comes out of the lag phase.[43] These synthetic reactions within the plastid appear to be under photocontrol of protochlorophyll(ide) in both *Euglena* and higher plants. Among the processes induced by this red-blue absorbing system is transcription of plastid DNA to form plastid proteins on plastid ribosomes.[45] Oxygen appears to act as an inducer of similar events in the mitochondrion during the development of this organelle.[8, 9] In *Euglena* it appears that induction by the plastid photoreceptor system of plastid development can bring about induction of the nonplastid system, even when the nonplastid system has not been directly induced by light.[43] To explain this (FIGURE 4), we suppose that on photoinduction of the plastid photoreceptor system, a mediator molecule is synthesized that travels to the nonplastid compartments, including

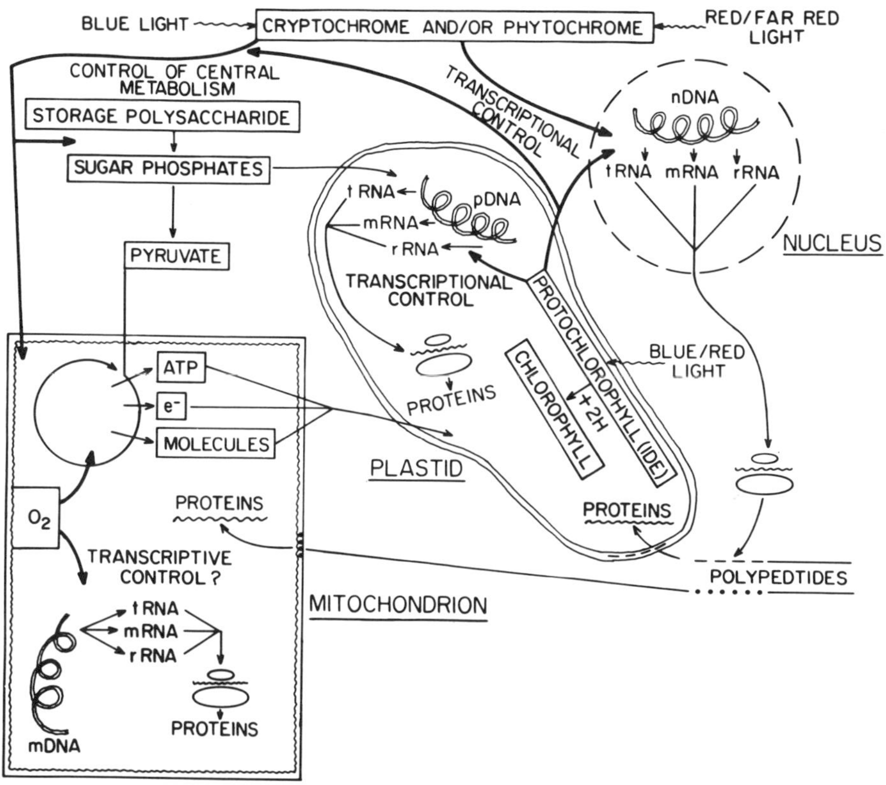

FIGURE 4. Summary of nutritional and informational interactions among organelles during chloroplast development in *Euglena* and other organisms. The DNAs of either the mitochondrion or the chloroplast code for that organelle's tRNAs and rRNAs and for certain proteins that are translated on the ribosomes of that organelle and remain within the organelle. Nuclear DNA codes for rRNAs and tRNAs of the cytoplasm and for proteins that are translated on cytoplasmic ribosomes. Some of these remain in the cytoplasm; others bearing signal polypeptides recognize and enter the appropriate organelle, losing the signal portion enroute. Light-induced breakdown of storage carbohydrate via glycolysis feeds mitochondrial respiration which supplies small molecules, energy and reducing power to the developing chloroplast. Blue light via cryptochrome or red/far-red light via phytochrome is thought to control the non-chloroplast aspects of plastid development including the central metabolism of the cell and nuclear transcription of genes coding for plastid proteins. Blue and red light (a protochlorophyll(ide)-type of photoreceptor) is thought to control plastid transcription and chlorophyll(ide) formation. Since those plastid enzymes that are nuclear-coded show protochlorophyll(ide)-type action spectra for induction, it is suggested that illumination of the plastid receptor results in a regulatory signal that passes to the nucleus and cytoplasm to activate processes ordinarily activated by the blue light receptor. Oxygen apparently controls mitochondrial development, particularly in yeast.[1]

the nucleus, to bring about the induction of the nonplastid systems when they themselves have not been photoinduced directly. Although proteins seem to move easily from one cellular compartment to another, no firm evidence for the movement of nucleic acids across organelle boundaries (other than the nucleus) has been obtained. All of the nucleic acids used within an organelle appear to be coded in that organelle's DNA; nuclear-synthesized nucleic acids supply the cytoplasm.

The results obtained with modern methods of molecular biology confirm the earlier genetic information that organelle phenotypes are determined both by organelle genes and by nuclear genes, since genetic information for the construction of the proteins of an organelle originate in the organelle's own DNA or in nuclear DNA (FIGURE 4).

The scheme shown in FIGURE 4 is, of course, speculative. The consequences of light induction are far reaching and some means of amplification is necessary to broadcast the primary photoevent to the rest of the cell. One means is transcriptive control, as shown, based on experimental evidence. Another possibility that has been suggested (but for which there is as yet no firm evidence from plastid development) is a membrane localization of the photoreceptors, enabling light to control the movements of ions and other strategic molecules.

By analogy to prokaryotic systems, we might expect photocontrol to be exerted at several levels of cellular metabolism. For example, it is known that protochlorophyll(ide) controls chlorophyll formation by serving as an intermediate that undergoes photoreduction by light (FIGURE 9) during chlorophyll synthesis.[18-20] But protochlorophyll(ide) (or a pigment of similar absorption judging from the action spectrum) also mediates photocontrol of transcription of plastid DNA to form plastid rRNAs.[45] The means by which control is exerted by cryptochrome and phytochrome over the nonplastid aspects of plastid development remain obscure. However, the fact that transcription in both the plastid and nuclear/cytoplasmic compartments appears to be under control by light during plastid development has led us to propose the following model for this level of control.

We suppose that the transcriptive control system is similar to that operating in prokaryotic systems (FIGURE 5).[1, 42, 45, 49] A gene (*i*) codes for the production of a repressor protein that binds to the *o* region preventing transcription of the genes in that operon. Usually, the inducer is a small molecule (e.g., lactose in the case of β galactosidase induction in *E. coli*) that combines with the repressor thereby freeing the *o* region and allowing transcription of the genes in the operon. In the simplest case, we suppose that repressors exist that are pigmented and can absorb light instead of combining with a small molecule (FIGURE 5). Absorption of light would change the conformation of the repressor so it could no longer occupy the *o* site leading to induction of the transcription of the operon. (It's worth remembering that phytochrome changes conformation upon photoconversion from one form to the other).[40, 41] Of course other variations are possible. For example, a pigmented small molecule (a proinducer) may absorb light and thereby be converted to an inducer that can combine with the repressor preventing its binding to *o* (FIGURE 5). But the essential thing is that there might be pigmented control molecules that allow light to influence protein synthesis in much the same manner as small molecules do. Light, after all, is a substrate for photosynthesis and once we think of light as an inducing substrate, mechanisms immediately suggest them-

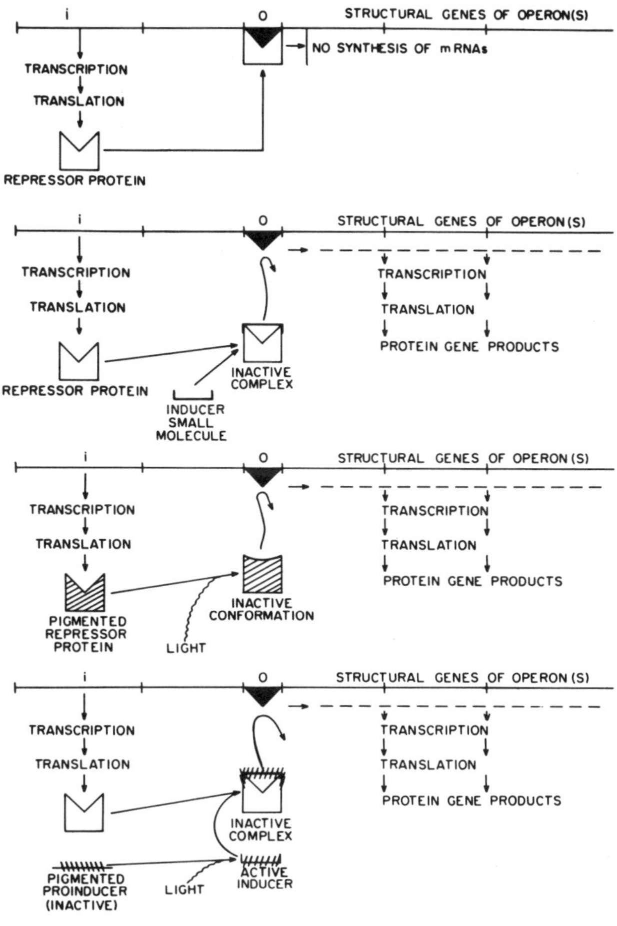

FIGURE 5. A hypothetical model for light as an inducing substrate for the control of transcription. In the first illustration the repressor protein is free and binds to the operator (*o*) site on the DNA preventing the transcription of mRNAs coded by the operon leading to an arrest of the synthesis of these protein gene products. In the second illustration, the usual mode of substrate induction is shown where a small molecule (the inducer) binds to the repressor preventing the repressor from binding to *o*, allowing transcription of mRNAs and the formation of the proteins coded by the operon. In the third illustration, we suppose that the repressor protein is pigmented and that absorption of light by the repressor converts it to a configuration that can no longer bind to *o*, inducing the formation of mRNAs and allowing the formation of the gene products. The fourth illustration is a variation where an inactive pigmented small molecule, or proinducer, is converted to an active inducer by light, allowing the active inducer to bind to the repressor and thereby prevent the repressor from binding to *o*; as a consequence mRNAs are transcribed and proteins are produced. These models apply to the transcription of any RNAs including tRNAs and rRNAs; mRNAs are shown as illustrations. The same repressor may bind to the operator of more than one operon leading to the control of formation of cadres of enzymes by light.

selves for substrate control of cellular synthesis at all levels including transcription, translation, and enzymatic catalysis.

To be adaptive over the course of evolution, these pigmented control molecules would come to have light absorptions similar to those of the processes they control. Thus the systems that control chloroplast development absorb in blue and/or red light, the same portions of the spectrum that are absorbed by chlorophyll in photosynthesis. It would be no more adaptive to have chloroplast development induced by light ineffective for photosynthesis than it would be for lactose to induce an enzyme for, say, arabinose utilization.

We must be prepared for the possibility, then, that photocontrol exerted at one level by a small molecule (like the control of chlorophyll formation through illumination of its precursor, protochlorophyll(ide)) may be mimicked at another level (such as transcription) by a similarly-absorbing control molecule, such as a pigmented repressor protein or associated pigment molecule. Of course, protochlorophyll(ide) itself could act as the associated molecule even at the level of transcription. Perhaps one of the reasons certain control molecules have been so elusive (e.g., the blue-absorbing cryptochrome) is that they act as repressors or in association with repressors. The concentrations of repressor molecules in cells such as *E. coli* are extremely low and the use of mutants making abnormally high amounts has been necessary to permit their detection and isolation.

Evolution of Plastid Pigments in Response to Available Natural Light

Considering the wavelengths of light available at the surface of the earth (Figure 6), it is paradoxical, at first sight, that chlorophyll *a*, the nearly universal photosensitizer of photosynthesis, has absorption in the blue and the red regions of the spectrum, not in the middle of the spectrum where a considerable amount of energy is available.[50] (In terms of number of available quanta, there is a constant plateau from about 550 to 800 nm with a steady decline from 550 to 400 nm.) One would think that life evolving under a yellow sun would evolve pigments to absorb light in the yellow-green region. Indeed, vision has evolved in this direction, where rhodopsin and other retinal pigments absorb maximally in the middle of the spectrum with small shifts to accommodate various ecological situations.[51] In seeking reasons for the selection of chlorophyll, we must consider both the requirements of photosynthesis and the environmental pressures that selected the pigment systems found in contemporary organisms.

All photosynthetic organisms contain plastid pigments other than chlorophyll *a* (or, in the case of the anoxygenic photosynthetic bacteria, bacteriochlorophyll *a*).[1] These pigments have absorptions that fill the gaps in the middle of the spectrum left by chlorophyll *a*. Studies of the photochemical relationship of these pigments to each other and to the reactions of photosynthesis have shown that the light energy absorbed by these "accessory" pigments can be transferred efficiently to chlorophyll *a* (or bacteriochlorophyll *a*, where appropriate).[52-56] For this to occur several conditions must be satisfied.

1. The energy transfer must be from pigments absorbing at shorter wavelengths to those absorbing at longer wavelengths (the absorptions under con-

sideration are those representing transitions from the ground state to the lowest excited singlet state). Since the energy in a photon or quantum of light is given by $E = hc/\lambda$ where h is Planck's constant, c is the speed of light and λ is the wavelength of light), it follows that if energy transfer is to occur it must do so in a manner that loses energy, since gaining energy from nowhere would violate the energy conservation laws. (This neglects acquisitions of thermal energy, which are small compared with photochemical quantities.) Thus in

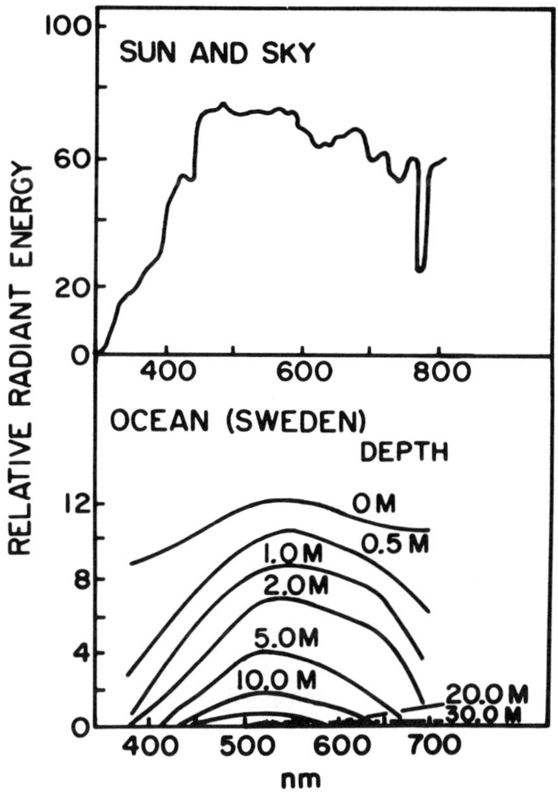

FIGURE 6. Distribution of spectral energy from the sun at the earth's surface and at various depths in the ocean on relative scales. To compare upper and lower halves of figure, note 0 depth curve in lower half.[1] (After [50, 62]).

being transferred, a quantum must stay the same or become smaller. If it becomes smaller, as it must do if the transfer process is less than 100% efficient, then the wavelength of the quantum must increase. Thus energy transfer is from shorter to longer wavelengths.

2. The two pigments involved must have energy states in common or, to put it another way, their absorption or fluorescence spectra must overlap.

3. The two pigments involved in energy transfer must be quite close to each other in space, usually less than 30–50 Å for efficient transfer to occur.

The thylakoid pigments satisfy these requirements and efficient energy transfer among these pigments has been measured in many systems.[52-54] Several basic patterns exist depending upon the specific pigments present, but transfer is always from pigments whose lowest excited states result from absorptions at shorter wavelengths, to those whose transitions result from absorptions at longer wavelengths (FIGURE 7 shows the basic schemes found in various groups). Although the series formulation shown in FIGURE 7 is usual, studies based on the isolation of light harvesting pigment-protein complexes from green and brown groups are beginning to provide evidence favoring parallel formulations.[57, 58] Thus carotenoids and chlorophyll *b* may transfer energy independently to chlorophyll *a* and fucoxanthin and chlorophyll *c* may transfer energy independently to chlorophyll *a* without transfer from carotenoids to chlorophylls *b* or *c* on the way to chlorophyll *a*.

In chlorophyll-containing organisms, the ultimate recipient of the light energy is chlorophyll *a*, serving as the reaction center of systems I and II of photosynthetic electron transport. In the photosynthetic bacteria, bacteriochlorophyll and bacteriopheophytin have been shown to be constituents of the reaction centers.[59, 60] The end result is that light energy absorbed in the region of the solar spectrum outside of the chlorophyll *a* absorptions themselves, is made available to the reaction centers and thus to the reactions of photosynthesis.

There are some clues as to how these accessory pigments have evolved. If we look at the way many of these groups of organisms are situated ecologically, for example, we find that they are sometimes stacked in a manner that takes advantage of energy transfer.[61] Some relationships among marine organisms which appear to result, at least in part, from light availability are shown in FIGURE 8. Land plants and the green algae occupying the upper layers in ponds and ocean contain chlorophylls *a* and *b* besides carotenoids. The organisms living below them (e.g., the brown algae) have evolved special brown carotenoids that absorb blue-green light not absorbed by the green algae.

Photosynthetic bacteria can use far-red light not absorbed by other organisms through the bacteriochlorophyll long-wavelength absorption. Blue-green and red algae are frequently the only photosynthetic organisms at very great depths. These organisms contain phycocyanin and phycoerythrin in phycobilisomes, which transfer energy to chlorophyll *a*. These absorptions are in the green and orange regions of the spectrum not absorbed by the organisms above them. Absorption of light by other organisms is not the only factor that limits light at great depths to the blue-green region of the spectrum (as you look down into a body of water, the light reflected back through the water is blue-green). Water itself and yellow impurities in the water absorb red and blue light and the longer the column of water above the organism, the greener the light (FIGURE 6).[62, 63]

The red and blue-green algae are well adapted, through the phycoerythrin absorption, to make use of this residual green light. It should also be noted that these two groups have another dimension of adaptation. Cells of many blue-green algae and red algae can regulate the proportions of phycocyanin and phycoerythrin through a process called complementary chromatic adaptation.[35, 36, 38, 39] Organisms growing in light rich in green wavelengths form more phycoerythrin; those growing in light rich in longer wavelengths form more phycocyanin. Chromatic adaptation of this type seems to be restricted to organisms containing phycobilisomes.[64] Perhaps the phycobilisome evolved as

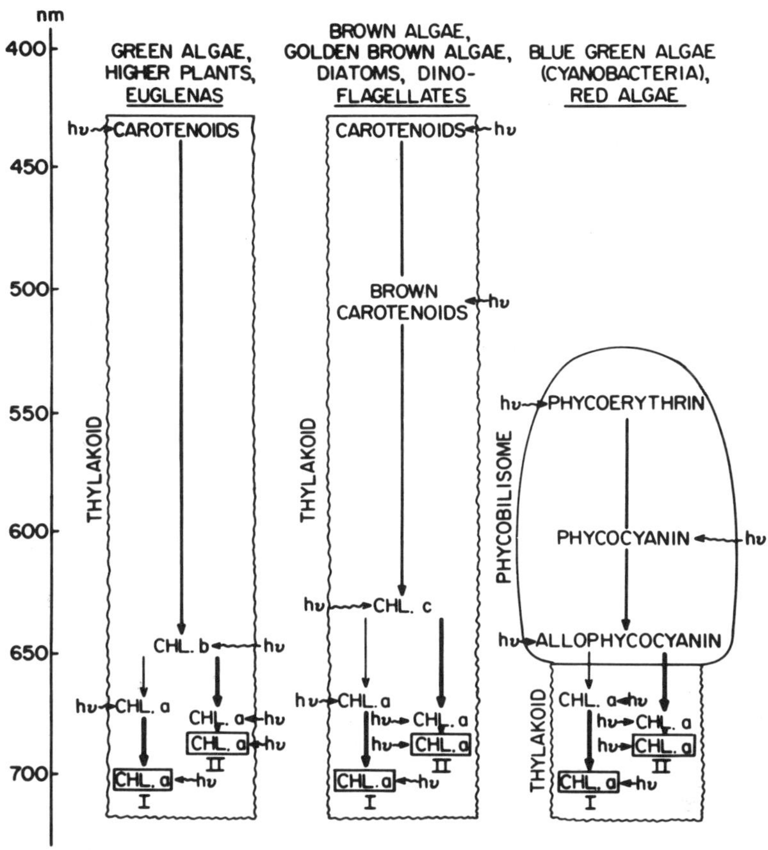

FIGURE 7. Diagrammatic representation of energy transfer in representative groups of photosynthetic organisms. Any pigment can absorb light at its absorption maxima but energy transfer is always from the lowest excited singlet state of a pigment to a pigment whose lowest excited singlet state is at longer wavelengths; i.e., energy transfer is always from the top of the diagram downwards as indicated by the vertical arrows. The thickness of the arrows represents, roughly, the extent of energy transfer, the thicker the arrow the greater the enegy transfer. The unboxed chlorophylls represent antenna pigments, the boxed chlorophylls represent the reaction centers of systems I and II which carry out the photodriven electron transport reactions with donors and receptors. All reactions are thought to occur in the thylakoids except for energy transfer among phycobilins in the red and blue green algae (cyanobacteria), which is thought to occur in phycobilisomes. Similar schemes involving bacteriochlorophylls can be written for energy transfer in the membranes of photosynthetic bacteria. The schemes shown are series formulations where energy transfer is from one pigment to the next; as discussed in the text, parallel formulations may be more realistic as more is known about the composition of pigment–protein complexes from thylakoids.[1]

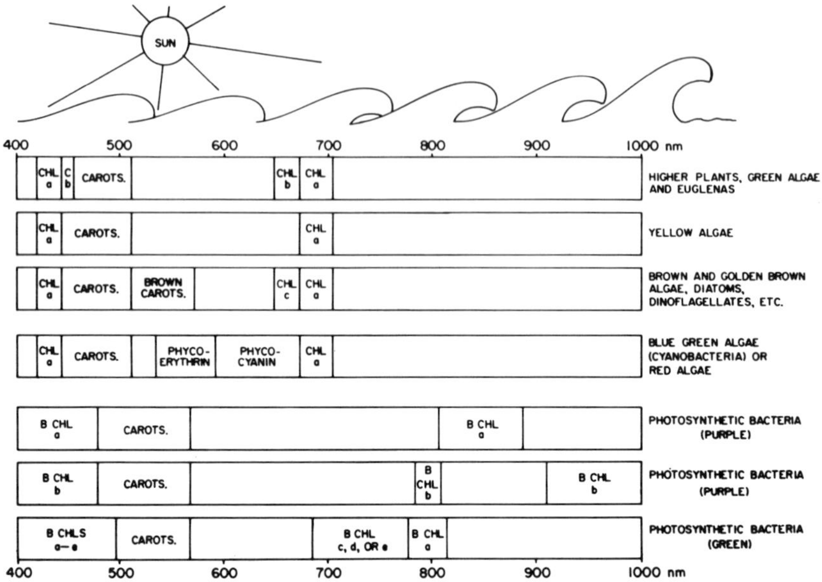

FIGURE 8. Diagrammatic representation of absorption by photosynthetic pigments in various groups of organisms in relation to depth in the ocean. Note "windows" where certain groups of organisms are transparent allowing light to penetrate to others that have pigments in the right absorption range. Remember, also, that water and substances in the water absorb light in the blue and red regions of the spectrum (FIGURE 6) leading to a predominance of green light as depth increases. The photosynthetic bacteria are found at interfaces between anaerobic environments such as mud and water where light (particularly far-red light) can penetrate but an anaerobic environment containing reductants such as H_2S and H_2 can be maintained.[1] Abbreviations: Chl a,b,c=chlorophyll *a,b,c;* Bchl a,b,c,d,e=bacteriochlorophyll *a,b,c,d,e;* Carots.=normal carotenoids (e.g., carotenes and xanthophylls); Brown Carots.= brown carotenoids (e.g. fucoxanthin).

a device to facilitate modulation of the levels of phycobilin accessory pigments without changing the composition of the thylakoids and their reaction centers. Even organisms that cannot regulate the various forms of pigments they contain are able to control the total amount of pigments in response to varying light levels. Thus one finds greater amounts of photosynthetic pigments in algae grown in dim light or in shaded leaves of higher plants.[65–69] As the available quanta decrease, more antenna pigments are added to increase the probability of photon capture. Thus both evolutionary adaptation and short-term regulatory mechanisms ensure efficient capture of light and transfer to the reaction centers of photosynthesis.

It appears, then, that one way evolution has adapted organisms to the light available is to provide accessory pigments in response to available light that transfer their energy to the chlorophylls of the reaction centers. But what can we add about the evolution of chlorophyll itself?

We begin with a consideration of the biosynthetic pathway for heme and chlorophyll (FIGURE 9). In animals (heme synthesis) and photosynthetic bacteria (bacteriochlorophyll synthesis) the pathway begins with an enzyme synthesizing aminolevulinic acid, ALA, from succinyl-CoA and glycine.[70-73] Although this system may participate in chlorophyll synthesis in cyanobacteria and algae, it has been extremely elusive in higher plants.[74, 75] Experiments in which labeled precursors are fed to algae and higher plants in which the utilization of ALA is blocked by administration of its analog, levulinic acid, the labeling of the accumulated ALA is consistent with compounds such as glutamate or alpha ketoglutarate being the precursors rather than succinyl-CoA and glycine; enzymatic investigations are in agreement.[75-78] In either case, condensation of two molecules of ALA yields the first pyrrole, porphobilinogen, and

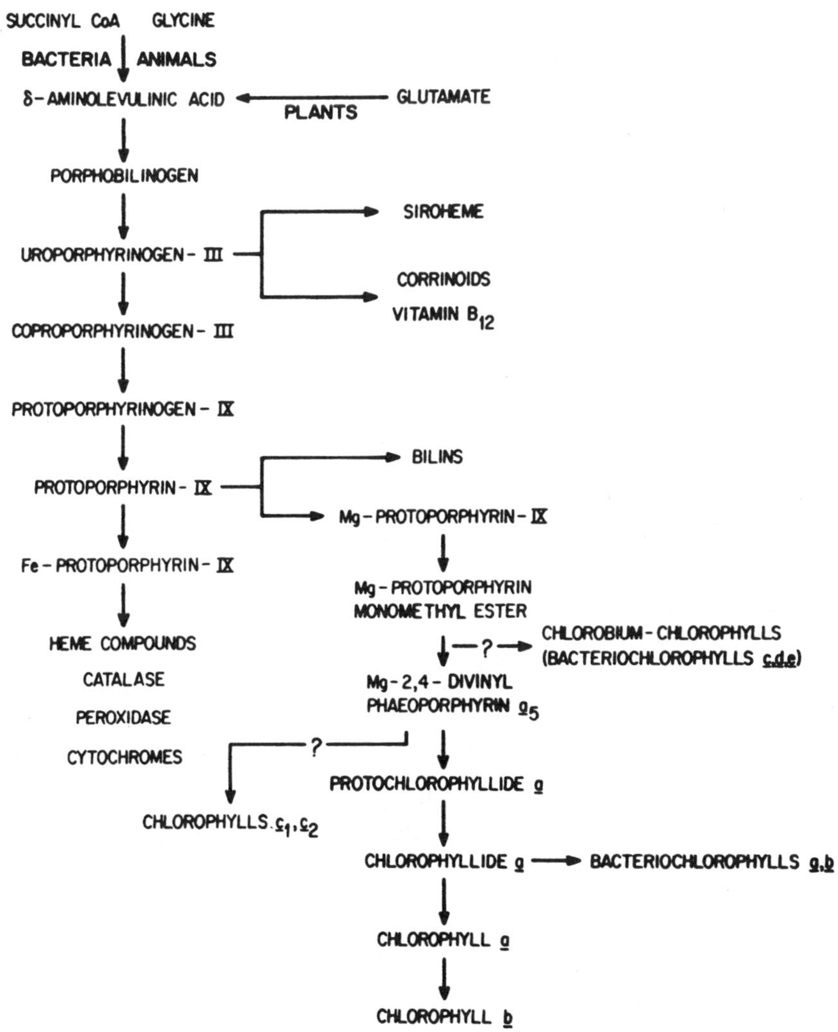

FIGURE 9. Summary of steps in heme and chlorophyll biosynthesis.[80]

further condensation of four pyrroles yields the tetrapyrroles. Protoporphyrin IX is the branch point at which either iron can be inserted to yield heme and, eventually, when necessary, the open-chain tetrapyrroles for phycocyanin and phycoerythrin formation by ring opening; or magnesium can be inserted to begin the synthetic leg leading to chlorophyll. Granick first pointed out that the precursors to chlorophyll leading from magnesium protoporphyrin IX to chlorophyll showed both increasing shifts towards the red region of the spectrum and increasing strength of absorption (FIGURE 10) of these bands.[79] He further proposed that during evolution, each of these pigments might have served, transiently, as the sensitizer of photosynthesis. Thus, an early photosynthetic pigment might have been Mg protoporphyrin IX. When mutation brought about the formation of enzymes for the conversion of Mg protoporphyrin IX to Mg vinyl pheoporphyrin *a*[5] or protochlorophyll(ide), this pigment was selected as the photosynthetic pigment and so on until the biosynthetic chain was lengthened to ultimately terminate in chlorophyll *a*. The longest wavelength of absorption of chlorophyll *a* is the farthest to the red of all of the precursors and this red band has the highest absorption coefficient of all of the precursors as well.

Now, if we combine all of these ideas, we can provide an explanation for the present pigment composition of photosynthetic organisms and its relation to available light. We must suppose that the evolution of chlorophyll for the reaction centers and the evolution of accessory pigments occurred concomitantly to adapt the organisms to the available light. If we adopt Granick's ideas, the long wavelength absorption of the early precursors to chlorophyll was more toward the yellow region of the spectrum where a great deal of light was available. As accessory pigments evolved, in response to each ecological situation (FIGURE 8), the absorption of the primary sensitizer of photosynthesis had to move toward longer wavelengths because it would have to serve as the sink for energy transfer from the accessory pigments. We must suppose that the chlorophyll type of pigment was selected early in evolution for its chemical properties, i.e., for its ability to lose an electron to an acceptor when excited by light of the available energy. Thus chlorophyll (or its precursors) became the converter of photon energy to chemical energy and provided the interface between light reactions and electron transport. As accessory pigments filled in the middle of the spectrum in various organisms, the absorption of the chlorophyll pigments was steadily pushed to the red to serve both as the ultimate acceptor of photons from energy transfer by accessory pigments and as the primary step in photosynthetic electron transport. Accessory pigments and the blue peak of chlorophyll itself exist to capture the available photons; the red peak captures photons and also functions as the lowest energy sink for all of this energy and its absorption lies in the region of maximum available quanta from sunlight. We must suppose that evolution has rendered the structure of chlorophyll optimal for this function and for losing electrons to and gaining electrons from the primary electron transport carriers. A similar argument can be made for bacteriochlorophyll, which has the added selective advantage of absorption at wavelengths so long that light is still available despite absorption by other organisms living above (FIGURE 8). Photosynthetic bacteria are anaerobic when they are carrying out photosynthesis and are frequently found at the interface between anaerobic mud and water, where both light and electron donors such as H_2S and H_2 are available. Since the photosynthetic bacteria lack system II of photosynthesis, which needs a quantum with enough energy to split water, the lowest

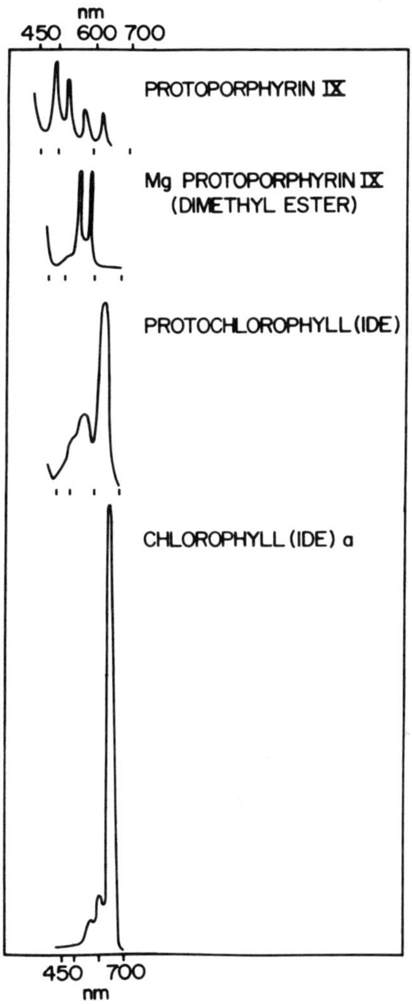

FIGURE 10. Absorption spectra of intermediates in chlorophyll biosynthesis drawn roughly to scale for comparison of wavelength and relative absorbance. Insertion of Mg^{+2} into protoporphyrin IX increases long wavelength absorbance (Mg protoporphyrin IX). On closure of the isocyclic ring and other modifications the long wavelength peak moves into the red with an increase in absorbance (protochlorophyll (ide)). On reduction of ring IV the peak moves further to the red with a further increase in absorbance (chlorophyll(ide) *a*).[1] (After[81, 82].)

excited state of bacteriochlorophyll could be pushed by ecological selection to longer wavelengths where the energy per quantum is still enough to drive photosystem I, but may not be great enough for splitting of water.

References

1. Schiff, J. A. 1980. Development, inheritance and evolution of plastids and mitochondria. *In* The Biochemistry of Plants. N. Edward Tolbert, Ed. Vol. **1:** 209–272. Academic Press, Inc. New York, N.Y.
2. Rhoades, M. M. 1946. Plastid mutations. Cold Spring Harbor Symp. Quant. Biol. **11:**202–207.
3. Margulies, L. 1970. Origin of eucaryotic cells. Yale University Press. New Haven, Conn.
4. Schiff, J. A. 1973. The development, inheritance and origin of the plastid in *Euglena. In* Advances in Morphogenesis. M. Abercrombie & J. Brachet, Eds. Vol. **10:**265–312. Academic Press, Inc. New York, N.Y.
5. Lewin, R. 1977. *Prochloron,* type genus of the prochlorophyta. Phycologia **16:**217.
6. Cohen-Bazire G., W. R. Sistrom & R. Y. Stanier. 1957. Kinetic studies of pigment synthesis by non-sulfur purple bacteria. J. Celi. Comp. Physiol. **49:** 25–68.
7. Padan, E. 1979. Facultative anoxygenic photosynthesis in cyanobacteria. Ann. Rev. Plant Physiol. **30:**27–40.
8. Watson, K., J. M. Haslam, B. Veitch & A. W. Linnane. 1971. Mitochondrial precursors in anaerobically-grown yeast. *In* Autonomy and Biogenesis of Mitochondria and Chloroplasts. N. K. Boardman, A. W. Linnane, R. M. Smillie, Eds.: 162–174. North-Holland Publ. Amsterdam.
9. Pearlman, P. S. & H. R. Mahler. 1974. Derepression of mitochondria and their enzymes in yeast: regulatory aspects. Arch. Biochem. Biophys. **162:**248–271.
10. Sager, R. 1958. The architecture of the chloroplast in relation to its photosynthetic activities. Brookhaven Symp. Biol. **11:**101–117.
11. Senger, H. & N. I. Bishop. 1972. The development of structure and function in chloroplasts of greening mutants of *Scenedesmus.* I. Formation of chlorophyll. Plant Cell Physiol. **13:**633–649.
12. Hudock, G. A. & H. Rosen. 1976. Formal genetics of *Chlamydomonas reinhardtii. In* Genetics of Algae. R. Lewin, Ed.: 29–48. University of California Press. Berkeley, Calif.
13. Kirk, J. T. O. & R. A. E. Tilney-Bassett. 1978. The Plastids (2nd Edition). W. H. Freeman. San Francisco.
14. Whatley, J. M. 1974. Chloroplast development in primary leaves of *Phaseolus vulgaris.* New Phytol. **73:**1097–1110.
15. Klein, S. & J. A. Schiff. 1972. The correlated appearance of prolamellar bodies, protochlorophyll(ide) species and the Shibata shift during development of bean etioplasts in the dark. Plant Physiol. **49:**619–626.
16. Klein, S., J. A. Schiff & A. Holowinsky. 1972. Events surrounding the early development of *Euglena* chloroplasts. II. Normal development of fine structure and the consequences of preillumination. Develop. Biol. **28:**253–273.
17. Gibbs, S. 1962. Nuclear envelope-chloroplast relationships in algae. J. Cell. Biol. **14:**433–444.
18. Lancer, H. A., C. E. Cohen & J. A. Schiff. 1976. Changing ratios of phototransformable protochlorophyll and protochlorophyllide of bean seedlings developing in the dark. Plant Physiol. **57:**369–374.
19. Cohen, C. & J. A. Schiff. 1976. Events surrounding the early development of *Euglena* chloroplasts. XI. Protochlorophyll(ide) and its photoconversion. Photochem. Photobiol. **24:**555–566.

20. Kindman, L. A., C. E. Cohen, M. H. Zeldin, Y. BenShaul & J. A. Schiff. 1978. Events surrounding the early development of *Euglena* chloroplasts. 12. Spectroscopic examination of the protochlorophyll(ide) phototransformation in intact cells. Photochem. Photobiol. **27**:787–794.
21. Horrum, M. A. & S. D. Schwartzbach. 1980. Nutritional regulation of organelle biogenesis in *Euglena*: repression of chlorophyll and NADP-glyceraldehyde-3-phosphate dehydrogenase synthesis. Plant Physiol. **65**:382–386.
22. App, A. A. & A. T. Jagendorf. 1963. Repression of chloroplast development in *Euglena gracilis* by substrates. J. Protozool. **10**:340–343.
23. Schiff, J. A., H. Lyman & G. K. Russel. 1971. Isolation of mutants from *Euglena gracilis. In* Methods in Enzymology. A. San Pietro, Ed. Vol. **23**:143–162. Academic Press, Inc. New York, N.Y.
(*For addendum see* Schiff, J. A., H. Lyman & G. K. Russel. 1980. *In* Methods in Enzymology. A. San Pietro, Ed. Vol. **69**:23–29. Academic Press, Inc. New York, N.Y.)
24. Clayton, R. 1964. Phototaxis in microorganisms. *In* Photophysiology. A. Geise, Ed. Vol. **2**:51–77.
25. Nultsch, W. 1970. Photomotion of microorganisms and interaction with photosynthesis. *In* Photobiology of Microorganisms. P. Halldal, Ed.: 213–251. John Wiley & Sons, Inc. New York, N.Y.
26. Diehn, B. 1969. Action spectra of phototactic responses in *Euglena*. Biochim. Biophys. Acta **177**:136–143.
27. Leedale, G. F. 1967. Euglenoid Flagellates. Prentice-Hall, Inc. Englewood Cliffs, N.J.
28. Buetow, D. E., Ed. 1968. The Biology of Euglena V. Academic Press, Inc. New York, N.Y.
29. Feinlieb, M. 1978. Photomovement in microorganisms. Photochem. Photobiol. **27**:849–854.
30. Senger, H., Ed. 1980. The Blue Light Syndrome. Springer-Verlag. Heidelberg, West Germany.
31. Howland, G. P. & M. E. Edwards. 1979. Photomorphogenesis of fern gametophytes. *In* The Experimental Biology of Ferns. A. F. Dyer, Ed. :394–427. Academic Press, Inc. New York, N.Y.
32. Briggs, W. 1964. Phototropism in higher plants. *In* Photophysiology. A. Giese, Ed. Vol. **1**:223–271. Academic Press, Inc. New York, N.Y.
33. Stockenius, W., R. H. Lozier & R. A. Bogomolni. 1979. Bacteriorhodopsin and the purple membrane of halobacteria. Biochim. Biophys. Acta **505**:215–278.
34. Dartnall, H. J. A., Ed. 1972. Handbook of Sensory Physiology. Vol. VII/I. Springer-Verlag. Heidelberg, West Germany.
35. Vogelman, T. C. & J. Scheibe. 1978. Action spectra for chromatic adaptation in the blue-green alga *Fremyella diplosiphon*. Planta **143**:233–240.
36. Bjorn, G. S. 1979. Action spectra for *in vivo* and *in vitro* conversions of phytochrome b, a reversibly photochromic pigment in a blue-green alga and its separation from other pigments. Physiol. Plantarum **46**:281–286.
37. Lazaroff, N. & J. A. Schiff. 1962. Action spectrum for developmental photoinduction of the blue-green alga *Nostoc muscorum*. Science **137**:603–604.
38. Fujita, Y. & A. Hattori. 1962. Photochemical interconversion between precursors of phycobilin chromoproteids in *Tolypothrix tenuis*. Plant Cell Physiol. **3**:209–220.
39. Haury, J. & L. Bogorad. 1977. Action spectra for phycobiliprotein synthesis in a chromatically adapting cyanophyte *Fremyella diplosiphon*. Plant Physiol. **60**:835–839.
40. Shropshire, W. 1977. Photomorphogenesis. *In* The Science of Photobiology. K. Smith, Ed.: 281–312. Plenum Publishing Corp. New York, N.Y.

41. SHROPSHIRE, W., JR. 1972. Phytochrome, a photochromic sensor. *In* Photophysiology. A. Giese, Ed. Vol. **7**:33–72.

42. HAUPT, W. & SCHONBOHM, E. 1970. Light-oriented chloroplast movements. *In* Photobiology of Microorganisms. P. Halldal, Ed.: 283–307. John Wiley & Sons. New York, N.Y.

43. SCHIFF, J. A. 1978. Photocontrol of chloroplast development in *Euglena. In* Chloroplast Development. G. Akoyunoglou J. H. Argyroudi-Akoyunoglou, Eds.: 747–767. Elsevier North-Holland, Inc. Amsterdam.

44. VIRGIN, H. 1972. Chlorophyll biosynthesis and phytochrome action. *In* Phytochrome. K. Mitrakos & W. Shropshire, Jr., Eds.: 371–406. Academic Press, Inc. New York, N.Y.

45. COHEN, D. & J. A. SCHIFF. 1976. Events surrounding the early development of *Euglena* chloroplasts. 10. Photoregulation of the transcription of chloroplastic and cytoplasmic ribosomal RNAs. Arch. Biochem. Biophys. **177**:201–216.

46. ELLIS, R. J. & R. BARRACLOUGH. 1978. Synthesis and transport of chloroplast proteins inside and outside the cell. *In* Chloroplast Development. G. Akoyunoglou & J. H. Argyroudi-Akoyunoglou, Eds.: 185–194. Elsevier North-Holland, Inc. Amsterdam.

47. DOBBERSTEIN, B., G. BLOBEL & N. CHUA. 1977. *In vitro* synthesis and processing of a putative precursor for the small subunit of ribulose-1,5-bisphosphate carboxylase of *Chlamydomonas reinhardtii.* Proc. Natl. Acad. Sci. USA **74:** 1082–1085.

48. SCHMIDT, G. & N. H. CHUA. 1978. *In vitro* synthesis, transport and assembly of ribulose-1,5-bisphosphate carboxylase subunits. *In* Photosynthetic Carbon Assimilation. H. W. Siegelman, Ed. :325–347. Plenum Press. New York, N.Y.

49. WATSON, J. D. 1970. Molecular Biology of the Gene. Benjamin. N.Y.

50. COLLINGBOURNE, R. H. 1966. General principles of radiation meteorology. *In* Light as an Ecological Factor. R. Bainbridge, G. C. Evans & O. Rockham, Eds.: 1–16. Blackwell. Oxford.

51. LYTHGOE, J. N. 1972. The adaptation of visual pigments to the photic environment. *In* Handbook of Sensory Physiology. H. J. Dartnall, Ed. Vol. **7**:566–603. Springer-Verlag. Berlin and New York.

52. BUTLER, W. L. 1978. Energy distribution in the photochemical apparatus of photosynthesis. Ann. Rev. Plant Physiol. **29**:345–378.

53. BLINKS, L. R. 1964. Accessory pigments and photosynthesis. *In* Photophysiology. A. C. Giese, Ed. Vol. **1**:199–222. Academic Press, Inc. New York, N.Y.

54. DUYSENS, L. N. M. 1952. Transfer of excitation energy in photosynthesis. Ph.D. Thesis, University of Utrecht. Utrecht, The Netherlands.

55. MYERS, J. & C. W. FRENCH. 1960. Evidences from action spectra for a specific participation of chlorophyll b in photosynthesis. J. Gen. Physiol. **43**:723–736.

56. FRENCH, C. S. 1960. The chlorophylls *in vivo* and *in vitro. In* Encyclopedia of Plant Physiology. A. Pirson, Ed. Vol. V/1:252–297. Springer-Verlag. Berlin and New York.

57. THORNBER, J. P., J. P. MARKWELL & S. REINMAN. 1979. Plant chlorophyll-protein complexes: recent advances. Proc. Natl. Acad. Sci. USA **74**:575–1216.

58. PRÉZLIN, B. B. & R. S. ALBERTE. 1978. Photosynthetic characteristics and organization of chlorophyll in marine dinoflagellates. Proc. Natl. Acad. Sci. USA **75**:1801–1804.

59. STRALEY, S. C., W. W. PARSON, D. C. MAUZERALL & R. K. CLAYTON. 1973. Pigment content and molar extinction coefficients of photochemical reaction centers from *Rhodpseudomonas spheroides.* Biochim. Biophys. Acta **305**:597–609.

60. ARNTZEN, C. & J. M. BRIANTAIS. 1975. Chloroplast structure and function. *In* Bioenergetics of Photosynthesis. Govindjee, Ed.: 51–115. Academic Press, Inc. New York, N.Y.

61. CRISP, D. J., Ed. 1971. European Marine Biology Symp. 4th. Cambridge University Press. Cambridge, England.
62. LEVRING, T. 1966. Submarine light and algal shore zonation. *In* Light as an Ecological Factor. R. Bainbridge, G. C. Evans & O. Rackham, Eds.: 305–318. Blackwell. Oxford.
63. KIRK, J. T. O. 1979. Spectral distribution of photosynthetically active radiation in some southeastern Australian waters. Aust. J. Mar. Freshwater Res. **30:** 81–91.
64. GRABOWSKI, J. & E. GANNT. 1978. Excitation energy migration in phycobilisomes: comparison of experimental results and theoretical predictions. Photochem. Photobiol. **28:**47–54.
65. HALLDAL, P. 1970. Photobiology of Microorganisms. Wiley (Interscience) New York, N.Y.
66. STEEMAN-NIELSEN, E. 1975. Marine Photosynthesis. Oceanography Ser. 13. American Elsevier. New York, N.Y.
67. RAMUS, J., S. I. BEALE & D. MAUZERALL. 1976. Correlation in changes in pigment content with photosynthetic capacity of seaweeds as a function of water depth. Marine Biol. **37:**231–238.
68. BJORKMAN, O. 1973. Comparative studies on photosynthesis in higher plants. *In* Photophysiology. A. Geise, Ed.: 1–64. Academic Press, Inc. New York, N.Y.
69. BOARDMAN, N. K. 1977. Comparative photosynthesis of sun and shade plants. Ann. Rev. Plant Physiol. **28:**355–377.
70. SHEMIN, D. & C. S. RUSSEL. 1953. δ aminolevulinic acid, its role in the biosynthesis of porphyrins and purines. J. Am. Chem. Soc. **75:**4873–4874.
71. SHEMIN, D., C. S. RUSSEL & T. ABRAMSKI. 1955. The succinate glycine cycle. I. The mechanism of pyrrole synthesis. J. Biol. Chem. **214:**613–626.
72. KIRUCHI, G., A. KUMAR, P. TALMAGE & D. SHEMIN. 1958. The enzymatic synthesis of δ-aminolevulinic acid. J. Biol. Chem. **223:**1214–1219.
73. GIBSON, L. D., W. D. LAVER & A. NEUBERGER. 1958. The purification and properties of δ aminolevulinic acid dehydrase. Biochem. J. **61:**618–629.
74. KLEIN, O. & H. SENGER. 1978. Two biosynthetic pathways to δ aminolevulinic acid in a pigment mutant of the green alga *Scenedesmus obliquus.* Plant Physiol. **62:**10–13.
75. HAREL, E. 1978. Initial steps in chlorophyll synthesis—problems and open questions. *In* Chloroplast Development. G. Akoyunoglou & J. H. Argyroudi-Akoyunoglou, Eds.: 33–44. Elsevier North-Holland, Inc. Amsterdam.
76. JURGENSON, J. E., S. I. BEALE, R. R. TROXLER, M. M. BARTOLF, M. P. FITZGERALD, J. RAMUS & J. A. SCHIFF. 1975. Biosynthesis of δ-aminolevulinic acid in *Cyanidium caldarium* and other algae. Biol. Bull. **149:**432.
77. BEALE, S. I. 1978. δ aminolevulinic acid in plants: Its biosynthesis, regulation and role in plastid development. Ann. Rev. Plant Physiol. **29:**95–120.
78. KANNANGARA, C. G. & S. P. GOUGH. 1978. Biosynthesis of Δ-aminolevulinate in greening barley leaves: glutamate-1-semialdehyde aminotransferase. Carlsberg Res. Commun. **43:**185–194.
79. GRANICK, S. 1949. The structural and functional relationships between heme and chlorophyll. Harvey Lect. Ser. **43:**220–245.
80. RAGAN, M. A. & D. J. CHAPMAN. 1978. A Biochemical Phylogeny of the Protists. Academic Press, Inc. New York, N.Y.
81. GRANICK, S. 1948. Magnesium protoporphyrin IX as a precursor of chlorophyll in *Chlorella.* J. Biol. Chem. **175:**333–342.
82. SMITH, J. H. C. & A. BENITEZ. 1955. Chlorophylls: Analysis in plant materials. *In* Moderne Methoden der Pflanzenanalyse. K. Paech & M. V. Tracey, Eds. Vol. **4:**142–196. Springer-Verlag. Berlin and New York.

DISCUSSION OF THE PAPER

DR. A. W. COLEMAN: I'm very interested in the ideas about constitutive chloroplasts and it occurs to me that we suffer from a bias, particularly in the alga, of looking at the wrong stage of the life cycle. A great many of these organisms make heavy-walled spores and we rarely look for chloroplasts or chloroplast pigments in those spores, and in fact they may have proplasts by nature.

DR. J. A. SCHIFF: Yes, I would agree.

DR. N. W. GILLHAM: First of all, one of the interesting things about Y mutations and *Chlamydomonas* is that they occur in very high frequencies. They also revert at very high frequencies. Second, prolamellar bodies in Y chloroplasts have been found. Third, with respect to looking for cryptochrome, it occurs to me that *E. coli* might not be the right organism to look for it in. You might want to look for it in a photosynthetic bacteria that you can do genetics on.

DR. SCHIFF: I didn't mean to imply that you would look for it in *E. coli* which has absolutely no photocontrol that I know about. I said to look at photocontrol in a chloroplast system using the prokaryotic model because after all a chloroplast is the residence for the prokaryote in the eukaryotic cell. One would think that eukaryotic control is built on a prokaryotic basis and indeed that turns out to be the case in most cases. Prolamellar bodies used to be thought of as precursors of the thylakoid membranes, they no longer are. They seem to be made of entirely different materials, they are not obligatory stages in chloroplast development, they've even been found in cyanophytes under certain conditions. I think the whole question of what a prolamellar body is, is another entire question and I wouldn't necessarily relate it to the problem of where a proplastid fits in the development.

DR. AARONSON (*Queen's College, Flushing, N.Y.*): From your discussion I had the sense that the prokaryotic photosynthetics which responded to photosynthetically active light might have to compete once eukaryotic alga evolved and they would therefore be in a better position to compete if they went directly to the light whereas the other algae in a competitive situation would have the opportunity to find some light underneath.

DR. SCHIFF: First of all, I forgot to mention one thing in the evolution of the photoreceptor and that is to be adaptable to the long haul of evolution pigmented repressors, or any of the systems if they exist, must be responsive to the ecological need. In other words there is no point in having chloroplast development induced by green light in an organism which uses chlorophyll for photosynthesis. Obviously you would want blue or red light to induce chloroplast development, the same light that is effective for photosynthesis. So carrying that into the ecological argument, I would say that I would expect the pigments that are used for development and for controlling chloroplast development or phototaxis would be those that are ultimately effective for photosynthesis in those organisms. And that's exactly what occurs. Cryptochrome absorbs in the blue, phytocytochrome in the red; they are both equally good for informing the organism that light appropriate for photosynthesis is available.

DR. L. MARGULIS (*Boston University, Boston, Mass.*): Besides the absorption spectrum of cryptochrome, what do you know about its distribution, composition, in fact, anything?

DR. SCHIFF: What we know about cryptochrome is the following: Cryptochrome, as you will gather, is a word expressing our ignorance about the composition of the pigment. It refers to a whole syndrome of blue light-induced effects that have been variously ascribed to carotenoids, flavoproteins, and unknowns. I think the last one probably comes closest to our current knowledge. Although there are cases where that might not be quite right.

What we know about those systems is that the action spectrum has one or more peaks in the blue and near-UV region of the spectrum and none beyond—no peaks in the green or the red, or far-red. They control morphogenic responses, organelle development, carotenogenesis, phototaxis, phototropism, and a large syndrome of other effects, and nobody has identified the photoreceptor. We don't know it's a protein. All we know, is its action spectrum. We know its absorption.

DR. ZECKENDORF (*Cornell University, Ithaca, N.Y.*): I think we are overlooking the work of Warren Butler.

DR. SCHIFF: No, not overlooking it, but I'd be glad to get into it.

DR. ZECKENDORF: Nice model systems in which isolated pigments are reducing cytochrome *b*, and may be responsible.

DR. SCHIFF: I was trying to avoid that and I'll tell you why. That requires a little history. In the early 1950s, Arthur Galston at Yale University fished out of a higher plant, a system which destroyed auxin indole acetic acid using blue light and he thought he had his hands on the phototropic response. Destroy auxin at the site of the plant in the light and you obviously have a redisposition of auxin, but of course we know that auxin is not destroyed, it's redistributed.

This system that Arthur Galston fished out can be brought about just by throwing in any old flavene off the shelf. Flavenes are very, very active and when you bust up a cell you release the flavoproteins and you can catalyze all sorts of photooxidations and photoreductions. I have always been very skeptical about such *in vitro* systems. Butler rediscovered this whole thing entirely over again. There's a large literature before him. Butler has done some excellent work, but he has not yet done the two critical experiments, nor has anybody, to verify that that system has anything whatever to do with the *in vivo* system—namely, an exact matching of the action spectrum of the *in vitro* response and the *in vivo* response.

Secondly, nobody has gotten mutants blocked in the primary photoresponse which are also blocked in the biological response. I would need those two pieces of information before I concluded that the pigment being measured in the test tube is the same as in the intact organism. I can show you any number of reactions that I can do in the test tube which are beautiful reactions, get huge optical density changes, but have nothing whatever to do with what goes on in the organism.

DR. R. A. LEWIN: One of the advantages of a meeting like this is it suggests for a number of us future experiments and other things to look for. Not only two diatoms, but as far as I know every heterotrophic photosynthetic diatom is capable of producing its photosynthetic pigments in darkness.

DR. MARGULIS: Would you like to say something about the status of heterotrophic diatoms? It's my understanding that there aren't any in nature.

DR. LEWIN: There are three permanently colorless diatoms with silica walls indicating that they're diatoms with no chlorophyll, no photosynthetic pigments at all. In addition to that, there are a lot of diatoms that can be grown in complete darkness on a medium containing glucose or lactate.

DR. SCHIFF: Organotrophic.

DR. MARGULIS: But all of that is laboratory stuff. Heterotrophy in diatoms was looked for in nature and my understanding is that they're obligate autotrophs.

THE CHLOROPLASTS OF SOME ALGAL GROUPS MAY HAVE EVOLVED FROM ENDOSYMBIOTIC EUKARYOTIC ALGAE

Sarah P. Gibbs

Department of Biology
McGill University
Montreal, P.Q. H3A 1B1
Canada

So far in this symposium it has been established that the chloroplasts of red algae arose during evolution from symbiotic blue green algae and that the chloroplasts of green algae arose from an ancestor of *Prochloron didemni,* which Lewin and his coworkers [27, 28, 34] have shown is a prokaryotic alga which lacks phycobilins and possesses both chlorophylls *a* and *b*. In this paper I present evidence that the red and green algae may be the only algae that acquired their chloroplasts directly from a prokaryotic symbiont.

What the chloroplasts of red and green algae have in common with each other and with the chloroplasts of green plants, which have evolved from green algae, is that they are surrounded solely by a double membrane. In the other groups of algae the chloroplasts are always enclosed by either three or four membranes (TABLE 1). The four membranes surrounding the chloroplasts of most groups of algae are easily explained; two are the chloroplast envelope and two are a layer of endoplasmic reticulum, termed chloroplast ER, which completely encloses the chloroplast.[13] The presence of a third membrane around the chloroplasts of *Euglena* and dinoflagellates was an enigma until I suggested recently that it could be the plasmalemma of a reduced eukaryotic symbiont.[15] Recently Greenwood [18] discovered that there is a small double-membrane-limited body, named a nucleomorph,[19] between the chloroplast envelope and the chloroplast ER in cryptomonads, and he [19] and Whatley [43] have suggested that this nucleomorph might be the vestigial nucleus of a chloroplast-containing eukaryotic symbiont. If nucleomorphs are vestigial nuclei, then not only the chloroplasts of cryptomonads but probably also those in all the other groups of algae containing chloroplast ER have arisen by the progressive reduction of a symbiotic eukaryotic alga.

At present the evidence for chloroplasts arising from the progressive reduction of a eukaryotic symbiont is, I believe, strongest in the case of *Euglena. Euglena* has long been a paradox both to beginning biology students and phycologists alike. On one hand, it contains chloroplasts that are very similar to those of green algae both in their ultrastructure and in their possession of chlorophylls *a* and *b*, so most phycologists have classified *Euglena* either as a green alga [5] or in a class of its own, the Euglenophyceae, which they placed adjacent to the Chlorophyceae.[32, 35] On the other hand, *Euglena* except for its chloroplasts bears almost no resemblance to green algae.[26] It has a flexible pellicle of interlocking proteinaceous strips, rather than a rigid polysaccharide-containing cell wall. Its storage product is paramylon, a β-1:3-linked glucose polymer, not starch which is α-1:4-linked. Paramylon grains are stored in the cytoplasm not in the chloroplast. The eyespot in *Euglena* lies free in the cyto-

0077–8923/81/0361–0193 $01.75/0 © 1981, NYAS

plasm, rather than in the chloroplast, and the photoreceptive pigment is found not in the cell membrane, as is believed to be the case in green algae,[1, 29] but in a small crystalline body located near the base of the long flagellum. Unlike green algae, the chromosomes of *Euglena* remain condensed throughout interphase and the nucleolus does not disperse during mitosis, but simply elongates and constricts in two.

I propose that the answer to this paradox—that *Euglena* has chloroplasts that are very similar to those of green algae yet in almost all its other ultrastructural characteristics is dramatically different from green algae—is that *Euglena* chloroplasts have arisen from symbiotic green algae. I believe that early in evolution green algae acquired their chloroplasts from chlorophyll *a* and *b*-containing prokaryotes, and much later small green algae invaded

TABLE 1

NUMBER OF MEMBRANES ENCLOSING THE CHLOROPLASTS OF ALGAE

Two (The chloroplast envelope)	Three (Two of chloroplast envelope plus postulated plasmalemma)	Four (Two of chloroplast envelope plus two of chloroplast ER)
Rhodophyceae (red algae)	Euglenophyceae (euglenoids)	Cryptophyceae (cryptomonads)
Prasinophyceae (green algae)	Dinophyceae (dinoflagellates)	Chloromonadophyceae
Chlorophyceae (green algae)		Prymnesiophyceae
		Chrysophyceae
		Xanthophyceae
		Eustigmatophyceae
		Bacillariophyceae (diatoms)
		Phaeophyceae (brown algae)

euglenoid flagellates. For some time the symbiosis persisted like that of *Chlorellae* in *Paramecium bursaria* today.[22] During this time the genetic material in the nuclei of the green algae, which controlled chloroplast characteristics, was transferred to the nucleus of the *Euglena*. Then since the chloroplasts were the useful part of the symbiosis to the *Euglena*, the symbiont became progressively reduced until all that remained of it was the plasmalemma and the chloroplast. Stewart and Mattox[36] have also suggested that euglenoid algae might have obtained their chloroplasts from symbiotic green algae.

Leedale,[26] however, believes that there is a simpler answer to this paradox, namely that *Prochloron*-type prokaryotes early invaded two very different types of algal groups—green algae and euglenoids. I have two main reasons for rejecting this simpler hypothesis. The first is that the structure and arrangement of the thylakoids in the chloroplasts of green algae and *Euglena* are similar to each other and different from *Prochloron*. In *Prochloron* the thylakoids are arranged predominantly in pairs with occasional larger stacks.[39, 41, 43, 44] In euglenoids and most green algae the thylakoids are arranged in lamellae of two to six tightly appressed thylakoids with lamellae of three appressed thylakoids predominating. In some green algae, however, thylakoids are arranged in grana. A distinctive ultrastructural feature of green algal chloroplasts is the presence of continuities between adjacent thylakoids at the ends of the lamellae. These continuities indicate that stacks of thylakoids are formed either by invaginations into existing thylakoids or by a thylakoid evaginating and growing back over itself. Similar continuities between adjacent thylakoids have been seen in *Euglena gracilis*.[14] Although connections between adjacent thylakoids have been found in all green plants, they have not been observed in any other algal group besides green algae and *Euglena*. Nor have I observed any continuities between adjacent thylakoids in the published micrographs of *Prochloron*. Thus green algal and *Euglena* chloroplasts have, in their ultrastructure, evolved considerably from their *Prochloron*-like ancestors and now display a "green" type of thylakoid arrangement that differs from that in all other algal classes.[14] I do not believe that this "green type" of chloroplast structure evolved independently in green algae and euglenoids. Thus I do not believe a *Prochloron*-like prokaryote invaded euglenoids and green algae separately.

My second reason for ruling out the possibility of a direct prokaryote ancestor for *Euglena* chloroplasts is the invariable presence of a third membrane around *Euglena* chloroplasts. FIGURES 1 and 2 are sections of the chloroplasts of *Euglena gracilis* var. *bacillaris* and it can be seen clearly that they are enclosed by three membranes. In FIGURE 1 the three membranes are evenly spaced, but in FIGURE 2 the inner two membranes which are the two membranes of the chloroplast envelope have at one place infolded a short distance into the chloroplast whereas the third membrane follows the contour of the chloroplast. In this figure there appears to be a fourth membrane around the chloroplast (*arrow*), but this membrane is the limiting membrane of a paramylon grain which lies close to the chloroplast. FIGURES 3 and 4, for comparison, illustrate the chloroplast boundary in algae whose chloroplasts are enclosed in four membranes, two of the chloroplast envelope and two of chloroplast ER. In FIGURE 3, which is of the chrysophyte alga *Ochromonas danica*, the chloroplast lies close to the nucleus and the two membranes of the chloroplast ER are continuous with the outer membrane of the nuclear envelope (*arrow*). Thus the chloroplast of *Ochromonas* lies in a double-membraned sac part of which is nuclear envelope and part of which is chloroplast ER. FIGURE 3 also shows that a layer of tubules and vesicles named the periplastidal reticulum by Falk and Kleinig [12] is present between the chloroplast envelope and the nuclear envelope. Periplastidal reticulum or equivalent vesicles is always present, usually at a characteristic location over the chloroplast, in all the algae that have chloroplast ER.[17] Also, chloroplast ER always has ribosomes present on the cytoplasmic surface of its outer membrane. These ribosomes were not preserved in the cell illustrated in FIGURE 3, for this cell was fixed simultaneously in glutaraldehyde and osmium tetroxide, a method which removes most cytoplasmic ribosomes. FIGURE 4 is a section of a cell of *Cryptomonas* sp. fixed by the conventional glutaraldehyde followed by osmium tetroxide method. The outer membrane of chloroplast ER appears studded with ribosomes.

Originally I believed that the three membranes surrounding *Euglena* chloroplasts arose by the fusion of two of the membranes of the chloroplast envelope—chloroplast ER complex,[14] and this view is presented in Bisalputra's [3]

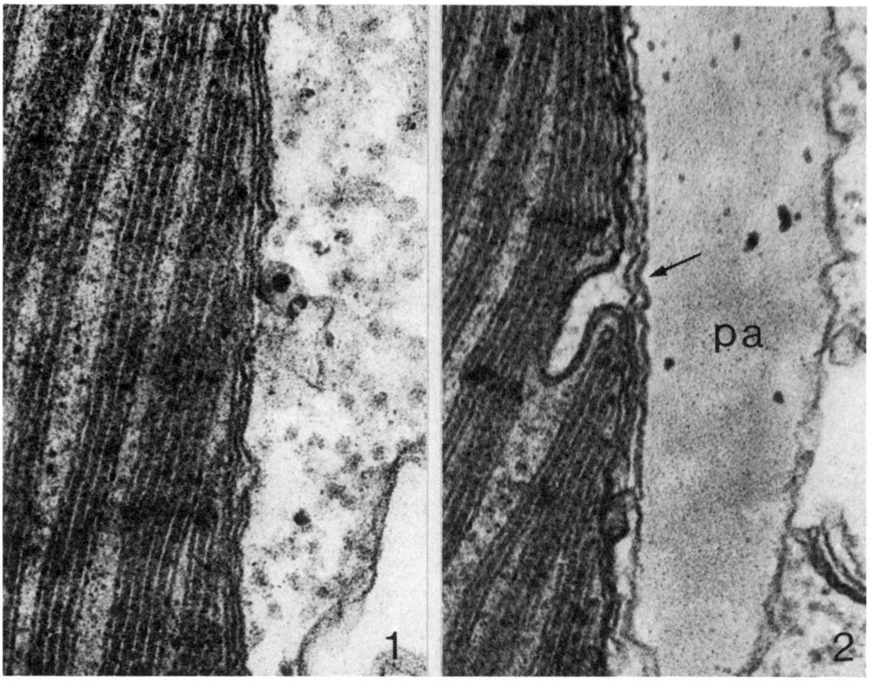

FIGURE 1. Section of a chloroplast of *Euglena gracilis* var. *bacillaris.* Note that the chloroplast is enclosed by three membranes. Ribosomes are present in the adjacent cytoplasm, but none are attached to the outermost membrane enclosing the chloroplast. ×128,000.

FIGURE 2. Section of a chloroplast of *Euglena gracilis* var. *bacillaris.* In this micrograph a membrane-bound paramylon grain (*pa*) lies adjacent to the chloroplast. At the arrow, one sees in succession the inner membrane of the paramylon grain, the third membrane surrounding the chloroplast, and the two membranes of the chloroplast envelope which have infolded a short distance into the chloroplast. ×103,000.

recent review of chloroplast structure. However, the view that the third membrane encircling *Euglena* chloroplasts is an ER membrane is no longer tenable, for several reasons. First, the membrane never has ribosomes attached to it as FIGURES 1 and 2 demonstrate. Sometimes, however, cytoplasmic ribosomes line up a short distance from the chloroplast in *Euglena* (FIGURE 3a in Reference 15), but they never touch the outer membrane. Bisalputra [3] has argued that there are ribosomes on the third membrane surrounding euglenoid chloroplasts, but in the micrograph of *Phacus* he uses to illustrate his argument there is a distinct space between most of the cytoplasmic ribosomes and the outer membrane of the chloroplast, with only an occasional ribosome actually appearing to make contact with the membrane. Secondly, there are never any connections between the outer chloroplast membrane of *Euglena* and the rough ER of the cytoplasm as is common in the other algae that have chloroplast ER.[15, 17] Nor do the two outer membranes of *Euglena* chloroplasts ever make contact with the outer membrane of the nuclear envelope.[15] However, algae with one nucleus and many chloroplasts frequently do not have a connection between chloroplast ER and the nuclear envelope,[17] so I would not expect to find such a connection in *Euglena*. A third argument against the outermost membrane being an ER membrane is that there are never any vesicles and tubules present between any of the membranes surrounding *Euglena* chloroplasts. Since these vesicles are present in all algae with chloroplast ER and since they play a vital role in the transport of plastid-bound proteins from their site of synthesis on the polysomes of chloroplast ER to the chloroplast matrix,[16] their absence, like the absence of ribosomes on the outer membrane, strongly argues against the third membrane of *Euglena* chloroplast being an ER membrane.

Thus I now believe that the third membrane surrounding the chloroplasts of *Euglena* is derived from the plasmalemma of the original green algal symbiont.

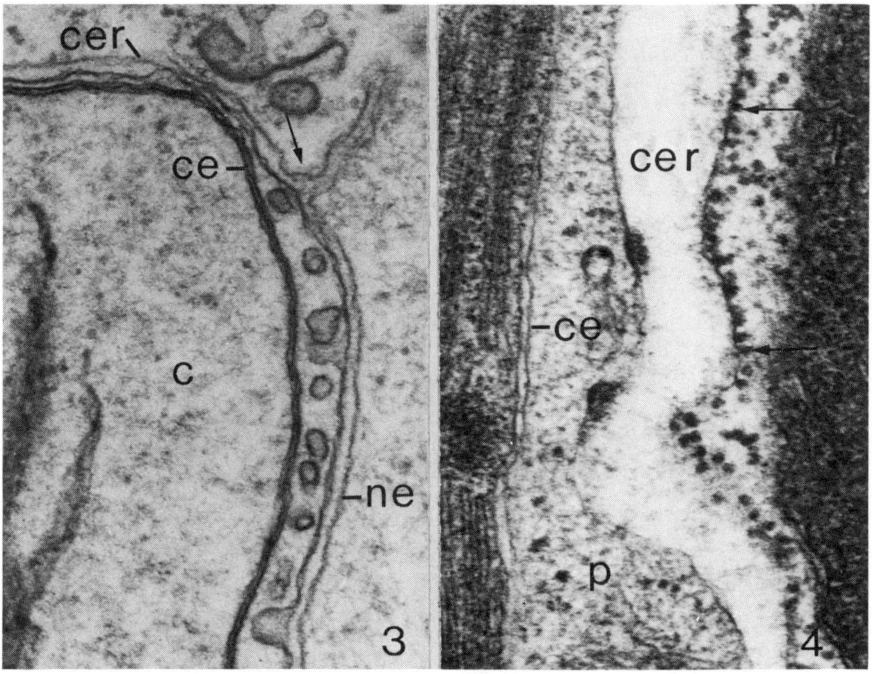

FIGURE 3. Section through the chloroplast (*c*) and adjacent nucleus of a greening cell of *Ochromonas danica*. The chloroplast is surrounded by four membranes, two of the chloroplast envelope (*ce*) and two of the chloroplast ER (*cer*). The chloroplast ER is continuous with the outer membrane of the nuclear envelope (*ne*) at the arrow. A layer of vesicles and tubules, the periplastidal reticulum, lies between the chloroplast and nucleus. ×73,200.

FIGURE 4. Section through the chloroplast and adjacent periplastidal compartment (*p*) of *Cryptomonas* sp. (R. R. L. Guillard's isolate ϕ). The chloroplast ER (*cer*) is moderately swollen and has conspicuous ribosomes (*arrows*) on the cytoplasmic surface of its outer membrane. Scattered ribosomes are also present in the periplastidal compartment. *ce*, chloroplast envelope. ×85,400.

When first captured, each green alga would have, of course, been separated from the cytoplasm of the host cell by the phagocytic vacuole membrane of the *Euglena*, its cell wall, and its plasmalemma. Why then do I believe it is the original cell membrane and not the phagocytic vacuole membrane that is the third membrane surrounding present day *Euglena* chloroplasts? Mainly because in order to survive a symbiont needs its cell membrane, whereas both cell walls and phagocytic vacuole membranes of symbionts can easily be dispensed with. As one modern day example, I cite the case of the symbiotic bacteria that inhabit *Paramecium aurelia.* Lambda and sigma symbiotic bacteria are still enclosed in a vacuole membrane whereas kappa and mu particles lie free in the cytoplasm.[2]

Earlier in this symposium and in a recent paper, Whatley [42, 43] has presented a somewhat different scheme for the origin of *Euglena* chloroplasts. She agrees with me that *Euglena* chloroplasts are derived from green algal chloroplasts, but she suggests that euglenoids captured these chloroplasts as chloroplasts much as the sacoglossan opisthobranch *Elysia viridis* captures chloroplasts of the green alga *Codium fragile.*[37] According to Hawes [20] each captured chloroplast is surrounded by a vacuole membrane of the host. Whatley, therefore, suggests that the third membrane surrounding *Euglena* chloroplasts is this phagocytic vacuole membrane. All the arguments I have presented in this paper for the origin of *Euglena* chloroplasts from the progressive reduction of a green algal symbiont could equally well be used to support Whatley's argument that *Euglena* chloroplasts are captured green algal chloroplasts. I favor my theory for I feel it is easier to imagine a whole cell symbiont becoming permanently integrated into a *Euglena* cell than it is to imagine a chloroplast becoming integrated in a similar manner. A naked chloroplast would be more susceptible to digestive attack. More importantly, if the chloroplast of the primitive green alga had lost any of its genetic information to the cell's nucleus during evolution, the *Euglena* host cell would not have the genetic information necessary for the growth and replication of the chloroplast. When a whole green algal cell is the symbiont, information from the symbiont's nucleus as well as from its chloroplast could be transferred with time to the host's nucleus.

There are other arguments that give indirect support to the theory that *Euglena* chloroplasts are reduced green algal symbionts or captured green algal chloroplasts. *Euglena* is the only organism that can be induced to lose its plastids experimentally. This can be done by heat, ultraviolet light, streptomycin, or several other drugs. Usually a small colorless proplastid persists, but in some cells all plastids have been lost.[15, 23] Also there are many colorless genera of euglenoids and most of these completely lack plastids.

The dinoflagellates are a second group of algae where there is suggestive evidence that the chloroplasts have arisen from the progressive reduction of a symbiotic eukaryotic alga. The main evidence is that the chloroplasts of almost all dinoflagellate species studied are, like those of euglenoids, enclosed in three membranes.[6, 14] The third membrane cannot be considered an ER membrane for it never has ribosomes attached to it, nor is it continuous with either the nuclear envelope or rough ER of the cytoplasm, and there is no trace of periplastidal reticulum between any of the three membranes enclosing the chloroplast. I believe that here, too, the third membrane represents the plasmalemma of an original eukaryotic symbiont. But in the case of dinoflagellates it is not obvious what group of algae were the symbionts. On the basis of chloroplast ultrastructure the dinoflagellates are most like the Prymnesiophyceae in that the chloroplast lamellae usually consist of three thylakoids and no girdle lamellae are present.[14] However, most dinoflagellate chloroplasts contain a unique carotenoid pigment peridinin, so if the eukaryotic invader were a prymnesiophycean alga, then it is necessary to postulate that peridinin evolved after the acquisition of chloroplasts from prymnesiophytes.

Recently it has been brought to my attention by M. Vesk that the chloroplasts of the dinoflagellate *Prorocentrum micans* are surrounded by only two membranes. A re-examination of the literature confirms her observations on *P. micans* (Figure 9, in Reference 24, and Figure 2, in Reference 8) and shows that two membranes are present around the chloroplasts of three other dinoflagellates (Plate 5B, in Reference 11; Plate 5A, in Reference 9; and Plate IIIB, in Reference 7). The absence of the third membrane in some dinoflagellate species does weaken the argument that the chloroplasts have evolved from eukaryotic symbionts, but on the other hand the third membrane could have originally been present in these species and have been lost.

Additional evidence that the chloroplasts of dinoflagellates could have evolved from symbiotic eukaryotic algae comes from the discovery by Tomas and Cox [40] that the dinoflagellate *Peridinium balticum* contains a large endosymbiotic chrysophyte alga that is separated from the host cell solely by the plasmalemma of the symbiont. In this dinoflagellate all the chloroplasts belong to the symbiont and this led Tomas and Cox [40] to suggest as early as 1973 that the chloroplasts of other photosynthetic dinoflagellates could have arisen from symbiotic chrysophyte algae. There is a second dinoflagellate species that today harbors a chloroplast-containing chrysophyte symbiont, *Peridinium foliaceum*,[21] as well as a third species in which there is a small eukaryotic symbiont but all the chloroplasts belong to the dinoflagellate.[15, 30]

It is interesting that in all three dinoflagellate species which at present contain an intact eukaryotic symbiont, there are remnants of a former eukaryotic symbiont. In the third species, which may be a *Gymnodinium* or a *Peridinium*, the remnants are the triple-membrane bounded chloroplasts in the dinoflagellate's cytoplasm. In *P. balticum* and *P. foliaceum* the dinoflagellate cytoplasm contains a large eyespot which is a modified chloroplast and is surrounded by three membranes.[10, 40] Presumably these two species of *Peridinium* once harbored a number of chloroplasts that had evolved from eukaryotic symbionts, but with time all the chloroplasts except the modified eyespot-containing one was lost. Relatively recently these species have been invaded by a chloroplast-containing chrysophyte which persists to this day.

The third group of algae where there is evidence that the chloroplast evolved from a symbiotic eukaryotic alga is the cryptomonads. The cryptomonads are a small class of flagellated algae that contain phycobilin pigments as well as chlorophylls *a* and *c* and have their chloroplast enclosed in a sac of chloroplast ER which is continuous with the nuclear envelope.[14] In all the other classes of algae with chloroplast ER, the space between the chloroplast envelope and the chloroplast ER, termed the periplastidal compartment, is very narrow or at places obliterated and contains no structures other than the periplastidal reticulum which is usually located over a specific region of the chloroplast.[17] In the cryptomonads, however, the periplastidal compartment is always very wide along the inner surface of the chloroplast and contains a variety of structures. Most conspicuous are numerous starch grains (Figure 5) which are usually present over the pyrenoid and along the inner surfaces of the chloroplast. Scattered vesicles and tubules (Figures 5 and 6) are also present in the com-

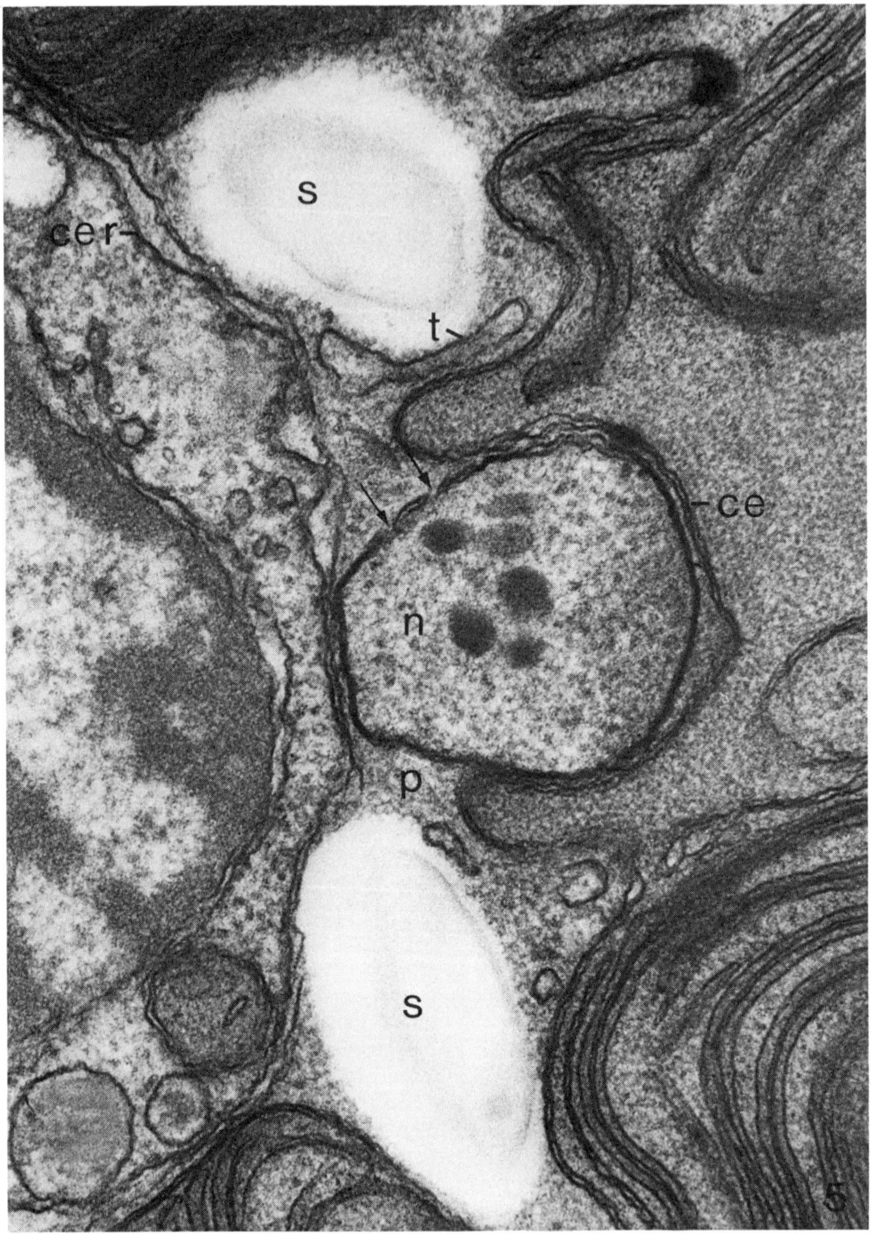

FIGURE 5. Section through the chloroplast, periplastidal compartment (*p*), and adjacent cytoplasm of *Chroomonas salina.* The periplastidal compartment lies between the chloroplast envelope (*ce*) and the chloroplast ER (*cer*) and contains starch grains (*s*), scattered vesicles and tubules (*t*) and a nucleomorph (*n*) which is slightly embedded in the pyrenoid region of the chloroplast. The nucleomorph is bounded by a double membrane which is interrupted by apparent pores (*arrows*). The nucleomorph contains dense granules and has a dense region of fibrillogranular material lying against the inside surface of the nucleomorph adjacent to the chloroplast. ×74,000.

partment. These vesicles and tubules do not have the organized structure of the periplastidal reticulum, i.e., a network of tubules with vesicles in the interstices; nonetheless, they are analogous to the periplastidal reticulum of other classes, for vesicles have been seen to pinch off the inner membrane of the chloroplast ER and to fuse with the outer membrane of the chloroplast envelope.[17] Thus I presume that in the cryptomonads the vesicles of the periplastidal compartment also transport plastid-bound proteins from the lumen of the chloroplast ER to the lumen of the chloroplast envelope. A third striking feature of the

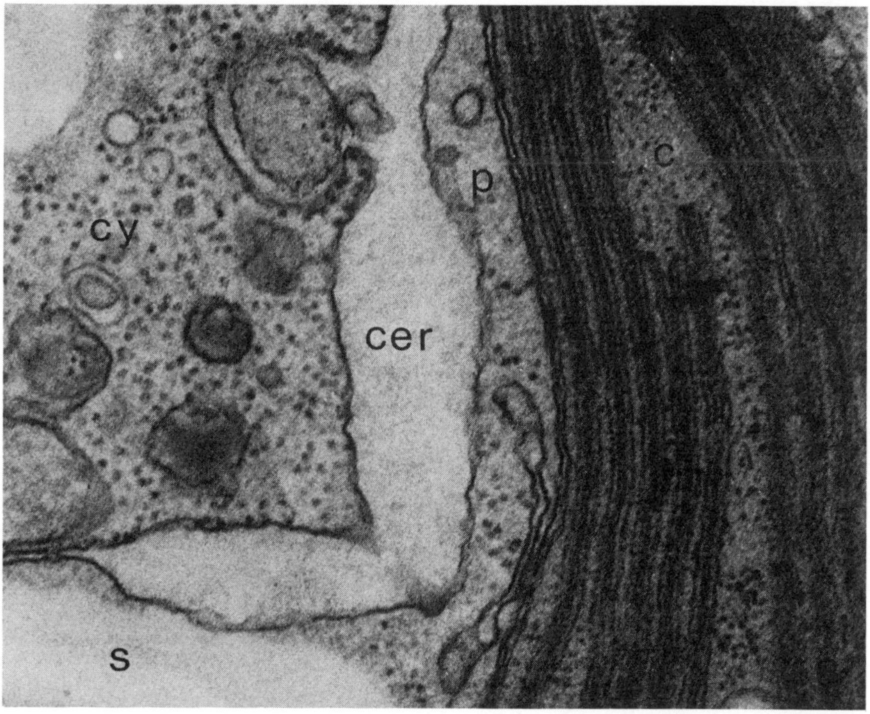

FIGURE 6. Section through the chloroplast and adjacent cytoplasm of *Cryptomonas* sp. (Guillard's isolate 3c). The ribosomes of the periplastidal compartment (*p*) are the same size as those in the cytoplasm (*cy*) and are distinctly larger than those in the chloroplast (*c*). Tubules and vesicles and a starch grain (*s*) are also present in the periplastidal compartment. *cer*, chloroplast ER. ×75,000.

periplastidal compartment in cryptomonads is the presence of ribosomes.[13] FIGURE 6 shows that these ribosomes are more sparsely distributed than either the cytoplasmic or chloroplast ribosomes. These periplastidal ribosomes measure the same size as the cytoplasmic ribosomes and are noticeably smaller than the chloroplast ribosomes (FIGURE 6). Sepsenwol[33] has measured the ribosomes in the periplastidal compartment of the colorless cryptomonad *Chilomonas paramecium* and found that they were the same size as the cytoplasmic ribosomes (220 Å) and distinctly larger than the 160 Å ribosomes of the cell's colorless plastid.

The most interesting structure in the periplastidal compartment of cryptomonads is a small double-membrane-limited body measuring 0.5 μm in width and 1.0 to 1.2 μm long, which Greenwood [19] has named a nucleomorph. The structure of the nucleomorph is illustrated in FIGURES 5, 7, 8 and 9. In cross section the double membrane limiting the organelle appears to be interrupted at places by pores (FIGURES 5 and 8, *arrows*), but a favorable tangential section through the limiting membranes of the nucleomorph shows that the pores are in fact slits (FIGURE 7, *arrows*).

Nucleomorphs invariably possess two characteristic inclusions. The first are dense granules, 300 to 500 Å in diameter which may be spherical or elongated (FIGURES 5, 7, 8, and 9). The second is a dense region of fibrillogranular material (FIGURES 5, 8, and 9) that Greenwood [18] noted looked like a nucleolus, but could also be a region of DNA and protein. FIGURE 8 is of a cell that was stained *en bloc*, i.e., just after fixing in osmium tetroxide, with 2% uranyl acetate at pH 4.0; whereas in the other micrographs, thin sections were stained with uranyl acetate and lead citrate. It can be seen that both the granules and the dense mass of fibrillogranular material are much more densely stained by uranyl acetate *en bloc*. Staining with uranyl acetate *en bloc* is known to increase the contrast of most cell components, but it particularly increases the contrast of DNA-containing structures, so it is possible that the dense fibrillogranular body or the granules may contain DNA.

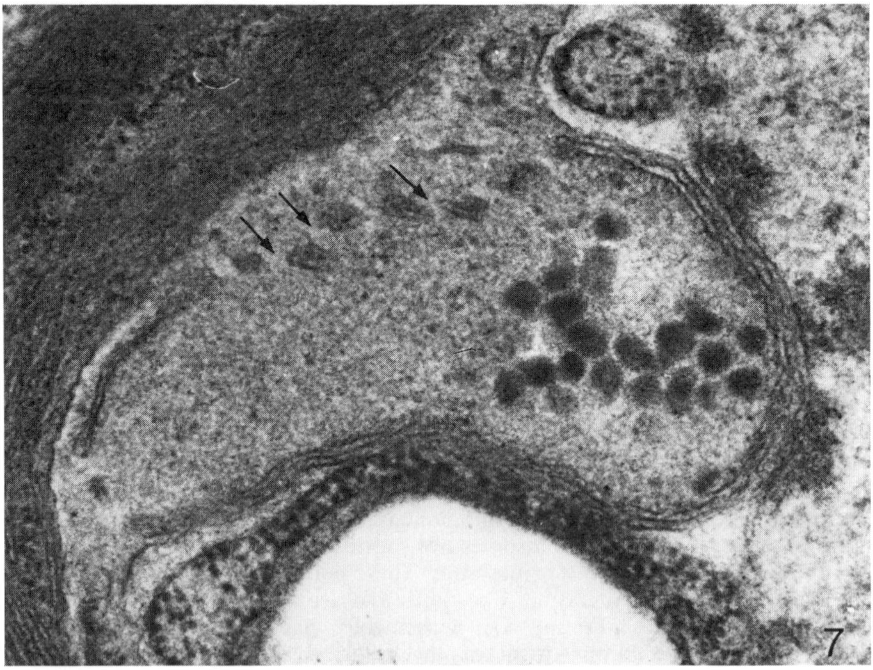

FIGURE 7. Nucleomorph of *Cryptomonas* sp. (Guillard's isolate ϕ). Slits (*arrows*) are present in the double-membraned envelope of the nucleomorph. ×95,700.

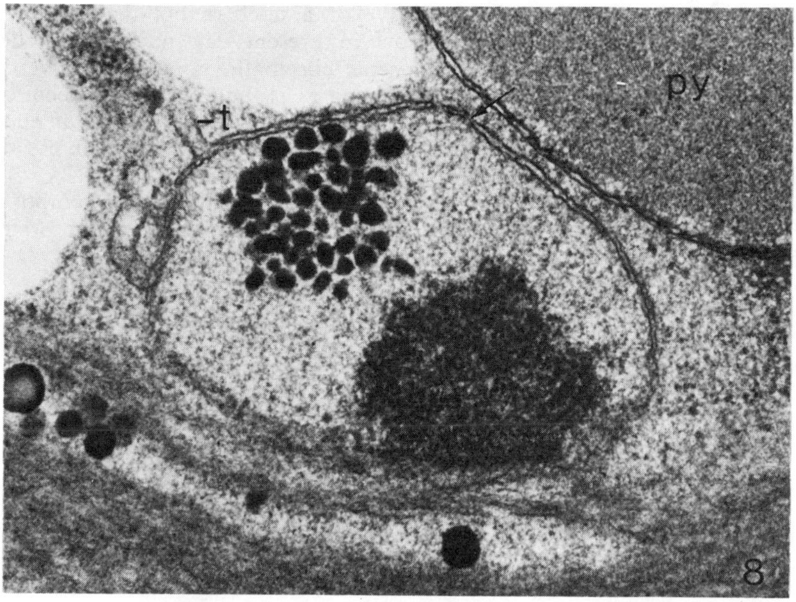

FIGURE 8. Nucleomorph of *Cryptomonas* sp. (Guillard's isolate ϕ). This cell has been stained *en bloc* with uranyl acetate. Note how much more darkly stained the dense granules and the fibrillogranular region of the nucleomorph are in this figure than they are in FIGURES 5, 7, and 9. In the latter figures only the sections have been stained with uranyl acetate. Note that the outer membrane of the nucleomorph is continuous with tubules (*t*) of the periplastidal compartment. The nucleomorph envelope is interrupted at the arrow. *py*, pyrenoid. ×61,650.

There is only a single nucleomorph per cell and this nucleomorph divides in two prior to prophase. Oakley and Dodge [31] have observed that in metaphase cells of *Chroomonas salina* there is a nucleomorph near each pole of the spindle. However, we have no information as yet on how nucleomorphs divide. We have studied nucleomorphs in six species of cryptomonads and they can be seen in the published micrographs of almost all species of cryptomonads that have been studied.[17] Depending on the species, the nucleomorph has a characteristic location in the cell. In a number of species, the nucleomorph sits in a groove in the pyrenoid (FIGURES 5 and 9). In one species, *Cryptomonas cryophila,* it sits in a groove in the nucleus but still inside the periplastidal compartment (Figure 17, in Reference 38). In other species, the nucleomorph lies along the inner surface of the chloroplast (FIGURES 7 and 8). In those species where the nucleomorph lies embedded in the pyrenoid, there is always a microbody in the cytoplasm lying close to the nucleomorph but separated from it by the two membranes of chloroplast ER (FIGURE 9).

Nucleomorphs are thus very distinctive organelles. What we are interested in is whether or not they represent the vestigial nucleus of a former eukaryotic symbiont. Some features suggest a nuclear origin; for example, the presence of only one per cell and its division at a specific time of the cell cycle. The double-membraned envelope is suggestive of a nucleus, but the slits in this envelope are unlike anything we know in present day nuclei. The dense fibrillogranular material possibly represents chromatin or perhaps a vestigial nucleolus. The dense granules are an enigma as is the fact that nucleomorphs may be associated with a microbody. Clearly, before we can decide if nucleomorphs are vestigial nuclei we need to know something about their chemical composition, particularly whether or not they contain DNA and RNA. Since their small size would make isolation and purification of nucleomorphs extremely difficult, we are presently employing electron microscopic autoradiography to determine whether nucleomorphs contain DNA and RNA.

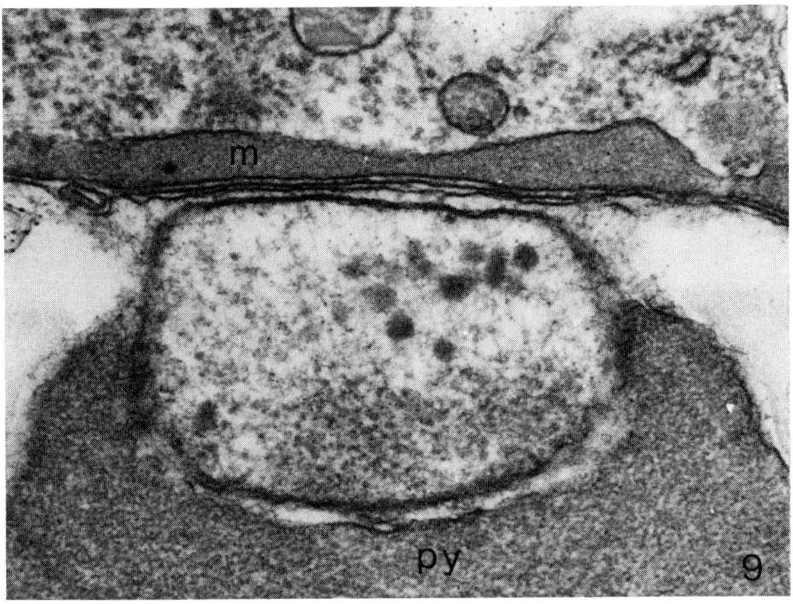

FIGURE 9. Nucleomorph of *Chroomonas salina.* A microbody (*m*) is present in the cytoplasm close to the nucleomorph separated from it by the two membranes of the chloroplast ER. Note that no ribosomes are present on the outer membrane of chloroplast ER where it lies close to the microbody. *py*, pyrenoid. ×61,650.

If nucleomorphs can be shown to be vestigial nuclei, then the chloroplast of cryptomonads would have evolved from the progressive reduction of a eukaryotic red alga. In contrast with *Euglena* and dinoflagellates where only the plasmalemma of the original symbiont remains, in cryptomonads all of the components of the periplastidal compartment would have once belonged to the symbiont—the nucleomorph, the ribosomes, the starch grains which in red algae today are located in the cytoplasm, and perhaps also the enclosing layer of the endoplasmic reticulum. Whatley [43] and Lee,[25] however, suggest that the inner membrane of the chloroplast ER is the plasmalemma of the original symbiont.

It is probable that all the other classes of algae with chloroplast ER have evolved from cryptomonads. In addition to possessing chloroplast ER, these algae like the cryptomonads possess chlorophyll *c* as well as chlorophyll *a* and it is unlikely that chlorophyll *c* evolved twice in evolution. There are several exceptions, some chrysophycean algae have only chlorophyll *a* as do the very few species of the Eugstigmatophyceae whose pigments have been examined,[4] but presumably these species have lost chlorophyll *c*. Thus if cryptomonad chloroplasts prove to be reduced red algal symbionts, then one can postulate that the chloroplasts in all the classes of algae with chloroplast ER have also evolved from reduced eukaryotic symbionts.

Acknowledgment

I wish to thank Dr. Marcelle Gillott for her astute suggestions and for a number of the micrographs in this paper.

References

1. Arnott, H. J. & R. M. Brown. 1967. Ultrastructure of the eyespot and its possible significance in phototaxis of *Tetracystis excentrica*. J. Protozool. **14:** 529–539.
2. Beale, G. H., A. Jurand & J. R. Preer. 1969. The classes of endosymbiont of *Paramecium aurelia*. J. Cell Sci. **5:**65–91.
3. Bisalputra, T. 1974. Plastids. *In* Algal Physiology and Biochemistry. W. D. P. Stewart, Ed.: 124–160. University of California Press. Berkeley, Calif.
4. Bold, H. C. & M. J. Wynne. 1978. Introduction to the Algae Structure and Reproduction. Prentice-Hall, Inc. Englewood Cliffs, N.J.
5. Christensen, T. 1964. The gross classification of algae. *In* Algae and Man. D. F. Jackson, Ed.: 59–64. Plenum Press. New York, N.Y.
6. Dodge, J. D. 1973. The Fine Structure of Algal Cells. Academic Press, Inc. New York, N.Y.
7. Dodge, J. D. 1974. A redescription of the dinoflagellates *Gymnodinium simplex* with aid of electron microscopy. J. Mar. Biol. Assoc. U.K. **54:**171–177.
8. Dodge, J. D. 1975. A survey of chloroplast ultrastructure in the Dinophyceae. Phycologia **14:**252–263.
9. Dodge, J. D. & B. T. Bibby. 1973. The Prorocentrales (Dinophyceae) I. A comparative account of fine structure in the genera *Prorocentrum* and *Exuviaella*. Bot. J. Linn. Soc. **67:**175–187.
10. Dodge, J. D. & R. M. Crawford. 1969. Observations on the fine structure of the eyespot and associated organelles in the dinoflagellate *Glenodinium foliaceum*. J. Cell Sci. **5:**479–493.
11. Dodge, J. D. & R. M. Crawford. 1971. A fine-structural survey of dinoflagellate pyrenoids and food-reserves. Bot. J. Linn. Soc. **64:**105–115.
12. Falk, H. & H. Kleinig. 1968. Feinbau und Carotinoide von *Tribonema* (Xanthophyceae). Arch. Mikrobiol. **61:**347–362.
13. Gibbs, S. P. 1962. Nuclear envelope—chloroplast relationships in algae. J. Cell Biol. **14:**433–444.
14. Gibbs, S. P. 1970. The comparative ultrastructure of the algal chloroplast. Ann. N.Y. Acad. Sci. **175:**454–473.
15. Gibbs, S. P. 1978. The chloroplasts of *Euglena* may have evolved from symbiotic green algae. Can. J. Bot. **56:**2883–2889.
16. Gibbs, S. P. 1979. The route of entry of cytoplasmically synthesized proteins into chloroplasts of algae possessing chloroplast ER. J. Cell Sci. **35:**253–266.

17. Gibbs, S. P. The chloroplast endoplasmic reticulum: structure, function and evolutionary significance. Int. Rev. Cytol. (In press.)
18. Greenwood, A. D. 1974. The Cryptophyta in relation to phylogeny and photosynthesis. Abstr. 8th Int. Cong. E.M., Canberra, vol. 2: 566–567.
19. Greenwood, A. D., H. B. Griffiths & U. J. Santore. 1977. Chloroplasts and cell compartments in Cryptophyceae. Br. Phycol. J. **12:**119.
20. Hawes, C. R. 1979. Ultrastructural aspects of the symbiosis between algal chloroplasts and *Elysia viridis*. New Phytol. **83:**445–450.
21. Jeffrey, S. W. & M. Vesk. 1976. Further evidence for a membrane-bound endosymbiont within the dinoflagellate *Peridinium foliaceum*. J. Phycol. **12:** 450–455.
22. Karakashian, S. J., M. W. Karakashian & M. A. Rudzinska. 1968. Electron microscopic observations on the symbiosis of *Paramecium bursaria* and its intracellular algae. J. Protozool. **15:**113–128.
23. Kivic, P. A. & M. Vesk. 1974. An electron microscope search for plastids in bleached *Euglena gracilis* and in *Astasia longa*. Can J. Bot. **52:**695–699.
24. Kowallik, K. 1969. The crystal lattice of the pyrenoid matrix of *Prorocentrum micans*. J. Cell Sci. **5:**251–269.
25. Lee, R. E. 1977. Evolution of algal flagellates with chloroplast endoplasmic reticulum from the ciliates. S. Afr. J. Sci. **73:**179–182.
26. Leedale, G. F. 1978. Phylogenetic criteria in euglenoid flagellates. BioSystems **10:**183–187.
27. Lewin, R. A. 1977. *Prochloron*, type genus of the Prochlorophyta. Phycologia **16:**217.
28. Lewin, R. A. & N. W. Withers. 1975. Extraordinary pigment composition of a prokaryotic alga. Nature (Lond.) **256:**735–737.
29. Melkonian, M. & H. Robenek. 1979. The eyespot of the flagellate *Tetraselmis cordiformis* Stein (Chlorophyceae): Structural specialization of the outer chloroplast membrane and its possible significance in phototaxis of green algae. Protoplasma **100:**183–197.
30. Mignot, J. P. 1970. Remarques sur le développement du reticulum endoplasmique et du système vacuolaire chez les Gymnodiniens. Protistologia **6:** 267–281.
31. Oakley, B. R. & J. D. Dodge. 1976. The ultrastructure of mitosis in *Chroomonas salina* (Cryptophyceae). Protoplasma **88:**241–254.
32. Round, F. E. 1965. The Biology of Algae. Edward Arnold. London.
33. Sepsenwol, S. 1973. Leucoplast of the cryptomonad *Chilomonas paramecium*. Exp. Cell Res. **76:**395–409.
34. Schulz-Baldes, M. & R. A. Lewin. 1976. Fine structure of *Synechocystis didemni* (Cyanophyta: Chroococcales). Phycologia **15:**1–6.
35. Silva, P. C. 1962. Classification of algae. *In* Physiology and Biochemistry of Algae. R. A. Lewin, Ed.: 827–837. Academic Press, Inc. New York, N.Y.
36. Stewart, K. D. & K. R. Mattox. 1975. Comparative cytology, evolution, and classification of the green algae with some consideration of the origin of other organisms with chlorophylls *a* and *b*. Bot. Rev. **41:**104–145.
37. Taylor, D. L. 1968. Chloroplasts as symbiotic organelles in the digestive gland of *Elysia viridis* (Gastropoda: Opisthobranchia). J. Mar. Biol. Assoc. U.K. **48:**1–15.
38. Taylor, D. L. & C. C. Lee. 1971. A new cryptomonad from Antarctica: *Cryptomonas cryophila* sp. nov. Arch. Mikrobiol. **75:**269–280.
39. Thinh, L.-V. 1979. *Prochloron* (Prochlorophyta) associated with the ascidian *Trididemnum cyclops* Michaelsen. Phycologia **18:**77–82.
40. Tomas, R. N. & E. R. Cox. 1973. Observations on the symbiosis of *Peridinium balticum* and its intracellular alga. I. Ultrastructure. J. Phycol. **9:**304–323.
41. Whatley, J. M. 1977. The fine structure of *Prochloron*. New Phytol. **79:** 309–313.

42. WHATLEY, J. M. 1980. Chloroplast evolution: ancient and modern. Ann. N.Y. Acad. Sci. This volume.
43. WHATLEY, J. M., P. JOHN & F. R. WHATLEY. 1979. From extracellular to intracellular: the establishment of mitochondria and chloroplasts. Proc. R. Soc. Lond. B. **204:**165–187.
44. WITHERS, N., W. VIDAVER & R. A. LEWIN. 1978. Pigment composition, photosynthesis and fine structure of a non-blue-green prokaryotic algal symbiont (*Prochloron* sp.) in a didemnid ascidian from Hawaiian waters. Phycologia **17:**167–171.

DISCUSSION OF THE PAPER

DR. J. A. SCHIFF: Is there any evidence from your work or from Jean Whatley's, from histochemistry or biochemistry, that differentiates the extra-membranes from the endoplasmic reticulum (ER) of the host?

Can the outermost membrane be distinguished in any way from the normal endoplasmic reticulum of *Euglena*?

DR. S. P. GIBBS: The main thing that distinguishes it from endoplasmic reticulum is that it never has ribosomes on it. Has there been any evidence that distinguishes it from smooth ER? No, there hasn't been.

DR. J. M. WHATLEY: About *Prochloron*, did you say that the membranes didn't overfold, or you hadn't seen them overfolding there?

DR. GIBBS: Yes I said that.

DR. WHATLEY: I think they do, I can send you some pictures if you like where I think the folding turns up.

DR. R. A. LEWIN: I believe that *Euglena gracilis* variety bacillaris is the only *Euglena* from which one can eliminate chloroplasts by treatment with streptomycin or heat. I believe it's the one that most people use from which they generalize to all of the *Euglena* phyta. Could I ask whether you have then looked for three membranes around the chloroplasts of other species of *Euglena*?

DR. GIBBS: I haven't myself but it's been done.

DR. J. SECKBACH (*Hebrew University, Jerusalem, Israel*): You propose that red algae invaded *Cryptomonads*. How can you explain that the plastids of *Cryptomonads* are not similar to the red algae?

DR. GIBBS: Red algae and Cryptomonal chloroplasts are different from each other. The big difference is that the phycobiliproteins are not on the thylakoids but in the lumen of the thylakoids, and also the thylakoids have paired. However, the *Cryptomonads* are the only group that has this predominance of paired thylakoids that puts them evolutionally close to the reds because reds have single thylakoids and then *Cryptomonads* have paired thylakoids, and then stacking becomes more complex as you go up the evolutionary tree. So I would say that chloroplasts have evolved once they got into the *Cryptomonads*.

DR. SECKBACH: How about the fact that the *Cryptomonads* don't have any phycobilisomes while the reds have?

DR. GIBBS: The *Cryptomonads* have phycobiliproteins inside thylakoids. I've heard recently that these phycobiliproteins are arranged inside the thylakoids in discrete structures that might be analogous to phycobilisomes.

DR. S. S. COHEN (*State University of New York, Stony Brook, N.Y.*): The usual answer to that question would be that modern red algae which you say have this kind of phycobilisome are not the kinds of red algae that Dr. Gibbs has within the *Cryptomonads*. In other words, there's a big evolutionary change. There's been an interesting development in polyamine biochemistry which will help to clarify some things about *Euglena*. I mentioned that certain thermophils and organisms that live in very acid waters have new kinds of polyamines and these have been exclusive of the old kinds, but *Euglena* is now found to have both, and within the last year several groups have reported that *Euglena* contains both thermine and spermine and the lower homologs of both of these compounds. It's quite possible that this is going to be revealed as a chloroplast or green algal function and this may help to separate these kinds of relationships.

DR. K.-P. KOLLER (*Harvard University, Cambridge, Mass.*): If you are talking about biliproteins you should really point out that the cryptophytes have a different kind of biliprotein than both the red and blue-greens. They differ remarkably in size and in the aggregation properties. This is a real difference between both groups.

THE APPLICATION OF DNA METHYLATION STUDIES TO THE ANALYSIS OF CHLOROPLAST EVOLUTION

Ruth Sager

Sidney Farber Cancer Institute
Boston, Massachusetts 02115

According to current hypotheses, chloroplast functions arose early in biological evolution, and for that reason alone, a detailed analysis of chloroplast DNA and the control of its expression should provide centrally important clues in the unraveling of early evolutionary history and phylogenetic relations. One mechanism of regulation that has recently become amenable to comparative studies is the methylation of DNA. This paper summarizes some of the current evidence about methylation functions, and describes the role of a coupled methylation-restriction system in the inheritance of chloroplast DNA. In addition, a new method we have developed for the detection of methylated residues in DNA is discussed.[1]

The most detailed analysis of the role of DNA methylation in a biological process comes from the work of Werner Arber and his colleagues on the degradation of phage DNA in some strains of bacteria, a process originally discovered by Luria[2] and called host modification and restriction. In a series of elegant genetic and biochemical studies from Arber's laboratory[3] it was established that bacteria contain specific restriction endonucleases capable of attacking DNAs at particular short sequences called recognition sites, unless the sequences are protected by methylation. In *E. coli,* pairs of enzymes that detect the same recognition site in DNA, one a restriction enzyme that cleaves the DNA at the site and the other a methyl transferase, act either to degrade DNA or to protect it by a secondary nonhereditary modification (i.e., methylation). The system operates, for example, to degrade phage DNA as it infects a bacterial cell by cleaving the DNA at a site that is methylated and therefore protected in the host, but not in the phage DNA.

In studies to be summarized in this paper, we have discovered in the eukaryotic alga *Chlamydomonas* a parallel system that is responsible for the maternal inheritance of chloroplast DNA. The restriction and methylation events occur in the chloroplast, which might be considered a prokaryotic-type organelle, but the system is regulated by the mating-type locus which is nuclear. Thus, a methylation-restriction system, previously described only in bacteria, is tightly integrated into both nuclear and chloroplast genomes in this eukaryotic microorganism.[4-8]

It seems likely that coupled methylation-restriction systems are present and will be found widely distributed not only in prokaryotes but also in eukaryotes, in specific situations in which particular DNA molecules are destroyed in the presence of homologous DNAs which are retained. The elimination of chromosomes from one parent but not from the other, which occurs in somatic and germ-line cells of particular genera of insects, is an example of a process that may be regulated by methylation and restriction as we have previously discussed.[9] Whether coupled methylation-restriction systems were significant in

0077–8923/81/0361–0209 $01.75/0

early evolution, and whether they may now provide clues to evolutionary relationships is clearly worth investigating.

In addition to coupled methylation-restriction systems, methylation *per se* was first proposed by Scarano [10] as a mechanism for regulation of transcription and thereby of developmental processes. In bacteria and some algae [11] both 6-methyl adenine (m^6A) and 5-methyl cytosine (m^5C) are present,[12] whereas in higher eukaryotes, m^5C is the only base that is methylated. Typically about 1–5% of bases are methylated in eukaryotic nuclear DNAs.[13] Chloroplast DNAs were reported as nonmethylated prior to our studies.

Reports of the m^5C content of mitochondria vary, presumably because of technical difficulties in the assays which, until recently, were performed by chemical methods or by radioisotope labeling *in vivo,* followed by isolation and purification of mitochondrial DNA. In mammalian cells, values of 1.5–3.0 mole % m^5C have been reported for mitochondrial DNA; [14] these values are in the same range as those for nuclear DNAs. More recently using the *Msp*I/*Hpa*II method described below, no methylated CCGG sequences were detected in mitochondrial DNAs from yeast, *Neurospora*, rat, or calf.[15] Also, no m^5C was detected in mitochondrial DNAs from human, frog, or *Paramecium*.[16, 17]

Recently, a simple method has been developed to scan isolated DNA fragments on gels to look for methylation at a particular sequence, CCGG, which is cleaved by the restriction enzyme *Msp*I [18] whether the site is methylated or not, whereas it is not cleaved by the isoschimeric enzyme *Hpa*II when methylated in the form C^mCGG.[19, 20] By comparing a pair of DNA preparations, one digested with *Msp*I and the other with *Hpa*II, it is easy to detect missing bands in the *Hpa*II digest that result from methylation. With this method DNAs from various tissues or developmental stages can be compared to look for differences in methylation of individual sites.[21–24]

This method, though simple and powerful, is limited to examining those recognition sites for which pairs of isoschimeric enzymes are available. This is an important limitation, as shown for instance by the fact that only some 20% of the methylated sites in the satellite I DNA of calf thymus are CCGG sites.[1] Because of this limitation, we were motivated to seek a more general method for the detection of *all* methylated bases in DNA regardless of sequence. The new method we have developed meets this requirement.

This method utilizes antibodies raised in rabbits against m^5C by a procedure first developed by Erlanger and Beiser.[26] They showed that specific antibodies could be developed against individual bases in DNA, and we have confirmed their results with antibodies directed against m^5C that do not recognize the nonmethylated cytosine.[1] In our method, DNA fragments that have been transferred from gels to nitrocellulose paper [27] or to DBM paper [28] are treated with antibody under conditions that promote antibody binding. Subsequently, the DNA-antibody complexes are further reacted with ^{125}I-labeled goat anti-rabbit IgG to enhance the sensitivity of the detection.[1]

The DNA fragments containing m^5C to which antibodies have bound are then visualized by autoradiography. The sensitivity of the method is such that individual m^5C residues can be detected in as little as 40–100 ng of DNA. Thus, the method makes possible the identification of individual methylated sites in DNA in the size range of single genes. The method was developed for eukaryotes and therefore is directed to the identification of m^5C, but for prokaryotes, the same method could be developed with antibodies directed against 6-methyl adenine, which is the major methylated base in bacteria. In addition,

m^5C (but not m^6A) can be detected in DNA sequencing gels[29] prepared by the Maxam and Gilbert method.[30]

In bacteria, some regulatory functions of DNA methylation have been identified and others proposed primarily from studies of *dam* mutants of *E. coli,* deficient in adenine methylation. Their DNA contains single-stranded breaks.[31] The *dam* gene codes for a methylating enzyme,[32] and strains carrying this mutation are hypermutable and recombination rates are high.[33] To explain this finding, it has been proposed that the methylation of adenine acts to discriminate between old and newly synthesized DNA strands in a repair mechanism which has been called methylation-instructed mismatch repair.[33] Methylation does not appear to be involved in UV-induced damage and repair however, since *dam* mutants are not more sensitive to UV than are wild-type cells.[33] In the single-stranded phage ϕX174, a single methylated cytosine serves as a signal during replication for a restriction endonuclease which cuts the newly synthesized DNA into fragments of genomic length.[34] This methylated base is present in virion DNA but not in the double-stranded replicative form (RF).[35]

In eukaryotes, m^5C is found abundantly in repetitive DNAs of mammals,[36] but not *Drosophila.*[37] Variations in the m^5C content of DNA has been reported from tissue to tissue and in different developmental stages.[38] Evidence that methylation levels affect gene expression has come, for example, from growing erythroleukemic Friend cells in the presence of ethionine, dimethylsulfoxide, or butyrate, which leads to hypomethylation, and stimulates the expression of the globin genes.[39]

Methylation has been implicated in the regulation of animal virus gene expression. In *Herpesvirus saimiri* infection of lymphoid cells, for example, viral DNA sequences present in cell DNA and as isolated episomes are extensively methylated, as shown by the *Msp*I/*Hpa*II method, in some cell lines which do not produce detectable virus, but unmethylated in virus-producing lymphoid lines.[40] Adenovirus type 12 (Ad12) DNA sequences integrated into nuclear DNA of transformed hamster cells are extensively methylated, whereas Ad12 DNA from virions is not methylated detectably using the *Msp*I/*Hpa*II method. Regions of the integrated viral genome that are being expressed as mRNA are undermethylated compared with adjacent viral segments (late regions) not being expressed.[41]

Maternal Inheritance

In higher plants, chloroplast and mitochondrial DNAs are transmitted to progeny only (or mainly) from the female parent in the sexual life cycle, a phenomenon called maternal inheritance. In the past it was widely assumed that the mechanism of maternal inheritance was simply the presence of abundant cytoplasm containing organelles in the female (egg) parent, and the absence of male cytoplasm in the sperm entering the egg at the time of fertilization. However, when maternal inheritance of chloroplast genes was discovered in *Chlamydomonas reinhardi,*[42] an isogamous species contributing equal cytoplasm from the two parental gametes in zygote formation, it became evident that a biochemical mechanism of maternal inheritance was required, at least for this species.

The first clues to the nature of this mechanism came from studies in which the buoyant densities of chloroplast DNAs from vegetative cells, gametes, and zygotes were examined in CsCl gradients in the analytical ultracentrifuge.[5] We made two key observations: (1) Chloroplast DNA from zygotes underwent a density shift of about 0.005 g/cm^3, becoming lighter than chlDNA from vegetative cells. (2) In reciprocal crosses between parental cells which had been prelabeled to distinguish the DNAs from the two parents by density ($^{15}N \times ^{14}N$ and $^{14}N \times ^{15}N$), the chlDNA of male (mt$^-$) origin was preferentially lost within the first few hours after zygote formation. We suggested that the density shift could be explained by methylation, but the direct evidence was not forthcoming until improved methods became available to detect DNA methylation.

The first direct evidence that chloroplast DNA of female (mt$^+$) origin is methylated in zygotes, whilst that of male origin is not methylated but rather is degraded, came from studies in which parental cells were prelabeled with [^{3}H]deoxycytidine in pairs of reciprocal crosses: female labeled × male unlabeled and the reverse. As previously reported[6] we found that [^{3}H]m^{5}C of female origin could be recovered from zygote DNA that was extracted, hydrolyzed, and bases separated on an Aminex-A6 column by high pressure liquid chromatography (HPLC). However, when the male parent was prelabeled, little or no [^{3}H]m^{5}C was found. These results directly confirmed our methylation hypothesis.

Further evidence about the degradation of the chlDNA of male origin came from pairs of reciprocal crosses in which one of the parents was prelabeled with [^{3}H]thymidine. It may be recalled that thymidine is incorporated into DNA of the *Chlamydomonas* chloroplast but not into nuclear DNA, providing a very convenient label for chloroplast events.[43] The results of such an experiment are shown in FIGURE 1, representing DNAs recovered from CsCl gradients. At T_0 just after gametes have fused to form zygotes, chlDNAs of both parents are present. After six hours, however, the chlDNA of male origin is virtually gone, whereas that of female origin is still present in approximately the same amount. However, as previously noted, it has undergone a density shift to a lighter buoyant density in CsCl.[6]

At this point it was already clear that both methylation and restriction had occurred, but with the availability of more refined methods to examine methylation, we pursued the investigation further. ChlDNAs were extracted from vegetative cells, from gametes, and from zygotes, and their restriction fragment patterns were compared after digestion with the enzymes *Msp*I and *Hpa*II. The results of those experiments, previously published[7] confirmed the earlier results and provided further detailed information. We found no differences in restriction fragment patterns of *Msp*I and *Hpa*II digests from vegetative cells of the two mating types, male and female, indicating no detectable methylation at CCGG sites in vegetative cells.

The male gametes showed little evidence of methylation but the female gametes were heavily methylated, as judged by the few restriction bands seen after *Hpa*II digestion. Most of the chlDNA was near the top of the gel, poorly cut by *Hpa*II, but the *Msp*I digest of the same DNA resembled that of the vegetative cells. Thus, we found that considerable CCGG methylation was occurring in gamete differentiation in the female but not in the male. ChlDNA from 6-hour zygotes proved to be more highly methylated than that of female gametes, as shown by the fact that the zygote DNA was refractile to cleavage not only by *Hpa*II, but also by *Msp*I, by *Eco*RI and by *Bam*HI. Thus, it appears that there are two waves of methylation, one in the female gametes and a second one after zygote formation.

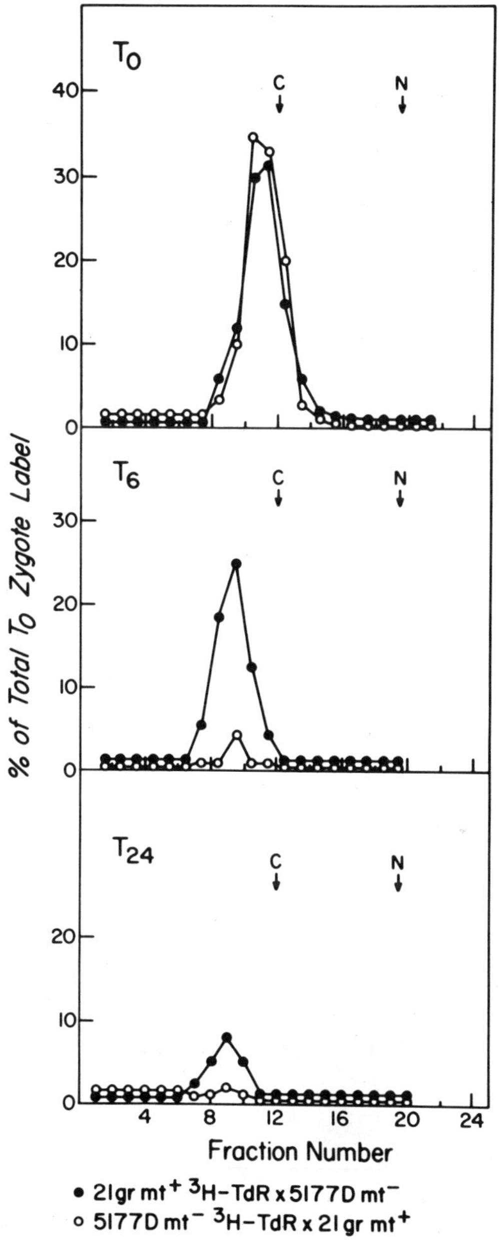

FIGURE 1. Buoyant density shift and loss of male (mt⁻) chloroplast DNA in zygotes. DNA was extracted from replicate cultures in which strain 21gr (female) was prelabeled with [^{3}H]thymidine and fused with unlabeled 5177D (male) (●); and the reciprocal experiment in which 5177D chloroplast DNA was prelabeled (○). Zygotes were sampled immediately after fusion (Top), 6 hr later (Middle), and 24 hr later (Bottom). Carrier cells were added, and DNA was extracted and fractionated in two successive CsCl gradients. Absorbance tracing of carrier DNA (not shown) from second CsCl gradient identifies position of chloroplast (C) and residual nuclear (N) peaks and provides the basis for estimating buoyant densities.

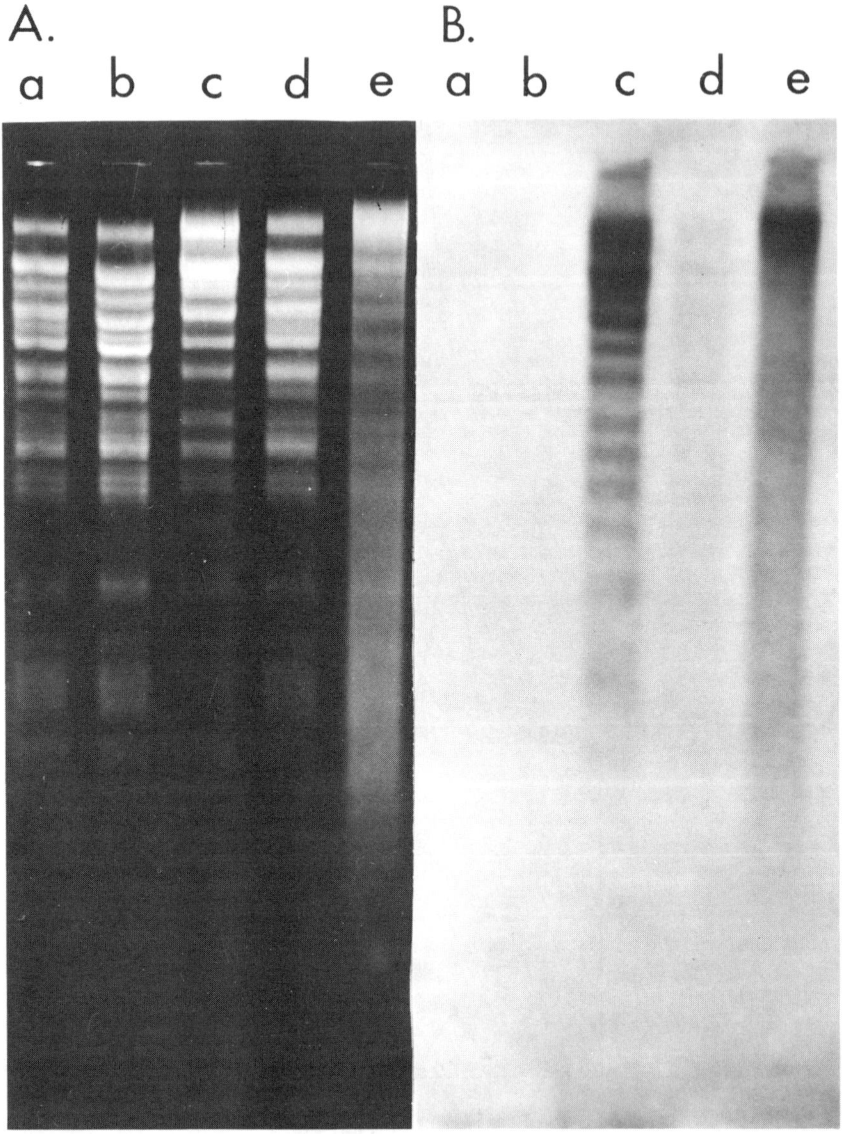

FIGURE 2. 50 μl reaction mixtures containing 0.1 M Tris-HCl, pH 7.6, 50 mM NaCl, 5 mM $MgCl_2$, 2 μg DNA and 25 units *Eco*RI were incubated at 37° C for 6 hr. Samples were electrophoresed at 100 V for 17 hr. DNAs were visualized by ethidium bromide staining (A) and m^5C was identified by antibody-binding (B). (a) DNA from female vegetative cells, (b) DNA from male vegetative cells, (c) DNA from female gamete cells (d) DNA from male gamete cells, and (e) DNA from zygote cells.

A question not resolved by these experiments concerns the distribution of the methylated bases. Are they restricted to localized regions containing runs of CCGG? To answer this question, as well as others, our new methylation method [1] described above was applied to this problem. As in the previous work, chlDNAs were prepared from vegetative cells, from gametes, and from 6-hour zygotes, digested with *Eco*RI and electrophoresed on agarose gels. The DNA fragments were then transferred to nitrocellulose paper by the Southern method [27] and incubated with rabbit antibody directed against m^5C and then with ^{125}I-labeled goat-anti-rabbit antibody.[1]

An example of the results is shown in FIGURE 2A, in which the DNAs were visualized by ethidium bromide staining of the gel; and in FIGURE 2B, in which the locations of m^5C residues were detected by autoradiography of the same preparation after transfer to nitrocellulose paper and incubation with ^{125}I-labeled antibody. No methylation was seen in the vegetative cells of either mating type, and only traces of methylation were found in the chlDNA from male gametes. In the chlDNAs from female gametes and from zygotes, however, the DNA was very heavily methylated.

These results confirm and extend our previous findings, showing (1) that many if not all *Eco*RI fragments are already methylated in the female gametes; and (2) that further methylation occurs in the zygote. The *Eco*RI fragments have been previously mapped.[44] Therefore, we are able to conclude that both waves of methylation, in the female gametes and in the zygotes, affect DNA sequences located all around the genome, rather than being restricted to a particular region.

Concluding Remarks

Let us turn now to a general consideration of DNA methylation in organelle evolution. As yet, the literature contains only hints of the multiple roles of this regulatory mechanism in present-day organisms, let alone in evolution. We now have available simple and precise new methods for the detection of DNA methylation, and a rapidly growing list of biological processes in which methylation is being shown to play a role. With the current explosion of new ideas and new techniques in the field of molecular genetics, it is now possible to ask new sets of questions and explore evolutionary and phylogenetic relationships in novel ways. It is to be anticipated that the application of these methods to key organisms will contribute to unraveling some of the puzzles posed but not resolved in this Symposium.

References

1. Sano, H., H.-D. Royer & R. Sager. 1980. Proc. Natl. Acad. Sci. USA (In press).
2. Luria, S. E. 1953. Cold Spring Harbor Symp. Quant. Biol. **18**:237.
3. Arber, W. 1974. Prog. Nucleic Acid Res. Mol. Biol. **14**:1–35.
4. Sager, R. 1972. Cytoplasmic Genes and Organelles. Academic Press, Inc. New York, N.Y.
5. Sager, R. & D. Lane. 1972. Brookhaven Symp. Biol. **23**:495–505.
6. Burton, W., C. T. Grabowy & R. Sager. 1979. Proc. Natl. Acad. Sci. USA **76**:1390–1394.

7. Royer, H.-D. & R. Sager. 1979. Proc. Natl. Acad. Sci. USA **76:**5794–5798.
8. Sager, R. 1979. Methylation and restriction of chloroplast DNA: The molecular basis of maternal inheritance in Chlamydomonas. *In* Extrachromosomal DNA. D. J. Cummings, P. Borst, I. B. Dawid, S. M. Weissman & C. F. Fox, Eds. Vol. **15:**97–111. Academic Press, Inc. New York, N.Y.
9. Sager, R. & R. Kitchin. 1975. Science **189:**426–433.
10. Scarano, E. 1967. *In* Comprehensive Biochemistry. M. Florkin & E. H. Stotz, Vol. **28:**55. Elsevier American Publishing Co., Inc. New York, N.Y.
11. Hattman, S., C. Kenny, L. Berger & K. Pratt. 1978. J. Bacteriol. **135:**1156–1157.
12. Pakhomova, M. W., G. N. Zaitseva & A. N. Belozerskii. 1968. Dok. Akad. Nauk SSSR **182:**712–714.
13. Burdon, R. M. & R. L. P. Adams. 1969. Biochem. Biophys. Acta **174:**322–329.
14. Kirnos, M. D. & B. F. Vanyushin. 1976. Biokhimiya **41:**68–74.
15. Groat, G. S. P. & A. M. Kroon. 1979. Biochem. Biophys. Acta **564:**355–357.
16. Dawid, I. B. 1974. Science **184:**80–81.
17. Cummings, D. J., A. Tait & J. M. Goddard. 1974. Biochim. Biophys. Acta **374:**1–11.
18. Waalwijk, C. & R. A. Flavell. 1978. Nucleic Acids. Res. **5:**3231–3236.
19. Garfin, D. E. & H. M. Goodman. 1974. Biochem. Biophys. Res. Comm. **59:** 108–116.
20. Mann, M. B. & H. O. Smith. 1977. Nucleic Acids Res. **4:**4211–4221.
21. Singer, J., J. Roberts-Ems & A. D. Riggs. 1979. Science **203:**1019–1021.
22. Singer, J., J. Roberts-Ems, F. L. Luthardt & A. D. Riggs. 1979. Nucleic Acids Res. **7:**2369–2385.
23. Kaput, J. & T. W. Sneider. 1979. Nucleic Acids. Res. **7:**2303–2321.
24. Waalwijk, C. & R. A. Flavell. 1978. Nucleic Acids Res. **5:**4631–4641.
25. McGhee, J. D. & G. D. Ginder. 1979. Nature. **280:**419–420.
26. Erlanger, B. F. & S. M. Beiser. 1964. Proc. Natl. Acad. Sci. USA **52:**68–74.
27. Southern, E. M. 1975. J. Mol. Biol. **98:**503–517.
28. Alwine, J. C., D. J. Kemp, B. A. Parker, J. Reiser, J. Renart, G. R. Stark & G. M. Wahl. 1979. Methods Enzymol. **68:**220–242.
29. Ohmori, H., J. Tomizawa & A. M. Maxam. 1978. Nucleic Acids Res. **5:**1479–1486.
30. Maxam, A. M. & W. Gilbert. 1977. Proc. Natl. Acad. Sci. USA **74:**560–564.
31. Marinus, M. G. & N. R. Morris. 1974. J. Mol. Biol. **85:**309–322.
32. Brooks, J. E. & S. Hattman. 1978. J. Mol. Biol. **126:**381–394.
33. Glickman, B., P. Van Den Elsen & M. Radman. 1978. Molec. Gen. Genet. **163:**307–312.
34. Friedman, J., A. Friedman & A. Razin. 1977. Nucleic Acids Res. **4:**3483–3496.
35. Razin, A., J. W. Sedat & R. L. Sinsheimer. 1970. J. Mol. Biol. **53:**251–259.
36. Vanyushin, B. F., S. G. Tkacheva & A. N. Belozersky. 1970. Nature (London) **225:**948–949.
37. Rae, P. M. M. & R. E. Steele. 1979. Nucleic Acids Res. **6:**2987–2995.
38. Vanyushin, B. F., A. L. Mazin, V. K. Vasilyev & A. N. Belozersky. 1973. Biochim. Biophys. Acta **299:**397–403.
39. Christman, J. K., P. Price, L. Pedrinan & G. Acs. 1977. Eur. J. Biochem. **81:**53–61.
40. Desrosiers, R. C., C. Mulder & B. Fleckenstein. 1979. Proc. Natl. Acad. Sci. USA **76:**3839–3843.
41. Sutter, D. & W. Doerfler. 1980. Proc. Natl. Acad. Sci. USA **77:**253–256.
42. Sager, R. 1954. Proc. Natl. Acad. Sci. USA **40:**356–363.
43. Swinton, D. C. & P. C. Hanawalt. 1972. J. Cell Biol. **54:**592–597.
44. Rochaix, J. D. 1978. J. Mol. Biol. **126:**597–617.

Discussion of the Paper

FLOOR: How do you make the antibody?

DR. R. SAGER: By immunizing rabbits with 5-methyl-C. Then we use a two-layer system of ^{125}I-labeled goat anti-rabbit IgG, and probably if one did a third level you could get even more sensitivity. But I think it's going to be very powerful and very useful.

DR. DAVIS (*University of Rochester, Rochester, N.Y.*): I noticed you had a peak wave of uracil on the graphs of your cell.

DR. SAGER: I'm not sure whether it's uracil or not, it's something that we have been wanting to follow up.

DR. DAVIS: I've looked at nuclear DNA several times and it has both methyl *c* in small amounts and methyl *a*. It seems to be strain-dependent but reproducible from strain to strain.

DR. SAGER: There are several reports in the literature of that. Our strain doesn't seem to have it.

DR. J. LOCKER: You have two DNAs that were heavily methylated by the antibody method, one of which cut and one which didn't cut with *E. coli.* It would seem to me that you would have to conclude that the thing that's inhibiting the *E. coli* one is not the methylation.

DR. LOCKER: One of the methylated DNAs cuts quite nicely with *E. coli*-1.

DR. SAGER: The other one is the zygote DNA which I discussed previously, so we have a lot of evidence about the fact that *Eco* RI does not cut chloroplast DNA, that is zygote DNA.

DR. LOCKER: You also showed inhibition of *Msp* I in some of these chloroplast preparations.

DR. SAGER: That's right and *Eco* RI and *Bam* HI and every enzyme that we've looked at.

DR. LOCKER: But *Msp* I as I understand it does not inhibit, it cuts the methylated sites.

DR. SAGER: No, it cuts only CCGG sites and the point that I was trying to make was that in addition to CCGG sites, cytosine is present in a large number of other sites which are also methylated. Other enzymes in which one hasn't previously described inhibition by methylation are in fact inhibited by methylation.

DR. LOCKER: I think that's very likely and the data are very clear that there is methylation. What I'm wondering is do you have other phenomena that are also inhibiting restriction enzymes besides simply the methylation?

DR. SAGER: If you are talking about the presence of protein or something like that, I can tell you we think we've gotten rid of it all.

DR. W. M. BROWN: If the mechanism is really a restriction in the nuclease type of mechanism in *Chlamydomonas* it has to be a non–site-specific kind of cleavage because otherwise you'd see a specific degradation pattern and you don't. You don't see discrete bands being produced.

DR. SAGER: I think it's both. I think that initially it is site specific and then, subsequently those fragments are rapidly chewed up. I think that one of the things that distinguishes our strain from some other strains in which there are many more recombinations seen, is that probably there are strains in which that later degradation is slower. In ours it's so fast we can't catch the fragments. And we're looking for strains in which it's slower.

DR. N. W. GILLHAM: Before Dr. Sager published her recent paper, we thought that the methylation that we were seeing in this special strain, which is streptomycin-dependent and in mating-type minus, might be related to a question of maternal inheritance. We now think however, that the methylation pattern is one of these other examples of methylation that is not related directly to the maternal inheritance pattern that Dr. Sager is talking about, in zygotes. In other words it's a different sort of methylation.

DR. F. CASTORA: If the *Msp* cuts in methylated sequence, what happens if one of the C's is methylated, if two of the C's are methylated, or if the whole thing is methylated?

DR. SAGER: It's only CC-GG, it's the internal C, the methylation of the internal C which has been studied. I don't actually know of any studies of methylation of the first C. It's the internal C that's the important one that's blocked in Hepa-II.

ON THE FOLDING AND ORGANIZATION OF CHLOROPLAST DNA IN *CHLAMYDOMONAS REINHARDTII* *

Kwen-Sheng Chiang, Edward Friedman, Michael J. Malavasic, Jr., Ming-hsien Lin Feng, Eva M. Eves, Teng-Yung Feng, and David C. Swinton

Departments of Biophysics and Theoretical Biology, Biology, and the Committee on Genetics
The University of Chicago
Chicago, Illinois 60637

Introduction

Certain molecular-biochemical approaches have been utilized fruitfully to elucidate the origin and evolution of intracellular organelles. Direct sequence homology analysis of specific proteins, certain easily obtainable RNA species, and even the total mitochondrial DNA genome have generated considerable incisive data concerning the evolutionary origins of organelles. Such data constitute, in fact, one of the major foci of this conference. We feel that another useful molecular-biochemical approach to the investigation of the evolutionary history of organelles is to study the folding and organization of the cytoplasmic DNAs. Packaging of a long DNA strand into a relatively compact structure is a universal biological phenomenon. The sophistication involved in such folding may have advanced in parallel with the trend, throughout the course of evolution, towards an increase in DNA sequence complexity.

The primary folding unit of eukaryotic nuclear DNA is the nucleosome, whose octet structure has been the subject of much investigation.[7, 15, 22] Higher orders of organization of the primary folding by sequential helical winding have also been postulated for both interphase chromatin and the metaphase chromosomes of higher eukaryotes.[1, 16, 38] In prokaryotes, the genomic DNA is also condensed into an easily recognizable, compact *in situ* structure, the nucleoid. In the case of the best-understood prokaryotic nucleoids, i.e., the *E. coli* membrane free nucleoid, little protein other than RNA polymerase, is found to be associated with the nucleoid. Under certain isolation conditions the *E. coli* DNA binding protein, HU, can also be found associated with the nucleoid.[35] This protein has been shown to organize SV40 viral DNA *in vitro* into a beaded structure.[44] A similar beaded structure of condensed *E. coli* DNA has also been detected under certain specific experimental conditions.[21]

These results raise the question of the exact relationship between the familiar protein-poor nucleoid[31] and the beaded fiber structure of condensed *E. coli* DNA as isolated by Griffith.[21] Whatever the answer to this question may ulti-

* Supported by the U.S. Public Health Service (Grant HD 05804). Partially supported by the National Science Foundation (Grant 78–09815, Department of Biophysics and Theoretical Biology Electron Microscopy Facility), and by the National Institutes of Health (Grants GM–07183, M.J.M., Jr.'s traineeship and GM–00090, E.M.E.'s traineeship).

0077–8923/81/0361–0219 $01.75/0 © 1981, NYAS

mately be shown to be, it seems reasonable to conclude that the overall folding and higher order organization of nuclear DNA in present day eukaryotes are distinctly different from those of present day prokaryotes. Very little, however, is known about the folding and organization of organelle DNA except that a DNA-binding protein has been isolated from yeast mitochondria.[6, 23]

In this paper, we would like to report on some new observations concerning the folding and organization of the chloroplast DNA in *Chlamydomonas reinhardtii*. It is hoped that these observations will shed some light on the structure of the organelle nucleoid and the evolution of eukaryotic intracellular organelles.

BACKGROUND ON *C. reinhardtii* CHLOROPLAST DNA

A DNA species with a buoyant density in CsCl distinctly lower than the nuclear (nc-) DNA was first shown to have been enriched in a crude chloroplast preparation of *C. reinhardtii* and was assumed to be the chloroplast (cp-) DNA.[37] This same chloroplast enriched DNA species, $\rho_{CsCl} = 1.695 \pm 0.002$ gm/cm^3, also appears as the largest distinct satellite in a CsCl gradient fractionation of whole cell DNA and has been referred to as the β-satellite.[10] The β-satellite has been shown to replicate semiconservatively and out of phase with the replication of nc-DNA.[8]

Since it is difficult to isolate intact and uncontaminated *C. reinhardtii* chloroplasts in quantities suitable for the biochemical-molecular characterization of the cp-DNA, most of the subsequent studies of cp-DNA in this organism have been done using the β-satellite isolated from whole cell DNA by repeated density gradient centrifugation. The β-satellite isolated in this manner has, up to this time, been generally accepted as the sole constituent of the chloroplast genome, while the DNA species that constitute the major band, α, in the CsCl gradient, i.e., those species having a peak density of 1.723 ± 0.002 gm/cm^3, were assumed to comprise the nc-DNA *in toto*. It has recently been shown, however, that DNA extracted from a relatively pure chloroplast preparation isolated by a substantially improved chloroplast isolation technique, contains two additional small, yet distinct, satellites that have buoyant densities in CsCl higher than that of β-satellite.[13] At this moment, it is no longer certain whether the β-satellite is the only DNA species present in the chloroplast. Our preliminary evidence suggests that two minor DNA species, $\epsilon(\rho_{CsCl} = 1.733)$ and $\tau(\rho_{CsCl} = 1.722)$, which we have found to be associated with the β-satellite, when isolated by our improved isolation procedure, may also be present in the chloroplast.[17] The verification of this suggestion awaits confirmation and will require the development of a chloroplast isolation procedure that insures the removal of all possible extrachloroplast DNA contamination from the chloroplast preparation.

Few circular cp-DNA molecules have been observed and no photographs have been published, although the observation of a limited number of 65 μm circular cp-DNA molecules has been reported.[3] The questions of whether the great majority of the cp-DNA molecules exist as such naked DNA circles *in vivo,* and whether the β-satellite is the sole constituent of the cp-genome remain to be answered. Recently, a circular restriction map has been constructed for the β-satellite using DNA samples purified by two sequential CsCl density gradient centrifugations.[34] Hence, it may even be possible that the two minor satellites, ϵ and τ, if they indeed exist inside the chloroplast, might be related to the packaging or folding of the *in situ* chloroplast nucleoid (see CONCLUDING REMARKS).

DNA CONTENT AND DIMENSIONS OF THE CHLOROPLAST NUCLEOID, *In Situ*

Ris and Plaut [33] were the first to demonstrate the presence of one or more DNA containing bodies in the chloroplast of *C. moewusi.* They accomplished this by staining with Feulgen reagents and acridine orange and by electron microscopy on thin sections.

The highly specific DNA fluorochrome, 4′-6′-diamidino-2-phenylindole (DAPI), has been successfully used to stain mitochondrial DNA in yeast [42] and chloroplast DNA *in situ* in a number of algal species.[11, 12] Using a procedure specifically adapted to *C. reinhardtii* grown under our experimental conditions, we have determined the number and dimensions of the non-nuclear, deoxyribonuclease (DNase)-sensitive, ribonuclease (RNase)-resistant DAPI fluorescent bodies. Since there are several mitochondria per cell and the mitochondrial DNA content of *C. reinhardtii* is less than 5% of the cp-DNA content,[32] these intensely DAPI-fluorescent large bodies are presumed to represent the chloroplast nucleoids (ND). The number of NDs per cell varies among different strains and within a given strain.[14] Some of these variations may be related to the growth stage within the cell cycle or to the developmental stage within the life cycle. The details of these variations will be reported elsewhere. The haploids and heterozygous diploids average 4 NDs per vegetative cell (FIGURE 1), while a homozygous diploid averages about 8 NDs per cell.[14] The apparent size of the *in situ* DAPI-stained cp-ND also varied considerably (from 0.3–0.7 μm). This was not surprising since a number of factors, such as the stage of a particular cp-ND in its divisional or metabolic cycle, the geometry and the degree of flatness of the fixed cell, the saturation of the stain, the excitation and emission conditions of the stain, the exposure duration, and other photographic conditions can alter the photographic image of the cp-ND. Some of these variables would probably cause an underestimation of the *in vivo* dimensions of the cp-ND while the others an overestimation. For the sake of simplicity, only those groups with the larger sizes will be used in the later discussion. The average larger cp-ND observed under our experimental conditions measured roughly 0.55 μm in diameter. It is not yet clear whether the cp-ND adapts a spherical or discoidal form *in vivo.*

In order to compare the size of the cp-ND with that of the chromosomes, cells were grown under the same conditions as utilized for the DAPI staining experiments and were stained with Giemsa stain. As shown in FIGURE 2, the *in situ* Giemsa-stained metaphase chromosomes appeared as discs rather than rods and had diameters ranging from 0.45 to 1.05 μm. There is general agreement in the literature that the *Chlamydomonas* metaphase chromosomes in the light microscope are of a disclike morphology.[4, 14, 26] This point has been further confirmed by our recent scanning electron microscopic observations of isolated metaphase chromosomes of *C. reinhardtii.*[24] Hence, it is apparent that the cp-ND and the metaphase chromosome have dimensions that are of the same order of magnitude *in vivo.*

Our earlier results have shown that *C. reinhardtii* contains 1.23×10^{-7} μg or 7.40×10^{10} daltons of DNA per cell, of which nuclear DNA constitutes

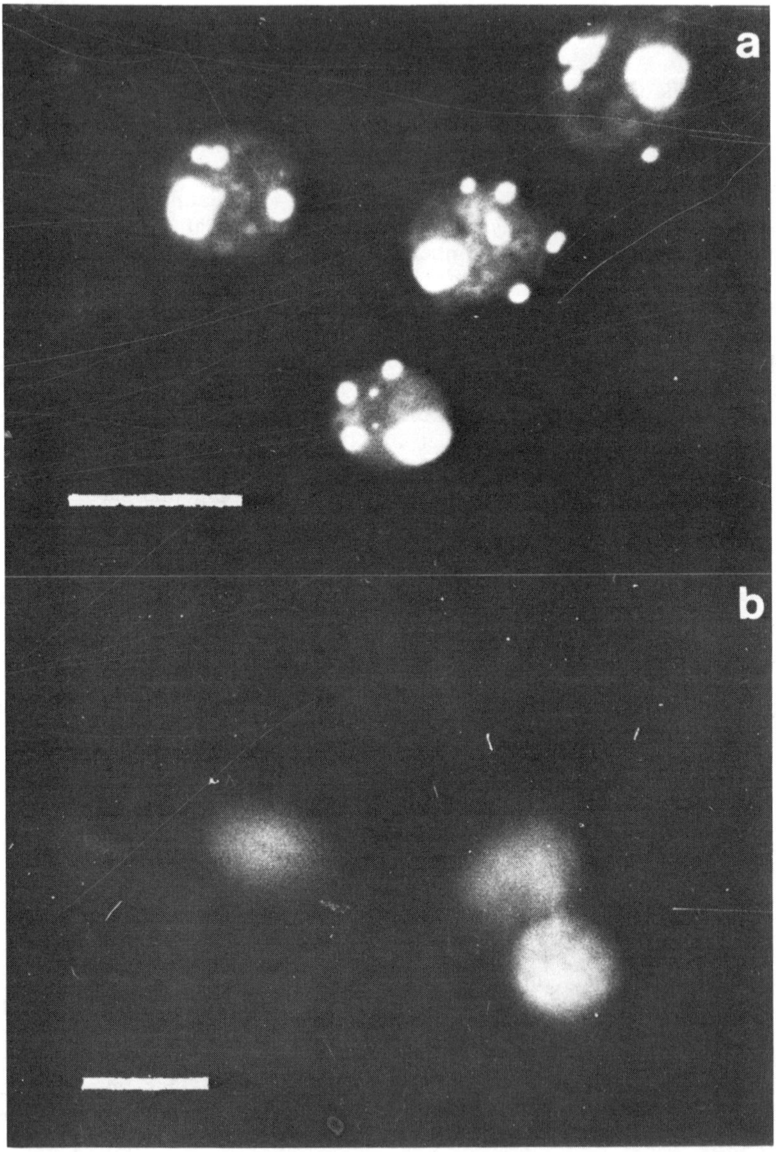
a
b

approximately 85% and the cp-DNA or the β-satellite about 14%.[8] Utilizing rate constants from cp-DNA reassociation reactions,[2, 41] the genomic complexity of the β-satellite has been estimated to be 1.94×10^8 daltons. The electron microscopic[3] and restriction mapping[34] data indicate that the basic unit of the β-satellite is approximately 65 μm in length. This converts to a molecular weight of 1.3×10^8 daltons which is in reasonable agreement with the values obtained from DNA reassociation kinetics data. TABLE 1 summarizes pertinent comparisons between the nc-chromosome and cp-ND systems in *C. reinhardtii.*

It is obvious from the data presented in TABLE 1 that the cp-DNA must be highly compacted, perhaps by a factor of the same order of magnitude as that of the metaphase chromosomes, since the *in situ* dimensions and DNA content of the chloroplast nucleoids are comparable to those of the metaphase chromosomes in this organism. Implicit in this argument is the idea that with the proper experimental conditions one should be able to isolate and identify a compact and tightly folded chloroplast DNA structure.

THE STRATEGY AND GENERAL EXPERIMENTAL APPROACH

In order to study the folding and organization of the cp-DNA, it is imperative that it be isolated under a set of experimental conditions that satisfy the requirements listed below. The basic idea is to keep the isolated cp-ND in a state as close to the "native state," i.e., the *in vivo* structure, as possible. (1) All the endogenous nuclease activity must be inhibited. Such activities may come not only from *bona fide* nucleases but also from the associated and/or reverse reaction activities of other nucleic acid metabolic enzymes. (2) All possible structural constituents of the cp-ND must be protected from degradation. In addition to protecting the DNA, RNA, and protein components of the cp-ND, the protection of other possible polymeric constituents, e.g., lipids or polysaccharides, must also be considered when devising an isolation procedure. (3) The covalent and noncovalent binding as well as other types of interactions among the structural elements of the cp-ND must be preserved or at least isolated in a state that is not significantly altered.

FIGURE 1. Morphology and size of chloroplast nucleoids *in situ*. *C. reinhardtii* cells were fixed in methanol:acetic acid (3:1) at room temperature for 30 minutes and were dropped onto slides. The cells were then stained with DAP1 (4′-6′-diamidino-2-phenylindole dihydrochloride) (Boehringer-Mannheim) at a concentration of 0.2 μg/ml in distilled water for 10 minutes at room temperature. The slides were rinsed once with distilled water and air dried. After moistening the slides with a drop of 0.2 M Na_2HPO_4 solution (pH 11.0) they were sealed by wax. A Leitz Orthoplan microscope, equipped with a 100× oil immersion NPL fluotar, an epi-illuminating objective (n.a.=1.31–0.60), a 200W mercury arc lamp, a broad band UV filter (System A), TK400 dichroic mirrors, and an Orthomat-W automatic camera, was used for observation and photography. Exposed Kodak Tri-X film was developed in Kodak D-76:water (1:1) for 10 minutes at 20° C. Original magnification 414X. (a) A group of DAP1-stained 137C plus mating-type vegetative cells. Bar represents 10 μm. (b) Deoxyribonuclease control. Fixed cells on slides were digested with deoxyribonuclease I (Worthington code D at 0.25 mg/ml in 2 mM $MgSO_4$ pH 6.0) at 35° C for 4 hours and rinsed with distilled water prior to staining. Bar represents 5 μm.

The optimization of a set of conditions favorable to the isolation of "native state" cp-ND is laborious and elusive for lack of any *a priori* experimental parameters that might serve as a basis for such a task. It is, therefore, necessary to rely upon an empirical approach to strike an acceptable compromise between purity and "nativeness." We found achieving an acceptable compromise to be a

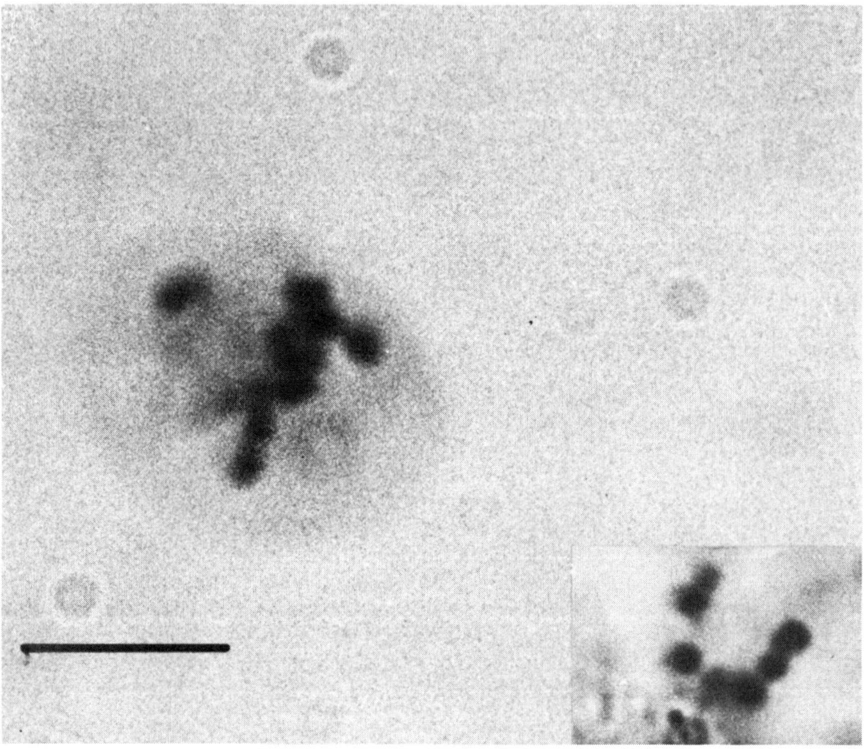

FIGURE 2. Morphology and size of metaphase chromosomes. *C. reinhardtii* cells fixed on glass slides as described in FIGURE 1 were stained with Gurr's Giemsa (1:10 in McIlvaine's Buffer pH 6.8) at room temperature for 4–5 minutes. Chromosomes are so identified and photographed only when their arrangement in a cell conformed to Buffaloe's[4] description of metaphase in *Chlamydomonas*. Photographic conditions were the same as described in FIGURE 1 except phase and bright field optics instead of the epi-fluorescent system were employed, i.e., with a NPL Fluotor Phaco 3 100× oil immersion objective (n.a.=1.32), and a 12V-100W tungsten halogen illumination lamp. The insert at lower right shows another frequently observed arrangement of metaphase chromosomes. Original magnification 414×. Bar represents 5 μm.

particularly acute problem since our initial control experiments revealed the presence in *C. reinhardtii* of powerful and previously unsuspected endogenous nuclease activities, which appeared to be tightly bound to the folded cp-DNA structure. For instance, a single wash of pelleted *C. reinhardtii* cells in cold standard saline citrate solution (SSC) reduced by more than 50% the amount

TABLE 1

A COMPARISON OF COMPACTION AND SOME OTHER PROPERTIES AMONG CHLOROPLAST PROKARYOTIC AND EUKARYOTIC NUCLEAR DNAs

DNA species / Parameters	*E. coli* (Stationary phase)	*Chlamydomonas reinhardtii*		Human Nucleus
		Chloroplast	Nucleus	
Relative cellular DNA content, %	–	~14	~85	–
Total DNA content, pg	4.15×10^{-3}	1.73×10^{-2}	1.04×10^{-1}	5.50
Total DNA content, daltons	2.5×10^{9}	1.04×10^{10}	6.29×10^{10}	3.31×10^{12} *
Kinetic complexity of unique DNA sequence, daltons	2.5×10^{9}	1.3×10^{8}	4.5×10^{10}	–
Presence of repetitive DNA sequences	no	no	yes	yes
Presence of snap-back DNA sequences	yes	yes	yes	yes
Reiteration of unique DNA molecules per cell or organelle	1	80	1	2
Number of nucleoids/metaphase chromosomes per cell	1	4 †	8	23×2
DNA molecule per nucleoid chromosomes or chromatids	1	20	1	1
DNA content per nucleoid/chromosome or chromatid, daltons	2.5×10^{9}	2.6×10^{9}	7.86×10^{9} ‡	7.12×10^{10} * §
Length of naked DNA fiber per nucleoid/chromosome or chromatid, μm	1,250	1,300	3,930 ‡	35,600 * §
Estimated *in situ* length of the longest dimension of nucleoid/chromosome or chromatid, μm	~0.45 ¶	~0.55 †	~0.77 **	~3.5 * §
Apparent DNA fiber contraction factor	2.77×10^{3}	2.36×10^{3}	5.62×10^{3} ‡	1.02×10^{4} §

* See Sedat and Manuelidis.[38]
† Only those larger and darker stained ones are counted.
‡ The DNA content of all eight chromosomes was assumed to be the same.
§ For chromosome 1 only.
¶ Estimated from Ryter and Chang.[36]
** Average value from the eight chromosomes.

of [³H]dThymidine-labeled cp-DNA recovered (see below). Due to the presence of these endogenous nuclease activities, a radical modification of the conventional DNA or nucleoid isolation procedures was found to be necessary in order to preserve the folding and organization of the cp-DNA in *C. reinhardtii.*

After a great deal of trial and error, the following basic experimental approach was developed: (1) Efficient nuclease inhibitors and other agents, which prevent significant disturbance of the binding between cp-DNA and other cp-ND constituents, are added to the cultures prior to harvesting and lysing of the unwashed cell pellet. (2) The cp-DNA is separated from the soluble fraction of the lysate immediately after lysing the cells in the presence of nuclease inhibitors and other agents to prevent the dissociation of condensed cpDNA fibers. (3) All subsequent isolation and characterization of the cp-DNA are performed as quickly as possible after the initial fractionation.

Extreme Lability of Some Chloroplast DNA Sequences or Species

Since exogenously supplied thymidine (dThd) is selectively incorporated into the β-satellites, but not into α DNA during the vegetative cell cycle of *C. reinhardtii*,[9, 40] [³H]-labeled dThd was used to label the cp-DNA in order to serve as a readily measurable indicator of cp-DNA recovery. This technique allows the quantitation of cp-DNA recovery during the various stages of the isolation procedure and under many different experimental conditions. It was an unexpected finding that the recovery of [³H]-labeled alkaline-resistant acid precipitable counts, which were taken to represent cp-DNA molecules, varied widely with different experimental conditions. For instance, under an identical DNA isolation procedure, a routine washing of the pelleted cells by SSC prior to resuspension and lysis would reduce the recovery of [³H]dThd by about 50% of unwashed values (Figure 3). Utilizing the DNA isolation conditions described in the legend to Figure 3, it was found that in addition to the familiar β-satellite, a comparable amount of [³H]dThd-labeled DNA appeared in a peak with a density near that of the α DNA species traditionally attributed to the *nc*-DNA. This [³H]dThd-labeled DNA could be converted *in vitro,* by shearing, to a species with a buoyant density in CsCl of normal β-satellite (Figure 4). Whether this *in vitro* converted DNA species is identical to the traditional β-satellite is currently under investigation. The available evidence suggests that the species of [³H]dThd-labeled DNA with a buoyant density near the α DNA peak is a cp-DNA species with an unusual conformation, possibly a partially unfolded cp-ND. The results presented in Figure 4 serve as the first indication that modification of the conventional DNA isolation procedure will allow the isolation of cp-DNA containing entities, as identified by [³H]dThd label, in configurations different from the naked cp-DNA molecule(s).

Perhaps even more striking was the fact that a significant increase in [³H]dThd-labeled DNA recovery was obtained by implementing drastic alterations in the conventional Proteinase K DNA isolation procedure (Figures 5a and 5b). It was found that by eliminating the washing step and incorporating nuclease inhibitors, sulfhydryl blocking agents, and high salt concentration from the very beginning of the cp-DNA isolation procedure, the recovery of [³H]-dThd-labeled DNA increased several fold (Figures 5c and 5d). It is of interest

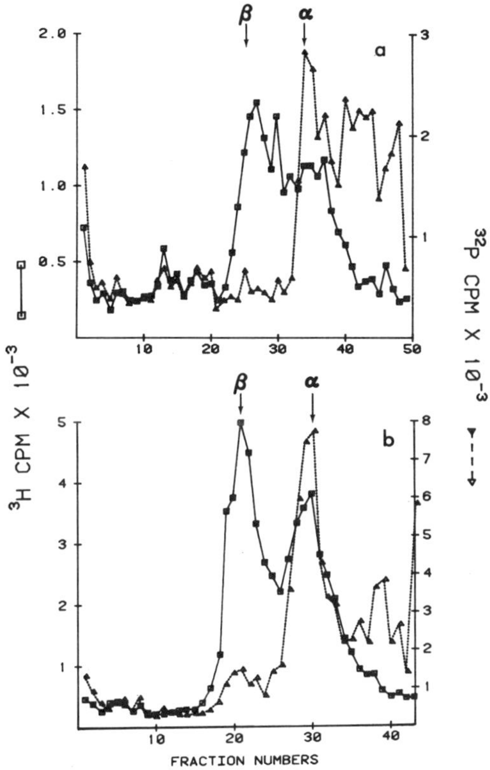

FIGURE 3. The effect of cell-wash on the recovery of [^{3}H]thymidine-labeled chloroplast DNA. (a) A 100 ml vegetative culture of a cell wall-less strain CW15 was grown (in a 125 ml Elenmyer flask) with a low-phosphate minimal medium (Wilson and Chiang[43]) at 25° C with constant stirring and aeration (0.5% CO_2 in air) under an alternating light (at an intensity of 135 μE m^{-2} s^{-1}) and dark regime of 12 hours each. At inoculation (4–8 $\times$ 10^4 cells/ml) this culture was labeled with methyl-^{3}H-thymidine (sp. act.=0.36 Ci/mM) and orthophosphate-^{32}P (carrier-free) at a final activity of 72 μCi/ml and 1.5 μCi/ml, respectively. When this synchronized culture reached 2.21 $\times$ 10^6 cell/ml, a 5 ml aliquot was harvested at L1, i.e., one hour after the onset of the light period, by centrifugation (2000 $\times g$, 5 min. at 4° C). The pelleted cells were resuspended in cold SSC (0.15 M NaCl, 0.015 M sodium citrate, pH 7.0) and then centrifuged as before. After this SSC wash step the cell pellet was resuspended in 0.14 ml of RS-1 (10 mM Tris-HCl, pH 8.0, 10 mM EDTA-Na_2, 50 mM N-ethyl-maleimide 0.25 M Sucrose, and 100 μg/ml of proteinase K) and incubated for 60 min. at 0° C. Triton X-100, Sarkosyl N.L. 97, and Sodium dodecyl sulfate were then added to a final concentration of 2.5, 2, and 2% respectively. After achieving a complete lysis (within a few minutes) the lysate was incubated at 8° C for 16 hours. Solid CsCl (4.2 g) and a diluting solution, DS-1 (10 mM Tris-HCl, 10 mM EDTA-Na_2, pH 8) were added to the lysate to adjust the refractive index to 1.4010–1.4015 with a final volume of approximately 4.45 ml. The density gradient centrifugation was effected by sedimenting this solution (in a 12 ml Oak Ridge tube) in a Sorvall OTD-2 ultracentrifuge with a Beckman 65 rotor at 40,000 RPM for 40 hours at 20° C. After collecting, each fraction was first treated with 1 N NaOH at 80° C for 60 min. The labeled DNA was then precipitated with 1 N HCl (final concentration) in the presence of carrier DNA. Following collection and drying on a Whatman GF/C filter the radioactivity of the DNA was determined in a scintillation counter. (b) Same as (a) except the SSC wash step was omitted.

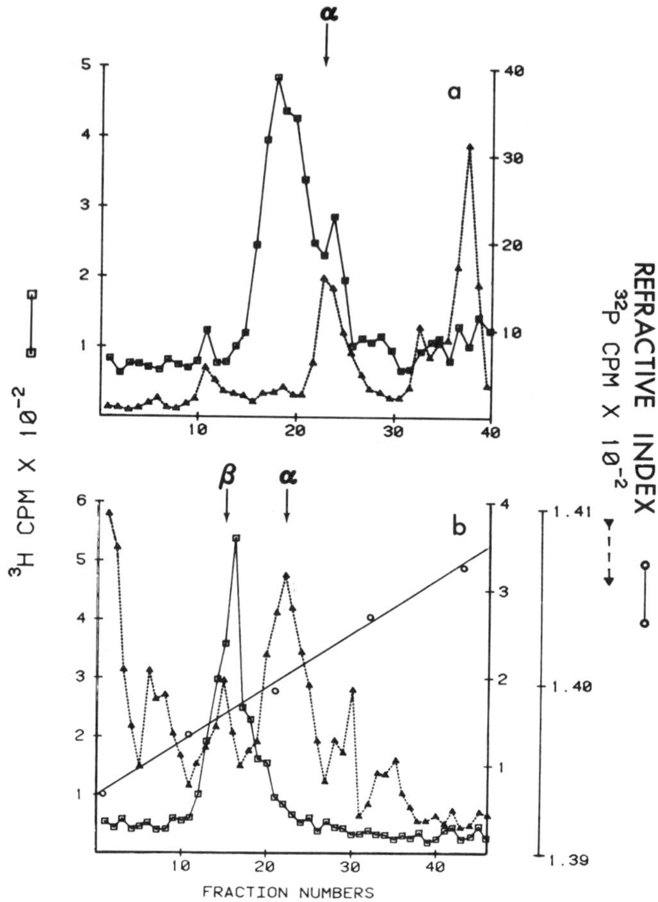

FIGURE 4. *In vitro* conversion of a higher density [³H]thymidine labeled DNA species to β satellite. (a) The DNA and supramolecular DNA species were isolated and fractionated in CsCl gradient under the same experimental condition as FIGURE 5c except that 10 mM dithiothreitol instead of p-hydroxymercuribenzoate was present in the cell pellet resuspension solution and that cell density at harvest was 2.8×10^6 per ml. Only a portion of the even numbered fractions was precipitated and counted as described in FIGURE 3a. (b) Fractions 24 to 30 of (a) were pooled, sheared (30 passes through a 2½ inch long 30 gauge needle) at room temperature and rerun in a CsCl gradient.

to note that this exceptionally high recovery of the [³H]dThd-labeled DNA did not result from the increased recovery of [³H]dThd-labeled β-satellite but rather from the recovery of a previously unencountered [³H]dThd-labeled species. This new species is presumably a cp-DNA containing entities with a very high buoyant density in a CsCl gradient (>1.76 g/cm³). This high density entity, as practically defined by its alkaline-resistant, acid precipitable counts (FIGURE 5c and 5d) will hereafter be referred to as the [³H]dThd-labeled high density components of cp-DNA or HDC. For a given experiment and under otherwise identical conditions, the recoverable amount of [^{3}H]dThd-labeled HDC varied dramatically depending primarily upon the pH and composition, i.e., the presence of nuclease inhibitor and sulfhydryl reagents, of the initial cell resuspension buffer (FIGURE 5, 6, and 7 legends). The growth stage within the cell cycle was another factor that led to variation in [^{3}H]dThd-labeled HDC recovery. The details of this variability and some of the properties of HDC will be documented below and elsewhere.[17] For the purpose of our present discussion, it is sufficient to point out that certain [^{3}H]dThd-labeled and presumably cp-DNA species or sequences were extremely labile (see below). Unless a combination of unconventional measures were taken during DNA isolation such DNA species could not be detected. The most likely explanation

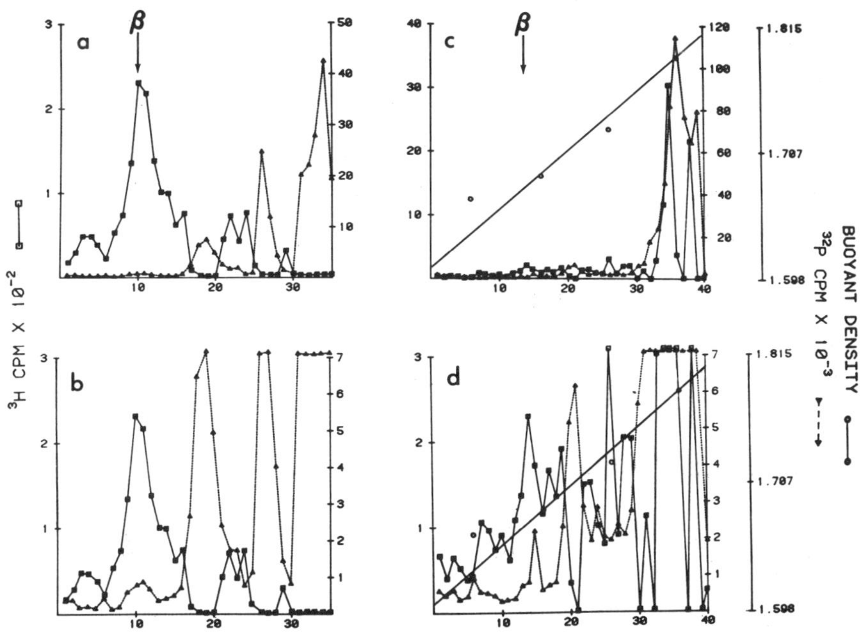

FIGURE 5. Recovery of [^{3}H]thymidine-labeled chloroplast DNA as a function of isolation conditions. (a) The growth, labeling, and harvesting conditions were the same as described in FIGURE 3b (no SSC wash step) except that lower [^{3}H]thymidine (0.5×) and higher ^{32}P (3.34×) were used. The pelleted cells from a 2 ml aliquot of a culture at 2.72×10^6 cell per ml was resuspended in RS-2 (0.1 M Tris-HCl, 0.01 M EDTA-Na$_2$, 0.25 M sucrose, and 100 μg/ml Proteinase K, pH 8.12) and immediately lysed by the addition of Triton X-100 Sarkosyl and sodium dodecyl sulfate to 2.5, 2, and 2%, respectively. The subsequent incubation of the lysate, CsCl gradient fractionation, and radioactivity counting were the same as described in FIGURE 3a. (b) A replot of (a) with expanded scale for ^{32}P. (c) Same as (a), except the cell pellet resuspension Solution RS-3 (10 mM Tris-HCl, 10 mM EDTA-Na$_2$, 0.25 M sucrose, 5 mM aurin tricarboxylic acid (ATA), 10 mM p-hydroxymercuribenzoate pH 6.17) and the CsCl gradient diluting solution DS-3 (100 mM Tris-HCl, 10 mM EDTA-Na$_2$, 5 mM aurin tricarboxylic acid, pH 8.07) were different. (d) A replot of (c) with expanded scales for both ^{3}H and ^{32}P.

for this phenomenon is that our newly developed DNA/ND isolation procedures somehow prevent or retard the spontaneous degradation of certain [^{3}H]dThd-labeled cp-DNA species by endogenous nuclease activities during the normal DNA isolation process. It is not known at present whether the exogenously supplied [^{3}H]dThd was incorporated randomly into all the available deoxythymidine (T) sites of the cp-DNA or whether a bias might exist that would favor certain special sequences. Possible bias in dThd incorporation and the nature of the endogenous nuclease activities are currently under investigation.

A [^{3}H]Thymidine-Labeled High Density Chloroplast-DNA Containing Component

As seen in Figure 5, under the experimental conditions that gave the highest recovery of alkaline-resistant acid-precipitable [^{3}H]-dThd counts, the majority of such [^{3}H]dThd-labeled DNA-containing component was not located at the β-satellite density region of a CsCl gradient but at an unusually high density, i.e., $\rho_{CsCl} > 1.78$ g/cm^3. The nature of this [^{3}H]dThd-labeled HDC was briefly examined. Since the HDC banded near the bottom of specially designed CsCl gradients which possessed unusually high starting densities (Figure 6a), its buoyant density in CsCl cannot be accurately determined. Cesium sulfate gradient centrifugation was then employed to fractionate a [^{3}H]dThd-labeled HDC sample isolated from the [^{3}H]dThd-labeled HDC peak taken from the highest density region obtainable from a fractionated CsCl gradient (Figure 6a). The Cs_2SO_4 gradient was so designed that after sedimentation equilibrium had been reached, the concentration of Cs_2SO_4 near the bottom of the gradient approached saturation under the experimental conditions. It was striking that the [^{3}H]dThd-labeled HDC again banded near the bottom of such a gradient and had an apparent density in Cs_2SO_4 greater than 1.69 gm/cm^3 (Figure 6b). Since no presently known biological macromolecule, in any conformation or shape, possesses a density in Cs_2SO_4 close to HDC, this observation was rather surprising. At least two possible explanations for the HDC behavior in Cs_2SO_4 gradients exist: (1) the HDC possesses a unique and exceedingly compact conformation, and/or (2) the high concentration of Cs_2SO_4 or CsCl required to prepare the density gradients caused the HDC to precipitate out of solution. To ascertain whether the buoyant density of HDC was related to its inherent conformation, the intercalating dye, propydium iodide (PI), was incorporated in the CsCl equilibrium gradient analysis of HDC. As shown in Figure 7a, the density of the major [^{3}H]dThd-labeled DNA containing component was dramatically reduced in the presence of PI; i.e., from a $\rho > 1.78$ without the dye to 1.60 g/cm^3 in CsCl with the dye added. When the fractions of the PI-CsCl gradient containing the major [^{3}H]dThd peak were pooled, subjected to dye removal by isopropanol/isoamyl alcohol extraction, and rebanded in a CsCl gradient, the usual behavior of HDC reappeared, i.e., it virtually pelleted to the bottom of the gradient (Figure 7b). It is evident from the above data that the buoyant density of the HDC species is at least partially related to their conformation.

The sedimentation behavior of [^{3}H]dThd-labeled HDC in isokinetic sucrose gradients is also rather unusual. As seen in Figure 8, at low DNA concentration, i.e., $\sim$1 μg whole cell DNA in a centrifuge tube with a 25 mm diameter, the [^{3}H]HDC exhibited an exceedingly high apparent sedimentation coefficient

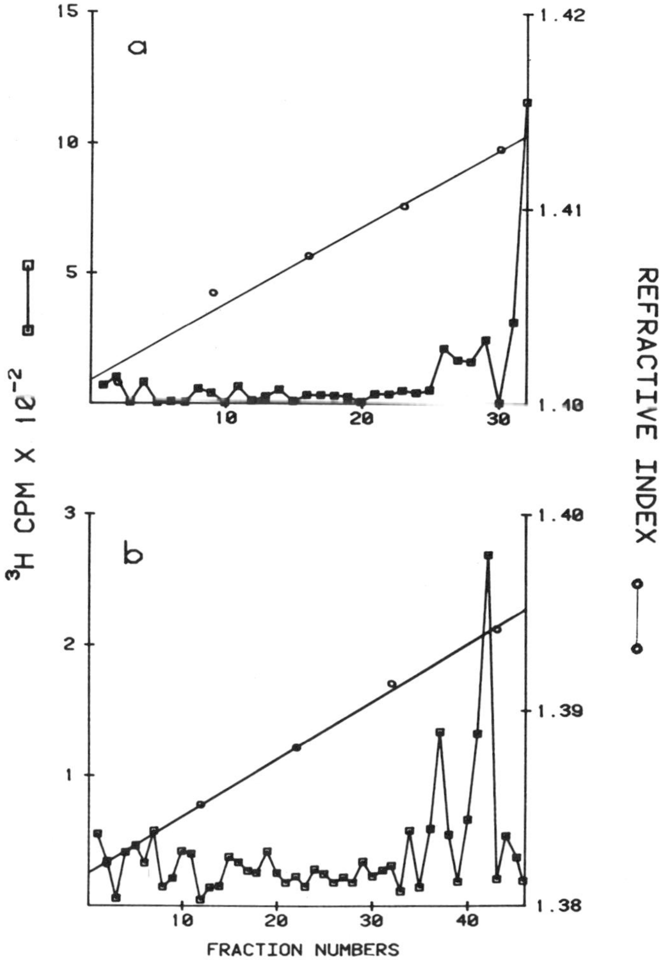

FIGURE 6. Fractionation of [^{3}H]thymidine-labeled supramolecular chloroplast DNA as in a cesium sulfate gradient. (a) The growth and labeling conditions were the same as described in FIGURE 3a. At L1, 0.16 ml of 0.25 M ATA was added to a 40 ml aliquot of culture (3.5×10^6 per ml), which was then harvested by centrifugation ($2000 \times g$, 5 min, 22°C). The pelleted cells were resuspended in 0.27 ml of RS-4 (10 mM Tris-HCl, 10 mM EDTA-Na_2, 0.25 M sucrose, 2 mM ATA, and 100 μg/ml Proteinase K) and lysed by adding 0.03 ml of 20% sarkosyl at room temperature. After an addition of 0.9 ml of a saturated CsCl solution (in DS-1) the lysate was incubated at 37°C for 60 min. Conditions for CsCl gradient fractionation and radioactivity determination were the same as described in FIGURE 3a. Only a portion of each collected fraction was precipitated and counted. (b) Fractions 31 and 32 from (a) were pooled and 1/50 volume was rerun in a Cs_2SO_4 gradient with a starting density of 1.68 (i.e., at a refractive index of 1.3875). Other conditions of the Cs_2SO_4 gradient fractionation were the same as described in FIGURE 3a for CsCl gradient.

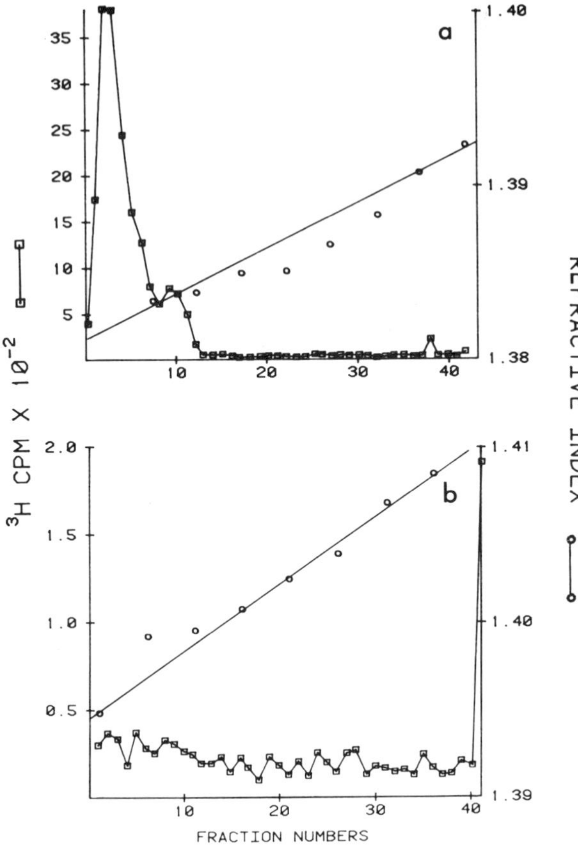

FIGURE 7. Fractionation of [³H]thymidine pulse-labeled supramolecular chloroplast DNA by propydium iodide-cesium chloride density gradient centrifugation. (a) The growth conditions were the same as in FIGURE 3a. At the L0 (onset of the light period) [³H]thymidine was added to a 50 ml culture with a cell density of 8.58×10^5 per ml. This pulse-labeled culture was harvested at L6 by centrifugation $4000 \times g$, 10 min., 25°C). The pelleted cells were resuspended in RS-3 and lysed as describable in FIGURE 6a. The conditions for CsCl gradient fractionation were the same as in FIGURE 3a except propydium diiodide was included in the 7 ml gradient at a final concentration of 0.5 mg/ml and the refractive index was adjusted to 1.3824. (b) Fractions 0 to 4 from (a) were pooled and extracted exhaustively with isopropanol and isoamyl alcohol to remove propydium iodide. Solid CsCl and DS-1 were then added. The mixture was then subjected to a regular CsCl gradient fractionation as described in FIGURE 3a.

of approximately 30,000 *s*. It is not clear at present whether the possible solubility problem of HDC was in any way related to this extremely high sedimentation coefficient obtained under our experimental conditions. This possibility is not very likely, however, due to the fact that our preliminary experiments have indicated that the sedimentation behavior of the [³H]dThd-labeled HDC could be altered by the addition of a low concentration of ethidium bromide (2.0–5.0 μg/ml) to the sucrose gradient.[17]

The [^{3}H]dThd-labeled HDC can be converted *in vitro* to an apparent β-satellite species in a CsCl gradient by RNase treatment, although considerable loss of the [^{3}H]dThd label accompanies this conversion process. Prolonged storage of the HDC in the cold will also convert a portion of the [^{3}H]dThd counts from HDC to an apparent β-satellite species in a CsCl gradient. Here, again, there is a considerable loss of [^{3}H]dThd counts. Additionally, we have found during the course of our evaluation of the recovery of [^{3}H]dThd-labeled DNA that both the recovery of ^{3}H-labeled HDC and the ratio of recovered ^{3}H-labeled HDC to β-satellite showed fluctuations at different times during the synchronized cell cycle.[17] We are currently investigating the reasons for and the nature of these losses and fluctuations in recovery of [^{3}H]dThd DNA and HDC counts. It is possible that such fluctuations could have resulted from alterations in the susceptibility of [^{3}H]dThd-labeled cp-DNA species or certain sequences therein to endogenous nuclease activities or changes in the endogenous nuclease activities themselves during the different stages of the cell cycle. It is also possible that these two factors might have a synergistic effect upon one another or that they may have risen or fallen in concert with changes in cp-DNA metabolism during the cell cycle.

Structural Studies by Electron Microscopy

The [^{3}H]dThd-labeled HDC isolated under our experimental conditions appeared to possess a combination of properties that are distinctly different

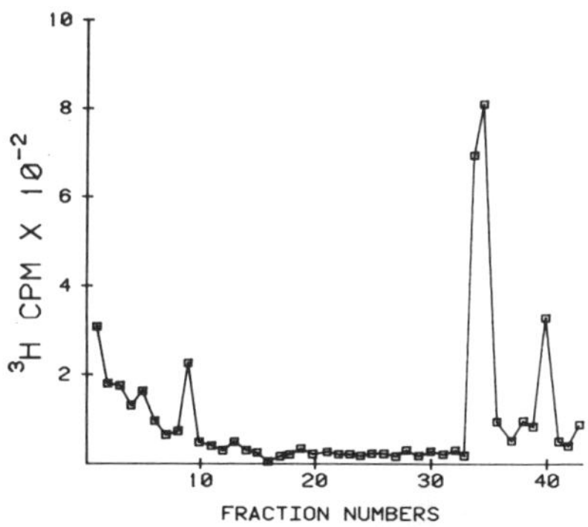

FIGURE 8. Isokinetic sucrose gradient sedimentation of [^{3}H]thymidine-labeled supramolecular folded chloroplast DNA. The growth and labeling conditions were similar to those described in FIGURE 3a except ^{32}P was not used. At L1 an aliquot of a culture containing 8.12×10^6 cells was pelleted ($12000 \times g$, 2 min. at 4°C) and resuspended in 1 ml of RS-5 (10 mM Tris-HCl, pH 8.0, 10 mM EDTA-Na_2, 0.15 M NaCl, 0.25 M sucrose, 100 μg/ml proteinase K and 1 mM 2-mercaptoethanol). Cells were lysed by the sequential addition of 0.1 ml 25% Triton$\times$100, 0.1 ml 20% sarkosyl, and 0.1 ml 20% sodium dodecyl sulfate with gentle mixing following each addition. After the complete lysis was achieved the lysate was gently poured onto a preformed 34 ml isokinetic sucrose gradient with a starting concentration of 10% w/w (McCarthy *et al.*[25]). This gradient was run in a Sorvall OTD-2 ultracentrifuge with a Beckman SW-27 rotor at a speed of 6,000 RPM for 60 min. at 4°C. After collecting the fractions, the conditions for alkaline treatment, acid precipitation, and counting were the same as described in FIGURE 3a.

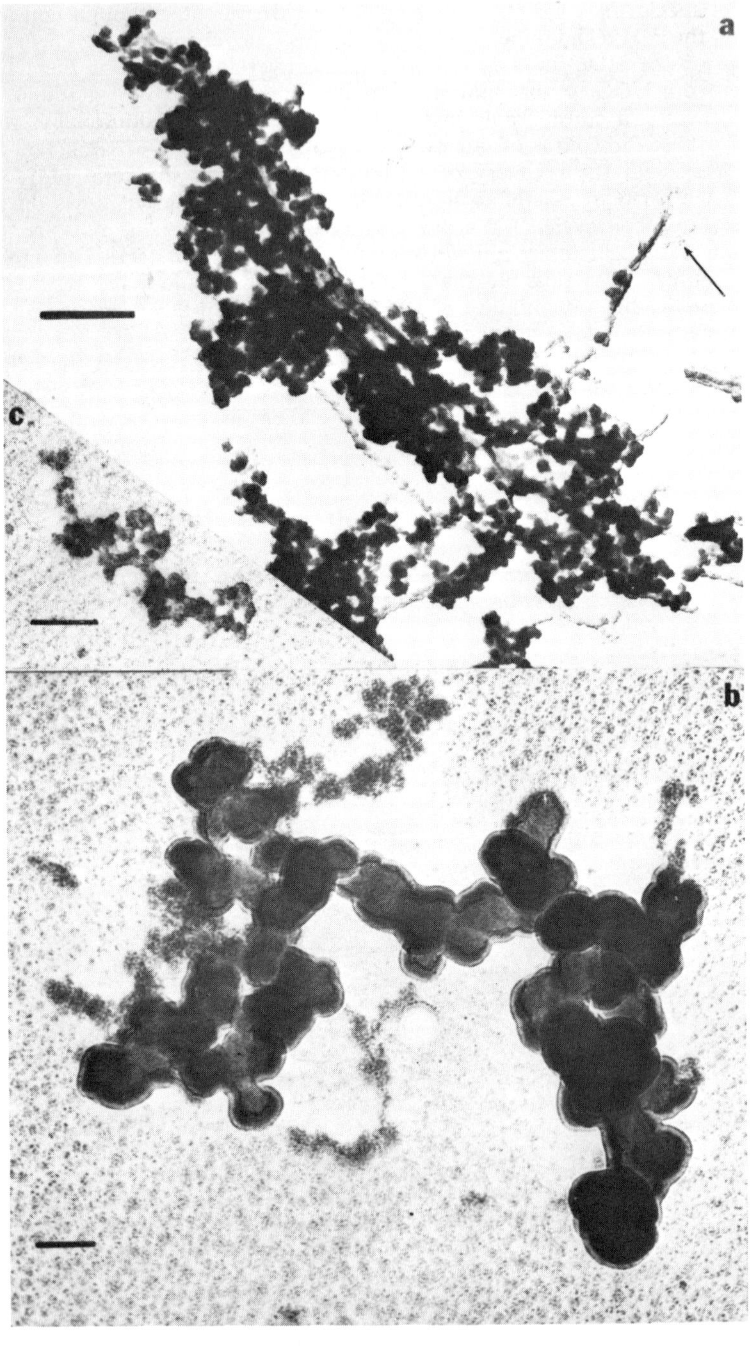
a
c
b

from the well-characterized naked cp-DNA or the β-satellite, yet the β-satellite and certain unidentified and extremely labile cp-DNA species were shown to have been contained in the HDC. These and other data, particularly data concerning the unusual, intercalating dye-modifiable conformations of HDC suggest that HDC might represent the cp-ND in its "native state" or a remnant of the "native state" structure that has survived the isolation and fractionation processes. To ascertain whether this was indeed the case, a structural analysis of HDC was carried out utilizing transmission and scanning electron microscopy. A synchronized *C. reinhardtii* culture was pulse-labeled with [^{3}H]dThd during the cp-DNA replication period, i.e., for the initial 6 hours of an alternating 12-hour light and 12-hour dark regime,[19] and then harvested at the completion of such replication (during nuclear interphase). HDC was prepared and fractionated on a PI-CsCl gradient (FIGURE 7a). The major (fraction 3) and the minor (fraction 10) [^{3}H]dThd peaks were isolated and spread on thin cytochrome *c* films (Kleinschmidt technique as modified by Paulson and Laemmli[29]). The great majority of the structures observed in the transmission electron microscope can be classified into four different types. Type A (FIGURE 9a) and Type B (FIGURE 9b) were found under the minor [^{3}H]dThd peak, while Type C (FIGURE 9c) and Type D (FIGURE 10a) were found under the major [^{3}H]dThd peak (see TABLE 2).

Several distinct sizes of fibers, "beads," and sheath or tubelike structures are observed in Type A structures: (1) ~120-Å diameter fibers with smooth surfaces that appear in short stretches interspersed between thicker fibers and tubes; (2) ~360-Å diameter beads that are distributed in between the ~120-Å diameter fibers; (3) ~360-Å diameter smooth fibers that appear in short stretches in between much thicker fibers; (4) ~750-Å diameter fibers with beaded surface structures that protrude out as short end pieces; (5) ~1100-Å diameter long fibers with beaded surfaces that often run as parallel arrays in between clusters of very dense ball-like structures with a diameter of ~1300 Å; (6) 1800–2200-Å diameter tube-like structures with an unsmooth surface covered with large dense "balls" of varying diameter (500 Å–1200 Å). One terminus of such a tubelike structure appears to have split open, revealing the

FIGURE 9. Transmission electron micrograph of type A structure (interphase nuclear chromatin). (a) A 10 μl sample of fraction 10 from the PI-CsCl gradient shown in FIGURE 7a was first diluted to 100 μl with DS-1 and added to a conical microcentrifuge tube containing 10 μl of fresh prepared cytochrome *c* solution (1 mg/ml in 5M ammonium acetate, pH 6.5). After gentle mixing, a small amount of precipitate appeared and was removed by centrifugation. The supernatant was then spread on an ammonium acetate hypophase, stained, and rotary-shadowed according to Paulson and Laemmli.[29] Samples were viewed in a Phillips EM300, photographed on Kodak electron-image film (4463), and developed in D19. Arrow shows ~360-Å diameter beads on ~120-Å diameter fiber. Original magnification 8,200×. Bar represents 1 μm. (b) Transmission electron micrograph of Type B structure. Experimental conditions were identical to FIGURE 9a. Original magnifications 55,000×. Bar represents 0.1 μm. (c) Transmission electron micrograph of Type C structure. A 10 μl sample of fraction 2 from the PI-CsCl gradient shown in FIGURE 7a was diluted to 100 μl with DS-1 and mixed with 10 μl of cytochrome *c* solution. After gentle mixing, this total mixture was spread, stained, shadowed, and viewed as described in FIGURE 9a. Original magnification 55,000×. Bar represents 0.1 μm.

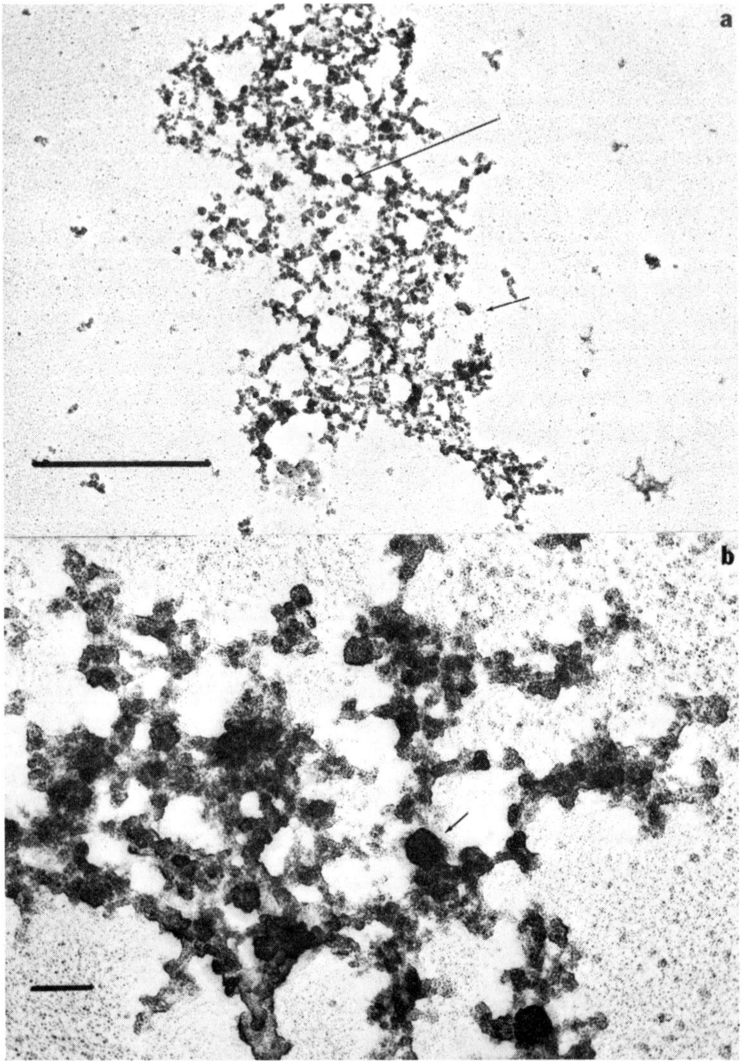

FIGURE 10. Transmission electron micrographs of Type D structure. (a) Experimental conditions were identical to FIGURE 9c. Short arrow shows a loop of ~185 Å diameter repeating beads with a periodicity of ~185 Å in between each pair of beads. Long arrow shows a 650–740 Å hexagonal particle unique to Type D structure. Original magnification 20,000×. Bar represents 1 μm. (b) Higher magnification transmission electron micrograph of the Type D structure. Experimental conditions were identical to FIGURE 9c. Arrow shows a 650–740 Å hexagonal knob. Original magnification 70,000×. Bar represents 0.1 μm.

~360 Å-diameter beads on the ~120 Å-diameter fiber extending outward (see arrow in FIGURE 9a). Most of these structures outlined above closely resemble structures found in higher eukaryotic interphase chromatin.[28, 38] This similarity and the massive size of the Type A structure are indicative of its nuclear origin.

Type B structures, as shown in FIGURE 9b, also exhibited several different sizes of fibers and beads, the most prominent being a thick, twisted fiber with a beaded appearance, having a diameter of approximately 1010 Å (ranging from 875–1150 Å for different component beads). From the periphery of this thick fiber emerges a smaller beaded string with an average diameter of 370 Å for the beads and about 170 Å for the interconnecting string. Very short stretches of a thin fiber measuring 40–45 Å in thickness are also observed. Most of these structural features resemble that of the interphase nuclear chromatin or the Type A structure as shown in FIGURE 9a. It is unclear at the moment whether the Type B structure is a part of the interphase nuclear chromatin having a different *in vitro* conformation from that of the Type A structure or whether the Type B structure represents a distinct subset of nuclear chromatin that may be in a different metabolic *milieu* than that represented by the Type A structure. Other large nuclear chromatin-like structures, which would not spread well with the Kleinschmidt technique, have been observed with the use of scanning electron microscopy. They appear to be greatly swollen, membraneless nuclei and dissociated fragments of swollen nuclei. A detailed characterization of the *C. reinhardtii* nuclear chromatin structure will be documented elsewhere.[24]

It is not clear why Type A and Type B structures were found under the minor [^{3}H]dThd peak, i.e., fraction 10 in FIGURE 7a, which was pulse labeled during nuclear interphase between L0 to L6. At least two possibilities exist: (1) the [^{3}H]dThd counts came from a newly synthesized minor organelle DNA species that had a distinct conformation and thus a different bouyant density from the majority of the newly replicated organelle DNA contained in major [^{3}H]dThd-labeled peaks; and (2) contrary to the great majority of nuclear DNA, a minute quantity of DNA present in the nucleus was synthesized and capable of utilizing supplied [^{3}H]dThd during the normal nuclear interphase. At present, these possible alternatives are being evaluated experimentally.

The Type C and Type D structures were found under the major [^{3}H]dThd peak, i.e., fraction 3 in FIGURE 7a, and are distinctly different from the Type A and Type B structures found under the minor [^{3}H]dThd peak. The Type C structure appears to be a supercoiled circular beaded string. The thickness of the thin fiber is approximately 40 Å, while the beads measure about 160 Å on the average (ranging from 130 to 200 Å). These beads are distinctly smaller than the beaded structure seen in Type A and Type B. Partial overlapping of many beads in the Type C structure, especially at the twisted regions, however, gives the structures the appearance of being considerably larger in diameter than may actually be the case. An accurate measure of the contour length of Type C structure is difficult to obtain due to its supercoiled conformation. A rough estimation gives an average value of approximately 0.80 μm. If one assumes that the contraction factor for packaging the naked DNA fiber into the ~160 Å beads is not too far from that of the condensed bacteriophage λDNA, i.e., ~6.5 [21] or the nucleosome, i.e., 7,[38] then the naked DNA fiber of the Type C structure should have a contour length of roughly 5.6 μm. This is of the same order of magnitude as the mitochondrial DNA of *C. reinhardtii,* which is approximately 5 μm in length and can be isolated as supercoiled closed-circular forms.[20, 32] This tentative suggestion awaits more direct experimental confirmation, which is currently in progress. Irrespective of the outcome of such studies, the DNA contour length of the Type C structure is much too short to comprise even a single molecule of the β-satellite or cp-DNA.

Tentative Indentification of the Folded Chloroplast DNA

The Type D structure is the predominant object found in the [^{3}H]dThd-labeled major HDC peak as shown in Figure 7a. A typical Type D structure is presented in Figure 10a. A micrograph taken at a higher magnification showing some of the structural components in greater detail is presented in Figure 10b. The structure of Type D appears to be a three-dimensional network composed of four basic structural elements. The largest of these is a dense, roughly hexagonal structure that measures about 650–740 Å across and appears to be scattered throughout the network (see the long arrow in Figure 10a). The shape, dimension, and degree of electron opaqueness of these hexagonal knobs are unique to the Type D structure. Interconnected round and ellipsoidal beads with diameters ranging from 410 to 430 Å are the predominant substructure of the network. A running fibrous structure connecting these hexagonal knobs and ellipsoidal beads has a width of between 150 and 260 Å (Figures 10a and 10b). Small round beads with a quite consistent diameter of ~185 Å are seen (Figure 10a) to form a circular pattern that has about equally distant gaps of 185 Å between each pair of beads. This structure is seen protruding from one of the many branch termini of the network (see the short arrow in Figure 10a). Round or slightly ellipsoidal beaded structures with the same diameter can also be found on the network proper.

Figure 11 shows a partially lysed Type D structure. The various levels of compaction of the constituent DNA are directly visible in a fortuitously positioned partially blocked circle that appears at the periphery of the structure. The thinnest line on that incomplete circle measures approximately 30 Å (see the long black arrow), while a short section of the incomplete circle measures about 150 Å in thickness. On the same circle, an even more compacted structure composed of two interconnected beads having a diameter of roughly 370 Å and a thickness of about 148 Å in between the two beads is clearly visible. The thinner fibers appear to have been unwound from both ends of this beaded structure (see the short black arrow). This result is suggestive, although it does not prove, that the 150 Å and 370 Å structures are products of two consecutive DNA fiber compactions. A number of small circles composed of 150 Å-diameter fibers are also visible. These circles have a contour length of about 0.65 μm (range of 0.60 to 0.85 μm). Some of these small circles appear to be interlocked with parts of the partially lysed network (see the long white arrow). Numerous fibers having a thickness close to that of naked DNA fibers can be seen running in a crisscross manner with respect to the network. Some electron opaque, round particles still remain relatively undeformed and clustered several to a given site. From the relative abundance, distribution, and electron opaqueness of these clustered dense particles, it is likely that they may have been degenerated from those extremely electron dense hexagon-like 650–

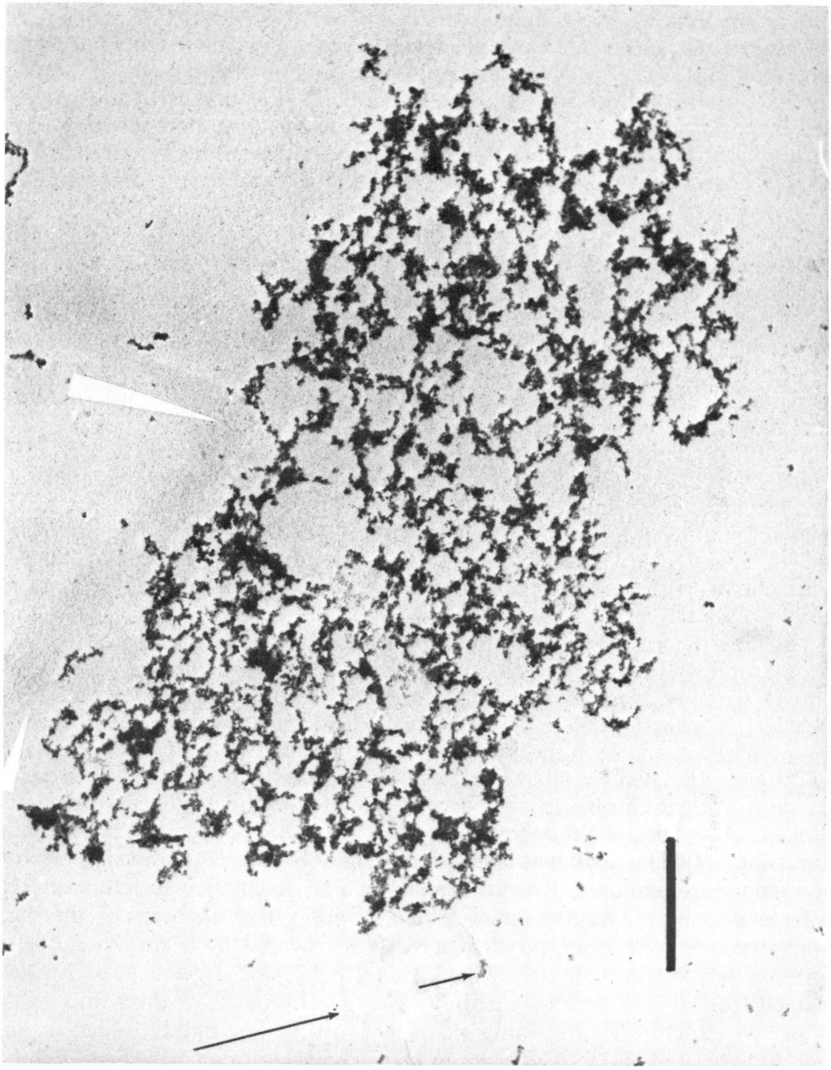

FIGURE 11. Transmission electron micrograph of a partially dissociated Type D structure. Experimental conditions were the same as in FIGURE 9c. Short black arrow shows a partially dissociated particular structure from which a circular ~30-Å diameter fiber emerges (long black arrow). Long white arrow shows one of many submicron DNA circles observed at the periphery of the Type D structure. Short white arrow shows one of several suspected single-stranded nucleic acid-containing regions extending outward from the partially lysed Type D structure. Original magnification 10,000×. Bar represents 1 μm.

740 Å structures seen in FIGURE 10a and 10b. These observations may imply that the hexagonal structure may be composed of an assembly of self-complimentary subunits whose binding to the compacted DNA structures is considerably more resistant to the ionic detergent-induced lysis than are some of the other "foldings" of the DNA fibers. Although it is slightly out of focus, a network of naked DNA and/or RNA, is visible at one corner of this partially lysed Type D structure (see the short white arrow). A more complete lysis will probably certainly result in the conversion of the entire Type D structure into a massive network of naked DNA that presumably also contains RNA fibers.

Since no other naked DNA or folded DNA species, other than structure C, were found in the major [^{3}H]dThd-labeled HDC peak, structure D appears to be the major species that incorporates [^{3}H]dThd label during the first six hours of the light period under an alternating 12-hour light 12-hour dark cell synchronization regime. It has been well established that the cp-DNA or the β-satellite is the only major DNA species that is undergoing replication during this period.[8, 19] This and other results presented above suggest that structure D is the most likely candidate to have been derived from the "native state" form of the folded chloroplast DNA. Due to the rapid degradation of the [^{3}H]dThd-labeled DNA molecules of the Type D structure during further manipulation, some technical problems remain to be resolved before more rigid proof of this tentative identification can be obtained. Hence, we searched for another independent isolation procedure of the [^{3}H]Thd-labeled, high density, high molecular weight structure to reinforce our tentative identification of it as the partially unfolded cp-ND.

Utilizing the same basic isolation premises described earlier in this paper, a new set of experimental conditions different from those used to isolate the Type D structure were developed to prepare and fractionate the [^{3}H]dThd-labeled HDC (see legend to FIGURE 12a). Particular emphasis was placed on keeping at least some of the cp-NDs intact, or nearly so, in order that they could be fractionated by molecular-sieve chromatography. Due to their larger size, most nuclear chromatin structures were eluted in the excluded volume of a Bio-Rad A150 agarose gel column. In the fractionation range of the column, a prominent [^{3}H]dThd peak was isolated and aliquots taken for scanning electron microscopic examination. FIGURE 12a shows a representative structure of what we believe to be the nearly intact folded cp-ND. The diameter of the large round structure, which appeared to have been covered by a sheath, measures approximately 0.5 to 0.56 μm. This diameter is in good agreement with values obtained from *in situ* DAPI-stained cp-NDs (see FIGURE 1). From this tightly compacted structure a three-dimensional network of less tightly folded or partially disassociated cp-DNA appears to be extruded. FIGURE 12b shows another partially dissociated cp-ND structure with the typical three-dimensional network appearance. The spherical particles seen in FIGURES 12a and 12b have a diameter of between 360 to 440 Å, while a few larger and more angularly shaped particles measure approximately 600 to 800 Å across. Those fibers that appear in between these small round and larger angular particles measure 110–160 Å in thickness. All these structural features are roughly equivalent to those observed in the Type D structure described by transmission electron microscope studies (FIGURES 10a, 10b, and 11 and TABLE 2).

TABLE 2

APPROXIMATE PHYSICAL PARAMETERS OF CONDENSED ORGANELLE AND NUCLEAR DNA FIBERS OBSERVED IN *Chlamydomonas reinhardtii* *

Figure No.	Structure Type	EM Magnification × 1000	Dimensions of Observed Substructures, Å: Bar Length μm	Smooth Fiber (thickness)	Beads/ Particles (diameter)	Beaded Thick Fibrils (thickness)	Others	Interpretation
9a	A	8.2		~120	~360	~750 ~1100	–	Interphase nuclear Chromatin
			1.0	~360	500–1200	1800–2200		
9b	B	55	0.1	40–45 ~170	~370	~1010	–	Interphase nuclear chromatin (a subset of Type A?)
9c	C	55	0.1	~40	~160 (130–200)	–	–	(Condensed mitochondrial DNA?)
10a	D	20	1.0	150–260	~185 ~420		hexagonal knobs 650–740	putative chloroplast nucleoid (dissociated form)
10b	D	70	0.1	"	"	–	"	"
11	D	10	1.0	~30 ~150	~370	–	0.65μ circles with 150 Å smooth fiber	partially lysed putative chloroplast nucleoid
12a	(SEM) †	50	0.2	110–160	360–440	–	Angular (particles) 600–800 Sheath covering the nucleoid (?)	Nearly intact putative chloroplast nucleoid (compact form)
12b	(SEM) †	50	0.2	"	"	–	Angular particles 600–800	Three-dimensional network of putative condensed chloroplast DNA (dissociated form)

* The increase in diameter or thickness of particles and fibers due to added metal was not corrected.

† Similar to type D but detected in scanning electron microscope.

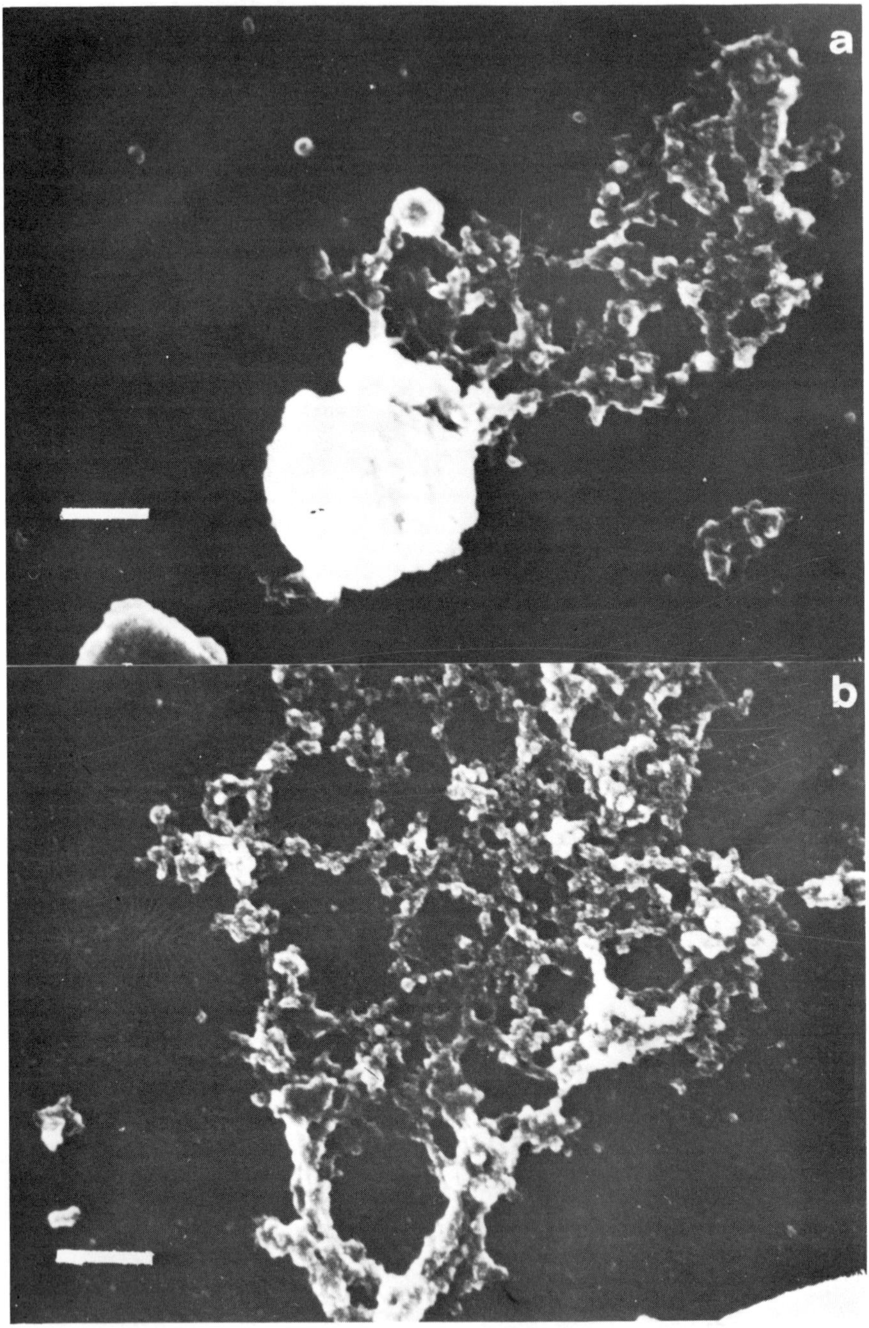
a
b

Interpretations and Implications of Our Observations

Utilizing two different isolation/fractionation principles and two different electron microscopic procedures, a tightly compacted structure has been tentatively identified, *in vitro,* as the folded cp-DNA. The dimensions and shape of this isolated cp-ND from *C. reinhardtii* are consistent with those of the cp-ND as seen by DAPI staining *in situ.* As expected, this cp-ND was labeled by a [^{3}H]dThd pulse during the well established cp-DNA replication period in the synchronized cell cycle. The intact cp-ND appeared to be enclosed in a sheath of unknown composition. A three-dimensional network of folded cp-DNA fibers apparently with varying degrees of organization and consequently of different dimensions and configurations was seen to protrude from the intact ND. The gross general pattern of cp-DNA folding and organization is distinctly different from its nuclear and putative mitochondrial counterparts. At least three levels of cp-DNA compaction were detected in partially lysed cp-NDs. The 30 Å thick naked DNA fiber appears to have been unraveled from two classes of beads as well as beaded fibrous structures, which measure 150 to 185 Å and 370 to 430 Å in diameter, respectively. A striking 650–740 Å hexagonal knob-like structure, which appears to be unique to the cp-ND, may represent a group of clustered smaller subunits possibly of nonDNA nature. Numerous tiny circles with submicron contour lengths appear to be associated with the three-dimensional network of the folded cp-DNA (Table 2). Possible

Figure 12. Scanning electron micrographs of a structure tentatively identified as the folded chloroplast DNA structure. The growth and labeling conditions were the same as described in Figure 3a. Towards the later part of the nuclear S phase at D 4.5 the cells were harvested, resuspended, lysed, and incubated as described in Figure 6a. Instead of running in CsCl gradient the crude lysate was fractionated on a Bio-Rad A150M (50–100 mesh) column (18×280 mm), equilibrated and eluted with a TENA buffer (10 mM Tris HCl, 10 mM EDTA, 0.15 M NaCl, and 1 mM ATA, pH 8.0). In the fractionation range of the column, which had been precalibrated by ^{14}C-labeled *E. coli* DNA (average size $3–5 \times 10^7$ dalton) and ^{3}H-labeled water, a prominent [^{3}H]thymidine-labeled DNA-containing peak (fraction 56) was identified by the alkaline-resistant, acid-precipitable counts of all eluted fractions. Large molecular weight polymers contained in fraction 56 were then pelleted onto 10 mm-diameter aluminum discs by centrifugation at maximum speed for 20 min. in an Eppendorf Model 3200 microcentrifuge. Fixation and dehydration procedures listed below were modified from that previously given.[29] (1) Fixed for 4 hours at 22°C in 0.1% glutaraldehyde in SEM Buffer (10 mM cacodylate, 10 mM NaCl, 5 mM $MgCl_2$, 0.1% NP-40, 0.5% deoxycholate, 1.0 mM ATA, pH 7.6); (2) Postfixed in 0.2% OsO_4 in SEM buffer for 1 hour at 22°C; (3) Dehydrated in 100% EtOH overnight at 4°C followed by two changes of 100% EtOH; (4) Critical-point dried; and (5) Coated with gold/palladium (60/40). Samples were viewed in a Hatachi HFS-2 field emission microscope and photographed with Polaroid positive/negative film ($4 \times 5''$ Type 55). The negatives were fixed in Na_2SO_3 (220 gm/l) and hardened in acid hardener (250 ml of 28% acetic acid and 1.6 gm KA1 $(SO_4)_2 \cdot 12 H_2O/\ell$). Original magnification 50,000×.
(a) A complete view of an almost intact chloroplast nucleoid with only a limited extrusion of the chloroplast supramolecular DNA fibrils. Bar represents 0.2 μm.
(b) A close-up view of the three-dimensional network of supramolecular DNA fibrils that extruded from the chloroplast nucleoid. Bar represents 0.2 μm.

interlocking between these circles and the network also is observed. Currently, we are striving to ascertain whether these tiny circles are related in any way to the two minor satellites, ϵ and τ, which we have discovered recently (see BACKGROUND section and Friedman and Chiang[17]).

As shown in FIGURE 5, under our experimental conditions the preservation of the HDC, from which the putative cp-ND was identified, appears to be dependent upon the integrity of certain exceedingly labile cp-DNA sequences or species. The protection of these sequences or species requires the presence of potent nuclease inhibitors that apparently counter the endogenous nuclease activity. The nature of these [^{3}H]dThd-labeled labile DNA species or sequences and the mechanism of their rapid degradation are at present unknown.

From a speculative standpoint, it is possible that these labile DNA sequence(s) or species may be so located as to be topologically pivotal to the integrity of the entire cp-ND and that certain nuclease activity-containing DNA metabolizing enzyme(s) may be closely associated with such pivotal sites. A shift in equilibrium and/or the reversal of the normal forward *in vivo* reaction of such enzyme(s) would, of course, result in the collapse and destruction of the cp-ND pivotal sites as well as those sequences that lie next to these sites. The final result would be the disintegration of the entire cp-ND. Those subunits that contain the hexagonal granules that dissociate more slowly than their surrounding folded cp-DNA "beads" (FIGURE 10) should be of particular interest in evaluating this possibility experimentally. If the cp-DNA sequence(s) or species at those postulated topologically "pivotal regions" are particularly rich in deoxythymidine, then, at least tentatively, this whole scenario could very well explain all of our experimental results, i.e., the data regarding the extreme lability of certain [^{3}H]dThd-labeled cp-DNA, the unusual difficulty in isolating the cp-ND *in vitro*, and the differential lysis-resistant behavior of the putative cp-DNA structural components. The experimental evaluation of this scenario is currently in progress in our laboratory. In this regard, it is of particular interest to note that a deoxythymidine site-preferred single-strand endonuclease has been isolated from *C. reinhardtii*.[5]

It is worth emphasizing that our investigation of the folding and organization of cp-DNA in *C. reinhardtii* is only in its initial stages and, as such, we are only just beginning to catch a glimpse of a complex and fascinating phenomenon. Since the preparation and fractionation methods used in this present study were distinctly different from the methods in conventional usage, it was possible to observe a few orders of consecutive packaging of the cp-DNA and a characteristic form of compacted cp-DNA organization in three-dimensional space which have not previously been appreciated. A number of problems, however, remain and some of them are currently under investigation. Foremost among these is to secure more direct evidence confirming our tentative identification of the native structure of the cp-ND. In addition, we are refining the techniques for quantity isolation of the putative cp-ND and investigating other crucial aspects that include (1) the identification and characterization of the cp-DNA binding proteins; (2) the characterization of certain species, tentatively identified as RNA, that are apparently indispensible in maintaining the integrity of the folded cp-genome; (3) the structural details and biochemical properties of the initial folding of the naked cp-DNA fiber into the 130–185 Å "beads on a string"-like structure. Although the data supporting the selective incorporation of the exogenously supplied [^{3}H]dThd label into cp-DNA[9, 40] are quite convincing, there does remain a small uncertainty in assuming, as we have throughout this study, that by following the [^{3}H]dThd label as an indicator for cpDNA-containing entities during the isolation and fractionation processes *in vitro* would unequivocally lead us to all or most of the constituents of the complete chloroplast genome *in vivo.* The particulate and periodic morphology of the condensed cp-DNA observed in the putative cp-ND has a superficial resemblance to the interphase chromatin of higher eukaryotes. This resemblance may not be very surprising in view of the fact that immediately after lysis, *E. coli* DNA can be found arranged in fibers having physical parameters very similar to those found in eukaryotic chromatin.[21]

Concluding Remarks

From the data we have thus far obtained, the following facts have become apparent: (1) In contrast to what has been generally assumed, cp-DNA molecules do not exist as naked DNA *in vivo,* they appear instead to take the form of thick supramolecular granules and fibrils. (2) The cp-DNA molecules are progressively folded and packaged into a few structurally distinct entities. These include a 130–185-Å diameter "beads on a string"-like structure, 260–400 Å granular fibril, and a 650–740 Å hexagonal particle (Table 2). A further compaction of the three-dimensional network of the 260–400 Å diameter granular fibrils, which are dotted with the hexagonal particles, results in the formation of an exceedingly compact chloroplast nucleoid that measures about 0.55 μm in diameter. (3) The two predominant structural entities of the cp-DNA fibers are similar, morphologically, to the 90–130 Å nucleofilament and the 300–500 Å chromatin fiber of the eukaryotic nucleus. The larger 650–740 Å hexagonal particles appear to be unique to the cp-ND.

An interesting dichotomy seems to exist when one examines the folding and organization of the cp-DNA. The cp-ND appears to contain certain features that are characteristic of eukaryotic nuclear DNA organization and certain features that are characteristic of prokaryotic organization. The cp-DNA of *C. reinhardtii* contains approximately 10% snap-back sequences,[18, 39] but is devoid of any detectable repetitive sequences.[2, 41] The total compaction of the DNA fiber in the cp-ND is by a factor roughly of 2.36×10^3 (Table 1). These characteristics are definitely prokaryotic in nature. On the other hand, the folded cp-DNA structures appear to possess ultrastructural characteristics similar to the two basic components of eukaryotic chromatin, i.e., the nucleofilament and the chromatin fiber. The higher orders of compaction found in eukaryotic chromatin are lacking, however.

How this apparent dichotomy, which seems to exist within the chloroplast DNA organization, will eventually contribute to the unravelling of the evolutionary history of the chloroplast genome is not clear at this moment. It does seem, however, that as we begin to understand the architecture of the chloroplast nucleoid, at least with respect to the packaging of the DNA molecules, a greater distinction may exist between the folding and organization of the mitochondrial and chloroplast DNAs than between the chloroplast and present-day eukaryotic nuclear DNA.

ACKNOWLEDGMENTS

The authors wish to acknowledge Dr. Rosann Farber for the use of a Leitz Orthoplan microscope and the Scanning Electron Microscope User's Laboratory at the University of Chicago, in which part of these results were obtained. The Laboratory is operated on a resource grant from the National Institute of Health.

REFERENCES

1. BAK, A. L., J. ZEUTLIEN & F. C. CRICK. 1977. Proc. Natl. Acad. Sci. USA **74:**1595–1599.
2. BASTIA, D., K. S. CHIANG, H. SWIFT & P. SIERSMA. 1971. Proc. Natl. Acad. Sci. USA **68:**1157–1161.
3. BEHN, W. & R. G. HERRMANN. 1977. Molec. Gen. Genet. **157:**25–30.
4. BUFFALOE, N. D. 1958. Bull. Torrey Botanical Soc. **85:**157–178.
5. BURTON, W. G., R. J. ROBERTS, P. A. MYERS & R. SAGER. 1977. Proc. Natl. Acad. Sci. USA **74:**2687–2691.
6. CARON, F., C. JACQ & J. ROUVIERE-YANIV. 1979. Proc. Natl. Acad. Sci. USA **76:**4265–4269.
7. CHAMBON, P. 1977. Cold Spring Harbor Symp. Quant. Biol. **42:**301–312.
8. CHIANG, K. S. & N. SUEOKA. 1967. Proc. Natl. Acad. Sci. USA **57:**1506–1513.
9. CHIANG, K. S., E. EVES & D. C. SWINTON. 1975. Devel. Biol. **42:**53–63.
10. CHUN, E. H. L., N. H. VAUGHAN, JR. & A. RICH. 1963. J. Mol. Biol. **7:**130.
11. COLEMAN, A. W. 1978. Exp. Cell Res. **114:**95–100.
12. COLEMAN, A. W. 1979. J. Cell Biol. **82:**299–305.
13. DRON, M., G. ROBREAU & Y. LEGAL. 1979. Expt. Cell Res. **119:**301–305.
14. EVES, E. M. 1979. Mendelian and non-Mendelian Genetics of Diploids and Anenploids of *Chlamydomonas reinhardtii*. Ph.D. dissertation. University of Chicago. Chicago, Ill.
15. FELSENFELD, G. 1978. Nature (London) **271:**115–122.
16. FINCH, J. T. & A. KLUG. 1976. Proc. Natl. Acad. Sci. USA **73:**1897–1901.
17. FRIEDMAN, F. & K. S. CHIANG. 1980. (Manuscript in preparation.)
18. GELVIN, S. B. & S. H. HOWELL. 1979. Molec. Gen. Genet. **173:**315–322.
19. GRANT, D., D. C. SWINTON & K. S. CHIANG. 1978. Planta **141:**259–267.
20. GRANT, D. & K. S. CHIANG. 1980. Plasmids. **4:**82–96.
21. GRIFFITH, J. D. 1976. Proc. Natl. Acad. Sci. USA **73:**563–567.
22. KORNBERG, R. D. 1977. Ann. Rev. Biochem. **46:**931–954.
23. KUROIWA, T., S. KAWANO & M. HIZUME. 1976. Exp. Cell Res. **97:**435–440.
24. MALAVASIC, M. J., JR., E. FRIEDMAN & K. S. CHIANG. 1980. (Manuscript in preparation.)
25. MCCARTHY, K. S., D. STAFFORD & O. BROWN. 1968. Anal. Biochem. **24:**314–329.
26. MAGUIRE, M. P. 1976. Genetica **46:**479–502.
27. MARSDEN, M. P. F. & U. K. LAEMMLI. 1979. Cell **17:**849–858.
28. OLINS, A. L. 1977. Cold Spring Harbor Symp. Quant. Biol. **42:**325–329.
29. PAULSON, J. R. & U. K. LAEMMLI. 1977. Cell **12:**817–828.
30. PETTIJOHN, D. E. & J. O. CARLSON. 1979. Cell Biology: A Comprehensive Treatise. **2:**2–57.
31. PETTIJOHN, D. E. 1976. C. R. C. Critical Reviews in Biochemistry **4:**175–202.
32. RYAN, R. P., D. GRANT, K. S. CHIANG & H. SWIFT. 1978. Proc. Natl. Acad. Sci. USA **75:**3268–3272.
33. RIS, H. & W. PLAUT. 1962. J. Cell Biol. **13:**383–391.
34. ROCHAIX, J. D. 1978. J. Mol. Biol. **126:**597–617.
35. ROUVIERE-YANIV, J., M. YANIV & J. E. GERMOND. 1978. Cell **17:**265–274.

36. RYTER, A. & A. CHANG. 1975. J. Mol. Biol. **98:**797–810.
37. SAGER, R. & M. A. ISHIDA. 1963. Proc. Natl. Acad. Sci. USA **50:**725–729.
38. SEDAT, J. & L. MANUELIDIS. 1977. Cold Spring Harbor Symp. Quant. Biol. **42:** 331–350.
39. SIU, C. H., K. S. CHIANG & H. SWIFT. 1975. J. Mol. Biol. **98:**369–391.
40. SWINTON, D. C. & P. HANAWALT. 1972. J. Cell Biol. **54:**592–597.
41. WELLS, R. & R. SAGER. 1971. J. Mol. Biol. **58:**611–622.
42. WILLIAMSON, D. H. & D. T. FENNELL. 1975. Methods Cell Biol. **12:**335.
43. WILSON, R. E. & K. S. CHIANG. 1977. J. Cell Biol. **72:**470–481.
44. GERMOND, J. E., J. ROUVIERE-YANIV, M. YANIV & D. BRUTLAG. 1979. Proc. Natl. Acad. Sci. USA **76:**3779–3783.

MOLECULAR SEQUENCE DATA INDICATING AN ENDOSYMBIOTIC ORIGIN FOR PLASTIDS *

W. Ford Doolittle and Linda Bonen

Department of Biochemistry
Dalhousie University
Halifax, Nova Scotia B3H 4H7
Canada

Introduction

Photosynthetic mechanisms in cyanobacteria (and presumably in prochlorophytes [1]) are so similar to those in eukaryotic photosynthesizers that it would be foolish to suggest that they evolved independently.[2] If eukaryotes did not now segregate photosynthetic functions and the DNA encoding some of these functions into discrete, membrane-bounded organelles (plastids), then we would all accept what Margulis [3] has called the "botanical myth;" the notion that plants arose from eukaryotic algae (specifically chlorophytes) and that these in turn arose in some straightforward way from oxygenic-photosynthetic prokaryotes, of which we now know only two kinds—the cyanobacteria and the prochlorophytes. Unfortunately, eukaryotic photosynthesizers *do* segregate photosynthetic functions and the DNA responsible for some of these functions.

There are really only two ways to explain this. One (the "botanical myth" or, less pejoratively, the direct filiation,[4] autogenous origin,[5] or developmental [6] hypothesis) assumes that nuclear and organellar genomes diverged, intracellularly, after or during the formation of the first "protoeukaryote," which was itself capable of oxygenic photosynthesis. The other (the endosymbiont [3] or xenogenous [7] origin hypothesis) assumes that the nuclear genome arose in a nonphotosynthetic protoeukaryote, while plastid genomes are of foreign origin, the descendants of the genomes of once free-living oxygenic-photosynthetic prokaryotic endosymbionts.

The question then is one of phylogenetic relationships between genomes. If plastid genomes prove to be more closely related to those of the nucleus of the cells in which they reside then they are to those of surviving oxygenic-photosynthetic prokaryotes, then autogenous origins are likely. If, on the other hand, they show stronger affinities to free-living prokaryotes, then xenogenous origins are likely.

All the genomes in question have diverged so markedly that meaningful measures of *overall* homology cannot be obtained. But we *can* look at conservative parts of these genomes. The ribosomal RNA (rRNA) genes are conservative parts of genomes, and have the further singular advantage of being present on all nuclear, organellar, and prokaryotic genomes.[8] We know of no *proteins* with identifiable homologs [3] coded for by each of these sorts of genome, and the plastid proteins whose sequences have been used to bolster xenogenous explanations are in most if not all cases actually the products of nuclear genes.[8]

* Supported by the Medical Research Council and the Natural Sciences and Engineering Research Council of Canada.

0077–8923/81/0361–0248 $01.75/0 © 1981, NYAS

METHODS

Homologies between rRNA genes can be directly determined from the sequences of the rRNAs which are their products. We have concentrated on the 16S rRNA of cyanobacteria and organelles, and their 18S (nuclear gene-coded) cytoplasmic homologs, because (i) not all organelles contain a recognizable 5S homolog, and this molecule is so small and conservative that only complete sequence determinations are of phylogenetic value,[8, 9] (ii) prokaryotic and organellar 23S rRNAs and their "28S" cytoplasmic homologs are too large, and often too labile,[10] for convenient sequence analysis, and (iii) Woese and his collaborators have assembled a truly massive amount of data (much of it unpublished) on bacterial 16S sequences, with which our data can be compared.

Complete sequencing of 16S and 18S rRNAs is now feasible, but remains tedious, and is unnecessary. Ample information can be obtained from the technique called *T1-oligonucleotide cataloging.*[11–13] This involves the complete digestion of ^{32}P-*in vivo*-labeled 16S or 18S rRNA with ribonuclease T1, which cleaves after each guanylate (Gp) residue, to yield some 300–500 oligonucleotides that range in size from one residue (Gp) to about 20 and all but one of which (that derived from the 3′-terminus of the molecule) will inevitably terminate in 3′-Gp. The oligonucleotides are resolved by two-dimensional electrophoresis on DEAE (diethylaminoethyl-cellulose) paper, separately eluted therefrom, and completely sequenced with the aid of the adenylate (Ap)-specific enzyme, ribonuclease U_2 and the pyrimidine residue-specific enzyme, pancreatic ribonuclease A.[11–13] All determined sequences are tabulated to form what we call a T1-oligonucleotide catalog. When oligonucleotides of less than five residues are eliminated (because they are inevitably present in almost all rRNAs) we have a set of some 100–150 oligonucleotides, each of which can be used as an individual character in assessing the homology between any two rRNAs. The mathematics of the situation are such that one expects, by chance, some 10–12 pentamers and some 2–4 hexamers to be common to any two 16S rRNAs even if they are completely unrelated (or random) in sequence, while one expects no chance coincidence in sequence between oligonucleotides of greater length.[14–16] One always observes *many* more than this number of oligonucleotide sequence coincidences between bacterial and cyanobacterial 16S rRNAs, and the observation of sequence coincidences between oligonucleotides of seven or more residues is unquestionable evidence for homology. For instance, the probability of obtaining by chance the three coincidences in oligonucleotide sequence observed[14] between octamers for the 16S rRNA of the cyanobacterium *Synechococcus* 6301 and the 16S rRNA of the rhodoplast of *Porphyridium cruentum* is approximately 0.00001.

The quantitative comparison of catalogs is essentially an exercise in numerical taxonomy, and neatly avoids the problems of alignment which plague measurements of homology between completely determined sequences of proteins or RNAs. We use a similarity coefficient, S_{AB}, defined[17] as follows: $S_{AB} = 2N_{AB}/(N_A + N_B)$, where N_A = total number of residues represented by oligonucleotides present in catalog A, N_B = total number of residues represented by oligonucleotides present in catalog B, and N_{AB} = total number of residues represented by oligonucleotides common to both catalog A and catalog B. Initially,[14, 15] we included for such calculations all oligonucleotides of five or more residues, but we have now limited our consideration to oligonucleotides of six or more residues, to bring our values into conformity with those of Woese,

Fox, and collaborators, and to eliminate almost all of the contribution from random sequence coincidences.

Two RNAs of identical sequence will show an S_{AB} of 1.0. Two RNAs that are of the same size and overall base composition as prokaryotic 16S molecules but completely unrelated in sequence will show an S_{AB} of about 0.05, reflecting the residual random contribution from hexamers. This does not mean, however, that RNAs showing an S_{AB} in this range are necessarily completely unrelated. S_{AB} and true sequence homology (as might be measured from complete sequences of intact molecules) are not linearly related. Homology becomes undetectable (indistinguishable from random) when homologous RNAs have diverged in sequence by some 40%.[15] All organellar and prokaryotic rRNAs (other than those of archaebacteria[18]) appear from our data and those of Woese and collaborators to have diverged by considerably less than this amount.

Results

We have obtained catalogs for eight cyanobacterial species. Three of these (6714, 6701 and 6308) are members of the unicellular group termed *Synechocystis* by Rippka *et al.*,[19] and nearly span the range of chromosomal G+C contents (34.7–48.0 mol%) shown by this group.[20] Three other unicells—6301 (55.1 mol% G+C), 7502 (39.0 mol% G+C), and *Agmenellum quadruplicatum* (49.0 mol% G+C)—are from the group Rippka *et al.*[19] call *Synechococcus,* although *Agmenellum* appears from our data to be a *Synechocystis* closely related to 6714 (47.4 mol% G+C). The remaining two are the filamentous species *Nostoc* (strain MAC) and *Fischerella ambigua,* which are classically considered members of different orders (Nostocales and Stigonematales). Plastid 16S rRNA catalogs are available from our laboratory for the rhodophyte *Porphyridium cruentum,*[14] and from Woese's laboratory for *Euglena gracilis*[21] and *Lemna* (duckweed, C. R. Woese, unpublished). Cytoplasmic 18S rRNA catalogs are available for *Porphyridium* and *Lemna* (C. R. Woese, unpublished).

Complete oligonucleotide catalogs for the 16S rRNAs of all of the cyanobacteria and the plastids of *Porphyridium* and *Euglena* have been presented recently elsewhere.[17] Table 1 shows the S_{AB} values calculated from all possible pairwise comparisons between them, while Figure 1 shows a dendrogram constructed from these values by average linkage (between the merged groups) clustering.[22] What Table 1 and Figure 1 do *not* show is comparable data published previously for other, noncyanobacterial prokaryotes or for the cytoplasmic 18S rRNA species. If they did, one would see that all pairwise comparisons between cyanobacterial 16S rRNAs or between cyanobacterial and plastid 16S rRNAs give S_{AB}'s higher than does any comparison between any cyanobacterial or plastid 16S and the 16S of *any* noncyanobacterial prokaryote. That is, the cyanobacteria and plastids form a monophyletic assemblage distinct from the rest of the prokaryotes (with the possible exception of *Prochloron,* which cannot yet be subjected to this sort of analysis). Furthermore, and more relevant here, one would see that there is no homology detectable by this method between the cytoplasmic 18S and plastid 16S rRNA of *Porphyridium* ($S_{AB} \simeq 0.05$), or in fact between the former RNA and any cataloged prokaryotic or plastid 16S molecule. Similar statements can be made for the cyto-

TABLE 1

VALUES OF S_{AB}

	Nostoc	Fischerella	Agmenellum	6714	6701	6308	Porphyridium	6301	7502	Euglena
Nostoc										
Fischerella	0.71									
Agmenellum	0.52	0.50								
6714	0.54	0.51	0.63							
6701	0.52	0.49	0.50	0.54						
6308	0.51	0.46	0.51	0.48	0.81					
Porphyridium	0.40	0.41	0.40	0.44	0.39	0.38				
6301	0.41	0.42	0.45	0.47	0.39	0.36	0.31			
7502	0.46	0.42	0.46	0.47	0.39	0.38	0.38	0.41		
Euglena	0.27	0.27	0.27	0.26	0.26	0.25	0.30	0.30	0.25	

Values of the association coefficient S_{AB} relating the 16S rRNAs of eight cyanobacteria and two plastids. See legend to FIGURE 1 for species designations.

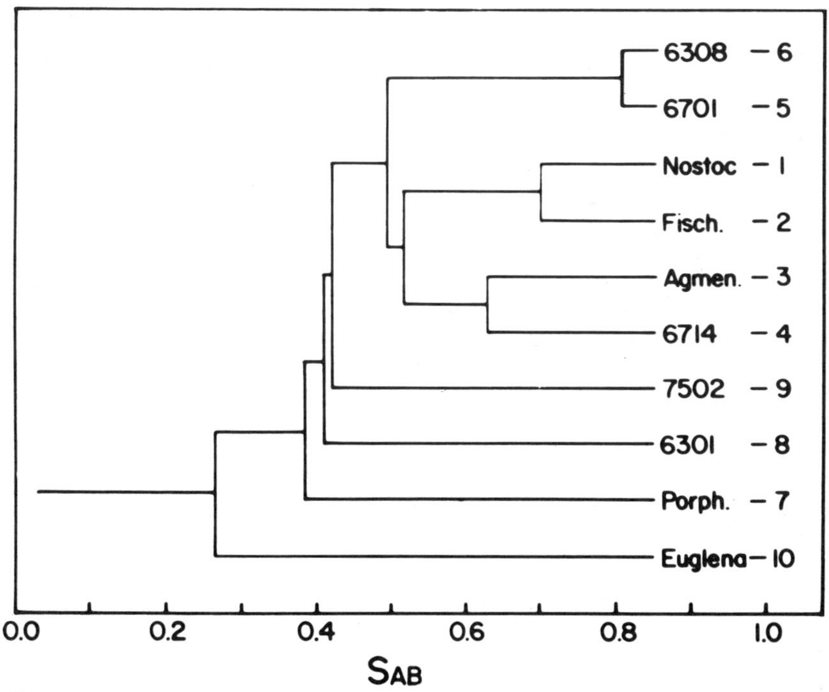

FIGURE 1. A dendrogram relating the 16S rRNAs of eight cyanobacteria (*Synechococcus* spp. 6301, and 7502, *Synechocystis* spp. 6308, 6701, and 6714, *Agmenellum quadruplicatum, Fischerella ambigua* and *Nostoc* sp. strain MAC) and the plastid 16S rRNAs of *Porphyridium cruentum* and *Euglena gracilis.* Reproduced by permission of the National Research Council of Canada from the *Canadian Journal of Biochemistry* **57:**879–888. 1979.

plasmic 18S and plastid 16S rRNAs of *Lemna* (C. R. Woese, personal communication). There is no comparable data for *Euglena* cytoplasmic 18S.

FIGURE 2 is a *schematic* representation of the relationships revealed by this sort of analysis. It includes also the results of an examination, performed with R. S. Cunningham and M. W. Gray,[16] of the homolog of 16S rRNAs found in wheat embryo mitochondria (which shows it to be of fundamentally prokaryotic, and hence likely endosymbiotic, origin) and typical results for non-cyanobacterial prokaryotes analyzed by Woese and collaborators.[18] There are two reasons why this representation is schematic. (i) Homologies between cytoplasmic 18S and prokaryotic or organellar 16S molecules are at or below

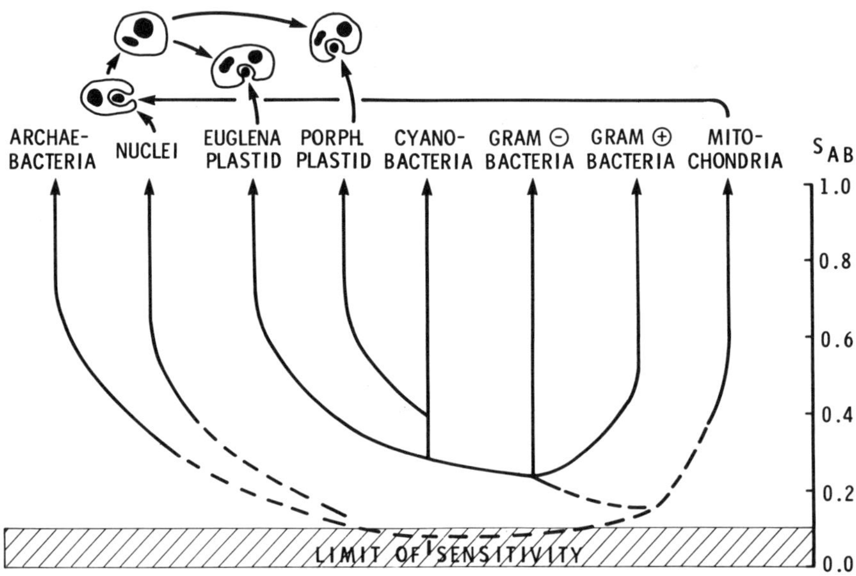

FIGURE 2. A *schematic* phylogenetic tree, based on 16S and 18S rRNA catalogs relating cyanobacteria, plastids, bacteria, mitochondria (wheat) and the nuclear-cytoplasmic components of eukaryotic cells. Data taken from sources cited in text and from C. R. Woese and G. E. Fox (personal communications).

the levels detectable by this method, so no proper S_{AB} value can be assigned to them. (ii) The mitochondrial rRNA is larger than its prokaryotic or plastid homologs, and this perturbs calculations of S_{AB} in ways that can only be dealt with arbitrarily.[16]

IMPLICATIONS

The Endosymbiont Hypothesis Appears Valid

Plastid 16S rRNAs from *Porphyridium* and *Euglena* are more closely related to the 16S rRNAs of cyanobacteria than they are to those of any other prokaryotes (again, *Prochloron* has not been examined). *Porphyridium* rhodoplast 16S rRNA is especially closely related to cyanobacterial 16S rRNAs. It is not unlikely that further analysis of cyanobacteria will reveal some species (for instance from the high G+C cluster of *Synechococcus*[20]) that are less closely related to the cyanobacteria so far examined than is *Porphyridium* plastid, thus enclosing this organelle (phylogenetically) *within* the cyanobacteria. On the other hand, there is no detectable homology between *Porphyridium* plastid 16S rRNA and the 18S rRNA molecule resident within the same cell. This result (and comparable results for *Lemna*) is fully consistent with xenogenous (endosymbiotic) explanations for plastid origins. It can be reconciled with autogenous origin hypotheses *only* by assuming that, after the intracellular divergence of plastid and nuclear rRNA genes, the latter have experienced rapid evolutionary sequence divergence. This assumption is *ad hoc,* otherwise unmotivated, and seems more liable to excision by Occam's razor than is its alternative. That alternative is that the dissimilarities in plastid 16S and cytoplasmic 18S rRNAs reflect a divergence of the genomic lineage leading to plastid genomes and the genomic lineage leading to nuclear genomes that is considerably more ancient than the divergence of cyanobacterial and plastid genomic lineages. There are reasons other than rRNA dissimilarities to believe in such an ancient divergence. Principle among them are the strikingly different ways in which genetic information is organized and expressed in eukaryotic nuclei and prokaryotes, recently stressed by us[23] and by Darnell.[24]

Plastids Are Polyphyletic

The 16S rRNA of the plastid of *Porphyridium* is more closely related to each of the cyanobacterial 16S rRNAs than it (or any cyanobacterial 16S) is to the 16S rRNA of the plastid of *Euglena.* Unless we suppose that cyanobacteria descend from plastids, the last common ancestor of cyanobacteria and *Porphyridium* plastids must have been a free-living prokaryote. If so, then the last common ancestor of *Porphyridium* and *Euglena* plastids must also have been a free-living (oxygenic-photosynthetic) prokaryote, and the two plastids derived independently from separate prokaryotes; plastids are *polyphyletic.* There is other evidence for the independent prokaryotic origin of plastids of other eukaryotic algal groups. For instance, both rhodophytes and cryptophytes contain the phycobiliproteins (phycocyanin and/or phycoerythrin) found otherwise only in the cyanobacteria, but available primary and higher-order structural information on these proteins suggests that the divergence of rhodophytan plastids and cyanobacteria is more recent than the divergence of either of these and crytophytan plastids.[2, 25] Chlorophyte and higher-plant plastids differ in important ways from those of other algal groups, and Lewin and coworkers suggest that the most likely ancestor for them is that prokaryotic lineage whose only free-living survivor is the newly discovered oxygenic-photosynthetic prokaryote, *Prochloron.*[1] Thus the earlier suggestion of Raven[26] and others[2, 3] that plastids arose on many different occasions from different oxygenic-photosynthetic prokaryotes, not all of which we would now call cyanobacteria, seems likely to be correct.

It is most difficult to accommodate such evidence for multiple plastid origins with autogenous or direct filiation hypotheses. To do so, we must assume that all of the other features that various eukaryotic algal groups share *as eukaryotes* (histones, chromosomal organization, endomembrane, and microtubular systems, 80S ribosomes, etc.) arose *independently* in each of the separate, originally prokaryotic, lineages, none of whose surviving free-living members show any of these traits. As Taylor [5] notes "the parallelisms invoked seem excessive."

Times of Divergence

Phylogenetic trees derived from macromolecular sequence data cannot be calibrated against time without independent evidence from the fossil record. Trees based on protein sequences for "higher" eukaryotes are reasonably congruent with phylogenies based on the fossil record, and molecular clocks for individual proteins appear to run at reasonably constant (albeit different) rates.[27] We can only *assume* that this is also true for prokaryotes. Their fossil record, although extensive, is likely never to yield a convincing independent phylogeny. Some information can be derived from the cyanobacterial fossil record. The origin of the Stigonematales (presumably from the Nostocales) has been estimated from the fossil record to have occurred between 0.4 and 1 billion years ago, while the divergence of the Nostocales from unicells is likely much more ancient (more than two billion years [28, 29]). The 16S rRNA catalog data summarized in TABLE 1 and FIGURE 1 indicate that these filamentous cyanobacteria are indeed more recently diverged than are many unicellular lineages, a fact consistent with the great range of chromosomal G+C contents (35–71 mol%) exhibited by chroococcalean species and the relatively restricted range of chromosomal G+C contents shown by all filamentous species (38–59 mol%).[20] If the cladistic relationships implied by FIGURE 1 are even approximately valid, then the divergence of prokaryotic lineages giving rise to *Porphyridium* and *Euglena* plastids is indeed very ancient. They might not be valid if for some reason plastid (and especially *Euglena* plastid) rRNA genes have diverged rapidly in sequence since their entrapment in the protoeukaryotic cytoplasm. This is of course possible, although the analysis of sequences which (by their nearly universal presence in plastid and prokaryotic 16S rRNAs) must be assumed to be under strong functional constraint suggests that plastid ribosomes have not diverged *functionally* from those of free-living prokaryotes.[29]

What Else Can Be Done?

The endosymbiont hypothesis for plastids will only finally be proven when the prokaryotic genomic lineage which gave rise to the nuclear genome has been identified and shown to be unrelated to oxygenic-photosynthetic prokaryotic lineages. For this reason, one hopes that the claims made for *Thermoplasma* [30] and/or the halobacteria [31] prove true. If, on the other hand, Fox and Woese [18] are correct in dividing the living world into three distinct lineages—that now represented by nuclei, that now represented by the "archaebacteria" (including *Thermoplasma* and the halobacteria), and that represented by all other prokaryotes, plastids, and mitochondria—then such a proof may never be possible. Future data will then simply reinforce the notion that nuclei and plastids operate in different ways, and, with the various ancillary *ad hoc* assumptions mentioned above, autogenous explanations can always be defended.

One thing is clear; it is no longer reasonable to suppose that *both* plastids and mitochondria arose autogenously. Homologies between the 16S rRNAs and cytochromes [32] of plastids and cyanobacteria are very strong, and can be rationalized by autogenous schemes only if we assume that eukaryotes arose from cyanobacteria. Homologies between cytochromes of mitochondria and purple nonsulfur bacteria and their presumed nonphotosynthetic aerobic derivatives like *Paracoccus denitrificans* are also very strong,[32, 33] and can be rationalized by autogenous schemes only if we assume that eukaryotes arose from members of this latter group. The last common ancestor of the purple nonsulfur bacteria and cyanobacteria is more ancient than either the last common ancestor of cyanobacteria and plastids or the last common ancestor of purple nonsulfurs and mitochondria. Uzzell and Spolsky [34] must be wrong in suggesting that "an entirely plausible phylogeny of eukaryotes can be constructed by beginning with the common ancestor of blue greens and aerobic bacteria" (the latter being derived by loss of photosynthesis from the purple nonsulfur lineage). Either mitochondria arose autogenously in protoeukaryotes derived from purple nonsulfur bacteria and plastids were acquired by symbiosis, or plastids arose autogenously in protoeukaryotes derived from cyanobacteria and mitochondria were acquired by symbiosis. These are mutually exclusive hypotheses, and *both* of them cannot be true. We suspect that *neither* of them is.

Acknowledgments

We thank C. R. Woese and G. E. Fox for communicating unpublished results.

References

1. Whithers, N. W., R. S. Alberte, R. A. Lewin, J. P. Thornber, G. Britton & T. W. Goodwin. 1978. Photosynthetic unit size, carotenoids, and chlorophyll-protein composition of *Prochloron* sp., a prokaryotic green alga. Proc. Natl. Acad. Sci. USA **75:**2301–2305.
2. Stanier, R. Y. 1974. The origins of photosynthesis in eukaryotes. Symp. Soc. Gen. Microbiol. **24:**219–243.
3. Margulis, L. 1970. Origin of Eukaryotic Cells. Yale University Press. New Haven, Conn.
4. Allsopp, A. 1969. Phylogenetic relationships of the Prokaryota and the origin of the eukaryotic cell. New Phytol. **68:**591–612.
5. Taylor, F. J. R. 1976. Autogenous theories for the origin of eukaryotes. Taxon **25:**377–390.
6. Dodson, E. O. 1979. Crossing the procaryote-eucaryote border: endosymbiosis or continuous development? Can. J. Microbiol. **25:**652–674.
7. Taylor, F. J. R. 1974. Implications and extensions of the serial endosymbiosis theory for the origin of eukaryotes. Taxon **23:**229–258.
8. Gillham, N. R. 1978. Organelle Heredity. Raven Press, Inc. New York, N.Y.
9. Sogin, S. J., M. L. Sogin & C. R. Woese. 1971. Phylogenetic measurements in prokaryotes by primary structural characterization. J. Mol. Evol. **1:**173–184.
10. Doolittle, W. F. 1980. The cyanobacterial genome, its expression and the control of that expression. Adv. Microbiol. Physiol. **20:**1–103.
11. Pechman, K. J. & C. R. Woese. 1972. Characterization of the primary structural homology between the 16S ribosomal RNAs of *Escherichia coli* and *Bacillus megaterium* by oligomer cataloging. J. Mol. Evol. **1:**230–240.
12. Uchida, T., L. Bonen, H. W. Schaup, B. J. Lewis, L. Zablen & C. Woese.

1974. The use of ribonuclease U_2 in RNA sequence determination. J. Mol. Evol. **3:**63–77.
13. WOESE, C., M. SOGIN, D. STAHL, B. J. LEWIS & L. BONEN. 1976. A comparison of the 16S ribosomal RNAs from mesophilic and thermophilic *Bacilli*: some modifications in the Sanger method for RNA sequencing. J. Mol. Evol. **7:** 197–213.
14. BONEN, L. & W. F. DOOLITTLE. 1975. On the prokaryotic nature of red algal chloroplasts. Proc. Natl. Acad. Sci. USA **72:**2310–2314.
15. BONEN, L. & W. F. DOOLITTLE. 1976. Partial sequences of 16S rRNA and the phylogeny of blue-green algae and chloroplasts. Nature **261:**669–673.
16. BONEN, L., R. S. CUNNINGHAM, M. W. GRAY & W. F. DOOLITTLE. 1977. Wheat embryo mitochondrial 18S ribosomal RNA: evidence for its prokaryotic nature. Nucleic Acids Res. **4:**663–671.
17. BONEN, L., W. F. DOOLITTLE & G. E. FOX. 1979. Cyanobacterial evolution: results of 16S ribosomal ribonucleic acid sequence analysis. Can. J. Biochem. **57:**879–888.
18. WOESE, C. R. & G. E. FOX. 1977. Phylogenetic structure of the prokaryotic domain: the primary kingdoms. Proc. Natl. Acad. Sci. USA **74:**5088–5090.
19. RIPPKA, R., J. DERUELLES, J. B. WATERBURY, M. HERDMAN & R. Y. STANIER. 1979. Generic assignments, strain histories and properties of pure cultures of cyanobacteria. J. Gen. Microbiol. **111:**1–61.
20. HERDMAN, M., M. JANVIER, J. B. WATERBURY, R. RIPPKA, R. Y. STANIER & M. MANDEL. 1979. Deoxyribonucleic acid base composition of cyanobacteria. J. Gen. Microbiol. **111:**63–71.
21. ZABLEN, L. B., M. S. KISSIL, C. R. WOESE & D. E. BUETOW. 1977. Phylogenetic origin of the chloroplast and prokaryotic nature of its ribosomal RNA. Proc. Natl. Acad. Sci. USA **72:**2418–2422.
22. ANDERBERG, M. R. 1973. Cluster Analysis for Applications. Academic Press, Inc. New York, N.Y.
23. DOOLITTLE, W. F. 1978. Genes-in-pieces: were they ever together? Nature **272:**581–582.
24. DARNELL, J. E. 1978. Implications of RNA:RNA splicing in evolution of eukaryotic cells. Science **202:**1257–1260.
25. GLAZER, A. N. 1976. Phycocyanins: structure and function. *In* Photochemical and Photobiological Reviews. D. C. Smith, Ed. Vol. **1:**71–115. Plenum Publishing Corp. New York, N.Y.
26. RAVEN, P. H. 1970. A multiple origin for plastids and mitochondria. Science **169:**641–646.
27. WILSON, A. C., S. S. CARLSON & T. J. WHITE. 1977. Biochemical evolution. Ann. Rev. Biochem. **46:**573–639.
28. SCHOPF, J. W. Paleobiology of the Precambrian: the age of blue-green algae. *In* Evolutionary Biology. T. Dobzhansky, M. L. Hecht, W. C. Steere, Eds. Vol. **7:**1–43. Plenum Publishing Corp. New York, N.Y.
29. BONEN, L. & W. F. DOOLITTLE. 1978. Ribosomal RNA homologies and the evolution of the filamentous blue-green algae. J. Mol. Evol. **10:**283–291.
30. SEARCY, D. C., D. B. STEIN & G. R. GREEN. 1978. Phylogenetic affinities between eukaryotic cells and a thermophilic mycoplasma. BioSystems **10:**19–28.
31. BAYLEY, S. T. & R. A. MORTON. 1978. Recent developments in the molecular biology of extremely halophilic bacteria. C.R.C. Crit. Revs. Biochem. **6:** 151–205.
32. SCHWARTZ, R. M. & M. O. DAYHOFF. 1978. Origins of prokaryotes, eukaryotes, mitochondria, and chloroplasts. Science **199:**395–403.
33. ALMASSY, R. J. & R. E. DICKERSON. 1978. *Pseudomonas cytochrome* c_{551} at 2.0 Å resolution: enlargement of the cytochrome *c* family. Proc. Natl. Acad. Sci. USA **75:**2674–2678.
34. UZZELL, T. & C. SPOLSKY. 1974. Mitochondria and plastids as endosymbionts: a revival of special creation? Amer. Scientist **62:**334–343.

Discussion of the Paper

DR. J. A. SCHIFF: It struck me that your sequencing data really are very much in line with the DNA base-composition data that we got many years ago also on the blue-greens and *Euglena*. *Euglena* is very far out, it's 75% AT and very far from any of the blue-green alga base composition. Similarly the red algae some Japanese workers have shown are again very far, very high in AT, not as high as *Euglena* but again toward *Euglena* and away from the blue-greens to some extent. So actually it's amazing that your relatedness state is very, very close to the base-composition data and I wonder, is there any necessary reason that you can think of why that might be so?

DR. W. F. DOOLITTLE: Well there's no necessary reason except that base composition is a rough indication of phylogenetic distance. In fact you could construct our tree for the blue-greens for which there's a lot of base-composition data that agrees very well with the base-composition data.

DR. SCHIFF: Your sequencing data within the blue-greens are astonishingly like the base composition.

DR. L. BOGORAD (*Harvard University, Cambridge, Mass.*): I think you should be congratulated on trying to bring rationality to faith. I agree that the main problem is to try to decide where a gene has been by what it is at the end and whether it stayed within a single cell and whether the evolution continued in both of those places. I think the polyphylectic arguments are helpful but I think one of the frustrations of this whole business is a difficulty of trying to measure what happened in between. No matter which position one would be interested in taking, and I really have no strong feelings in either case, it's the evidence that is almost impossible to obtain about what happened in between times. As you yourself alluded to, the kinds of arguments one can make about relative rates of evolution with regard to compartments (or many of the other things that one sees again and again), which is somewhat disturbing, is a taking on faith. Taking the arguments on faith permits frequently building one piece of faith on another piece of faith and these are really, it seems to me, the kinds of issues we have to keep ourselves aware of.

DR. DOOLITTLE: I think the only kind of acceptable proof will be established when we can identify that prokaryotic lineage which gave rise to the nuclear component and show that it is not cyanobacterial. I personally think that the eukaryotic lineage diverged very long ago and we may never see prokaryotic affinities in it.

DR. T. UZZELL (*Academy of Natural Sciences, Philadelphia, Pa.*): As you say it's very easy to read a dendrogram as representing phylogeny and you know very well if you put the nuclear RNA's catalogs on the same dendrogram you showed us you'd have that divergent outside of blue-greens and of *Porphyridium* and *Euglena*. You draw the conclusion from this that all the blue-greens are much more recently evolved and the prokaryotes follow the eukaryotes in origin? I'm only trying to caution you on making that interpretation of the dendrogram. You wind up having the blue-greens evolving much more rapidly and much more recently because they are more similar, and eukaryotes preceding blue-greens, because they are outside of it.

DR. DOOLITTLE: In fact, I believe the implication is that the nuclear lineage diverged from the prokaryotic lineages very early. That doesn't mean that the eukaryotes that we see now are as divergent as the blue-greens because the survivors of the nuclear lineage may represent a very small spectrum.

DR. S. S. COHEN: It would be very nice if we had more information on the lineage within the group of cyanobacteria. Most of us tend to lump them all together and then pick on *Anacystis* which, it turns out, does prove to be fairly close to the plastids. But one could think of metabolic parameters or chemical parameters that might be of value in classifying these organisms and giving us some fix as to what the most primitive ones were.

We think of blue-greens as the organisms that first learned to evolve molecular oxygen. We think that this has been important in producing the oxidized state of cupric ion which becomes available; anaerobes generally do not contain copper. We find very few copper enzymes but we do apparently turn up a copper enzyme in a blue-green namely, plasticyanin, though I must say, to my knowledge nobody has ever identified the copper in a plasticyanin of a blue-green very rigorously.

DR. DOOLITTLE: The proteins have been sequenced but I don't know whether it's copper.

DR. COHEN: So we are now in a circular state, namely a plasticyanin is required for the sequence that makes oxygen, then there must have been perhaps a primitive blue-green which was minus copper prior to that event of generating this cupric-containing enzyme. One wonders if in classifying blue-greens, we might not look for some primitive characters for which this copper might be thought of as one, and developing some markers which would permit us to date the blue-greens. I would suggest that it's quite possible that we're going to turn up some blue-greens, perhaps without copper, and that we might be able to use this as one such parameter.

DR. DOOLITTLE: That might be a good parameter. As you know the blue-greens show an alarming degree of metabolic monotony, which makes it very difficult to classify them on any other than morphological ground at present. I think that there was a common oxygenetic photosynthetic ancestor to *Prochloron* and the Cyanobacteria which was not a Cyanobacteria, and presumably didn't have phycobilin proteins. And presumably also to the *Cryptomonad.*

DR. G. G. BROWN (*State University of New York, Stony Brook, N.Y.*): The idea is that the molecules that you are comparing, in this case the small ribosomal RNA molecule, have to be evolving at a relatively constant rate and divergently. I was going to ask, do you have any evidence that this is the case?

DR. DOOLITTLE: I don't I think that they are a good candidate for one that would be evolving that way because the function of the ribosomal RNA has been constant probably for the last 3½ billion years, so that I think the changes we see are nonconservative, so that our sequence is not so important a function because they all share a conservative core. The large fraction of the homology is this conservative core. But the only extensive ribosomal RNA cataloging that has been done has been done among prokaryotes in which there is no independent way of estimating age.

DR. MARGULIS: This question really relates to what Dr. Chiang has said in a previous talk as well as Dr. Doolittle, and that has to do with the organization of the genome in plastids and chloroplasts and in free-living blue-greens. If I understood Dr. Chiang correctly, he claimed to show things called nucleosomes. Nucleosomes are structures that are DNA-connected with type 2-A and B and

3 and 4-histones wrapped in a very precise way with the eight-fold symmetry that makes a structure that is really very precisely defined. If someone is claiming that such a structure exists either in blue-greens or chloroplasts, I really would like to see evidence of that. As far as I know they don't exist in blue-greens.

CHLOROPLAST ORIGINS: INFERENCES FROM PROTEIN AND NUCLEIC ACID SEQUENCES *

R. M. Schwartz and M. O. Dayhoff

National Biomedical Research Foundation and Georgetown University Medical School Washington, D.C. 20007

Introduction

The endosymbiotic theory for the origin of the eukaryote cell pictures the mitochondria and the chloroplasts as once having been free-living prokaryotes whose symbioses with a protoeukaryote evolved into their current roles as cellular organelles.[1] Assuming that the phenotypes of the organelles and their prokaryote ancestors resemble each other in important ways, the chloroplasts are thought to have evolved from blue-greens (or cyanobacteria), whereas the mitochondria are thought to have evolved from aerobic bacteria. The available molecular data, including plastocyanin, ferredoxin, and cytochrome c_6 sequences,[2] as well as transfer RNA (tRNA) sequences,[3] and 16S ribosomal RNA catalogues,[4] describe a close relationship between chloroplasts and blue-greens. 5S ribosomal RNA (5S rRNA) and tRNA sequences describe an evolutionary history for the eukaryote host that is quite distinct and separate from that of this organelle. By combining information from different types of protein and nucleic acid sequences, we have shown that the eukaryote organelles share a recent common ancestry with bacteria and that this ancestry is indeed quite separate from the eukaryote host line.[2]

In this paper, we pursue from a slightly different perspective the question of whether or not the eukaryote chloroplast was originally a blue-green. Because new sequence data are available, we can examine phylogenetic trees based on ferredoxin and cytochrome c_6 sequences in greater detail than was previously possible. Branches leading to eukaryote chloroplasts are intermixed with the branches leading to blue-greens on these trees. The simplest explanation of this intermixing of branches is the endosymbiotic theory. The alternatives are that the eukaryote cell arose more than once or that at least some blue-greens evolved from eukaryote chloroplasts. Neither of these alternatives is reasonable. In examining evolutionary histories based on ferredoxins and cytochrome c_6 sequences, we also address very interesting questions concerning the details of contemporary blue-green phylogeny and their relationships to eukaryote chloroplasts.

The evolutionary trees presented here are based on complete nucleic acid or protein sequences and were derived by the least-squares matrix method that is described in our other paper in this volume.

* Supported by National Aeronautics and Space Administration (Contract NASW-3317) and the National Institutes of Health (Grants GM-08710 and RR-05681).

0077–8923/81/0361–0260 $01.75/0 © 1981, NYAS

5S Ribosomal RNA

5S rRNAs are relatively short molecules, about 120 nucleotides long. They are associated with the large ribosomal subunit and may be involved in non-specific tRNA binding during protein synthesis.[5] Figure 1 pictures an evolutionary tree based on 5S rRNA sequences elucidated by a number of workers, including T. A. Dyer, R. N. Nazar, and C. R. Woese.[6-8] The base of the tree (or point of earliest time) is located on the clostridial line in conformance with

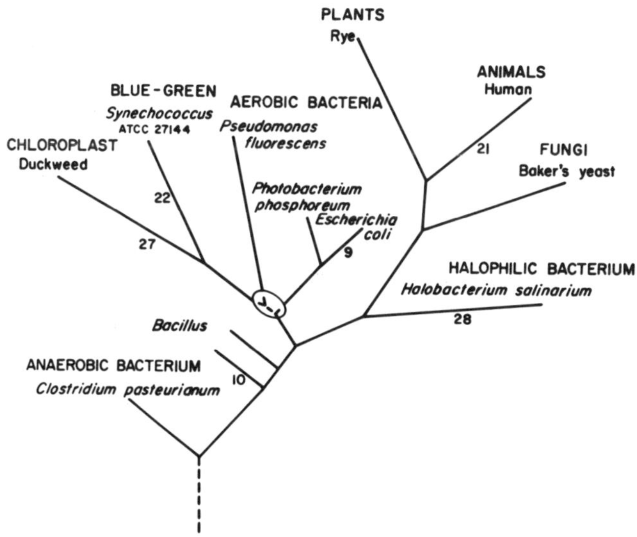

Figure 1. Phylogenetic tree based on 5S ribosomal RNA sequences. Branches are drawn proportional to the amount of inferred evolutionary change; branch lengths are given in accepted point mutations per 100 residues (PAMs). The order of divergences near the divergence of the line leading to *Pseudomonas fluorescens* and that leading to *Escherichia coli* and *Photobacterium phosphoreum* is not clearly resolved. The shortest topology is shown. One that pictures the *Synechococcus*–duckweed chloroplast branch as diverging from a branch that leads to *Ps. fluorescens*, on one hand, and *E. coli–P. phosphoreum*, on the other, has nearly as short an overall length. The first *Bacillus* divergence leads to *B. stearothermophilus* and the second to *B. licheniformis* (5S rRNA 1). *Synechococcus* ATCC 27144 is also called *Anacystis nidulans*. *H. salinarium* is called *H. cutirubrum* by the researchers who sequenced this molecule.

a tree based on bacterial ferredoxins and the strong evidence for a gene doubling in those sequences, particularly well conserved in the clostridial ferredoxins.[2] This tree develops the framework for the evolution of the prokaryotes in which we are interested. Two branches of the tree are particularly noteworthy. The branch leading to animals, plants, and fungi is based on sequences from cytoplasmic ribosomes and, in our opinion, represents the evolution of the eukaryote cell that was the host for the symbioses that led to the establishment of the mitochondria and chloroplasts. The other branch to note is the one leading to the sequences from *Synechococcus* ATCC 27144 (a *Synechococcus* from subgroup 2[9]) and from the chloroplast ribosome from duckweed; this branch represents the evolution of the blue-greens including the ancestor to the chloroplast of duckweed and the other higher plants. Clearly, the higher plants are composed of bacteria with two very different evolutionary histories—those that were ancestral to the eukaryote host and those that led to the chloroplasts. We expect plant mitochondrial 5S rRNA sequences will lie on yet a third branch, possibly diverging from the *Pseudomonas* line. The evolution that we will discuss in the remainder of this article concerns the ancestry of eukaryote chloroplasts and can be pictured as having occurred in the neighborhood of the *Synechococcus*–duckweed chloroplast divergence on this tree.

2Fe-2S Ferredoxins

Ferredoxins are a class of iron-sulfur proteins that act as electron carriers in a number of biochemical processes, including nitrogen fixation, nitrite reduction, and photosynthetic electron transport.[10] The 2Fe-2S ferredoxins have been isolated from the bacterium *Halobacterium halobium,* blue-greens, and the chloroplasts of eukaryotes, although chloroplast ferredoxins are translated on cytoplasmic ribosomes.[11] Thus it appears that ferredoxin, like mitochondrial cytochrome *c*, is coded in the nucleus but reflects the evolutionary history of an organelle. In the endosymbiotic theory, this is explained by postulating the transfer of genes from the organelle to the nucleus. Ferredoxin sequences were done by a number of workers, including H. Matsubara and K. T. Yasunobu.[12–20]

In *H. halobium*, ferredoxin functions as the electron acceptor in the oxidation of alpha-ketoglutarate, pyruvate, and alpha-ketobutyrate.[21] The large discrepancy in the distances between the divergences of *Bacillus* and *Halobacterium* on the line leading to the blue-green in trees based on 5S rRNAs (Figure 1) and ferredoxins[20] could be explained by either a gene duplication on the line leading to the 2Fe-2S ferredoxins or the different *Halobacterium* species on the two trees.

Figure 2 shows an alignment of the amino-terminal 23 residues of the ferredoxins for which complete sequences have been determined. The double sequences from horsetail and pokeweed are the result of recent gene duplications in each of those lines. The two sequences from *Aphanothece* are a different matter; they may be the result of a very ancient duplication. In sequence I from *Aphanothece,* there are deletions at positions 11 and 15 with respect to sequence II. (We believe that these are deletions in gene I rather than insertions in gene II because the sequence from *H. halobium* can be aligned best with sequence II without introducing gaps into the amino-terminal portion of either sequence.) The amino-terminal sequences of two ferredoxins from *Nostoc* MAC have been determined[22] and are consistent with this gene duplication having preceded the divergence of *Nostoc* and *Aphanothece*. Further evidence that this duplication preceded the divergences of the extant blue-green lines comes from reports of two ferredoxins in both *Spirulina maxima*[23] and *Nostoc muscorum*.[24]

Because this duplication preceded the blue-green divergences, we can place the point of earliest time on the ferredoxin tree near that shown in Figure 3. This is extremely important because it provides a temporal ordering to all the divergences on the tree.

	1	2	3	4	5	6	7	8	9	10	11 ↓	12	13	14	15 ↓	16	17	18	19	20	21	22	23
1 Aphanothece sacrum II	-	A	T	Y	K	V	T	L	I	N	E	E	E	G	I	N	A	I	L	E	V	A	D
2 Mastigocladus laminosus	-	A	T	Y	K	V	T	L	I	N	E	A	E	G	L	N	K	T	I	E	V	P	D
3 Spirulina platensis	-	A	T	Y	K	V	T	L	I	N	E	A	E	G	I	N	E	T	I	D	C	D	D
4 Spirulina maxima	-	A	T	Y	K	V	T	L	I	S	E	A	E	G	I	N	E	T	I	D	C	D	D
5 Nostoc muscorum	-	A	T	F	K	V	T	L	I	N	E	A	E	G	T	K	H	E	I	E	V	P	D
6 Cyanidium caldarium	-	A	S	Y	K	I	H	L	V	N	K	D	Q	G	I	D	E	T	I	E	C	P	D
7 Porphyra umbilicalis	-	A	D	Y	K	I	H	L	V	S	K	E	E	G	I	D	V	T	F	D	C	S	E
8 Aphanothece sacrum I	-	A	S	Y	K	V	T	L	K	T	-	P	D	G	-	D	N	V	I	T	V	P	D
9 Synechococcus lividus	-	A	T	Y	K	V	T	L	V	R	-	P	D	G	-	E	T	T	I	D	V	P	E
10 Scenedesmus quadricauda	-	A	T	Y	K	V	T	L	K	T	-	P	S	G	-	D	Q	T	I	E	C	P	D
11 Taro	-	A	T	Y	K	V	K	L	V	T	-	P	S	G	-	Q	Q	E	F	Q	C	P	D
12 Wheat	-	A	T	Y	K	V	K	L	V	T	-	P	E	G	-	E	V	E	L	E	V	P	D
13 Alfalfa	-	A	S	Y	K	V	K	L	V	T	-	P	E	G	-	T	Q	E	F	E	C	P	D
14 Spinach	-	A	A	Y	K	V	T	L	V	T	-	P	T	G	-	N	V	E	F	Q	C	P	D
15 Leucaena glauca	-	-	A	F	K	V	K	L	L	T	-	P	D	G	-	P	K	E	F	E	C	P	D
16 Pokeweed I	-	A	T	Y	K	V	T	L	V	T	-	P	S	G	-	T	Q	T	I	D	C	P	D
17 Pokeweed II	A	A	S	Y	K	V	T	F	V	T	-	P	S	G	-	T	N	T	I	T	C	P	A
18 Horsetail I	-	-	A	Y	K	T	V	L	K	T	-	P	S	G	-	E	F	T	L	D	V	P	E
19 Horsetail II	-	-	A	Y	K	V	T	L	K	T	-	P	D	G	-	D	I	T	F	D	V	E	P

FIGURE 2. Alignment of the amino-terminal residues from blue-green and chloroplast ferredoxin sequences.

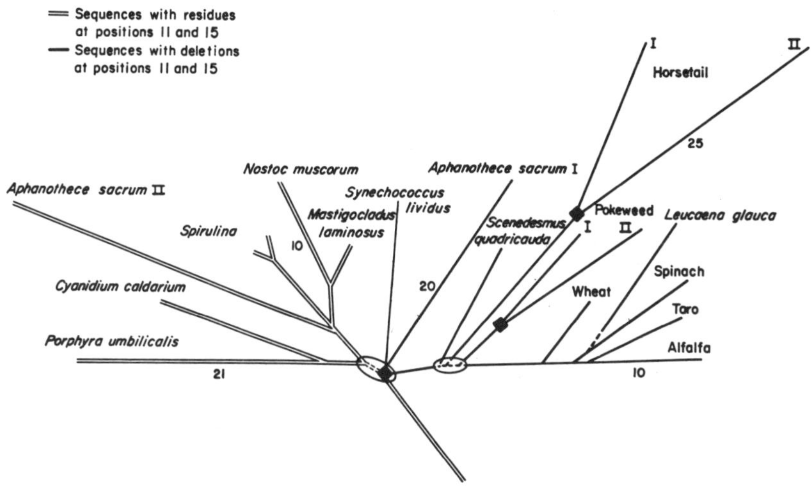

FIGURE 3. Evolutionary tree derived from 2Fe-2S ferredoxins. Branches are drawn proportional to the amount of inferred evolutionary change; selected branch lengths are given in PAMs. Double lines indicate sequences with residues at positions 11 and 15 in the alignment (FIGURE 2); single lines indicate those with deletions at positions 11 and 15. Diamonds indicate gene duplications. The *Leucaena glauca* chloroplast branch could also have diverged with either the wheat or taro branches; all three configurations are nearly equal in overall length. The order of divergence of the *Scenedesmus quadricauda*, horsetail, and pokeweed branches could not be resolved. The precise order of branching in the neighborhood of the gene duplication in *Aphanothece sacrum* is not clearly resolved computationally; the configuration among those with the shortest overall length that also segregates proteins on the basis of positions 11 and 15 in their alignment is shown.

In the ferredoxin tree (FIGURE 3), the two types of 2Fe-2S ferredoxin segregate as one would expect if a gene duplication preceded the blue-green divergences. The topology of the tree in the immediate vicinity of the ancient gene duplication is not clearly resolved; however, the improbability of deletions at positions 11 and 15 occurring in two separate lines makes the topology shown the most reasonable choice. Because the chloroplasts of *Scenedesmus* and the higher plants show no evidence of having the ancestral sequence without the two deletions, we presume that their ancestor lost the gene for this sequence.

If one scales branches on the right side of the tree so that the two *Aphanothece* branches are of equal length, the two halves of the tree can be super-

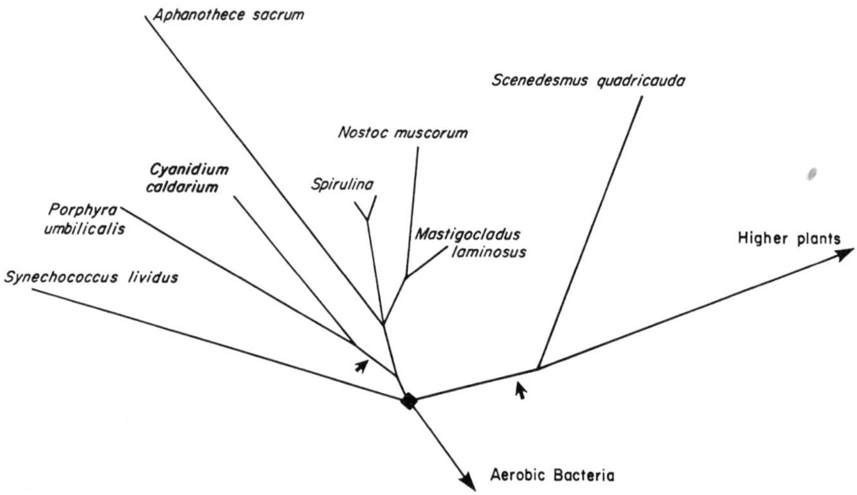

FIGURE 4. Phylogenetic tree based on 2Fe-2S ferredoxin sequences. This tree was deduced by scaling the branches for sequences with the two amino-terminal deletions so that they would be comparable to those for sequences without deletions at positions 11 and 15, and then superimposing the two halves of the tree. Arrows indicate branches that independently gave rise to eukaryote chloroplasts.

imposed to provide a description of the evolutionary relationships among the organisms containing the sequences. Depending on the exact location of the gene duplication, three phylogenies are possible. FIGURE 4 pictures one alternative. The earliest divergence on this tree was the one between prokaryotes that were ancestral to *Synechococcus,* to *Aphanothece,* and to the chloroplasts of higher plants and green algae such as *Scenedesmus.* This last prokaryote may have been related to *Prochloron,* a photosynthetic prokaryote containing chlorophylls *a* and *b.*[25] Although we cannot be sure the chloroplasts of the green algae and the higher plants are monophyletic, there was at least one symbiosis on this line. On this tree at least one additional symbiosis led to the chloroplasts of *Porphyra,* a red alga, and *Cyanidium,* a eukaryote alga of uncertain affinity. The ancestor of these chloroplasts was also ancestral to *Aphanothece* and *Synechococcus* and might, therefore, be expected to resemble contemporary blue-greens. The radiation of blue-greens subsequent to the divergence of the line leading to the chloroplasts of *Porphyra* and *Cyanidium* suggests that an evolutionary schema that pictures the origin of filamentous blue-greens as a simple bifurcation of the coccoid and filamentous types is overly simplified. *Aphanothece* is coccoid, whereas *Spirulina* and *Nostoc* are filamentous. Alternative explanations of this tree would require either two independent origins for the eukaryote cell or an evolutionary descent from a eukaryote chloroplast for a wide variety of contemporary blue-greens, including *Aphanothece, Spirulina, Nostoc,* and *Mastigocladus.*

A second possibility can be inferred if the gene duplication is placed nearer to the *Porphyra–Aphanothece* II divergence (FIGURE 3). If the two halves of this tree are superimposed, it is possible to place all of the eukaryote chloroplasts on a single branch. However, this requires that we assume the two chloroplast lines share an evolutionary history subsequent to their divergence from *Synechococcus.* Without any species in common, the superposition of the *Porphyra–Cyanidium* and the *Scenedesmus*–higher plant branches seems unwarranted. This configuration is consistent with a monophyletic origin for the chloroplasts. If the chloroplast branches are not superimposed this tree is essentially equivalent to the first alternative we described.

The third possibility results from placing the gene duplication at the *Porphyra–Aphanothece* II divergence. The base of the tree lies on the *Porphyra–Cyanidium* chloroplast line. This requires that the *Porphyra–Cyanidium* chloroplast line diverged prior to the gene duplication and predated the divergences of all the blue-greens pictured here. In other respects, this alternative is similar to the first. It also requires at least two symbiotic events leading to the chloroplasts on the tree.

CYTOCHROME c_6

Cytochrome c_6, also referred to as soluble cytochrome *f* or cytochrome $c_{552-554}$, is an electron transport protein that has been shown to substitute for plastocyanin in the electron transport chain between photosystems 1 and 2 depending upon the availability of copper to the organism.[26] Sequences have been determined by a number of workers, including A. Aitken, R. Ambler, and M. V. Laycock.[19, 27, 28] An evolutionary tree based on these sequences is shown in FIGURE 5. The point of earliest time on the tree cannot be inferred from the sequences but was placed to conform with the gene duplication in the 2Fe-2S ferredoxins shown in FIGURE 4. This tree has branches leading to *Spirulina maxima* and *Synechococcus lividus* in common with the ferredoxin tree; moreover, the branches leading to the chloroplasts of *Porphyra umbilicalis* and *P. tenera* on the two trees probably share a long evolutionary history, including the symbiotic event leading to their current status as organelles. *Synechococcus* ATCC 27167, like *Synechococcus* ATCC 27144 on the 5S rRNA tree, is classified on the basis of DNA base ratio and genome size as belonging to subgroup 2.[9]

In its overall configuration, this tree resembles the ferredoxin tree quite closely. The divergence of the branches leading to *Spirulina, Plectonema,* and the chloroplasts of *Euglena* and *Monochrysis* probably corresponds to the radiation of blue-green types on the ferredoxin tree that included *Spirulina, Aphanothece,* and *Nostoc.* Regardless of the exact position of the ferredoxin duplication and, by inference, the base of this tree, the intermixing of eukaryote and prokaryote branches here requires a minimum of two symbioses: one leading to the chloroplasts of *Alaria* and *Porphyra* and the other leading to those of *Euglena* and *Monochrysis.* The divergence of *Monochrysis* and *Euglena* being

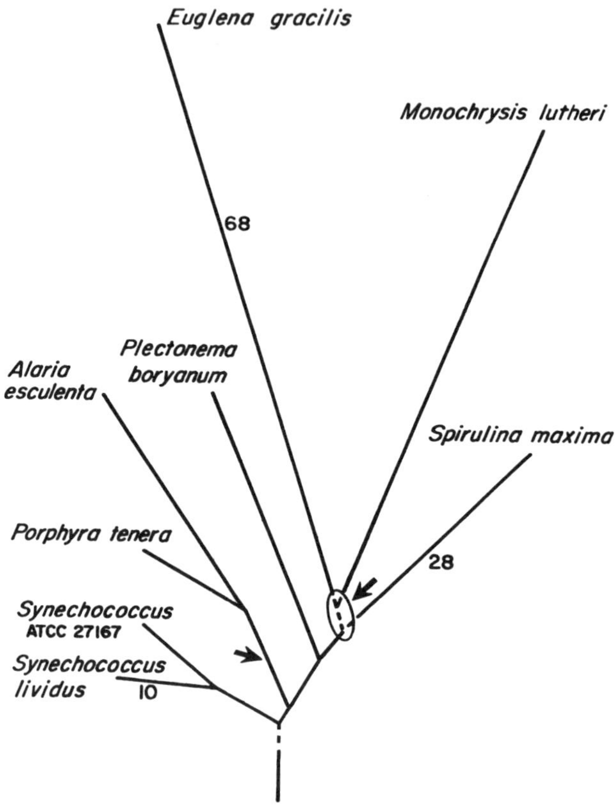

FIGURE 5. Phylogenetic tree derived from cytochrome c_6 sequences. Branches are drawn proportional to the amount of inferred evolutionary change; selected branch lengths are given in PAMs. The point of earliest time was placed to conform with the duplication in the 2Fe-2S ferredoxin tree in FIGURE 4. Arrows indicate branches that independently gave rise to eukaryote chloroplasts. The order of divergence of the branches leading to *Euglena, Monochrysis,* and *Spirulina* could not be resolved.

so close to what we believe was a radiation of blue-green types suggests that these chloroplasts may have arisen from different blue-greens; however, both were derived from a blue-green ancestor that was also common to *Plectonema* and *Synechococcus*.

At present, there is an unfortunate lack of biological and biochemical data for organisms such as *Alaria* and *Monochrysis*. It has been suggested that *Alaria* and *Monochrysis* are more closely related than the topology of this tree depicts.[29] The chloroplasts of both organisms contain chlorophyll *c* in contrast to chlorophyll *b* found in the chloroplast of *Euglena* and the bilin pigments found in the blue-greens and the chloroplast of *Porphyra*. This may mean that the ancestor of *Spirulina* and *Synechococcus* had a wider range of photosynthetic pigments than any of the descendants on this tree. It has also been suggested that the chloroplast of *Euglena* is the result of a symbiosis between a nonphotosynthetic protist and a eukaryote green alga.[30] This tree does not address that suggestion (for which the morphological evidence is quite appealing). If the chloroplast of *Euglena* was most recently a free-living, eukaryote alga, the history of that alga's chloroplast is represented here. Clearly it was a blue-green, sharing an ancestor with *Spirulina* and *Plectonema*.

Conclusion

Protein and nucleic acid sequence data offer distinct advantages in classifying prokaryotes and understanding their evolutionary relationships. In this area, where morphological characteristics, for example, have proved very inadequate, these gene products can provide unique insights. In this regard, what is true of free-living prokaryotes is even truer of their close relatives, the chloroplasts and mitochondria.

We have constructed evolutionary trees based on 5S rRNA, 2Fe-2S ferredoxin, and cytochrome c_6 sequences. The 5S rRNA tree provides an overview of the bacterial evolution in which we are interested; it pictures the evolution of 5S sequences from chloroplasts and free-living blue-greens as quite separate from that of sequences from cytoplasmic ribosomes. This clearly supports the endosymbiotic origin of the chloroplasts.

The intermixing of branches leading to chloroplasts and blue-greens in trees based on 2Fe-2S ferredoxins and cytochrome c_6 sequences is also most reasonably explained by the endosymbiotic theory. Alternative explanations require assuming multiple origins of the eukaryotes or chloroplast ancestors for contemporary genera of blue-greens.

A gene duplication in the 2Fe-2S ferredoxins places the point of earliest time in that tree and provides a temporal ordering for its divergences. Near this point of earliest time three lines diverged. One leads to the chloroplasts of *Scenedesmus* and the higher plants; a second to *Synechococcus lividus*, a thermophilic blue-green; and the third to *Aphanothece sacrum*, another blue-green. The branch leading to the chloroplasts of *Porphyra* and *Cyanidium* also diverged in this portion of the tree.

The tree based on cytochrome c_6 sequences is consistent with this evolutionary history. In addition it places the chloroplast from *Alaria,* a brown alga, on a branch with that from *Porphyra.* The chloroplasts from *Euglena,* a protist, and *Monochrysis,* a golden-brown alga, are located on another chloroplast branch that arose subsequent to the divergence of *Plectonema* from the line leading to *Spirulina.* A minimum of two and probably three independent symbioses leading to chloroplasts are required by the topologies of the two trees; there may have been more. The elucidation of additional sequence data will undoubtedly provide exciting new details in this area.

Acknowledgments

We would like to thank Drs. D. O. Hall and E. Margoliash for supplying us with sequences in advance of their publication, and K. Lawson for drafting the figures.

References

1. Margulis, L. 1970. Origin of Eukaryotic Cells. Yale University Press. New Haven.
2. Schwartz, R. M. & M. O. Dayhoff. 1978. Science **199:**395–403.
3. Chang, S. H., C. K. Brum, M. Silberklang, U. L. RajBhandary, L. I. Hecker & W. E. Barnett. 1976. Cell **9:**717–723.
4. Doolittle, W. F. & L. Bonen. 1980. Ann. N.Y. Acad. Sci.
5. Monier, R. 1974. *In* Ribosomes. M. Nomura, A. Tissieres & P. Lengyel, Eds.: 141. Cold Spring Harbor Laboratory. Cold Spring Harbor, N.Y.
6. Dyer, T. A. & C. M. Bowman. 1979. Biochem. J. **183:**595–604.
7. Nazar, R. N., A. T. Matheson & G. Bellemare. 1978. J. Biol. Chem. **253:** 5464–5469.
8. Schwartz, R. M. & M. O. Dayhoff. 1979. *In* Atlas of Protein Sequence and Structure. M. O. Dayhoff, Ed. Vol. **5**(Suppl. 3): 327–336. National Biomedical Research Foundation. Washington, D.C. 5S rRNA sequences and our alignment of them are collected here.
9. Rippka, R., J. Deruelles, J. R. Waterbury, M. Herdman & R. Y. Stanier. 1979. J. Gen. Microbiol. **111:**1–61.
10. Rao, K. K. & D. O. Hall. 1977. *In* The Evolution of Metalloenzymes, Metalloproteins and Related Materials. G. J. Leigh, Ed.: 39–65. Symposium Press. London.
11. Armstrong, J. J., S. J. Surzycki, B. Moll & R. P. Levine. 1971. Biochemistry **10:**692–701.
12. Hase, T., K. Wada & H. Matsubara. 1977. J. Biochem. **82:**267–276.
13. Hase, T., S. Wakabayashi, K. Wada & H. Matsubara. 1978. J. Biochem. **83:** 761–770.
14. Hase, T., S. Wakabayashi, K. Wada, H. Matsubara, F. Juttner, K. K. Rao, I. Fry & D. O. Hall. 1978. FEBS Lett. **96:**41–44.
15. Hase, T., S. Wakabayashi, H. Matsubara, K. K. Rao, D. O. Hall, H. Widmer, J. Gysi & H. Zuber. 1978. Personal communication.
16. Takruri, I., B. G. Haslett, D. Boulter, P. W. Andrew & L. J. Rogers. 1978. Biochem. J. **173:**459–466.

17. Wakabayashi, S., T. Hase, K. Wada, H. Matsubara, K. Suzuki & S. Takaichi. 1978. J. Biochem. **83:**1305–1319.
18. Takruri, I. & D. Boulter. 1979. Biochem. J. **179:**373–378.
19. Borden, D. & E. Margoliash. 1979. Personal communication.
20. Schwartz, R. M. & M. O. Dayhoff. 1979. *In* Atlas of Protein Sequence and Structure. M. O. Dayhoff, Ed. Vol. **5**(Suppl. 3): 45–55. National Biomedical Research Foundation. Washington, D.C. Ferredoxin sequences and our alignment of them are collected here.
21. Kerscher, L. & D. Oesterhelt. 1977. FEBS Lett. **83:**197–201.
22. Hutson, K. G., L. J. Rogers, B. G. Haslett, D. Boulter & R. Cammack. 1978. Biochem. J. **172:**465–477.
23. Cammack, R., K. K. Rao, C. P. Bargeron, K. G. Hutson, P. W. Andrew & L. J. Rogers. 1977. Biochem. J. **168:**205–209.
24. Hase, T., K. Wada, M. Ohmiya & H. Matsubara. 1976. J. Biochem. **80:**993–999.
25. Lewin, R. A. 1980. Ann. N.Y. Acad. Sci.
26. Wood, P. M. 1978. Eur. J. Biochem. **87:**9–19.
27. Aitken, A. 1979. Eur. J. Biochem. **101:**297–308.
28. Schwartz, R. M. & M. O. Dayhoff. 1978. Atlas of Protein Sequence and Structure. M. O. Dayhoff, Ed. Vol. **5**(Suppl. 3):28–43. National Biomedical Research Foundation. Washington, D.C. Cytochrome c_6 sequences and our alignment of them are collected here.
29. Ragan, M. A. & D. J. Chapman. 1978. A Biochemical Phylogeny of the Protists: 211–232. Academic Press, Inc. New York, N.Y.
30. Gibbs, S. P. 1980. Ann. N.Y. Acad. Sci.

Discussion of the Paper

Dr. R. Sager: I would like to know where the *Chlamydomonas* chloroplast fits into all of this? Because there are a number of lines of evidence available about similarities of chloroplast ribosomes and bacterial ribosomes which have always intrigued me. Could somebody tell me about that?

Dr. R. M. Schwartz: There are several things that I'd like to see done. For example, I'd like to see a *Prochloron* sequence either of a cytochrome c_6 or a ferridoxin.

Dr. Sager: Well let me try to sort of spur somebody on to do this because so much has been done on chloroplast ribosomes and on their antibiotic resistance compared to bacteria. It seems to me it's a rather pivotal position.

Dr. L. Neward (*University of Hartford, Hartford, Conn.*): There are two proteins in photosynthetic eukaryotes. Are they nuclear coded or are they chloroplast coded? Because if they are nuclear coded you are looking at nuclear phylogenetic trees and you are not looking at chloroplast phylogenetic trees. Then you don't need an endosymbiont to come in three times.

Dr. Schwartz: I believe that both are very important to the photosynthetic apparatus. I believe they are chloroplast coded but I'm not absolutely sure.

Dr. Neward: Then how can you say that there's an endosymbiont coming in?

DR. SCHWARTZ: If it were nuclear coded one could argue that there had been transfer of genetic information between the chloroplast and the nucleus, which is you know, sort of an excuse, but I'm really not sure it's known where those two proteins are coded.

DR. T. UZZELL: I'd like to ask you about the ferridoxin sequences. You showed us evidence of a rather old gene duplication in the blue-greens.

DR. SCHWARTZ: The evidence I know of indicates that all of the blue-greens seem to have these two ferridoxins. I don't know whether *Cyanidium* has both

DR. UZZELL: If very early eukaryotes actually retained both loci as the blue-greens do, could you imagine losing one of these by a deletion in some of the eukaryotes and losing the other one in the rest of your eukaryotes so that you had to appeal to no endosymbiosis whatsoever?

DR. SCHWARTZ: You could certainly have lost them; you also could have drawn the tree around another way with the eukaryote lines first and there was gene duplication which led to the double pair in the blue-greens.

DR. UZZELL: Yes, but you could have the blue-greens in, you could have the gene duplication in the blue-greens as you suppose that it is or even in the ancestor of the blue-greens.

DR. SCHWARTZ: If you go back down the tree you do hit a ferridoxin from halobacteria, which is very clearly a prokaryote. It has one ferridoxin sequence and that sequence is continuous; it does not have the deletions.

DR. N. GILLHAM: With respect to the ferridoxins, I believe Sam Wildman showed some years ago that by making inter-specific crosses, ferridoxin in tobacco shows Mendelian inheritance.

DR. SCHWARTZ: You think they are coded in the nucleus?

DR. GILLHAM: Those pieces of evidence would suggest that.

DR. J. SECKBACH: If I'm not mistaken there is a Japanese work where this group claims that *Cyanidium caldarium* has a typical red alga type ferridoxin.

DR. M. MÜLLER (*Rockefeller University, New York, N.Y.*): Would you conclude that *Cyanidium caldarium* might be a red alga and forget about the endosymbiotic speculative hypothesis?

DR. SCHWARTZ: Then you have to explain why the relationship is so close to, for example, *Synechococcus* ferridoxins and why those branches are intermixed with ferridoxins from *Spirulina*.

DR. J. F. FREDRICK: Why can't you assume that *Cyanidium* is a bridge or transition form between the blue-greens and the rhodophytes?

DR. SECKBACH: How about the Japanese work which states quite specifically that *Cyanidium caldarium* is a rhodophyte with red alga type rhodopsin?

DR. SCHWARTZ: Oh, it certainly looks like it in sequence. That's why they show up so closely on the tree.

DR. SECKBACH: So closely and yet separated?

DR. SCHWARTZ: There is some separation. They are not identical. The group in Japan said about it that it's really quite closely related. I've seen trees that they draw and they're very similar to the one that I've shown here. They said that this very close relationship may indeed reflect an endosymbiontic origin.

DR. DUSING (*Yale University, New Haven, Conn.*): Clearly any endosymbiotic event requires finally the integration of the two genomes and loss of genes from the chloroplast. If you have two endosymbiotic events, the likelihood of both of those endosymbionts losing exactly the same genes to the nucleus would give us some understanding of whether or not the events occurred.

DR. SCHWARTZ: It's just very fortuitous.

DR. DUSING: What is our current understanding of the degree of gene similarity in terms of the numbers of groups that are being coded for by chloroplast genomes of one type versus another type?

DR. SCHWARTZ: I simply don't know.

DR. N. WEEDEN (*University of California, Davis, Calif.*): Most of the chloroplast proteins in the higher green plants are coded by the nucleus; very little work has been done on red algae or green algae as far as I know. The small subunit showed the same distribution in most cases but the other proteins, such as the ribosomal proteins, just haven't been studied nor have the enzymatic proteins.

DR. UZZELL: There are two alternatives at least for the ferridoxins for explaining the mixture of prokaryotic and eukaryotic sequences. One is an endosymbiotic and the other is not.

DR. SCHWARTZ: There is one problem with that and that is why the *Synechococcus* from subgroup two appears very closely related to *Synechococcus lividus* and to the chloroplasts of various high organisms on one hand, and on the other it seems to be very distantly related on the 5S ribosomal RNA tree. It's an entirely different branch.

DR. M. V. SIMPSON: It seems to me that there are two general problems in discussing evolution, particularly and making trees with new methods available. We ought to start addressing them. There is still a raging conflict amongst professional evolutionists about the constancy of evolution. The constancy of rate of evolution is a problem that biochemists or biochemical techniques have not really attacked.

The other problem is the question of convergence which nobody has addressed. If it is involved to any serious extent it's going to cause lots of trouble. One perhaps needs some kind of system where one knows an awful lot about the sequences of different genes and about the metabolic characteristics of such a system so that one can be sure that certain kinds of characteristics could not possibly arise from converging evolution. There may be other ways of doing this but I think this particular meeting is, in a way, the first of its kind to start thinking very seriously about how to attack these two problems.

DR. J. LOCKER: The discussion on ferridoxins reminded me that in Dr. Dayhoff's paper, cytochrome *c* was one of the essential proteins that was sequenced to construct this phylogeny, and of course cytochrome *c* is definitely a nuclear gene. What occurs to me is that if one presumes that a symbiosis occurred the protoeukaryote that didn't have a mitochondria must have had an energy transducing system. Quite likely it had a fairly elaborate electron transport system of some variety and so there was not simply a completely new set of genes being introduced to the cell, but in many cases there were duplicate genes in both genomes. Cytochrome *c* in particular is found in a large variety of bacteria that have quite different solutions to electron transport. So the crucial test has to be examination of other kinds of protein sequences. Cytochrome *c* is probably the most widely sequenced protein, it makes it very convenient but it's quite important to compare cytochrome *c* for instance, with a gene that might be much more closely associated then with DNA metabolism, like thymidine kinase, or for that matter with a gene which is clearly mitochondrial like cytochrome *b*. Until we have this kind of sequence data and also until a lot of genera are filled in in the trees, there has to be a kind of ambiguity. The trees are obviously a valid analysis of the data, the big problem comes with the interpretation.

ADAPTATIONS TO ENDOSYMBIOSIS IN GREEN HYDRA *

Clayton B. Cook

Department of Zoology
The Ohio State University
Columbus, Ohio 43210

Introduction

Most of the papers in this symposium have examined properties of organelles with the implicit objective of understanding the evolutionary changes that have influenced these properties. Much of the current research on endosymbiosis has followed this approach, focusing on genetic function and evolutionary changes in organelles and symbionts. Less attention has been given to the evolutionary implications of endosymbiosis for host cells. In this paper I compare some cellular characteristics of green hydra (*Hydra viridis* = *Chlorohydra viridissima*) with some of a nonsymbiotic species, *Hydra littoralis,* in the context of adaptations to endosymbiosis.

Nonsymbiotic hydra species never possess algal symbionts, and do not form stable symbioses with them, in contrast with the aposymbiotic hydra which are derived from symbiotic strains.

The hydra comparison is particularly appropriate for this purpose. Aposymbiotic (free of algal symbionts) strains of *H. viridis* are essentially laboratory constructs, genotypically identical to parental symbiotic strains. Aposymbiotic green hydra are very rarely found in nature. Such rarity suggests reduced fitness of aposymbionts relative to both symbiotic hydra and nonsymbiotic hydra; this has been demonstrated in experimental studies.[1] Given the assumption that symbiotic hydra evolved from nonsymbiotic ancestors, any unique feature of *H. viridis* that reduces fitness of aposymbiotic hydra, but that favors symbiotic hydra, must be an adaptation that was acquired since the formation of the symbiosis. Other specializations of *H. viridis* cells may represent "pre-adaptations" that favored the establishment of the association and that have been retained in subsequent evolutionary history. In the ensuing discussion I shall distinguish between these alternatives when possible.

Adaptations Favoring Formation of the Symbiosis

The green hydra symbiosis can be artificially re-established by the injection of aposymbiotic hydra with appropriate algae.[2, 3] "Infection" of hydra digestive cells involves the endocytosis of algae and the subsequent retention of the symbiont population. One characteristic of cells from both symbiotic and aposymbiotic *H. viridis* is the rapid phagocytosis of potential symbionts. Acceptable algae stimulate the formation of endocytic microvilli (Figure 1A), resulting from immunological interactions between receptors on the

* Supported in part by the National Science Foundation (PCM 75 03380) and by the Graduate School of The Ohio State University.

0077-8923/81/0361-0273 $01.75/0 © 1981, NYAS

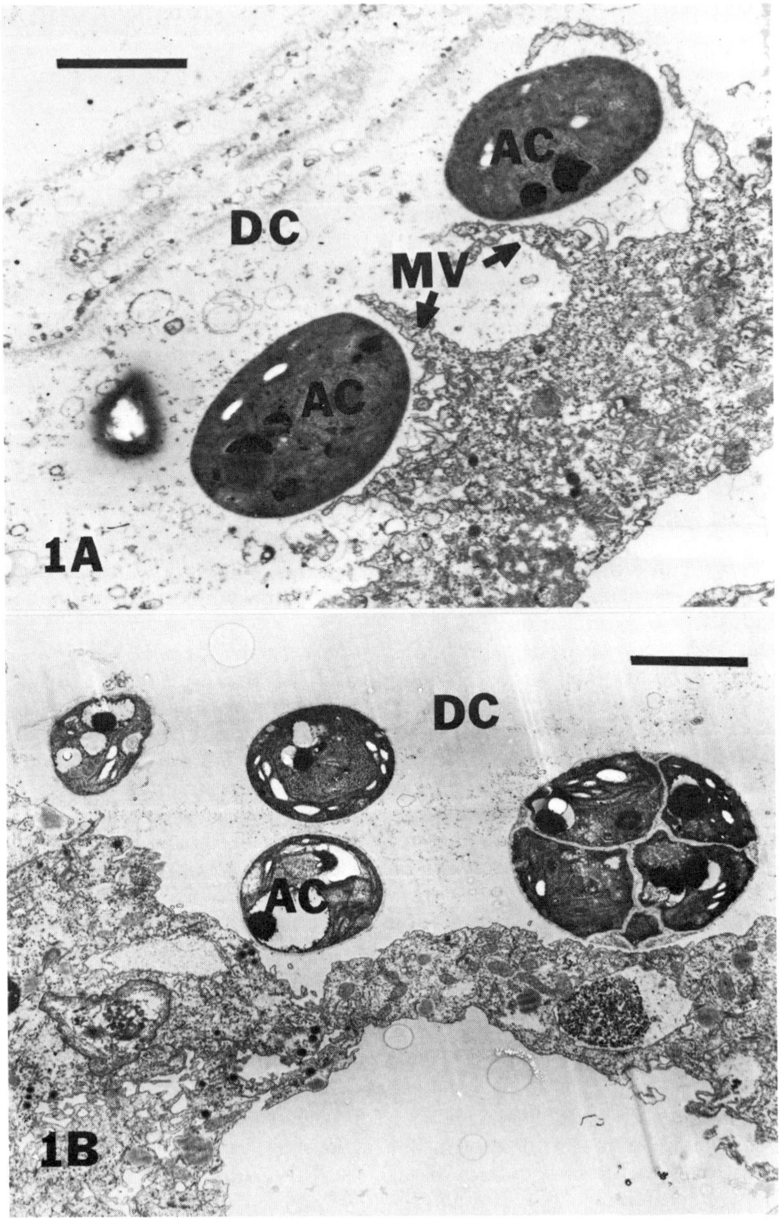

FIGURE 1. Electron micrographs of aposymbiotic *Hydra viridis* injected with symbionts from Florida strain green hydra (1A, top) and with a strain of free-living *Chlorella vulgaris* (1B, bottom). Both hydra were fixed 10 minutes after the injection of algae into the digestive cavity (DC). Endocytic microvilli (MV) have formed only in response to symbiotic algae. AC, algal cells. Scale bars: 1A, 2 μm; 1B, 5 μm.

hydra cell and components, probably glycoproteins, on the algal cell surface.[4] Unacceptable algae do not stimulate microvillar formation (FIGURE 1B). There is a corresponding spectrum of uptake specificity: competent symbionts > incompetent symbionts > free-living algae [see Reference 5 for review]. In contrast, digestive cells from nonsymbiotic hydra have an extremely low affinity for symbiotic algae (TABLE 1) or for *Chlorella*-type algae in general (Cook, unpublished).

The digestive cells in all hydras are phagocytes which take up food materials (cell particulates and free molecules) from the digestive cavity. Kinetic studies [6, 7] and morphological studies [8, 9] indicate that food uptake mechanisms are identical in *H. viridis* and *H. littoralis*. The differences in algal uptake appear to result from the absence of appropriate receptors in *H. littoralis*, or perhaps from the failure of receptors to trigger the membrane events involved in endocytosis.

TABLE 1

UPTAKE OF SYMBIOTIC ALGAE FROM "OHIO" STRAIN HYDRA BY DIGESTIVE CELLS OF APOSYMBIOTIC *Hydra viridis* (OHIO) AND NONSYMBIOTIC *H. littoralis*

Strain (n)	Algae per Cell *	% Cells with Algae (n) **
H. littoralis (6)	0.003 ± 0.007	0.22% ± 0.53 (500)
H. viridis (6)	2.960 ± 0.41	78.00% ± 5.50 (50)

* Includes uninfected cells.

** n = number of digestive cells assayed per hydra.

Algae in cell macerations counted under incident fluorescent excitation (BG-12 exciter, 0–590 barrier) and transmitted phase-contrast illumination. Hydra were fixed 2 hours after injection of algae. Data expressed as $\overline{X} \pm$ S.D.

The second phase in the establishment of the green hydra symbiosis is the maintenance of endocytosed symbionts. Muscatine and Pool [10] and Cook [5] have summarized the problems of digestive cell function that must be circumvented if the algae are to be retained as hereditary endosymbionts. These include exocytosis, degradation by host cell hydrolases, and the regulation of host and symbiont cell division. Digestive cells in *H. viridis* utilize cytoskeletal elements to segregate symbionts from regions of digestion and exocytosis.[3, 5, 11] Muscatine and Pool [10] have suggested mechanisms by which algal populations could be controlled. Because the fate of the few algae endocytosed by digestive cells of nonsymbiotic hydra is not known (except that they are not retained as symbionts), it is not clear if these mechanisms are specific adaptations of *H. viridis* cells or are general features of hydra digestive cells.

METABOLIC ADAPTATIONS

The specializations of *H. viridis* cells that relate to the establishment of the symbiosis involve structural elements of the cells (receptor molecules, microtubules, microfilaments). Following the establishment of the symbiosis, metabolic adjustments may occur that facilitate the exchange of metabolites between host cell and symbiont. Adaptations in carbohydrate metabolism of the symbionts have been reviewed elsewhere.[12, 13] Metabolic adaptations of host cells include uptake of dissolved nutrients and transfer of nutrients to algal endosymbionts.

Nutrient Uptake

Recent work on marine algae-invertebrate associations indicates that host cells have mechanisms for the uptake of dissolved nutrients useful to algal endosymbionts (phosphate,[14] nitrate,[15] and ammonium [15]); these compounds are normally excreted by metazoan cells. Evidence is accumulating that the nitrogen compounds are assimilated by symbiotic zooxanthellae, and that products of these compounds are translocated to hosts (sea anemones and corals).[16]

Cells of *H. viridis* have a greater affinity for dissolved inorganic sulfate than cells of nonsymbiotic hydra (TABLE 2).[17] This has been found in both symbiotic and aposymbiotic strains, although the presence of algae enhances uptake. Like ammonium, sulfate is a metabolite that normally is excreted by animal cells. The sulfate-sulfur derived from inorganic sulfate is incorporated into protein by green hydra tissue, probably in the form of sulfate-reduction products of the algae. This incorporation does not occur in aposymbiotic hydra tissue.[18] Over a 5-day period, symbiotic *H. viridis* retain 75% of the sulfur which they obtain as SO_4^{2-}; aposymbiotic hydra gradually excrete this sulfur (FIGURE 2). The small amount of sulfate-sulfur taken up by nonsymbiotic hydra is retained, in part as sulfated mucopolysaccharide.[18]

When starving hydra are exposed to exogenous sulfate for extended periods, it becomes apparent that high levels of sulfate are toxic in hydra tissue lacking algae. Sulfate stimulates bud production in symbiotic hydra (1 millimolar > 0.1 millimolar > controls), while 1 millimolar sulfate decreases the survival of aposymbiotic hydra (TABLE 3). In contrast, 1 millimolar sulfate has no effect on the survival of *H. littoralis* (TABLE 4). The different effects of sulfate on

TABLE 2

UPTAKE OF RADIOSULFATE BY SYMBIOTIC AND NONSYMBIOTIC HYDRA

Strain	dpm per Hydra *	dpm per μg Protein *
H. viridis (symbiotic)	35978 ± 4518 (3)	3707 ± 435 (4)
H. viridis (aposymbiotic)	14924 ± 4141 (5)	1984 ± 289 (4)
H. littoralis	7755 ± 1633 (5)	121 ± 36 (5)

* Including algae and attached buds. Sample size in parentheses.

Each strain was incubated in $Na_2{}^{35}SO_4$ (10 $\mu c \cdot ml^{-1}$) for 24 hours, followed by a 24 hour chase in sulfate-free medium. Results are expressed as $\bar{X} \pm$ S.D.; sample size in parentheses. Data from Cook, 1976.

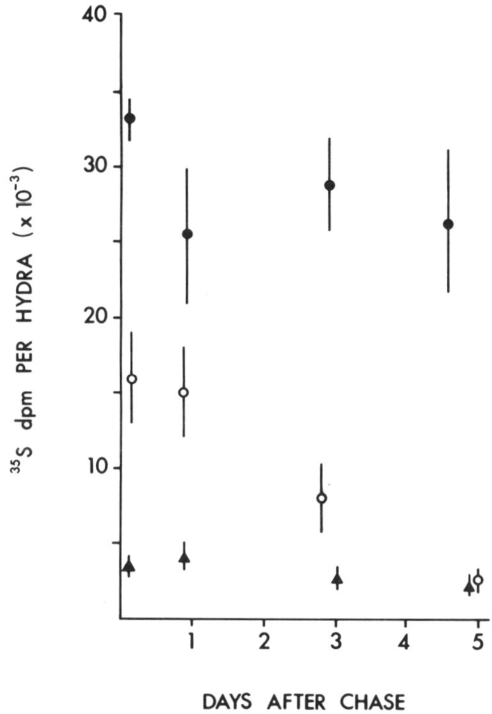

FIGURE 2. The retention of radiosulfur by symbiotic (●) and aposymbiotic (○) *H. viridis*, and by *H. littoralis* (▲). The hydra were incubated with $Na_2{}^{35}SO_4$ for 24 hours (see TABLE 2 for details), and transferred to sulfate-free culture solution. Each point is the mean of five samples, including algae and attached buds. Vertical bars indicate ± one standard deviation.

aposymbiotic and nonsymbiotic hydra are probably due to the greater accumulation of sulfate by the aposymbionts.

Nutrient Transfer to Symbionts and Tissue Turnover

Host cell metabolites obtained from ingested food serve as nutrient sources for symbionts.[19] Studies of the physiology of starvation in symbiotic, aposymbiotic, and nonsymbiotic hydra have indicated that *H. viridis* tissue is adapted for the rapid turnover of cellular constituents, probably acting to either increase the flow of nutrients to the algae, or to make optimal use of translocated algal products. During starvation, aposymbiotic hydra have reduced survival and show a rapid loss of behavioral coordination in comparison to both symbiotic *H. viridis* and nonsymbiotic hydra.[1, 7] In aposymbionts, glycogen and lipid stores are rapidly depleted and total tissue pro-

TABLE 3

SURVIVAL OF FASTED SYMBIOTIC AND APOSYMBIOTIC *H. viridis* (OHIO STRAIN) IN CULTURE SOLUTION WITH ADDED Na_2SO_4

Strain	Days	[SO_4^{2-}]	Number of Hydra
Symbiotic *H. viridis*	0	0	10.0 ± 0.0
		10^{-4} M	10.0 ± 0.0
		10^{-3} M	10.0 ± 0.0
	9	0	14.2 ± 2.4
		10^{-4} M	15.2 ± 1.5 (n.s.)
		10^{-3} M	17.1 ± 1.6 **
	20	0	15.1 ± 3.4
		10^{-4} M	16.3 ± 2.7 (n.s.)
		10^{-3} M	17.7 ± 1.6 *
Aposymbiotic *H. viridis*	0	0	10.0 ± 0.0
		10^{-4} M	10.0 ± 0.0
		10^{-3} M	10.0 ± 0.0
	9	0	10.0 ± 1.3
		10^{-4} M	9.4 ± 1.6 (n.s.)
		10^{-3} M	9.8 ± 1.9 (n.s.)
	15	0	6.5 ± 3.4
		10^{-4} M	5.1 ± 1.9 (n.s.)
		10^{-3} M	2.7 ± 2.0 **

* $P < 0.05$
** $P < 0.01$
Initial hydra cultures contained five hydra, each with one bud. Data expressed as $\overline{X} \pm$ S.D. Data analyzed with analysis of variance; significance levels refer to differences between experimentals and sulfate-free controls. Each point is the result of 8–10 replicates.

TABLE 4

SURVIVAL OF FASTING NONSYMBIOTIC *H. littoralis* IN THE PRESENCE OF ADDED INORGANIC SULFATE

Days	SO_4^{2-}	Number of Hydra
0	0	10.0 ± 0.0
	10^{-4} M	10.0 ± 0.0
	10^{-3} M	10.0 ± 0.0
9	0	11.1 ± 1.5
	10^{-4} M	10.0 ± 0.6 (n.s.)
	10^{-3} M	10.0 ± 0.0 (n.s.)
15	0	10.9 ± 1.2
	10^{-4} M	9.7 ± 0.8 (n.s.)
	10^{-3} M	9.7 ± 0.5 (n.s.)

Experimental and statistical procedures as in TABLE 3.

tein is much reduced in comparison to the other two strains.[20] These differences are reflected in ultrastructural features. Starving aposymbionts show decreases in the amount of epithelial tissue and increased numbers of autophagous vacuoles in comparison to similarly starved symbiotic hydra or *H. littoralis* (FIGURES 3–5).[21]

Other Metabolic Adaptations

Other biochemical differences between *H. viridis* and nonsymbiotic hydra have not been examined. One adaptation could be the occurrence of maltase

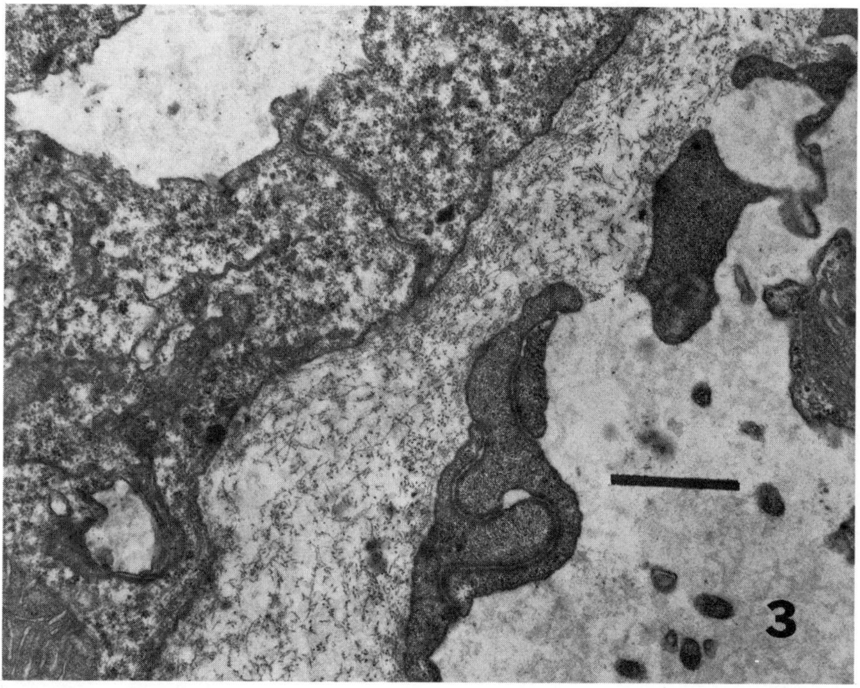

FIGURE 3. Electron micrograph of a one-week starved aposymbiotic *H. viridis*. Note the space between epidermal cells in the ectoderm; ectodermal cells are indicated by arrows. MG, mesoglea; GD, gastrodermal cell layer. Scale bar = 1 μm. (Micrograph courtesy of M. O. Kelty.)

(α-glucosidase) activity in symbiotic hydra tissue. Muscatine[22] detected maltase activity in extracts of tissue from *H. viridis;* the symbionts associated with these hydra released maltose as the principal translocated photosynthetic product. Maltase activity has not been examined in either aposymbiotic or nonsymbiotic hydra, so it is not known if this represents a specific adaptation of green hydra cells. Differences in the activity of catabolic enzymes between the two species have also not been investigated.

ENDOSYMBIOTIC ADAPTATIONS OF HYDRA: PRE-ADAPTATIONS?

It should be clear that some of the adaptations of *H. viridis* cells discussed above must have evolved after the formation of the symbiosis with algae. These adaptations (sulfate-uptake mechanisms, rapid catabolism) increase fitness of symbiotic green hydra during starvation, but reduce the survival of starving aposymbionts compared to both symbiotic and nonsymbiotic hydra. Specializations of *H. viridis* cells that favor the endocytosis and retention of algal symbionts could represent "pre-adaptations", i.e., characteristics of non-symbiotic cells that permitted the establishment of the symbiosis. The major selective force for the retention of these specializations (especially endocytic recognition) is probably the restitution of the association. Aposymbiotic *H. viridis* are occasionally produced during sexual reproduction,[3, 7] and the endocytosis and sequestering of appropriate algal symbionts serve as "insurance" that aposymbiotic cells will be re-infected.

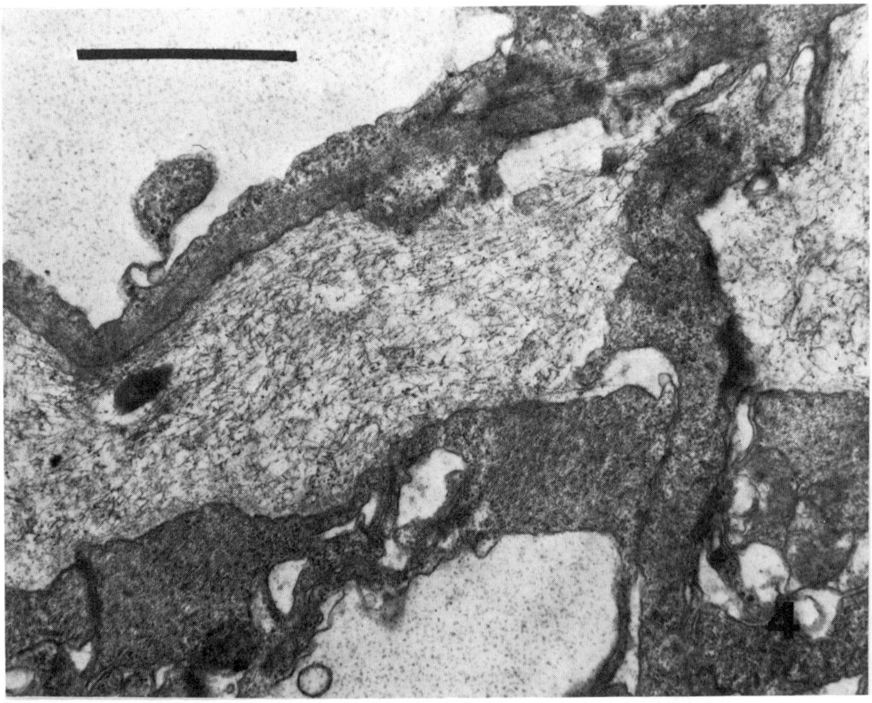

FIGURE 4. Electron micrograph of a one-week starved *H. littoralis*. The section is taken through a region of the body column comparable to FIGURE 3. Few spaces occur between epidermal cells (EC) in the ectoderm. Scale bar = 1 μm. (Micrograph courtesy of M. O. Kelty.)

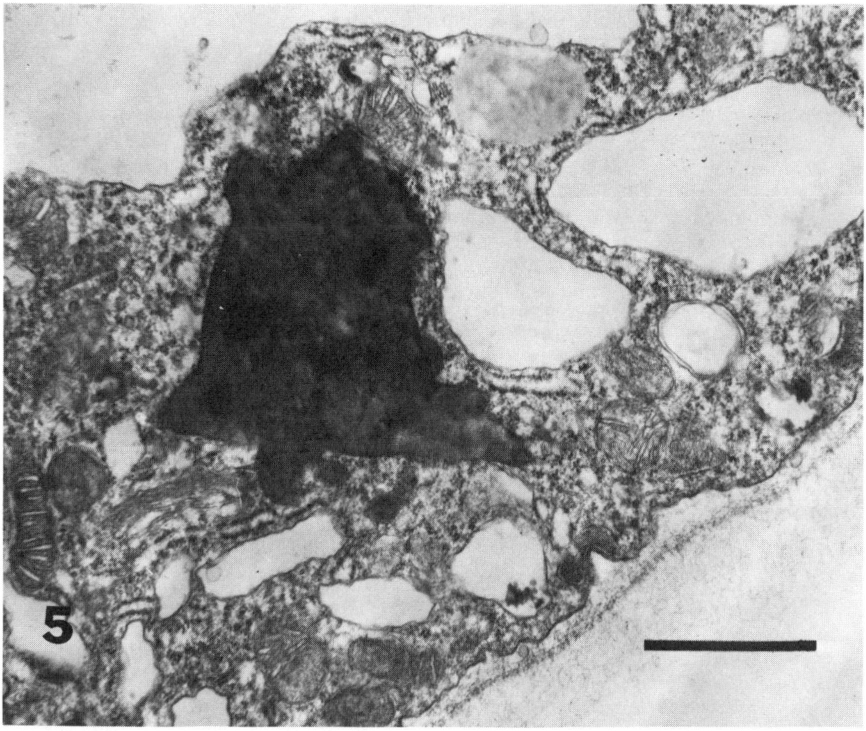

FIGURE 5. An autophagous vacuole in the ectoderm of a one-week starved aposymbiotic hydra. Scale bar = 1 μm. (Micrograph courtesy of M. O. Kelty.)

ACKNOWLEDGMENTS

I am most grateful to one of my former students, Marcia Kelty, for her contributions to this study, including the electron micrographs in FIGURES 3–5. FIGURE 1 represents material fixed while I was an N.I.H. Postdoctoral Trainee with Dr. Leonard Muscatine in the Department of Biology, University of California at Los Angeles.

REFERENCES

1. KELTY, M. O. & C. B. COOK. 1976. Survival during starvation of symbiotic, aposymbiotic and nonsymbiotic hydra. *In* Coelenterate Ecology and Behavior. G. O. Mackie, Ed.: 409–414. Plenum Press. New York, N.Y.
2. PARDY, R. L. & L. MUSCATINE. 1973. Recognition of symbiotic algae by *Hydra viridis*. A qualitative study of the uptake of living algae by aposymbiotic *H. viridis*. Biol. Bull. **145**:565–579.
3. MUSCATINE, L., C. B. COOK, R. L. PARDY & R. R. POOL. 1975. Uptake, recog-

nition and maintenance of symbiotic *Chlorella* by *Hydra viridis.* Symp. Soc. Exp. Biol. **29:**175–203.
4. Pool, R. R. 1979. The role of antigenic determinants in the recognition of potential algal symbionts by cells of *Chlorohydra.* J. Cell Sci. **35:**367–79.
5. Cook, C. B. 1980. Infection of invertebrates with algae. *In* Cellular Interactions in Symbiosis and Parasitism. C. B. Cook, P. W. Pappas & E. D. Rudolph, Eds.: 46–73. The Ohio State University Press. Columbus, Ohio.
6. Lenhoff, H. M. 1961. Digestion of protein in *Hydra* as studied using radioautography and fractionation by differential solubilities. Exp. Cell Res. **23:**335–353.
7. Muscatine, L. & H. M. Lenhoff. 1965. Symbiosis of hydra and algae. II. Effects of limited food and starvation on growth of symbiotic and aposymbiotic hydra. Biol. Bull. **129:**316–328.
8. Slautterback, D. B. 1967. Coated vesicles in absorptive cells of *Hydra.* J. Cell Sci. **2:**563–572.
9. Cook, C. B., C. F. D'Elia & L. Muscatine. 1978. Endocytic mechanisms of the digestive cells of *Hydra viridis.* I. Morphological aspects. Cytobios **23:** 17–31.
10. Muscatine, L. & R. R. Pool. 1979. Regulation of numbers of intracellular algae. Proc. R. Soc. Lond. Ser. B. **204:**131–139.
11. Cooper, C. G. & L. Margulis. 1978. Delay in migration of symbiotic algae in *Hydra viridis* by inhibitors of microtubule polymerization. Cytobios **19:**17–20.
12. Smith, D. C., L. Muscatine & D. H. Lewis. 1969. Carbohydrate movement from autotrophs to heterotrophs in parasitic and mutualistic symbiosis. Biol. Rev. **44:**17–90.
13. Muscatine, L. 1974. Endosymbiosis of algae and cnidarians. *In* Coelenterate Biology: Reviews and New Perspectives. L. Muscatine & H. M. Lenhoff, Eds.: 359–395. Academic Press, Inc. New York, N. Y.
14. D'Elia, C. F. 1977. The uptake and release of dissolved phosphorus by reef corals. Limnol. Oceanogr. **22:**301–315.
15. Muscatine, L. & C. F. D'Elia. 1978. The uptake, retention and release of ammonium by reef corals. Limnol. Oceanogr. **22:**725–734.
16. Muscatine, L. 1980. Uptake, retention and release of dissolved inorganic nutrients by marine alga-invertebrate associations. *In* Cellular Interactions in Symbiosis and Parasitism. C. B. Cook, P. W. Pappas & E. D. Rudolph, Eds.: 229–244. The Ohio State University Press. Columbus, Ohio.
17. Cook, C. B. 1976. Sulfate utilization in green hydra. *In* Coelenterate Ecology and Behavior. G. O. Mackie, Ed.: 415–422. Plenum Press. New York, N. Y.
18. Cook, C. B. 1980. Sulfur metabolism in the green hydra symbiosis. *In* Endosymbiosis and Cell Research. W. Schwemmler & H. Schenk, Eds. (In press.)
19. Cook, C. B. 1972. Benefit to symbiotic zoochlorellae from feeding by green hydra. Biol. Bull. **142:**236–242.
20. Kelty, M. O. & C. B. Cook. 1980. Food levels in starving hydra: the effects of algal symbiosis. (Submitted to J. Exp. Zool.)
21. Kelty, M. O. & C. B. Cook. 1980. In preparation.
22. Muscatine, L. 1965. Symbiosis of hydra and algae. III. Extracellular products of the algae. Comp. Biochem. Physiol. **16:**77–92.

Discussion of the Paper

Dr. R. Sheinin (*University of Toronto, Toronto, Canada*): Did I understand you to infer that the algae provide either energy or pathways for synthesis of the sulfate in mucopolysaccharides by this beast?

Dr. C. B. Cook: No, I didn't say that. We've hydrolyzed that insoluble stuff and there are sulfate-labeled amino acids in the hydrolysate which implies, unless the animal cell is doing it, that the host cell is receiving sulfate reduction products from the algae. It turns out fixed carbon from the algae winds up essentially in all the major biochemical fractions of host tissue. It's translocated as carbohydrate, but as you might expect, it goes everywhere.

METHODICAL ANALOGIES BETWEEN ENDOCYTOBIONT AND DNA-ORGANELLE RESEARCH

Werner Schwemmler and Bart Luuring

Institut für Pflanzenphysiologie und Zellbiologie
Freie Universität Berlin
1000 Berlin 33
Federal Republic of Germany

1. Introduction

The small cicada *Euscelidius variegatus* K. exhibits two types of bacterial, intracellular symbionts (so-called endocytobionts). In accordance with Müller,[1] these have been named primary symbiont *a* and auxiliary symbiont *t* (Figure 1). Only type *a* can occur alone in the cicadas. Type *t* appears, if at all, only in association with type *a*. It has been proposed that both symbiont species belong to the recently described prokaryotic group of the Protoplastoides.[2, 3] For transfer to the next generation, they are inserted between the egg envelope and egg cell of the host, in the form of a symbiont ball, consisting of extracellular infectious stages.[4] In the further course of egg development, these forms are taken up into special host cells (bacteriocytes) that merge in the abdomen to paired lateral symbiont organs (bacterioms). There, each differentiates into its respective intracellular reproductive forms, the vegetative stages.[5]

The cicada endocytobionts show close analogies to DNA-containing cell organelles (plasts) such as mitochondria and plastids (review: Reference 6). These exhibit, as do the cicada endocytobionts, several substructure forms. The mitochondria exist in orthodox and condensation forms; the plastids exist as proplastids, amyloplasts, leukoplasts, etioplasts, chromoplasts, and chloroplasts (Figure 10). Both plasts and infectious endocytobionts are enclosed in a double membrane. The symbiotic vegetative stages even have a triple membrane. The inner membrane stems from the prokaryotic system, the outer membrane from the eukaryotic system. Both have a more or less reduced protein-biosynthesis

Figure 1. Two different forms of fine structure can be distinguished in *a*- and *t*-endocytobionts. The smaller and optically denser infectious form consists of diffuse nucleus material covered by a double membrane, highly granulated plasm, and abundant cell organelles, such as ribosomes, inclusion bodies, mesosomes, and membrane layer bodies (type *a*). The larger and optically more transparent vegetative form has condensed nucleus material, a triple membrane, a less granulated plasm, glycogen inclusions, vacuoles, and a few organelles, such as ribosomes and inclusion bodies.[5] A_1–large, amorphous inclusion body, A_2–small, amorphous inclusion body, Gly–glycogen inclusion, Gra–granula, I_1–infectious division stage, I_2–infectious stage, K_1–spherical, paracrystal inclusion body (q=lateral, l=longitudinal), K_2–hexagonal, paracrystal inclusion body, M_1–symbiontic cytoplasm membrane, M_2–symbiontic membrane with cell wall rudiments, M_3–bacteriocytic cytoplasm membrane of the symbiont, Mes–mesosome, Mi–mitochondrion, Ms–membrane layer body, R–ribosome, TS–division septum, V_1–vegetative division stage, V_2–vegetative stage, and Va–vacuole (b–arising, r–peripheral).

0077-8923/81/0361-0284 $01.75/0 © 1981, NYAS

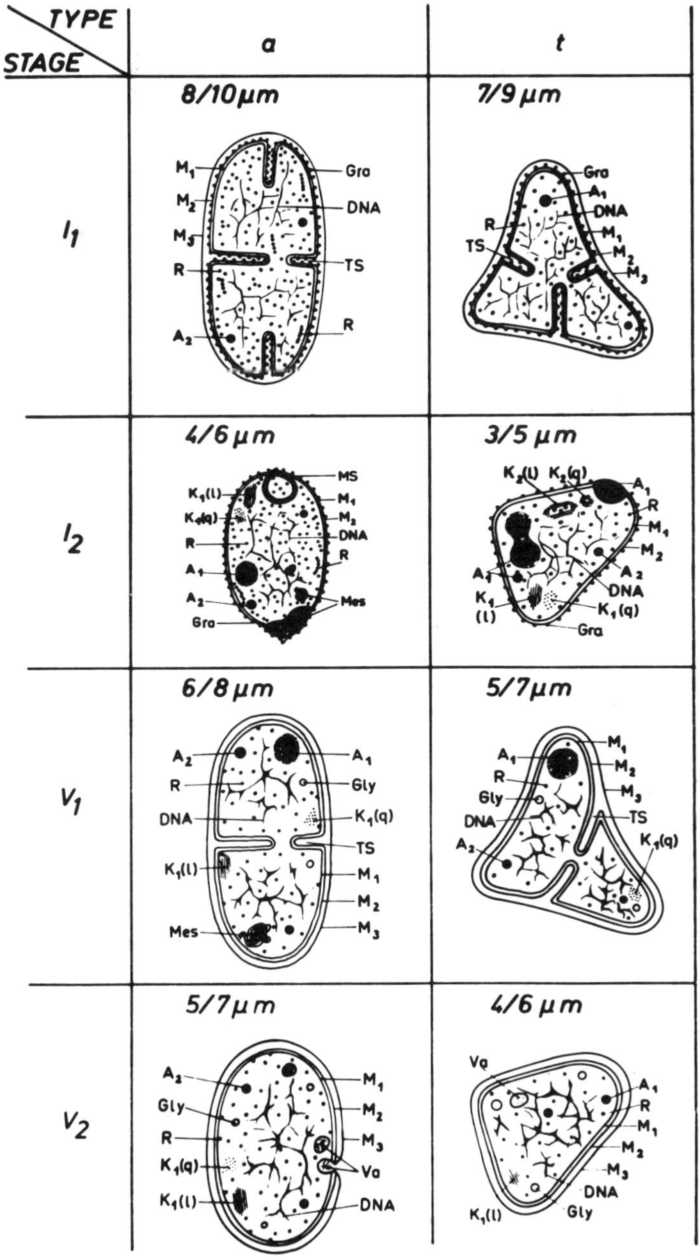
TYPE
STAGE
a
t
I1
I2
V1
V2
8/10 µm
7/9 µm
4/6 µm
3/5 µm
6/8 µm
5/7 µm
5/7 µm
4/6 µm
M1
M2
M3
R
A1
A2
Gra
DNA
TS
MS
K1(l)
K1(q)
Mes
Gly
Va

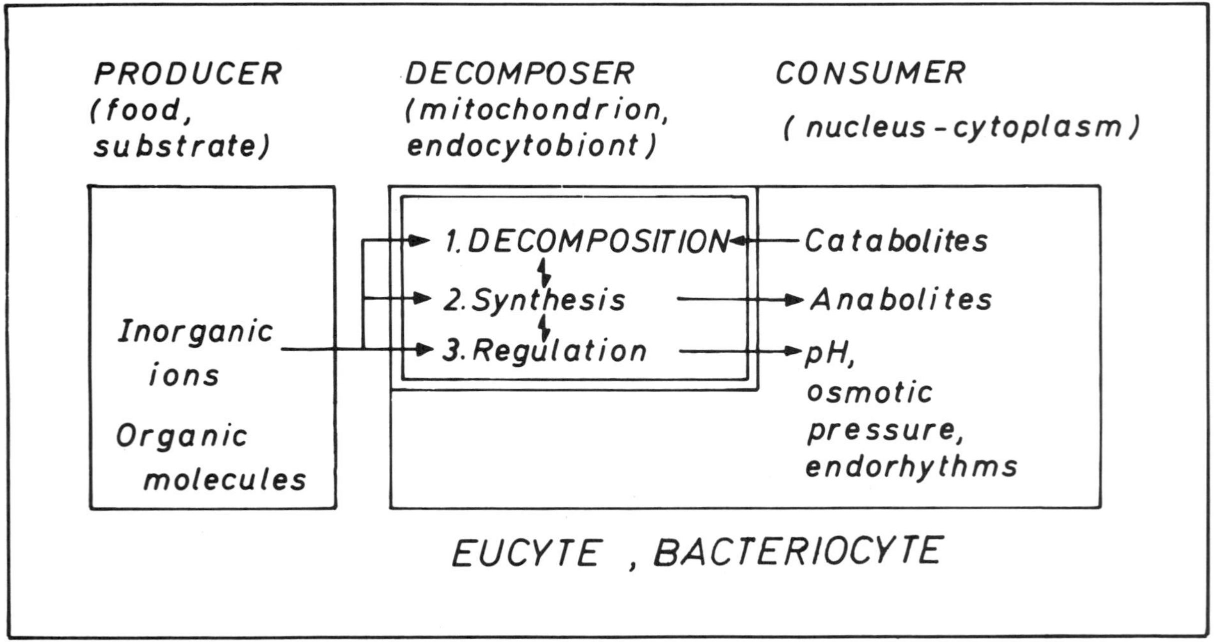

FIGURE 2. Analogous functions of DNA-containing cell organelles and cicada endocytobionts (*a*, *t*), formulated as intracellular ecosystem: Both serve as destruents of food (producer) or eukaryotic cell (consumer) waste products, as synthesizers of essential nutrients, and as regulators of pH, osmotic pressure, and certain endogenous rhythms of the eukaryotic cytoplasm–nucleus system.[6]

machinery, resembling that of the prokaryotes and differing from that of the eukaryotes. The supposed DNA content of the endocytobionts, maximally 10^8 daltons, is only slightly higher than that of the mitochondria, with 10^7 daltons, and approximately equal to that of the plastids. Finally, endocytobionts and plasts use eukaryotic catabolites (i.e. urea and uric acid) to synthesize essential metabolites for the eukaryotic cells (i.e. amino acids and vitamins) (FIGURE 2). In doing so, they also regulate the pH values, osmotic pressure, and certain endogenous rhythms of the eukaryotic cell. They function, in a sense, as intermediaries, between the eukaryotic cytoplasm–nucleus system (consumer) on the one hand, and its food (producer) on the other. They are therefore thoroughly integrated into the eukaryotic cell and are not attacked by its defense mechanisms. Neither the cytoplasm–nucleus system, nor the plasts or the endocytobionts are capable of existing on their own.

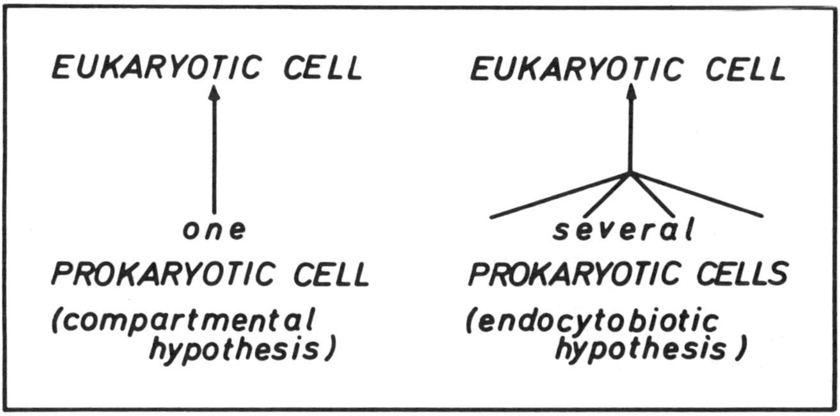

FIGURE 3. The alternative hypothesis of eukaryotic cell origin (details see text and Reference 6).

Two contradictory hypotheses have been proposed to explain the analogies between plasts and cicada endocytobionts (FIGURE 3). The endosymbiontic, or better endocytobiontic, hypothesis accounts for the analogies as follows: *various* prokaryotic endocytobionts, having been integrated into a eukaryotic precursor, carried on an active gene transfer with this precursor, and thereby evolved in the course of millions of years of common evolution into plasts.[6,7] According to the compartmental hypothesis, on the other hand, the plasts are attributed to intracellular differentiation and compartmentation of a *single* prokaryotic cell. The extranuclear DNA originates either from genome duplications of the primitive prokaryotic form, followed by diverging further development,[7,8] or from the migration of plasmids or episomes into membrane vesicles.[9]

2. Materials and Methods

The question to be resolved is, in effect, whether or not DNA-containing cell organelles are endocytobionts and whether certain endocytobionts are DNA organelles. For the experimental investigation of this question, symbiontic methods should be used in conjunction with cell biological methods. *In vitro* experiments with extracellular plastids will be described. These methods have been developed in cicada symbiosis research and have yielded, in this case, indications of the extent of autonomy of the plastids from the nucleus.

Wheat plants (*Triticum vulgare*) that were kept in the fully automatic climate chamber under the standard conditions as the cicada culture [10, 11, 27] are used to isolate sterile chloroplasts. The 15–20 cm long leaves are cut off approximately two weeks after planting and are surface-sterilized for 10 min with 5% sodium hypochlorite solution (ACE) plus a few drops of Tween 20 to reduce surface tension. 100–150 g of this treated leaf material is then washed several times with double-distilled water using sterile procedure and homogenized with a mortar or mixer (i.e. Virtis-Mixer, Stomacher, ball mill) in phosphate buffer (11.73 g Na_2HPO_4, 10.32 g KH_2PO_4, 2 g BSA, 90 g sucrose in 1 l double-distilled water, pH 7.3 and 410 mosmol). The homogenate is cleaned by means of repeated cloth filtrations (Nylon, ø 31 μm) and centrifugation at $900 \times g$ for 7 min. It is passed through a chromatographic column at a running speed of 1 ml/min.[15] Sephadex G 25–50 (coarse) is a suitable column material. Controlled-pore glass (CPG-10, Serva) can also be used, whereby its high adsorption rate can be reduced by using 0.05 M glycerin. CPG-10 has the advantage of being suitable for hot sterilization together with the column. The identification of the sterile chloroplast eluate was by means of fluorescent microscopy (chlorophyll fluoresces red in blue light). The percentage of intact chloroplasts can be determined in a phase-contrast microscope using a counting cell (Thoma chamber). The number of plastids in the eluate can be ascertained in the same manner, following treatment with 0.1 M citric acid and 0.1% crystal violet. The chloroplast substructure is checked by electron-optical means [16, 17] for damage (FIGURE 4).

Pea (*Pisum sativum*) and bean (*Vicia faba*) seeds are employed to obtain sterile etioplast fractions (protochloroplasts). They are superficially sterilized by a short immersion in 70% ethyl alcohol and immersion for ten min in 15% ACE plus one drop Tween 20. Following repeated washings in distilled water the seeds are placed in autoclaved, aired, glass containers filled with soil, using sterile procedures. Cultivation is in a climate chamber in the dark under standard conditions. After 14 days the etiolized leaves are harvested, weighed, and again surface-sterilized (see above) under sterile conditions in a laminar-air box at 4° C, using a green safety light. Approximately 10–20 g of the leaf material per series is processed further; the epidermis is removed and the leaves are cut into small strips. This is followed by plasmolysis in a cell–protoplast washing medium, i.e. CPW (27.2 mg KH_2PO_4, 101 mg KNO_3, 1,480 mg $CaCl_2 \cdot 2\ H_2O$, 246 mg $MgSO_4 \cdot 7\ H_2O$, 0.16 mg KI, 0.016 mg $CuSO_4$, and 6.5 g mannitol in 1 l double-distilled water; pH ~5.7, ~760 mosmol).[18] Next is enzyme treatment with a CPW solution containing 13% mannitol, 0.5% cellulose, and 0.1% maceroenzyme, conducted for 16 hours under gentle shaking.[19] After centrifugation, the bottom fraction is floated for 10 min at $50 \times g$ in CPW plus 20% sucrose. The supernate is again plasmolysed. The protoplast fraction obtained is checked in the suspension medium of Rathnam and Edwards,[18] as to the number and condition of the protoplasts, using a light microscope. Then the protoplasts are osmotically shocked (at approx. 400 mosmol) by adding dithiothreitol (DTT); this causes rupture, allowing the freed etioplasts to be isolated by cloth filtration (Nylon, ø 20 μm) and centrifugation (300 × *g* for 5 min). Here, sterility is a problem that has not yet been solved to our complete satisfaction. The identification of the etioplasts can be achieved by measurement of the absorption spectrum of the protochlorophyll. The electron microscope evaluation is done according to the usual methods in agar blocks (FIGURE 4).[15]

The isolated chloroplasts or etioplasts can then be further analyzed *in vitro*. The synthetic cell culture medium developed for the cicada cells and endocytobionts is used.[19] The additions of 0.1% antibacterial lysozyme and the growth hormone auxin to the plastid medium are significant modifications of the original composition. The exact composition, as well as the pH and osmotic values of the medium, has been expressed in terms of constant and variable

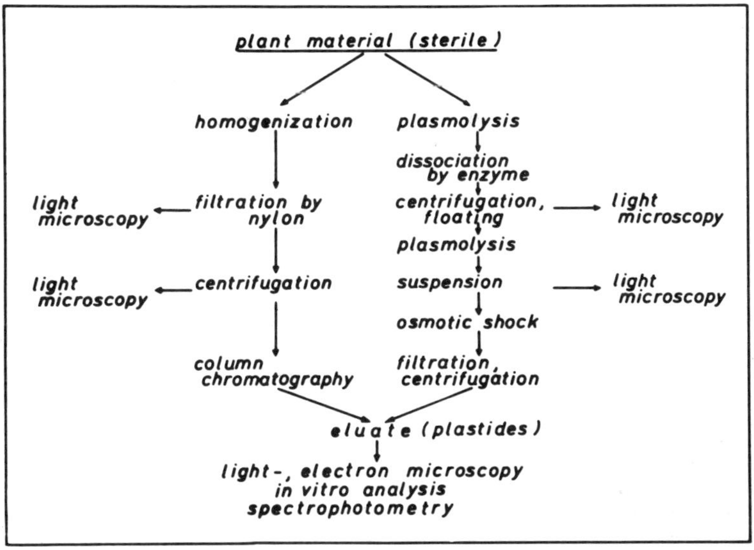

FIGURE 4. Working diagram for the *in vitro* analysis of plastids (details see text).

fractions and is found in TABLE 1. The plastids are conserved in a test tube at 27° C by mixing 2 ml eluate with 2 ml medium and sealing with paraffin. The conservation can also be carried out in hollow ground slides in hanging drops or in special micro-continuous flow chambers.[5, 20] Plastid-free protoplasts or algae or animal cell cultures can then be "reinfected" by means of centrifugation or phagocytosis with the isolated plastids.[21] Intracellular redifferentiation (for example of etioplasts to chloroplasts), as well as growth and propagation, of the plastids after reinfection may be seen as proof of successful *in vitro* vital conservation. Furthermore, an advantage of the *in vitro* method is that morphological changes of the plastids following directed applications (for example, for the purpose of optimizing the medium) may be directly observed with the Invertoskop (Zeiss, Wetzlar) and recorded in microphotographs and microfilms. These observations should in any case be checked by electron microscope analysis (FIGURE 5.4) and biochemical tests, such as those for chlorophyll content, Hill reaction, phosphorylation, CO_2 fixation, and O_2 consumption.[22]

FIGURE 4 briefly summarizes the various steps involved in plastid isolation, *in vitro* analysis, and evaluation.

3. RESULTS

Of all DNA-containing cell organelles, plastids show the greatest genetic autonomy. Their molecular weight of 10^8 daltons almost equals that of the primitive bacteria. They are therefore most suited for *in vitro* experiments using synthetic medium. By adding precursors of essential plastid proteins [23] as a kind of "chemical crutch," these plastids can be extracellularly, vitally conserved for more than two weeks (FIGURE 5.2). This is in agreement with the results of Ridley and Leech [24] and Giles and Sarafis [25] as well as with those

TABLE 1

PLASTID *In Vitro* MEDIUM [27]

Inorganic Components *		
Constant Fraction		
KH_2PO_4	20 mg	
KCl	60 mg	
$CaCl_2 \cdot H_2O$	30 mg	
$MgSO_4 \cdot 7\ H_2O$	110 mg	
NaH_2PO_4	20 mg	
NaCl	180 mg	
Variable Fraction		
Fe^{2+}-EDTA	1.5 mg	
Zn^{2+}-EDTA	1 mg	
Mn^{2+}-EDTA	1 mg	
Cu^{2+}-EDTA	0.5 mg	
$NaHCO_3$	~5 mg	→ pH 7
Organic Components *		
Constant Fraction		
gelose or agar	200 mg	
Bacto-Peptone (Difco, USA)	200 mg	
Yeastolate (Difco, USA)	200 mg	
glucose	400 mg	
ficoll	250 mg	
chicken extract (Lesquendieu, France)	20 mg	
fetal calf serum (Gibco, USA)	20 mg	
bovine serum albumin	1 mg	
Variable Fraction		
DNA/RNA from blue-green algae (1:3) (General Biochem., USA)	50 mg	
ATP, NAD, DPN, CoA, Cytochrome (Serva, BRD)	each 5 mg	
auxin	10 mg	
lysozyme	50 mg	
sorbitol	4,000 mg	→ mosmal 350

* Amounts are mg/100 ml.

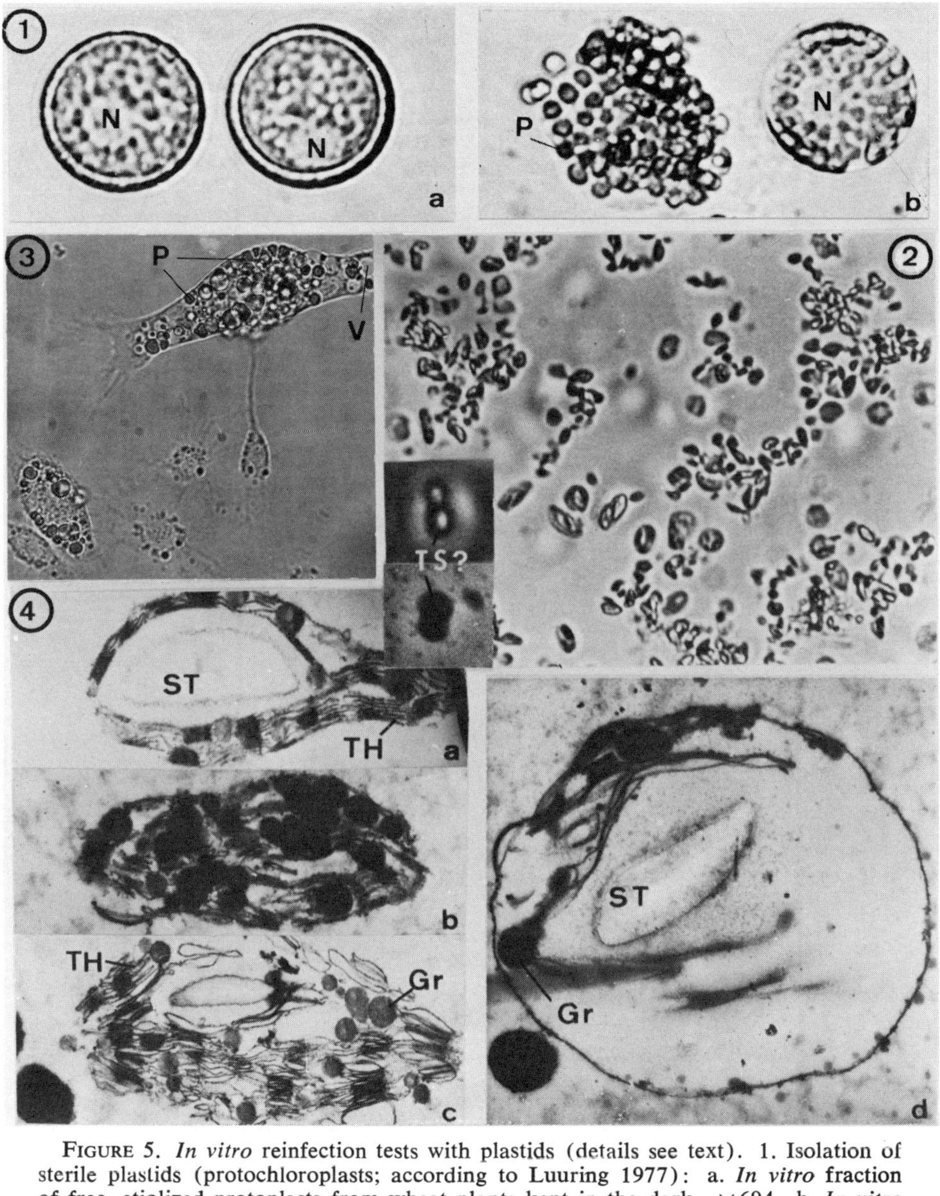

FIGURE 5. *In vitro* reinfection tests with plastids (details see text). 1. Isolation of sterile plastids (protochloroplasts; according to Luuring 1977): a. *In vitro* fraction of free, etiolized protoplasts from wheat plants kept in the dark. ×694. b. *In vitro* fraction of free plastids (protochloroplasts) after rupture of the protoplast cell membrane. ×694. 2. *In vitro* conservation of wheat plastids (chloroplasts) in a synthetic tissue culture medium for one week with division stages (×515, detail ×893). 3. "Infection" of animal cultures with the plastids kept *in vitro* (×340). 4. Analysis of the fine structure of plastids kept *in vitro* for different long periods (chloroplasts): All transitions from intact (a) and poor conservation (b) to total degeneration (c,d) are shown (×8,500). Gr–granula, N–nucleus, TH–thylakoid, TS–division stages, ST–starch particle, and P–plastid.

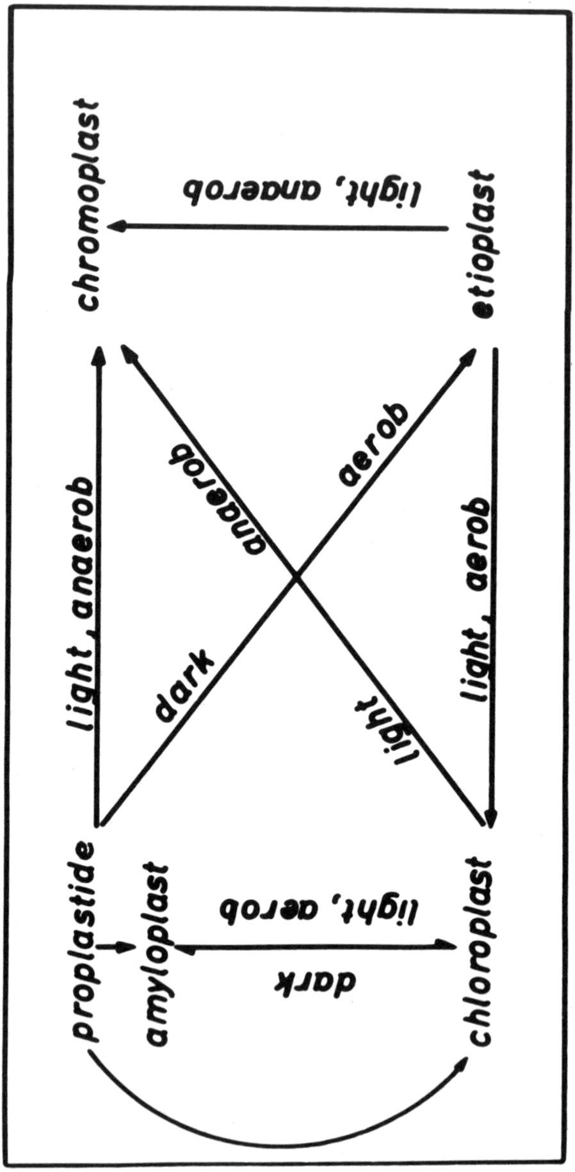

FIGURE 6. Diagram of the various plastid types which, according to medium or substrate conditions, change from one form to another. The different substructure types of the plastids should therefore be the expression of the different metabolic types on which they are based.

of Edmunds and Savillo[26] (see also Reference 15). These authors also report occasional extracellular, *in vitro* division of plastids without growth, which seems to be verified in our experiments by microfilms and counts made before and after *in vitro* procedure (FIGURE 5.2).[27]

Plastids can also be reimplanted into plastid-free plant cells and transplanted into animal cells (FIGURE 5.3). Since we have only recently begun the reinfection experiments, only a few substantial statements can be made. It is known, however, that chloroplasts that have been transplanted into mouse fibroblasts have remained photosynthetically active for weeks.[28] Chloroplasts extracted from green algae survive in cultures of gut cells from certain sea slugs.[29, 30] It should therefore be possible to investigate the autonomy from the nucleus or the vitality of *in vitro*, extracellularly changed plastids after their reimplantation into an eukaryotic system, with the help of reinfection tests. Division or differentiation, for example from etioplast to chloroplast, is taken as proof of vitality and can be checked light microscopically (FIGURE 6).

The reinfection tests were developed using cicada endocytobionts.[20] Like the plastids, they can be vitally conserved *in vitro* outside their bacteriocytes for a limited time only and exhibit, if any, only one division stage (FIGURE 7.6). The vitality of the symbionts kept *in vitro* can be effectively demonstrated by reinfecting the bacteriocytes with them (FIGURE 7.5).

DISCUSSION

Because of the close analogies between DNA-containing organelles and certain endocytobionts, methods of symbiontic research and cell biology can be employed so as to mutually supplement each other. The consistent use of combined methods of *in vitro* analysis should help to elucidate the autonomy or dependence of plastids and endocytobionts from their nucleus–cytoplasm system, with regard to structure, function, and information.

The combination of *in vitro* analysis with enzyme tests could clear up another important problem, i.e., the intrinsic correlation between substructural changes and metabolic changes in plastids or endocytobionts. For example, certain blue-green algae switch from anaerobic fermentation in the dark to aerobic photosynthesis in light by way of specific substructural transformation (e.g. synthesis of thylakoids and photopigments). Since the plastids are today generally considered to be of the same origin as blue-green algae,[6, 31] it is probable that the different substructural types of plastids (amyloplasts, leukoplasts, proplastids, etioplasts, chromoplasts, chloroplasts) also indicate different metabolic types such as fermentation, respiration, photosynthesis, etc. (FIGURE 6).[6]

The distribution of special substructural differences between aerobic, infectious and anaerobic (?) vegetative cicada endocytobionts may also be attributable to significant metabolic differences (FIGURE 1).[27] The enzyme tests that have begun to apply to the elucidation of this question make such an assumption probable. However, the nature of these metabolic differences is as yet uncertain.[12, 13, 14]

Such investigations could provide a definite answer to the question posed at the beginning of this paper. Are the DNA-containing cell organelles of an endocytobiontic nature and are some endocytobionts of an organelle nature? The research that has resulted from the search for answers to this question has

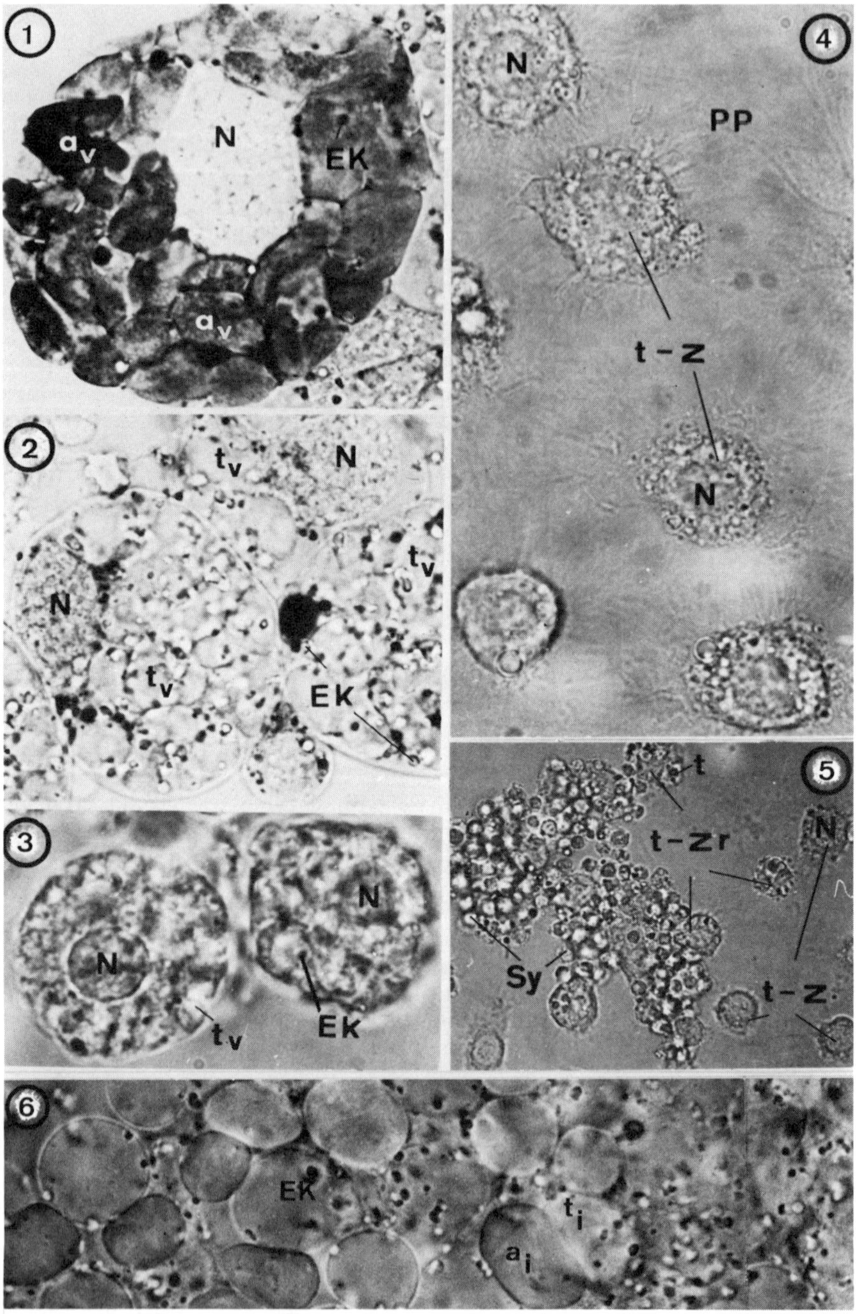
1
N
a_v
EK
a_v
2
t_v
N
t_v
N
t_v
EK
3
N
N
t_v
Ek
4
N
PP
t-Z
N
5
t
t-Zr
N
Sy
t-Z
6
EK
t_i
a_i

FIGURE 7. *In vitro* reinfection tests with cicada endocytobionts.[20] 1. Vitally stained *a*-bacteriocyte with *a*-endocytobiont *in vitro* (×893). 2. Vitally stained *t*-bacteriocyte with *t*-endocytobiont *in vitro* (×893). 3. Two-week old *in vitro* culture of *t*-bacteriocytes carrying symbionts (×893). 4. Two-week old culture of noninfected *t*-cells (prospective *t*-bacteriocytes) (×850). 5. Reinforced *t*-cell cultures with *in vitro* conserved endocytobionts (see FIGURE 6) (×298). 6. *In vitro* conservation of infectious endocytobionts of the symbiont ball for one week (×893). $a_{i,\ v}$–primary symbiont *a* (infectious, vegetative), EK–symbiontic inclusion body, HS–heterochromatin, N–nucleus, PP–pseudopodia, Sy–symbiont, $t_{i,\ v}$–auxiliary symbiont *t* (infectious, vegetative), and t-Z(r)–t-cell (reinfected).

led to the establishment of endocytobiology as a synthesis of recent research between symbiosis and cell biology.[31]

REFERENCES

1. MÜLLER, H. J. 1949. Zur Systematik und Phylogenie der Zikaden-Endosymbiosen. Biol. Zbl. **68:**343–368.
2. SCHWEMMLER, W. 1971. Intracellular symbionts: a new type of primitive prokaryotes. Cytobiologie **3:**427–429.
3. SCHWEMMLER, W., G. HOBOM & M. EGEL-MITANI. 1975. Isolation and characterization of leafhopper endosymbiont DNA. Cytobiologie **10:**249–259.
4. KÖRNER, H. 1978. Intraovarially transmitted symbionts of leafhopper. Zoolog. Beitr. **24:**59–68.
5. SCHWEMMLER, W. 1974. Studies on the fine structure of leafhopper intracellular symbionts during their reproductive cycles (Hemiptera: Deltocephalidae). Appl. Ent. Zool. **9:**215–224.
6. SCHWEMMLER, W. 1979. Mechanismen der Zellevolution—Grundriss einer modernen Zelltheorie. Walter De Gruyter Verlag. Berlin.
7. PARTHIER, B. 1975. Zur Evolution von Chloroplasten und Mitochondrien. Nova Acta Leopoldina N.F. **42:**223–239.
8. CAVALIER-SMITH, T. 1975. The origin of nuclei and of eukaryotic cells (review). Nature **256:**463–468.
9. RAFF, R. A. & H. R. MAHLER. 1972. The non symbiotic origin of mitochondria. Science **177:**575–582.
10. SCHWEMMLER, W., J.-L. DUTHOIT, G. KUHL & L. VAGO. 1973. Sprengung der Endosymbiose von *Euscelis plebejus* F. und Ernährung aposymbiontischer Tiere mit synthetischer Diät (Hemiptera, Cicadidae). Z. Morph. Tiere **74:** 297–322.
11. ZABEL, U. & W. SCHWEMMLER. Darstellung der Insekten-Embryogenese am Beispiel der Kleinzikade *Euscelidius variegatus* K. (Jassidae, Cicadina, Homoptera). Biologie in unserer Zeit. **10:**120–123.
12. KNY, U. 1976. Histochemische Untersuchungen infektiöser und vegetativer Zikadenendocytobionten. Diplomarbeit (Master's thesis), Freie Universität Berlin :1–45.
13. GLAUERT, A. M. & R. H. GLAUERT. 1958. Araldite as an embedding medium for electron microscopy. J. Biophys. Biochem. Cytol. **4:**191–195.
14. MONNERON, A. & W. BERNHARD. 1966. Action de certaines enzymes sur des tissues inclus en epon. J. Meroscop. **5:**697–713.
15. WELLBURN, A. R. & F. A. M. WELLBURN. 1971. A new method for the isolation of etioplasts with intact envelopes. J. Exp. Botany **22:**972–979.

16. Burstone, M. S., Ed. 1962. Enzyme histochemistry. Academic Press, Inc. New York, N.Y.
17. Geyer, G. 1973. Ultrahistochemie. Gustav Fischer Verlag. Stuttgart.
18. Rathnam, C. K. M. & G. E. Edwards. 1976. Protoplasts as a tool for isolating chloroplasts from leaves. Plant Cell Physiol. **17:**177–186.
19. Gosch, G., Y. P. S. Bajaj & J. Reinert. 1975. Isolation, culture and induction of embryogenesis in protoplasts from cell-suspensions of *Atropa belladonna*. Protoplasma **86:**405–410.
20. Schwemmler, W. 1973. In vitro Vermehrung intrazellulärer Zikadensymbionten und Reinfektion asymbiontischer Myzetozytenkulturen. Cytobios **8:**63–73.
21. Bajaj, Y. P. S. 1974. Potentials of protoplast culture work in agriculture. Euphytica **23:**633–649.
22. Renger, G., D. Difiore, B. Luuring & P. Gräber. 1979. The variation of the electrochromic difference spectrum at various stages of the chloroplast development. Z. Naturforsch. **34c:**120–124.
23. Berthold, M. 1979. Analyse der Interaktionen zwischen Kern- und Chloroplastengenom (Review). Staatsexamensarbeit (Master's Thesis), Freie Universität Berlin :1–185.
24. Ridley, S. D. & R. M. Leech. 1970. Division of chloroplasts in an artificial environment. Nature **227:**463–465.
25. Giles, K. L. & V. Sarafis. 1971. On the survival and reproduction of chloroplasts outside the cell. Cytobios **4:**61–74.
26. Edmunds, L. & R. L. Savillo. 1974. Long-term in vitro culture of cell-free preparations of chloroplast isolated from *Euglena gracilis*. *In* Proceedings of the Third Internatl. Congress on Photosynthesis. M. Avron, Ed.:1719–1730. Elsevier Scient. Publ. Co. Amsterdam.
27. Schwemmler, W. 1974. Analyse elementarer Zusammenhänge von Struktur, Funktion und Evolution der Zikadensymbiose zur Ableitung eines allgemeinen Endosymbiose-Konzepts. Habilitationsschrift, Freie Universität Berlin :1–400.
28. Nass, M. M. K. 1969. Uptake of isolated chloroplasts by mammalian cells. Science **165:**1128–1131.
29. Trench, R. K. 1969. Chloroplasts as functional endosymbionts in the mollusc *Tridachia crispata* (Bergh.) (Opisthobranchia, Sacoglossa). Nature **222:**1071–1072.
30. Trench, R. K., R. W. Greene & B. G. Bystome. 1969. Chloroplasts as functional organelles in animal tissues. J. Cell Biology **42:**404–417.
31. Schwemmler, W. M. & H. E. A. Schenk, Eds. 1980. Endocytobiology. Endosymbiosis and Cell Biology. Proc. Intl. Coll. on Endosymbiosis and Cell Research. Tübingen, April 1980. Walter de Gruyter. Berlin.

Discussion of the Paper

Dr. K.-S. Chiang: If I heard correctly you mentioned a mitochondrial species, the DNA of which is 10^8 daltons. Do you know anything more about that?

Dr. Müller (*translating for Dr. Schwemmler*): Would you repeat that question slowly?

Dr. Chiang: Oh, sorry. I think he mentioned a mitochondria species with a genome kinetic complexity of 10^8 daltons. Does he know anything else about this piece of DNA?

Dr. Müller: This is a value which was rounded up. And he thinks that that gives you the answer. The cyclomitochondria have a higher DNA content, a higher molecular weight than the plant mitochondria do.

Dr. Chiang: Yes, I understand but still 10^8 daltons was probably the largest ever known perhaps.

Dr. Müller: He agrees with your point.

Dr. Chiang: May I ask how this kinetic work was performed?

Dr. Müller: First of all with radioactive labeling, sedimentation analysis, and conformation analysis. They got circular DNA and therefore they feel that the values correctly determined.

Ms. B. Davis (*University of Massachusetts, Boston, Mass.*): What kind of cell of the cicada is involved in the plastid infection and does it play an active part?

Dr. Müller: Plastid chloroplast? [Yes] These were those cells which normally contain the bacteria.

Ms. Davis: And do they phagocytize actively?

Dr. Müller: The question is still not answered but they go through an active phagocytosis through a lysis of the cell membrane. He has preliminary evidence for just phagocytosis.

MOLECULAR APPROACHES TO THE ENDOSYMBIOTIC HYPOTHESIS

Derek O. Phillips and Noel G. Carr

Department of Biochemistry
University of Liverpool
Liverpool L69 3BX
England

Background

After the initial speculation on the origins of chloroplasts around the turn of the century, relatively little attention was paid to these suggestions which so disturbed accepted biological opinion. With the advent of modern biology, the last thirty years has seen an increasing array of techniques applied to the relationship of eukaryotic organelles to free-living prokaryotes. More recently the emphasis of the questions asked has changed, with increasing attention being paid to the nature of the recipient, proto-eukaryotic cell and the co-evolution of eukaryotic cell and organelles. In a broad sense, attempts to understand the origin of chloroplasts and mitochondria have proceeded on three fronts, which succeeded and grew from each other. The physicochemical characterization of the machinery of protein synthesis and of gene structure supplied immediate comparisons between organelles, prokaryotes, and eukaryotic cytoplasm; the 70S and 80S ribosome distribution being the most vivid indication that structures of crucial importance had occurred in common. Examination of the shape and size of macromolecules was supplemented in due course by the understanding of chemical constitution and arrangement. The procedures of nucleic acid hybridization, and later, of nucleic acid sequencing, developed rapidly and have been central in the formation of current opinion on the origin of organelles. This third front, started much later and still as yet much less defined, asks questions as to the evolution of organelles and eukaryotic cells and to how, at what stage, and to what extent these integrate and coordinate their metabolism and growth.

By the mid-seventies, DNA/RNA hybridization[1] and direct sequencing (oligonucleotide cataloging) studies[2,3] had shown unequivocally that chloroplast and eubacterial ribosomal RNAs (rRNAs) belong in a closely knit family of sequences that share minimal homology with the cytoplasmic rRNAs of eukaryotes. Justifiably, these facts were taken as clear-cut evidence of an endosymbiotic origin for chloroplasts. Yet one or two lingering doubts remained: perhaps the size differences between prokaryotic and eukaryotic rRNAs rendered them less than directly comparable; perhaps also, the lack of sequence homology between eubacterial and cytoplasmic rRNAs reflected not the antiquity of their divergence, but rather, resulted from some hidden event intimately connected with the origin of the eukaryotic condition itself. To try and overcome these problems, and conclusively distinguish between the direct filiation and endosymbiotic hypotheses for the evolutionary origin of chloroplasts, we directed our attention to one rRNA species (the 5S RNA) and one modified rRNA sequence (the two adjacent dimethyladenosine residues and their flank-

0077–8923/81/0361–0298 $01.75/0 © 1981, NYAS

ing nucleotides located near the 3′-ends of 16S and 18S rRNAs) which both appeared to bridge the prokaryote/eukaryote discontinuity in somewhat less equivocal manners. We asked two questions: in a single photosynthetic eukaryote is the chloroplast 5S rRNA more similar to its cytoplasmic counterpart or to the 5S rRNA of some free-living photosynthetic prokaryote?; and, did the dimethylation of two adjacent adenosine residues in 16S/18S rRNAs arise before or after the divergence of the cyanobacteria from the main path of microbial evolution? Definitive answers to both these questions have been obtained. Yet, if anything, the basic enigma of the early development of the eukaryotic condition today seems more puzzling than ever.

Questions with Answers

Chloroplast 5S rRNA: Is It More Similar to Prokaryotic or Eukaryotic 5S rRNAs?

The large subunits of all ribosomes, except those of animal and fungal mitochondria, contain a small ribonucleic acid, about 120 nucleotides long, termed 5S rRNA.[4] To date, chloroplast 5S rRNAs have been detected in several flowering plants,[5] and in the photosynthetic protists, *Chlorella pyrenoidosa,*[6] *Euglena gracilis,*[7] and *Chlamydomonas reinhardii.*[8] Two techniques have been successfully used to determine the prokaryotic or eukaryotic affinities of chloroplast 5S rRNAs: DNA/RNA hybridization and complete RNA sequence analysis.

We chose the technique of DNA/RNA hybridization. Our experiments[9] involved the isolation of 5S rRNAs (by sequential phenol extraction, DEAE-cellulose chromatography, Sephadex gel filtration, and preparative polyacrylamide gel electrophoresis) from seven microorganisms: four cyanobacteria—*Anabaena variabilis, Nostoc* sp. MAC, *Synechococcus* AN (*Anacystis nidulans*), and *Synechocystis* 6308 (*Gloeocapsa alpicola*)—a single aerobic eubacterium, *Escherichia coli,* and the protists, *Euglena gracilis* * and *Euglena* W_{15}ZHL (a chloroplast DNA-free mutant strain of *Euglena gracilis* Z). Each 5S rRNA was labeled *in vitro* to high specific activity with [^{125}I]iodide and allowed to hybridize with denatured (single-stranded) *Euglena gracilis* chloroplast DNA bonded to nitrocellulose membrane filters using standard procedures.

Figure 1 shows the plots of a series of hybridizations of ^{125}I-labeled 5S rRNAs with *Euglena gracilis* chloroplast DNA. For each 5S rRNA the degree of hybridization increased linearly with increasing RNA input; none had any tendency to saturate the chloroplast DNA. Each line is therefore drawn as the statistically best fit through its points, and represents the median of the regression of *y* on *x* and the regression of *x* on *y*. Linearity was expected, as the maximum level of hybridization attained in each case—even by total *Euglena gracilis* 5S rRNA (0.00895%)—fell well below the predicted saturation value

* *Euglena gracilis* total 5S rRNA proved impossible to separate into cytoplasmic and chloroplast components by preparative polyacrylamide gel electrophoresis—even in 12.5% gels.

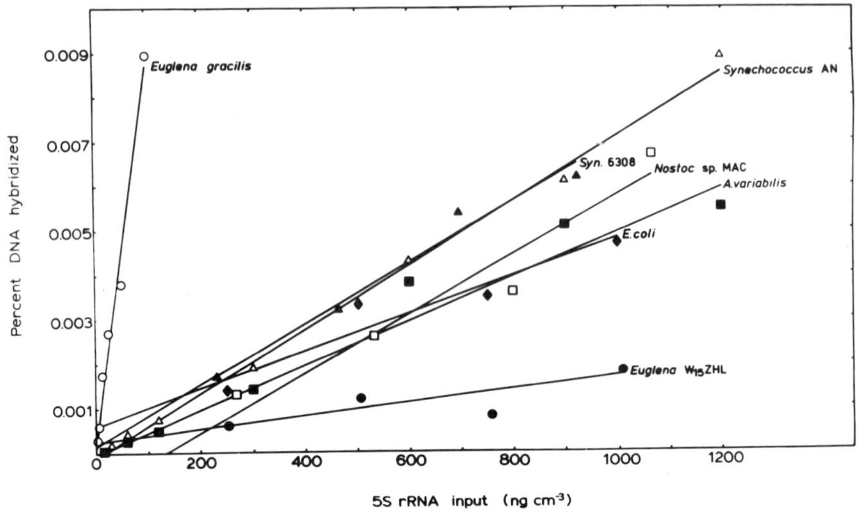

FIGURE 1. Hybridization of 5S rRNAs with *Euglena gracilis* chloroplast DNA. Filters containing 5μg of *Euglena gracilis* chloroplast DNA were annealed individually with increasing concentrations of ^{125}I-labeled 5S rRNAs from *Euglena gracilis* (○), *Synechococcus* AN (△), *Synechocystis* 6308 (▲), *Nostoc* sp. MAC (□), *Anabaena variabilis* (■), *Escherichia coli* (◆), and *Euglena* W_{15}ZHL (●) as described.[19]

for *Euglena gracilis* chloroplast 5S rRNA alone (0.13%),† so all points would have remained on the initial linear portion of the titration curve.

The hybridization of *Euglena gracilis* 5S rRNA must have been due almost entirely to a fraction lacking in the 5S rRNA from *Euglena* W_{15}ZHL, its electrophoretic counterpart: this fraction is concluded to have been chloroplast 5S rRNA. The high efficiency of the homologous hybridization of the chloroplast 5S rRNA contrasted markedly with the low efficiencies of the heterologous hybridizations. For example, an input concentration of 1200 μg/ml of *Synechococcus* AN 5S rRNA was required to reach the same hybridization level attained by *Euglena gracilis* chloroplast 5S rRNA at an input of less than 25 μg/ml—assuming that chloroplast 5S rRNA accounted for about one-quarter of the total *Euglena gracilis* 5S rRNA.[12, 13] Yet despite their comparative inefficiencies the heterologous hybridizations displayed a definite gradation; moreover, all prokaryotic 5S rRNAs showed more homology with *Euglena gracilis* chloroplast DNA, and consequently its chloroplast 5S rRNA, than did the cytoplasmic 5S rRNA of *Euglena* itself.

In TABLE 1 all hybridizations are compared at a common arbitrary DNA/RNA ratio of 5/1. The values do differ slightly from those reported earlier,[9]

† The predicted value of 0.13% for the saturation of *Euglena gracilis* chloroplast DNA by its homologous 5S rRNA is based on: (1) a molecular mass of 92×10^6 daltons for *Euglena gracilis* chloroplast DNA;[10] (2) estimates of three tandemly repeated compound rRNA transcription units per molecule of *Euglena gracilis* chloroplast DNA;[11] and (3) an approximate molecular mass of 40,000 daltons for *Euglena gracilis* chloroplast 5S rRNA.

due to slight modifications—hopefully, improvements—in the statistical methods applied. By the new methods, the 5S rRNAs of the two unicellular cyanobacteria, *Synechococcus* AN and *Synechocystis* 6308, share the greatest—almost equal—homology with *Euglena gracilis* chloroplast DNA, the 5S rRNAs of the two filamentous cyanobacteria being slightly less related. Among the prokaryotes, *E. coli* 5S rRNA has the minimum homology, but the cytoplasmic 5S rRNA of *Euglena* is by far the least related of all.

Thermal analysis confirmed that the hybridization results were genuine and not due to nonspecific interactions. The thermal stability profiles are shown in FIGURE 2. Incubation at 50° C for 10 min removed between 20 and 50% of the total radioactivity bound on each filter. This was probably due to the melting of short oligoribonucleotides produced by ribonuclease treatment. Plateaus were reached around 55° C. The counts released above this temperature in the case of the prokaryotic 5S rRNAs (plus *Euglena gracilis* 5S rRNA) show the characteristic S-shaped curves that result from the melting of longer sections of genuine hybrid. The 5S rRNAs of *Synechococcus* AN (FIGURE 2B) and *E. coli* (FIGURE 2F) produced the sharpest profiles, with mean hybrid melting temperatures (placing the baselines at 55° C) of 80.5° C and 79.5° C, respectively. If in each case these profiles reflected the release of a single oligoribonucleotide specifically annealed with *Euglena gracilis* chloroplast DNA, calculations would suggest that they resulted from the melting of 12–14 nucleotides in the former, and 14–16 nucleotides in the latter.

The existence of long stretches of sequence homology between chloroplast and cyanobacterial 5S rRNAs has recently been confirmed.[14] In FIGURE 3 Dyer and Bowman's complete sequence of *Lemna minor* (duckweed) chloroplast 5S rRNA is compared with the 5S rRNAs from *Lemna* cytoplasm and *Synechococcus* AN. The picture tells its own story. The overall quantity of homology between the chloroplast and cyanobacterial 5S rRNAs (68.8%) far exceeds that between the chloroplast and cytoplasmic 5S rRNAs of *Lemna* itself (49.6%). Flowering plant chloroplast 5S rRNAs are slightly more conserved than their cytoplasmic counterparts, but only an endosymbiotic origin for chloroplasts can adequately explain these results.[14]

TABLE 1

RELATIVE HYBRIDIZATION OF 5S rRNAS WITH *Euglena gracilis* CHLOROPLAST DNA

Sources of 5S rRNA	% of DNA hybridized ± one standard error of estimate at DNA/RNA ratio of 5/1 (×10⁴)
Euglena gracilis	871.5±3.2
Synechococcus AN	71.8±1.9
Synechocystis 6308	70.1±2.9
Nostoc sp. MAC	57.8±5.2
Anabaena variabilis	49.5±3.0
Escherichia coli	48.1±3.5
Euglena W_{15}ZHL (cytoplasm)	17.2±2.9

After Phillips & Carr [9]

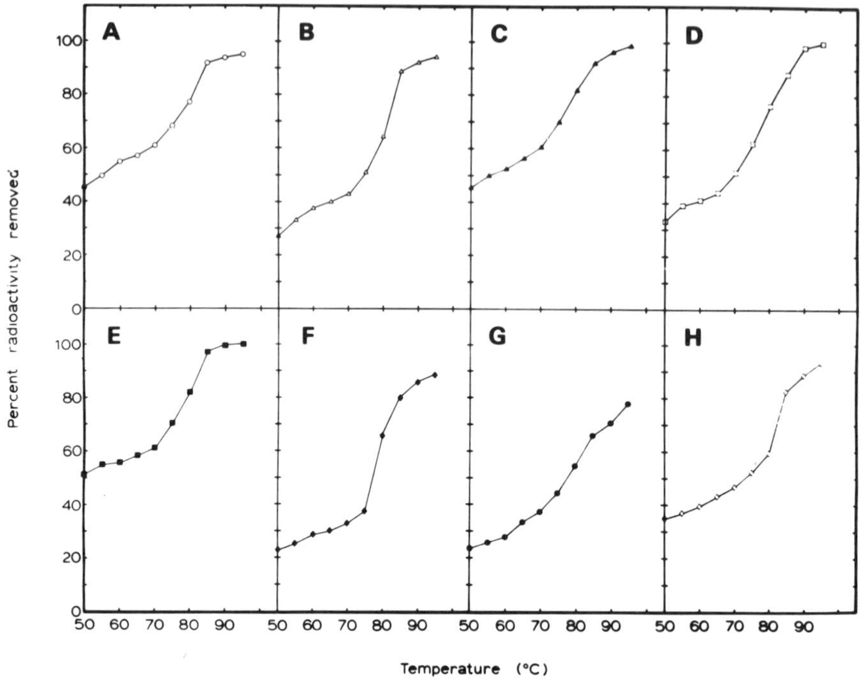

FIGURE 2. Thermal stability profiles of *Euglena gracilis* chloroplast DNA-5S rRNA hybrids. Sources of 5S rRNA: A, *Euglena gracilis;* B, *Synechococcus* AN; C, *Synechocystis* 6308; D, *Nostoc* sp. MAC; E, *Anabaena variabilis;* F, *Escherichia coli;* G and H, *Euglena* $W_{15}ZHL$. A–G were hybridized with *Euglena gracilis* chloroplast DNA of buoyant density 1.685 g cm^{-3}. H was hybridized with *Euglena gracilis* DNA of mean buoyant density 1.700 g cm^{-3}. Hybrids were formed and melted as described.[19]

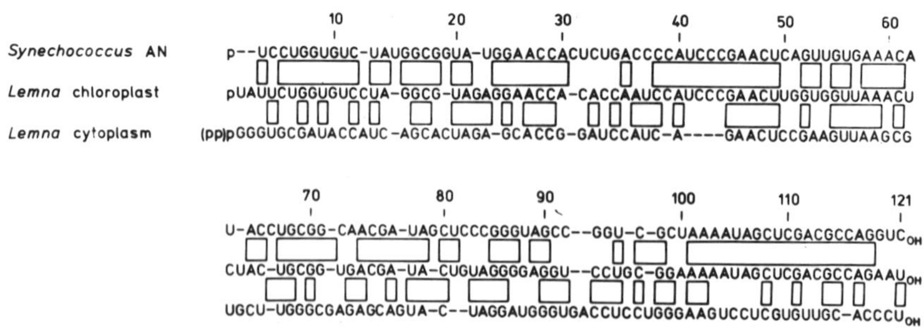

FIGURE 3. Alignment comparison of 5S rRNA sequences. Hyphens denote gaps introduced to obtain maximum sequence homology. After Dyer and Bowman.[14]

Are There Phylogenetic Implications in Kasugamycin Sensitivity and Resistance?

To establish the endosymbiotic hypothesis it is necessary to show that organelles and prokaryotes share derived character states—that is, that they share features that would not have been present in the protoeukaryote host cell before the symbiosis was established. Alternatively, the elimination of those hypotheses which infer that eukaryotes arose by direct filiation from a cyanobacterial-like ancestor requires an identification of character states—common to certain prokaryotes and eukaryotic cytoplasm but lacking from cyanobacteria and chloroplasts—that developed after the divergence of cyanobacteria from the main path of microbial evolution.

Modified nucleotides have been strongly conserved during evolution [15] though their functions remain largely unknown.[16] As a result they are particularly useful to evolutionary investigations. Post-transcriptional modifications are introduced into all high molecular weight ribosomal RNA molecules during their processing from larger precursors. These involve the specific addition of methyl groups and the conversion of certain uridine residues to pseudouridine. The number and location of the methyl groups within a mature rRNA is characteristic of the evolutionary position of the organisms from which it derives. The rRNAs of prokaryotes mainly have methyl groups attached to positions on the purine and pyrimidine bases, whereas in the cytoplasmic rRNAs of eukaryotes methylations occur predominantly at the 2′-*O* position of ribose residues.

In any phylogenetic study based on rRNA, coincidences in post-transcriptionally modified sequences are individually more significant than those involving normal (unmodified) oligonucleotides, because in such cases not only are the sequences common, but the modifying enzyme systems and certain local nucleic acid conformations must be very similar if not identical as well.[17]

The phylogenetic distribution of the tetramethylated sequence -$Gm_2^6Am_2^6$AC- is particularly relevant to the evolutionary origin of chloroplasts. It has been shown [18] that the adjacent N^6,N^6-dimethyladenosine residues in the *E. coli* 16S rRNA sequence -$Gm_2^6Am_2^6$ACCUG- are required for the action of the antibiotic kasugamycin. The presence or absence of this sequence in the small subunit rRNA of an organism parallels its sensitivity or resistance to the bacteriostatic effects of this antibiotic. The aim of the results presented, therefore, was to extend the known phylogenetic distribution of the late methylated sequence -$Gm_2^6Am_2^6$AC- through use of kasugamycin; and, by combining these results with the direct sequence data of others, to establish whether this character evolved before or after the divergence of cyanobacteria from the main path of microbial evolution.

A wide range of prokaryotes, and a lesser number of eukaryotes, were tested for kasugamycin sensitivity.[19] This was defined as 40% inhibition of growth by 100 μg/ml of the antibiotic. All cultures were examined in the exponential growth phase. The results of this survey are included in TABLE 2, together with a compilation of the presence or absence of the -$Gm_2^6Am_2^6$AC- sequence; these have been taken from the literature and are a result of direct sequence analysis.

Both the results of the kasugamycin sensitivity and direct sequence data clearly show that the small subunit ribosomal RNA sequence -$Gm_2^6Am_2^6$AC- is a primitive rather than derived character state: it is found across the entire

TABLE 2

PHYLOGENETIC DISTRIBUTION OF THE LATE METHYLATED SEQUENCE -G-m_2^6A-m_2^6A-C- IN PROKARYOTIC AND CHLOROPLAST 16S rRNAS AND EUKARYOTIC 18S rRNAS

SEQUENCE PRESENT IN	Type of Evidence		SEQUENCE ABSENT IN	Type of Evidence	
PROKARYOTES	Sequence	Kasugamycin	PROKARYOTES	Sequence	Kasugamycin
Aerobacter aerogenes	+		*Alcaligenes faecalis*	—	
Erwinia carotovora		+			
Escherichia coli	+	+	*Sporolactobacillus inulinus*	—	
Proteus vulgaris	+				
Yersinia pestis	+		*Bacillus cereus*		—
			Bacillus megaterium	—	
Pseudomonas aeroginosa	+		*Bacillus pasteurii*	—	
Pseudomonas coronofaciens		+	*Bacillus pumilus*	—	
Pseudomonas fluorescens		+	*Bacillus stearothermophilus*	—	
Pseudomonas lachrymans		+	*Bacillus subtilis*		—
Azotobacter sp.		+	*Sporosarcina ureae*	—	
Micrococcus luteus		+	*Staphlococcus aureus*		—
Paracoccus denitrificans	+		*Agmenellum quadruplicatum*	—	
			Anabaena cylindrica		—
Rhodomicrobium vannielii		+	*Anabaena variabilis*		—
Rhodopseudomonas sphaeroides	+	+	*Chlorogloeopsis fritschii*		—
Rhodospirillum rubrum		+	*Fischerella ambigua*	—	
			Gloeotrichia sp.		—
Chlorobium thiosulfatophilum		+	*Nostoc entophytum*		—
			Nostoc muscorum		—

Clostridium perfringens	+	+	*Nostoc* sp. MAC	–	(+)*
Clostridium septicum		+	*Synechococcus* AN	–	–
			Synechococcus 7502	–	
Streptococcus faecalis		+	*Synechocystis* 6308	–	–
			Synechocystis 6701	–	
Leptospira icterohaemorrhagiae		+	*Synechocystis* 6714	–	–
Acholeplasma laidlawii		+	*Halobacterium halobium*	–	–
Methanobacterium arbophilicum	+		*Methanobacterium mobile*	–	
Methanobacterium formicium	+		*Methanobacterium barkeri*	–	
Methanobacterium ruminantium	+		*Methanospirillum hungatii*	–	
Methanobacterium thermoautotrophicum	+				
EUKARYOTES			EUKARYOTES		
Chlamydomonas reinhardi	+	–	*Euglena gracilis*		
Euglena gracilis		+	chloroplast	–	
Euglena W_{15}ZHL		+	*Porphyridium* sp.		
Porphyridium sp.	+	–	chloroplast	–	
Pyricularia oryzae		+	*Lemna minor*		
Saccharomyces carlsbergensis	+	–	chloroplast	–	
Lemna minor	+				
Triticum vulgare	+				
Xenopus laevis	+				
Chick embryo	+				
HeLa cell	+				

* The effect of kasugamycin in this case was somewhat aberrant. The antibiotic was bacteriocidal at concentrations as low as 2μg cm^{-3} rather than bacteriostatic. (From Phillips.[19])

evolutionary spectrum—from methanogens to Man. In those organisms in which it is absent, the lack probably reflects a secondary loss not a primary deficiency. This permits only limited phylogenetic conclusions to be drawn from its distribution. Nevertheless, apart from the methanogens, which because of their extreme antiquity are a special case, the presence or absence of this primordial sequence does appear to be consistent within any single, major group of organisms. For instance, while it is totally lacking from bacilli, and cyanobacteria and chloroplasts—the latter pair constitute a single evolutionary family —it is present throughout the anaerobic photosynthetic bacteria and enteric-vibrio species. Moreover, it is probably universal in the cytoplasmic 18S rRNAs of eukaryotes.

These facts alone are sufficient to suggest that the chloroplasts and cytoplasm of any photosynthetic eukaryote are of independent evolutionary origin. To provide a specific example: hybridization results [1, 9] have indicated a pronounced similarity between the rRNAs of *Synechocystis* 6308 and the chloroplast rRNAs of *Euglena gracilis*. Since the sequence -$Gm_2^6Am_2^6$AC- is absent in the 16S rRNAs from both of these sources, yet apparently present in *Euglena gracilis* 18S rRNA, the most plausible explanation for this phenomenon must be that the chloroplasts and cytoplasm of this protist are of temporally separate origin. Of course, it could still be argued, as Uzzell and Spolsky [20] have done, that the protoeukaryote arose shortly after the divergence of the cyanobacteria from the main line of bacterial evolution, and hence could have possessed the sequence -$Gm_2^6Am_2^6$AC- at that time. But such a view is untenable: both the electron transport chains of mitochondria and the primary and tertiary structures of their cytochrome *c* molecules are most like those of *Paracoccus denitrificans* and the purple non-sulphur bacteria [21, 22]—organisms considered to be only distantly related to cyanobacteria on the basis of 16S rRNA sequence data.[23]

Features of the Eukaryotic Protein Synthesizing System Designed to Prevent the Translation of Prokaryotic Messenger RNAs on Cytoplasmic Ribosomes

Alteration in the Function of 5S rRNA: Appearance of 5.8S rRNA

All sequenced prokaryotic 5S rRNAs share one feature in common: the presence of the hexanucleotide -CCGAAC- around position 45.[24] The function of this conserved sequence is to assist in the binding of aminoacyl-tRNAs at the acceptor (A)-site of the 70S ribosome via complementary base-pair interaction with the conserved sequence -GT(or U)ΨCR- present in loop IV of prokaryotic elongator tRNAs.[25, 26] The binding of the initiator formylmethionyl-$tRNA^{fMet}$ and peptidyl-tRNAs at the 70S ribosomal puromycin-releasible (P)-site does not require complex formation with 5S rRNA.[25, 27]

The function of eukaryotic 5S rRNA is unclear. Only one thing is certain: its role is different from that of prokaryotic 5S rRNA. Suggestions have been made that eukaryotic 5S rRNA is located at the P-site of the 80S ribosome and involved in the initiation phase of cytoplasmic protein biosynthesis; [4, 28] or alternatively, its function is concerned with the reversible association of ribosomal subunits.[29] Presently, a limited amount of data can be mustered to support both these speculative ideas.

Eukaryotic cytoplasmic ribosomes contain, in addition to 5S rRNA, a second low molecular weight rRNA species: the 5.8S rRNA.[30, 31] No definite function has yet been ascribed to 5.8S rRNA. But several workers[32-34] have made an informed guess on the basis of circumstantial evidence that its role is similar to that of 5S rRNA in prokaryotes, i.e., it assists in binding aminoacyl-tRNAs at the (A)-site of the 80S ribosome. This seems very plausible, especially since eukaryotic 5S rRNA (at least in fungi and animals) is incapable of fulfilling such a role itself. In fact, all sequenced 5.8S rRNAs[24] can potentially base-pair with the -GTΨCR- sequence in loop IV of elongator tRNAs: this may have been the one function to have been preserved during their evolutionary transmogrification from bacterial 5S rRNA.[35, 36]

Specific Deletion of the Functionally Important Sequence -CCUCC- During the 16S → 18S rRNA Transition

Close to the 3′-ends of the 16S rRNAs of many, if not all, prokaryotes (eubacteria, archaebacteria, and chloroplasts) is located the polypyrimidine sequence -CCUCC-, contained within a terminal sequence of the form -G/AUCACCUCCU$N_{X\text{-OH}}$ (N_X stands for 0–5 additional pyrimidine residues, sometimes—as in *E. coli*—terminating in -A_{OH}).[37, 38] During the initiation of protein biosynthesis on *E. coli* ribosomes a base-pairing occurs, that involves at least two nucleotides in the sequence -CCUCC- and a purine-rich region preceding the initiation (-AUG-) codon in the mRNA.[39] The available 3′-terminal sequences of eukaryotic 18S rRNAs share considerable homology with the 3′-end of *E. coli* 16S rRNA,[40, 41] yet in every case examined the entire -CCUCC- sequence is absent—during the prokaryote to eukaryote transition it has been specifically deleted. The developing cytoplasmic ribosome seems to have been forced to evolve an alternative mechanism for specific recognition of mRNA initiator regions that does not involve base-pair formation between the mRNA 5′-noncoding sequences and the 3′-terminus of the 18S rRNA.

Addition of 'Caps' to Eukaryotic Messenger RNAs

Many eukaryotic mRNAs are modified at their 5′-termini by a cap structure consisting of 7-methylguanosine in 5′-to-5′ linkage with the first encoded base. Translation experiments with various eukaryotic cell-free protein synthesizing systems indicate that the presence of the cap moiety is required for efficient translation of these mRNAs. Several prokaryotic mRNAs have been translated in both mammalian and wheat germ cell-free systems to yield accurate and functional protein products. However, the translational efficiencies of prokaryotic mRNAs are significantly lower than those of most eukaryotic mRNAs; they are, in fact, translated at levels equivalent to those obtained with eukaryotic mRNAs lacking cap structures (for references to the above, see Reference 42). It has now been shown that if certain prokaryotic mRNAs are modified at their 5′-termini by the addition of a cap structure *in vitro,* they are translated in a wheat germ cell-free system at least as efficiently as eukaryotic mRNAs.[42] The presence or absence of a cap appears sufficient to distinguish prokaryotic and eukaryotic mRNAs at the initiating 80S ribosome.

The three different features of the protein synthesizing systems of prokaryotes and eukaryotes outlined above appear specifically designed to prevent the translation of prokaryotic mRNAs on cytoplasmic ribosomes. In our opinion, the acquisition of an aerobic prokaryotic endosymbiont (the protomitochondrion) by a similarly-functioning prokaryotic host caused a crisis: how were mRNAs from the two sources to be distinguished? The problem was overcome by a series of subtle changes in the initiation mechanism of the host cell. Unlike others,[43, 44] we do not believe that the intricacies of the eukaryotic protein synthesizing system—such as, larger ribosomes, more initiation factors, capped messengers, genes-in-pieces, mRNA splicing, etc.—are relics of an inefficient, primordial protein synthesizing system. Yet, it is indeed difficult, if not impossible, to imagine how so many changes could have been achieved rapidly in a coordinated manner. Alas! we can never go back and see what really happened.

Modification to Cyanelle Genomes

The kinetics of renaturation, whereby thermally separated, single strands of DNA reanneal to produce the double-stranded DNA molecule, may be measured conveniently by decrease in extinction at 260 nm. The rate of renaturation can be used to calculate the degree of complexity of the original genetic material;[45] put simply, small genome molecules such as those of bacteriophage renature quickly, compared to large DNA molecules such as comprise eukaryotic chromosomes. The development of this procedure has allowed the size of a considerable number of bacterial genomes to be determined; all these fall into size range $1.0–3.6 \times 10^9$ daltons. It has recently been shown[46] that the genomes of most unicellular cyanobacteria are within this range whilst, most interestingly, the pleurocapsalean and filamentous cyanobacteria possess genomes approximately double or triple this size.

The DNA of chloroplasts has been characterized by both chemical and physical means. There is general agreement between chloroplast DNA sizes from diverse organisms. The chloroplast genome size is small compared to that of prokaryotes, and there is considerable duplication of genome copies within the organelle (Table 3). Some physiological consequences of the extensive duplication of the rather small unit of chloroplast DNA may be imagined. As the organelle DNA relinquishes the coding of many gene products to the nuclear DNA, its size may be reduced.[47] The large number of copies of chloroplast DNA achieves a much greater degree of gene amplification than found elsewhere in prokaryotes. This may be a necessary prerequisite for the massive accumulation of certain proteins that occurs in chloroplasts.

Cyanophora paradoxa is a biflagellate unicellular eukaryote, whose taxonomic position has been the subject of some dispute. It contains several clearly discernible organelles that contain phycobiliprotein pigments. These "cyanelles" possess a fine structure comparable to that of unicellular cyanobacteria. This has led to the description of the organelle of *Cyanophora paradoxa* as an endosymbiotic cyanobacterium. The distinction of these intracellular organelles from chloroplasts is emphasized by their lysozyme sensitivity, indicating the presence of a muramic acid external layer,[48] which is typical of prokaryotes. Herdman and Stanier[49] have measured the genetic complexity of *Cyanophora paradoxa* organelle DNA by renaturation kinetics and found that it was

1.17×10^8 daltons, well within the range of chloroplast DNA (TABLE 3). There were approximately 60 copies of the genome in the organelle and Herdman and Stanier concluded that the cyanelle of *Cyanophora paradoxa* should, on these grounds, be considered a photosynthetic organelle rather than a symbiotic cyanobacterium. These results offer some interesting insights into the possible modes of integration of organelles within eukaryotic cells. The possession of phycobiliprotein pigments and a muramic acid cell surface certainly suggests that the organelle of *Cyanophora paradoxa* has undergone fewer modifications than have most chloroplasts. It is tempting to think of it as a relative early and therefore unmodified result of an endosymbiotic association. If this is correct it would imply that a necessary event in the establishment of

TABLE 3

GENOME SIZES AND CONTENTS OF CHLOROPLASTS, *Cyanophora paradoxa* CYANELLES AND UNICELLULAR CYANOBACTERIA

Organism	Genome size (daltons $\times 10^{-8}$)	Number of copies of genome
Chloroplasts of:		
Chlamydomonas	1.94– 2.0	24–52
Chlorella	2.0 – 2.3	20–40
Euglena	0.93	9–72
Higher plants	0.82– 0.91	35–100
Cyanelle of:		
C. paradoxa	1.17	60
Unicellular Cyanobacteria:		
Synechocystis spp.	18.9 –23.9	1–few
	17.6 –23.0	
Synechococcus spp.	29.0 –39.2	1–few
	16.4 –30.6	1–few
	23.2 –25.3	

After Herdman & Stanier.[49]

such an association would lie in the modification of the organelle genome to the chloroplast pattern, i.e., large duplication of a DNA molecule with a low degree of genetic complexity. The possibility that the mode of integration of organelles should be examined at the level of alterations induced in genetic structure and expression should yield experimental approaches to the study of organelle integration.

REFERENCES

1. PIGOTT, G. H. & N. G. CARR. 1972. Science **175:**1259–1261.
2. ZABLEN, L. B., L. BONEN, R. MEYER & C. R. WOESE. 1975. Proc. Natl. Acad. Sci. USA **72:**2418–2422.
3. BONEN, L. & W. F. DOOLITTLE. 1976. Nature **261:**669–673.

4. Erdmann, V. A. 1976. *In* Progress in Nucleic Acid Research and Molecular Biology. W. E. Cohn, Ed. Vol. **18:**45–90. Academic Press, Inc. New York, N.Y.
5. Dyer, T. A., C. M. Bowman & P. I. Payne. 1977. *In* Nucleic Acids and Protein Synthesis in Plants. L. Bogorad & J. H. Weil, Eds.: 121–133, Plenum Publishing Corp. New York, N.Y.
6. Galling, G. & B. R. Jordan. 1972. Biochimie **54:**1257–1265.
7. Avadhani, N. G. & D. E. Buetow. 1974. Biochem. J. **140:**73–78.
8. Rochaix, J. D. & P. Malnoe. 1978. Cell **15:**661–670.
9. Phillips, D. O. & N. G. Carr. 1975. FEBS Lett. **60:**94–97.
10. Manning, J. E. & O. C. Richards. 1972. Biochim. Biophys. Acta **259:**285–296.
11. Rawson, J. R. Y., S. R. Kushner, D. Vapnek, N. K. Alton & C. L. Boerma. 1978. Gene **3:**191–209.
12. Cohen, D. & J. A. Schiff. 1976. Arch. Biochem. Biophys. **177:**201–216.
13. Chelm, B. K., P. J. Hoben & R. B. Hallick. 1977. Biochemistry **16:**782–786.
14. Dyer, T. A. & C. M. Bowman. 1979. Biochem. J. **183:**595–604.
15. Klagsbrun, M. 1973. J. Biol. Chem. **248:**2612–2620.
16. Poldermans, B., C. P. J. J. Van Buul & P. H. Van Knippenberg. 1979. J. Biol. Chem. **254:**9090–9094.
17. Zablen, L. B. & C. R. Woese. 1975. J. Mol. Evol. **5:**25–34.
18. Helser, T. L., J. E. Davies & J. E. Dahlberg. 1972. Nature New Biol. **235:** 6–9.
19. Phillips, D. O. 1981. Ph.D. Thesis, University of Liverpool. Liverpool, England.
20. Uzzell, T. & C. Spolsky. 1974. Am. Sci. **62:**334–343.
21. John, P. & F. R. Whatley. 1975. Nature **254:**495–498.
22. Dickerson, R. E. 1980. Nature **283:**210–212.
23. Gibson, J., E. Stackebrandt, L. B. Zablen, R. Gupta & C. R. Woese. 1979. Curr. Microbiol. **3:**59–64.
24. Erdmann, V. A. 1979. Nucl. Acids. Res. **6:**r29–r44.
25. Sprinzl, M., T. Wagner, S. Lorenz & V. A. Erdmann. 1976. Biochemistry **15:**3031–3039.
26. Schwarz, U., H. M. Menzel & H. G. Gassen. 1976. Biochemistry **15:**2484–2490.
27. Dohme, F. & K. H. Nierhaus. 1976. Proc. Natl. Acad. Sci. USA **73:**2221–2225.
28. Cox, R. A. 1977. Prog. Biophys. Mol. Biol. **32:**193–231.
29. Azad, A. A. 1979. Nucl. Acids Res. **7:**1913–1929.
30. Pene, J. J., E. Knight & J. E. Darnell. 1968. J. Mol. Biol. **33:**609–623.
31. Rubin, G. M. 1973. J. Biol. Chem. **248:**3860–3875.
32. Nishikawa, K. & S. Takemura. 1974. FEBS Lett. **40:**106–109.
33. Ginsburg, D. & J. A. Steitz. 1975. J. Biol. Chem. **250:**5647–5654.
34. Ulbrich, N., A. Lin & I. G. Wool. 1979. J. Biol. Chem. **254:**8641–8645.
35. Phillips, D. O. & N. G. Carr. 1977. Taxon **26:**3–42.
36. Wrede, P. & V. A. Erdmann. 1977. Proc. Natl. Acad. Sci. USA **74:**2706–2709.
37. Balch, W. E., L. J. Magrum, G. E. Fox, R. S. Wolfe & C. R. Woese. 1977. J. Mol. Evol. **9:**305–311.
38. Schwarz, Zs. & H. Kössel. 1979. Nature **279:**520–522.
39. Steitz, J. A. & K. Jakes. 1975. Proc. Natl. Acad. Sci. USA **72:**4734–4738.
40. Alberty, H., M. Raba & H. J. Gross. 1978. Nucl. Acids Res. **5:**425–434.
41. Hagenbüchle, O., M. Santer, J. A. Steitz & R. J. Mans. 1978. Cell **13:**551–563.
42. Paterson, B. M. & M. Rosenberg. 1979. Nature **279:**692–696.
43. Darnell, J. E. 1978. Science **202:**1257–1260.
44. Doolittle, W. F. 1978. Nature **272:**581–582.
45. Wetmur, J. G. & N. Davidson. 1968. J. Mol. Biol. **31:**349–370.
46. Herdman, M., M. Janvier, R. Rippka & R. Y. Stanier. 1979. J. Gen. Microbiol. **111:**73–85.

47. ELLIS, R. J., P. E. HIGHFIELD & J. SILVERTHORNE. 1977. *In* Proceedings of the Fourth International Congress on Photosynthesis. D. O. Hall, J. Coombs & T. W. Goodwin, Eds.: 497–506. The Biochemical Society. London.
48. AITKEN, A. & R. Y. STANIER. 1979. J. Gen. Microbiol. **112:**219–223.
49. HERDMAN, M. & R. Y. STANIER. 1977. FEMS Microbiol. Lett. **1:**7–12.

DISCUSSION OF THE PAPER

DR. J. F. FREDRICK: On the general size of the cyanelles, is there any data on the size of the cyanelles of *Glaucocystis*?

DR. N. G. CARR: Not that I now know of.

DR. F. J. R. TAYLOR (*University of British Columbia, Vancouver, B.C.*): These contemporary examples of blue-greens that might be turning into chloroplasts always worried me a little bit because if this type of process is taking place now, and if what I would expect is that this is a counter-process that could have been taking place over the whole history of eukaryotes, why don't we have an enormous range of chloroplast-like things all of which have come in at different times over the whole span? It seems to me very peculiar that we would have a whole lot of these events taking place a long time ago and then nothing at all for a long time, or other than these kind of secondary things that Jean Whatley talked about, and then have some happening now.

DR. CARR: There's not an awful lot of them around these indeterminate stages. Perhaps one answer to your question would be that we are seeing the results of a very infrequent occurrence over a large period of time. So maybe there are not many, and what we really need is much wider microbial knowledge of eukaryotic protists.

A MYCOPLASMA-LIKE ARCHAEBACTERIUM POSSIBLY RELATED TO THE NUCLEUS AND CYTOPLASM OF EUKARYOTIC CELLS

Dennis G. Searcy, Diana B. Stein, and Karen B. Searcy

Zoology Department
University of Massachusetts
Amherst, Massachusetts 01003

INTRODUCTION

Thermoplasma acidophilum is a freeliving microorganism that is without a rigid cell wall. It has been isolated only from burning refuse piles associated with coal mines.[1, 2] These piles are both hot and rich in sulfuric acid; it is not surprising, then, that optimal culture conditions were found to be 59° C and pH 1 to 2.[1] No one knows where Thermoplasma lived before there were coal mines.

T. acidophilum was originally reported to be a thermophilic mycoplasma. However, its resemblance to other mycoplasmas is apparently superficial, and on the basis of both its 16S ribosomal RNA [3] and its unusual membrane lipids,[4] it has now been classified with the Archaebacteria. In contrast, by the same criteria, the typical parasitic mycoplasmas are closely related to the true bacteria. Thus, *T. acidophilum* has had a long evolutionary history separate from typical bacteria, and has a number of unusual features.

As advocated by Margulis,[5] eukaryotic cells may have arisen from a symbiotic association between an aerobic bacterium and an amoeboid host cell, presumably about 2 billion years ago. Since several of the features of *T. acidophilum* are highly suggestive of this hypothetical host cell, we have advanced the hypothesis that they may indeed have been related.[6] A possible evolutionary scheme incorporating this idea is shown in FIGURE 1.

However, in trying to identify any modern cell with an evolutionary event that occurred so long ago, one must take into account that in all probability it has undergone considerable evolution in the meantime. In this regard, evidence is accumulating that prokaryotes and eukaryotes have evolved in separate adaptive directions.

For example, as pointed out by Reanney [7] and Doolittle,[8] a reasonable argument can be made that the apparent size and redundancy of eukaryotic cells is actually a primitive feature, especially in regard to the repetitive DNA sequences. A similar point has been made by Darnell,[9] who has suggested that messenger RNA splicing, with the elimination of intervening sequencies, may also be a primitive feature that has been retained in eukaryotic cells. According to these hypotheses, modern bacterial cells are the product of selection for regulatory economy and rapid growth rates, this being achieved through the elimination of redundant or unnecessary genetic material. For example, in bacteria the structural genes have become grouped together into operons, with several genes being regulated by a single control element, whereas in eukaryotes, possibly representing a more primitive condition, the genes are scattered and individually regulated. Thus, primordial cells may have been larger and more redundant

0077–8923/81/0361–0312 $01.75/0 © 1981, NYAS

than are modern prokaryotic cells. Indeed, living bacteria are often smaller than Precambrian microfossils.[10, 11]

T. acidophilum has presumably followed this same pattern of reduction, becoming today smaller, simpler, and more highly specialized than its common ancestor with eukaryotic cells. In fact, possibly as a result of its extreme environment, *T. acidophilum* has now acquired the smallest genome of any known freeliving organism.[12] Eukaryotic cells, on the other hand, have followed the alternative evolutionary strategy of increased size and complexity. Thus, *T. acidophilum* and eukaryotes have taken different evolutionary paths. Nevertheless, if there is indeed an evolutionary relationship between them, the signs of this relationship should still persist.

The purpose of this paper is to review our observations on *T. acidophilum* since we first described this hypothesis.[6] Our research has continued to focus

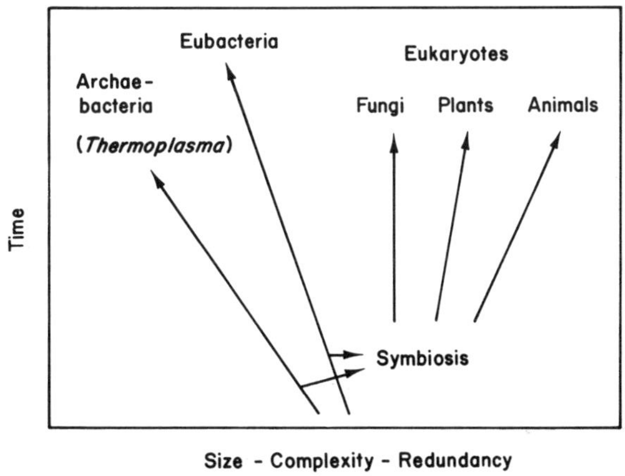

FIGURE 1. A hypothetical scheme for the evolution of the major kingdoms of life. The eukaryotic kingdoms originated as a symbiotic association between an organism such as Thermoplasma and an aerobic Eubacterium. On the *x*-axis, prokaryotes are shown as having evolved in the direction of reduction, whereas eukaryotes have generally evolved in the direction of complexity.

on three major areas; the nucleoprotein, cytoskeleton, and respiratory metabolism.

The Nucleoprotein

One of the better studied aspects of *T. acidophilum* is a small basic protein (HTa) that is associated with the DNA. In its size and amino acid composition, HTa resembles eukaryotic histones.[13] Furthermore, like eukaryotic histones,[14] it condenses DNA into globular particles, except that in *T. acidophilum* the particles are smaller and simpler than in eukaryotic cells.[15] In *T. acidophilum,* each nucleoprotein particle is composed of a loop of 40 base pairs of DNA coiled around a core of 4 molecules of HTa. However, in *T. acidophilum,* the actual function of this protein is probably not to condense the DNA, but to stabilize it against thermal denaturation.[15]

Since the eukaryotic histones are among the most slowly evolving of all known proteins, it has been of particular interest to compare the primary structure of HTa with that of the histones. As shown in FIGURE 2, the N-terminal sequence of HTa is identical to histone H2A in 6 out of 24 residues. Taking into account the amino acid composition of these sequences, the probability of such a match is less than 0.005. Furthermore, of the remaining 18 nonidentical amino acids in this sequence, 13 are potentially related by single base substitutions in the genetic code.

This particular sequence in H2A is an unusually interesting one because, as pointed out by Reeck *et al.*,[17] it is one of a set of homologous sequences that can be identified in each of the eukaryotic histones, except for H1. Thus, this sequence appears to be a highly conserved vestige of an ancestral protein common to all of these histones. The sequence occurs twice in histone H2A, so that HTa can be aligned also with the N-terminal region of H2A, resulting in 5 identical residues, which is also statistically significant ($P \approx 0.02$). Another alignment can be made with an internal segment in H3, again producing 5 identical residues. Additional alignments may also be made with each of the other sequences identified by Reeck *et al.*, but these were of lower statistical significance.

As previously pointed out by Reeck *et al.*,[17] the amino acid sequences that we are considering are not particularly basic, and so these similarities are not likely to have arisen from convergence due to direct interactions of the peptides with the DNA. Thus, we believe that these data are evidence of a genuine evolutionary relationship.

In *Escherichia coli,* another small basic protein (HU) has been described.[18] In size and amino acid composition, HU is similar to HTa. However, the amino acid sequence of HU has been reported to show no homology to eukaryotic histones.[19, 20] Nevertheless, HU does show some relationship to HTa, 5 residues matching out of the N-terminal 24. Thus, HTa may be an evolutionary link between eukaryotic histones and the bacterial proteins, which otherwise would show little evidence of kinship to each other.

However, there are also important differences between HU and HTa. For example, when *E. coli* nucleoprotein was isolated using physiological ionic

	(1)				(5)					(10)		
HTa	met	val	gly	ile	ser	glu	leu	ser	lys	glu	val	ala-
	○	○	○	●	○	●	●	○			○	●
H2A	-thr	ala	glu	ile	leu	glu	leu	ala	gly	asn	ala	ala-

			(15)					(20)				
HTa	-lys	lys	ala	asn	thr	thr	gln	lys	val	ala	arg	thr-
	○			○	○	●	○	○	○	○	●	
H2A	-arg	asp	asn	lys	lys	thr	arg	ile	ile	pro	arg	his-

FIGURE 2. The N-terminal amino acid sequence of the *T. acidophilum* histonelike protein (HTa) compared with a sequence in calf histone H2A. (●)–identical amino acids; (○)–amino acids that can potentially be related by single base changes in the genetic code. Sources: HTa–R. J. DeLange (personal communication); Calf H2A–Isenberg,[41] starting from residue #59.

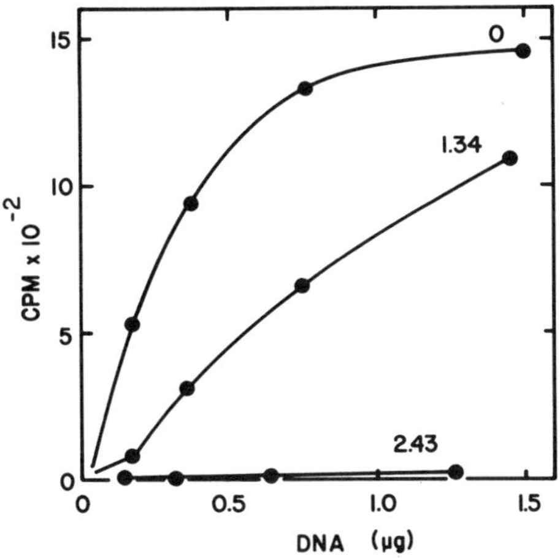

FIGURE 3. The effect of HTa upon the template activity of calf DNA. The DNA was reconstituted as described by Stein and Searcy [16] with increasing amounts of HTa. The ratio (w/w) of HTa to DNA is shown in Figure. Each assay (0.25 ml) contained 10 μg of *E. coli* RNA polymerase, 30 mM Tris-Cl, 4 mM $MgCl_2$, 1 mM $MnCl_2$, 12 mM 2-mercaptoethanol, 0.4 mM each nucleoside triphosphate, 10 Ci/mole [^{3}H]ATP, pH 7.5, and increasing amounts of template shown on the *x*-axis. Incubation was for 10 minutes at 38° C, and then was stopped by the addition of 10% tricholoroacetic acid. Insoluble material was collected on glass fiber filters, and its radioactivity measured. In the absence of DNA template, background radioactivity was 31 counts per minute (CPM), which has been subtracted from each point.

conditions, HU was not associated with the DNA, nor was the DNA stabilized against thermal denaturation.[13] HU has been reported to form globular particles with DNA,[21] but only at ionic strengths lower than physiological. These particles each contained 16–20 molecules of HU plus one loop of DNA, in contrast to *T. acidophilum* nucleoprotein, where there are 4 molecules of HTa per loop of DNA,[15] and to eukaryotic cells, where there are 8 molecules of histone per approximately 2 loops of DNA.[14]

Another important, but less well understood, aspect of histones is their effect upon RNA synthesis. Eukaryotic histones, at least *in vitro,* appear to inhibit transcription.[22, 23] A phage histonelike protein has also been shown to clearly inhibit transcription.[24] In contrast, *E. coli* HU stimulates transcription.[18] According to Berthold and Geider,[25] this stimulatory effect might be explained by a high affinity for RNA, so that the protein is helping to remove the transcript from the template. If so, then HU might not be a DNA-associated protein at all, but rather, is associated with RNA in living cells.

Thus, it has been of interest to examine the influence of *T. acidophilum* HTa on RNA synthesis *in vitro.* As shown in FIGURE 3, it was inhibitory, like eukaryotic histones. In the reconstituted nucleoproteins shown, the amounts of DNA directly associated with HTa in each sample were about 0%, 60%, and 100%, as measured by nuclease digestion [15] and thermal denaturation.[16] Thus, the inhibition of RNA synthesis was about proportional to the fraction of DNA that was complexed with the protein.

Therefore, in terms of binding affinity, DNA condensation, and effect upon transcription, HTa appears to be more similar to eukaryotic histones than to the bacterial protein HU. In amino acid sequence, it appears to be distantly homologous to each of these proteins, possibly being an evolutionary link that could relate them all to each other.

The Cytoskeleton

Since *T. acidophilum* is without a rigid cell wall, and in shape is often irregular and filamentous, as shown in Figure 4, it might be anticipated that there is some type of internal cytoplasmic skeleton. The lability of this cytoskeleton can be readily seen, since when the cells are cooled, they lose their characteristic filamentous shape and become spherical (Figure 5). A similar effect of cooling has already been described in certain animal cells,[26] and is thought to be a frequent characteristic of actomyosin-type cytoskeletal systems.

There are additional observations that may be related to the cytoskeleton of *T. acidophilum*. For example, growth of the organism is inhibited by cytochalasin B,[6] a drug that is thought to be a specific inhibitor of actin-related movement in eukaryotic cells.[26] In addition, as with eukaryotic cells,[26, 27] cytoplasmic extracts can be prepared from *T. acidophilum* that are capable of both gelation and contraction. These extracts were prepared in buffers that contained ethyleneglycol-tetraacetate (EGTA), which chelates calcium. Then, when calcium was added back, gelation occurred as shown in Figure 6. Birefringent strands of gel could also be made by combining calcium and the cell extract on a microscope slide, and then, with the addition of small amounts of ATP, these strands could then be made to contract. However, it should be mentioned that if too much ATP was added, the strands redissolved.

When these gels were examined by electron microscopy, filaments about 6 nm in diameter could be seen. (J. Pustell, personal communication). This is about the same size as eukaryotic actin filaments.[28] Similar filaments could also be seen in thin sections of whole cells, although less clearly. However, these filaments did not bind rabbit heavy meromyosin, which is a diagnostic feature of all types of actin filaments that have been examined so far.[28] Thus, if these filaments in *T. acidophilum* are indeed related to eukaryotic actin, further study will be required before any judgements about homology can be made. Nevertheless, it is clear that an internal cytoskeleton does exist in this organism, although it may have evolved a considerable distance from typical eukaryotic actomyosin.

Curiously, an actin-type cytoskeleton has also been described from certain parasitic mycoplasmas.[29, 30] These organisms contain actin-like filaments that have been reported to bind heavy meromyosin. However, as mentioned earlier, these organisms are not evolutionarily related to *T. acidophilum,* and instead, are better thought of as degenerate bacteria.[3] Since bacteria presumably do not contain actin,[6, 31] and since these mycoplasmas are parasitic upon eukaryotic cells, it is interesting to speculate that they may have recently acquired their actin genes directly from their eukaryotic hosts. This intriguing possibility will probably be resolved when the amino acid sequences of these proteins become available.

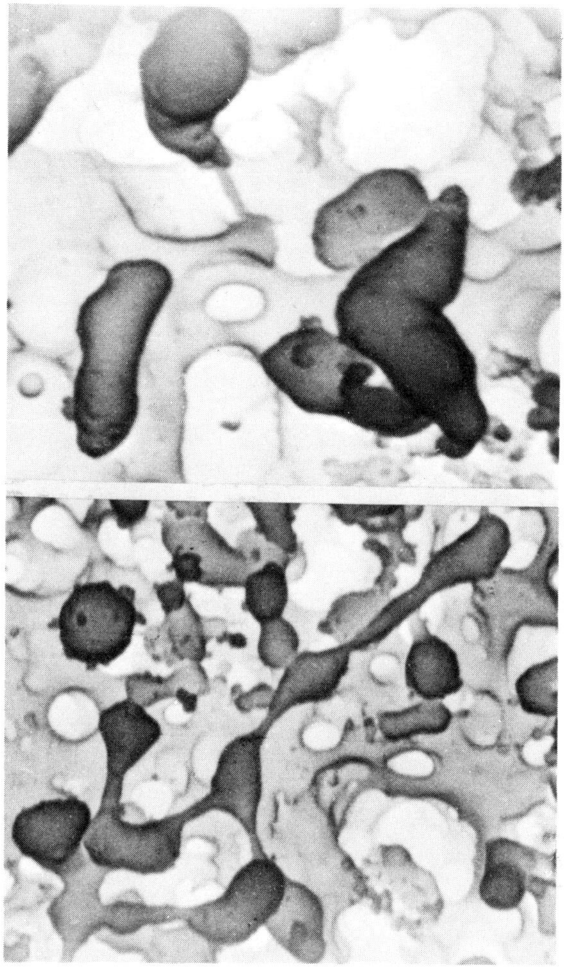

FIGURE 4. Scanning electron micrographs of *T. acidophilum*. Actively growing cultures were adjusted at 59° C to pH 3.5 with 2 M NH_4OH, and then fixed with 2% glutaraldehyde. The cells were collected on a membrane filter, and critical-point dried. Upper: 28,000×. Lower: 15,000×. This picture was taken by James Pustell.

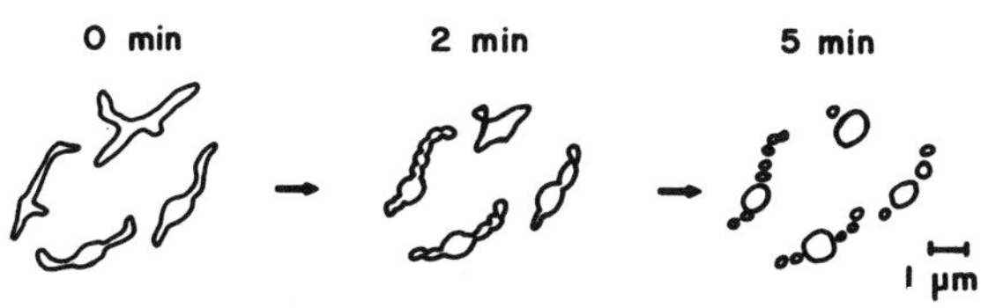

FIGURE 5. Diagrammatic representation of the shape of *T. acidophilum* after cooling from culture conditions (59° C) to room temperature.

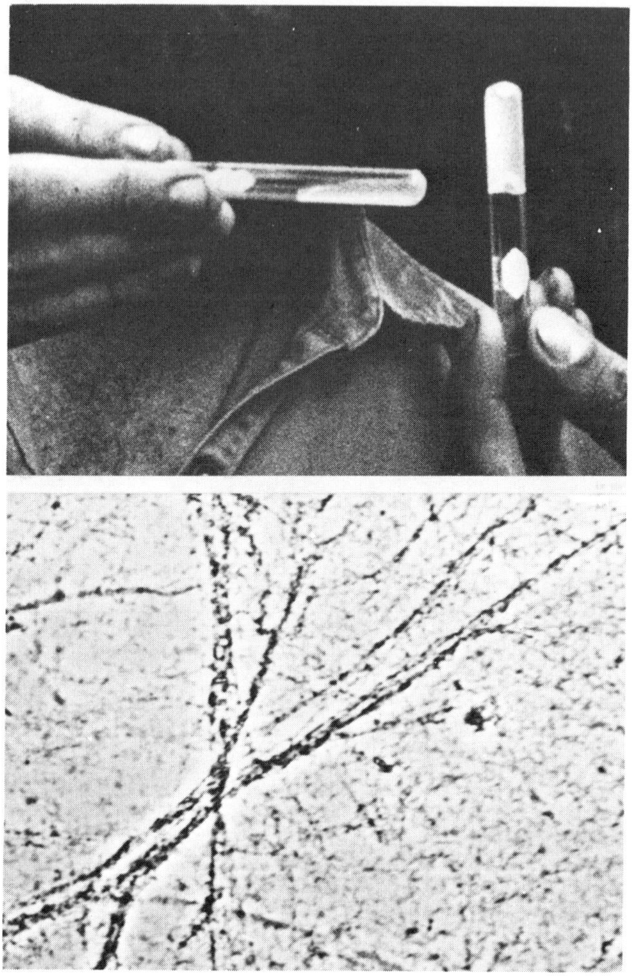

FIGURE 6. Gelation in cytoplasmic extracts of *T. acidophilum*. Upper: gelation in a test tube caused by the addition of calcium, compared to an ungelled control (left). Lower: strands of gel formed on a microscope slide, shown about 500×. The extract was prepared in 0.34 M sucrose, 1 mM ATP, 1 mM EGTA, and 10 mM imidazole chloride, pH 7. After centrifugation, the pH of the extract was adjusted to 5.5, and then gelled by the addition of 0.1 to 0.5 mM $CaCl_2$. Optimal conditions for gelation are still under investigation. This work is by Greg Williams.

THE RESPIRATORY METABOLISM

Concerning the metabolic pathways of *T. acidophilum*, our preliminary results indicate that this organism is at least partially fermentative, and may be lacking oxidative phosphorylation (D. Searcy and F. R. Whatley, in preparation). For example, radioactive glucose was metabolized to excretory products consisting of only acetic acid and carbon dioxide. The Embden-Myerhof and pentose-phosphate pathways seemed to account for this metabolism, since key enzymes and metabolic intermediates from both of these pathways could be detected.

However, *T. acidophilum* was unexpectedly insensitive to dicyclohexylcarbodiimide (DCCD). This drug is a general inhibitor of the H^+-translocating membrane ATPase.[32] For example, growth of *T. acidophilum* was not affected even by saturated solutions of DCCD, which are about 10^{-3} M. In contrast, growth of *Bacillus acidocaldarius*, a true bacterium that grows in similar environmental conditions, was inhibited by 10^{-5} M DCCD. Growth of *T. acidophilum* was also unaffected by oligomycin, another inhibitor of H^+-ATPase.

When purified membranes from *T. acidophilum* were examined, an ATPase activity was present. However, this ATPase was activated only by sulfate, and thus is likely to be a specific sulfate pump. Confirming the observations with whole cells, there was no indication of any DCCD-inhibitable ATPase associated with the purified membranes. In contrast, DCCD-inhibited ATPases have been easily detected in most other prokaryotes examined so far.[32, 33]

Thus, although not proven, it appears likely that *T. acidophilum* is without a H^+-ATPase. Since this enzyme is an essential component of oxidative phosphorylation, it follows that *T. acidophilum* can not synthesize ATP by this mechanism. Such a possibility would fit nicely with the observation that in eukaryotic cells oxidative phosphorylation is associated entirely with the mitochondria. A prediction of this hypothesis is that where H^+-ATPases do occur in the cytoplasmic membranes of eukaryotic cells, they should have been secondarily acquired, possibly from the genetic information of the mitochondria.

T. acidophilum is nevertheless obligately aerobic (Searcy and F. R. Whatley, in preparation). For example, when cultures were deprived of oxygen, an oxygen debt developed, and after 30 minutes the cells were killed. However, in cell homogenates, only about 10% of the cellular respiration was associated with the membrane fraction. These membranes contained a single *b*-type cytochrome plus a quinone, and presumably constitute a simple electron transport chain that can function to expel H^+ from the cell. The remaining 90% of the oxygen consumption in cell homogenates was associated with soluble cytoplasmic enzymes. Although none of these enzymes have yet been isolated, they appear from their aggregate absorbance spectrum to be primarily flavin oxidases.

Raff and Mahler[34] have pointed out that the ancestor of the cytoplasm of eukaryotic cells must have been aerobic; it is interesting that *T. acidophilum* appears to satisfy this requirement. It is also noteworthy that in respiring primarily by flavin oxidases, the metabolism of *T. acidophilum* resembles that of the microbodies of eukaryotic cells. This is consistent with the earlier suggestion of de Duve[35] that the microbodies of eukaryotes might be vestiges of primitive premitochondrial respiratory pathways.

The Superoxide Dismutase

Among the many differences between prokaryotic and eukaryotic cells, one is the type of superoxide dismutase that is present. This is an enzyme that removes the toxic free radical HO_2, which is apparently produced in all aerobic cells.[36] As shown in Table 1, prokaryotic superoxide dismutases (including those of mitochondria) typically contain either manganese or iron. In contrast, the enzyme found in the cytoplasm of eukaryotic cells contains two atoms of copper plus two of zinc. The dismutase from *T. acidophilum* is different from either of these, containing two atoms of iron plus two of zinc.

TABLE 1

SUPEROXIDE DISMUTASES: SOME TYPICAL FEATURES

	Protein subunit composition	Metal content	Activity after $CHCl_3$ treatment	Inhibited by 1 mM HCN	Amino acid composition: residues per subunit tryptophan	tyrosine
Prokaryotic I	$2\times 18{,}000$	Fe_1	No	No	4	6
Prokaryotic IIa	$2\times 22{,}000$	Mn_1	No	No	6	9
Prokaryotic IIb	$4\times 21{,}000$	Mn_2	No	No	6	10
Mitochondrial	$4\times 20{,}000$	Mn_2	No	No	5	8
Eukaryotic	$2\times 16{,}000$	Cu_2Zn_2	Yes	Yes	0	1
T. acidophilum	$4\times 20{,}000$	Fe_2Zn_2	Yes	No	4	9

Sources: Prokaryotic I, *E. coli* (Yost & Fridovich [37]); Prokaryotic IIa, *Streptococcus faecalis* (Britten *et al.*[38]); Prokaryotic IIb, *Thermus aquaticus* (Sato & Harris [39]); Mitochondrial and Eukaryotic, chicken liver (Weisiger & Fridovich [40]); *T. acidophilum*, Searcy and Searcy, in preparation.

There are additional differences between the prokaryotic and eukaryotic superoxide dismutases. For example, in amino acid composition, the prokaryotic forms are rich in tryptophan and tyrosine. In addition, the prokaryotic enzymes are not inhibited by cyanide. In these respects, the *T. acidophilum* dismutase resembles the prokaryotic enzymes.

However, one other difference between the types of dismutases is their resistance to denaturing agents, such as chloroform plus ethanol, sodium dodecylsulfate, urea, or heat. By these criteria, the *T. acidophilum* enzyme resembles the eukaryotic superoxide dismutase; it is not inactivated by any of these agents, and is active even in boiling water. It appears to be one of the toughest enzymes yet described.

In summary, the *T. acidophilum* superoxide dismutase is unique, but in some ways may be intermediate between the typical prokaryotic and eukaryotic forms of this enzyme.

Conclusion

Although *T. acidophilum* has some features of eukaryotic cells, certain other features, such as its extreme small size and simplicity, indicate that it has had a long evolutionary history of its own. During this time, it has clearly continued to remain a prokaryotic cell. It has now evolved into an extreme ecological niche, and by doing so, appears to have escaped from competition by more efficient and highly evolved organisms. However, this would seem to be an evolutionary dead-end. Thus, we do not expect that *T. acidophilum* has remained unchanged since the origin of eukaryotic cells. Nevertheless, some eukaryotic features are still recognizable, and so, through further study, we hope to learn more about the relationship of this organism to eukaryotic cells.

Acknowledgments

We thank Gregory Williams for the studies on gelation. James Pustell for the electron microscopy, and G. Ray Green for his frequent help and advice.

References

1. Darland, G., T. D. Brock, W. Samsonoff & S. F. Conti. 1970. A thermophilic, acidophilic mycoplasma isolated from a coal refuse pile. Science **170:**1416–1418.
2. Belly, R. T., B. B. Bohlool & T. D. Brock. 1973. The genus Thermoplasma. Ann. N.Y. Acad. Sci. **225:**94–107.
3. Woese, C. R., J. Maniloff & L. B. Zablen. 1980. A phylogenetic analysis of the mycoplasmas. Proc. Nat. Acad. Sci. USA **77:**494–498.
4. Tornabene, T. G. & T. A. Langworthy. 1979. Diphytanyl and dibiphytanyl glycerol ether lipids of methanogenic archaebacteria. Science **203:**51–53.
5. Margulis, L. 1970. The origin of eukaryotic cells. Yale University Press. New Haven, Conn.
6. Searcy, D. G., D. B. Stein & G. R. Green. 1978. Phylogenetic affinities between eukaryotic cells and a thermophilic mycoplasma. BioSystems **10:** 19–28.
7. Reanney, D. C. 1974. On the origin of prokaryotes. J. Theor. Biol. **48:**243–251.

8. Doolittle, W. F. 1978. Genes in pieces: were they ever together? Nature **272:**581–582.
9. Darnell, J. E., Jr. 1978. Implications of RNA-RNA splicing in evolution of eukaryotic cells. Science **202:**1257–1260.
10. Pflug, H. D. & H. Jaescke-Boyer. 1979. Combined structural and chemical analysis of 3,800-Myr-old microfossils. Nature **280:**483–486.
11. Schopf, J. W. 1970. Precambrian micro-organisms and evolutionary events prior to the origin of vascular plants. Biol. Rev. **45:**319–352.
12. Searcy, D. G. & E. K. Doyle. 1975. Characterization of *Thermoplasma acidophilum* deoxyribonucleic acid. Int. J. Syst. Bacteriol. **25:**286–289.
13. Searcy, D. G. 1975. Histone-like protein in the prokaryote *Thermoplasma acidophilum*. Biochim. Biophys. Acta **395:**535–547.
14. Kornberg, R. D. 1977. Structure of chromatin. Ann. Rev. Biochem. **46:** 931–954.
15. Searcy, D. G. & D. B. Stein. 1980. Nucleoprotein subunit structure in an unusual prokaryotic organism: *Thermoplasma acidophilum*. Biochim. Biophys. Acta **609:**180–195.
16. Stein, D. B. & D. G. Searcy. 1978. Physiologically important stabilization of DNA by a prokaryotic histone-like protein. Science **202:**219–221.
17. Reeck, G. R., E. Swanson & D. C. Teller. 1978. The evolution of histones. J. Mol. Evol. **10:**309–317.
18. Rouvière-Yaniv, J. & F. Gros. 1975. Characterization of a novel, low-molecular-weight DNA-binding protein from *Escherichia coli*. Proc. Nat. Acad. Sci. USA **72:**3428–3432.
19. Laine, B., P. Sautière, G. Biserte, M. Cohen-Solal, F. Gros & J. Rouvière-Yaniv. 1978. The amino- and carboxy-terminal amino acid sequences of protein HU from *Escherichia coli*. FEBS Lett. **89:**116–120.
20. Mende, L., B. Timm & A. R. Subramanian. 1978. Primary structures of two homologous ribosome-associated DNA-binding proteins of *Escherichia coli*. FEBS Lett. **96:**395–398.
21. Rouvière-Yaniv, J., M. Yaniv & J.-E. Germond. 1979. E. coli DNA binding protein HU forms nucleosome-like structure with circular double-stranded DNA. Cell **17:**265–274.
22. Hoare, T. A. & E. W. Johns. 1971. The inhibition of RNA synthesis by histones. Biochim. Biophys. Acta **247:**408–411.
23. Wasylyk, B., G. Thevenin, P. Oudet & P. Chambon. 1979. Transcription of *in vitro* assembled chromatin by *Escherichia coli* RNA polymerase. J. Mol. Biol. **128:**411–440.
24. Geiduschek, E. P. & C. Beard. 1978. Effect of the SPO1 DNA-binding protein, TF1, on promoter competition. J. Mol. Biol. **126:**481–488.
25. Berthold, V. & K. Geider. 1976. Interaction of DNA with DNA-binding proteins: the characterization of protein HD from *Escherichia coli* and its nucleic acid complexes. Eur. J. Biochem. **71:**443–449.
26. Pollard, T. D. 1976. Cytoskeletal functions of cytoplasmic contractile proteins. J. Supramol. Struct. **5:**317–334.
27. Condeelis, J. C. & D. L. Taylor. 1977. The contractile basis of amoeboid movement. V. The control of gelation, solation, and contraction in extracts from *Dictyostelium discoideum*. J. Cell Biol. **74:**901–927.
28. Korn, E. D. 1978. Biochemistry of actomyosin-dependent cell motility (A Review). Proc. Nat. Acad. Sci. USA **75:**588–599.
29. Neimark, H. C. 1977. Extraction of an actin-like protein from the prokaryote *Mycoplasma pneumoniae*. Proc. Nat. Acad. Sci. USA **74:**4041–4045.
30. Maniloff, J. & U. Chaudhuri. 1979. J. Supramol. Struct. **12:**299–304.
31. Rosenbusch, J. P., G. R. Jacobson & J.-C. Jaton. 1976. Does a bacterial elongation factor share a common evolutionary ancestor with actin? J. Supramol. Struct. **5:**391–396.

32. DOWNIE, J. A., F. GIBSON & G. B. COX. 1979. Membrane adenosine triphosphatases of prokaryotic cells. Ann. Rev. Biochem. **48:**103–131.
33. SONE, N., M. YOSHIDA, H. HIRATA & Y. KAGAWA. 1975. Purification and properties of a dicyclohexylcarbodiimide-sensitive adenosine triphosphatase from a thermophilic bacterium. J. Biol. Chem. **250:**7917–7923.
34. RAFF, R. A. & H. R. MAHLER. 1972. The nonsymbiotic origin of mitochondria. Science **177:**575–582.
35. DE DUVE, C. 1969. Evolution of the peroxisome. Ann. N.Y. Acad. Sci. **168:** 369–381.
36. FRIDOVICH, I. 1978. The biology of oxygen radicals. Science **201:**875–880.
37. YOST, F. J. & I. FRIDOVICH. 1973. An iron containing superoxide dismutase from *E. coli.* J. Biol. Chem. **248:**4905–4908.
38. BRITTON, L., D. P. MALINOWSKI & I. FRIDOVICH. 1978. Superoxide dismutase and oxygen metabolism in *Streptococcus faecalis* and comparisons with other organisms. J. Bacteriol. **134:**229–236.
39. SATO, S. & J. I. HARRIS. 1977. Superoxide dismutase from *Thermus aquaticus.* Isolation and characterization of Mn and Apo enzymes. Eur. J. Biochem. **73:**373–381.
40. WEISIGER, R. A. & I. FRIDOVICH. 1973. Superoxide dismutase. Organelle specificity. J. Biol. Chem. **248:**3582–3592.
41. ISENBERG, I. 1979. Histones. Ann. Rev. Biochem. **48:**159–191.

DISCUSSION OF THE PAPER

DR. J. F. FREDRICK: Dr. Searcy, one of the things that strikes me immediately as to the environmental niche that this *Thermoplasma* has, it is identical with *Cyanidium caldarium* both in pH and temperature optima. Now if as an endosymbiontist, we look for a host organism, wouldn't studies on the nuclear histones of *Cyanidium* compared with the *Thermoplasma* protein that you mentioned be of great value?

DR. D. G. SEARCY: Yes, I think it would. As it turns out there is another key factor that I think is important in understanding why thermoplasma has a histone, and that is that it doesn't have a cell wall and apparently is in osmotic equilibrium with its environment. It can't withstand a high pressure inside as do almost all bacteria. There are thermophylic bacteria that are known, but these invariably have a high salt concentration inside and that stabilizes the DNA. *Thermoplasma* has virtually no salt inside, in fact we've measured it down to 17 millimolar potassium which is the lowest known value for any organisms and this stabilizes the DNA. The absence of cations has made thermoplasma resort to the alternative mechanism of stabilizing the DNA with the histone.

DR. FREDRICK: I know that although the environmental conditions are similar, you don't find the *Thermoplasma* in the same environment that you do *Cyanidium.*

DR. SEARCY: That's an interesting point. I think the survey of histones should be extended to organisms such as that. In the past we've had difficulty doing that because whenever we started to make a survey, there would be another great leap forward in the technology and then we decided that we were doing it all wrong, and we'd go back and start over again. So we've been looking at the same dozen organisms over and over again by different criteria.

But it has still turned out that *Thermoplasma* is the only one that has a histone-like protein that really resembles eukaryotes with the possible exception of another archaic bacterium which does not have a rigid cell wall.

DR. AARONSON: I wonder if you have any information on the membrane of this creature and does it show any affinities with the mycoplasmas or with bacteria? And what, if any, is the association of this putative cytoskeleton with this membrane?

DR. SEARCY: I don't have any idea what the association of the cytoskeleton with the membrane is. We've just begun to study this. The membrane has been studied by Tom Langworthy at the University of South Dakota and it's an unusual membrane. The membrane has more rigidity and is an adaptation to heat. It is interesting that the membrane is loaded with squaline, and so the organism—although it does not have a sterol, and no prokaryote does—is only one step away from evolving cholesterol. What is the affinity to mycoplasma? *Thermoplasma* was originally classified as a mycoplasma because of the superficial resemblance. Recently some investigators have looked at the 16S ribosomal RNase and they place it with the bacteria and conclude that the mycoplasmas are actually degenerate bacteria, so they are on completely separate branches of evolution.

DR. S. S. COHEN: I'm glad you mentioned the presence of squalene. One distinguishes prokaryotes from eukaryotes in one part by the use of molecular oxygen and to go from squaline to cholesterol requires 5 molecules of molecular oxygen. There would be another criterion which would be important to look for, and that is conversion of tryptophan to niacin, which requires three molecules of molecular oxygen in the eukaryote. So there are a whole string of specific biochemical characters that are rather important.

DR. L. MARGULIS: It may be true that there are five steps from squalene to cholesterol but there's only one from squaline to lanosterol. So there really is only one step to make the fundamental steroid ring. It's likely that up to squalene is in the host anyway or in the nuclear cytoplasm and after that is mitochondrial and the eukaryotes. I think you have a very strong analogy. The methane oxidizing bacteria had up to 2% steroid and although the pathway, as far as I know has not been worked out, and it's probably different, that's the only exception to this generalization which you know is basically valid; that steroids are eukaryotic features not prokaryotic.

PROCHLORON AND THE THEORY OF SYMBIOGENESIS *

Ralph A. Lewin

Scripps Institution of Oceanography
La Jolla, California 92093

I can trace my ancestry back to a protoplasmal primordial atomic globule.
Pooh Bah, in *The Mikado*, by W. S. Gilbert

Marine biologists have known for many decades that certain tropical didemnids harbor zoochlorella-like symbionts. (Didemnids are ascidians that form encrusting colonies, often somewhat calcified, with a common cloacal system. Zoochlorellae are green unicellular algae associated with marine invertebrates; the intracellular symbionts of *Paramecium aurelia* and *Hydra viridis*, and the intercellular symbionts of the flatworm *Convoluta roscoffensis*, are well known examples.) The symbionts of didemnids are not eukaryotic green algae, however, but are prokaryotic, like cyanophytes. They contain the same two chlorophylls, *a* and *b*, as are found in chlorophytes. They lack the distinctive photosynthetic pigments of blue-green algae (or cyanophytes). Although some people still regard them as aberrant blue-green algae [1] or as (even more aberrant) green photosynthetic bacteria, we are now generally inclined to consider them as representing a separate algal division, the Prochlorophyta, with so far only a single genus, *Prochloron*. Despite some criticism (Chadefaud [2]), these taxa were recently established [3, 4] not so much as a matter of fact or faith as one of convenience. Other treatments, assigning such paradoxical cells to an older, relatively clearly defined taxon, would distort the sets of basic characteristics that distinguish the Cyanophyta from the Chlorophyta.

In considerations of organelle evolution, *Prochloron* has a special significance in relation to other photosynthetic organisms with or without chlorophyll *b*, and with or without internal cellular membranes. The cells are spherical, and range in diameter from 6 to 30 micrometers. (The size ranges of the algal cells in different didemnid hosts differ. We cannot yet say to what extent such differences may be attributable to specific or varietal differences in the algae, or to different physiological conditions on or in the host colonies.) They are clearly prokaryotic, as shown by electron microscopy. In this and in the mode of cell division they resemble cyanophytes. They show this affinity, too, in the fine structure and biochemical composition of the cell walls.

Like the cells of many blue-green algae, they contain polyhedral bodies (which, presumably, are composed of the protein ribulose-diphosphate carboxylase), but in *Prochloron* these bodies tend to occur among the thylakoids of the peripheral chromatoplasm, and not in the central nucleoplasm as they do in blue-green algae. Unlike cyanophytes, prochlorophytes have paired or stacked thylakoids, as in (eukaryotic) chlorophytes. The thylakoid membranes show no evidence of phycobilisomes, and the cells lack the blue or red water-soluble pigments (phycobiliproteins) which characterize photosynthetic cyanophytes. These and other biochemical and physiological data on the algae, either isolated or in association with the didemnid host, have been recently reviewed [5] and will

* Supported by the National Science Foundation (Grant DEB–7826857).

0077–8923/81/0361–0325 $01.75/0 © 1981, NYAS

not be further discussed here. We have still much to learn about prochlorophyte biology, not only because these algae have only recently come to the attention of microbiologists, plant physiologists, and biochemists, but also because we still do not know how to grow them in culture. Data can now be obtained only from studies carried out on living systems close to their native habitats—areas of coral reefs or mangrove swamps—or from analyses of the frozen, dried, or otherwise preserved algal cells.

Until they can be cultured in the laboratory, we have to study their biology *in situ*. So, to introduce didemnids and their algal symbionts to those who, like myself ten years ago, may be unaware of their existence, I am presenting some pictures illustrating their appearance and habitats (FIGURES 1–8). I hope that these will help to familiarize scientists with this fascinating group of creatures, until recently known to only a few specialists, e.g., Kott.[6] Note that they live mostly, perhaps exclusively, in the warm, shallow waters of tropical seas,[6, 7] which is not a bad kind of place to go hunting for the origins of organelles as important as chloroplasts, especially in weather like this!

The past or present existence of algae with prochlorophytic characteristics has been postulated ever since botanists suggested, as a possible origin of plastids, invading prokaryotic algae that became symbiotically associated with previously apochlorotic hosts. This idea was put forward by several biologists in the nineteenth century, and was perhaps most elaborately expounded by Mereshkovsky in a 1905 thesis on "symbiogenesis." Unfortunately, this work, being in Russian, received scant attention in the West. The German version, published in 1910,[8] was likewise generally ignored, largely because in those days one had almost no ways to test Mereshkovsky's hypothesis.* (Incidentally, Mereshkovsky indicated possible symbiogenetic origins not only of plastids but also of mitochondria, as indicated in his figure, reproduced here as FIGURE 9.) Subsequent studies provided information to support this originally speculative hypothesis, which was revived half a century later by Goksøyr,[9] Schnepf and Brown,[10] and others. Margulis [11] has enthusiastically compiled these data, and has reviewed the evidence for symbiotic origins of plastids and other organelles (see also reviews by Flavell [12] and others).

Prokaryotes presumably evolved some 3,500 million years ago; possibly among them there were prochlorophytes as well as cyanophytes. Since it has been shown that *Prochloron* can be silicified in the laboratory to produce fossil-like artifacts,[13] some of the Precambrian fossils now identified as cyanophyte remains may in fact represent prochlorophytes, though this would be hard to prove. Eukaryotes are believed to have evolved some 1,200 million years ago; possibly some of the prochlorophytes at that time gave rise to the chlorophytes, either by direct intracellular differentiation or by symbiogenesis. Theorization along these lines is based on the speculative hypothesis that biosynthesis of

* Mereshkovsky [8] wrote something to this effect: "The cyanophytes developed from the bacteria (among which there were already many different colored forms: red, yellow, green, etc.) by an increase in pigment formation. They invaded various amoebae and flagellates that had already been formed from the symbiosis of bacteria with nucleus-less Monera, and thereby initiated the development of several (6–9) independent branches of the plant kingdom. Such a highly polyphyletic origin of the plant kingdom, which can now be accepted as finally established, seems to have resulted from different cyanophytes (green, brown and red) invading morphologically different flagellates with one flagellum, two similar flagella, or two dissimilar flagella."

FIGURE 1. Various encrusting invertebrates, etc., on intertidal roots of a mangrove (*Rhizophora*), including grey colonies of *Didemnum* sp. bearing green patches of *Prochloron* cells. Isla San José, B.C., México; January 1967.

FIGURE 2. White colonies of *Didemnum* sp., bearing green patches of *Prochloron* cells. Approx. 1 m depth. Punta Espinosa, Isla Isabella, Galápagos Isles, Ecuador; February 1978. Scale in mm.

FIGURE 3. Grey, calcified colonies of *Lissoclinum* sp. (?), containing *Prochloron* cells inside common cloacal system. Approx. 1 m depth. Dravuni Is., Fiji; August 1978. Snorkel (ca.2.5 cm diam.) indicates scale.

FIGURE 4. Green colonies of a different species of *Lissoclinum*, containing *Prochloron* cells inside common cloacal system. Palau, W. Caroline Is.; September 1979. Scale as in FIGURE 2.

FIGURE 5. Colonies of a large fleshy species, *Lissoclinum patella*, collected on shore just above mean low-water level. Pulau Salu, Singapore: April 1977. (Photo: W. Vidaver)

FIGURE 6. Colony of *L. patella*. Pulau Salu, Singapore; April, 1977. Finger indicates scale. (Photo: W. Vidaver)

FIGURE 7. *Prochloron* cells expressed from *L. patella*. Pulau Salu, Singapore; May 1980. Scale bar = 50 micrometers. (Photo: P. N. Avadhani)

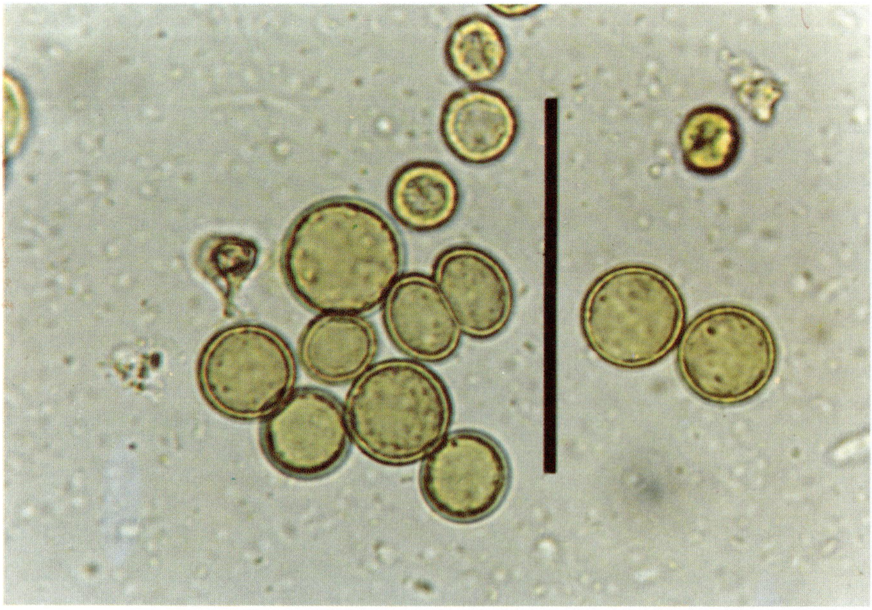

FIGURE 8. *Prochloron* cells expressed from *L. patella*. Pulau Salu, Singapore; May 1980. Scale bar = 50 micrometers. (Photo: P. N. Avadhani)

chlorophyll *b*, which is an extraordinarily conservative trait, was "invented" only once in the evolution of life on Earth.

So far, prochlorophytes are known almost exclusively in symbiotic association with didemnids. These animals, usually characterized by their skeletal elements (calcareous spiculospheres), are known from as far back as the Upper Cambrian. In that epoch, when oxygen concentrations in the sea or atmosphere may have been lower than they are today, prochlorophytes were perhaps commoner, but in subsequent epochs they almost died out except in symbioses

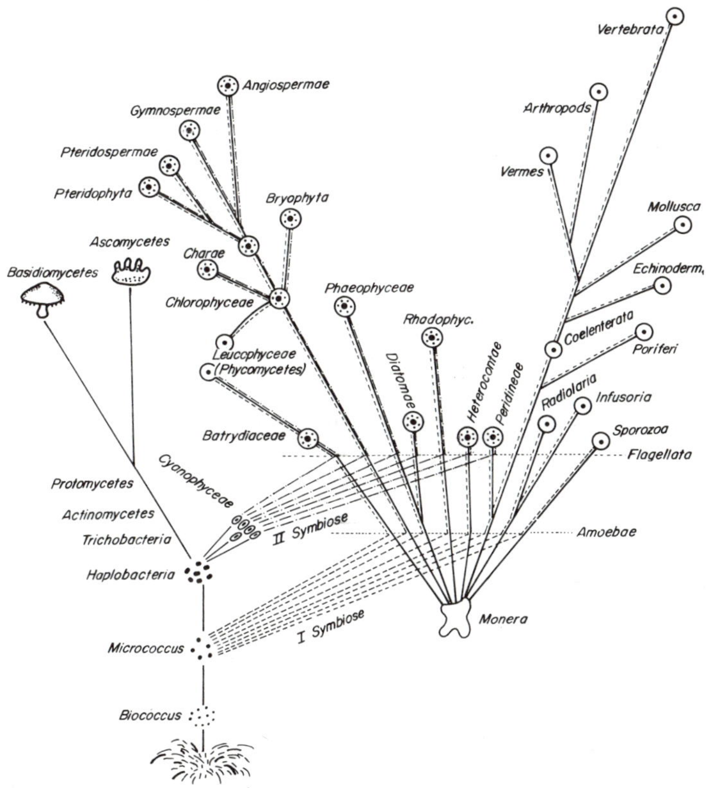

FIGURE 9. Phylogenetic tree indicating origins of eukaryotes by symbiogenesis according to Mereshkovsky.[8]

where ambient oxygen levels are lowered biologically. But these are mere speculations.

However, now that we know of the contemporary existence of prochlorophytes, we recognize the need for biological, physiological, and biochemical studies of these hitherto overlooked algae. Although the hypothesis of symbiogenesis is still far from being confirmed, certain of its assumptions can now be examined with data from *Prochloron*. We need in particular a critical comparison of the nucleotide sequences in its nucleic acids with corresponding sequences in chloroplasts of green algae and angiosperms. We may find close parallels, such as those that have recently been shown to exist between the nucleic acids in the plastids of rhodophytes and those of certain blue-green algae—but we may find no such interrelationships. The years ahead are full of promise; its fulfillment will come earlier if we can overcome the obstacles that have hitherto prevented us from growing prochlorophytes in culture.

Acknowledgments

I gratefully acknowledge helpful discussions on this topic with Drs. Lanna Cheng and Joan W. Stewart.

References

1. Antia, N. J. 1977. A critical appraisal of Lewin's Prochlorophyta. British Phycol. J. **12:**271–276.
2. Chadefaud, M. 1978. Sur la notion de prochlorophytes. Rev. Algol. **13:**203–206.
3. Lewin, R. A. 1976. Prochlorophyta as a proposed new division of algae. Nature **261:**697–698.
4. Lewin, R. A. 1977. *Prochloron,* type genus of the Prochlorophyta. Phycologia **16:**217.
5. Lewin, R. A. 1980. Prochlorophytes. *In* The Prokaryotes. M. P. Starr, H. Stolp, H. G. Trueper, A. Balows & H. G. Schlegel, Eds. Springer-Verlag. Berlin. (In press.)
6. Kott, P. 1977. Algal supporting didemnid ascidians of the Great Barrier Reef. Proc. 3rd Int. Coral Reef Symp. (Miami, Fla.) pp. 616–622.
7. Lewin, R. A. 1978. Distribution of symbiotic didemnids associated with prochlorophytes. Proc. Int. Symp. Marine Biogeography. July 17–20, New Zealand Dept. of Scientific and Industrial Research. Auckland, N.Z. pp. 365–369.
8. Mereschkowsky, C. 1910. Theorie der zwei Plasmaarten als Grundlage der Symbiogenesis, einer neuen Lehre von der Entstehung der Organismen. Biol. Centralbl. **30:**278–303, 321–347, 353–367.
9. Goksøyr, J. 1967. Evolution of eukaryotic cells. Nature **211:**1161.
10. Schnepf, E. & R. M. Brown. 1969. On relationships between endosymbionts and the origin of plastids and mitochondria. *In* Origin and Development of Cell Organelles. W. Ursprung & C. Reinert, Eds. Springer-Verlag. Berlin.
11. Margulis, L. 1970. Origin of Eukaryotic Cells. Yale University Press. New Haven, Conn.
12. Flavell, R. 1972. Mitochondria and chloroplasts as descendants of prokaryotes. Biochem. Genetics **6:**275–291.
13. Francis, S., E. S. Barghoorn & L. Margulis. 1978. On the experimental silicification of microorganisms. Implications of the preservation of the green prokaryotic alga *Prochloron* and other coccoids for interpretation of the microbial fossil record. Precambian Res. **7:**377–383.

Discussion of the Paper

Dr. D. A. Shoenberg (*Lehigh University, Bethlehem, Pa.*): I have three questions which pertain to the natural history of this symbiosis. First, are those two the only reports of zoochlorellae in your chordates? Second, have you found any aposymbiotic *dideminids* of the species? And third, could you possibly speculate on *why* tunicates and these particular ones?

DR. R. A. LEWIN: First of all a number of people have reported on them. Eldrige pointed out the particularly interesting thing that the symbiotic species, when they produce larvae, equip the larvae with a little capital so they can start a new business when they go off. So they've got little pockets next to their tail with algae on them.

Secondly, have we found aposymbiotic ones? I think not, there are plenty of species in which the majority of species do not have symbiotic algae associated with them. The species which do, I believe nearly always do. The one exception is the one that I showed you first, the one with the material on the outside and that very often occurs aposymbiotically, much more so than associated with the algae.

The last question, do I have any idea why ascidians and not other things? They have some funny things about their excretory mechanism, they have some species that have acid cells in their blood, some of them accumulate valadium or niobium, we're looking into these possibilities.

DR. J. A. SCHIFF: I was interested to see the paired thylakoids because I always sort of run on the assumption that the reason the thylakoids don't pair in the cyanophytes and the red algae is because the phycobilisomes don't permit it. The phycobilisomes of course achieve the same end providing the energy transfer that would ordinarily be enhanced by stacking. Perhaps these things do have regulation of energy transfers through stacking just like the higher chloroplasts and therefore, then fall in the line of the chlorophytes in that sense also.

DR. S. P. GIBBS: It's a slight simplification to say that green algae chloroplasts have paired thylakoids. Paired thylakoids can occur in green algae chloroplasts. They're not common. Three to 6 in a group are more common. Grana are sometimes present and also the thylakoids at the edge of the band in green algae continue with each other often. They are not in *Prochloron*. So there are very many ultrastructural differences in thylakoid arrangements between greens and *Prochloron*.

DR. LEWIN: Somebody asked about the chlorophyll–protein complex. I don't remember the figures exactly, but there is a figure given for the usual ratio of chlorophyll to chlorophyll–protein complex (the free to the associated) and this figure is different in the green algae, the blue-green algae, and in *Prochloron* it's intermediate.

THE ESTABLISHMENT OF MITOCHONDRIA: *PARACOCCUS* AND *RHODOPSEUDOMONAS*

F. R. Whatley

University of Oxford
School of Botany
Oxford OX1 3RA
England

The theory of serial endosymbiosis [1] proposes *inter alia* the uptake of bacteria by a primitive eukaryote to yield mitochondria. In support of this view, several features present in prokaryotes can be identified in mitochondria, e.g. biochemical components, circular DNA, properties of DNA and protein synthesis. An outstanding feature of the mitochondrion is its ability to carry out oxidative phosphorylation (electron transport to oxygen coupled to ATP synthesis). Many aerobic bacteria can also carry out electron transport to oxygen,[2] or to alternative terminal electron acceptors, e.g. *Paracoccus denitrificans, Escherichia coli, Rhodopseudomonas spheroides*, and a few have been shown experimentally to couple this transport to ATP synthesis, e.g. *E. coli* [3] and *P. denitrificans*,[4–6] while growth yield experiments clearly show that other bacteria must also do oxidative phosphorylation.[7] We have been interested, from our own experiments with *P. denitrificans* and from a survey of the literature, in identifying the points of similarity and difference between the details of electron transport and phosphorylation in bacteria on the one hand and mitochondria on the other. Although there are some recognizable differences between plant and animal mitochondria these differences are not sufficient at present to indicate a multiple origin for mitochondria in different groups of organisms (unlike chloroplasts). It is suggested that the original bacterium supposed to have been taken up endosymbiotically by the protoeukaryote host must have resembled in many ways *P. denitrificans* or *Rh. spheroides*, though care must be taken in thinking about the evolutionary sequence not to imagine that either of these modern forms was the actual bacterial ancestor of the mitochondrion.

The plasma membrane of *P. denitrificans* contains a constitutive respiratory chain which is almost identical to the electron transport chain of mammalian mitochondria.[8] The respiratory components present and the response to respiratory inhibitors are surprisingly similar in both *Paracoccus* and mitochondrion, as are the main physiological features of ATP formation. In addition to the constitutive chain, *P. denitrificans* also has many adaptive constituents (e.g. hydrogenase, formate dehydrogenase, nitrate and nitrite reductases) that enable the bacterium to grow with a large number of respiratory substrates compared with the "restricted diet" available to the mitochondrion.[9] If we examine the distribution of mitochondrial features among aerobic bacteria (facultative as well as obligate aerobes) we find hardly any feature (apart from the possession of the adenine nucleotide carrier and certain details of the carrier systems, both of which reflect the change of environment between free-living bacterium and mitochondrion) to be absent from *Paracoccus*.[8] On

0077–8923/81/0361–0330 $01.75/0

the other hand all the mitochondrial features can be found in other aerobic bacteria but not all together.

Consequently the majority of bacterial electron transport chains are very different from those of mitochondria. In the case of the photosynthetic bacterium *Rhodopseudomonas spheroides* the electron transport chain during aerobic growth in the dark [10] is also very similar to that of the constitutive chain in *Paracoccus* and in the mitochondrion. It may be that further extensive work on *Rhodopseudomonas* will demonstrate its similarity to *Paracoccus* in virtually all mitochondrial features, including ATP synthesis and respiratory control. A pseudomitochondrial chain is also a striking feature of the aerobe *Alcaligenes eutrophus.*[2]

A number of features of aerobic bacteria, including *Paracoccus* and *Rhodopseudomonas*, are of very wide distribution. These include NADH dehydrogenase and succinate dehydrogenase (feeding hydrogens into the system), the proton translocating ATPase (responsible for ATP synthesis), and the enzymes of the tricarboxylic acid cycle (required for processing of the carbon skeletons used as a source of hydrogen for the electron transport chain). In addition the NADH-NADPH transhydrogenase has frequently been recorded in aerobic bacteria.[2, 8]

Other mitochondrial features are of more restricted distribution among aerobic bacteria. For example, cytochrome aa_3 (cytochrome oxidase) is the terminal oxidase in *Paracoccus* [4, 5, 11] as well as in all mitochondria (FIGURE 1). While cytochrome oxidase is present in a number of aerobic bacteria, e.g. *Alcaligenes, Mycobacterium, Micrococcus lutea (inter alia)*, the terminal oxidase in other bacteria is either a mixture of cytochromes (cytochrome *o, a,* and *d* are present in *Actobacter vinelandii* and *Haemophilus para-influenzae*) or cytochrome *o* alone (*E. coli*).[2, 3, 12] *Rh. spheroides* grown aerobically in the dark has cytochrome aa_3.[10] *Paracoccus* contains two readily distinguishable cytochromes *b* when grown aerobically. This is also true for some other aerobic bacteria, e.g. *Micrococcus luteus* and *M. lysodeikticus* (which have cyt a/a_3) and *Escherichia coli* and *Klebsiella pneumoniae* (which lack cyt a/a_3). Other aerobic bacteria appear to have only one cytochrome *b* (e.g. *Alcaligenes* and *Pseudomonas ovalis*). Like mitochondria, *Paracoccus* has two cytochromes *c*, as have *Mycobacterium phlei, Azotobacter,* and *M. leuteus*; in other aerobic bacteria only one cytochrome *c* component has been found (e.g. *Alcaligenes* and *Rhodopseudomonas*) and there is no cytochrome *c* in *E. coli, Klebsiella,* or *Bacillus megaterium.*[2, 3, 10, 12]

Cytochromes *c* are small molecular mass proteins which are important constituents of the electron transport system. Their amino acid sequences have been widely studied. When the sequences are arranged according to the extent of their similarities it becomes apparent that the cytochromes *c* from a variety of eukaryotes are very similar and that they bear a close resemblance to the cytochromes c_2 from a group of bacteria which includes *Paracoccus* and the photosynthetic purple non-sulphur bacteria *Rhodopseudomonas spheroides* and *Rhodopseudomonas vannielii.*[13] The actual shapes of several cytochrome *c* molecules have been determined by X-ray crystallography, including cytochrome *c* from tuna mitochondria and cytochrome c_2 from *Paracoccus* (usually termed cytochrome 550). A great similarity in the shape of the two molecules is immediately apparent (FIGURE 2).[14] The main differences are the inclusion in the cytochrome c_{550} sequence of three short surface loops together with a long amino acid tail situated at the back of the molecule. An evolutionary deletion

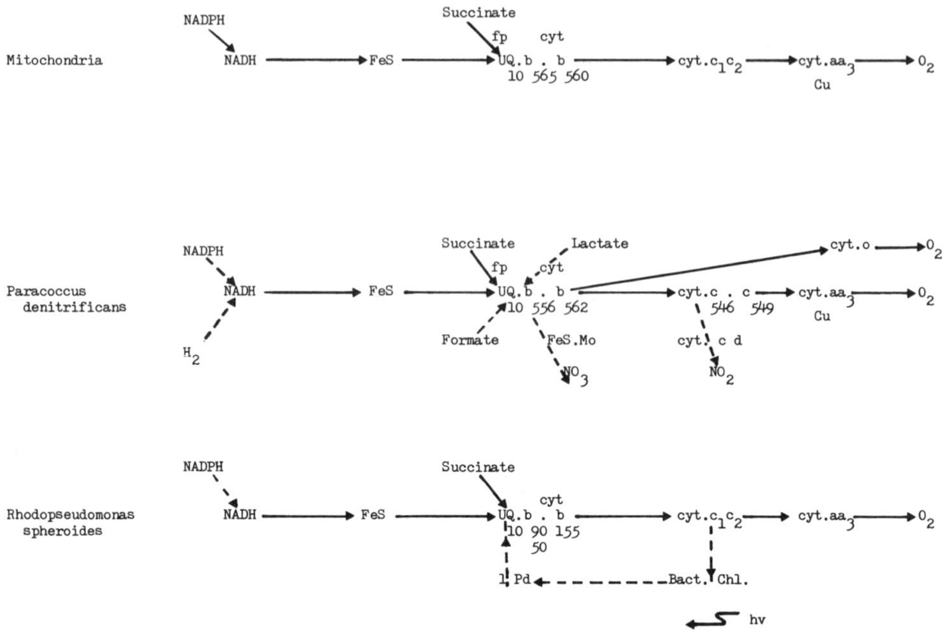

FIGURE 1. Electron transport chains in mitochondria, *Paracoccus* and *Rhodopseudomonas. Key:* NADPH, nicotinamide adenine dinucleotide phosphate; NADH, nicotinamide adenine dinucleotide; FeS, iron-sulphur centers with g=1.94 signal; fp, flavoprotein; UQ_{10}, ubiquinone-10; cyt, cytochrome; Pd, "photoredoxin;" and Bact.Chl, bacteriochlorophyll.

of the loops and the tail would convert cytochrome c_{550} of *Paracoccus* into a very respectable mitochondrial type of cytochrome *c*.

Gram-negative bacteria have either ubiquinone-10 or ubiquinone-8 as constituents of their electron transport chain while mitochondria, *Paracoccus*, and *Rhodopseudomonas* have only ubiquinone-10. By contrast the gram-positive aerobic bacteria use naphthoquinone.[2, 12] Experiments so far indicate that aerobic bacteria are largely insensitive to antimycin or rotenone, whereas *Paracoccus* has a typical mitochondrial sensitivity to these two inhibitors. Both mitochondria and *Paracoccus* particles show respiratory control, which is released by the addition of ADP or by uncouplers.[8, 15] This type of control is rarely seen in other aerobic bacteria, though one should note that very few organisms have been tested for this and the picture is complicated by the fact that only inside-out particles (whose sidedness is the reverse of the normal) can use ADP. The extent of respiratory control must depend on the ratio of inside-out particles, which can respire and accumulate ATP, to right-side out particles, which can only respire.

Implicit in the current ideas of oxidative phosphorylation is the pumping of protons, whose return to the inside of the bacterial cell (or organelle) is accompanied by ATP synthesis. The number of protons produced between the span NADH→O_2 has been determined as six per oxygen atom in mitochondria and *Paracoccus*.[2,3,9] In many other aerobic bacteria the ratio is consistently two or four.[2,3] Finally one notes that the membranes contain phosphatidyl choline as a major constituent in mitochondria and *Paracoccus* whereas it is a rare constituent in other aerobic bacteria. Most aerobic bacteria contain branched chain and cyclopropane fatty acids whereas in mitochondria and *Paracoccus* all the fatty acids are straight chained; a mixture of saturated and unsaturated types are found. (See Reference 12.)

In addition to these observations there are a number of experiments showing additional mitochondrial features which have been looked for and found in *Paracoccus*, although often they have not yet been sought in other aerobic bacteria. These include features of the respiratory chain and of the ATPase. For example the NADH dehydrogenase segment of the respiratory chain was shown to contain FeS centers 1–4 when examined by electron paramagnetic resonance at 15° K.[16] Similar FeS centers are also reported in *Rhodopseudomonas*.[10] A rapid oxygen-induced antimycin-dependent reduction of *b*-type cytochrome characteristic of mitochondria is also seen in *Paracoccus* [17] (indicating a great degree of similarity of structural arrangement between the two). Oxidation of succinate is inhibited by carboxin [18] and thenoyltrifluoroacetone [11] and the oxidation of NADH is sensitive to piericidin [19] just as in mitochondria. It has also been demonstrated that *Paracoccus* ATPase, like mitochondrial ATPase, contains tightly bound nucleotides which are exchangeable on energization,[20] that the ATP synthesis is inhibited by aurovertin but not ATP hydrolysis,[21] and that the ATPase in inhibited by venturicidin [21] and also by 7-chloro-4-nitrobenzo-2-oxal-1,3-diazole, the latter observation suggesting the importance of a tyrosine group in the active center of the ATPase in both mitochondria and *Paracoccus*.[22]

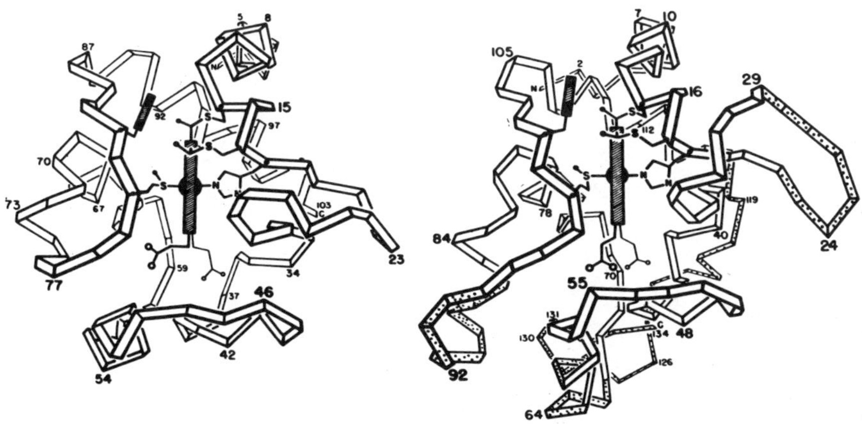

FIGURE 2. Front views of cytochromes *c*. Left, mitochondrial cytochrome *c* from tuna; right, *Paracoccus* cytochrom c_{550}. The additional tail and three loops in cytochroma c_{550} are indicated by stippling. (From Almassy & Dickerson,[14] 1978.)

No significant feature of the mammalian respiratory chain has been shown to be absent from *Paracoccus* and this fact has been used to suggest that the evolutionary transition from an aerobic bacterium resembling *Paracoccus* to mitochondrion would be biochemically a simple one (FIGURE 3). In this transition the ATPase and the constitutive components of the respiratory chain would be kept, the adaptive components of the respiratory chain would be lost and the transport properties of the bacteria plasma membrane would be modified so that an effective integration of the metabolism of the organelle with that

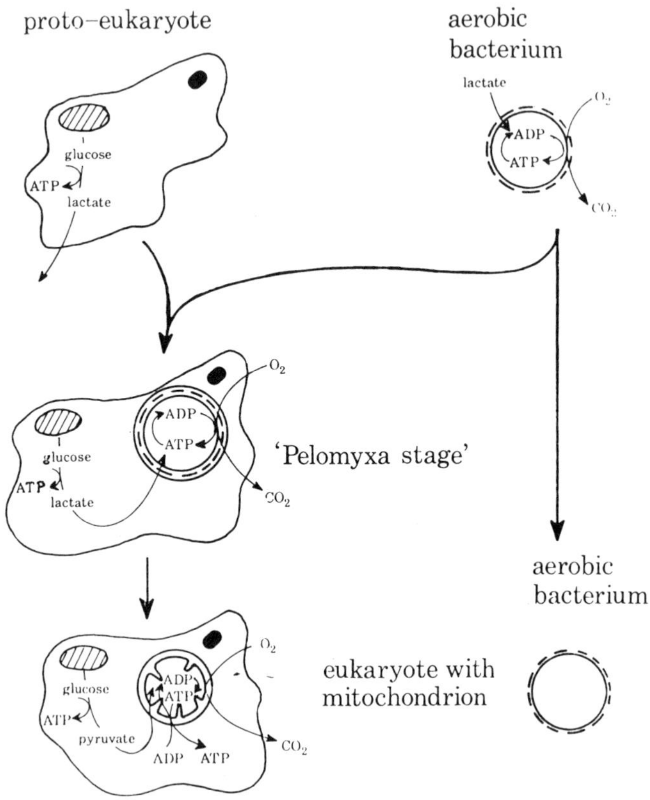

FIGURE 3. An evolutionary origin for mitochondria. (Redrawn from John & Whatley,[8] 1975.)

of the surrounding cell was finally attained.[8] One could easily imagine that the adaptive components of the respiratory chain of *Paracoccus,* which would not be needed by a mitochondrion operating in the controlled environment of the cytoplasm, could readily be dispensed with. For example, we know in *Paracoccus* that synthesis of the respiratory nitrate reductase is easily repressed in the presence of oxygen.[23] In addition many of the specific carriers required for the uptake of a large number of metabolites from the environment only appear when the specific metabolite is present; the mitochondrion of course has a very restricted number of carriers and "lives" on a highly simplified "diet." The only new component necessary for an evolutionary transition from the plasma membrane of a bacterium like *Paracoccus* to the mitochondrial membrane is the adenine nucleotide carrier, which appears to be absent from free-living bacteria but present in all mitochondria. The ATP synthesized by the endosymbiont would become available to the host only after the acquisition of this carrier. When the bacterium has acquired the carrier it may be said to have become, metabolically speaking, a mitochondrion.[8]

The mitochondrial adenine nucleotide carrier might have been thought of as imposed by the host on the bacterium as it changed into a mitochondrion. However this is not necessarily true. Winkler has shown that the endoparasitic bacterium, *Rickettsia prowazeki*, has a nucleotide carrier which can exchange an ATP for an ADP in a one-to-one ratio across its boundary membrane [24] and in this way has the possibility of picking up ATP from its host (a sort of mitochondrion in reverse). The carrier is reversible and it may be that the supposed mitochondrial ancestor could actually have provided its own adenine nucleotide carrier. However the observation that the mitochondrial adenine nucleotide carrier is sensitive to the inhibitors atractyloside and bongkrekic acid, whereas that of *Rickettsia* is not, may perhaps be taken not to support this speculation.

Mitochondrial cytochrome *c* is coded for by the nuclear DNA. Obviously cytochrome *c* in free-living bacteria is coded for by the bacterial DNA. In an evolutionary transition there must have been transfer of information from the bacteria to the nuclear genome if the endosymbiotic origin of the mitochondrion is correct. There has not yet been any conclusive demonstration of genetic transfer from a prokaryotic cell to a eukaryotic nucleus. However, experiments have shown that the prokaryote *Agrobacterium tumefaciens* contains a plasmid which can be transferred to a plant cell.[25, 26] Following transfer of the plasmid the plant cell develops into a crown gall; the information for this morphogenetic change is provided by the plasmid. It is not known at present whether the transferred plasmid joins the main nuclear genome, whether it goes into the satellite genome of the chloroplast or mitochondrion, or whether it remains a free plasmid. If the plasmid were shown to enter the chloroplast or the mitochondrion we would only be seeing the equivalent of a plasmid transfer from one prokaryote to another. If it merely remains free in the cytoplasm that is essentially uninteresting from the evolutionary point of view! If however the plasmid could be shown to join the nuclear genome then we would have a demonstration of genetic transfer from a prokaryotic cell to a eukaryotic nucleus, the type of transfer required by the evolutionary theory of endosymbiosis.

A present-day model for a transitional stage, which illustrates how mitochondria might have evolved from bacteria, is provided by the giant amoeba *Pelomyxa palustris*.[27] This multinucleate amoeba is a primitive eukaryote whose highly vacuolated cytoplasm lacks mitochondria and Golgi bodies (FIGURE 4). However two types of endosymbiotic bacteria are present within the amoeba throughout its entire life cycle. The larger bacteria are enclosed within tightly fitting vacuoles whose bounding membrane is provided by the host cell. These bacteria thus lie not in the cytoplasm itself but within the endomembrane system which traverses the host cell and are therefore topologically outside the cell, an arrangement expected from the endosymbiotic theory. When *Pelomyxa* is about

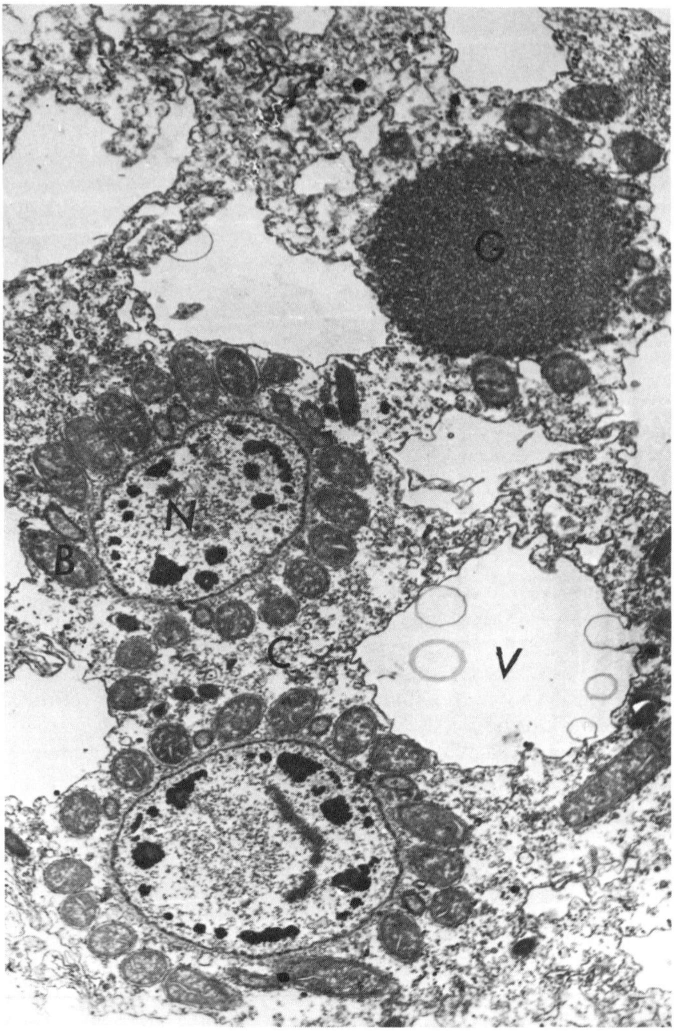

FIGURE 4. *Pelomyxa paulustris* and its endosymbiotic bacteria. (×5,000) *Key:* B, endosymbiotic bacterium; C, cytoplasm; G, glycogen body; N, nucleus; and V, structural vacuole. Electron micrograph supplied by Jean M. Whatley.

to divide the (large) bacteria aggregate around the nuclei; when cell division is completed each daughter amoeba contains one to three nuclei together with their associated bacteria. The symbiont population within the daughter cell is maintained not only by migration of the bacteria to the vicinity of the nucleus at cell division but also by the formation at the time of division of tubular connections between the outer nuclear membrane and the membranes of the vacuoles in which the bacteria are situated.[28] The tubules effectively tie the bacteria to the nuclei, since the nuclear membrane does not disappear during mitosis, as commonly happens in higher organisms. Thus when the daughter nuclei are formed by pinching following mitosis the appropriate population of bacteria is carried off to the daughter cells.[27]

A new apparently permanent symbiotic association was experimentally produced recently in Jeon's laboratory. Cells of *Amoeba proteus* became infected by alien bacteria in 1967, at which time they caused the amoeba to become sick and to grow at a lower rate than the uninfected cells. However the relationship between the amoebae and the bacteria became established to the extent that by 1976 it was shown that the amoebae had become dependent on the presence of the bacteria and were no longer viable without them.[29] Development of an endosymbiotic relationship need not be a prolonged process, and in fact there cannot be a prolonged delay in any of the steps leading to the integration.

An unexpected addition to the range of circular DNA-containing organelles found in eukaryotic cells is the hydrogenosome; unlike mitochondria and chloroplasts, it has a fermentative mode of ATP synthesis. The recent work of Lindmark & Müller and of Cerkasovova and coworkers on *Tritrichomonas*

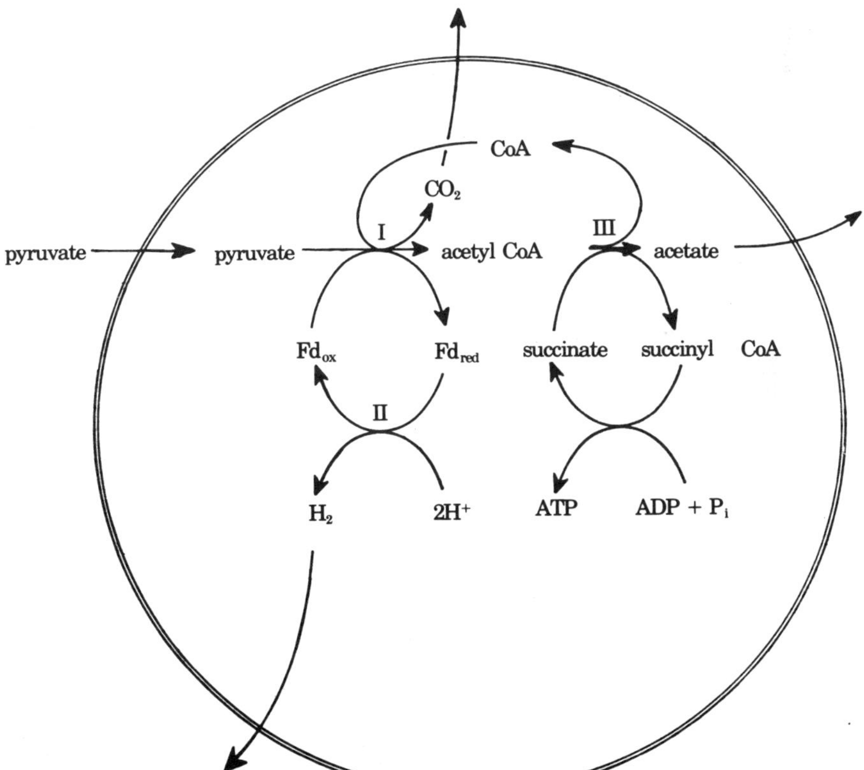

FIGURE 5. Metabolism of a hydrogenosome. (After Müller,[32] 1975.) *Key:* I, pyruvate synthase; II, hydrogenase; and III, acetate thiokinase.

foetus suggests that the hydrogenosome may be viewed as having had a free-living ancestor, probably resembling a present-day species of *Clostridium.*

Tritrichomonas foetus is an aerotolerant flagellate which inhabits the vagina of the cow. It lacks organelles resembling mitochondria, and has neither the tricarboxylic acid cycle nor cytochromes. Carbohydrates are fermented anaerobically with the production of H_2, CO_2, and succinate. The carbohydrate is catabolized in the cytosol via the glycolytic pathway with the production of pyruvate. The pyruvate is then oxidized to acetate in the microbody-like organelles called hydrogenosomes, as illustrated in FIGURE 5. This reaction occurs via acetyl CoA, and a coupled substrate-level phosphorylation yields ATP. Ferredoxin is reduced during the oxidation and is regenerated by donating its electrons to protons under the influence of hydrogenase; the resulting H_2 is discarded.[30]

Ferredoxin-linked pyruvate synthase and hydrogenase are enzymes normally found only in anaerobic prokaryotes like clostridia. These two enzymes are not known from aerobic eukaryotes. Thus it seems possible that the hydrogenosome had an exogenous, prokaryotic evolutionary origin. Further features of the hydrogenosome which support this notion are: the possession of a superoxide dismutase which is CN^--insensitive; [30] a high content of cardiolipin; [31] and, of even greater significance, the occurrence of circular DNA [31] in isolated hydrogenosomes.

It is not clear whether the hydrogenosome is bounded by a single or a double membrane. The membrane obviously provides an effective physiological limit to the hydrogenosome, since all of the enzymatic activities assayed in the isolated hydrogenosome (except for hydrogenase) are stimulated by treating hydrogenosomes with the detergent Triton.[32] It would be of particular interest to know more about the permeability properties of this membrane, particularly with respect to ADP and ATP.

The available evidence strongly suggests that the hydrogenosome has evolved from a free-living prokaryote resembling present-day clostridia. Intermediates in the evolution may well be represented by the endosymbiotic bacteria of those ciliates which flourish in anaerobic, sulphur-rich habitats.[33]

The theory of endosymbiosis receives obvious support from the existence of many present-day symbioses between prokaryotic or eukaryotic symbionts and eukaryotic hosts. When such a union has been arranged there is good evidence for modification of metabolism in both the contracting parties. A few bacteria, notably *Paracoccus denitrificans,* and perhaps to a lesser extent *Rhodopseudomonas spheroides* and *Alcaligenes eutropha*, have cell membranes containing mitochondrial-like electron transport chains and the ability to do oxidative phosphorylation. Continuing studies, especially on *Paracoccus*, have extended the spectrum of similarities. A principal remaining problem is the necessary transfer of genetic information from endosymbiont to host; work on *Agrobacterium tumefaciens* may well provide definitive examples this transfer, and relevant information may also result from modern techniques of DNA and RNA sequencing. We have therefore in *Paracoccus* and a few other bacteria a present-day model that contains a great many of the mitochondrial features to be expected of the ancestral bacterium that supposedly gave rise to the mitochondrion.[34]

Summary

Many aerobic bacteria (both facultative and obligate) possess a number of those biochemical features of mitochondria which are concerned with energy metabolism. However, only a restricted number, notably *Paracoccus denitrificans* and *Rhodopseudomonas spheroides*, have the majority of these features. The theory of endosymbiosis proposes that a primitive eukaryote took up bacteria to yield mitochondria. The present-day *Paracoccus* then resembles the ancestral bacterium in many respects the primitive amoeba, *Pelomyxa palustris*, which lacks mitochondria but contains a permanent population of unique symbiotic bacteria, has many of the characteristics of a present-day transitional form. The evolution of mitochondria from endosymbiotic bacteria would involve their integration with the host cell both biochemically and structurally: a number of the intermediate steps are discussed. Attention is drawn to the existence in some ciliates of hydrogenosomes, which function as anaerobic mitochondria.

References

1. Margulis, Lynn. 1970. Origin of Eukaryotic Cells. Yale University Press. New Haven, Conn.
2. Jones, C. W. 1977. Symp. Soc. Exptl. Microbiol. **27:**23–59.
3. Haddock, B. A. & C. W. Jones. Bacteriol. Rev. **41:**47–99.
4. Imai, K., A. Asano & R. Sato. 1968. Biochim. Biophys. Acta **143:**462–467.
5. John, P. & F. R. Whatley. 1970. Biochim. Biophys. Acta **216:**342–352.
6. Knobloch, K., M. Isaque & M. I. H. Aleem. 1971. Arch. Microbiol. **76:**114–124.
7. Stouthamer, A. H. 1977. Symp. Soc. Exptl. Microbiol. **27:**285–315.
8. John, P. & F. R. Whatley. 1975. Nature (London) **254:**495–498.
9. John, P. & F. R. Whatley. 1977. Biochim. Biophys. Acta **463:**129–153.
10. Dutton, P. L. & D. F. Wilson. 1974. Biochim. Biophys. Acta **346:**165–212.
11. Scholes, P. B. & L. Smith. 1968. Biochim. Biophys. Acta **153:**363–375.
12. John, P. & F. R. Whatley. 1977. Adv. Bot. Res. **4:**51–115.
13. Ambler, R. P., T. E. Meyer & M. Kamen. 1976. Proc. Natl. Acad. Sci. USA **73:**472–475.
14. Almassy, R. J. & R. E. Dickerson. 1978. Proc. Natl. Acad. Sci. USA **75:**2674–2678.
15. John, P. & W. A. Hamilton. 1971. Eur. J. Biochem. **23:**528–532.
16. Meijer, E. M., R. Wever & A. H. Stouthamer. 1977. Eur. J. Biochem. **81:** 267–275.
17. John, P. & S. Papa. 1978. F.E.B.S. Lett. **85:**179–182.
18. Tucker, A. N. & T. T. Lillich. 1974. Antimicrob. Agents Chemother. **6:**572–578.
19. Lawford, H. G., J. C. Cox, P. B. Garland & B. A. Haddock. 1976. F.E.B.S. Lett. **64:**369–374.
20. Harris, D. A., P. John & G. K. Radda. 1977. Biochim. Biophys. Acta **459:** 546–559.
21. Ferguson, S. & P. John. 1977. Biochem. Soc. Trans. **5:**1525–1527.
22. Ferguson, S., P. John, W. J. Lloyd, G. K. Radda & F. R. Whatley. 1974. Biochim. Biophys. Acta **357:**457–461.
23. Pichinoty, F. 1965. Ann. Inst. Pasteur, Paris. **109:**248–255.
24. Winkler, H. H. 1976. J. Biol. Chem. **251:**389–396.
25. Drummond, M. 1979. Nature (London) **281:**343–347.
26. Schell, J., M. van Montagu, M. de Beuckeleer, M. de Block, A. Depicker,

M. DE WILDE, G. ENGLER, C. GENETELO, J. P. HERNALSTEENS, M. HOLSTERS, J. SEURINCK, B. SILVA, F. VAN VLIET & R. VILLARROEL. 1979. Proc. Roy. Soc. Lond. B **204:**251–266.
27. CHAPMAN-ANDRESEN, C. 1971. Ann. Rev. Microbiol. **25:**27–48.
28. WHATLEY, JEAN M. 1976. New. Phytol. **76:**111–118.
29. JEON, K. W. & M. S. JEON. 1976. J. Cell Physiol. **89:**337–344.
30. LINDMARK, D. G. & M. MÜLLER. 1973. J. Biol. Chem. **248:**7724–7728.
31. CERKASOVA, A., J. CERKASOV, J. KULDA & J. REISCHIG. 1976. Folia Parasitol., Prague **23:**33–37.
32. MÜLLER, M. 1975. Ann. Rev. Microbiol. **29:**467–483.
33. FENCHEL, T., T. PERRY & A. THANE. 1977. J. Protozool. **24:**154–163.
34. RICHMOND, N. H. & D. C. SMITH, Eds. 1979. The Cell as a Habitat. Proc. Roy. Soc. Lond. **B204:**115–286.

DISCUSSION OF THE PAPER

DR. J. M. OLSON: In view of the tree published by Schwartz and Dayhoff in *Science*, and which I believe you showed, I would think that the original endosymbiont from which mitochondria have developed might have been a photosynthetic organism and the great similarity between *Paracoccus* and the mitochondrion is convergent evolution.

DR. F. R. WHATLEY: Yes I think that's one of the points which I perhaps should have made more firmly, that *Rhodopseudomonas* is just as good a candidate as *Paracoccus*. It is simply that we know somewhat less about it than *Paracoccus*. As you say, we may be suffering from convergent evolution. It would certainly make sense to look at *Rhodopseudomonas* from this particular biased view. It seems, the bias, if you like, in our presentation is that we know a great deal more about *Paracoccus* and can't fault it. Equally, we know less about *Rhodopseudomonas* but we cannot fault it. Is that an adequate answer?

DR. D. E. GREEN: I wonder if the similarity goes to the specific phospholipid that's required for energy coupling namely, cardiolipin. Do you find cardiolipin in these organisms?

DR. F. R. WHATLEY: I do believe the answer is yes, and it's similar to the hydrogenosome which has cardiolipin. The hydrogenosome is in fact an anaerobic mitochondrion, if you like to define it as such. This does substrate phosphorylation and in fact doesn't have any requirement for cardiolipin to do oxidative phosphorylation.

DR. GREEN: Does it have cardiolipin?

DR. F. R. WHATLEY: It does have cardiolipin. I showed that.

DR. GREEN: I was anxious recently to find out whether there were any cases of mitochondria-like systems that were deficient in cardiolipin.

DR. F. R. WHATLEY: I think that this particular organism, *Paracoccus*, does have it. I don't think this information is known for other organisms.

CHLOROPLASTS: PRESUMPTIVE AND *DE FACTO* ORGANELLES *

Robert K. Trench

Department of Biological Sciences and the
Marine Science Institute
University of California, Santa Barbara
Santa Barbara, California 93106

INTRODUCTION

> . . . it is the possession of plastids which, at the cellular level, more than anything else distinguishes the Plant Kingdom from the Animal Kingdom. In fact, taxonomically speaking, it might be reasonable to make the presence or absence of plastids the basis for defining an organism as plant or animal.
>
> J. T. O. Kirk and R. A. E. Tilney-Bassett

Until quite recently the chloroplast was thought to be an organelle that was located and functioned exclusively within the plant cell. Hence the *presence* of functional chloroplasts was thought to be a criterion for distinguishing between plants and animals.[1] In this article, I shall briefly recall evidence that clearly demonstrates sustained plastid function in animal cells. The phenomenon of "foreign organelle retention" is consistent with the concept of the possible origin of some chloroplasts by endocytosis and subsequent symbiosis.

Because chloroplasts possess their own genetic and protein synthesizing apparatus, they were thought to be autonomous entities. However, the vast body of evidence now available supports the concept that in their developmental continuity, chloroplasts are semiautonomous and rely to a very large extent on genetic information and biosynthetic processes that are extraplastidal. Hence, in plastid development, it is the coordinated intercompartmental cooperation between the plastome and plastid stroma, the genome and the cell cytoplasm, that eventually produces the mature differentiated chloroplast. But how autonomous is the differentiated chloroplast? To what extent is the mature plastid dependent on the plant cell nucleus and cytoplasm? Studies of mature chloroplasts away from the sphere of influence of the plant nucleus and cytoplasm were made possible by the discovery that certain animals sequestered plant chloroplasts and that such chloroplasts remained functional for very extended periods.[2]

To obtain insight into the possible symbiotic origin of chloroplasts, better experimental systems are provided by certain protozooans that harbor intracellular cyanobacteria. Based on our current understanding of chloroplast structure and function, it is virtually impossible to resolve whether the photosynthetic entities found in organisms such as *Cyanophora paradoxa* and *Glaucocystis nostochinearum* are indeed symbionts or *de facto* chloroplasts.

* Supported by the National Science Foundation (Grant PCM 78–15209).

0077–8923/81/0361–0341 $01.75/0 © 1981, NYAS

A. *Retention of Functional Chloroplasts by Animal Cells*

Certain marine slugs such as *Tridachia crispata* and *Elysia viridis* feed on siphonaceous seaweeds such as *Codium.* The slugs feed by puncturing a hole in the algal cell wall and sucking out the cytoplasm. Through some sorting mechanism, at present unknown, the plant cell nuclei, mitochondria, and other organelles, with the exception of the chloroplasts, are voided. The chloroplasts only are actively phagocytosed by certain digestive cells (FIGURE 1). Initially the phagocytosed plastids are in vacuoles interpreted as phagosomes,[2] but in many instances, the phagosome membrane soon disappears, and the plastids come to lie freely in the animal cell cytoplasm (FIGURE 2). The general consensus is that the plant chloroplasts are sequestered by the animal cells free of any contaminating plant cytoplasm. All studies to date have concluded that chloroplasts sequestered by slug cells are not inherited by subsequent generations of slugs.

In the animal cells, sequestered chloroplasts demonstrate oxygenic photosynthesis which is inhibited by DCMU (5×10^{-6} M). Photosynthetic CO_2 fixation rates were found to be comparable to intact *Codium* plants (TABLE 1), and was also DCMU sensitive. Continued photosynthetic function by slugs after removal from a source of plastid replenishment varies but *Tridachia crispata, Elysia viridis,* and *Placobranchus ocellatus* are able to maintain high chloroplast photosynthetic function for periods of up to three months [2] (Dunlap, M. personal communication).

Comparative studies of photosynthetic carbon fixation patterns by chloroplasts in the plant cell, the animal cells, and *in vitro* [3, 4] indicate that in terms of primary carbon fixation, chloroplasts in the slug cells produced the same spectrum of compounds including sugars, sugar phosphates, and amino acids as chloroplasts in the plant cell or in artificial media (TABLE 2 and FIGURE 3). The transport of primary photosynthate from the plastids to the cytoplasm of the animal cells also proceeded unimpaired.[2, 5]

Although the evidence indicates that chloroplasts in slug cells are capable of all aspects of photosynthesis and transport, studies of macromolecular biosynthesis have indicated minimum turnover or *de novo* synthesis of proteins or nucleic acids by such chloroplasts. Trench *et al.*,[4] demonstrated the absence of any synthesis of chlorophylls *a* and *b* by chloroplasts in slug cells. They also demonstrated the absence of synthesis of the three chloroplast-specific glycolipids, monogalactosyldiglyceride, digalactosyldiglyceride and sulfoquinovosyldiglyceride (FIGURE 4). Trench and Ohlhorst [6] demonstrated the absence of synthesis of ribulosediphosphate carboxylase and of chloroplast membrane proteins by plastids in slug cells.

The evidence cited above is consistent with the interpretation that mature differentiated plastids are autonomous with respect to the function of the photosynthetic apparatus which can continue to operate for extensive periods without concomitant macromolecular turnover. This would imply remarkable stability of chloroplast photosynthetic components and hence a long lifespan or slow turnover of such organelles. In slug cells, genetic information coding for chloroplast components are not likely to be present, but it is possible to conceive that if some apoplastidal cell (i.e., a cell that once had plastids) should incorporate by some mechanism plastids from another source, the complete maintenance and replication of such foreign chloroplasts might indeed be feasible.

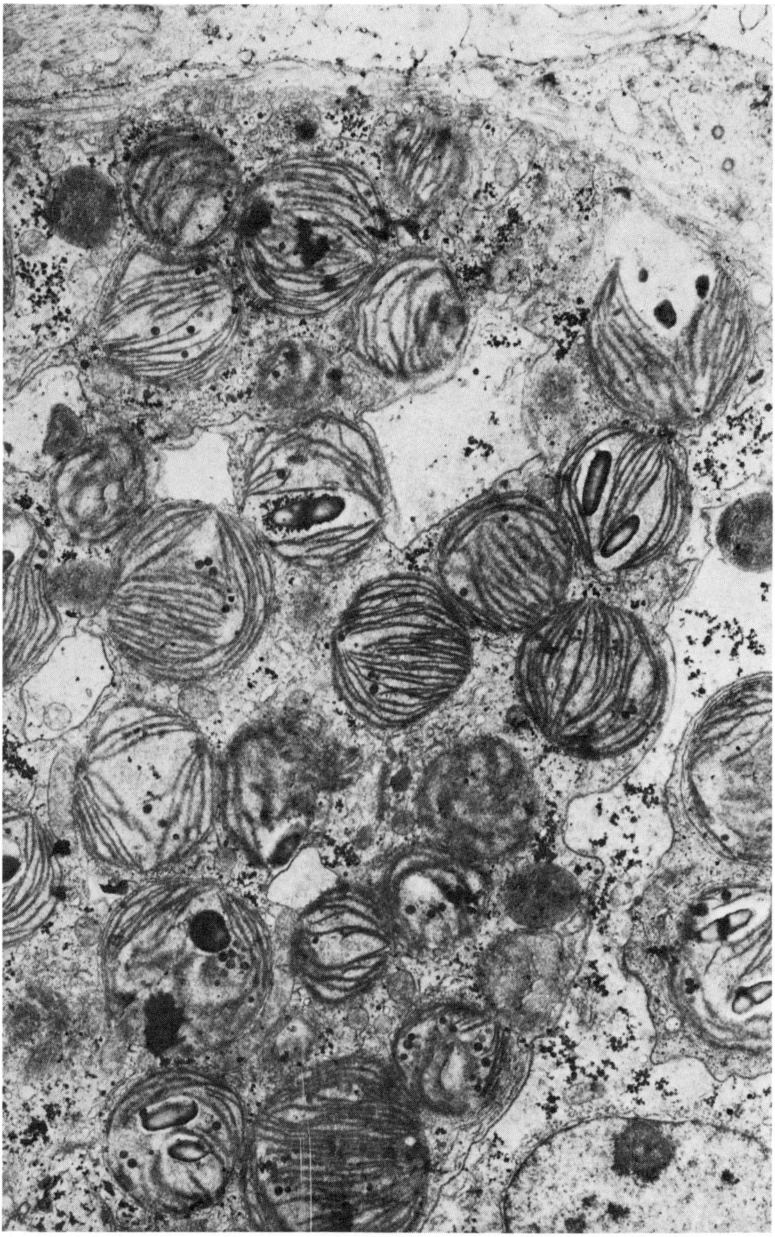

FIGURE 1. Chloroplasts from *Codium fragile* in a digestive cell of the mollusc *Elysia viridis*. Reproduced by permission from the Royal Society. Magnification approximately 8,000×.

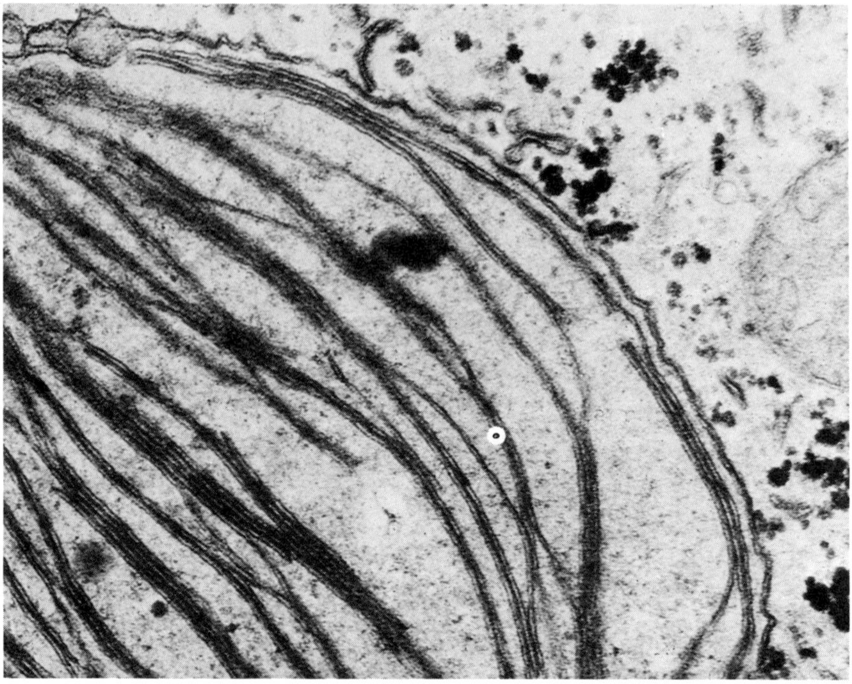

FIGURE 2. Electron micrograph of a plastid in the digestive cell of *E. viridis* showing the chloroplast lying free in the cytoplasm. The phagosome membrane is absent. Reproduced by permission of the Royal Society. Magnification approximately 78,300×.

TABLE 1

COMPARISON OF RATES OF CO_2 FIXATION IN CHLOROPLASTS FROM *Codium fragile* IN THEIR NATURAL ENVIRONMENT, THE PLANT CELL, IN AN ARTIFICIAL MEDIUM AND, IN CELLS OF *E. viridis*

		CO_2 fixation rate μ mol CO_2 mg^{-1} chlorophyll hr^{-1}	
	Temp. °C	Light	Dark
Intact *C. fragile*	18	11.9–12	–
Intact *E. viridis*	18	7.9	–
Isolated *Codium* plastids	18	18.0–23	–
Intact *C. fragile*	12	6.4	0.14
Intact *E. viridis*	12	–	0.13
Isolated *Codium* plastids	12	8.2	0.16

Data from Trench *et al.*[3, 4] By permission of the Royal Society of London.

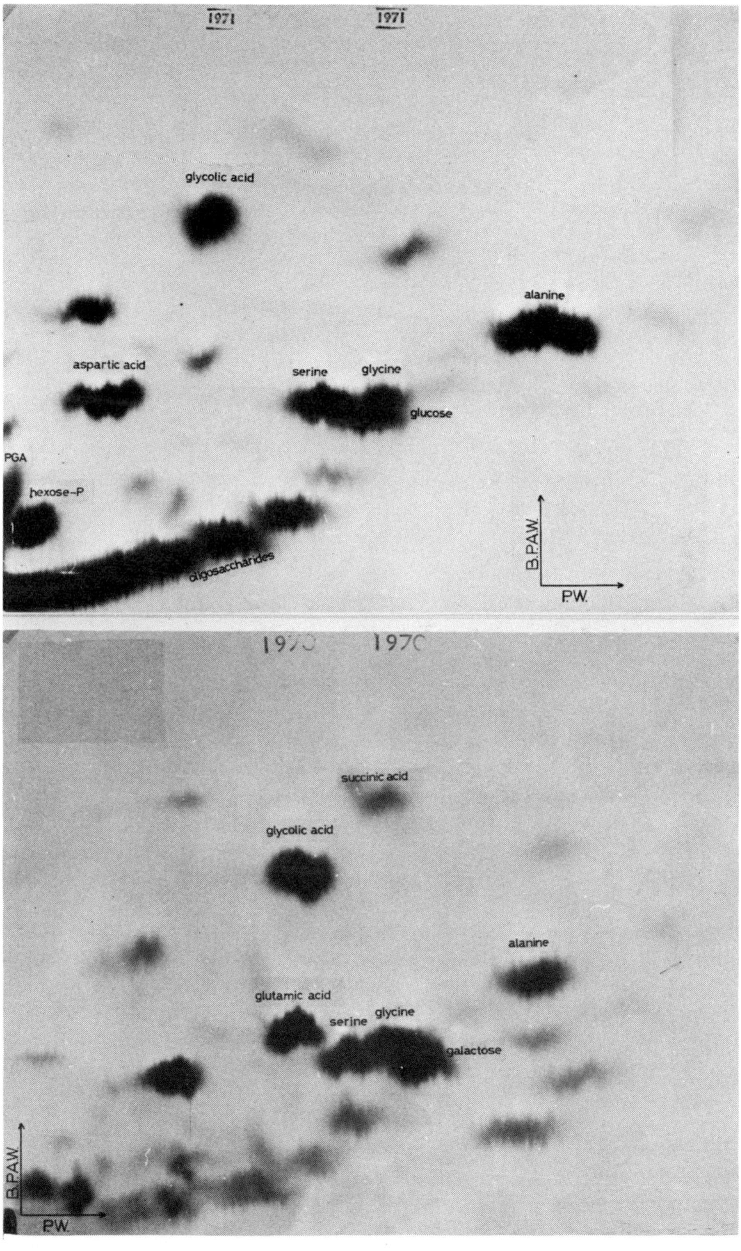

FIGURE 3. a) Radiochromatogram of methanol extract of products of photosynthetic ^{14}C fixation by chloroplasts isolated from *C. fragile.* P.W., Phenol-water; B.P.A.W., Butanol-propionic acid-water. b) Radiochromatogram of methanol extract of products of photosynthetic ^{14}C fixation by chloroplasts in *E. viridis.* Both figures are reproduced by permission of the Royal Society.

TABLE 2

COMPARISON OF THE CHARACTERISTICS OF PHOTOSYNTHETIC ^{14}C FIXATION BY *Codium* CHLOROPLASTS IN DIFFERENT ENVIRONMENTS

		Codium	intact *Elysia*	crushed *Elysia*	isolated *Codium* chloroplasts
Net rate of carbon fixation at 18° C and 21,500 lm m^{-2} illumination (μ mol C mg^{-1} chlorophyll h^{-1})		12	8	20	18 to 23
^{14}C incorporated during photosynthesis for 1 hr at 21,500 lm m^2 and 18° C	sucrose	++	—	—	—
	glucose	+	trace	++	trace
	galactose	—	++	—	—
	glycolicaoid	+	+	+	++
	carotene+xanthophyll	+	+		
	chlorophyll	+	—		
	galactolipid	+	—		
% fixed ^{14}C released from chloroplasts		?	36	40	2

++, abundant; +, present; —, absent.
Data from Trench *et al.*[4] By permission of the Royal Society of London.

B. *Presumptive Chloroplasts: Cyanelles*

Cyanelles are phototrophic cyanobacteria, and several occur as intracellular or intercellular symbionts associated with a wide range of host taxa (FIGURE 5). Several such prokaryote-eukaryote associations have been regarded as extant examples of prokaryotic endosymbionts becoming organelles (chloroplasts). Indeed, there is great confusion among phycologists over whether organisms such as *Cyanophora paradoxa, Glaucocystis nostochinearum, Glaucosphera vacuolata, Gloeocheate wittrockiana, Paulinella chromatophora,* and *Cyanidium caldarium* represent a symbiosis between cyanobacteria and apochlorotic eukaryotic hosts or *de facto* red algae with authentic chloroplasts.[7–10] Several investigators have suggested that such associations as those cited above are actually intermediate stages or "bridges" between symbiotic associations and true plant cells with chloroplasts.[11–13] Such concepts might imply a smooth transition from a once autonomous cytobiont to a semiautonomous organelle. The extent to which such ideas are valid appears to be a direct function of the emphasis placed on morphological and biochemical analogies and homologies between cyanobacteria and chloroplasts.[14]

Hall and Claus,[15, 16] after ultrastructural analysis of *Cyanophora* and *Glaucocystis* concluded that the lamellated inclusions in these organisms were cyanobacteria, and in fact went so far as to suggest taxonomic binomials for them. By contrast, Schnepf and Brown [7] proposed a "radical alternative," namely to regard the cyanelles as chloroplasts. These investigators listed several criteria that support their view, and I shall elaborate on the same topics adding pertinent evidence which has become available since their article was published.

1. *Wall—A Typical Cyanophycean Cell Wall is Absent.*[15, 17]

In the case of *G. nostochinearum* it appears that the peptidoglycan wall is absent, or at least has not been resolved with the electron microscope. However, in *C. paradoxa* a reduced wall has been resolved at the ultrastructural level.[18, 19] It should be emphasized that the reduction or complete loss of cell walls by "plant" endosymbionts is a feature repeatedly encountered in unrelated symbiotic associations,[14, 20, 21] involving organisms that are totally autonomous.

2. *Diaminopimelic Acid—This Substance, a Typical Constituent of the Cell Wall of Bacteria and Cyanophycean Algae, is Absent.*[22]

In *G. nostochinearum* Holm-Hansen *et al.*[22] did not find diaminopimelic acid, but in *C. paradoxa* Aitken and Stanier[13] found diaminopimelic and muramic acids associated with the isolated cyanelles. Nonetheless, they concluded that the cyanelles were actually chloroplasts, the peptidoglycan wall being regarded as vestige of their recent evolutionary past.

3. *"Wall Remnant"—A Thin Nonconvoluted Electron-Dense Layer Sometimes Is Seen Between the Two Unit Membranes Enclosing the Cyanelles. Sometimes in* Glaucocystis *These Membranes Are Separated by an Enlarged Space.*

All the cyanelles studied to date are found within 'host' cell vacuoles. This

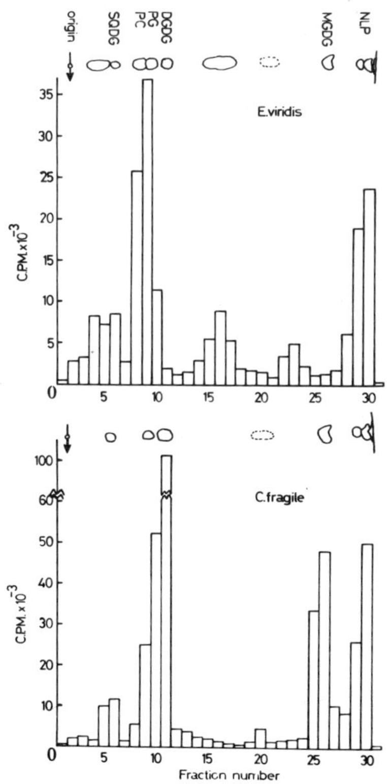

FIGURE 4. Distribution of radioactivity on one-dimensional thin-layer chromatograms of lipids extracted from *E. viridis* and *C. fragile* after 10 h incubation in the light in ^{14}C. NLP, neutral lipids and pigments; MGDG, monogalactosyl-diglyceride; DGDG, digalactosyl-diglyceride; PG, phosphatidyl glycerol; PC, phosphatidyl choline; SQDG, sulfoquinovosyldiglyceride. Reproduced by permission of the Royal Society.

is the case in *C. paradoxa* and *G. nostochinearum* as well as *P. chromatophora* and *G. wittrockiana.*[23, 24] Chloroplasts are not usually found within vacuoles. The concept that the outermost chloroplast envelope membrane is homologous with the host vacuolar membrane is little more than speculation, even though the outer and inner chloroplast envelope membranes have distinct properties.

4. *Thylakoid Arrangement—The Thylakoids Are Separate and Arranged Concentrically.*

The thylakoid arrangement in the cyanelles of *C. paradoxa* and the presumed chloroplasts of *C. caldarium* is essentially identical (FIGURES 6 and 7). In both cases they are concentrically arranged.

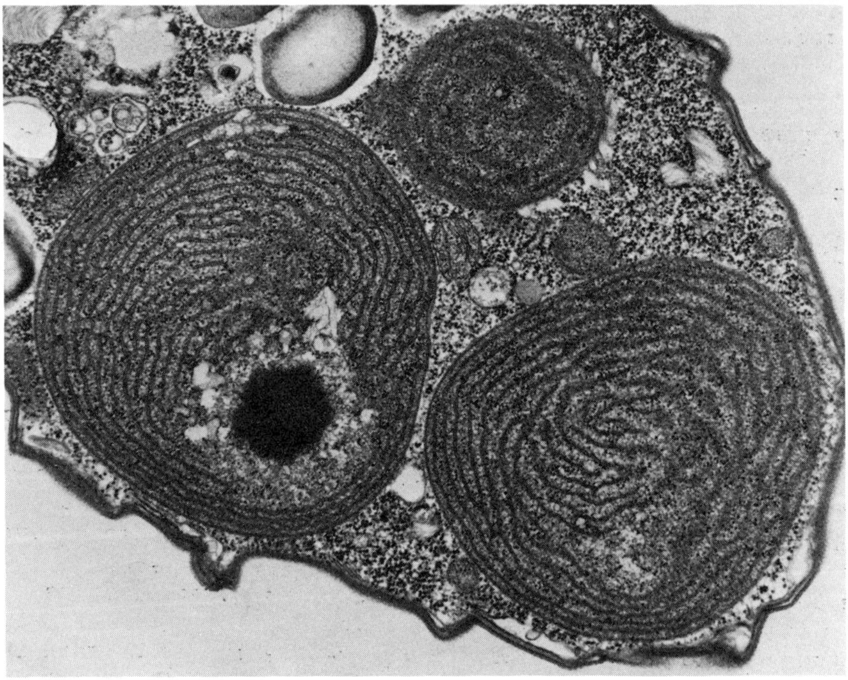

FIGURE 5. Electron micrograph of *Cyanophora paradoxa* demonstrating the intracellular location of the symbiotic cyanobacteria or presumed chloroplasts. Magnification approximately 22,500×.

5. *Phycobilisomes—Granules With a Diameter of About 35 nm, Attached to the Thylakoids Containing Phycobilins, Are Present.*

In *C. paradoxa, G. nostochinearum,* and *C. caldarium* the phycobilisomes are interthylakoidally attached. The same is true for several free-living cyanobacteria, e.g., *Synechococcus,*[25] and for the chloroplasts in *Porphyridium cruentum.*[26]

6. *Pigments—C- and R-Phycocyanin Are Present; the Common Cyanophycean Xanthophylls, Echinenone and Myxoxanthophyll, Are Absent.*

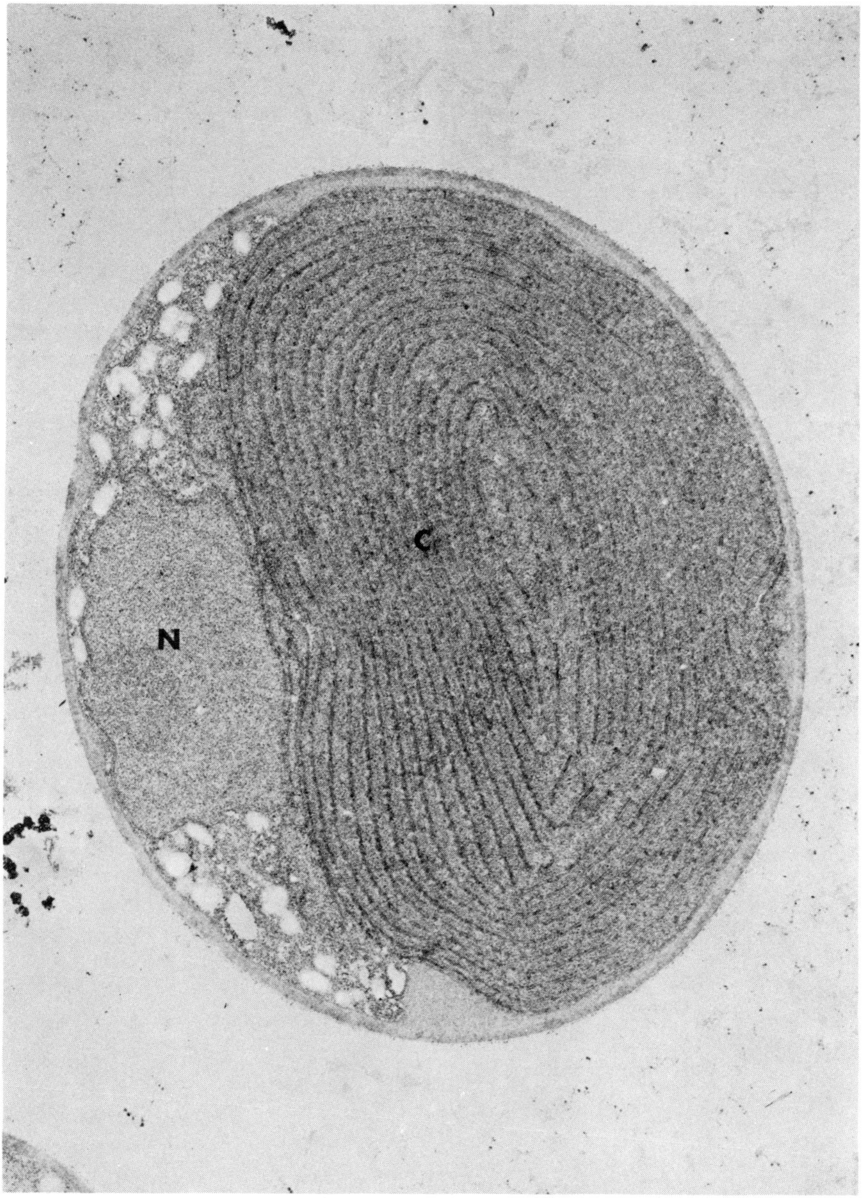

FIGURE 6. Electron micrograph of *Cyanidium caldarium* fixed with $KMnO_4$. N, nucleus; C, presumed chloroplast or cyanobacterium. Photograph kindly provided by Dr. M. Edwards. Magnification approximately 27,000×.

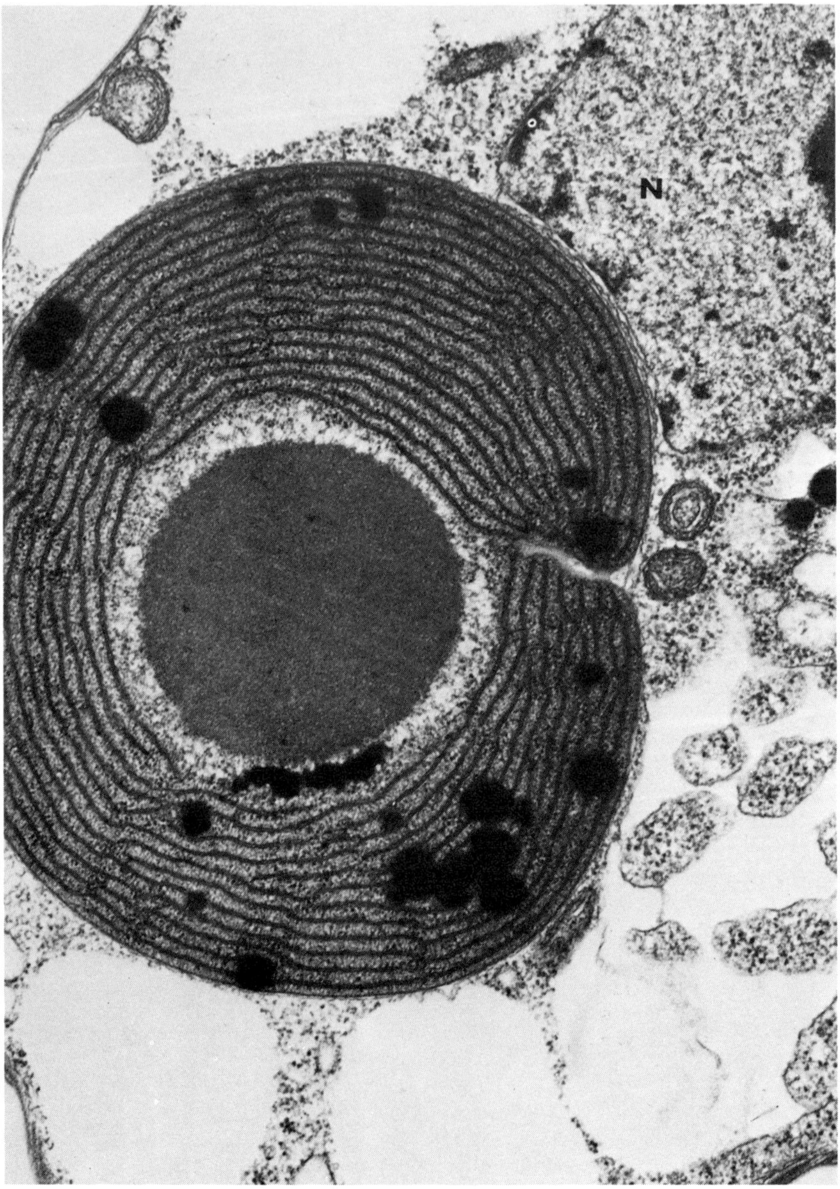

FIGURE 7. Electron micrograph of *C. paradoxa* showing details of the cyanelle. N, host nucleus. Magnification approximately 31,500×.

The cyanelles in *C. paradoxa* have been shown to contain C-phycocyanin.[27, 28] Although regarded as a red alga, *C. caldarium* also possesses C-phycocyanin (Trench, unpublished),[22] identical in its electronic adsorption and fluorescence excitation and emission spectra to the pigment from *C. paradoxa.*

7. *DNA—A) The Staining Properties with Acridine Orange are Similar to Chloroplasts and not to the Centroplasm of Free-Living Cyanophyceae. B) The G-C Ratio of* Cyanophora *DNA (Actually the DNA of the Cyanelles) Is Similar to that of Chroococcalian Cyanophyceae.*[29]

Edelman *et al.*[29] reported a mol % (G + C) value of 57 for the DNA of the cyanelles in *C. paradoxa.* They also reported a buoyant density of 1.716 g/cm^3. These values have been corroborated by L. Vernon (personal communication). However, Herdman and Stanier [9] reported a buoyant density of 1.695 g/cm^3 and a mol % (G + C) value of 35.7. These values are identical with that reported for DNA from the chloroplasts of *Chlamydomonas reinhardii.*[14] In addition, Herdman and Stanier,[9] using DNA renaturation kinetics to assay genome complexity, came to the conclusion that the cyanelles in *C. paradoxa* had a genome size of 0.117×10^9 daltons. This value is about 5–10% of that of free-living cyanobacteria or about the size of the chloroplast genome.

In discussions of the possible origin of chloroplasts, much emphasis in the past has been placed on morphological and biochemical homologies. For example, the same spectrum of photosynthetic pigments exist in cyanobacteria and chloroplasts of red algae [8] and indeed there is remarkable homology between cyanobacterial phycobilins and those of red algal chloroplasts.[30–33] In addition, chloroplast DNA and cyanobacterial DNA occur as covalently closed circular molecules.[34, 35] Chloroplast protein synthesis occurs on 70s ribosomes as does protein synthesis in cyanobacteria, and in both, synthesis is sensitive to the same spectrum of antibiotics. The initiation of polypeptide synthesis in both systems requires formylmethionine-tRNA.[14, 42] All these similarities effectively support both the serial endosymbiosis hypothesis [36–39] and the alternative cluster-clone hypothesis [40, 41] for the origin of chloroplasts, particularly when one considers the possibility of evolutionary convergence.

In our laboratory we have been placing the emphasis on *functional* analogies between symbiotic cyanobacteria and chloroplasts, using *C. paradoxa* as the model system. It is known that the developing chloroplast is dependent on the nuclear-cytoplasmic protein synthesizing apparatus for the synthesis of many plastid components.[14, 42] It is therefore reasonable to ask to what extent are the presumed chloroplasts in *C. paradoxa* dependent on the 'host's' nuclear-cytoplasmic machinery.

Several studies on developing plastids have demonstrated that the synthesis of chlorophyll and chloroplast rRNA are inhibited when cytoplasmic protein synthesis is inhibited.[4, 43–46] Trench and Siebens [47] found that the incorporation of photosynthetically fixed ^{14}C into chlorophyll *a* by the cyanelles in *C. paradoxa* in log phase of growth was inhibited by cycloheximide more than by chloramphenicol or rifampicin. However, they also observed that the incorporation of ^{33}P into cyanellar rRNA was totally insensitive to cycloheximide, but was inhibited by rifampicin and chloramphenicol. The data are consistent with the interpretation that the cyanelles in *C. paradoxa* are dependent on some process which is inhibited by cycloheximide in the 'host' cytoplasm with respect to chlorophyll synthesis but are independent with respect to rRNA synthesis.

Trench [14] reported that the synthesis of the large subunit of C-phycocyanin in *C. paradoxa* was chloramphenicol sensitive, and that synthesis of the small subunit was cycloheximide sensitive. Although this observation might suggest that synthesis of the small subunit polypeptide occurs on 'host' ribosomes, the evidence available does not really warrant such a conclusion.

Observations on the degradation patterns of cyanellar rRNA [48] suggest close similarity to chloroplast rRNA. When chloroplast rRNA is degraded, the 23s subunit (1.1×10^6 daltons) usually cleaves toward the center of the molecule yielding fragments of approximately 0.67 and 0.53×10^6 daltons.[49] By contrast, 23s rRNA from free-living cyanobacteria cleaved toward the end of the molecule yielding fragments of 0.9 and 0.2×10^6 daltons.[50, 51] The rRNA from the cyanelles of *C. paradoxa* yielded fragments of 0.61 and 0.43×10^6 daltons on degradation.

The chloroplast genome is about $0.094–0.230 \times 10^9$ daltons, and the plastid chromosome has a contour length of about 44 μm. This quantity of DNA is thought to be able to code for the synthesis of about 126 proteins of 40,000 daltons. We know that this is not adequate to direct the synthesis of all platsid components. If in fact the genome size of the cyanelles in *C. paradoxa* is 0.117×10^9 daltons, then one would expect that the cyanellar genome would be fairly close to 44 μm in contour length and be able to code for the synthesis of about 126 proteins.

The evidence presented above, taken as a whole implies strongly that the cyanelles in *C. paradoxa* are more autonomous than chloroplasts are. Indeed, preliminary measurements of the contour length of isolated DNA from the cyanelles of *C. paradoxa* indicate that it is appreciably longer than 44 μm (Trench, unpublished).

It is indeed apparent that mature plastids are biochemically autonomous entities with respect to the function of the photosynthetic apparatus. It is plausible that an apochlorotic cell with the genetic information coding for plastid components could maintain through generations plastids derived from some other source. If mouse fibroblasts [52] and slug cells [2] can provide a cellular environment that sustains chloroplast function, how much more so could a cell that once had its own plastids?

The use of symbioses involving cyanobacteria and eukaryotic cells as model systems to test the serial endosymbiosis hypothesis appears to hold much promise, but I feel that we should exercise much more caution before conceptual leaps are made rendering cytobionts as organelles.

Acknowledgment

I am grateful to Dr. M. Edwards for kindly providing the electron micrograph of *Cyanidium caldarium*.

References

1. Kirk, J. T. O. & R. A. E. Tilney-Bassett. 1967. The Plastids. W. H Freeman & Co. London and San Francisco.
2. Trench, R. K. 1975. Of 'leaves that crawl': functional chloroplasts in animal cells. Soc. Exp. Biol. **29**:229–265.
3. Trench, R. K., J. E. Boyle & D. C. Smith. 1973. The association between chloroplasts of *Codium fragile* and the mollusc *Elysia viridis*. I. Characteristics of isolated *Codium* chloroplasts. Proc. R. Soc. Lond. (B) **184**:51–61.

4. TRENCH, R. K., J. E. BOYLE & D. C. SMITH. 1973. The association between chloroplasts of *Codium fragile* and the mollusc *Elysia viridis*. II. Chloroplast ultrastructure and photosynthetic carbon fixation in *E. viridis*. Proc. R. Soc. Lond. (B) **184:**63–81.
5. HINDE, R. 1978. The metabolism of photosynthetically fixed carbon by isolated chloroplasts from *Codium fragile* (Chlorophyta: Siphonales) and by *Elysia viridis* (Mollusca: Sacoglossa). Biol. J. Linn. Soc. **7:**161–171.
6. TRENCH, R. K. & S. OHLHORST. 1976. The stability of chloroplasts of siphonaceaus algae in symbiosis with marine slugs. New Phytol. **76:**99–109.
7. SCHNEPF, E. & R. M. BROWN, JR. 1971. On the relationships between endosymbiosis and the origins of plastids and mitochondria. *In* Origin and Continuity of Cell Organelles. J. Reinert & H. Ursprung, Eds.: 299–322. Springer-Verlag. New York, Heidelberg & Berlin.
8. CHAPMAN, D. J. 1973. Biliproteins and bile pigments. *In* The Biology of Blue-green Algae. N. G. Carr & B. A. Whitton, Eds.: 162–185. Blackwell Scientific. Berkeley & Los Angeles, Calif.
9. HERDMAN, M. & R. Y. STANIER. 1977. The cyanelle: chloroplast or endosymbiotic prokaryote? F.E.M.S. Lett. **1:**7–12.
10. KREMER, B. P., G. B. FEIGE & H. A. W. SCHNEIDER. 1978. A new proposal for the systematic position of *Cyanidium caldarium*. Naturwissenschaften **65:**157.
11. IKAN, R. & J. SECKBACH. 1972. Lipids of the thermophilic alga *Cyanidium caldarium*. Phytochemistry **11:**1077–1082.
12. FREDRICK, J. F. 1976. *Cyanidium caldarium* as a bridge alga between Cyanophyceae and Rhodophyceae: Evidence from immunodiffusion studies. Plant and Cell Physiol. **17:**317–322.
13. AITKEN, A. & R. Y. STANIER. 1979. Characterization of peptidoglycan from the cyanelles of *Cyanophora paradoxa*. J. Gen. Microbiol. **112:**219–223.
14. TRENCH, R. K. 1980. Cyanelles. *In* Origins and Evolution of Chloroplasts. R. Y. Stanier, J. A. Schiff & H. Lyman, Eds. (In press.)
15. HALL, W. T. & G. CLAUS. 1963. Ultrastructural studies on the blue-green algal symbiont in *Cyanophora paradoxa* Korschikoff. J. Cell Biol. **19:**551–563.
16. HALL, W. T. & G. CLAUS. 1967. Ultrastructural studies on the cyanelles of *Glaucocystis nostochinearum* Itzigsohn. J. Phycol. **3:**37–51.
17. SCHNEPF, E., W. KOCH & G. DEICHGRÄBER. 1966. Zur Cytologie und taxonomischen Einerdnung von *Glaucocystis*. Soch. Mikrobiol. **55:**149–174.
18. PICKETT-HEAPS, J. 1972. Cell division in *Cyanophora paradoxa*. New Phytol. **71:**561–567.
19. TRENCH, R. K., R. R. POOL, JR., M. LOGAN & A. ENGELLAND. 1978. Aspects of the relation between *Cyanophora paradoxa* (Korschikoff) and its endosymbiotic cyanelles *Cyanocyta korschikoffiana* (Hall and Claus). I. Growth ultrastructure, photosynthesis and the obligate nature of the association. Proc. R. Soc. Lond. (B) **202:**423–443.
20. MUSCATINE, L., R. R. POOL, JR. & R. K. TRENCH. 1975. Symbiosis of algae and invertebrates: Aspects of the symbiont surface and the host-symbiont interface. Trans. Amer. Micros. Soc. **94:**450–469.
21. TRENCH, R. K. 1979. The cell biology of plant-animal symbiosis. Ann. Rev. Plant Physiol. **30:**485–531.
22. HOLM-HANSEN, O., R. PRASAD & R. A. LEWIN. 1965. Occurrence of ϵ-diaminopimelic acid in algae and flexibacteria. Phycologia **5:**1–14.
23. KIES, L. 1974. Elektrovenmikroskopische untersuchungen an *Paulinella chromatophora* Lauterborn, einer Thekamöbe mit blaugrünen Endosymbionten (cyanellen). Protoplasma **80:**69–89.
24. KIES, L. 1976. Untersuchungen zur Feinstruktur und taxonomischen Einordung von *Gloeochaete wittrockiana*, einer apoplastidalen capsalen Alge mit blaugrünen Endosymbionten (cyanellen). Protoplasma **87:**419–446.
25. EDWARDS, M. & E. GANTT. 1971. Phycobilisomes of the thermophilic blue-green alga *Synechococcus lividus*. J. Cell Biol. **50:**896–899.

26. Gantt, E. & S. F. Conti. 1965. The ultrastructure of *Porphyridium cruentum.* J. Cell. Biol. **26:**365–381.
27. Chapman, D. J. 1966. Pigments of the symbiotic algae (cyanomes) of *Cyanophora paradoxa* and *Glaucocystis nostochinearum* and two Rhodophyceae, *Porphynidium aerugineum* and *Asterocytis ramosa.* Arch. Mikrobiol. **55:**17–25.
28. Trench, R. K. & G. S. Ronzio. 1978. Aspects of the relation between *Cyanophora paradoxa* (Korschikoff) and its endosymbiotic cyanelles *Cyanocyta Gorschikoffiana* (Hall and Claus). II. The photosynthetic pigments. Proc. R. Soc. Lond. (B) **202:**445–462.
29. Edelman, M., D. Swinton, J. A. Schiff, H. T. Epstein & B. Zeldin. 1967. Deoxynibonucleic acid of the blue-green algae. Bacteriol. Rev. **31:**315–331.
30. Berns, D. S. 1967. The immunochemistry of biliproteins. Plant Physiol. **42:** 1569–1586.
31. Bogorad, L. 1965. Studies of phycobiliproteins. Rec. Chem. Prog. **26:**1–12.
32. Glazer, A. N., G. Cohen-Bazire & R. Y. Stanier. 1971. Comparative immunology of algal biliproteins. Proc. Natl. Acad. Sci. USA **68:**3005–3008.
33. Glazer, A. N., G. S. Apell, C. S. Hixon, D. A. Bryant, S. Rimon & D. M. Brown. 1976. Biliproteins of cyanobacteria and Rhodophyta: Homologous family of photosynthetic accessory pigments. Proc. Natl. Acad. Sci. USA **73:**428–431.
34. Herrmann, R. G., H. J. Bohnert, K. V. Kowallik & J. M. Schmitt. 1975. Size conformation and purity of chloroplast DNA of some higher plants. Biochem. Biophys. Acta **378:**305–317.
35. Lau, R. H. & W. F. Doolittle. 1979. Covalently closed circular DNA in closely related unicellular cyanobacteria. J. Bacteriol. **137:**648–652.
36. Margulis, L. 1970. Origin of Eukaryotic Cells. Yale University Press. New Haven, Conn.
37. Margulis, L. 1976. Genetic and evolutionary consequences of symbiosis. Exp. Parasitol. **39:**277–349.
38. Raven, P. H. 1970. A multiple origin for plastids and mitochondria. Science **169:**641–646.
39. Taylor, F. J. R. 1974. Implications and extensions of the serial endosymbiosis theory for the origin of eukaryotes. Taxon **23:**229–258.
40. Uzzell, T. & C. Spolsky. 1974. Mitochondria and plastids as endosymbionts: a revival of special creation? Am. Sci. **62:**334–343.
41. Bogorad, L. 1975. Evolution of organelles and eukaryotic genomes. Science **188:**891–898.
42. Kirk, J. T. O. & R. A. E. Tilney-Bassett. 1978. The Plastids. 2nd Edit. Elsevier/North Holland. Amsterdam, New York, Oxford.
43. Ingle, J. 1968. Synthesis and stability of chloroplast r-RNA's. Plant Physiol. **43:**1448–1454.
44. Gassman, M. & L. Bogorad. 1967a. Control of chlorophyll production in rapidly greening bean leaves. Plant Physiol. **42:**774–780.
45. Gassman, M. & L. Bogorad. 1967b. Studies on the regeneration of protochlorophillide after brief illumination of etiolated bean leaves. Plant Physiol. **42:**781–784.
46. Detchon, P. & J. V. Possingham. 1973. Chloroplast ribosomal ribonucleic and synthesis in cultured spinach leaf tissue. Biochem. J. **136:**829–836.
47. Trench, R. K. & H. C. Siebens. 1978. Aspects of the relation between *Cyanophora paradoxa* (Korschikoff) and its endosymbiotic cyanelles *Cyanocyta korschikoffiana* (Hall and Claus). IV. The effects of rifampicin, chloramphenicol and cycloheximide on the synthesis of ribosomal ribonucleic acids and chlorophyll. Proc. R. Soc. Lond. (B) **202:**473–482.
48. Siebens, H. C. & R. K. Trench. 1978. Aspects of the relation between *Cyanophora paradoxa* (Korschikoff) and its endosymbiotic cyanelles *Cyanocyta korschikoffiana* (Hall and Claus). III. Characterization of ribosomal ribonucleic acids. Proc. R. Soc. Lond. (B) **202:**463–472.

49. LEAVER, C. J. & J. INGLE. 1971. The molecular integrity of chloroplast ribosomal ribonucleic acid. Biochem. J. **123:**235–243.
50. SZALAY, A., D. MUNSCHE, R. WOLLGIEHN & B. PARTHIER. 1972. Ribosomal ribonucleic acid in *Anacystis nidulans.* Biochem. J. **129:**135–140.
51. GRIERSON, D. & H. SMITH. 1973. The synthesis and stability of ribosomal RNA in blue-green algae. Eur. J. Biochem. **36:**280–285.
52. NASS, M. M. K. 1969. Uptake of isolated chlorophlasts by mammalian cells. Science **165:**1128–1131.

DISCUSSION OF THE PAPER

DR. AARONSON: Dr. Whatley described the amoeba that had the bacteria with which it traveled in the course of reproduction. I wonder what would happen if you were to induce lysosomal activity with some of the drugs that are apparently available for induction or prevention of lysosomal activity, to the endosymbionts under those conditions?

DR. R. K. TRENCH: I can answer that from the point of view of the slug system. You don't have to bother to induce it. The chloroplast in the slugs remains functional for a fairly long time. Apparently, and this is an interpretation on my part, this is consistent with the observation once the chloroplasts that are free in the cytoplasm stop functioning, they're enveloped by a new membrane assumed to be an autophagic vacuole. At that point lysosomes move to it, fuse, enter the contents, and clobber it.

DR. AARONSON: It would be interesting to see what happens to the amoeba under the same set of circumstances where you induce that.

DR. C. B. COOK (*Ohio State University, Columbus, Ohio*): A question for Dr. Aaronson if I may. What are these drugs you are talking about sir? I've not heard of them.

DR. AARONSON: There are some associated with rheumatic fever; there are a number of situations where lysosomal induction occurs as a result of rheumatic fever and you can simulate some of these. There are some drugs that will simulate some of the lysosomal effects that are induced in rheumatic fever.

MICROTUBULES, UNDULIPODIA AND *PILLOTINA* SPIROCHETES *

Lynn Margulis, Leleng P. To

Department of Biology
Boston University
Boston, Massachusetts 02215

and

David Chase

Sepulveda Veterans' Administration Hospital
Sepulveda, California 91343

This short review discusses in outline form two mutually exclusive concepts of the origin of microtubule systems of eukaryotic cells: the endogenous nonsymbiotic concept[26] and the spirochete symbiont or exogenous hypothesis.[24] It also indicates where most of the experimental work upon which each is based, often in disparate literatures and presented with differing terminologies and underlying assumptions, has already been published. These must be made explicit if an understanding of the evolutionary origin of microtubules, mitosis, meiosis, and hence Mendelian genetics is ever to be achieved. Because of inconsistent use of terminology, a word list is included in this paper.

The endogenous hypothesis suggests that microtubules and their organizing centers developed by differentiation within ancestral cells under selection pressures to evenly distribute genetic material. This differentiation led to the evolution of mitosis. It took place in ancestors to some eukaryotic microorganisms, perhaps to red algae. Only subsequent to the origin of mitosis in immotile cells did the (9+2) organelles of motility evolve. That is, the evolution of undulipodia (cilia, eukaryotic flagella) followed the evolution of mitosis.

The exogenous hypothesis postulates that tubulin proteins evolved first in an extremely limited group of prokaryotic microbes including certain spirochetes. Spirochetes, by definition, are slender helical motile heterotrophic bacteria which have flagella lodged within the spaces between the cell wall and the membrane. In spirochetes, the flagella lie in the periplasm and are attached to the cell via typical plate-shaped basal structures. Certain spirochetes containing tubulin (in addition to flagellin) developed a series of symbioses with a variety of hosts. Some of these associations became motility symbioses. From motility symbioses between heterotrophic protists and tubulin-containing spirochetes undulipodiated nonmitotic protists evolved: the first mastigotes. With time symbiotic spirochetes became undulipodia. Motile protists of many varieties evolved including those in which mitosis eventually appeared. The genetic material and motility systems of the symbiotic spirochetes were deployed inside the cell: the nucleic acid and protein of the symbionts were used in the elaboration of microtubule organizing centers including centrioles, centromeres,

* Supported by The Guggenheim Foundation, National Aeronautics and Space Administration grant NGR 004–025, and the National Science Foundation Cell Biology program.

0077–8923/81/0361–0356 $01.75/0 © 1981, NYAS

and kinetosomes. The kinetosomes of all protists, animals, and plants are analogous to spirochete attachment sites. The acquisition of spirochetes and the formation of motility symbioses preadapted the symbiotic partners for the origin of the mitotic apparatus and eventually meiosis. The details of this hypothesis occupy several chapters in a forthcoming book.[20]

For those dealing routinely with both prokaryotic and eukaryotic microorganisms the terminological ambiguity is intolerable. The term flagellum has been used interchangeably for the small solid structure composed of flagellin protein and the longer entirely different, nonhomologous eukaryotic (9+2) structure composed of tubulins. The word cilium has been used for the shorter (9+2) eukaryotic organelle of motility. The cilium and eukaryotic flagellum are not only homologous, they are virtually identical. Thus the adoption of the term, first suggested by Smagina (1948) and later by Kuznicki *et al.*[15] "undulipodia" for the eukaryotic structure is in order.

This issue (discussed in Margulis [21]) is summarized by the following definitions:

Basal Structures—Circular insertion structures of flagella. Limited to prokaryotes. Also called basal apparatus or basal bodies.

Centrioles—The (9+0) structure, extremely similar or identical to kinetosomes but lacking the axoneme (Wilson, 1925; Cleveland, 1957). Found at the poles in animal and many protist mitoses.

Flagella—Solid protruding organelles. Their presence is correlated with bacterial motility. They are composed of flagellin and intrinsically nonmotile. Rotary locomotion is generated at the points of insertion of the organelle into the cell wall.[2] Limited to bacteria, not surrounded by membrane but may be sheathed. Diameter: $\simeq$150–250 Å. Includes the axial filaments or axial fibrils of spirochetes, bacteria in which the flagella are situated in the periplasm.

Kinetosomes—The (9+0) basal structure from which all undulipodia develop. This structure contains RNA and protein.[9, 12, 13] Often called basal bodies in the zoological literature.

MTOC—Microtubule-organizing center.[25] Refers to several types of structures (from well-organized centromeres to amorphous granulofibrosal material) only visible with electron microscopy from which microtubules are seen to emanate. Also called MC's (microtubule centers).

Undulipodia—Cilia and eukaryotic flagella. Intrinsically motile organelles composed of the (9+2) arrangement of 240 Å diameter microtubules. They consist of the axoneme [the (9+2) shaft] surrounded by plasma membrane. Movement involves dephosphorylation of ATP to ADP and calcium ions.

Mastigotes—It is suggested that this term, already familiar to protistologists replace the term "flagellates" for eukaryotic microorganisms with a small number of undulipodia. It is already in use for example in the case of polymastigotes such as *Trichomonas* or *Devescovina* or hypermastigotes such as *Trichonympha*.

Pillotinas—Large spirochetes of the family Pillotaceae [14] (see To *et al.*[37] and Margulis *et al.*[22] for illustration).

Although the terminology itself is consistent with either evolutionary hypothesis of the origin of undulipodia, the concepts themselves are mutually exclusive. The following questions are immediately brought to mind:

What is the current status of the evidence for each of the two hypotheses, the endogenous and the exogenous?

What is the major difference between the two hypotheses?

What observations must be comprehended by the hypotheses if they are to be judged adequate?

How is each to be strongly inferred (if not proved)?

The exogenous and endogenous hypotheses for the origin of the microtubule system has recently been reviewed in the context of the transition from prokaryotes to eukaryotes.[10] That paper should be consulted for a status report on varying concepts of the origin of all the major eukaryotic organelles (e.g. plastids, mitochondria, nuclei).

The major evidence for the endogenous "mitosis-first" concept of the origin of microtubules is the discovery of the presence of microtubules associated with the wall structure in two species of *Nostoc*, a cyanobacterium.[3] Pickett-Heaps[26] suggests that microtubular systems in ancestors of red algae (such as blue greens) led to those in red algae in which the selection pressure was far greater efficiency in the distribution of genetic material. A difficulty here is that it has not been shown that the smaller wall tubules of these *Nostoc* have any homology with eukaryotic tubulin tubules. Neither is it clear how red algae evolved all of their nonplastid features (e.g. mitochondria, chromosomes, endoplasmic reticulum, and so forth) if they differentiated directly from the prokaryotic blue greens.

The exogenous hypothesis is supported only by weak evidence as well. Microtubules that may be made of tubulin have been seen in certain large spirochetes of the family Pillotaceae.[23] Tubulin is absent in most prokaryotes studied; that the putative tubulin protein in pillotina spirochetes is homologous to the proteins of eukaryotic microtubules awaits verification by primary amino acid sequence determination (Fracek and Laursen, in progress). Since the alpha tubulin of mammalian brain has been almost entirely sequenced[28, 29] the probability that comparative sequence data from eukaryotic tubulins and putative spirochete tubulins will eventually be obtained is high.

Other evidence relevant to the spirochete hypothesis is indirect: it is argument by analogy. Associations between spirochetes and host cells, whether tissue cells of animals or single protists are being studied. Most of the associations that have recently been described are those within the complex microbial communities harbored in the hypertrophied hindguts of termites and wood-eating cockroaches (*Cryptocercus*). Both the prokaryotic and eukaryotic microorganisms of *Cryptocercus* as well as members of the Rhinotermitids (subterranean termites) and Kalotermitids (dry wood termites) tend to be highly motile. Many sorts of associations in the crowded gut are seen; virtually none of them has been analyzed in biochemical or genetical detail. Several types of associations between spirochetes and protist hosts have been observed by microcinematography in live material and by electron microscopy. They can be organized into a series from less to greater degrees of association between partners; such a series is shown diagramatically in FIGURE 1. Unidentified spirochetes and protists can be seen in typical hindgut profusion at low magnification in FIGURE 2. It is usual that spirochetes of various diameters and lengths are found abundantly among the regular-sized undulipodia (FIGURE 2A). Spirochetes, apparently "feeding," are coursing up and down in between the undulipodia of live protists (FIGURE 2B). Independent spirochetes temporarily associate with each other and generate conspicuous waves (FIGURE 2C).[22, 37] The similarity between the coordinated beating in hindgut spirochetes and the metachronal waves of the undulipodia of *Opalina*, a well-known protist found as a commensal in frog intestine was pointed out to us by A. T. W. Cheung who kindly supplied FIGURE 2D.

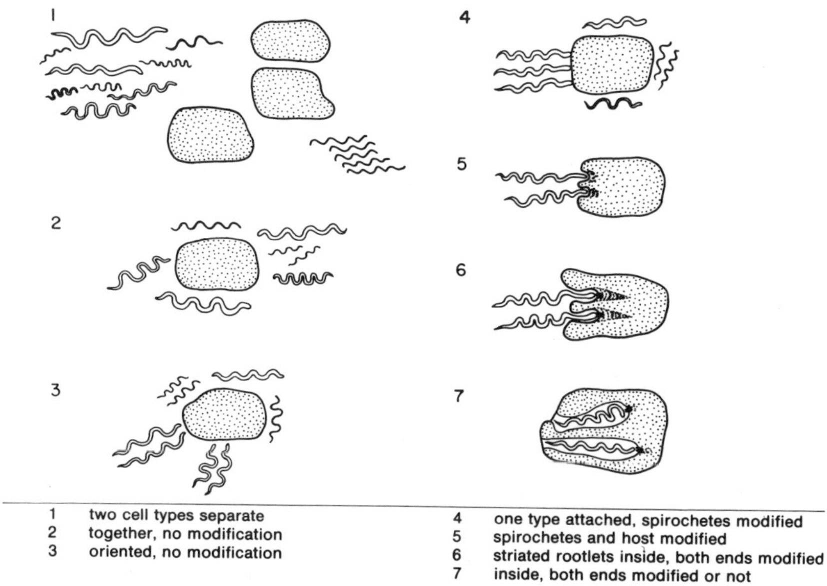

FIGURE 1. Continuum of association between spirochetes and protist hosts (diagrammatic).

Many "free-living" (in the sense that they are not attached to each other, to protist hosts, or to the gut epithelial cells) spirochetes are found in the hindguts of wood-eating insects. Some of these are shown in FIGURE 3. Some (FIGURES 3A & B) are unmodified. Some (FIGURES 3C & D) are slightly modified, perhaps for attachment at certain stages or under certain conditions. The host may show no modification for the spirochete attachment or they may show extensive morphological alteration at the site on the cortex where the spirochete attaches (FIGURES 4 & 5). In some cases only a single end of the spirochete seems modified, in others both ends are elongated or in some way distinguishable, in still other cases there are conspicuous differences between the modifications at each end (for example, FIGURE 4b in Margulis *et al.*[24]). Particularly obvious and best studied[24, 33] are the flattened ends of the spirochete attachment sites to *Pyrsonympha* and the correlated internal differentiations of the *Pyrsonympha* host (FIGURES 5A & B).

Organisms such as *Pyrsonympha* and *Metadevescovina* (FIGURE 4) may have none, few, or many spirochetes attached to them, whereas organisms such as *Mixotricha paradoxa*[8] regularly have a complex arrangement of surface spirochetes. In the former cases the attachment is presumed to be gratuitous to the host and of selective advantage to the spirochetes that possibly derive food from the protists; in the latter case, the spirochetes which constitute a portion of the species description of the organisms, determine the motility of the protist hosts. The posterior clear zone region of many of the wood-ingesting protists is apparently sensitive. Contact with wood particles leads to ingestive activity in that zone whereas contact with wood particles at the anterior rostral region of the cell has no effect. Often spirochetes are seen close to the sensitive wood-ingesting region suggesting that they are able to feed there. One might imagine that the motility of the spirochetes attached near the ingesting zone increases the probability of contact with wood particles and hence enhances their rate of entry into the protist.

Spirochetes attached in bundles to undulipodial protists, apparently propelling their host cells forward have been seen (FIGURE 6c, Margulis *et al.*[24]). They form a continuum between the casual associations of *Pyrsonympha* and the well-developed motility symbioses such as the spirochete one studied by Cleveland and Grimstone [8] and the flagellated bacterial symbiosis described by Tamm.[36] One can easily imagine an analogous transition in an ancestral population of spirochetes and hosts from free-living partners to motility organelles. These sorts of associations are especially frequent in organic rich anaerobic or microaerophilic environments.

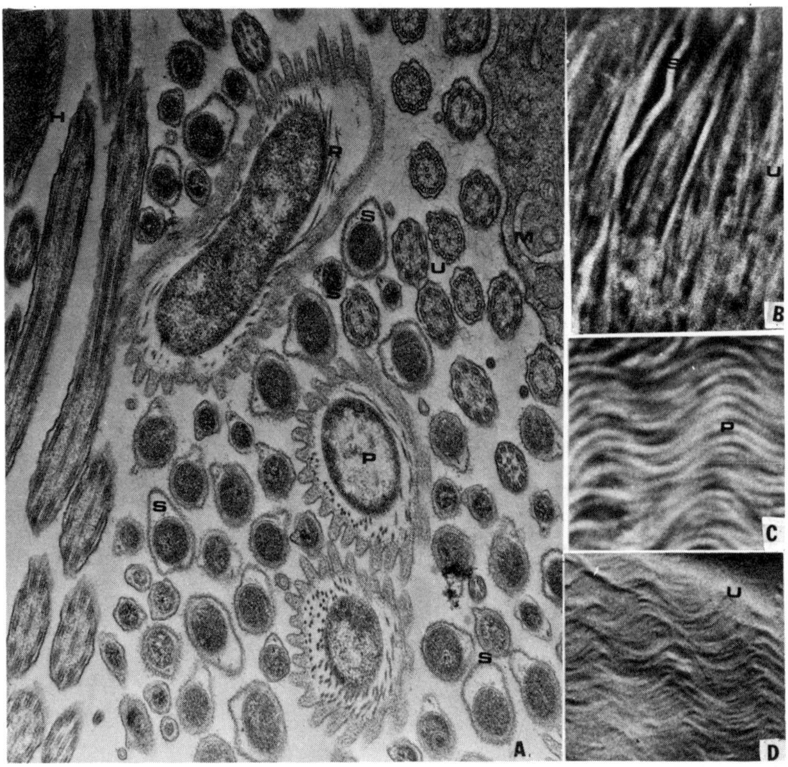

FIGURE 2. Comparison of spirochetes and undulipodia. A. Thin section of spirochetes (S) and undulipodia (U) from the hindgut microbial ecosystem of *Reticulitermes hesperus.* M=unidentified mastigote, P=*Pillotina* sp. spirochetes, H=*Hollandina* sp. spirochetes, S=unidentified spirochetes of various kinds. (×22,500) B. Live spirochetes (S) coursing between undulipodia (U) of *Coptotermes formosanus* (From 16 mm film, ×2,250.)[32] C. Live pillotina spirochetes (P) moving in synchronous waves from *Pterotermes occidentis* (From 16 mm film, ×2,250.) D. Undulipodia (U) moving in synchronous waves from *Opalina* (Opalinids are variously classified as ciliates or flagellates; they are placed in Phylum Zoomastigina, Kingdom Protoctista in Schwartz and Margulis.[33] From 16 mm film of A. T. W. Cheung, photo courtesy of Dr. Cheung.)

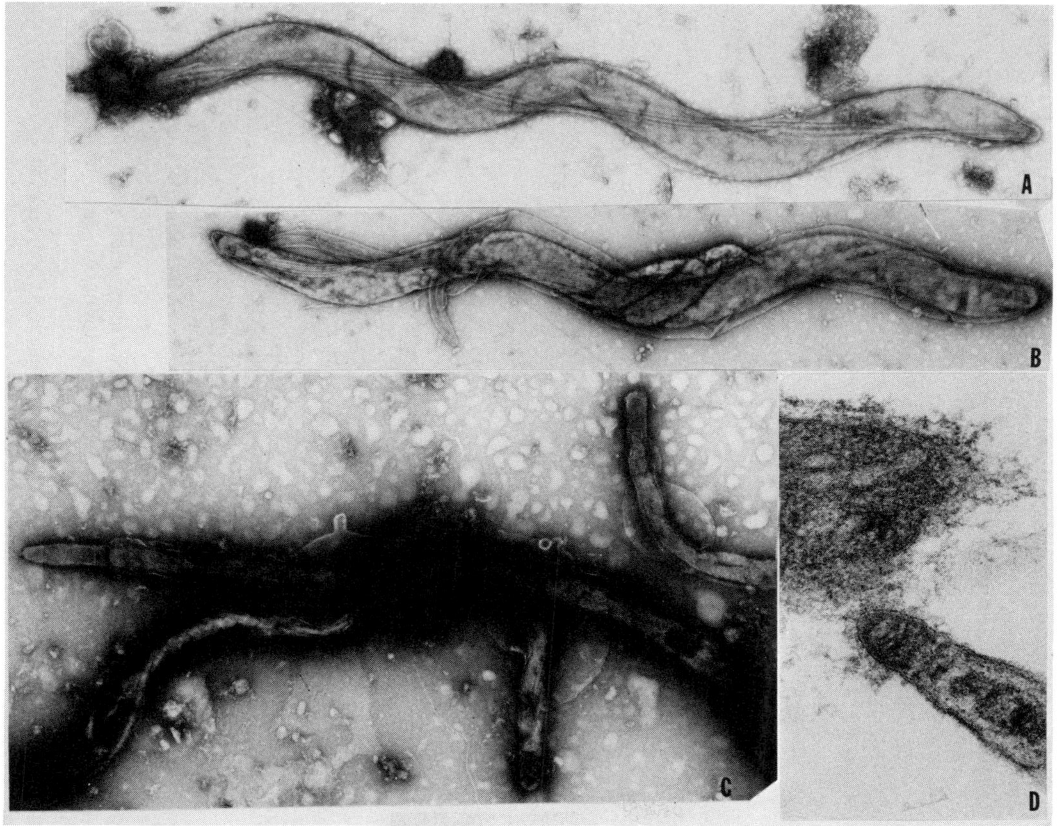

FIGURE 3. Unmodified spirochetes and those modified for attachment. A. Unmodified spirochete with four flagella, from the termite *Reticulitermes hesperus* from Joshua Tree National Monument. (Negative stain, electron micrograph ×20,625) B. Unmodified spirochetes probably bearing eight flagella from *Reticulitermes tibialis* (Negative stain, ×13,750.) C. Three different spirochetes with modified ends, unidentified, from *Reticulitermes hesperus.* (Negative stain, ×27,500.) D. Unidentified microbe (spirochete?) attached to unidentified mastigote showing slight modification of both partners. From *Reticulitermes hesperus.* (Negative stain, ×82,500.)

In summary the evidence for neither endogenous nor exogenous (spirochete) hypothesis is compelling. In fact there are various endogenous hypotheses in which the undulipodia differentiate within the primitive eukaryotic cell. The differences between the endogenous and exogenous hypotheses have already been summarized (TABLE 1, p. 8, Margulis [18]). A major difference is that no relation between any spirochetes and the undulipodial system of eukaryotes is expected on the endogenous hypothesis whereas the proof of such a relationship is mandatory on the exogenous hypothesis. Neither are these the only possibilities: for example, pillotina spirochetes may show a relationship to the microtubule system because the entire eukaryotic cell itself evolved from pillotina spirochetes. The observations of Searcy *et al.*[31] on the similarities between *Thermoplasma* and the nucleocytoplasm make this possibility unlikely but do not definitely exclude it. Furthermore it is the resemblance of attached surface spirochetes to undulipodia that prompted the symbiotic hypothesis in the first place. In addition the nucleocytoplasm itself is not any more similar to pillotina spirochetes than to other prokaryotes. Another possibility is the

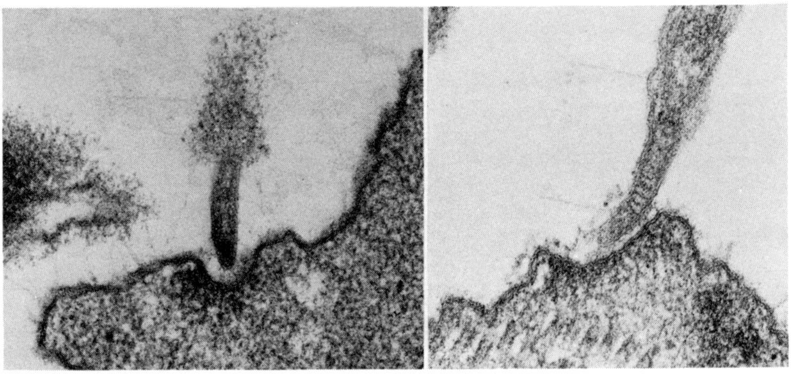

FIGURE 4. Two examples of thin sections of spirochetes modified for attachment to the protist *Metadevescovina* sp. from the termite *Incisitermes (Kalotermes) minor*. Variations in form have been seen even on spirochetes attached to the same host species. Note indentation and thickening of the host surface at points of attachment. (Both ×75,000.)

viral or plasmid transfer origin of the undulipodium. If the tubulin genes were plasmid or viral borne they may have originated endogenously in eukaryotes and were subsequently transferred to those spirochetes that are commonly associated with undulipodiated eukaryotes in the hindgut (as in FIGURE 2A). Many possibilities exist.

Adequate proof of the symbiotic spirochete concept may never be available but the following would constitute strong inference:

1. Identification of proteins homologous to α tubulin in association with ATP-mediated motility systems in free-living and host-associated spirochetes.
2. Discovery of tubulin-containing spirochetes in environments devoid of eukaryotes and thus in which the transfer of genes from undulipodiated eukaryotes to spirochetes would be deemed improbable.

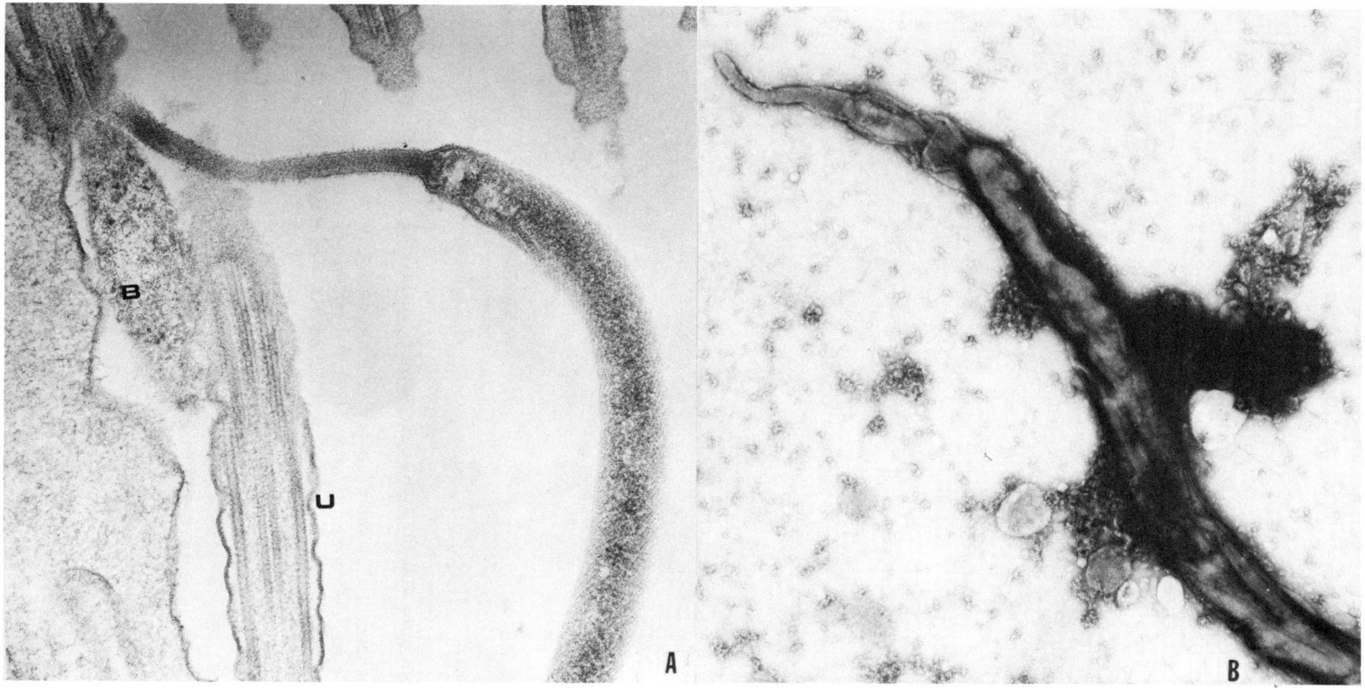

FIGURE 5. Thin section (A) and negative stain (B) of spirochetes highly modified at one end for attachment to protist host. Both the host protist and the spirochete are unidentified. From the Sonoran desert dry wood termite, *Pterotermes occidentis.* (A) Thin section electron micrograph, B=apparently another bacterium, U=undulipodia of protist host, ×45,000. (B) Negative stain electron micrograph, note the striations in the highly pointed end. (×30,000)

3. Variation in tubulin protein and in the arrangement and size of microtubules in populations of motile prokaryotes including pillotina spirochetes.
4. Discovery of a nine-fold symmetrical arrangement of microtubules, namely an internal structure comparable to the arrangement of tubules in undulipodia, in free-living spirochetes.
5. Discovery of distinctive ribosomal or cell membrane proteins in spirochetes that are restricted to tubule-containing spirochetes and eukaryotes.

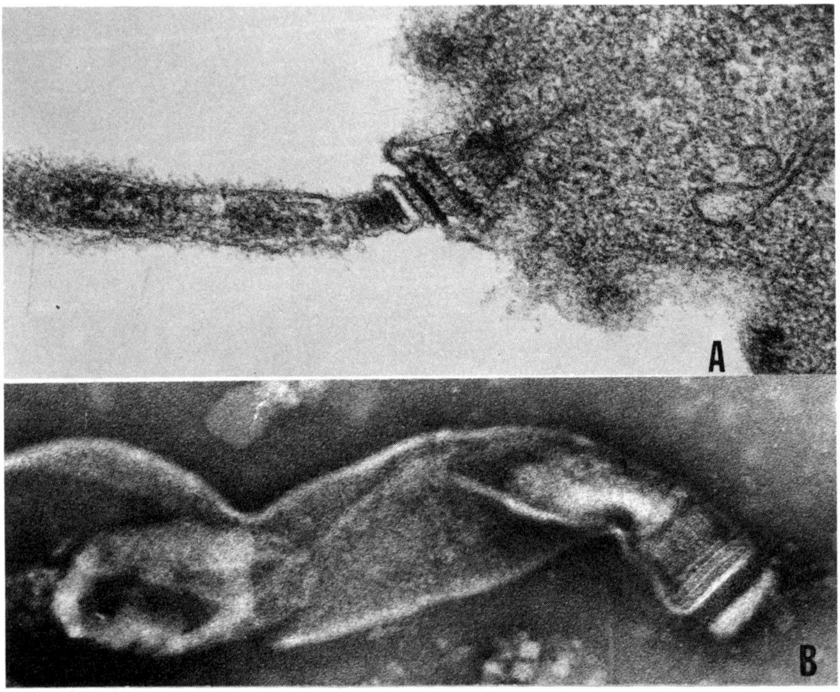

FIGURE 6. Spirochetes highly modified for attachment to the protist *Pyrsonympha* sp. from the termite *Reticulitermes tibialis.* These flat-ended spirochetes are recovered attached (A, thin section ×67,500) and unattached (B, negative stain ×101,250).

6. Discovery of spirochete attachment sites that resemble kinetosomes (centrioles).
7. Discovery of plasma membrane loss in recognizable spirochete–host associations.
8. Absence of tubulin and eukaryotic-style microtubules in any photosynthetic prokaryotes.
9. Discovery of sequence homology between the RNA known to be present in the lumen of kinetosomes [9, 12] and DNA sequences in tubulin-containing spirochetes.
10. Discovery of sequence homology between centromeric (chromosomal) DNA and DNA from tubulin-containing spirochetes. (For further

explanation of those steps thought to lead from spirochete motility symbioses to the origin of the mitotic apparatus see chapters 9 and 10 in Margulis.[20]

The endogenous hypothesis would predict no greater tendency of spirochetes to have tubulin than other prokaryotes and in fact would expect the presence of tubulin protein in photosynthetic prokaryotes, especially in those blue greens thought to be ancestral to the red algae.

It also should be mentioned briefly that on any hypothesis of the evolutionary origin of undulipodia the following observations at least must be taken into consideration.

1. The pattern of arrangement of the kinetosomes and microtubules (for example, of the ciliate cortex) may be inherited independently of the nuclear genes.[1]

2. No undulipodium forms without the prior presence of the kinetosome. Kinetosomes and undulipodia can form from amorphous granulo-fibrosal material (MTOCs).

3. MTOCs vary widely in ultrastructure.

4. Undulipodia often vary in length and in number but they never vary in width or size of component microtubules. Structures composed of undulipodia (cirri, membranelles) are larger on larger protists (e.g. ciliates) by becoming composed of a greater number of standard-sized undulipodia.

5. No red algae show a tendency towards motility by undulipodia whereas all studied show normal eukaryotic, mitotic, and meiotic patterns.

6. Although mitotic patterns vary a great deal in protists such as dinoflagellates and amoeboflagellates undulipodia show a great constancy in structure and function. Undulipodiated organisms that lack standard mitosis are known (dinoflagellates) and mitotic organisms that lack undulipodia are known (e.g. fungi, *senso strictu*). Many protists lack undulipodial motility during the mitosis process but after cell division undulipodia return (*Naeglaria*, certain dinoflagellates).

7. There has been no homology reported between flagellin (prokaryotes) and tubulin proteins of eukaryotes.

8. Those spirochetes in which microtubules have been seen also contain flagella. Some mycoplasms (e.g. spiroplasms) are highly motile and yet lack flagella or any other obvious mechanisms of motility; whether or not tubulin proteins are present is unknown.

9. Tubulin proteins of the mitotic apparatus and of undulipodia are homologous if not identical. No movement of chromosomes occurs unless microtubules are formed.

Clearly, although the evolutionary origin of the microtubule system is an unsolved problem, because of the incredible conservatism of biological systems, it is approachable by experiment and inference. It is hoped that the next decade will provide the data upon which one can base a definitive answer to the question of the origin of undulipodia: from symbiotic spirochetes or not?

Acknowledgments

We are grateful for helpful discussion to H. Ponstingle, S. Fracek, R. Laursen, K. H. Nealson, R. Ober, J. Stolz, H. Hartman, and A. T. W. Cheung. We are also grateful to A. T. W. Cheung for his generous donation of micrographs.

REFERENCES

1. BEISSON, J. & T. M. SONNEBORN. 1965. Cytoplasmic inheritance of the organization of the cell cortex in *Paramecium aurelia*. Proc. Natl. Acad. Sci. USA **53:**265–282.
2. BERG, H. C. 1975. Bacterial behaviour. Nature **254:**389–92.
3. BISALPUTRA, T., B. C. OAKLEY, D. C. WALKER & C. H. SHIELDS. 1975. Microtubular complexes in blue-green algae. Protoplasma **86:**19–28.
4. BLOODGOOD, R. A. & T. P. FITZHARRIS. 1976. Specific association of prokaryotes with symbiotic flagellate protozoa from the hindgut of the termite *Reticulitermes* and the wood-eating cockroach *Cryptocercus*. Cytobios **17:**103–122.
5. BREZNAK, J. A. 1973. Biology of nonpathogenic host-associated spirochaetes. Crit. Rev. Microbiol. **2:**457–489.
6. CHEUNG, A. T. W. & T. L. JAHN. 1975. Helical nature of the continuous ciliary beat of *Opalina*. Acta Protozool. **14:**111–25.
7. CHEUNG, A. T. W. & L. MARGULIS. 1977. Microbiota from *Pterotermes occidentis*. 16 mm film. California Institute of Technology. Pasadena, California.
8. CLEVELAND, L. R. & A. V. GRIMSTONE. 1964. The fine structure of the flagellate *Mixotricha paradoxa* and its associated microorganisms. Proc. R. Soc. Lond. B **157:**668–683.
9. DIPPELL, R. V. 1976. Effects of nuclease and protease digestion on the ultrastructure of *Paramecium* basal bodies. J. Cell Biol. **69:**622–637.
10. DODSON, E. 1979. Crossing the procaryote-eucaryote border: endosymbiosis or continuous development? Can. J. Microbiol. **25:**651–674.
11. GROSOVSKY, B. D. D. & L. MARGULIS. 1980. Termite microbial community: heat, starvation and antibiotic induced alterations. *In* Microbial Ecology. R. Burns, Ed. Blackwells Scientific Publishers. Oxford.
12. HARTMAN, S., T. GURNEY & T. PUMA. 1974. Evidence for the association of RNA with the ciliary basal bodies of *Tetrahymena*. J. Cell Sci. **16:**241–259.
13. HEIDEMANN, S. R., G. SANDER & M. W. KIRSCHNER. 1977. Evidence for a functional role of RNA in centrioles. Cell **10:**337–350.
14. HOLLANDE, A. & I. GHARAGOZLOU. 1967. Morphologie infrastructurale de *Pillotina calotermitidis* nov.gen., nov.sp. Spirochaetale de l'intestin de *Calotermes praecox*. C. r. hebd. Seanc. Acad. Sci. Paris **265:**1309–1312.
15. KUZNICKI, L., T. L. JAHN & J. R. FUMESCA. 1970. Helical nature of the ciliary beat of *Paramecium multimicronucleatum*. J. Protozool. **17:**16–24.
16. LAANE, M. M. & F. B. HAUGLI. 1974. Division centres in mitotic nuclei of *Physarum polycephalum plasmodia*. Norweg. J. Botany **21:**309–318.
17. LUDUEÑA, R. F. & D. O. WOODWARD. 1975. α- and β-tubulin: separation and partial sequence analysis. Ann. N.Y. Acad. Sci. **253:**272–283.
18. MARGULIS, L. 1975. Microtubules and evolution. *Intern. Symp. on Microtubules and Microtubule Inhibitors.* (Beerse, Belgium) M. DeBrabander & M. Borgers, Eds.: 3–18. North Holland Publishing Co. Amsterdam.
19. MARGULIS, L. 1976. The theme (mitotic cell division) and the variations (protists): implications for higher taxa. Taxon **25:**391–403.
20. MARGULIS, L. 1981. Symbiosis in Cell Evolution. W. H. Freeman & Co. San Francisco, Calif.
21. MARGULIS, L. 1980. Flagella, cilia and undulipodia. BioSystems. **12:**105–108.
22. MARGULIS, L., L. P. TO & D. CHASE. 1981. Large spirochaetes symbiotic in termites: the genera *Pillotina, Hollandina, Diplocalyx,* and *Clevelandina*. *In* The Prokaryotes: A Handbook on Habitats, Isolation and Identification of Bacteria. M. P. Starr, H. Stolp, H. G. Trupper, A. Balows & H. G. Schlegel, Eds. Springer-Verlag. New York, N.Y.
23. MARGULIS, L., L. TO & D. CHASE. 1978. Microtubules in prokaryotes. Science **200:**1118–1123.

24. MARGULIS, L., D. CHASE & L. TO. 1979. Possible evolutionary significance of spirochetes. Trans. R. Soc. Lond. (Ser. B) **204:**189–198.
25. PICKETT-HEAPS, J. D. 1971. The autonomy of the centriole: fact or fallacy. Cytobios **3:**205–214.
26. PICKETT-HEAPS, J. D. 1974. The evolution of mitosis and the eukaryotic condition. BioSystems **6:**37–48.
27. PICKETT-HEAPS, J. D. 1975. Aspects of spindle evolution. Ann. N.Y. Acad. Sci. **253:**352–61.
28. PONSTINGLE *et al.* Approaches to the sequence of tubulin. *In* Methods in Peptide and Protein Sequence Analysis. C. Birr, Ed.: 225–234. Elsevier/North Holland Biomedical Press.
29. PONSTINGLE, H., M. LITTLE, E. KRAUHS & T. KEMPF. 1980. Carboxy-terminal amino acid sequence of α tubulin from porcine brain. Nature. (In press.)
30. RIS, H. 1975. Primitive mitotic systems. Symposium on the evolution of mitosis in eukaryotic micro-organisms (see entire November issue.) BioSystems **7:** 298–304.
31. SEARCY, D. G., D. B. STEIN & G. R. GREEN. 1978. Phylogenetic affinities between eukaryotic cells and a thermoplasmic mycoplasm. BioSystems **10:** 19–28.
32. SCHAADT, J. G. & L. MARGULIS. 1976. Spirochaetes. 16 mm film. Boston University. Boston, Massachusetts.
33. SCHWARTZ, K. V. & L. MARGULIS. 1980. Guide to Life on Earth: Phyla of the Five Kingdoms. W. H. Freeman & Co. (In press.)
34. SMITH, H. E., H. E. BUHSE & S. J. STAMLER. 1975. Possible formation and development of spirochaete attachment sites found in the surface of symbiotic polymastigote flagellates of the termite *Reticulitermes flavipes.* BioSystems **7:**374–9.
35. SONNEBORN, T M. 1970. Determination, development and inheritance of the structure of the cell cortex. Symp. Soc. Dev. Biol. **9:**1–13.
36. TAMM, S. 1978. Membrane specializations associated with the attachment of prokaryotes to a eukaryote cell. J. Cell Biol. **79:**216a.
37. TO, L., L. MARGULIS & A. T. W. CHEUNG. 1978. Pillotinas and hollandinas: distribution and behaviour of large spirochetes symbiotic in termites. Microbios **22:**103–133.
38. YOUNGER, K. B., S. BANERJEE, J. K. KELLEHER, M. WINSTON & L. MARGULIS. 1972. Evidence that the synchronized production of new basal bodies is not associated with DNA synthesis in *Stentor coeruleus.* J. Cell Sci. **11:**621–637.

DISCUSSION OF THE PAPER

DR. D. RUIZ (*Rockefeller University, New York, N.Y.*): In the slide you showed you postulated three separate periods of symbiotic invasion of these protomitochondria, chloroplasts, and flagella. What I was curious about was, you showed the invasion by the mitochondria first, followed by invasion by these flagella, and then in some subset, chloroplasts. What's the basis for postulating a prior invasion by mitochondria rather than flagella? Especially since you mentioned at one point that there are these organisms which have the flagella and have no mitochondria.

DR. MARGULIS: Organisms that have undulipodia and also have no mitochondria, have relatives that are pretty close among standard protozoa with mitochondria, so one assumes they went back. In other words, the organisms I was showing you, organisms that had classical undulipodiated ancestors or the ciliated type ancestors, lost their mitochondria when they went back into tremites, into anaerobic environments. The reason for that is they look like organisms that do have mitochondria. So I think it's a secondary loss. They acquired spirochetes after that.

DR. RUIZ: The reason I would suspect that if this was the case it might indeed have been prior to mitochondria leads into my second question. We've been hearing in the last couple of days about the relationship between a number of bacteria and either mitochondria or chloroplasts and I think we're all very impressed by the association of these spirochetes with bacteria. But the question is, what degree of molecular analogy can you make between the spirochetes and the forms of undulipodia which are normally found?

DR. MARGULIS: I have to say quite frankly that in *Pelomyxa* there has been a report that it has a nonfunctional 9-plus-2 structure in it, and although it has no mitosis at all, it has some cortical microtubules. It does not have mitochondria and one expects that's a primitive state with that organism. If that's the case, then it's probably correct that the 9-plus-2 condition or the association with spirochetes preceded the mitochondria. One feels that chloroplasts came later because there aren't any cells that have chloroplasts. But there is no hard evidence as far as mitochondria versus the undulipodia.

Your second question about what chemistry can you do that shows a relation between spirochetes and undulipodia, is a subject of one of my graduate student's thesis. He's looking for tubulin protein and going to try to first get it out and sequence the tubulin protein that is making up the walls of the tubules that we see morphologically. We think if we get the protein sequence on that, that probably would be strong evidence.

DR. AARONSON: There are an awful lot of associations between all kinds of microorganisms and a variety of environments. I'm wondering if in any of these you know of any association between a spirochete or even a bacterium and the eukaryote which would indicate at least a third or a half of its length be covered by the eukaryotic plasma membranes, which is the state of the flagella.

DR. MARGULIS: Any of them that are inside the cell are covered.

DR. AARONSON: In your case it would be a depression in which the symbiont is surrounded by a significant amount of plasma membranes.

DR. K. H. NEALSON: This association is in the hind-gut of a eukaryotic organism loaded with other dead and dying eukaryotes, all possessing the genetic information for tubulin. So mechanisms for transfer probably exist there. What are the criteria?

DR. MARGULIS: We are aware of this problem. We have a spirochete that's free-living, that has been isolated. It is growing in culture from Baja California muds in which the scene is overwhelmingly prokaryotic. It's really hard to believe that some eukaryote came in there and transferred to prokaryotic organisms. We have a putative tubulin positive result with fluorescent antibody made against authentic brain tubulin. These spirochetes indeed glow with an antifluorescent-tagged antibody made against tubulin. It's that kind of situation in pure culture in a prokaryotic scene where you'd have to get the protein out to answer that question.

ENDOSYMBIONTS, BIOGENIC AMINES, AND A HETERODYNE HYPOTHESIS FOR CIRCADIAN RHYTHMS *

M. Levandowsky

Haskins Laboratories of Pace University
New York, New York 10038

INTRODUCTION

Much of the evidence for the endosymbiont hypothesis of the origin of eukaryote cells arises from their spatial complexity: the heterogeneous array of distinct organelles suggests very naturally a colonial evolutionary history. Less studied, in this context, is the rich *temporal* structure of eukaryote physiology. Beyond the traditional four parts of the cell cycle, a fine structure of sequential metabolic events can now be seen. In yeasts and several flagellates, enzymes appear during the cycle in an invariant, orderly succession of "peaks" and "steps," and temperature-sensitive mutants of these can serve as markers in the genetic analysis of temporal organization. In the yeast, *Saccharomyces cereviseae*, about 30 of these biochemical events have been pinpointed.[1]

In addition to the *qualitative* temporal organization seen in such orderly sequences, there are also many indications of a *quantitative* structure—i.e., regular metabolic oscillations. Well-known examples are the glycolytic oscillations in many cells,[2, 3] and the regular pulses of cAMP secretion by certain cells during cellular slime-mold aggregation.[4, 5, 6] Many examples of such oscillations are now known and various models have been proposed to account for them. In recent years the realization has grown that they must be the result of very nonlinear processes and therefore are probably *limit cycles*. That is, after perturbation, the oscillation always returns asymptotically to a stable cycle of given amplitude, wave form, and period.[7] Two important features of most of these biochemical oscillations are a relatively short period, of the order of several minutes, and a strong temperature dependence—e.g., a Q_{10} of about 4 in the case of the glycolytic oscillation of yeasts.

Perhaps the most striking feature of eukaryotic temporal architecture, however, is the much slower, temperature-insensitive circadian cycle–the biological clock.[8] Since this phenomenon is usually considered a peculiarity of eukaryotes, it seems reasonable to enquire into its possible bearing on the endosymbiont theory of eukaryote origins.†

* Supported in part by a U. S. Public Health Service Grant No. FR-05596 to A. M. Liberman, Haskins Laboratories.

† A few reports have appeared of circadian rhythms in prokaryotes. Recently Sturtevant,[9] using statistical tools, reported circadian cycles in rather noisy growth-rate data of two bacterial species but interpretation of these are made difficult by the diurnal periodicity of the illumination that was used; subsequent work on this appears to be lacking. If a bacterial circadian rhythm could be demonstrated, then the problem of a special relation to eukaryote architecture would need to be reexamined. In that case, though, it would seem odd that the rhythm has not been noted elsewhere in the immense literature on bacterial growth, particularly in the many precise studies with continuous cultures.

0077-8923/81/0361-0369 $01.75/0 © 1981, NYAS

Difficult Problems Posed by Biological Clocks

As noted, circadian rhythms differ greatly from other known biochemical oscillations in at least two basic ways: their low frequency and their virtual temperature-insensitivity. Additional features that must be addressed by any model are their persistence in enucleated cells (in *Acetabularia*[10]) and the lack of effect of a variety of standard metabolic poisons and protein-synthesis inhibitors.[8]

Concerning insensitivity to both temperature and metabolic inhibitors, recent work suggests that part of the explanation may be that the circadian oscillation is a membrane phenomenon. That is, experiments with selective ionophores and other agents show that the rhythm can be altered by changes in potassium permeability.[11] Which particular membranes are involved—whether the outer cell membrane itself, interior membranes, or both—is not yet clear, nor how potassium permeability is used. Sweeney[12] and others have also noted circadian changes in the ultrastructure of membranes, but whether these changes are causal, or merely an effect of the primary rhythm (the "hands of the clock" rather than the clock itself) is also not yet known.

At any rate the circadian rhythm is apparently not linked intimately to the Krebs cycle or glycolysis, but does involve membrane permeability to K^+ somewhere in the cell, and now it seems probable that a membrane potential coupled with a gated K^+ channel may be a part of the basic oscillator. Temperature-insensitivity suggests a curious lack of enzymic properties in the oscillator: are the putative K^+ gates and pumps (proteins, presumably) temperature-insensitive? Models have been proposed in which two processes with opposite types of temperature sensitivity precisely balance each other, but arguments of this nature seem terribly *ad hoc* and evidence for them is lacking. On the other hand physical properties of the lipid bilayer itself would probably be rather constant over broad temperature ranges, except for phase transition points, and would not be affected by metabolic blockers such as cyanide, azide, dinitrophenol, etc.

Membrane-based oscillations are known in neurobiology. These are limit cycles of the relaxation-oscillation type—a relatively slow phase in which membrane potential builds to a threshold, followed by rapid discharge in response to permeability change at the threshold voltage. These phenomena have been modeled fairly successfully. Known relaxation-oscillations of this type have high frequencies, however, and this presents severe difficulties in attempts to model a circadian rhythm.

Sweeney[13] and others have suggested that a slow component in the cycle could be obtained by introducing a spatial diffusion process, in which a molecule or ion diffused slowly, either across a membrane or laterally along it. However, if one considers the diffusion rate of even heavy molecules in a variety of solvents, and the tiny subcellular distances in such rhythmic eucells as *Chlamydomonas*, it seems very unlikely that diffusion could provide much delay on this scale—it is virtually instantaneous from the standpoint of a circadian rhythm. Furthermore, diffusion rates are temperature dependent.

A Possible Solution to the Low Frequency Problem

One way to get a slow oscillation is to couple two fast ones, with slightly different frequencies. This leads to a resultant oscillation whose frequency, $\Delta\omega$, is the difference between the 2 primary frequencies, ω_1, ω_2.

This is an obvious idea, and has been suggested before in connection with the biological clock.[14] A fairly serious drawback with it is a certain lack of stability. This arises from the fact that, if $\Delta\omega$ is small enough, it may be close to the noise levels of ω_1 and ω_2, and small statistical fluctuations in these latter may give rise to large proportional changes in $\Delta\omega$. This is a quantitative aspect, and rather difficult to evaluate on our level of speculation; nevertheless, some general comments can be made. Certainly, as the very name implies, circadian rhythms are not precise oscillations but rather very noisy, with approximate periods in the range 23–25 hours. If the underlying primary oscillations are "stiff" limit cycles and recover rapidly from perturbations, then statistical noise in the slow resultant oscillation could be kept within this range (White[15] discusses random fluctuations of limit cycles).

For some people, a qualitative drawback of a coupled oscillation model will be the necessity of postulating two primary oscillations. This is essentially an esthetic criticism however, based on Occam's razor. On the other hand, in relation to the endosymbiont, or colonial, hypothesis such models become rather appealing: we might suppose that the two primary oscillations are properties of distinct partners in the original symbiosis—host and symbiont, perhaps, or two different symbiotic elements. By oscillating at slightly different frequencies, they could achieve a resultant slow endogenous cycle, attuned roughly to the diurnal rhythm of the environment. Fine tuning and phase-setting would then come from external cues, or *zeitgebers*.

Temperature-independence of the resultant slow oscillation can also be found in this model: if the nearly identical frequencies of the two primary oscillations change by the same amount with temperature then their *difference*—the frequency of the slow oscillation—will be temperature-independent.

Biogenic Amines

Now we must consider the mechanism of interaction between the two putative primary oscillators. Since the oscillators themselves are hypothetical, it will not be surprising that our suggestion is completely speculative (those who dislike speculation must surely have stopped reading by now!).

Since we are postulating membrane potentials as primary oscillators, it is plausible to think of membrane receptors as part of the interaction mechanism. This in turn would imply the presence of chemical stimuli, and we come to the question of possible chemical signals between the two oscillators.

The biogenic monoamines (quaternary amines, catecholamines, indoleamines, certain amino acids, and purines) are widely distributed, perhaps ubiquitous in eukaryotes. Their roles as intercellular messengers, or neurohormones in metazoa are well known. They have been found wherever they were diligently sought, and their functions in certain parasitic members of the lower phyla (e.g., trematodes) are now of great clinical interest as possible drug targets.[16] In the metaphyta they are less well known, with exception of the auxins (indoleamines), though a number of scattered examples, such as the curious presence of acetylcholine in potatoes, suggest a wide distribution and yet to be discovered functions. Among single-celled eukaryotes, there are also reports, some controversial, of biogenic monoamines and/or their receptors, but their possible function has seemed rather obscure.[17] The best-studied case, and that only recently, is the ciliate *Tetrahymena*.

This organism can synthesize both serotonin and epinephrine,[18] and responds behaviorally and metabolically to low levels of these and other amines.[19] The possible biological role of such molecules in a single-celled organism appeared problematic to many however, and such observations have tended to be dismissed as laboratory curiosities. Blum[20] however suggested a role in metabolic regulation, on the basis of experiments with various known pharmacological agents. Recently, this notion was put on firmer ground with the elucidation of the function of epinephrine in the metabolic response to glucose availability (FIGURE 1).

From this line of research, just beginning, it is clear that epinephrine is critically involved in at least one control system of *Tetrahymena* metabolism in

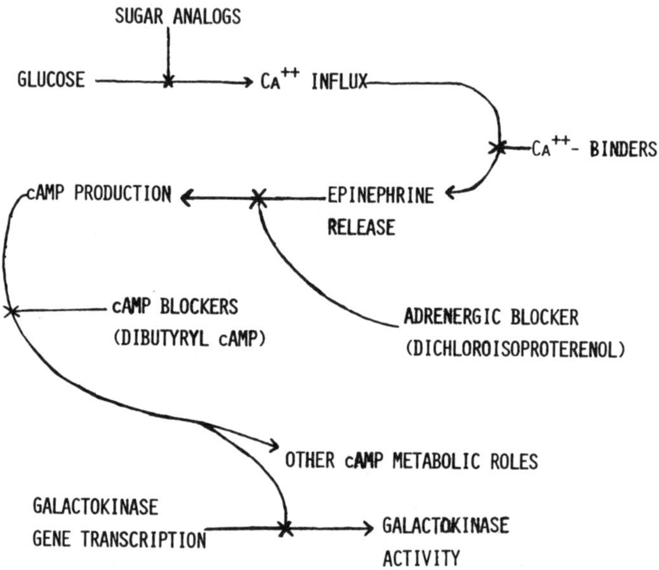

FIGURE 1. The role of epinephrine in *Tetrahymena* galactokinase regulation. Recent studies[24, 25] with auxotrophic mutants and blocking agents indicate that catecholamines are intimately involved in the transduction linking the glucose receptor to galactokinase suppression.

a manner reminiscent of (though far from identical to) its function in, e.g., mammalian liver cells. We may reasonably anticipate that serotonin may prove to have a similar function, and all this is probably the tip of a biochemical iceberg of other roles for such molecules in the single-celled eukaryotes.

Among prokaryotes, however, there is little evidence for a biochemical role for these monovalent cations. Here and there we find organisms that produce such substances as acetylcholine (e.g., *Lactobacillus plantarum*[21]), but this is far from generally true, apparently. However, most of the biochemical machinery for synthesis of the major monoamines would appear to be widely available: synthesis of catecholamines, indoleamines, and quaternary amines can be viewed as mere variations of the metabolism of tyrosine and catechol, tryptophan and indole, and choline, respectively. A very intriguing recent observation is that *Escherichia coli* and other bacteria respond behaviorally to these amines, and in some cases the response is blocked by well-known pharmacological agents.[22, 23]

Whether this is a biochemical accident—a preadaptation if one believes in such things—or a sign that we have overlooked a functional role for these molecules in bacteria is not clear, and there appears to be very little work on this interesting question at the moment. However, the fact that prokaryotes *do* respond to monoamines, and that in many species these could presumably be produced with relatively minor additions or changes in existing pathways, is of interest for the present speculation.

Let us suppose, then, that two prokaryote colonists of the putative *ur*-eukaryote cell oscillated biochemically, with slightly different short periods, emitting pulses of, say, epinephrine and serotonin respectively. Then a longer, circadian period might depend on the coincidence of the two signals—a heterodyne model, essentially. For such a scheme to work, the pulsed signals must also be quickly degraded, as happens, e.g., in neural synapses or the cAMP pulsations of aggregating slime molds.

It would be interesting to know whether there are any psychoactive substances, monoamine-blockers in other systems, that have an effect on biological rhythms in the single-celled eukaryotes.

[**Note added in proof:** Hardeland (1980. *Comp. Biochem. Physiol.* **66c**:53–58) reports blocking of circadian rhythm in a dinoflagellate by the adrenergic antagonist propranolol, but suggests this may be due to an effect on membrane fluidity.]

References

1. Hartwell, L. H. 1974. *Saccharomyces cereviseae* cell cycle. Bacteriol. Rev. **38**:164–198.
2. Winfree, A. T. 1972. Oscillatory glycolysis in yeast: the pattern of phase resetting by oxygen. Bioch. Biophys. **149**:388–401.
3. Chance, B., R. W. Estabrook & A. Ghosh. 1964. Damped sinusoidal oscillations of cytoplasmic reduced pyridine nucleotide in yeast cells. Proc. Natl. Acad. Sci. USA **51**:1244–1251.
4. Shaffer, B. M. 1962. The Acrasina, Part 1. Adv. Morphogen. **2**:109–182.
5. Robertson, A., D. J. Drage & M. H. Cohen. 1972. Control of aggregation in *Dictyostelium discoideum* by an external periodic pulse of cyclic adenosine monophosphate. Science **175**:333–335.
6. Gerisch, G & U. Wick. 1975. Intracellular oscillations of cyclic AMP from *Dictyostelium* cells. Biochem. Biophys. Res. Commun. **65**:364–370.
7. Minorsky, N. 1962. Nonlinear Oscillations. Van Nostrand. Princeton, N.J.
8. Bünning, E. 1973. The Physiological Clock. Springer-Verlag. Heidelberg.
9. Sturtevant, R. P. 1973. Circadian variability in *Klebsiella* demonstrated by cosinor analysis. Int. J. Chronobiol. **1**:141–146.
10. Schweiger, E., H. G. Wallraff & H. G. Schweiger. 1964. Endogenous circadian rhythm in cytoplasm of *Acetabularia*: influence of the nucleus. Science **146**:658–659.
11. Njus, D., V. O. Gooch, D. Mergenhagen, F. Sulzman & J. W. Hastings. 1976. Membranes and molecules in circadian systems. Fed. Proc. **35**:2353–2357.
12. Sweeney, B. M. 1976. Evidence that membranes are components of circadian oscillators. *In* The Molecular Basis of Circadian Rhythms. J. W. Hastings & H. G. Schweiger, Eds.:267–282. Dahlem Konferenzen. Berlin.

13. SWEENEY, B. M. 1974. A physiological model for circadian rhythms derived from the *Acetabularia* rhythm paradoxes. Int. J. Chronobiol. **2**:25–33.
14. TYSON, J. J., S. G. A. ALIVISATOS, F. GRÜN, T. PAVLIDIS, O. RICHTER & F. W. SCHNEIDER. 1976. Mathematical background, group report. *In* The Molecular Basis of Circadian Rhythms. J. W. Hastings & H. G. Schweiger, Eds.:86–108. Dahlem Konferenzen. Berlin.
15. WHITE, B. S. 1977. The effects of a rapidly varying random environment on systems of interacting species. SIAM J. Applied Math. **32**:666–693.
16. BUEDING, E. & J. BENNETT. 1972. Neurotransmitters in trematodes. *In* Comparative Biochemistry of Parasites. H. Van den Bossche, Ed.:95–100. Academic Press, Inc. New York, N.Y.
17. LEVANDOWSKY, M. & D. C. R. HAUSER. 1978. Chemosensory responses of swimming algae and protozoa. Int. Rev. Cytol. **53**:145–210.
18. JANAKIDEVI, K., V. DEWEY & G. KIDDER. 1966. The biosynthesis of catecholamines in two genera of protozoa. J. Biol. Chem. **241**:2576–2578.
19. CSABA, G. & T. LANTOS. 1973. Effects of hormones on protozoa. Cytobiologie **7**:361–365.
20. BLUM, J. J. 1967. An adrenergic control system in *Tetrahymena*. Proc. Natl. Acad. Sci. USA **58**:81–88.
21. STANASZEK, P. M., J. F. SNELL & J. J. O'NEILL. 1977. Isolation, extraction and measurement of acetylcholine from *Lactobacillus plantarum*. Appl. Envir. Microbiol. **34**:237–239.
22. MARUYAMA, H. B. & H. AZUMA. 1976. Regulation of bacterial motility by cyclic nucleotides and effect of biogenic amines. J. Cyclic Nucleotide Res. **2**:129–138.
23. AZUMA, H. & H. B. MARUYAMA. 1974. Biochemically active substances from actinomycetes and other organisms II. Effects of various compounds on motilometry. J. Antibiotics **27**:185–191.
24. NANDINI-KISHORE, S. G. & G. THOMPSON, JR. 1978. Increased levels of adenosine 3′,5′-cyclic monophosphate in *Tetrahymena* stimulated by glucose and mediated by Ca^{2+} and epinephrine. Proc. Natl. Acad. Sci. USA **76**:2708–2711.
25. ROBERTS, C. T. & D. E. MORSE. 1978. Genetic regulation of galactokinase by cyclic AMP, glucose, and epinephrine. Proc. Natl. Acad. Sci. USA **75**:1810–1814.

DISCUSSION OF THE PAPER

DR. S. S. COHEN: Might you mention the word endosymbiosis in here someplace?

DR. M. LEVANDOWSKY: The idea is that the bacterial prokaryotes, both gram-negative and gram-positive, are shown to have what appear to be sensitive specific receptors to certain of these compounds. It's known that some bacteria will produce these compounds which isn't surprising, after all indoleamines are just one step away from common compounds as are catecholamines. Acetylcholine is easily made from choline produced by many bacteria. So they can produce them, and they have receptors for them. Why they have receptors for them is another question which I really have no answer for. But they seem to, so in terms of the symbiotic hypothesis, it seems reasonable that they might be using these messages to sort of "talk" to each other and regulate their total integration.

DR. K. H. NEALSON: There are lots of oscillators known in bacteria but they're all quite temperature-sensitive and in general, in higher systems these biological clocks aren't temperature sensitive, so how do you fit that in with your model of chemical oscillators which by their nature ought to be temperature sensitive?

DR. LEVANDOWSKY: First of all there are a few old reports that bacteria might have circadian oscillators, this is the sort of thing we don't like to think about because it ruins the pattern. There were a couple of papers published with two bacterial species and there was evidence of a certain circadian oscillation which is something you wouldn't see normally and perhaps that's why it's not seen by bacteriologists, if it's there. There's so much "noise" but if you analyze statistically, circadian rhythm seems to come out of that pattern. That work hasn't been, as far as I know, repeated or analyzed or studied. As to the temperature insensitivity, you are right. I tried to make the point that the chemical oscillations that I started mentioning are indeed quite temperature sensitive. It's thought now that the insensitivity of the biological clock has to do with its location in the membrane's permeability. A different opening and closing of channels for example, for ionic permeability might be a lot less temperature sensitive as can be seen in neurophysiology. Their temperature effects in membranes usually involve things like phase transitions within the membrane itself. If you are not in a region of a phase transition, there are broad reaches of the membrane that don't seem to have any great change with temperature. The suggestion is that the rhythm involved here involves receptors in membranes and flux of ions across membranes.

THE LOCKED-CELL HYPOTHESIS: ON THE ORIGIN OF MITOCHONDRIA AND THE TRANSITION FROM PROKARYOTIC TO EUKARYOTIC CELLS

Ezzatollah Keyhani *

Laboratory for Cell Biology and Biochemistry
University of Tehran
Institute of Biochemistry and Biophysics
P.O. Box 314-1700
Tehran, Iran

One of the most intriguing and still unresolved problems in modern biology is the evolutionary discontinuity between prokaryotic and eukaryotic cells. During the past decade two distinct hypotheses have been proposed to account for the origin of intracellular organelles. According to the endosymbiotic hypothesis, mitochondria, chloroplasts, and centrioles and related 9 + 2 structures originated, respectively, from an invasion of a bacterium-like, a blue-green algae-like, and spirochete-like organism into protoeukaryote anaerobic-type cells.[1-7] The symbiont progressively transferred most of its genetic information into host cells during the course of evolution, and so evolved to the present-day status of mitochondria, chloroplast, or flagellum in eukaryotic cells.[1-7] A second hypothesis is that the intracellular organelles of eukaryotic cells arose by a gradual and continuous evolutionary process.[8-14]

Most evidence presented in favor of the symbiotic origin of mitochondria is derived from comparative studies of the structural and functional analogy between mitochondria and bacteria. Like bacterial protein synthesis, mitochondrial protein synthesis is inhibited by chloramphenicol but is insensitive to cycloheximide, which specifically inhibits cytoplasmic protein synthesis. Another argument of the proponents of symbiotic hypothesis has been the similarity between the structure and function of bacterial and mitochondrial protein synthesis machinery, but recent observations show a great number of dissimilarities between the two synthesis machineries (for review, see References 9–11). Both the structure and base composition of DNA and tRNA of mitochondria exhibit characteristics distinct from DNA and tRNA of bacterial origin.[15, 16]

A well-established fact is that the great majority of mitochondrial proteins are coded by nuclear genes, synthesized on cytoplasmic ribosomes, and subsequently transported into mitochondria. Present evidence indicates that the proteins of the outer mitochondrial membrane and the mitochondrial matrix, most of the proteins of the inner membrane, and most of the proteins involved in the replication, transcription, and translation of mitochondrial DNA are coded by nuclear DNA.[17] The genome of the mitochondrion has coding capacity for ribosomal RNA, tRNA, and a dozen polypeptides; among these are three large subunits of cytochrome *c* oxidase,[17] four subunits of ATPase,[18] and one subunit of cytochrome *b*.[19] Mitochondria thus have a limited autonomy in regard to the biosynthesis of their own proteins.

* Mailing address: Johnson Research Foundation G4, University of Pennsylvania School of Medicine, Philadelphia, Pa. 19104.

0077–8923/81/0361–0376 $01.75/0 © 1981, NYAS

Numerous reviews have been published in recent years, accounting for the argument proposed against the symbiotic hypothesis of mitochondrial origin. Of particular relevance are those of Raff and Mahler,[9-11] Uzzell and Spolsky,[12, 13] Bell,[8] Cavalier-Smith,[14] and Avadhani *et al.*[20] In this paper, only one aspect relevant to the argument in favor of nonsymbiotic origin of mitochondria will be developed.

The sequence for the symbiotic origin of the mitochondrion is that a gram-negative eubacterium containing an aerobic Krebs cycle invaded a fermentative anaerobic cell.[4] The stabilization of this once-predatory association led to the formation of the mitochondrion.[4] The nucleus and other endoplasmic membranes evolved autogenously after the steroid biosynthesis pathway was made available by the presence of this aerobic eubacterium.[4] However a convincing set of evidence is now available, that indicates that the host cell was aerobic and contained enzymatic systems for the metabolism of oxygen and several biochemical pathways requiring the presence of molecular oxygen.

1. The existence of respiratory pigments and oxidative phosphorylation in the nuclear membrane of a wide variety of cells is the best indication of the aerobic nature of the "protoeukaryotic" cells. TABLE 1 shows the enzymic properties of nuclear membrane.[21-46] Many enzymes present in the mitochondria are also found in the nucleus, including Krebs cycle enzymes[23, 24, 26-28] and cytochrome *c* oxidase.[25b, 33, 35, 37-39, 42-46] In addition, cytochrome *c* purified from nucleus and from mitochondria of calf thymus, were found to be indistinguishable from each other in spectral properties, redox potential, electrophoretic mobilities, amino-acid composition, and peptide maps.[34]

2. Molecular oxygen is required for the synthesis of sterols such as cholesterol and steroid hormones. It is also necessary for the formation of polyunsaturated fatty acids. These molecules play a key role in the metabolism of eukaryotes.

3. Cytoplasmic catalase and superoxide dismutase are necessary for enzymatic protection against oxygen toxicity. Catalase and superoxide dismutase are ubiquitous cytoplasmic enzymes in all aerobic organisms and play a vital role in the prevention of oxygen toxicity. These enzymes protect the cell against the potential degradation by the superoxide radical O^{2-}, which is generated in a variety of enzymic and nonenzymic reactions that utilize molecular oxygen.[47]

These observations indicate that protoeukaryotic cells were aerobic and that the basic argument of symbiotic hypothesis concerning the anaerobic nature of protoeukaryotes does not stand. It scarcely seems useful for an aerobic system to have an aerobic symbiont. On the other hand, the arguments for the origin of mitochondria by evolution of the components of the protoeukaryotic cell, are more compelling.[8-14]

Origin of Mitochondria From Intracellular Differentiation

Various cytoplasmic organelles were believed to be involved in the origin of the mitochondria. The formation of mitochondria from the nuclear envelope has been suggested by Hoffman and Grigg,[48] Brandt and Papas,[49] Bell and Muhlethaler,[50] Pannese,[51] and more recently, Keyhani.[52] Other organelles believed to be involved in the origin of mitochondria include the formation from microbodies,[53-55] plasma membrane,[56] endoplasmic reticulum (for review,

TABLE 1

OXIDATIVE PHOSPHORYLATION AND RESPIRATORY ENZYMES IN NUCLEAR MEMBRANE

Tissue	Enzymes	Inhibitors	Reference
Calf thymus	N.L.*	KCN, NaN_3, Dinitrophenol	21
Rat thymus spleen intestinal mucosa bone marrow lymph node	N.L.*	N.S.	22
Rat liver, Pig kidney, and Ox brain	Isocitrate dehydrogenase, Malate dehydrogenase, and α-G-P-dehydrogenase	N.S.	(23 and 24)
Rat liver	Succinate oxidase	Dinitrophenol	25a
kidney	Cytochrome *c* oxidase, DPN cyt. *c* reductase, and DPNH neotetrazolium reductase	Anoxia, NaN_3, Chloropromazine, and Dicoumarin	25b
Calf thymus	Isocitrate dehydrogenase, Succinate dehydrogenase, and Malate dehydrogenase	KCN, NaN_3, Dinitrophenol, Amytal, DNAse, Fluoroacetic acid, and Dehydroacetic acid	26–28
Rat thymus	N.L.*	Rotenone	29
Rat liver	N.L.*	Dinitrophenol, KCN, Antimycin A, and Amytal	30
Rat liver	Cytochromes *c*, b_5, and Flavin	Antimycin A	31
Rat thymus	N.L.*	DNase, Rotenone, Amytal, Oligomycin, Dinitrophenol, Antimycin A, and CO	32
Rat thymus	Cytochromes aa_3, *b*, *c*	N.S.	33
Calf thymus	Cytochrome *c*	N.S.	34
Calf thymus	Cytochromes aa_3, *b*, *c*	N.S.	35
Rat liver	Cytochrome *c*	N.S.	36

Rat liver	Cytochrome *c* oxidase, Glutamate dehydrogenase, NADH cyt. *c* reductase, NADPH cyt. *c* reductase, ATPase, NADH oxidase, and succinate cyt. *c* reductase	Oligomycin, Dinitrophenol, and Gramidicin	37
Calf thymus	Cytochromes aa_3, c, c_1, b, b_5	CO	38
Bovine liver	NADH cyt. *c* reductase, NADPH cyt. *c* reductase, succinate cyt. *c* reductase, NADH oxidase, and Cytochrome *c* oxidase	DNAse, Rotenone, Amytal, Antimycin A, and Piericidin A	39
Bovine liver	Cytochromes aa_3, b_5	N.S.	40
Rat liver	DPNH cyt. *c* reductase (rotenone insensitive), cytochrome b_5, and TPNH cyt. *c* reductase	N.S.	41
Bovine liver	Cytochrome *c* oxidase, NADH oxidase, NADH cyt. *c* reductase, and Mg^{2+}-stimulated ATPase	N.S.	42
Calf thymus	Cytochrome *c* oxidase, Cytochromes c, c_1, and *b*-type	N.S.	43
Rat liver thymus †	Cytochromes aa_3, c, c_1 b, Succinate oxidase (trace), Succinate PMS reductase (trace), Rotenone sensitive NADH, Cytochrome *c* reductase, Rotenone insensitive NADH, and Cytochrome *c* reductase	Rotenone, CO, KCN, NaN_3, Sulfide hydroxylamine	44
Rat liver	Cytochrome *c* oxidase and NADH cyt. *c* reductase	Rotenone	45

N.L.=Not localized. N.S.=Not studied.
* The authors' studies refer to nuclear phosphorylation.
† Rotenone-sensitive NADH cytochrome *c* reductase is absent in the nuclear membrane of thymus.

see Reference 51), and Golgi complex.[57] Most of these observations were done in the period when the techniques of preparation, as well as the resolution of electron microscopy, were inadequate and so have been subjected to numerous criticisms. In order to clarify this point, the association between mitochondria and nuclei in the spermatocytes of the Mexican Axolotl *Ambystoma mexicanum* and the yeast *Candida utilis* were examined by electron microscopy using newer preparation techniques and higher resolution than heretofore.

Spermatocyte of the Mexican Axolotl Ambystoma Mexicanum

The association between the nucleus and mitochondria can be visualized in all stages of spermatocyte development. FIGURE 1 shows, at low magnification, portions of a spermatocyte containing an abundant number of mitochondria, usually in clusters throughout the cytoplasm. Some of the clusters of mitochondria (arrow) are directly apposed to the nuclear membrane. Morphologically, these associations can be classified in three categories. 1) The mitochondria are tightly in contact with the nucleus, in such a way that it is not possible to determine where the mitochondria stop and nucleus starts (FIGURE 2). Such an association gives the impression that the mitochondria are budding from the nucleus. In other instances, despite such a tight association it is possible to delineate the limits of nuclear and mitochondrial membranes (FIGURES 3 & 4). 2) Mitochondrion-like structures are located between the outer and inner nuclear membranes (FIGURE 3). In a well-oriented and sectioned specimen the outer nuclear membrane is continuous with the outer mitochondrial membrane (FIGURE 4, solid arrow), or fusion between the two outer membranes can be seen (FIGURE 4, open arrow). And finally, 3) in some cells a relatively large mitochondrion is in part linked to the nuclear membrane, with the suggestive image of the formation of a "pore" between the two organelles (FIGURES 5 & 6).

FIGURES 7 through 9 (see also DIAGRAM 1) illustrate the suggestive sequence of the formation of a mitochondrion from the nuclear membrane: 1) evagination of the nuclear membrane (FIGURE 7); 2) internalization (FIGURE 8), in which the inner nuclear membrane encloses a space within the evagination, thus forming the matrix of the mitochondria; 3) formation of cristae (FIGURE 9) and maturation; and 4) externalization refers to the process separating mitochondria from nucleus, and is characterized by the appearance of a "pore" between mitochondria and nucleus before complete separation.

Association of Nucleus and Mitochondria in the Aerobic Yeast Candida Utilis

FIGURES 10, 11, and 12 show various stages in the formation of the mitochondrion from the nuclear membrane of *Candida utilis* yeast cells. FIGURE 10 shows the early formation of mitochondria from nuclear membrane. In FIGURE 10 the early stage, a clear continuity between outer nuclear membrane and future mitochondria is observed. FIGURE 11 shows the formation of cristae; the outer nuclear membrane (arrow) is still in continuity with outer mitochondrial membrane. FIGURE 12 probably shows the final stage of separation of mitochondria from nuclear membrane.

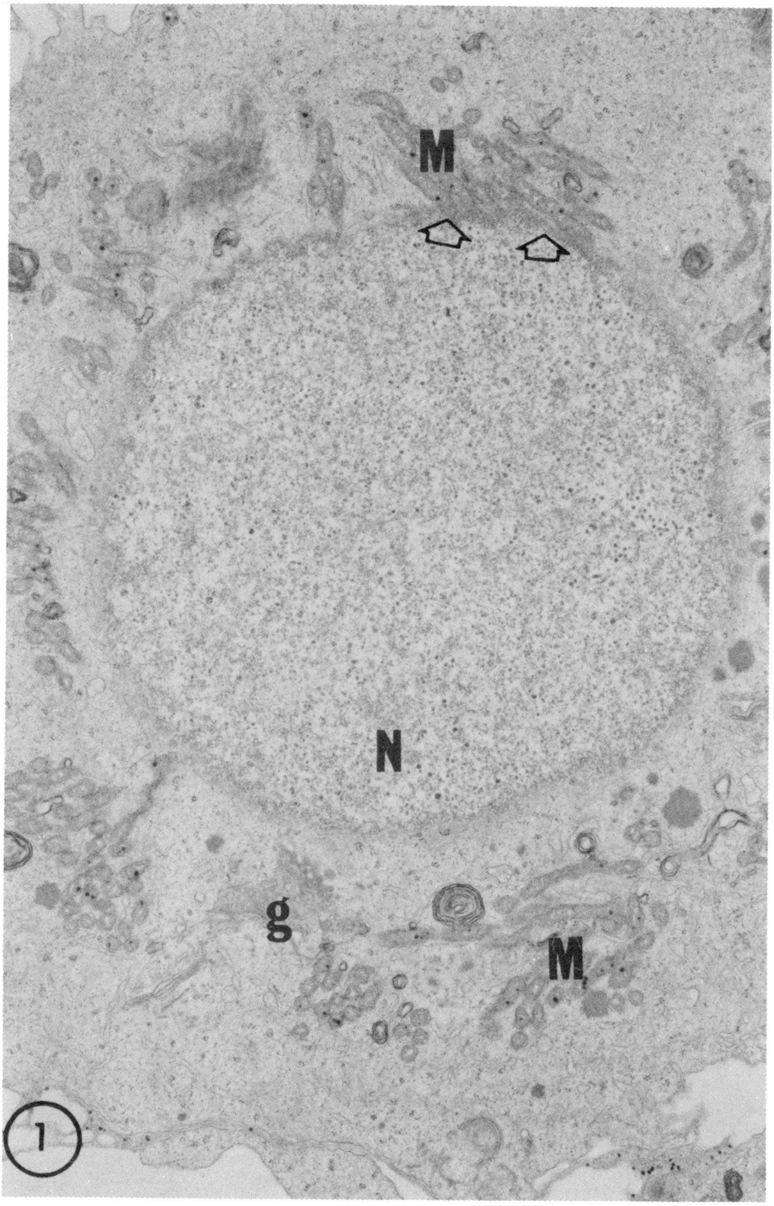

FIGURE 1. Spermatocyte of Mexican Axolotl, *Ambystoma mexicanum* at low magnification. The mitochondria (M), usually in clusters, are scattered throughout the cytoplasm. Arrows indicate a cluster of mitochondria in contact with nucleus (N). g : Golgi apparatus. Magnification ×12,150.

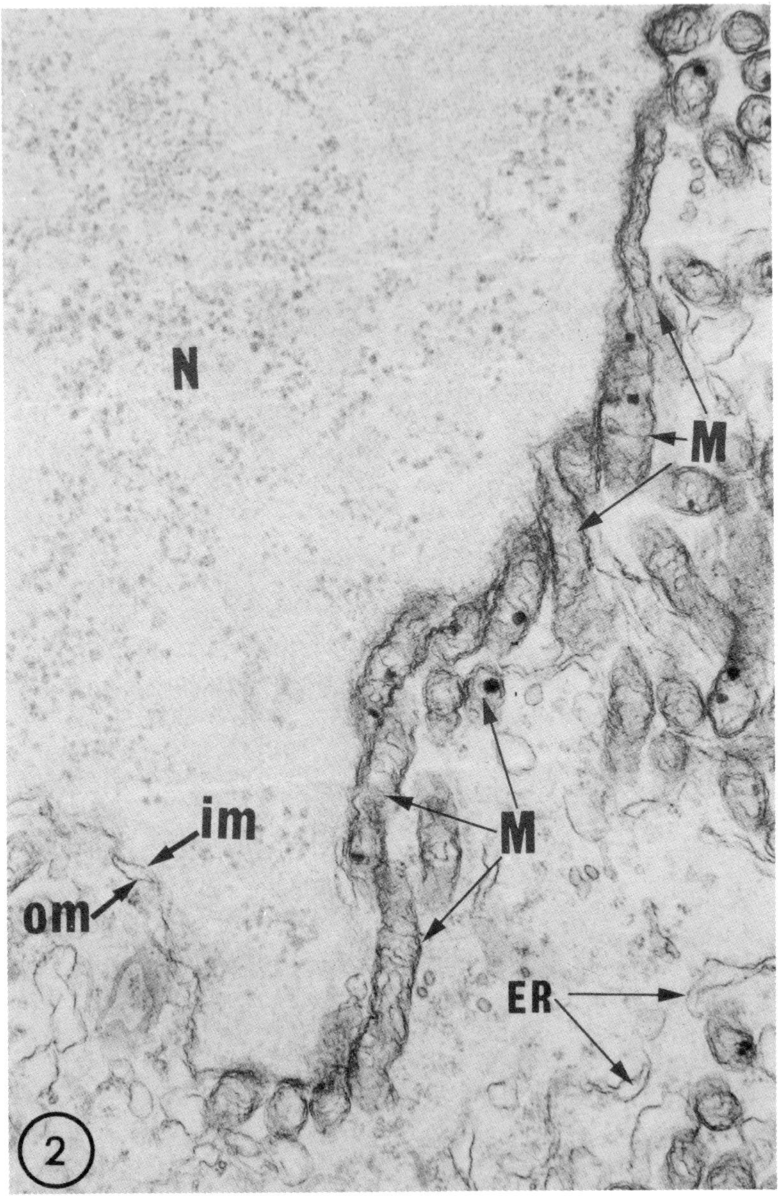

FIGURE 2. This figure shows a portion of nucleus (N) of a spermatocyte and its associated mitochondria (M). The mitochondria in this electron micrograph give the impression of budding from nucleus. om : outer nuclear membrane; im : inner nuclear membrane; ER : endoplasmic reticulum. Magnification ×27,900.

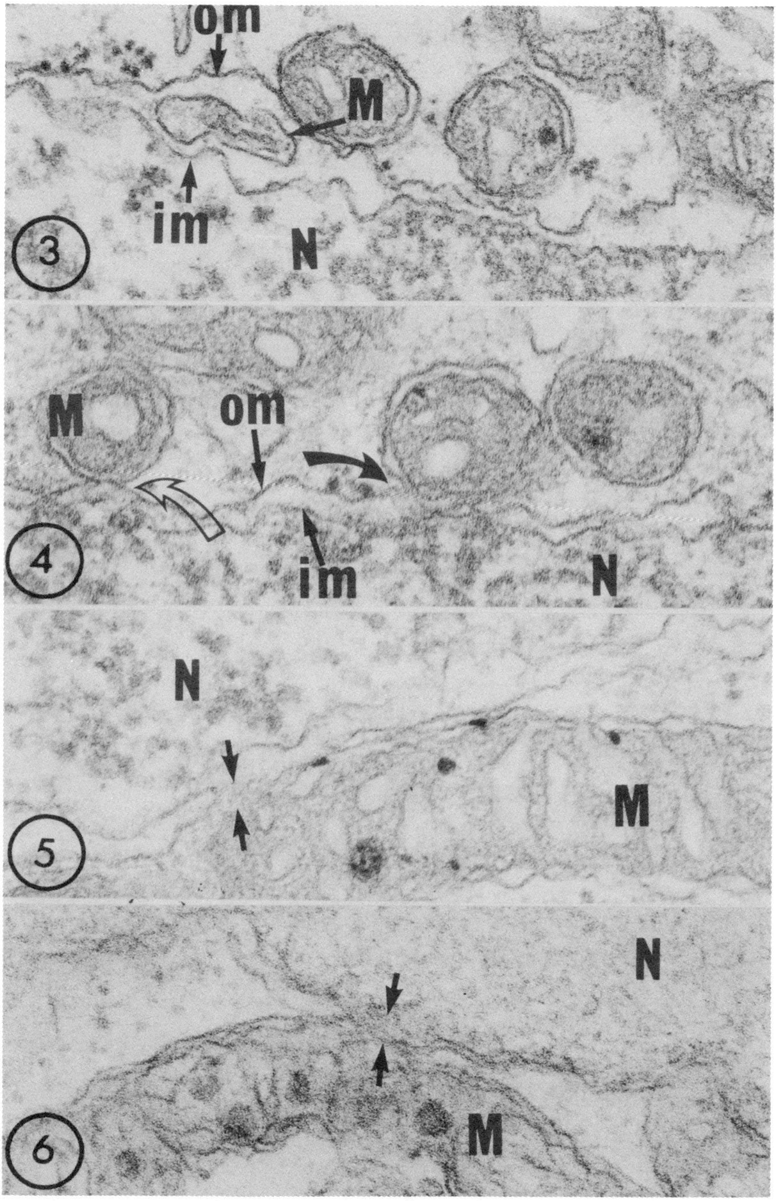

FIGURE 3–6. Diverse modes of association between nucleus and mitochondria in spermatocyte. FIGURE 3. A mitochondrion-like structure (M) between the outer nuclear membrane (om) and inner nuclear membrane (im). N : nucleus. Magnification ×68,400. FIGURE 4. The fusion between outer nuclear membrane (om) and outer mitochondrial membrane (open arrow). On the middle right side of the picture the outer nuclear membrane (om) is continuous with outer mitochondrial membrane (plain arrow). im : inner nuclear membrane; N : nucleus. Magnification ×91,800. FIGURES 5 and 6. The formation of pore (arrows) between nucleus (N) and mitochondria (M) in spermatocyte. Magnification ×91,800.

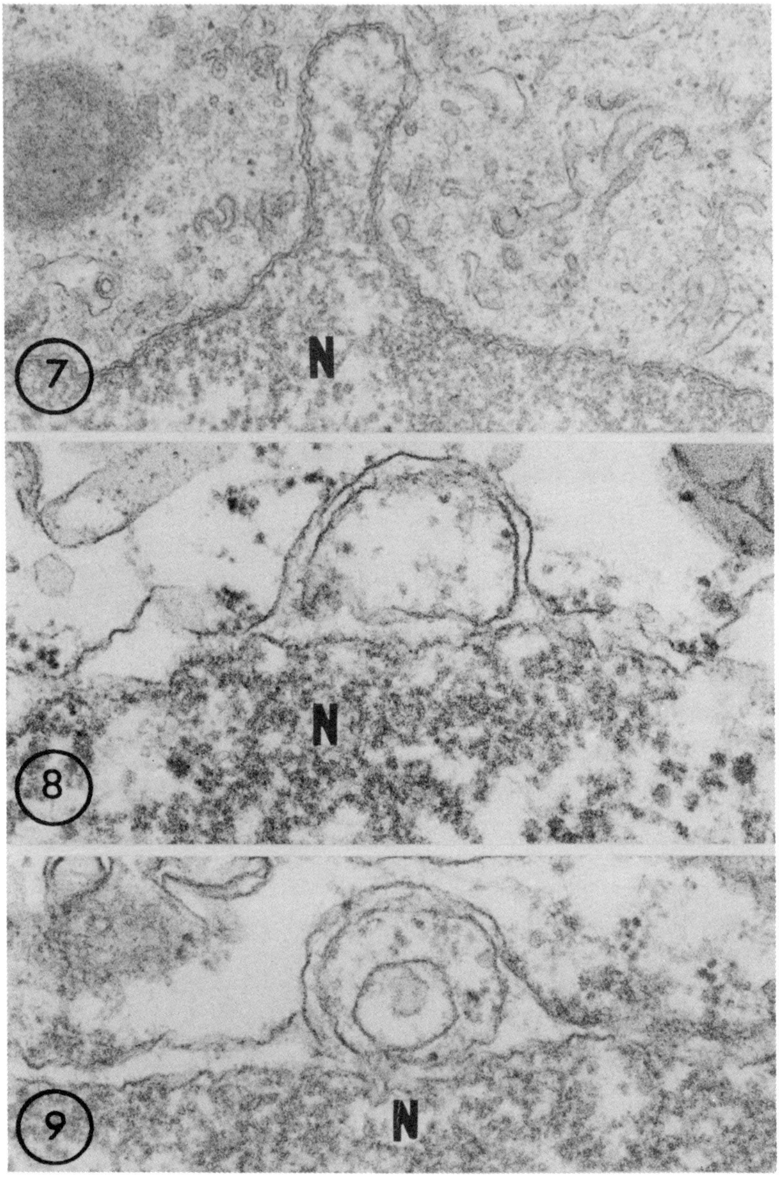

FIGURES 7–9. Sequence of the formation of mitochondria from nucleus in spermatocyte. FIGURE 7. Evagination (Magnification ×27,900). FIGURE 8. Internalization (Magnification ×68,400). FIGURE 9. Maturation (Magnification ×68,400).

SIGNIFICANCE OF THE ASSOCIATION BETWEEN NUCLEUS AND MITOCHONDRIA

The associations between nucleus and mitochondria raise three fundamental questions.

1) Are associations the product of artefacts? The intimate contact between nuclear and mitochondrial membranes bears little resemblance to an adventitious contact due to mechanical response, including cell shrinkage due to the fixation procedure. Although the possibility exists that pathological or other factors may have induced a condition favorable for tight contact between nuclear membrane and mitochondria, the spermatogonial tissues appear to have been healthy and free of preparative artefacts. Alternatively it may be argued that the illustrated images are the product of the deregulation of mitosis, so that occasionally a mitochondrion is "caught up" with the nuclear membrane. Perhaps a change in the nuclear-cytoplasmic volume ratio could be a more likely causal influence in the contact between nucleus and mitochondria. However the presence of cytoplasmic regions unoccupied by any organelles, the presence of mitochondria between the nuclear membranes, and apparent fusion

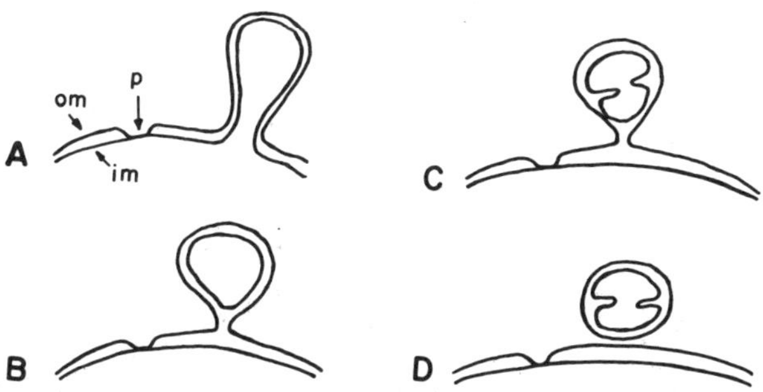

DIAGRAM 1. Formation of mitochondria from nuclear membrane in the spermatocyte of Mexican Axolotl *Ambystoma mexicanum* and in the yeast *Candida utilis*. A : Evagination; B : internalization; C : maturation and D : externalization. OM : outer nuclear membrane; IM : inner nuclear membrane; P : nuclear pore.

of others with the nuclear membrane are against such possibilities. In eleven male Axolotl studied for several years, such contact was found to be a natural phenomenon. Presumably these observations can be interpreted as supporting the hypothesis that the mitochondria arise by intracellular differentiation.

2) Are such associations an indication of the involvement of mitochondria in the biogenesis or metabolism of the nucleus? It is very unlikely that mitochondria have any participation in the genesis of the nucleus. A survey of the literature indicates the absence of any metabolic reason(s) for the association of the two organelles. Of course it is possible to suggest several interpretations to this purpose, e.g. mitochondria in this way provide energy in the form of ATP for the nucleus. But the transfer of these small molecules does not necessitate such an extensive physical link to the nucleus.

3) Are such associations an indication of the involvement of nucleus in the biogenesis of the mitochondria? Such a question cannot be answered with certitude at this stage of investigation. However it is possible to reconstruct the sequence of events leading to the formation of mitochondria from nuclear membrane as illustrated in FIGURES 7–9.

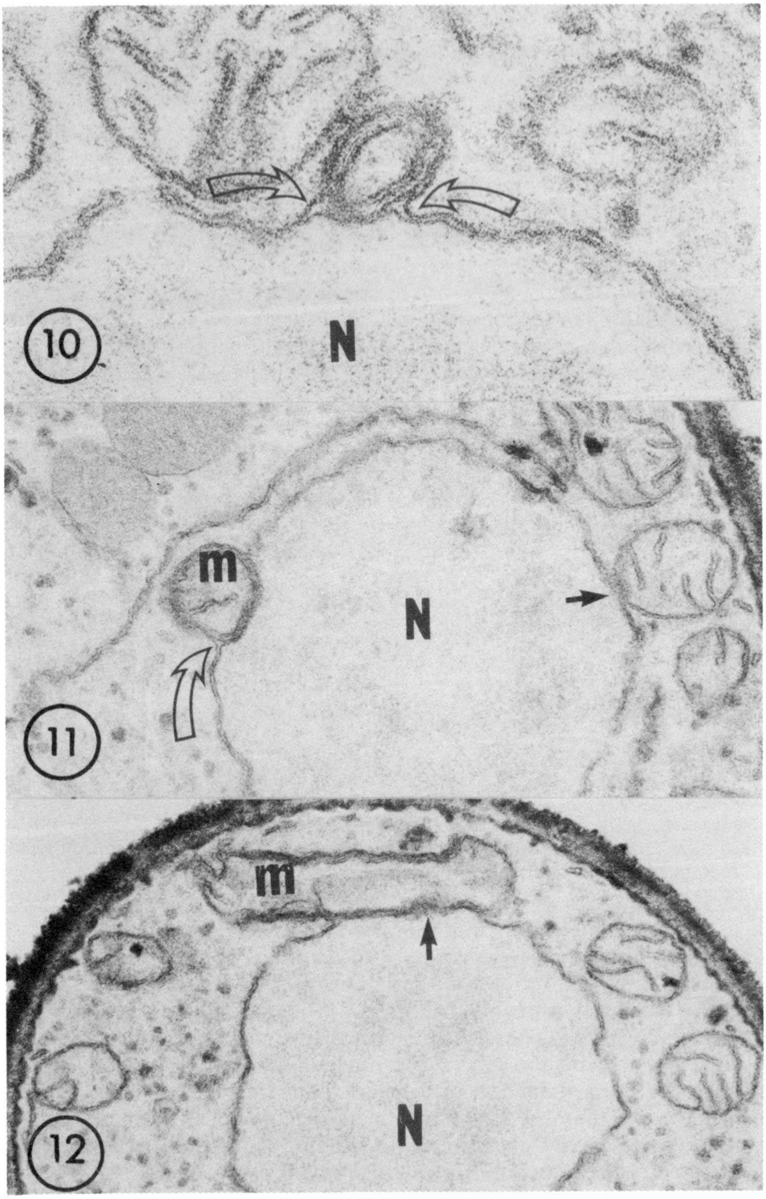

FIGURES 10–12. Association between nucleus and mitochondria in yeast *Candida utilis*. FIGURE 10. Early phase in the formation of mitochondria. Note the continuity between outer nuclear membrane and outer mitochondrial membrane. N : nucleus. Magnification ×74,700. FIGURE 11. Formation of cristae in the mitochondria (m) associated with nucleus (N). Magnification ×32,400. FIGURE 12. Pore (arrow) between mitochondria (m) and nucleus (N). Magnification ×28,800.

Nonsymbiotic Models (Diagram 2)

The Bell Model[8]

The transition from prokaryote to eukaryote could have taken place more or less instantaneously as a consequence of physical principles identical to those that probably led to the first cell. Assuming that the plasma membrane had been suddenly subjected to hypertonic conditions, the membrane would have been compressed and the cell shrunk. Assuming that the contents of the cell

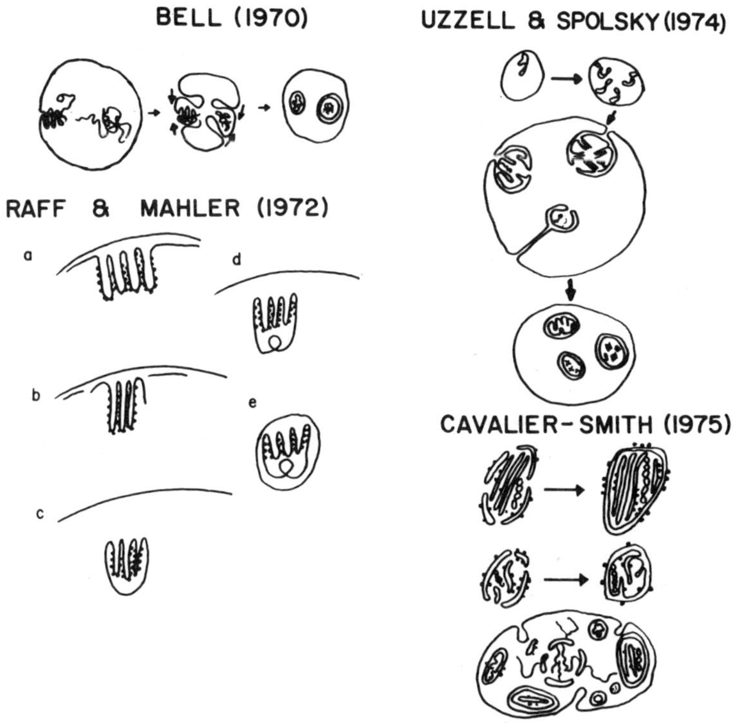

Diagram 2. Various models (see text) of the nonsymbiotic origin of mitochondria and chloroplasts. Redrawn after References 8, 9, 13, and 14.

were more viscous than the surrounding environment, a sphere or cylinder (surrounded by two layers of membrane) would then have been formed within prokaryotes as the cell surface area was reduced. Such collapse would presumably have arisen in the structurally simpler area, and followed the path of least resistance in the cell. The collapse on each side of the respiratory membrane system and the site of attachment of chromosomes to plasma membrane produced, respectively, the mitochondria and the nucleus. Ultimately these structures are surrounded by additional membrane, exhibiting the present day mitochondrion and nucleus.

The Raff and Mahler Model[9]

As the protoeukaryote becomes large, a more extensive respiratory surface becomes necessary. This necessity produces an invagination of the plasma membrane possessing respiratory functions. The invagination subsequently separates from the plasma membrane and forms a topologically closed respiratory organelle. Subsequent to this system, a plasmid containing genes for ribosomal components and some elements of the respiratory membrane becomes integrated. The final step is the acquisition of an outer membrane.

The Uzzell and Spolsky Model[13]

According to Uzzell and Spolsky, the initial step in the origin of the nuclear and organellar structure was genome duplication without cell division. The cell membrane then invaginated near the genome attachment site to form double membrane structures, with each genome attached to the inner surface of the inner membrane of such structures. Initially nuclear and organellar structures contained a complete genome; all may have had membrane-bound respiratory and photosynthetic activities. With time, increasing differentiation of function occurred; nuclear membranes lost photosynthetic and respiratory activities, which became sequestered in plastids and mitochondria, respectively. Organellar genomes lost many duplicated genes, while the nuclear genome through additional duplication increased in size and differentiated tremendously.

The Cavalier-Smith Model[14]

According to this author, the plastids, mitochondria, and nuclei evolved by fusion of DNA-associated thylakoids or endoplasmic reticulum cisternae. The plastids resulted from an association of photosynthetic and plasmid-associated thylakoids that formed compartments containing DNA, ribosomes, and Calvin-cycle enzymes. Mitochondria resulted from fusion of plasmid-associated respiratory thylakoids to form a compartment containing Krebs cycle and fatty acid metabolizing enzymes.

The Locked-Cell Hypothesis

There is ample indication that we will not convincingly resolve the origin of mitochondria in the near future and one can neither prove nor disprove the symbiotic or nonsymbiotic hypotheses. Therefore our present goal is restricted to arriving at an intellectually satisfying and biochemically acceptable account for the origin of this organelle.

It is possible that the first stage in formation of the eukaryotic cell was the organization of a well-defined nucleus surrounded by membrane and separated from plasma membrane by the cytoplasm. Once the cell had acquired a stable nuclear structure and surrounding cytoplasm, it evolved toward the present day eukaryotic cell by progressive refinement and differentiation from the preexisting structure.

It is possible to find the formation of the nucleus by examining the life of present day bacteria. The sporulation of bacterial cell may provide an insight into how the chromosome came to be separated by plasma membrane. FIGURE 13 shows an electron micrograph of the sporulating cell at stage III. At this stage morphologically the enclosure of the bacterial genome by plasma membrane gives a striking resemblance to present day nucleus of eukaryotic cell.

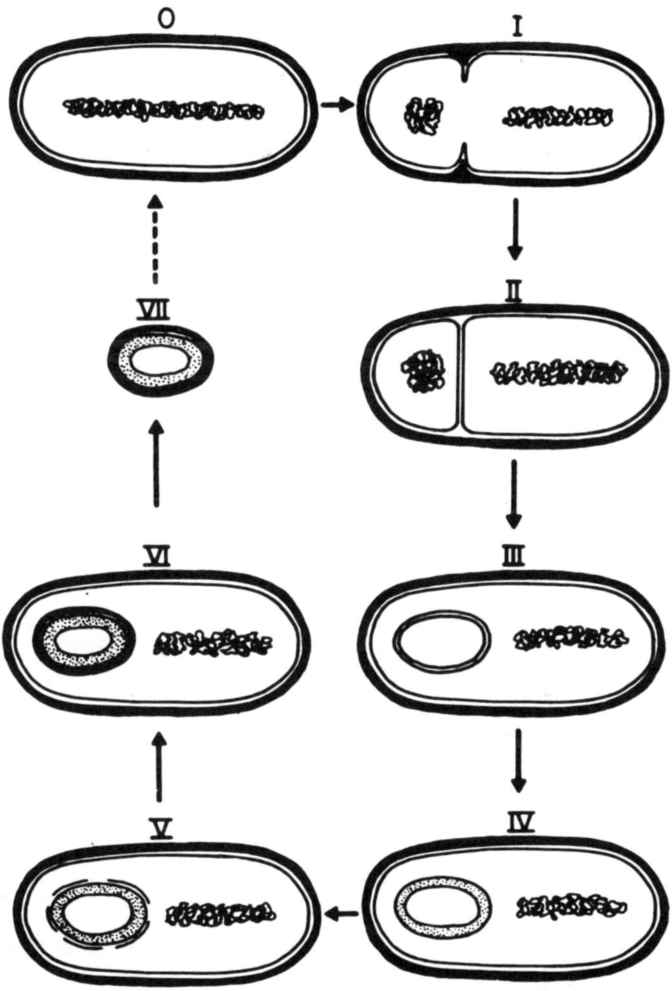

DIAGRAM 3. Various stages of sporulation, according to Dr. Hoch.[74] Reproduced with his permission.

The basic pattern of morphological sequence of events leading to spore formation seems to be similar in different species of sporulating bacteria.[58-63] Both strictly aerobic[58] and anaerobic[64] bacteria are subject to sporulation. DIAGRAM 3 shows the various stages of the sporulation and the mechanism by which hereditary information is enclosed by the plasma membrane of the bacteria. A similar process could have been used for the formation of the eukaryotic nucleus.

The first stage in the formation of a spore is the replication of bacterial chromosome (stage 0). This stage is followed by stage I, in which a replicate of chromosome migrates to one pole of the cell. At stage II there is a forespore septum formation involving the mesosome. In stage III, the double membrane of the septum leads to the formation of the prespore. At this stage the chromosome is surrounded by double membrane. Subsequently at stage IV, a cortex of mucopeptide material forms between the double membrane layers of the prespore, leading to stage V in which the spore coat is formed around the outer prespore membrane. After the maturation (stage VI), the spore formation is completed and the lysis (stage VII) produces the liberation of free spores.

At stages III and IV we are essentially in the presence of an eukaryotic-resembling cell (see also FIGURE 13) in which the genetic information is sequestered in a double membrane system originated from the plasma membrane and a cytoplasm, and is active by synthesizing diverse elements to complete formation of spore.

The biochemistry of the sporulating cell has some resemblance to that of eukaryotes and differs from that of the vegetative cell. There are diverse biochemical indications that the metabolism of vegetative cells has a pattern distinct from that of sporulating cells. In a number of sporulating aerobic bacteria, the tricarboxylic acid cycle (TCA) enzymes are absent from vegetative cells,[65–70] and TCA enzymes are not necessary for vegetative development of the cells. At the beginning of sporulation, the TCA enzymes are synthesized and are necessary for spore development. The TCA enzymes utilize the acetic acid and pyruvic acid formed from glycolysis during the vegetative period. Cells in which mutations affect the formation of Krebs cycle enzymes do not sporulate. Mutants deficient in aconitase[71] are unable to synthesize glutamate and have lost the capacity to sporulate even when the required amino acid was supplied to the medium at the beginning of the growth. Freese *et al.*[72] suggested that in citric acid cycle mutants, sporulation is prevented because not enough ATP is available to allow sporulation to take place. An interesting point concerning the metabolism of sporulating cells is the existence of a relatively stable mRNA.[71] Such characteristics of sporulating cells resemble those of present-day eukaryotic cells.

The genetic analysis of bacterial sporulation leads to the location of the genes involved in this process. In *Bacillus subtilis* there are approximately 700 genes involved in the process of sporulation.[73, 74] These genes are scattered along *Bacillus subtilis* chromosome, and some of them are clustered into an operon-like unit (~42). Many of these genes are structural genes and code for specific enzymes, but most of them are probably regulatory genes which may be involved in the control of the synthesis and function of the products of these structural genes. Various types of mutants have been obtained which showed a specific blockage of the stage of sporulation. The sporulation mutants blocked at any of the stages show no morphological structure characteristic of a stage later than the block. In addition when an early gene does not function, there is no biochemical expression of the genes later in the sequence.[74] For example, a blockage at stage III, in which the double membrane of forespore is completed, prevents the deposition of mucopeptide between the double membrane.

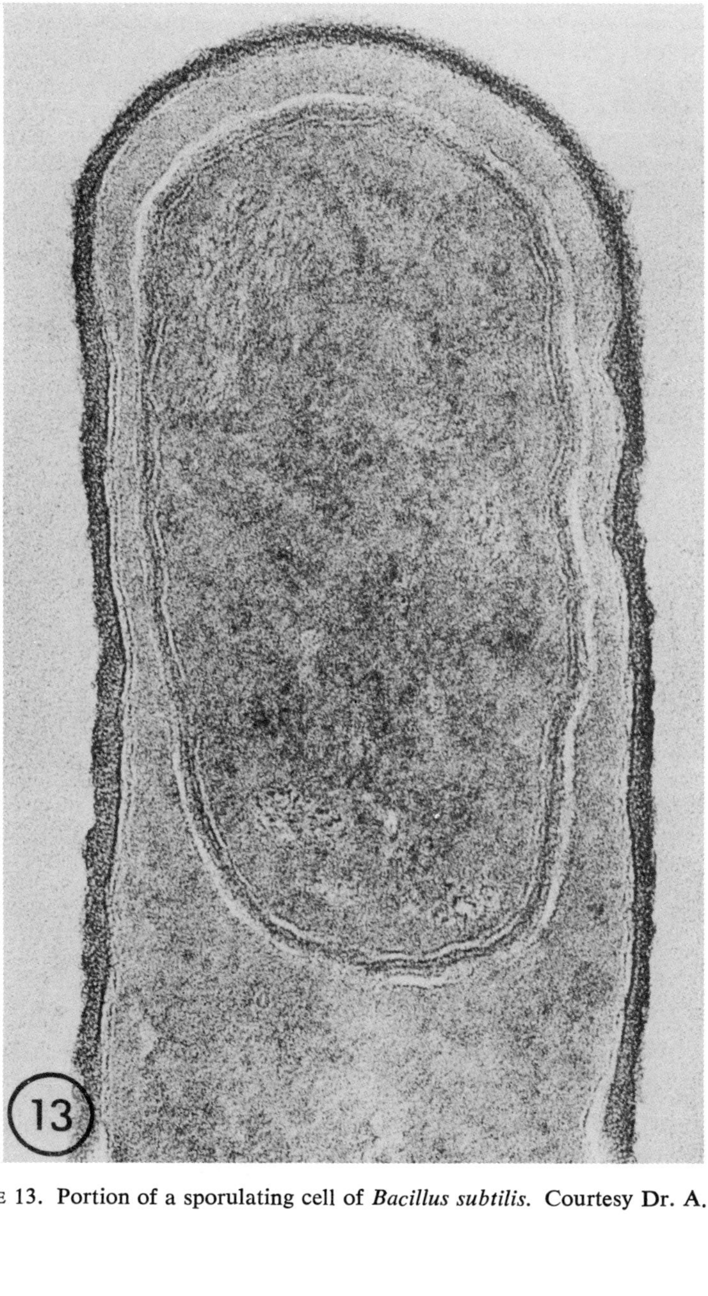

FIGURE 13. Portion of a sporulating cell of *Bacillus subtilis*. Courtesy Dr. A. Ryter.

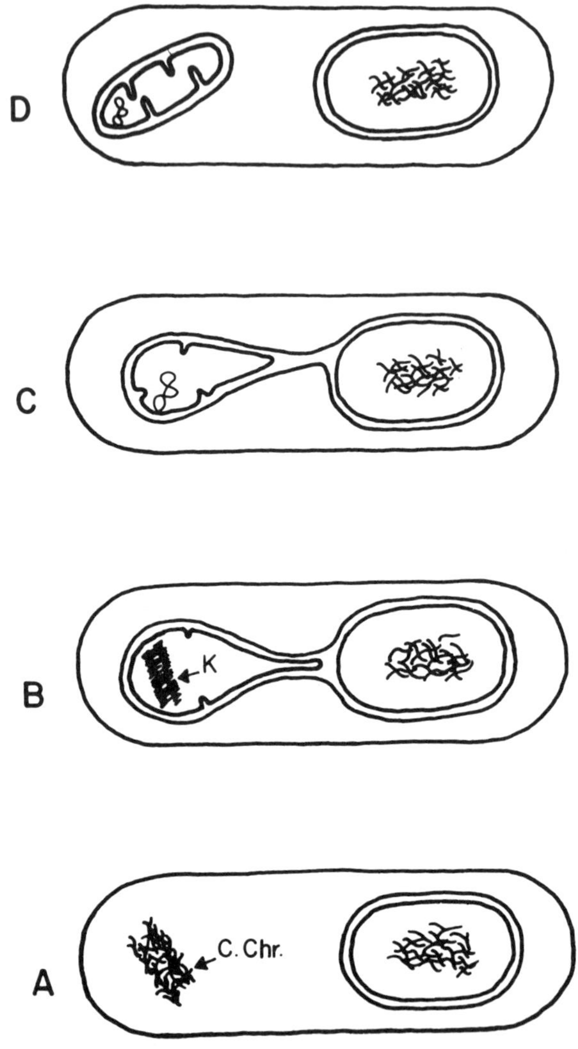

DIAGRAM 4. Hypothetical events leading to the formation of present-day eukaryotic cells. A : The locked cell exhibiting both nucleus and cytoplasmic chromosome (C.Chr.); B : formation of mitochondria by evagination of nuclear membrane enclosing the cytoplasmic chromosome which forms the kinetoplast (K); C : loss of kinetoplast, but preservation of physical link with nucleus; and D : loss of physical link with nucleus.

In essence the mutated cells have three possibilities: they can revert to the vegetative state, lyse, or remain in the blocked state. My view is that this blocked cell ("locked cell") provided a possible basis for future eukaryotic cells by gradual differentiation over millions of years. The gap between prokaryotes and eukaryotes exists because the locked cell either had reverted or evolved to the present day eukaryotic cell. Thus the locked sporulating cell had a nucleus with genetic information, a cytoplasm containing the remains of bacterial chromosome (normally destined for disintegration), a Krebs cycle enzymes system, and, most important, a volume several times larger than the bacteria from which it originated. This required an increasing fitness to the new situation, promoting specialization and giving rise to adaptation to a new form of metabolic activity in which both energy and nutritional requirements were satisfied. Without change, the cells would either have died or survived in a state unsuitable for the dynamism of evolution, since evolution is impossible without change and continuity. The diversity of genotype (and phenotype) that can be produced by the mutation of sporulating cells is almost unimaginable, thus providing an enormous pool for natural selection.

After formation of the locked cell, the cytoplasmic chromosome became enclosed by an extension of the nuclear membrane, possessing respiratory functions (DIAGRAM 4). Concomitantly, changes in the structure of some area of nuclear membrane produced nuclear pores, which allowed greater exchange between nucleus and its surrounding cytoplasm. An early division of labor between these two genetic systems (nuclear, and cytoplasmic enclosed in the membrane from nucleus) led to the properties of present-day mitochondria and nuclei. The biosynthesis of most mitochondrial proteins came under the control of nuclear DNA and the redundant genes present in chromosomes of mitochondria were eliminated. A residual system was maintained that provided the machinery necessary for the synthesis of hydrophobic polypeptides and perhaps regulatory proteins, repressors, and inducers for the control of mitochondrial protein synthesis. Such a system was subject to less stringent evolutionary pressure and so preserved most of its primitive properties, while the nucleus responded to the more stringent evolutionary pressure of the environment and so produced the modifications of properties seen in the present-day eukaryotes. The kinetoplast is reminiscent of this cytoplasmic chromosome which is lost in other eukaryotic cells; indeed the spontaneous loss of the kinetoplast has been shown in some hemoflagellates.[75]

Obviously the model implies that a copy of the mitochondrial genome is present in nuclear DNA. Such a master copy [76, 77] has been postulated, as a result of hybridization experiments [78, 79] but opinion is not unanimous concerning its presence in the nucleus.[80] More sensitive hybridization methods should eventually resolve this question.

REFERENCES

1. SAGAN, L. 1967. J. Theor. Biol. **14**:225–274.
2. MARGULIS, L. 1970. Origin of Eucaryotic Cells. Yale University Press. New Haven, Conn.
3. MARGULIS, L. 1971. Sci. Am. **225**(2):48–57.
4. MARGULIS, L. 1975. Symp. Soc. Exptl. Biol. **29**:21–28.
5. STANIER, R. Y. 1970. Symp. Soc. Gen. Microbiol. **20**:1–38.
6. TAYLOR, F. JR. 1974. Taxon **23**:229–258.

7. Cohen, S. S. 1970. Am. Sci. **58:**281–289.
8. Bell, P. R. 1970. Symp. Soc. Exptl. Biol. **24:**109–128.
9. Raff, R. A. & H. R. Mahler. 1972. Science **177:**575–582.
10. Raff, R. A. & H. R. Mahler. 1975. Symp. Soc. Exptl. Biol. **29:**41–92.
11. Mahler, H. R. & R. A. Raff. 1975. *In* International Review of Cytology. G. H. Bourne & J. F. Danielli, Eds.: **43:**1–124.
12. Uzzell, T. & C. Spolsky. 1973. Science **180:**516–517.
13. Uzzell, T. & C. Spolsky. 1974. Amer. Sci. **62:**334–343.
14. Cavalier-Smith, T. 1975. Nature (London) **256:**463–468.
15. Bernardi, G. 1978. Nature (London) **276:**558–559.
16. Miller, D. L., N. C. Martin, H. D. Pham & J. F. Donelson. 1979. J. Biol. Chem. **254:**11735–11740.
17. Schatz, G. & T. L. Mason. 1974. Ann. Rev. Biochem. **43:**51–87.
18. Tzagoloff, A., M. S. Rubin & M. F. Sierra. 1973. Biochim. Biophys. Acta **301:**71–104.
19. Weiss, H. & B. Ziganke. 1974. Eur. J. Biochem. **41:**63–71.
20. Avadhani, N. G., F. S. Lewis & R. J. Rutman. 1975. Sub-Cell. Biochem. **4:** 93–145.
21. Osawa, S., V. G. Allfrey & A. E. Mirsky. 1957. J. Gen. Physiol. **40:**491–513.
22. Creasey, W. A. & L. A. Stocken. 1959. Biochem. J. **72:**519–523.
23. Siebert, G. 1961. Biochem. Z. **334:**369–387.
24. Siebert, G., K. H. Bässler, R. Hannover, E. Adolff & R. Beyer. 1961. Biochem. Z. **334:**388–400.
25a. Rees, K. R. & G. F. Rowland. 1961. Biochem. J. **78:**89–95.
25b. Rees, K. R., F. R. Hazel & G. F. Rowland. 1962. Biochem. J. **83:**523–527.
26. McEwen, B. S., V. G. Allfrey & A. E. Mirsky. 1963. J. Biol. Chem. **238:** 758–766.
27. McEwen, B. S., V. G. Allfrey & A. E. Mirsky. 1963. J. Biol. Chem. **238:** 2571–2578.
28. McEwen, B. S., V. G. Allfrey & A. E. Mirsky. 1963. J. Biol. Chem. **238:** 2579–2586.
29. Betel, I. & H. M. Klouwen. 1963. Biochem. Biophys. Acta **76:**327–329.
30. Penniall, R., J. Saunders & S. M. Liu. 1964. Biochemistry **3:**1459–1465.
31. Conover, T. E. & G. Siebert. 1965. Biochem. Biophys. Acta **99:**1–12.
32. Betel, I. & H. M. Klouwen. 1967. Biochem. Biophys. Acta **131:**453–467.
33. Betel, I. 1967. Biochem. Biophys. Acta **143:**62–69.
34. Yamagata, S. & R. Sato. 1968. J. Biochem. (Tokyo) **64:**549–556.
35. Ueda, K., T. Matsuura, N. Date & K. Kawai. 1969. Biochem. Biophys. Res. Commun. **34:**322–327.
36. Davidian, N., R. Penniall & W. B. Elliott. 1969. Arch. Biochem. Biophys. **133:**345–358.
37. Zbarsky, I. B., K. A. Perevoshchikova, L. N. Delektorskaya & V. V. Delektorsky. 1969. Nature (London) **221:**257–259.
38. Conover, T. E. 1970. Arch. Biochem. Biophys. **136:**541–550.
39. Berezney, R., L. K. Funk & F. L. Crane. 1970. Biochem. Biophys. Res. Commun. **39:**93–98.
40. Berezney, R. & F. L. Crane. 1971. Biochem. Biophys. Res. Commun. **43:** 1017–1023.
41. Kasper, C. B. 1971. J. Biol. Chem. **246:**577–581.
42. Berezney, R., L. K. Macaulay & F. L. Crane. 1972. J. Biol. Chem. **247:** 5549–5561.
43. Matsuura, T. & K. Ueda. 1972. Arch. Biochem. Biophys. **150:**440–450.
44. Jarasch, E. D. & W. W. Franke. 1974. J. Biol. Chem. **249:**7245–7254.
45. Sikstrom, R., J. Lanoix & J. J. M. Bergeron. 1976. Biochim. Biophys. Acta **448:**88–102.
46. Zbarsky, I. B. 1978. *In* International Review of Cytology. G. H. Bourne & J. F. Danielli, Eds.: **54:**295–360.

47. Fridovich, I. 1974. *In* Horizons in Biochemistry and Biophysics. E. Quagliariello, F. Palmieri & T. P. Singer, Eds.: **1:**1–37. Addison-Wesley. London.
48. Hoffman, H. & G. W. Grigg. 1958. Exptl. Cell Res. **15:**118–131.
49. Brandt, P. W. & G. D. Pappas. 1959. J. Biophys. Biochem. Cytol. **6:**91–96.
50. Bell, P. R. & K. Mühlethaler. 1964. J. Cell Biol. **20:**235–248.
51. Pannese, E. 1966. J. Ultrastruct. Res. **15:**57–65.
52. Keyhani, E. 1976. J. Cell Biol. **70:**81a.
53. Belt, W. D. 1957. Anat. Rec. **127:**264.
54. Roth, L. E. 1957. J. Biophys. Biochem. Cytol. **3:**985–999.
55. Rouiller, C. & W. Bernhard. 1956. J. Biophys. Biochem. Cytol. **2**(Suppl.): 355–359.
56. Geren, B. B. & F. O. Schmitt. 1954. Proc. Natl. Acad. Sci. USA **40:**863–870.
57. Lever, J. D. 1956. J. Biophys. Biochem. Cytol. **2**(Suppl.):313–317.
58. Rytter, A., P. Schaeffer & H. Ionesco. 1966. Ann. Inst. Past. **110:**305–315.
59. Fréhel, C. & A. Rytter. 1969. Ann. Inst. Past. **117:**297–311.
60. Kay, D. & S. C. Warren. 1968. Biochem. J. **109:**819–824.
61. Walker, P. D. 1970. J. Appl. Bact. **33:**1–12.
62. Decker, S. & S. Maier. 1975. J. Bact. **121:**363–372.
63. Bechtel, D. B. & L. A. Bulla, Jr. 1976. J. Bacteriol. **127:**1472–1481.
64. Grain, J. & J. Senaud. 1976. J. Ultrastruct. Res. **55:**228–244.
65. Hanson, R. S., V. R. Srinivasan & H. O. Halvorson. 1963. J. Bacteriol. **86:** 45–50.
66. Hanson, R. S., J. Blicharska & J. Szulmajster. 1964. Biochem. Biophys. Res. Commun. **17:**1–7.
67. Halvorson, H. O. 1965. Symp. Soc. Gen. Microbiol. **15:**343–368.
68. Freese, E. & P. Fortnagel. 1967. J. Bacteriol. **94:**1957–1969.
69. Fortnagel, P. & E. Freese. 1968. J. Bacteriol. **95:**1431–1438.
70. Bulla, L. A., Jr., G. St. Julian & R. A. Rhodes. 1971. Canad. J. Microbiol. **17:**1073–1079.
71. Szulmajster, J. 1973. Symp. Soc. Gen. Microbiol. **23:**45–83.
72. Freese, E., P. Fortnagel, R. Schmitt, W. Klofat, E. Chappelle & G. Picciolo. 1969. *In* Spores. L. L. Campbell, Ed. Vol. **4:**82–101. American Society for Microbiology. Bethesda, Md.
73. Piggot, P. J. 1973. J. Bacteriol. **114:**1241–1253.
74. Hoch, J. A. 1976. *In* Advances in Genetics. E. W. Caspare, Ed. **18:**69–98. Academic Press, Inc. New York, N.Y.
75. Simpson, L. 1972. *In* International Review of Cytology. G. H. Bourne & J. F. Danielli, Eds. **32:**139–207.
76. Wilkie, D. 1963. J. Mol. Biol. **7:**527–533.
77. Whittacker, P. A., R. C. Hammond & A. A. Luha. 1972. Nature (London) **238:**266–268.
78. Starti, R. V. & J. H. Sinclair. 1974. Biochemistry **13:**4447–4455.
79. Wintersberger, E. & G. Vienhauser. 1968. Nature (London) **220:**699–702.
80. Fukuhara, H. 1970. Mol. Gen. Genetics **107:**58–70.

Discussion of the Paper

Dr. J. M. Olson: Where you outlined the evidence for respiratory function in nuclei—and there's always a question of, when you're looking at organelles, the level of contamination by other organelles—could you say anything about the level of contamination of the nuclear material by mitochondrial material in those experiments?

Dr. J. Keyhani: They indicated that they have the electron microscope and they did not find any contamination by the mitochondria. They also do gradient centrifugation so you can expect that they can isolate nucleus from the mitochondria.

Dr. D. E. Green: I think this problem of demonstrating electron transfer activity in the nuclear membrane is one that we live with in mammalian cells; there have been many reports. All that work has come to a standstill because the electron microscopic evidence has been shown to be inconclusive in that it is well known that different kinds of membranes will fuse so that one appears to be derived from the other. The only real evidence that one can rely upon is the isolation of the pure membrane and the demonstration that that pure membrane has in fact those activities, which of course hasn't been done in this particular test. So, I would say that one has to weigh the evidence of electron microscopy; it's interesting, suggestive, and sometimes it's more dramatic than it really is because if you take sarcoplasmic reticulum and show very beautiful associations with mitochondria (I've done that myself), you can almost deduce that sarcoplasmic reticulum arises from mitochondria or vice versa. That kind of evidence is very circumstantial.

Dr. Keyhani: I agree with you when you say it's suggestive evidence and I use it as suggestive evidence of course.

Dr. Chiang: I'm wondering whether you have analyzed whether the organisms you studied show some budding-type activities of the nuclear membrane to form possibly mitochondria? There are small circular DNAs in the nucleus. These are well known in eukaryote cells. Their small circular molecule is similar to those in mitochondria. My point was whether or not some of the small circular DNA that has been discovered in the nucleus can possibly account for precursor to mitochondria.

Dr. Keyhani: Yes. The model that I have presented shows that the mitochondrial genome is present in the nucleus. Such a copy, which has been called master copy, has been postulated as a result of hybridization experiments but the opinion is not unanimous concerning its presence in the nucleus. There are very good hybridization experiments which have been done to show that there is a copy of the mitochondrial gene in the nucleus.

Dr. R. Sheinin: I wonder if you considered two other possibilities in regard to your data with the axolotl and the yeast. One is simply a functional association between the mitochondria and the nucleus in terms of energy requirements. The other possibility is what is the chromosomal distribution of the nuclear genes which code for mitochondrial information in these two organisms? Are they localized to a single chromosome or a piece of a chromosome, and is there any possibility that in some of these organisms some of the genetic information contributing to mitochondrial DNA replication moves from the nucleus out in this way?

Dr. Keyhani: Yes I did consider the significance of the association. The thing that I saw is that mitochondria provide energy to the nucleus as you pointed out. When you say energy you say ATP, and ATP is a small molecule. I didn't find any justification, in my opinion of course, that such extensive association is required for a small molecule like ATP to go into the nucleus. The second question, it is not known where the genome is localized, at which kind of chromosome. Because the amount of the gene that is in the mitochondria is very small, there have been developed very sensitive hybridization experiments and these have been done, but some people don't believe it because the radioactivity that has been recovered is not very high.

SOME POSSIBLE CHEMOTAXONOMIC APPROACHES TO UNDERSTANDING CYANOBACTERIAL TAXONOMY AND PHYLOGENY

Raymond W. Holton

Department of Botany
The University of Tennessee
Knoxville, Tennessee 37916

INTRODUCTION

Although a few blue-green algae (now called cyanobacteria) had been studied by plant physiologists, particularly Fogg and his students in England, prior to 1955, no organism easily cultured and manipulated was available for biochemical studies until the isolation of *Anacystis nidulans* (=*Synechococcus*, ATCC 27144, PCC 6301 of Rippka *et al.*[1]) by Kratz and Myers.[2] Since then the availability of *Anacystis* and other cyanobacteria in axenic culture, coupled with new chromatographic and electrophoretic techniques that were being used with higher plants and animals, made possible the use of biochemical techniques to supplement and supplant the traditional botanical taxonomic approaches being used. Although the cyanobacteria are prokaryotic, they had been all but ignored by bacteriologists perhaps in part because the traditional bacteriological nutritional criteria for classification were not usable with these organisms, many of which are obligate photoautotrophs.

For the historical perspective it is of interest to point out the possible value and concomitant risk of the chemotaxonomic approaches by mentioning fatty acid compositional data obtained on *Anacystis nidulans* more than fifteen years ago in a study done to examine the effect of temperature on fatty acid composition.[3] The unexpected finding that only monounsaturated (and no polyunsaturated) fatty acids were present was the first such detailed report on an oxygen-evolving photosynthetic organism and contradicted the statement that appeared at the same time concerning *Anabaena variabilis*, which was said to have polyunsaturated acids, and led to a generalization that blue-greens were like higher plants in their fatty acid composition as well as oxygenic photosynthesis.[4] Generalizations from either *Anacystis* or *Anabaena* in this case gave erroneous inferences for the whole group but led me to a general interest in biochemical approaches to their taxonomy. These additional studies, when coupled with other published reports, led me to suggest that the complexity of the fatty acid patterns seemed to be correlated with the morphological complexity.[5] It has been most interesting to see the use of this property and other biochemical characters aid the impressive efforts of Stanier and his associates to bring greater order to cyanobacterial taxonomy.[1, 6]

The utilization of biochemical characters of the cyanobacteria for phylogenetic and taxonomic purposes has been of great interest in recent years as evidenced by the number of recent reviews,[7–10] although earlier ones [11, 12] were particularly influential and noteworthy in developing my own interest in these approaches. My intention here is to select primarily recent examples of biochemical studies or neglected older work that might have future utility. This

0077–8923/81/0361–0397 $01.75/0 © 1981, NYAS

review is biased towards selection of examples that emphasize the diversity of biochemistry among the cyanobacteria on the one hand but on the other make mention of some well-known and some less-known affinities with higher green algae and plants. Finally I will pay special attention to the unique qualities of *A. nidulans* among those of the cyanobacteria.

Nucleic Acids and Nucleases

I will first mention briefly several newer and widely known approaches, including RNA sequence studies and DNA compositional studies. The genome size of cyanobacteria of some 128 strains has been shown by Stanier and collaborators [13] to range from 1.6×10^9 to 8.6×10^9 daltons, which the authors point out is "a remarkable range of genetic complexity," and that the upper limit far exceeds the largest genome previously described in prokaryotes. Not only is the genome size range broad, but the DNA base composition is broad, ranging from 35 to 71 mol % G+C, which is nearly as great a range as in all the bacteria known.[14] A different approach based on DNA/DNA hybridizations show that the percent relatedness of *A. nidulans* to a *Plectonema boryanum* of the LPP group [1] is about 6%, which is about the same as *Plectonema's* relatedness to *Bacillus subtilis*.[15] In the same studies, no genotypical differences between LPP strains could be detected, i.e., they were ~100% related to each other.

Sequence studies on the 16S ribosomal ribonucleic acids by Doolittle and colleagues [16] are also impressive achievements. However, the results based on association coefficients derived from oligonucleotide catalogues show more diversity among the eight cyanobacteria than in all the *Bacillus* or enteric-vibrio species catalogued. Indeed *A. nidulans* is "remote" from other coccoid strains examined as well as from filamentous cyanobacteria. In fact the coccoid strains including *A. nidulans* were as closely related to the red alga *Porphyridium* chloroplasts as they were to the other cyanobacteria.

In a related type of study [17] involving a comparison of 5S RNA sequences, *A. nidulans* was the only example of the cyanobacteria included. It was shown to be somewhat distantly related to other prokaryotes and even farther from higher plants. But how representative this organism is of the cyanobacteria is questionable, although the difficulties of carrying out these kinds of analyses understandably makes it impossible to examine the number of different strains that would be desirable.

Determination of the RNase and DNase activities in extracts of 5 cyanobacteria showed very low activities for *A. nidulans* compared to that from filamentous strains.[18] Because the low activities of *A. nidulans* extracts is not typical of either bacteria or plants, additions of possible activators and attempts to remove possible inhibitors were made but all were ineffectual.

Photosynthesis and Carbohydrate Metabolism

The pathway of photosynthetic CO_2 fixation is also one of key importance to these phototrophic organisms. Although C_3 fixation would appear to be the primary pathway according to some investigators,[19] others have proposed that the C_4 pathway may occur at least under typical environmental CO_2 concentrations and that photorespiration does not occur except under elevated CO_2

conditions used in experimental situations.[20-22] Further studies with additional strains are needed here before any phylogenetic importance is clarified.

A most interesting development is the discovery of facultative anoxygenic photosynthesis in some cyanobacteria.[23] Approximately one-half of the 21 strains screened could photoassimilate CO_2 in the presence of S^{2-}, the PS II inhibitor, DCMU, and light absorbed only by PS I. Unfortunately, there was no obvious taxonomic distinction between those which could and could not as both filamentous and unicellular strains were in each group. However, *A. nidulans* (*Synechococcus* 6301) has not been examined, and only one heterocystous strain has been studied; but there is some hope that this physiological capacity will be of taxonomic use.[24]

Reserve polysaccharide and its metabolism in *Anacystis nidulans* has been studied by Wöber and colleagues. The branched α-D-glucan from *A. nidulans, Escherichia coli,* and *Arthrobacter* sp., sweet corn, and shell fish were treated with an α-amylase to selectively hydrolyze all (1 → 6)-α-D linkages and the resulting linear chains subjected to gel-permeation chromatography. The elution profiles for *A. nidulans* glycogen and phytoglycogen were very similar to each other but were only superficially similar to the two bacterial glycogens.[25] However, no other cyanobacteria appear to have been studied.

With relation to the biosynthesis of the glucan, Levi and Preiss[26] have demonstrated that ADP-glucose was the primary sugar nucleotide used for glycogen synthesis in *A. nidulans* and is thus like other bacteria rather than higher plants for which UDP-glucose is also active. However, the enzyme is activated by glycerate-3-phosphate and P_i, as in green algae and higher plants, but not by the compounds (pyruvate and fructose phosphates) that activate the bacterial enzymes. Interestingly enough, Fredrick had shown some years earlier that either ADP-glucose or UDP-glucose was active in the filamentous strain, *Oscillatoria princeps.*[27] Thus in *A. nidulans* polysaccharide metabolism has a range of properties many of which are like higher plants but some others like the bacteria. Studies on other more "advanced" cyanobacteria would be of interest in this regard.

Properties of another enzyme of carbohydrate metabolism recently characterized from *Anacystis nidulans* may be of taxonomic interest. A specific fructose 1,6-bisphosphatase (FBPase) was purified by Bishop,[28] and it did not fall into either the heterotrophic or the autotrophic class of FBPases known. It does not closely resemble the FBPase from the lithoautotrophic prokaryote, *Thiobacillus neapolitanus* either. Because it has characteristics of both auto- and heterotrophic FBPases, Bishop suggests it may have roles in both types of metabolism.

The fructose-1,6-diphosphate aldolase studies done some years ago have resolved this enzyme into two classes with the cyanobacterial enzyme being in Type II, which is more typical of bacteria.[10] What is puzzling is that chloroplast aldolase is Type I rather than the expected Type II, although the issue has yet to be clearly resolved. It should be noted that *Anacystis nidulans* aldolase has some properties that are distinctive from other prokaryotes, and indeed they resemble those of yeast aldolase.[29]

Lipids

Another old observation concerns quinone composition. As Ragan and Chapman[10] point out, the belief that the quinone composition of cyanobacteria was different from that of higher plants was due to misleading quinone data on *A. nidulans*, which lacks α-tocopherol and β-tocopherol quinones. These compounds are found in *Anabaena variabilis* and are implicated in the protection of polyunsaturated acids from peroxidation in membranes; *Anacystis* lacks polyunsaturated acids, so the absence of the tocopherols makes sense.[30, 31] *A. nidulans* is an "atypical" representative of the group. But it also is the only unicellular form thus far studied except for *Synechococcus lividus*, (Synechococcus PCC 61717),[1] in which plastoquinone complement was similar to that of other cyanobacteria, but other quinones were not investigated.[32] Because *S. lividus* lacks polyunsaturated fatty acids,[6] it would be expected to lack the tocopherol quinones as does *A. nidulans*, but further analyses are necessary to ascertain if useful relationships can be obtained from quinone data.

Other lipid compounds, including sterols and carotenoids, have been thought to be potentially useful in systematics, but as yet applications of them have not been rewarding.[10]

Electron Transport Proteins

Low molecular weight proteins have been used for phylogenetic considerations by comparison of the primary sequences. Ferredoxin is an example of one which has been isolated and characterized from several higher plants, cyanobacteria, and other prokaryotes. Not only are the cyanobacterial ferredoxins similar to those of higher plants, but they differ significantly from those of other prokaryotes.[10] In fact a recent sequence of ferredoxin from *Mastigocladus laminosus* (a branched heterocystous thermophile of Section V of Rippka *et al.*,[1] and called by them, *Fischerella*) is similar to that of other filamentous cyanobacteria.[33] It is of interest to note that the sequence of ferredoxin from the red alga, *Porphyra umbilicalis*, is more similar to that of the filamentous cyanobacteria, *Spirulina*, than *Spirulina* is to a unicellular cyanobacterium, *Aphanothece*.[34] The sequences are still few, but it is clear that the blue-greens have an affinity to the higher plants rather than other prokaryotes in this characteristic. The immunochemistry of the cyanobacterial ferredoxins confirm this closer relationship of the cyanobacteria to the higher algae rather than to the bacteria.[35]

Other electron transport proteins for which some use in taxonomy and phylogeny are possible include superoxide dismutase (SOD), cytochrome *c*, and plastocyanin. The SOD homologies in the amino terminal sequences seem to be great whether the metal is Fe or Mn, as in the prokaryotes and higher plants, or Cu or Zn, as in eukaryote animals and fungi.[36, 37] It would be of interest to have amino terminal sequence information of ferredoxins from unicellular cyanobacteria such as *Anacystis* to complement those known from two filamentous strains.

A discovery by Wood[38] that in algae, including cyanobacteria, a bound cytochrome *f* could be released from the membranes by appropriate treatment has shown this cytochrome to be distinct from the long-known water-soluble cytochrome *c*-553 (or -554). Cytochrome *c*-553 has been recently isolated from other cyanobacteria.[39] Additional work of Wood[40] demonstrated that the copper protein, plastocyanin, and the cytochrome *c*-553 are functionally interchangeable. Krogmann[41] has compared the effectiveness as electron donors of plastocyanin and cytochrome *c*-553 in their affinities for spinach P700. The compounds tested can be divided into three groups based on isoelectric point (IEP). Those with IEPs below 5 included four higher plant and higher algal electron donors, while the cyanobacteria occurred in two groups—one with IEPs of near 5 and another with IEPs near 8. Krogmann suggests that the IEP may be useful in classifying the nearness of cyanobacteria to the eukaryotes. *Spirulina maxima*, a filamentous organism of Section III of Rippka *et al.*[1] with helical trichomes, is the closest to spinach on this basis, while *Aphanizomenon flos-equae*, also a filamentous alga, is most distant. As with ferredoxin, the amino acid sequences of the three cyanobacterial cytochrome *c*-553's and the one red algal cytochrome *c*-553 examined show that the cytochrome from *S. maxima* is closer to that from the red alga, *Porphyra umbilicalis*, than it is to another cyanobacterium, *A. flos-aquae*.[41] Both reactivity and sequence studies may be instructive when additional data are available.

Another approach to comparing cytochromes *c*-553 (and *c*-554) is by measuring their reactivity to nitrite reductase of *Pseudomonas aeruginosa* and cow cytochrome oxidase.[42] Although *Spirulina platensis* cytochrome *c*-553 was reactive with similar relative rates to those of higher algae, reactivity of *Anacystis nidulans* cytochrome *c*-554 with the nitrite reductase was less than 20% that of *S. platensis*. None of the cyanobacterial or algal cytochromes reacted with the oxidase.

Of possible value for purposes of comparison is the presence of a low potential, carbon monoxide-binding *c*-type cytochrome. First isolated from *Anacystis nidulans*[43] it has recently been found in five other blue-greens, but in various amounts.[39] It is also known from the diatom, *Navicula pelliculosa*,[44] and recently from another diatom, *Phaeodactylum tricornutum*,[45] and from the green marine alga, *Bryopsis maxima*.[46] It also has been described from a marine *Nostoc*.[47] Studies of sequences and comparison to bacteria may be useful, and the possible involvement of this cytochrome in anoxygenic photosynthesis with S^{2-} has been suggested by Ho *et al.*[39]

Nitrogen and Sulfur Metabolism

The cyanobacteria possess interesting variations, of possible phylogenetic significance, in what little is known of their amino acid metabolism. Examples include the path of tyrosine synthesis in which an intermediate, pretyrosine, is found instead of 4-hydroxyphenylpyruvate, which is present in the pathway of eukaryotic organisms and nonphotosynthetic bacteria.[48, 49] These authors suggest that the cyanobacterial pathway is the more primitive in an evolutionary sense. On the other hand, it should be noted that the tryptophan biosynthesis pathway is similar in the bacteria and cyanobacteria in that they are present as discrete aggregates but in other organisms different aggregations are found.[10]

Another example from amino acid metabolism is that glutamine synthetase (GS) from *A. nidulans* differs from bacterial GS by not being adenylated nor membrane-bound.[50] In higher plants GS is not adenylated,[51] and thus control mechanisms must be by other means such as metabolite levels. The enzyme from *Anabaena* sp.[52] appears to be much like that from *A. nidulans*.

Sulfate reduction enzymes also offer the presence of divergent pathways. Schmidt[53, 54] has shown that there are two different kinds of sulfotransferases in the cyanobacteria. The sulfonucleotide, PAPS (3′-phosphoadenosine-5′-phosphosulfate), is required for one group which includes *A. nidulans* (*Syne-*

chococcus 6301), *Spirulina,* and *Synechocystis* and is also required in the bacteria. On the other hand, *Plectonema* has a requirement for APS which is like *Chlorella,* spinach, and other higher plants.

Photoreversibly Photochromic Pigments

A kind of physiological response known in higher algae and plants that has a counterpart in the cyanobacteria is based on the green-red light antagonism presumably moderated by a photoreversibly photochromic pigment, phytochrome, or something similar to it.[55] Not only is the phenomenon, chromatic adaptation, probably mediated by such a pigment, but also phycoerythrin synthesis and photomorphogenesis have action spectra suggesting a photoreversible pigment such as a phycobiliprotein. Isolation of these compounds has been attempted but is not yet to the stage of knowledge of phytochrome, which is a biliprotein. Further studies of isolated pigments will be necessary before detailed conclusions of homology can be drawn.

Isozymes

Isozymes have been investigated as possible tools for the study of taxonomy and phylogeny in a variety of organisms.[56] However, a 1973 review showed relatively few studies on algae in general or cyanobacteria in particular.[57] While there have been several reports of isozyme studies on blue-green algae since then, in the two primarily chemotaxonomic papers,[58, 59] only malic dehydrogenase, α-esterase, leucine aminopeptidase, and phosphatase have been investigated using starch gel and acrylamide techniques. Isozymes were found, but there was a lack of consistency of bands between one experiment and another on the same organism, so some questions must be raised about the reliability of the results in one study.[58] The possible effects of the developmental stage of the organism when present and of environmental conditions have not yet been intensively studied nor has a large enough number of samples been analyzed to give reliable statistical evaluations as has been done in other organisms.

On the positive side, several studies with isozymes have yielded interesting results. Fredrick's use of isozyme approaches has given interesting insights into relationships between the cyanobacteria and other algae groups. In a recent paper[60] he found several similar patterns of isozymes involved in starch metabolism in the cyanelles of *Glaucocystis nostochinearum* and in *Anacystis nidulans,* although the total numbers of proteins found in the cyanelle was less than in the cyanobacterium. Isoelectric focusing on acrylamide gel of crude extracts and staining for superoxide dismutase was used by Lumsden and Hall,[61] and three of the four filamentous cyanobacteria examples had a single enzyme with similar mobilities, while in *Spirulina platensis,* the enzyme had a different mobility.

It would be encouraging to be able to report studies on cyanobacteria that are as impressive as those done by Murphy and Guillard[62] in which clones of presumably the same diatom (*Thalassirosira*) isolated from waters surrounding different continents typically gave isozyme patterns that were the same or very similar. Cheney and Babbel[63] have assessed the degree of genetic differentiation among four species of the seaweed, *Eucheuma.* After analysis of the patterns for five isozymes, they were able to compare phenotypic and genetic identity statistics of nine populations of the four species collected. This technique permitted the authors to evaluate the morphological plasticity often seen in field collections. Unfortunately similar studies have not been done with the cyanobacteria, but because these prokaryotes are essentially haploid, the isozyme patterns are simplified since genic polymorphism, when present, cannot be accounted for by heterosis.

Although at this time, the use of isozyme approaches has been rare among the cyanobacteria, I believe that isozyme patterns may give some aid in taxonomic studies at the genus and species level. In addition, enzyme polymorphism may be useful in other ways. First, the complement of several isozymes might be used to monitor cell cultures and serve as "genetic signatures" as has been proposed for human cell cultures.[64] The various problems with the *Anacystis* cultures around the world points out the value of being able to identify organisms by properties other than morphology in addition to the more sophisticated biochemical approaches such as DNA base composition.

Another possible use of isozyme patterns is in evaluating the morphological variation seen under different growth conditions as in a recent study by Stan and Holleman of 25 cyanobacterial strains.[65] If it can be shown that the isozyme complement is maintained under different conditions and does not vary as the morphology does, isozyme patterns may be both useful in the field and the laboratory in identification of species or strains because the analyses are relatively easy and straightforward.

Conclusion

This brief review focuses on the biochemical diversity of the cyanobacteria, which has been recognized by those who have considered the question.[9,10] *Anacystis nidulans* has been used frequently to represent this group in comparative biochemical studies as well as in phylogenic relationships, but the evidence is abundant (Table 1) that it is an unusual or atypical member of the group. (As more strains are studied, the unusualness of *A. nidulans* may disappear!) In some of the biochemical studies cited, *A. nidulans* is closer to other nonphotosynthetic bacteria than it is to filamentous cyanobacterium such as *Spirulina*, which in turn sometimes has closer relationship to the chloroplasts of the red algae than to *A. nidulans*. It would seem that the affinities of the cyanobacteria to the bacteria on the one hand and the higher plants on the other might be better recognized by placing them in a separate phylum within the superkingdom, Prokaryota, as has been recently suggested by Whittaker and Margulis.[66]

Comparative biochemistry of the cyanobacteria should eventually lead to suggestions of the most likely candidates for endosymbiosis or even for rejection of that theory (see also Ragan and Chapman[10]). The recent evidence presented here does not bring us much closer to that goal but perhaps suggests some approaches that might be usefully pursued.

Acknowledgment

I thank Professor David W. Krogmann for giving me access to several of his unpublished manuscripts.

TABLE 1

UNUSUAL PROPERTIES OF *Anacystis Nidulans* (SYNECHOCOCCUS 6301)

System	Comment	Reference
DNA/DNA hybridization	~6% relatedness to DNA of two LPP strains	15
16S rRNA sequences	"remote" to other strains including unicells	16
RNase and DNase activity	low compared to filamentous strains	18
α-D-glucan fine structure	similar to corn but not to two bacteria	25
ADP-glucose pyrophosphorylase	its activator like green plants; its sugar nucleotide like bacteria	26
Fructose bisphosphatase	unusual combination of auto- and heterotrophic properties	28
Fructose-1,6-diP aldolase	Type II but with similarities to yeast enzyme	29
Quinones	lacks tocopherol quinones found in other strains	30
Cytochrome *c*-554 reactivity	~20% of other strains with *Pseudomonas* nitrite reductase	42
Glutamine synthetase	not adenylated as in bacteria; *Anabaena* enzyme similar	50

REFERENCES

1. RIPPKA, R., J. DERUELLES, J. B. WATERBURY, M. HERDMAN & R. Y. STANIER. 1979. Generic assignments, strain histories, and properties of pure cultures of cyanobacteria. J. Gen. Microbiol. **111**:1–61.
2. KRATZ, W. A. & J. MYERS. 1955. Nutrition and growth of several blue-green algae. Am. J. Bot. **42**:282–287.
3. HOLTON, R. W., H. H. BLECKER & M. ONORE. 1964. Effect of growth temperature on the fatty acid composition of a blue-green alga. Phytochemistry **3**:595–602.
4. ERWIN, J. & K. BLOCH. 1963. Polyunsaturated fatty acids in some photosynthetic microorganisms. Biochem. Z. **338**:496–511.
5. HOLTON, R. W., H. H. BLECKER & T. S. STEVENS. 1968. Fatty acids in blue-green algae: possible relation to phylogenetic position. Science **160**:545–547.
6. STANIER, R. Y., R. KUNISAWA, M. MANDEL & G. COHEN-BAZIRE. 1971. Purification and properties of unicellular blue-green algae (Order *Chroococcales*). Bacteriol. Rev. **35**:171–205.
7. LEWIN, R. A. 1974. Biochemical taxonomy. *In* Algal Physiology and Biochemistry. W. D. P. Stewart, Ed.:1–39. University of California Press. Berkeley and Los Angeles, California.
8. SMITH, A. J. & D. S. HOARE. 1977. Specialist phototrophs, lithotrophs, and methylotrophs: a unity among a diversity of procaryotes? Bacteriol. Rev. **41**:419–448.
9. STANIER, R. Y. & G. COHEN-BAZIRE. 1977. Phototrophic prokaryotes: the cyanobacteria. Ann. Rev. Microbiol. **31**:225–274.
10. RAGAN, M. A. & D. J. CHAPMAN. 1978. A Biochemical Phylogeny of the *Protists*. Academic Press, Inc. New York, N.Y.
11. AARONSON, S. & S. H. HUNTER. 1966. Biochemical markers and microbial phylogeny. Quart. Rev. Biol. **41**:13–46.
12. KLEIN, R. M. & A. CRONQUIST. 1967. A consideration of the evolutionary and

taxonomic significance of some biochemical, micromorphological, and physiological characters in the thallophytes. Quart. Rev. Biol. **42:**105–296.
13. HERDMAN, M., M. JANVIER, R. RIPPKA & R. Y. STANIER. 1979. Genome size of cyanobacteria. J. Gen. Microbiol. **111:**73–85.
14. HERDMAN, M., M. JANVIER, J. B. WATERBURY, R. RIPPKA, R. Y. STANIER & M. MANDEL. Deoxyribonucleic acid base composition of cyanobacteria. J. Gen. Microbiol. **111:**63–71.
15. STAM, W. T. & G. VENEMA. 1977. The use of DNA-DNA hybridization for determination of the relationship between some blue-green algae (Cyanophyceae). Acta Bot. Neerl. **26:**327–342.
16. BONEN, L., W. F. DOOLITTLE & G. E. FOX. 1979. Cyanobacterial evolution: results of 16S ribosomal ribonucleic acid sequence analysis. Can. J. Biochem. **57:**879–888.
17. HORI, H. & S. OSAWA. 1979. Evolutionary change in 5S RNA secondary structure and a phylogenic tree of 54 5S RNA species. Proc. Natl. Acad. Sci USA **76:**381–385.
18. NORTON, J. & J. S. ROTH. 1967. Some aspects of nuclease activity in *Anacystis nidulans* and other algae. Comp. Biochem. Physiol. **23:**361–371.
19. IHLENFELDT, M. J. A. & J. GIBSON. 1975. CO_2 fixation and its regulation in *Anacystis nidulans* (*Synechococcus*). Arch. Microbiol. **102:**13–21.
20. COLMAN, B. & J. R. COLEMAN. 1978. Inhibition of photosynthetic CO_2 fixation in blue-green algae by malonate. Plant Sci. Lett. **12:**101–105.
21. COLMAN, B., K. H. CHENG & R. K. INGLE. 1976. The relative activities of PEP carboxylase and RuDP carboxylase in blue-green algae. Plant Sci. Lett. **6:** 123–127.
22. DÖHLER, G. 1974. C_4-Weg der Photosynthese in der Blaualge *Anacystis nidulans*. Planta **118:**259–269.
23. GARLICK, S., A. OREN & E. PADAN. 1977. Occurrence of facultative anoxygenic photosynthesis among filamentous and unicellular cyanobacteria. J. Bacteriol. **129:**623–629.
24. PADAN, E. 1979. Facultative anoxygenic photosynthesis in cyanobacteria. Ann. Rev. Plant Physiol. **30:**27–40.
25. WEBER, M. & G. WOBER. 1975. The fine structure of the branched α-D-glucan from the blue-green alga *Anacystis nidulans*: comparison with other bacterial glycogens and phytoglycogen. Carbohydr. Res. **39:**295–302.
26. LEVI, C. & J. PREISS. 1976. Regulatory properties of the ADP-glucose pyrophosphorylase of the blue-green bacterium *Synechococcus* 6301. Plant Physiol. **58:**753–756.
27. FREDRICK, J. F. 1968. Biochemical evolution of glucosyl transferase isozymes in algae. Ann. N.Y. Acad. Sci. **151:**413–423.
28. BISHOP, R. H. 1979. Regulatory characteristics of a fructose bisphosphatase from the blue-green bacterium *Anacystis nidulans*. Arch. Biochem. Biophys. **196:** 295–300.
29. GIBBS, M., E. LATZKO, M. J. HARVEY, Z. PLAUT & Y. SHAIN. 1970. Photosynthesis in the algae. Ann. N.Y. Acad. Sci. **175:**541–554.
30. NICHOLS, B. W. 1973. *In* The Biology of Blue-Green Algae. N. G. Carr & B. A. Whitton, Eds.: 144–161. University of California Press. Berkeley and Los Angeles, California.
31. WOLK, C. P. 1973. Physiology and cytological chemistry of blue-green algae. Bacteriol. Rev. **37:**32–101.
32. SHERIDAN, R. P. 1972. A qualitative and quantitative study of plastoquinone A in two thermophilic blue-green algae. J. Phycol. **8:**47–50.
33. HASE, T., S. WAKABAYASHI, H. MATSUBARA, K. K. RAO, D. O. HALL, H. WIDMER, J. GYSI & H. ZUBER. 1978. The amino acid sequence of ferredoxin from the alga *Mastigocladus laminosus*. Phytochemistry **17:**1863–1867.
34. TAKRURI, I, B. G. HASLETT, D. BOULTER, P. W. ANDREW & L. J. ROGERS. 1978.

The amino acid sequence of ferredoxin from the red alga *Porphyra umbilicalis*. Biochem. J. **173:**459–466.

35. TEL-OR, W., R. CAMMACK, K. K. RAO, L. J. ROGERS, W. D. P. STEWART & D. O. HALL. 1977. Comparative immunochemistry of bacterial, algal and plant ferredoxins. Biochim. Biophys. Acta **490:**120–131.
36. LUMSDEN, J., L. HENRY & D. O. HALL. 1977. Superoxide dismutase in photosynthetic organisms. *In* Superoxide and Superoxide Dismutase. A. M. Michelson, J. M. McCord & I. Fridovitch, Eds.: 437–450. Academic Press. London.
37. HARRIS, J. I. & H. M. STEINMAN. 1977. Amino acid sequence homologies among superoxide dismutases. *In* Superoxide and Superoxide Dismutase. A. M. Michelson, J. M. McCord & I. Fridovitch, Eds.: 225–230. Academic Press. London.
38. WOOD, P. M. 1977. The roles of c-type cytochromes in algal photosynthesis. Eur. J. Biochem. **72:**605–612.
39. HO, K. K., E. L. ULRICH, D. W. KROGMANN & C. GOMEZ-LOJERO. 1979. Isolation of photosynthetic catalysts from cyanobacteria. Biochim. Biophys. Acta **545:**236–248.
40. WOOD, P. M. 1978. Interchangeable copper and iron proteins in algal photosynthesis. Eur. J. Biochem. **87:**9–19.
41. KROGMANN, D. W. 1979. Personal communication.
42. YAMANAKA, T., Y. FUKUMORI & K. WADA. 1978. Cytochrome c-554 derived from the blue-green alga *Spirulina platensis*. Plant Cell Physiol. **19:**117–126.
43. HOLTON, R. W. & J. MYERS. 1963. Cytochromes of a blue-green alga: extraction of a c-type with a strongly negative redox potential. Science **142:**234–235.
44. YAMANAKA, T., H. DE KLERK & M. D. KAMEN. 1967. Highly purified cytochromes c derived from the diatom, *Navicula pelliculosa*. Biochim. Biophys. Acta **143:**416–424.
45. SHIMAZAKI, K., K. TAKAMIYA & M. NISHIMURA. 1978. Studies on electron transport systems in the marine diatom *Phaeodactylum tricornutum*. I. Isolation and characterization of cytochromes. J. Biochem. **83:**1631–1638.
46. KAMIMURA, Y., T. YAMASAKI & E. MATSUZAKI. 1977. Cytochrome components of green alga, *Bryopsis maxima*. Plant Cell Physiol. **18:**317–324.
47. PULICH, W., JR. 1977. Cytochrome c_{548} in *Nostoc* sp. (Cyanophyceae): an electron acceptor from reduced NADP in the dark. J. Phycol. **13:**40–45.
48. STENMARK, S. L., D. L. PIERSON, R. A. JENSEN & G. I. GLOVER. 1974. Blue-green bacteria synthesize L-tyrosine by the pretyrosine pathway. Nature **247:**290–292.
49. JENSEN, R. A. & D. L. PIERSON. 1975. Evolutionary implications of different types of microbial enzymology for L-tyrosine biosynthesis. Nature **254:**667–671.
50. EMOND, D., N. RONDEAU & R. J. CEDERGREN. 1979. Distinctive properties of glutamine synthetase from the cyanobacterium *Anacystis nidulans*. Can. J. Biochem. **57:**843–851.
51. MIFLIN, B. J. & P. J. LEA. 1976. The pathway of nitrogen assimilation in plants. Phytochemistry **15:**873–888.
52. STACEY, G., F. R. TABITA & C. VAN BAALEN. 1977. Nitrogen and ammonia assimilation in the cyanobacteria: purification of glutamine synthetase from *Anabaena* sp. strain CA. J. Bacteriol. **132:**596–603.
53. SCHMIDT, A. 1977. Assimilatory sulfate reduction via 3′-phosphoadenosine-5′-phosphosulfate (PAPS) and adenosine-5′-phosphosulfate (APS) in blue-green algae. FEMS Microbiol. Lett. **1:**137–140.
54. SCHMIDT, A. & U. CHRISTEN. 1978. A factor-dependent sulfotransferase specific for 3′-phosphoadenosine-5′-phosphosulfate (PAPS) in the cyanobacterium (*Synechococcus* 6301). Planta **140:**239–244.
55. BJÖRN, L. O. 1979. Photoreversibly photochromic pigments in organisms: properties and role in biological light perception. Quart. Rev. Biophysics **12:** 1–23.

56. MARKERT, C. L., Ed. 1975. Isozymes. IV. Genetics and Evolution. Academic Press, Inc. New York, N.Y.
57. HOLTON, R. W. 1973. Electrophoresis and the taxonomy of algae. Bull. Torrey Bot. Club **100**:297–303.
58. KLEIN, S., D. J. CHAPMAN & E. D. GARBER. 1973. Chemotaxonomy of the *Oscillatoria-Phormidium* Complex. Biochem. System **1**:173–177.
59. SCHENK, H. E. A., I. HOFER & H. METZNER. 1973. Malat-Dehydrogenase Isoenzymbanden als potentielles chemotaxonomisches Kriterium fur Cyanophyceen-Species. Biochem. System. **1**:179–184.
60. FREDRICK, J. F. 1977. Protein and isozyme patterns of the cyanelles of *Glaucocystis nostochinearum* compared with *Anacystis nidulans*. Phytochemistry **16**: 1571–1573.
61. LUMSDEN, J. & D. O. HALL. 1975. Superoxide dismutase in photosynthetic organisms provides an evolutionary hypothesis. Nature **257**:670–672.
62. MURPHY, L. S. & R. R. L. GUILLARD. 1976. Biochemical taxonomy of marine phytoplankton by electrophoresis of enzymes. I. The centric diatoms *Thalassiosira pseudonana* and *T. fluviatilis*. J. Phycol. **12**:9–13.
63. CHENEY, D. P. & G. R. BABBEL. 1978. Biosystematic studies of the red algal genus *Eucheuma*. I. Electrophoretic variation among Florida populations. Marine Biology **47**:251–264.
64. O'BRIEN, S. J., G. KLEINER, R. OLSON & J. E. SHANNON. 1977. Enzyme polymorphisms as genetic signatures in human cell cultures. Science **195**:1345–1348.
65. STAM, W. T. & H. C. HOLLEMAN. 1979. Cultures of *Phormidium, Plectonema, Lyngbya* and *Synechococcus* (Cyanophyceae) under different conditions: their growth and morphological variability. Acta Bot. Neerl. **28**:45–66.
66. WHITTAKER, R. H. & L. MARGULIS. 1978. Protist classification and the kingdoms of organisms. BioSystems **10**:3–18.

DISCUSSION OF THE PAPER

DR. J. M. OLSON: I just want to make a comment on the use of isoelectric focusing in comparing proteins from various strains of organisms. What you are looking at there are isoelectric points and not molecular sizes and molecular weights. I think you could easily be misled. In general, I would say it would be better to look at the electrophoresis pattern on the gel rather than isoelectric focusing for comparing proteins.

DR. R. W. HOLTON: If you do both isoelectric focusing and electrophoresis, it gives you a little more validity.

DR. R. A. LEWIN: I think that the thing often called *Anacystis nidulans* is a unicellular form of a filamentous organism. If it's grown under certain conditions it will form filaments, if it is subjected to mutagenesis it readily gives filamentous forms.

DR. S. S. COHEN: Actually any unicellular organism, when subjected to the usual conditions of mutagenesis which affect DNA synthesis, will produce elongation without subsequent steps of DNA synthesis and division. You can get tremendous filaments from *E. coli* as Dr. Doolittle pointed out, and you can revert that rather readily to unicellular forms and these will occur in coccoid forms. I really find Dr. Lewin's statement to be very puzzling and I'm not at all sure that it holds up.

DR. HOLTON: I think lots of the other biochemical indicators have put *Anacystis nidulans* squarely in the cellular camp rather than in the flagellate camp.

DR. COHEN: There are many other characteristics of *Anacystis* that make it nice to work with and we can grow it very easily. It's very nice to have a medium in which most organisms won't grow because there really isn't an organic substrate. There isn't a lot of carbohydrate in the medium which gums up things in the isolation—there are a myriad of reasons why one picks this and still one may get into a good deal of trouble because you've concentrated on something with a very special kind of biochemistry.

DR. HOLTON: Well if you're going to put the blue-greens on a scale from 1 to 10 with 1 being the extreme and perhaps towards the other prokaryotes and 10 towards the higher plants, I would certainly say it was down towards a 1 rather than being in the middle.

ULTRASTRUCTURAL STUDIES OF *CYANIDIUM CALDARIUM*: CONTRIBUTION TO PHYLOGENESIS

Joseph Seckbach,* I. S. Hammerman, and J. Hanania

Department of Life Sciences
Bar-Ilan University
Ramat-Gan, Israel

We can say nothing with certainty about anything because the picture presented to us is not constant.

Philo, 20 B.C.

Cyanidium caldarium, a eukaryotic unicellular alga, has served as a model of an algal pioneer in the dispute concerning the evolutionary relationships among algae.[1–7] Indeed, this greenish-bluish fascinating genus has (or lacks) certain analogous homologous cellular phylogenetic markers which tempt its investigators to place this paradoxical organism between the prokaryotic blue-green and the primitive eukaryotic red algae.[3, 4, 8, 9] Others claim this organism to be an endosymbiont cyanophyte within an unknown host originally lacking chloroplasts.[2, 6, 7, 17] Such cyanophytic origin of plastids was first proposed nearly 100 years ago.[18]

My personal interest and involvement in this organism started about a dozen years ago. I was asked by Prof. W. F. Libby of the University of California at Los Angeles to study the possibilities of life on the planet Venus. Our aim was to demonstrate that terrestrial photosynthetic organisms are able to tolerate and survive extreme conditions such as were believed to exist on Venus polar caps.[10, 11] Such organisms had to indicate growth in an atmosphere of pure CO_2 and at elevated temperature and pressure, as well as in a very acidic media. During our initial investigation with several representatives of the plant kingdom, I received a letter from Prof. L. Bogorad who suggested that we try his "favorite organism *Cyanidium caldarium*" as a candidate for the Venusian project. William N. Doemel, then a graduate student of Prof. T. D. Brock (at Indiana University), sent us axenic cultures of *Cyanidium,* isolated from samples collected at Norris Geyser Basin of Yellowstone National Park. The primary results of algal growth and physiology in high concentration of CO_2, elevated temperature, and acidic media were reported to the Eleventh International Botanical Congress (Seattle, Wash., 1969) and elsewhere.[11, 12] At that Congress I met Prof. R. M. Klein (University of Vermont) who indicated strong interest and was quite excited with our electron micrographs of *Cyanidium caldarium.* It was he who inspired me to consider a publication on the ultrastructural observations with a taxonomic approach to this topic.[8] Later, in Jerusalem with Prof. R. Ikan, we studied the fatty acids and sterols of *Cyanidium* as chemosystematic tools for assessing its position among the algae.[9, 13] The present report and our recent observations at Bar-Ilan University on the micromorphological features of *Cyanidium* could not have taken place without the perpetual

* Present address: School for Overseas Students, The Hebrew University, Jerusalem 91905, Israel.

encouragement of Dr. J. F. Fredrick, who also supplied the algal cultures (Allen strain) by air mail (the cells survived the low temperatures during the long trans-Atlantic flight).

It seems that whoever is involved with this interesting genus draws additional attention to *Cyanidium*; indeed, a recent survey of the literature of this organism lists about 100 papers written during the past 20 years.[14]

Cyanidium—Who, What and Why

The initial, and now classical, detailed comprehensive review on the genus *C. caldarium* appeared in the Klein-Cronquist essay on the phylogeny of the Thallophyta.[1] Since then—and particularly during the past decade—additional papers, as well as major articles [15-17] or ultrastructural studies [8, 9, 19, 20]—were all devoted primarily to *C. caldarium* (see Fukuda's list [14]). It seems that *Cyanidium* is a rather primitive alga (Figure 1) that tolerates severe physiological conditions,[11, 12, 15-17, 21-24] such as 1 N H_2SO_4, grows at 50–60° C, and oxygen is not necessary for growth.[17, 23] This organism occupies a unique environmental niche of acidic thermal habitats throughout the world,[15-17] and thrives in a pure atmosphere of CO_2.[10-12, 23] Because of its morphology and cellular constituents (shared by this alga with other algal classes), it has been placed by several investigators in different taxonomic categories.[1-4, 8, 9, 30] At one time or another it has been considered to be a member of the Cyanophyta, Rhodophyta, Cryptophyta, or even Chlorophyta.[8, 15, 17] Yet the differences seem valid enough to remove *C. caldarium* from all of these groups. In order to solve this dilemma Klein and Cronquist,[1] Klein,[3] and Fredrick [4] proposed *C. caldarium* as a "bridge form" that occupies a transition status between the blue-green prokaryotes and the eukaryotic primitive red algae. This later proposition was rejected by the followers of the endosymbiotic hypothesis.[2, 6, 7, 19] In this paper we shall avoid discussion of the proposed possible capacity of early eukaryotes to harbor blue-green endosymbionts.[25-27] According to this interesting theory the eukaryotic chloroplasts have been in fact blue-green, free-living organisms (Cyanophyceae) and got incorporated in the past into a host where they function as organelles [2, 6, 7, 18, 25] or cyanelles. We prefer not to step into this "mine field," even in order to pick up some orchids. Margulis and others have already attempted to place their evidence in the context of the endosymbiotic hypotheses.[2, 26, 27]

The comparison of *Cyanidium caldarium* cellular features with that of other algal classes with which it has been classified is presented in Table 1. We placed *Cyanidium* morphology, cellular organelles as well as biochemical products (storage carbohydrates, cell wall, and sterols), side by side with these characteristics as reported for the other algal classes. The Rhodophyta seems to be the group most similar to *Cyanidium* in all the aspects compared, as shown in Table 1. The red algae are followed by the blue-green algae, which are closer to *Cyanidium* than the green algae, while the Cryptophyceae are the most remote group, at least for these features.

The Photosynthetic Apparatus

We focused attention upon the chloroplast organization and pigmentation of *Cyanidium* because this organelle is of great phylogenetic value. It is well known that the chloroplast fine structure and pigmental content is characteristic of the different algal groups.[28] TABLE 2 compares the photosynthetic apparatus—organization and pigments—of *Cyanidium* with the four algal classes involved in *Cyanidium* taxonomy. It seems that the chloroplasts of the Rhodophyta have the closest similarity with those of *Cyanidium*. The prokaryotic

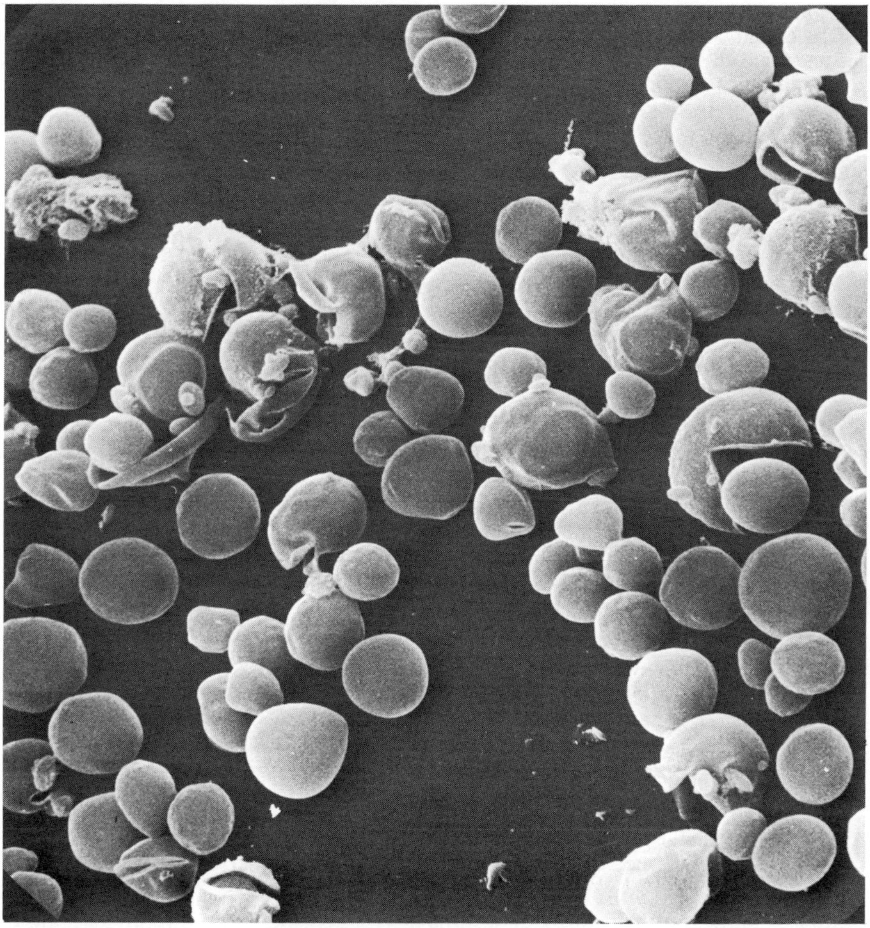

FIGURE 1. The appearance of *Cyanidium caldarium* cells as displayed by the scanning electron microscope. These cells look like *Chlorella*. They divide like *Chlorella* and others in forming four endospores. The larger cells have already divided and some stages of the mother cell wall opening are visible. Note the degree of cellular wall curling and that one is twisted into a shape like a flat bowl. ×5,700.

cells of the Cyanophyceae closely resemble the internal architecture of the red algal chloroplasts [1–4, 17, 18, 25, 26] but have less in common with the pigment content. Thus it is quite obvious (TABLE 2) that the photosynthetic apparatus structures of *Cyanidium*, and blue-green algae are very similar, which again indicates their phylogenetic relationships. The chloroplasts of the other algal

TABLE 1

FEATURES OF *Cyanidium caldarium* * AND OF DIFFERENT ALGAL GROUPS

Characteristic	*C. caldarium*	Rhodophyta	Cyanophyta	Chlorophyta	Cryptophyta
Morphology					
Unicellular	+	+	+	+	+
Spherical cell divides by forming autospores (daughter cells)	+	+	+	+	+
Sexual reproduction	—	+/—	—	+/—	+/—
Microstructure					
True nucleus, Mitochondrion Chloroplast	+	+	—	+	+
Golgi dictyosomes †	+	+/—	—	+	?
Pyrenoid	—	+/—	—	+	—/+
Storage granules	+	+	+	+	+
Vacuoles	+	+	—/+	+	+
Flagellum	—	—	—	+	+
Storage products					
Polyglucoside with many 1,6-glucosyl linkages	+	+	+	—	?
Branching enzyme active on amylose and amylopectin	+	+	+	—	?
Cell Wall					
High protein	+	—	—	—	?
Low cellulose	+	+/—	—	—	—
Hemicellulose	+	+	—	+	?
Diamino pimelic acid (DMP)	—	—	+	—	—
Amino sugars	—	—	+	+	—
Muramic acid	—	—	+	—/+	—
Sterol Content					
Ergosterol	+	—	—	+	?
β-sitosterol	+	+/—	—	+	?
Campesterol	+	+	—	—	?
Cholesterol	+	+	+	+/—	?

* After Seckbach and Ikan,[9] Brock.[17]

† Demonstrated in this paper.

Key: + = feature present, — = feature absent, +/— = feature variable in the group, and ? = presence of feature uncertain or not detected.

groups (Chlorophyceae and Cryptophyceae) have less in common with these features of *Cyanidium*. The score of all the characteristics as presented in both tables and the total "phylogenetic level" of *Cyanidium* as compared with these four algal groups (by giving a positive point for a similar feature) are shown in TABLE 3. The total score of the algal relationships as listed above indicates that there is a decreasing level of similarity in the total similar/dissimilar values between *C. caldarium* and the other algae. The results of TABLE 3 show that the *Cyanidium* affinities (expressed in percent) to the other algal classes is as follows:

Cyanidium	:	Rhodophyta	:	Cyanophyta	:	Chlorophyta	:	Cryptophyta
100%		87%		57%		47%		37%

The relation and resemblance of *Cyanidium* seems to be the strongest to the red algae (TABLE 3). We do not ignore the possibility that *Cyanidium* may have additional features as yet undiscovered, and that further details are needed concerning all the algal groups in order to settle the arguments and unriddle the confusion about the phylogeny and systematic position of *Cyanidium*.

RESULTS AND DISCUSSION

The present study was undertaken in order to examine additional fine-structural features of *C. caldarium* in relation to its uncertain taxonomic position among the algal groups. The appearance of these cells under light microscopy was recently published,[11, 12, 17] but very little information can be extracted from such magnifications. On the other hand, several investigators have examined *Cyanidium* with the electron microscope (EM) and revealed

TABLE 2

ORGANIZATION AND ARCHITECTURE OF THE PHOTOSYNTHETIC APPARATUS AND PIGMENTATION OF *Cyanidium caldarium* * AND OTHER ALGAE

Characteristic	*C. caldarium*	Rhodophyta	Cyanophyta	Chlorophyta	Cryptophyta
Chloroplast			plastid absent		
Number of thylakoids per band	1	1	1	2–6	2
Girdle bands or concentric thylakoids	+	+/— †	+	—	—
Parallel lamellation	+	+	+	—	—
Grana (stocked lamellae)	—	—	—	+	—
Degree of thylakoid assemblage in band	—	—	—	tight	loose
Branching of thylakoid (lamellar interconnection)	+	+	—	+	—
Phycobilisomes	+	+	+	—	—
Localization of phycobilins	adjacent to both thylakoid membranes (stroma side)			—	intra-thylakoidal lumen
Starch in cytoplasm	+	+	+	—	—
Pigments					
Chlorophyll *a*	+	+	+	+	+
Chlorophyll *b*	—	—	—	+	—
Phycocyanin	+	+	+	—	+
Phycoerythrin	—	+	+	—	+
α carotene	—	+	—	+	—
β carotene	+	+	+	+	+
Lutein	+	+	?	+	—
Zeaxanthin	+	+	—	—	+
Violoxanthin	+	+	—	+	—
Cryptoxanthin	+	—	—	—	+

* Modified after: Gibbs,[28] Brock[17] (see also Dodge[18]).
† In some species outermost thylakoid(s) encircle(s) the plastid.
Key: + = character present, — = character absent, and ? = character uncertain.

TABLE 3

THE "PHYLOGENETIC SCORE" OF *Cyanidium* IN COMPARISON TO OTHER ALGAL GROUPS *

Characteristic	*Cyanidium*	Rhodophyta	Cyanophyta	Chlorophyta	Cryptophyta
Morphology and microstructure	9	9	7	7	7
Storage products Cell wall+Sterols	12	10	3	6	?
Photosynthetic apparatus	9	9	8	1	1
Photopigments	10	7	5	5	7
Total score	40	35	23	19	15
Ratio	1.00	.87	.57	.47	.37

* Based on the total evaluation markers as presented in TABLES 1 and 2.

the internal fine structure, thus assisting evaluation of its systematic position.[8, 9, 17, 19, 20, 22, 29, 30] In general the *Cyanidium* eukaryotic cellular type has been established; its plastids have been considered "primitive," resembling the Cyanophyceae. Our knowledge about some eukaryotic components within this genus, such as the nature of the nucleus, Golgi bodies, microbodies (peroxisomes), and additional fine features (e.g., presence of flagellate, pyrenoid, basal body) have not been revealed until now. These characteristics are valuable as phylogenetic markers for the algal evolutionary relationships.

FIGURE 1 shows the appearance of *Cyanidium* cells as revealed with the scanning electron microscope (SEM). The life cycle of this alga is identical to *Chlorella* and these two algal species cannot be separated under the light microscope.[17] The mode of the asexual reproduction and the new daughter cells are shown in the second SEM photograph (FIGURE 2). It seems that the release of the autospores is associated with a cell wall "zipper-like" split. The SEM photographs reveal that newly discharged cells are still connected to one another. A different view of this interconnection is presented in FIGURE 3. Under some physiological conditions the *Cyanidium* mother cell has more than four spores (perhaps eight) and might release them while still connected in a "cord," one following the other. One of the cells seems quite artistically formed and its contents resemble the "historical"[29, 30] nature of this alga as seen in 1961/1962 (FIGURE 4). This electron micrograph displays a large chloroplast with parallel thylakoids, a mitochondrion and an organelle believed to be a nucleus. There are almost no additional details in such a section. A higher cellular organization of the *C. caldarium* fine morphology is shown in FIGURE 5. The individual cell sections show from one to six chloroplasts. Our previous examinations of *Cyanidium* ultrastructure[8, 9, 22] have shown new aspects of chloroplast architecture, pattern of lamellations, and the first demonstration of the phycobilisomes in this alga. The investigation of the phycobilisomes of this organism by Edwards and Mainwaring[19] showed defined and well-preserved bodies, while in several other studies[22, 29, 30] these bodies were not seen.

The cells examined in this study showed several features characteristic of higher algae; other aspects of fine structure were closely examined for the first time. Even at relatively lower EM magnification (FIGURE 5) it is clear that *Cyanidium* possesses rather complex cellular compartmentalization, chloroplasts, nuclei, mitochondria and dictyosomes (D), and granulated cytoplasm. Most of the chloroplastids have well-defined single thylakoids running in parallel bands across the plastid and are encircled by one, two, or more girdle lamellae adjacent to the plastid envelope (see FIGURE 5, cells 1,2,4,5,7,8). The cell designated as number 6 contains flat disks in parallel lines without any girdle external lamella as also reported for another red alga *Porphyridium cruentum*.[31] On the other hand chloroplasts such as in cell number 3 or in cells 2 and 8 (at their upper region) and in FIGURE 9 have concentric thylakoids which occupy most of the stroma space. In fact *Cyanidium caldarium* cells (FIGURE 5) resemble other red algal cells.[31, 32] The chloroplast of this alga is

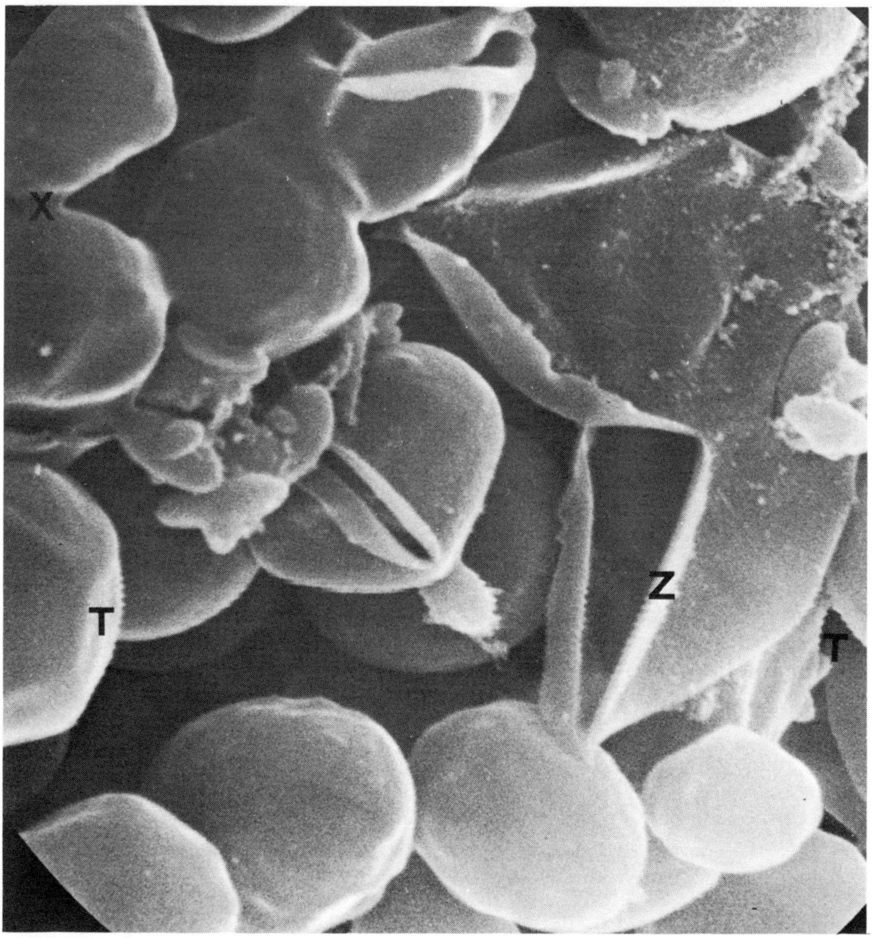

FIGURE 2. Scanning electron micrograph of cellular reproduction in *Cyanidium*. Note the opened "zipper" of the large cell while releasing the daughter cells (z); the central cell opens up its wall while some cells show a crown of "teeth" (T) before the spores discharge; two young cells seem to be interconnected (x). ×19,000.

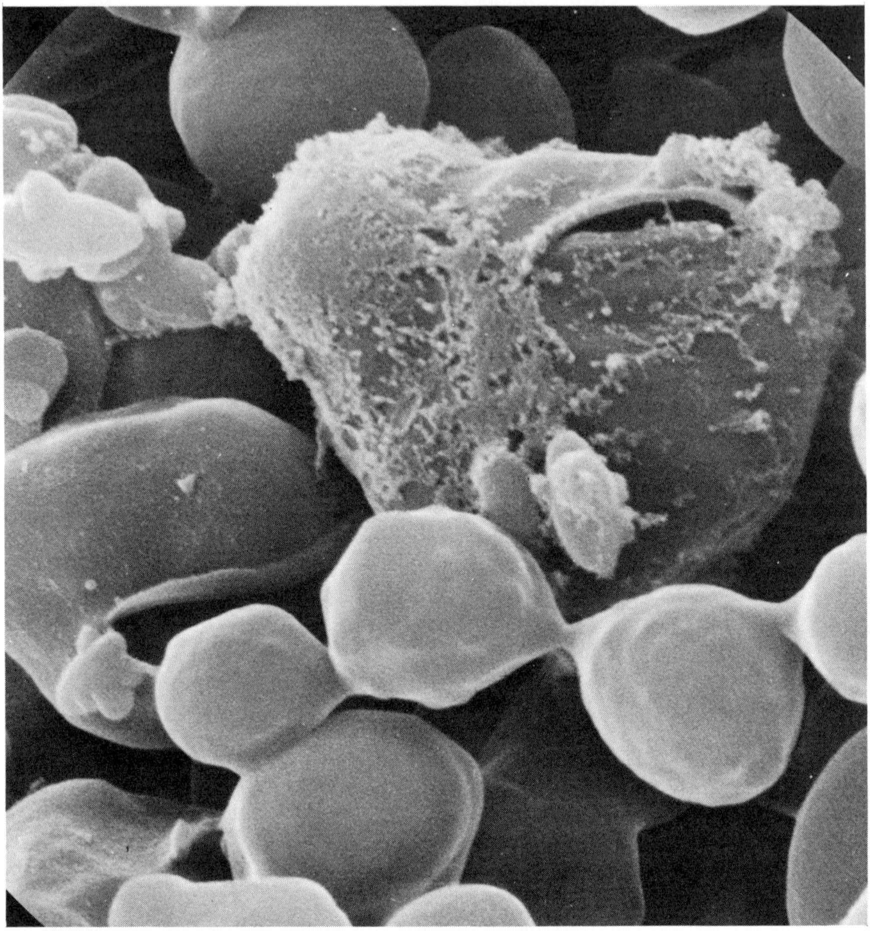

FIGURE 3. Another view of interconnected young *Cyanidium* cells after release from the mother cell. Note the wall opening of the two cells, the bigger one with unsmooth wall contains more than four autospores. Same culture as the previous scanning fine structures. Cells were not shaken during growth. ×19,500.

very similar to the rhodoplasts of the freshwater *Batrachospermum* [31] and *Lemanea* [32] which have a few concentric lamellae and a variable number of nonconcentric thylakoids, while another red alga was reported to have chloroplasts consisting entirely of concentrically arranged thylakoids.[31] All these structural differences in rhodoplasts examined in several red algae belonging to various taxonomic groups are observed in *Cyanidium* chloroplasts seen also in FIGURE 5. It seems to us that the three configurations of the plastids are developmental phases or the result of environmental effects on the individual organism, perhaps also the particular strain might "prefer" one of these thylakoid arrangements. For example, when the present study was performed with *Cyanidium caldarium* strain A (Allen) and under low illumination we noticed that the dominant chloroplast structure showed the "parallel lines-type" thylakoids, i.e., the outermost lamellae encircling the other nonconcentric and parallel-lying thylakoids. On the other hand, our previous observations [8, 9, 22] with strain B (Brock) of *Cyanidium caldarium* (freshly isolated from Yellowstone hot springs and cultured at high light intensity) presented several cells with much more extensive chloroplast development, and the whole cell was filled with the chloroplastids' concentric lamellation, as seen in FIGURE 9. A clearer cellular view of *C. caldarium* is demonstrated in FIGURE 6. In contrast to previous *Cyanidium* papers where only one chloroplast was reported,[33] this cell has five plastid profiles all with parallel-running thylakoids and one outermost girdle lamella. The central nucleus is connected through pores or

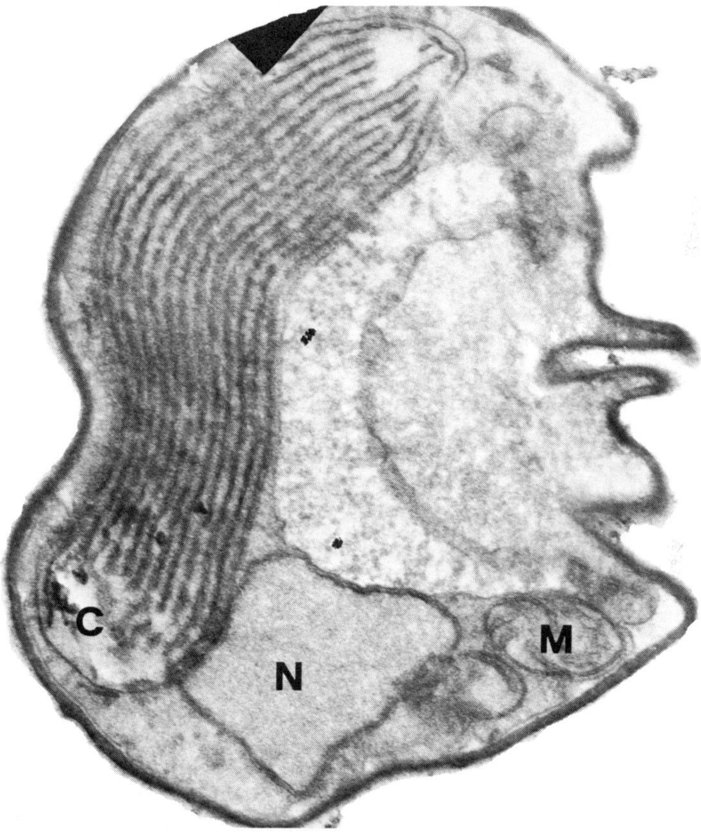

FIGURE 4. "Miss *Cyanidium*" represents a cell of light-grown culture which pictographically outlines a female head. This general electron micrograph resembles the first fine-structural photographs of this alga published in the early 1960s.[29, 30] This section shows one chloroplast (C) with parallel thylakoids, the nucleus (N) enclosed by a double membrane, mitochondria (M), cytoplasmic membranes, and vacuole (around the "face"). These organelles appear quite "primitive" as exhibited in this photograph and indeed some generalizations about *Cyanidium* fine features were misleading in the past 18 years. ×28,000.

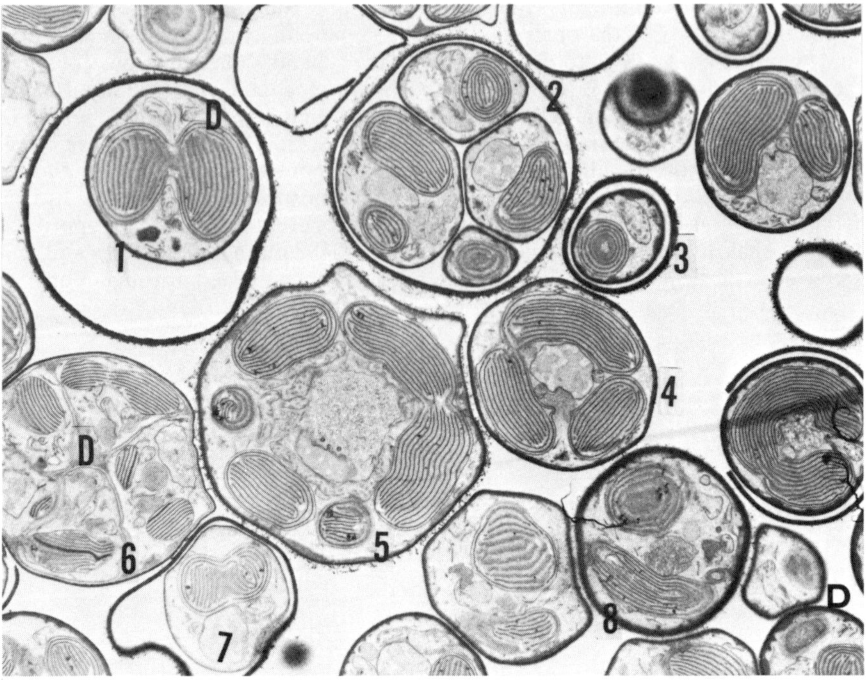

FIGURE 5. Lower magnification of *Cyanidium caldarium* cells. This electron micrograph presents a "higher" cellular status than the previous photo from this alga. Many cells show intensive dividing bodies (cells #1,2,4,5,6,7) as well as several chloroplast profiles. These *Cyanidium* cells seem to contain: a nucleus with chromatin or it appears evenly granulated (cells #2,4,5,7,8), several clear or long mitochondria (cells #2,3,5,7), and several dictyosome-like figures (D) are associated with the dividing cells. Note the thylakoid orientation and the outermost lamellae in many of these chloroplasts, most are parallel but some turn to form concentric thylakoids (cells #3 and 8). ×5,900.

membrane to the cytoplasm, and is surrounded by a ring of about ten mitochondria. In addition, there are a few bodies of rather simple dictyosomes at the cell periphery. The four single membrane-bounded spheres above the nucleus are presumably microbodies (peroxisomes); similar figures were reported to be in the green algae and were detected in the red unicellular alga *Porphyridium*.[18] These microbodies consist of a fine granular matrix 0.5–1.0 μm in diameter and are often situated adjacent to the mitochondria (FIGURE 6, as also elsewhere [18]). FIGURE 7 depicts the main cellular inclusions of *Cyanidium* as observed in this study as well as those published for other rhodoplasts from freshwater algae.[31, 32] In comparison to the fully developed organelles, the dividing cell of *Cyanidium caldarium* grown heterotrophically in dark is shown in FIGURE 8. There is a drastic reduction of the plastid dimensions which are smaller than the mitochondria; both nuclei are active in the reproduction and contain nucleolus and chromosomes, while the nuclear envelope seems to be associated with the cytoplasm; one microtubule is also visible in the nucleus. A clear separation line is being formed between both nuclei. This is the first demonstration of nuclear inclusions; indeed, it is interesting that Kikuchi *et al.*[33] predicted, on the basis of the inhibitory effect of cycloheximide on RNA synthesis, that *Cyanidium* had to possess a nucleolus.

In contrast to the petite plastids of the heterotrophically grown cell, FIGURE 9 represents a fully expanded chloroplast of *Cyanidium* (strain B) which has 15 circular lamellae filling all the cytoplasmic volume of this section and appear as annual rings of a vascular plant. The explanation of such dimorphic plastids as the result of the influence of the environmental conditions has been reported elsewhere.[17, 18, 22]

CONCLUSION

This study supplies additional cellular fine structural inclusions and stresses the photosynthetic apparatus. From the published data (presented in the Tables) as well as from the present SEM and EM observations, we are tempted to state that *Cyanidium caldarium* has a very close phylogenetic position with

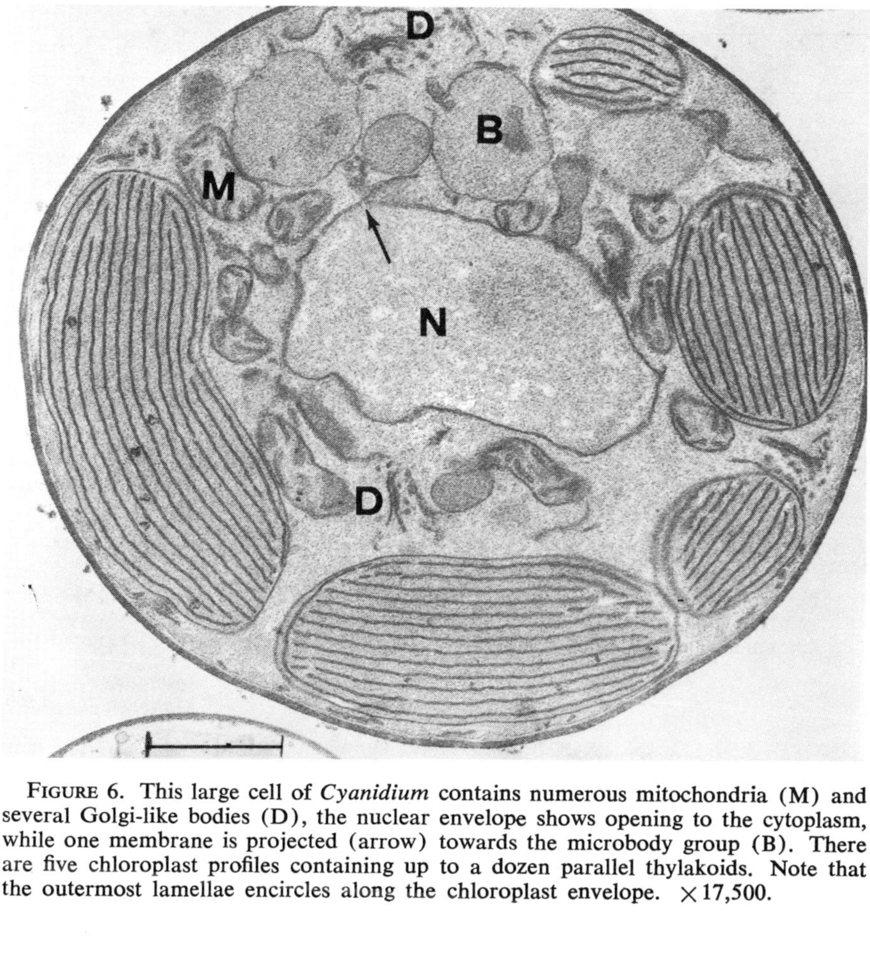

FIGURE 6. This large cell of *Cyanidium* contains numerous mitochondria (M) and several Golgi-like bodies (D), the nuclear envelope shows opening to the cytoplasm, while one membrane is projected (arrow) towards the microbody group (B). There are five chloroplast profiles containing up to a dozen parallel thylakoids. Note that the outermost lamellae encircles along the chloroplast envelope. ×17,500.

the rhodophytes. The proposed transitional link of *Cyanidium* between the blue-green prokaryotic algae and the primitive eukaryotic rhodophytes [1, 3, 4, 9] could be moved closer to the red algae. From this study it seems that the algal representative of the postulated Uralga,[1, 2] which serves as a bridge between the prokaryotes (blue-greens) and the eukaryotes [4] (Rhodophyta), is somewhat further advanced in its evolutionary qualities. With all the above phylogenetic and ultrastructure data in mind, we propose that *C. caldarium* is a blue-green rhodophyte. This is supported also by others [5, 13, 18] with different taxonomic approaches. In order to obtain additional evidence concerning the taxonomy of

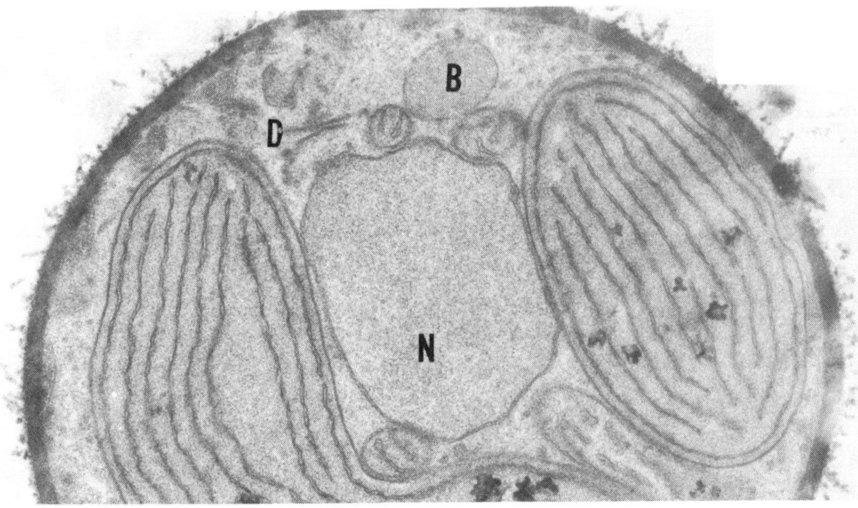

FIGURE 7. A section through a cell of *Cyanidium caldarium* showing the main cellular constituents: two chloroplasts with girdle bands around the plastid rim, the central nucleus is embraced with four mitochondria, a dictyosome-like membrane (D), and a microbody (B) are also present as in other rhodophytes.[18] ×27,200.

Cyanidium among the algal classes, one might search for some special chloroplast proteins such as phytoferritin [34, 35] which is present in many taxonomic groups, but so far not detected in the algae. We do hope that the problem with this fascinating alga will soon be settled, so that this black sheep among the algae will turn out to have been only an ugly duckling.

Our reply to the argument that *C. caldarium* differs from the rhodophytes is:

A woman can dye her hair but not change her character.

Immanuel
Mahberot, ca. 1300

[**NOTE ADDED IN PROOF:** Since the submission of this manuscript, new references relevant to this study have reached our attention. These additional sources have been incorporated into further ultrastructural, taxonomic and biochemical investigations of *Cyanidium caldarium* and published in *Origin of Life* (Proceedings of the Sixth International Conference on the Orgins of Life held in Jerusalem, June 1980). The rebuttal concerning the endosymbiotic origin, as suggested by others, for this enigmatic alga is being published elsewhere.]

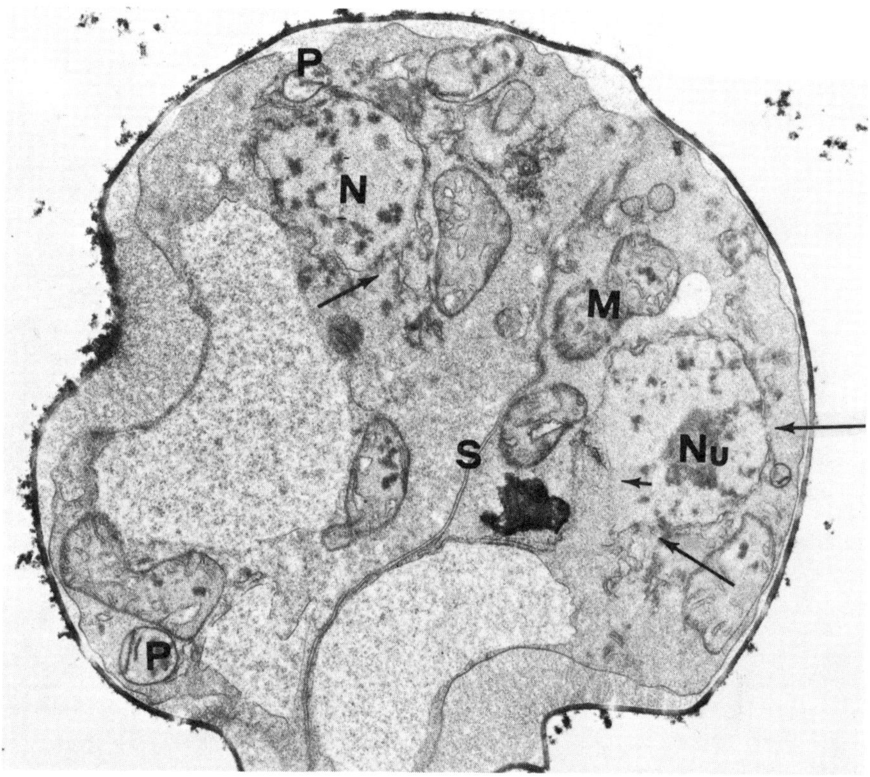

FIGURE 8. This cell was cultured heterotrophically in dark with glucose. There is a drastic reduction in the plastid size (P); they barely contain lamellation. It undergoes cellular division: the separating outline of the two daughter cells is already being formed (S) and it splits the cell into two regions. Both nuclei (N) contain chromosomes and the nucleolus is present (Nu), the nuclear envelope is not complete and it seems to open up to the cytoplasm (arrows), a microtubule is visible (arrow head). Some mitochondria are seen to divide in both regions (M). ×15,000.

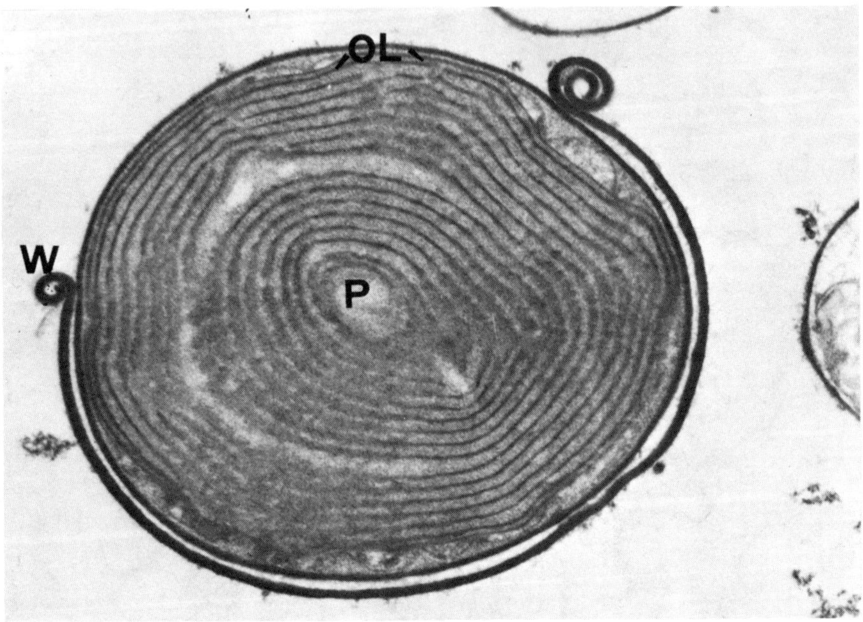

FIGURE 9. The chloroplast (P) of this cell (from strain B culture) occupies most of the cytoplasm volume and contains 15 concentric thylakoids. The scroll-like wall (W) occurs as the cell reproduces (see FIGURES 2, 3, as viewed by the SEM). The outermost lamella (OL) runs parallel with the plastid envelope. ×27,500. From Seckbach [22] with permission from the *Israel Journal of Botany*.

References

1. Klein, R. M. & A. Cronquist. 1967. A consideration of the evolutionary and taxonomic significance of some biochemical, micromorphological, and physiological characters in the Thallophytes. Quart. Rev. Biol. **42:**105–296.
2. Margulis, L. 1968. Evolutionary criteria in Thallophytes: A radical alternative. Science **161:**1020–1022.
3. Klein, R. M. 1970. Relationships between blue-green and red algae. Ann. N.Y. Acad. Sci. **175:**623–633.
4. Fredrick, J. F. 1976. *Cyanidium caldarium* as a bridge alga between Cyanophyceae and Rhodophyceae: Evidence from immunodiffusion studies. Plant & Cell Physiol. **17:**317–322.
5. Hase, T., S. Wakabayashi, K. Wada, H. Matsubara, F. Jüttner, K. K. Rao, I. Fry & D. O. Hall. 1978. *Cyanidium caldarium* ferredoxin: A red algal type? FEBS Letters **96:**41–44.
6. Raven, P. H. 1970. A multiple origin for plastids and mitochondria. Science **169:**641–646.
7. Raven, P. H., R. F. Evert & H. Curtis. 1976. Biology of Plants. 2nd edit. Worth Publishing, Inc. New York, N.Y.
8. Seckbach, J. 1972. On the fine structure of the acidophilic hot-spring alga *Cyanidium caldarium*: a taxonomic approach. Microbios **5:**133–142.
9. Seckbach, J. & R. Ikan. 1972. Sterols and chloroplast structure of *Cyanidium caldarium*. Plant Physiol. **49:**457–459.
10. Seckbach, J. & W. F. Libby. 1969. Life in pure CO_2. Studies with thermophilic algae grown at elevated temperatures and under one atmosphere of CO_2. Plant Physiol. Supp. **44:**15.
11. Seckbach, J. & W. F. Libby. 1970. Vegetative life on Venus? or investigations with algae which grow under pure CO_2 in hot acid media at elevated pressures. Space Life Sci. **2:**121–143.
12. Seckbach, J. & W. F. Libby. 1971. Life on Venus? *In* Planetary Atmospheres. Sagan, C., T. C. Owen & H. J. Smith, Eds.: 62–83. Reidel Publ. Dordrecht-Holland.
13. Ikan, R. & J. Seckbach. 1972. Lipids of the thermophilic alga *Cyanidium caldarium*. Phytochem. **11:**1077–1082.
14. Fukuda, I. 1979. A possible literature survey of a thermal alga *Cyanidium caldarium* Geitler (I). Sci. Univ. of Tokyo. Kagurazaka, Shinjuku-ku, Tokyo.
15. Doemel, W. N. 1970. The physiological ecology of *Cyanidium caldarium*. Ph.D. Thesis. Indiana University, Bloomington, Ind.
16. Doemel, W. N. & T. D. Brock. 1971. The physiological ecology of *Cyanidium caldarium*. J. General Microbiol. **67:**17–32.
17. Brock, T. D. 1978. Thermophilic microorganisms and life at high temperatures. Chapter 9: The genus *Cyanidium*. 255–302. Springer–Verlag. New York, Heidelberg, Berlin.
18. Dodge, J. D. 1973. The Fine Structure of Algal Cells. Academic Press. London.
19. Edwards, M. R. & J. D. Mainwaring. 1973. Ultrastructural localization of phycocyanin in the acidophilic, thermophilic alga, *Cyanidium caldarium*. *In* 31st Ann. Proc. Electron Microscopy Soc. Amer. C. J. Arceneaux, Ed. New Orleans, La.
20. Wollman, F. A. 1979. Ultrastructural comparison of *Cyanidium caldarium* Wild Type and III-C mutant lacking phycobilisomes. Plant Physiol. **63:**375–381.
21. Seckbach, J. 1971. Pigmentation of algae under pressure. Lim. Oceanogr. **16:** 567–572.
22. Seckbach, J. 1971. Size, composition and fine structure of *Cyanidium caldarium* cultured under pure CO_2. Israel J. Bot. **20:**302–310.

23. Seckbach, J., F. A. Baker & P. M. Shugarman. 1970. Algae thrive under pure CO_2. Nature **227**:744–745.
24. Seckbach, J., H. Gross & M. B. Nathan. 1971. Growth and photosynthesis of *Cyanidium caldarium* cultured under pure CO_2. Israel J. Bot. **20**:84–90.
25. Stanier, R. Y. 1970. Some aspects of the biology of cells and their possible evolutionary significance. *In* Symp. Soc. Gen. Microbiol., No. 20, Prokaryotic and eukaryotic cells. H. P. Charles & B. C. Knight, Eds. 1–38. Cambridge, England.
26. Fogg, G. E., W. D. P. Steward, P. Fay & A. E. Walsby. 1973. The blue-green algae. Academic Press. London and New York.
27. Cohen, S. S. 1970. Are/were mitochondria and chloroplasts microorganisms? Amer. Sci. **58**:281–289.
28. Gibbs, S. P. 1970. The comparative ultrastructure of the algal chloroplast. Ann. N.Y. Acad. Sci. **175**:454–473.
29. Rosen, W. G. & K. A. Siegesmund. 1961. Some observations on the fine structure of a thermophilic, acidophilic alga. J. Biophys. Biochem. Cytol. **9**:910–914.
30. Mercer, F. V., L. Bogorad & R. Mullens. 1962. Studies with *Cyanidium caldarium*. I. The fine structure and systematic position of the organism. J. Cell Biol. **13**:393–403.
31. Brown, D. L. & T. E. Weier. 1968. Chloroplast development and ultrastructure in the fresh water red alga *Batrachospermum*. J. Phycol. **4**:199–206.
32. Bergfeld, R. 1970. Die Feinstruktur und Entwicklung der Plastiden einiger Süsswasserrotalgen. Cytobiol. **1**:411–419.
33. Kikuchi, T., M. R. Ishida, T. Matsubara, G. Tsushimoto & N. Mizuma. 1973. Some feature of RNA and protein synthesis in thermophilic alga *Cyanidium caldarium*: The effects of antibiotics on *in vivo* incorporation of ^{14}C-uracil and ^{14}C-leucine. Ann. Rep. Res. Reactor Inst. Kyoto Univ. **6**:29–37.
34. Seckbach, J. 1971. Iron ferritin and plastid inclusions in leaf cells of iron treated *Xanthium* plants. Cytobios **4**:183–192.
35. Seckbach, J. 1972. Electron Microscopical observations of leaf ferritin from iron-treated *Xanthium* plants: Localization and diversity in the organelle. J. Ultrastr. Res. **39**:65–76.

Discussion of the Paper

Dr. R. K. Trench: I recently received a culture of *Cyanidium caldarium* from David Chapman's lab at U.C.L.A., it's a thing called 3D2 and we established the culture with it and we had it in glucose-containing medium. By accident, one of my students transferred one of the cultures into the light. I now have a very beautiful densely green culture of *Cyanidium* growing on glucose in the light. So with respect to this particular isolate, it can happen. The question I'd like to ask of you is, recently Bruno P. Kramer in Germany published an article in which he presented pretty good evidence that the pattern of photosynthetic [^{14}C] fixation in *Cyanidium* is essentially identical to that observed in his survey of Cyanobacteria and quite distinct from that that he also observed in red algae. Could you comment on that please?

Dr. J. F. Fredrick: Dr. Trench, Dr. Bogorad describes several mutants of *Cyanidium caldarium*. It's possible that the original culture had a mutant in it.

Dr. Trench: This is why I specifically said that this came from Chapman and it was called 3D2, I don't know precisely what the history of 3D2 is, however, if I could just elaborate on that possibility. Several of the other cultures he sent me are things that have been maintained in his lab in the dark for in excess of five years. I've got these now and spent three months putting these on nonglucose medium in the light, and they refuse to green up. Some of those are suposed to be wild-type.

DR. K.-P. KOLLER: All of the unicellular red algae have only one plastid and that plastid is really lobed. That means if you cut through the cell you get single different cuts which makes you feel that it has different plastids. I assume that you know that this thing has a few plastids, but again would that mean that you would have to form a new order for the *Cyanidium?*

DR. SECKBACH: No, because it doesn't really fit into the unicellular red algae. There are also many other objections. On the other hand, among every population some people are different than others, so also the algae.

DR. TRENCH: If my analysis is correct the phycocyanin from *Cyanidium* is identical in its absorption spectrum, its fluorescence excitation, and emission spectrum to that from *Cyanophora paradoxa.*

DR. SECKBACH: There is a red alga called *Rhodella* which looks red but that has phycocyanin. And also *Cyanidium* has phycocyanin. The only pigment or biliprotein that is really known to be only in red algae is the R-type of phycoerythrin and nothing else. So the posession of phycocyanin is no indication of any algae belonging to the red or green group.

THE BIOSYNTHESIS OF STORAGE GLUCANS IN PROKARYOTIC AND EUKARYOTIC ALGAE

Jerome F. Fredrick

The Research Laboratories
The Dodge Chemical Company
Bronx, New York 10469

The Embden-Meyerhof pathway, whereby the energy in the glucose molecule is released from the hexose, appears to be ubiquitous to practically all types of cells. It seems that this metabolic pathway is an ancient one, possibly one of the very first energy-releasing mechanisms developed during the course of evolution.[1]

The first cellular entities, evolving from the coacervates and microspheroids in the prebiotic "soup," obviously had no problem obtaining the necessary glucose, which was abiotically synthesized and available in the melange of chemicals dissolved in the waters of the primitive earth.[2] However, organisms with the Embden-Meyerhof (E-M) pathway soon depleted the supplies of this hexose. Thus, whereas glucose had probably entered the first primitive cell structures by a process of simple diffusion, the dwindling supplies now made necessary the expenditure of energy in order to transport these scarce molecules, against a shifting density gradient, across the newly developed cell membranes.[1]

Those primitive prokaryotes having the necessary enzymic pathways whereby the entering glucose molecules could be stored for later energy demands, in the form of polymers such as starch and glycogen, under the new conditions of glucose scarcity, would have had some selective advantage. These cells, with their internal supplies of stored glucose polymers, would no longer be dependent upon the outside concentrations of the abiotically formed sugar. Since the degradation of starch and glycogen to glucose-1-phosphate and its subsequent conversion to glucose-6-phosphate (whereby it enters the E-M pathway) does not require adenosine triphosphate (ATP), one such energy-rich ATP molecule per glucose molecule could have been conserved for use in other biochemical pathways.

Because of this, it would seem probable that those enzymes necessary for the polymerization of glucose to form storage glucans, probably in the form of large molecular weight, highly branched molecules,[3] were among the first catalytic proteins to appear. Indeed, since the E-M pathway is also present in nonphotosynthesizing bacteria, it seems probable that these enzymes evolved even long before photosynthetic pathways came into being.

The enzymes involved in the formation and degradation of storage glucans belong to three main groups. Two of these groups of glucosyltransferases, the *phosphorylases* (E.C. 2.4.1.1) and the *synthases* (E.C. 2.4.1.11) are capable of forming alpha-1, 4–glycosidic linkages between glucose molecules.[4,5] These enzymes can build up linear polymers such as *amylose* (actually, these alpha-1, 4–linked polymers exist as left-handed helices [6]). The third group of glucosyltransferases, known as *branching enzymes* (E.C. 2.4.1.18), can establish alpha-1, 6–glycosidic linkages between adjacent amylose chains.[7,8] Hence, these branching enzymes are ultimately responsible for the formation of such branched glucose polymers as phytoglycogen, floridean starch, and the amylopectin component of starch. The enzymes of all three groups of glucosyltransferases exist in multiple molecular forms (isozymes) in the algae [9] and in higher plants.[10–12]

There is much controversy as to whether the phosphorylases, which utilize glucose-1-phosphate as substrate, are involved in the biosynthesis of polyglucan,[13, 14] whether they only initiate "primer" formation,[15] or whether their function is exclusively degradative.[16] In the case of the synthases, which utilize the glucose nucleotides adenosine diphosphoglucose and uridine diphosphoglucose as substrates, there is general agreement that they are responsible for the addition of glucose residues in alpha-1,4–linkages to already-formed maltosaccharide primers.[17] Certainly, with regard to efficiency in forming alpha-1,4 linkages, the synthases appear to be superior to the phosphorylases.[18]

The branching isozymes apear to be of two main types. One type, known as the classical "Q" enzyme is able to insert alpha-1,6 branched linkages into linear amylose polymers, thereby forming amylopectin.[19, 20] The other type, *b.e.*, is able to insert further branched linkages into already-branched polymers such as amylopectin and other branched dextrins to form highly branched phytoglycogens.[21]

Since there was great advantage to the cell in storing energy-rich molecules such as glucose, the first polymers used for such storage were probably high molecular weight, branched glucans akin to the glycogens. In fact, the two groups of organisms in existence today which form the Prokaryota, the bacteria and the blue-green algae (Cyanobacteria), store glucose as glycogen and phytoglycogen. It is only among the higher eukaryotes that glucose is stored as a mixture of a moderately branched polymer, amylopectin, and a long unbranched molecule, amylose. The mixture is known as "starch" and probably represents a more evolutionary advanced form of storage glucan.[21] In fact, when phytoglycogens do appear in higher plants, such as maize and *Cecropia*,[22, 23] they are regarded as the result of a *regressive* mutation.[24]

It seems also conceivable that in the first primitive cells, instead of the three separate types of enzymes immunochemically related to the glucosyltransferases in existence today, the storage glucan was synthesized by a *single* polypeptide with the ability to form both alpha-1,4– and alpha-1,6–glucosyl linkages. Such a primordial enzyme has been postulated by Fredrick [25] as existing in the first cells to store glucan. During the course of evolution, the possible insertion of nonsense condons into the gene coding for this primitive polypeptide might have caused the synthesis of the two types of bonds to become functions of "separate" but adjacent genes.

A similar *modus operandi* exists in a eukaryote (yeast) and a prokaryote (*E. coli*) for the third and fourth steps of tryptophan biosynthesis. In yeast, two separate enzymes catalyze the two steps, while in the bacterium, both steps are brought about by a single polypeptide.[26, 27]

There appears to be substantial evidence that the genes corresponding to the three enzymes of glycogenesis are closely linked. In addition to immunological similarities,[25, 28] the amino acid sequences of the phosphorylated sites of phosphorylase and synthase are identical.[29, 30] In *E. coli* K-12, there is strong evidence that these genes probably constitute a single unit of transcription.[31]

The later development of photosynthesis made glucose molecules available intracellularly as the monomer. It would seem logical that the already-functioning glucosyltransferases, able to polymerize the monomer to storage forms, would now become closely associated with the sites of synthesis of the glucose, the developing photosynthesis structures.

SYNTHESIS OF STORAGE GLUCANS IN PROKARYOTES AND PRIMITIVE EUKARYOTES

In bacteria, as well as in the Cyanobacteria, the storage glucans formed are invariably highly branched polyglucans.[32] The primitive thylakoids in Cyanophyceae show alpha-granules (or more correctly, *glycogen* granules) between them.[33, 34] In both the cyanophytes, *Nostoc* and *Oscillatoria,* the photosynthetic thylakoids usually occur in the periphery of the cell and are not membrane-enclosed.[35] In both these cyanophytes, the storage glucans consist of highly-branched phytoglycogens and the branching isozymes are all of the *b.e.* type.[33, 36]

The enigmatic hot-springs alga, *Cyanidium caldarium* forms a storage glucan practically indistinguishable from that of cyanophytes.[37] The thylakoids of this primitive eukaryotic alga are gathered together into a membrane-enclosed primitive type of chloroplast.[38] Despite ultrastructural studies, it is not clear whether the limiting membrane surrounding the thylakoids is a single or a double membrane [39] as is found around the chloroplasts of chlorophytes.[40] The storage glucan, a type of phytoglycogen, although intimately associated with the chloroplast, is found *outside* of this structure (FIGURE 1).

Because *Cyanidium caldarium* resembles the chlorophyte *Chlorella* in its morphology and life cycle, it had been suggested by endosymbiont proponents that the alga represented a colorless *Chlorella* (possibly a *Prototheca*) that had acquired cyanobacteria and that these blue-green algae functioned as its chloroplast.[41] The location of storage glucan *within* the chloroplast in *Chlorella* and other chlorophytes would appear to contradict this view.[42]

Polyacrylamide gel electrophoretic studies of the glucosyltransferases of *Cyanidum* have revealed the presence of branching isozymes that are identical with those found in the cyanophyte, *Oscillatoria princeps.*[43] Disruption of the cell wall of *Cyanidium* and the isolation of its chloroplasts disclosed that the glucosyltransferases were not present in the chloroplast. Since the storage glucan is found in the cytoplasm, the polymer-forming enzymes are probably located there. Hence, although the synthesis of glucose occurs within the primitive chloroplast of this alga, the formation of storage glucan appears to take place only outside of the chloroplast.

In this respect, the situation is similar to that found in the red algae. In rhodophytes, the chloroplast is quite primitive [42] and the storage glucan, floridean starch, is found outside of the limiting membranes of the chloroplast [44] (FIGURE 2). The storage glucan is moderately branched and amylopectin-like.[45] While not as highly branched as the phytoglycogens of *Oscillatoria, Nostoc,* and *Cyanidium,* the floridean starch isolated from the red alga, *Rhodymenia pertusa,* for example, resembles phytoglycogen more than it does amylopectin.[45] Recent studies indicate that in another red alga, *Serratacardia maxima,* the glucosyltransferases responsible for the formation of the storage glucan are found only in the cytoplasm of the alga.[46] This appears to be the general situation in the Rhodophyceae.

The branching isozymes of red algae consist of two types, one type identical with the *b.e.* isozymes is found in the cyanophytes and in *Cyanidium caldarium,* and the other of the *Q* type is found in Chlorophycean algae and in higher plants.[45] The two types have been shown to be *charge isomers* of each other, probably differing by only a single amino acid residue substitution.[47]

Undoubtedly, the structure of the storage glucan of red algae, intermediate in degree of branching between the phytoglycogens of cyanophytes and the amylopectins of chlorophytes, reflects the influence of the *Q* isozymes acting in concert with the *b.e.* isozymes of these algae. It has recently been shown that in maize, the various types of branching isozymes determine the ultimate branching characteristics of the different glucans.[48]

The cyanophytes and the rhodophytes seem to be closely related in an evolutionary sense, particularly with regard to the enzymes of glycogenesis.

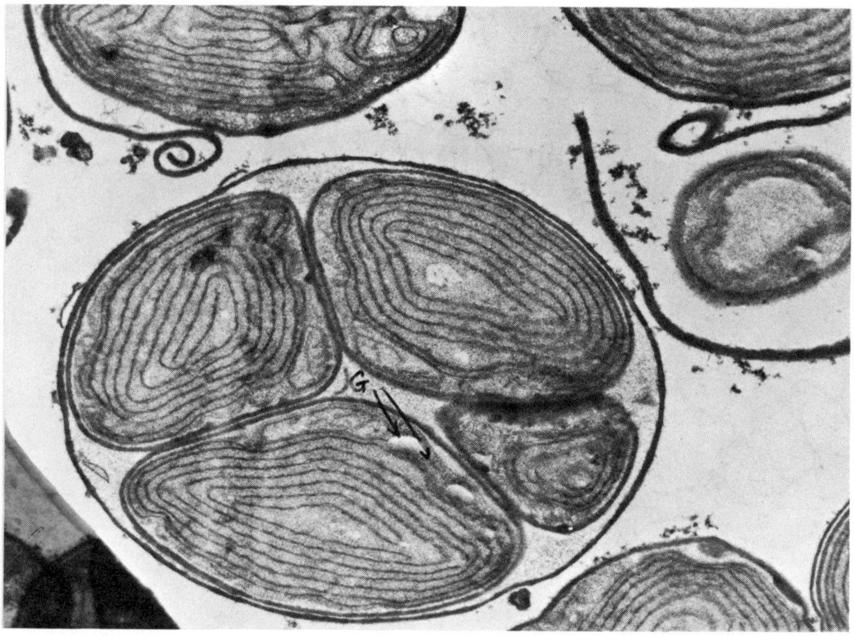

FIGURE 1. *Cyanidium caldarium.* Note the presence of storage glucan granules (G) outside of the chloroplastic membrane. The storage glucan is synthesized in the cytoplasm and not inside of the chloroplast. (×27,300). Courtesy of Dr. Joseph Seckbach.

Those enzymes that form storage glucans in these algae, appear to form an evolutionary progression from the Cyanophyceae to the Rhodophyceae. Intermediate between these two groups of prokaryotic and eukaryotic algae, the enigmatic *Cyanidium caldarium* fits as a "missing link," having properties of both groups. It may very well be *the* transition form between prokaryotic and eukaryotic algae.

Synthesis of Storage Glucans in Eukaryotic Algae

In green algae (Chlorophyceae) such as *Chlorella* and *Spirogyra,* the chloroplast is more structured, with the thylakoids forming grana and the chloroplast surrounded by a double membrane. The storage glucan formed by green algae and by higher plants is *true* starch, a mixture of the moderately branched amylopectin and the exclusively alpha-1,4–linked amylose. The starch granules are found within the chloroplast itself.

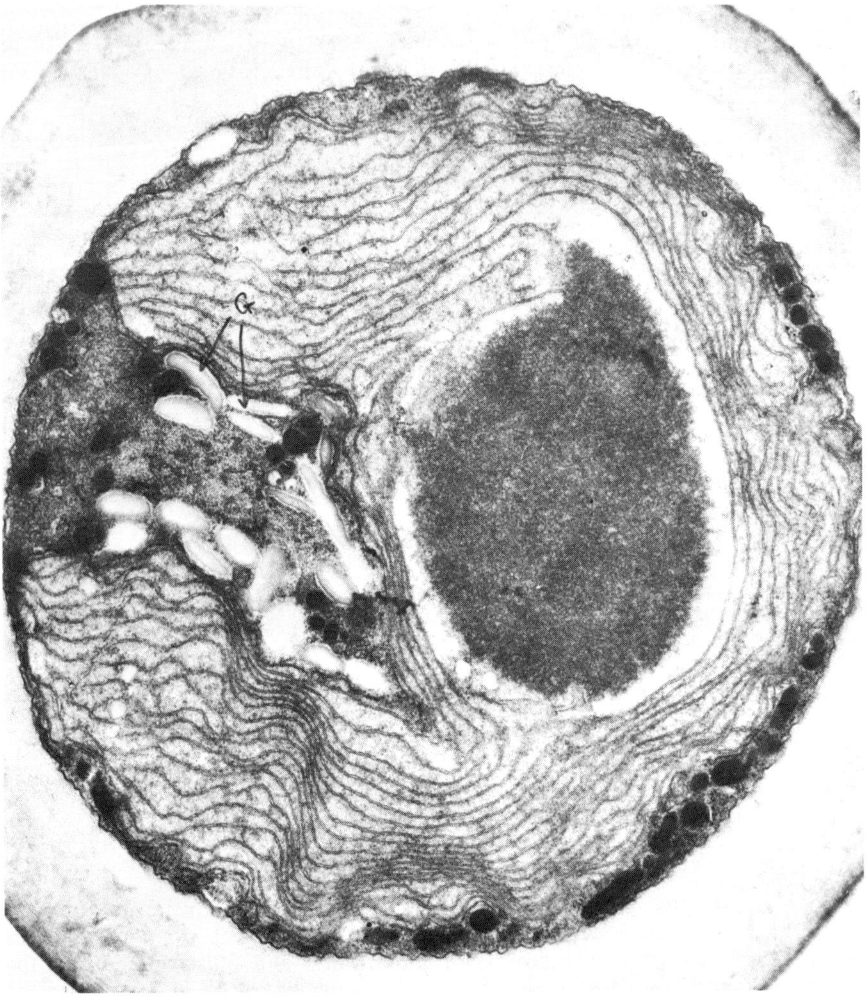

FIGURE 2. *Acrochaetium daviesii.* The granules of floridean starch (G) of this rhodophyte are also formed outside of the chloroplast. (×25,500). Courtesy of Dr. Hendrik H. van der Velde.

The pattern of glucosyltransferases obtained after two-dimensional polyacrylamide gel electrophoresis indicates the presence, exclusively, of the *Q* type of branching isozymes.[12] In *Chlorella pyrenoidosa,* as well as in *Spirogyra setiformis,* no *b.e.* type of branching enzyme has been reported. Amylopectin appears to be *the* only branched glucan formed by these chlorophytes. It is much less branched than the phytoglycogens found in blue-green algae and in *Cyanidium* or the floridean starch of red algae.

The unusual chlorophyte, *Glaucocystis nostochinearum,* has often been cited by endosymbiontists as an example of the acquisition of blue-green algae which act as chloroplasts ("cyanelles") for a eukaryotic alga.[49] Because of the brilliant blue coloration of these cyanelles, it was thought that unicellular cyanobacteria (even classified as a new genus of Cyanophyceae) formed the photosynthetic organelles of this alga.

Ultrastructural studies of *Glaucocystis* reveal the presence of chloroplasts which appear to be surrounded by a thin plasma membrane. This absence of a double membrane, and also of murein (a normal constituent of prokaryotic membranes [50]), has been explained by endosymbiontists as an adaptation to the symbiosis, even though double membranes are present around the chloroplasts of all other chlorophytes. Storage glucan granules are found in the cytoplasm of this alga and have not been reported within the cyanelles.[49] The storage glucan is a true starch, identical with that found in *Chlorella* and other members of the Chlorophyceae.[51]

Polyacrylamide gel electrophoretic studies have shown that the branching isozymes of this alga are all of the *Q* type.[51] (FIGURE 3). Disruption of the cell wall of the alga by treatment with Cellulysin, and the subsequent isolation of the cyanelles, has made possible the detection of phosphorylases, synthases, and *Q* branching isozymes in these structures. All three groups of glucosyltransferases appear to be present in *both* the cyanelles and the cytoplasm of this eukaryotic alga. This situation is rather unique in that glucose polymerizing enzymes exist in the chloroplasts or cyanelles but storage glucan is formed only in the cytoplasm. The lack of the usual double membrane around these photosynthetic organelles may allow diffusion of the enzyme molecules between the organelles and the cytoplasm.

If *Glaucocystis* represents a eukaryotic alga (chlorophyte) that has lost its chloroplasts and reacquired cyanobacteria,[49] then the branching isozymes of these cyanobacteria should cause the synthesis of a phytoglycogen-type of storage glucan rather than the true starch found in the cytoplasm of this alga.

Recent investigations on the colorless alga, *Prototheca zopfii* have shown that this alga, devoid of chloroplasts and chloroplastic DNA, nevertheless forms a true starch in its cytoplasm.[52] If *Prototheca* also represents a chlorophyte that has lost its chloroplasts, it too has retained the ability to form a true starch as its storage glucan. Therefore, the ability to form true starch may be the result of the biosynthesis of *Q* types of branching enzymes, by *nuclear* genes and not chloroplastic genes.

It seems posible that *Glaucocystis* may represent an intermediate stage in the evolution of the Chlorophycean chloroplast,[53] where the biosynthesis of storage glucan is still extrachloroplastic, and determined by genes residing in the nucleus.

Immunological studies have shown that the *Q* branching enzymes of Rhodophyceae may be charge isomers of the *b.e.* type of branching enzymes of Cyanophyceae. The mutation responsible for the conversion of the *b.e.*-coding gene to the *Q*-coding gene obviously occurred along the way from the Cyanophyceae to the Rhodophyceae; the *Q* type of branching isozymes first making their appearance in the rhodophytes. However, *b.e.* types of branching isozymes exist in red algae together with the *Q* types. This is not so in the

Chlorophyceae, where the branching isozymes are exclusively of the *Q* type. This has led to speculation that the red algae were the progenitors of the Chlorophyceae.[53, 54]

Recent studies, using the techniques of tandem-crossed immunoelectrophoresis, however, have indicated that the *Q* branching isozymes of red algae are more immunologically related to the *b.e.* isozymes of cyanophytes than to the

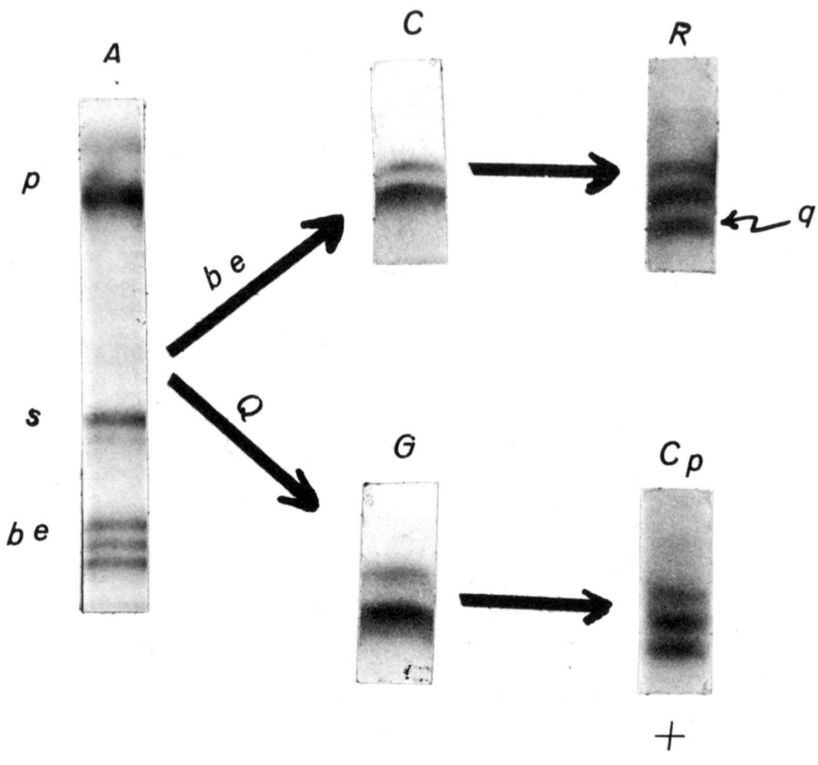

FIGURE 3. Gel electrophoresis showing possible dual line of evolution of glucan branching enzymes of algae. *p*, phosphorylases; *s*, synthases; *b.e.*, branching enzymes. *A* = whole gel pattern of glucosyltransferases of blue-green alga, *Anacystis nidulans.* The branching enzymes are at the bottom (anodal) end of the gel. All three are capable of branching amylopectin to glycogen. *C* = the branching enzymes of *Cyanidium caldarium.* All are of the same type (*b.e.*) as those of *A*. *G* = branching isozymes of the green alga, *Glaucocystis nostochinearum.* The two branching isozymes are of the classical *Q* type and can only branch amylose to amylopectin. *R* = the three branching enzymes of the red alga, *Rhodymenia pertusa.* Two are of the *b.e.* type, while the more anodal (+) is a *Q* type (arrow). *Cp* = the three branching isozymes of the green alga, *Chlorella pyrenoidosa* are all of the *Q* type.

Q isozymes of chlorophytes.[55] This would seem to support the thesis for a biphyletic evolution of the red algae and the green algae from a Cyanophycean ancestor.[56, 57]

It is, of course, hazardous to construct any type of phylogeny on the basis of a single trait. However, since it appears that the formation of storage glucans was of great importance in the survival of the first primitive cells, this particular characteristic may have had selective value during the course of evolution.

With this in mind, we can begin to ask certain questions as to the origins of the eukaryotic chloroplast. If this structure was the result of an endosymbiotic relationship, then the formation of storage glucan *outside* of the chloroplast in red algae and in *Cyanidium,* and the formation of storage glucan within the chloroplastic lamellae in green algae and in higher plants, seems unusual. Particularly one must account for the extrachloroplastic glucan always being highly branched (phytoglycogen, floridean starch, etc.) as compared with the intrachloroplastic glucan (amylose, amylopectin). To our knowledge, prokaryotic algae never form any detectable quantity of amylose, while the eukaryotic algae always do.

The intimate association of the glucosyltranferase enzymes responsible for the biosynthesis of storage glucans, with the photosynthesis structures, make it mandatory that these isozymes be taken into consideration whether one chooses to derive the origin of the chloroplast from endosymbiotic blue-green algae or from the cyanophytes via the accumulation of advantageous mutations.

Acknowledgments

We are indebted to the Dodge Institute for Advanced Studies for financial support of this study. Thanks are also due to Mrs. Margaret S. Boothe of the Cambridge Research Laboratories of the Dodge Chemical Company for technical assistance.

References

1. De Witt, W. 1977. Biology of the Cell. :340–357. W. B. Saunders Company. Philadelphia, Pa.
2. Calvin, M. 1969. Chemical Evolution. :144–161. Oxford University Press. Oxford, England.
3. Dodd, J. L. & D. A. McCracken. 1972. Mycologia **59:**1341.
4. Cori, C. F. & G. T. Cori. 1939. J. Biol. Chem. **131:**397.
5. Leloir, L. F., M. A. R. De Fekete & C. E. Cardini. 1961. J. Biol. Chem. **236:**636.
6. French, D. 1975. *In* Biochemistry of Carbohydrates. W. J. Whelan, Ed.: 267–335. Butterworth. London, England.
7. Kainuma, K. & D. French. 1972. Biopolymers **11:**2241.
8. Whelan, W. J. 1976. J. Jap. Soc. Starch Sci. **23:**101.
9. Fredrick, J. F. 1962. Phytochem. **1:**153.
10. De Fekete, M. A. R. 1968. Planta **79:**208.
11. Slabnik, E. & R. B. Frydman. 1970. Biochem. Biophys. Res. Commun. **38:**709.
12. Tsai, C. Y. & O. E. Nelson. 1968. Plant Physiol. **43:**103.
13. Badenhuizen, N. P. 1969. The Biogenesis of Starch in Higher Plants. :27–37. Appleton-Century-Crofts. New York, N.Y.
14. Mangat, B. S. & N. P. Badenhuizen. 1970. Starke **22:**329.
15. Frydman, R. B. & C. E. Cardini. 1967. Plant Physiol. **42:**628.
16. Nordin, J. H. & S. Kirkwood. 1965. Ann. Rev. Plant Physiol. **16:**393.
17. Badenhuizen, N. P. 1963. Nature **197:**464.
18. Ryman, B. E. & W. J. Whelan. 1971. Advances Enzymol. **34:**285.
19. Peat, S., W. J. Whelan & J. M. Bailey. 1953. J. Chem. Soc. (London): 1422.

20. BARKER, S. A., E. J. BOURNE, S. PEAT & I. A. WILKINSON. 1950. J. Chem. Soc. (London): 3022.
21. LAVINTMAN, N. 1966. Arch. Biochem. Biophys. **116:**1.
22. BLACK, R. C., J. D. LOERCH, F. J. MCARDLE & R. G. CREECH. 1966. Genetics **53:**661.
23. RICKSON, F. R. 1971. Science **173:**344.
24. NELSON, O. E. 1973. Ann. N.Y. Acad. Sci. **210:**113.
25. FREDRICK, J. F. 1973. Ann. N.Y Acad. Sci. **210:**254.
26. DE MOSS, J. A. 1965. Biochem. Biophys. Res. Commun. **18:**850.
27. CREIGHTON, T. E. & C. YONOFSKY. 1966. J. Biol. Chem. **241:**4615.
28. FREDRICK, J. F. 1977. Phytochem. **16:**55.
29. LARNER, J. & F. SANGER. 1965. J. Mol. Biol. **11:**491.
30. NOLAN, C., W. B. NOVOA, E. G. KREBS & E. H. FISCHER. 1964. Biochem. **3:**542.
31. CREUZAT-SIGAL, N., M. LATIL-DAMOTTE, J. CATTANEO & J. PUIG. 1972. *In* Biochemistry of the Glycosidic Linkage. R. Piras & H. Pontis, Eds.: 647–680. Academic Press, Inc. New York, N.Y.
32. MANNERS, D. J. 1968. *In* Control of Glycogen Metabolism. W. J. Whelan, Ed.: 83–100. Academic Press, New York, N.Y.
33. PANKRATZ, H. S. & C. C. BOWEN. 1963. Am. J. Bot. **50:**387.
34. GIESY, R. M. 1964. Am. J. Bot. **51:**387.
35. LANG, N. J. 1968. *In* Algae, Man and the Environment. D. F. Jackson, Ed.: 235–248. Syracuse University Press, Syracuse, N.Y.
36. FREDRICK, J. F. 1968. Phytochem. **7:**931.
37. FREDRICK, J. F. 1968. Phytochem. **7:**1573.
38. SECKBACH, J. 1972. Microbios **5:**133.
39. ENAMI, I., H. NAGASHIMA & I. FUKUDA. 1975. Plant Cell Physiol. **16:**221.
40. GIBBS, S. P. 1970. Ann. N.Y. Acad. Sci. **175:**454.
41. RAVEN, P. H. 1970. Science **169:**641.
42. GIBBS, S. P. 1962. J. Cell Biol. **14:**433.
43. FREDRICK, J. F. 1976. Plant Cell Physiol. **17:**317.
44. NAGASHIMA, H., S. NAKAMURA & K. NISIZAWA. 1968. Bot. Mag. (Tokyo) **81:**411.
45. FLEMING, I. D., E. L. HIRST & D. J. MANNERS. 1956. J. Chem. Soc. (London): 2831.
46. NAGASHIMA, H., H. OZAKI, S. NAKAMURA & K. NISIZAWA. 1969. Bot. Mag. (Tokyo) **82:**462.
47. FREDRICK, J. F. 1971. Physiol. Plant. **24:**55.
48. BOYER, C. D. & J. PREISS. 1978. Carbohydrate Res. **61:**321.
49. HALL, W. T. & G. CLAUS. 1967. J. Phycol. **3:**37.
50. SCHENK, H. E. A. 1971. Z. Physiol. Chem. **352:**321.
51. FREDRICK, J. F. 1977. Phytochem. **16:**1489.
52. FREDRICK, J. F. 1979. Phytochem. **18:**1823.
53. CHADEFAUD, M. & R. HEIM. 1976. C.R. Acad. Sci. (Paris) **283:**1029.
54. ALLSOPP, A. 1969. New Phytol. **68:**591.
55. FREDRICK, J. F. 1980. Phytochem. **19:**539.
56. KLEIN, R. M. 1970. Ann. N.Y. Acad. Sci. **175:**623.
57. FREDRICK, J. F. 1973. Plant Cell Physiol. **14:**443.

DISCUSSION OF THE PAPER

DR. J. M. WHATLEY: I would like to ask if you had ever looked at any cryptomonad starch.

DR. J. F. FREDRICK: No I haven't. The only report that I've seen of the cryptomonad starch is from a fellow at the University of Toronto, or University of Montreal, but he reports that's strictly a true starch made up of amylopectin and amylose.

TUMOR VIRUSES AS MODIFIERS OF THE NUCLEAR GENOME OF EUKARYOTIC CELLS

Rose Sheinin

Department of Microbiology and Parasitology
University of Toronto
Toronto, Canada M5S 1A1

INTRODUCTION

It is the purpose of this paper to explore the mechanism of action of tumor viruses as modifiers of host genome. At one extreme, such modification results in the conversion of normal into tumor cell and the death of the host organism; at the other it has given rise to successful chromosomal parasitism reflected in the co-evolution of tumor virus genomes with a number of animal species.

Agents capable of inducing cancer in wild and laboratory animals are found amongst six of the seven families of DNA-containing viruses of eukaryotic organisms (TABLE 1). In contrast, of the nine groups of RNA-containing viruses, only the *Retroviridae* have oncogenic members. The tumor viruses are distinguished from their nononcogenic analogues by one major property. Each is able to effect genetic transformation of a susceptible host cell, giving rise to a new entity in which viral and cellular genome have become one.[11] It is in this sense that the terms transforming virus, virus transformation, and transformed cells are used here.

Neoplastic transformation by a tumor virus is a two-step process (FIGURE 1). In the first, viral DNA becomes covalently linked to the chromosomal DNA of the host cell. This genetic modification, although an essential component of viral oncogenesis, is in itself inadequate to produce a tumor cell. This necessitates step II, expression of a specific, newly added DNA-sequence (the T gene in FIGURE 1), which is ultimately seen in the uncontrolled growth of a malignant cell.

If integration of tumor virus genome occurs only in somatic cells, protected immortality of the virus species is clearly short-lived, in the absence of an infectious virus intermediate which is horizontally transmitted from animal to animal. If, however, the tumor virus genome becomes a part of the germ line inheritance, immortality is assured for a very long time indeed, by vertical transmission from parent to progeny animals.

EVOLUTION OF TUMOR VIRUSES AS CHROMOSOMAL PARASITES

Speculation on the origin of viruses has invoked a variety of mechanisms including evolution from primordial organic matter, and evolution from cellular structures with some capacity for replication. Beyond this one can envisage a number of paths taken by the semi-autonomous microorganisms whose major biological function is reproduction, with or without concomitant destruction of the host cell (FIGURE 2), towards the successful chromosomal parasitism of tumor viruses. The first step would seem to require development of a mecha-

0077–8923/81/0361–0435 $01.75/0

TABLE 1

TUMOR VIRUSES OF EUKARYOTIC ORGANISMS

Family [1]	Example	Genome		Neoplastic disease induced in		Transmission		In vitro
		Nucleic acid	Molecular Weight	Nature	Laboratory	Hori-zontal	Germ cell	Trans-formation
	Hepatitis B [184]	DNA	1.6×10^6	Hepatocellular carcinoma	Hepatocellular carcinoma	+	−	+
Papovaviridae: Genus B	Polyoma, SV40 [2]	DNA	$\approx 3 \times 10^6$	Unknown	Multi-target [3]	+	−	+
Papovaviridae: Genus A	Bovine papilloma virus [4]	DNA	$\approx 5 \times 10^6$	Benign, malignant disease [5]	Benign, malignant disease	+	−	+
Adenoviridae	Human adenoviruses [6]	DNA	$\approx 20 \times 10^6$	Unknown	Multi-target	+	−	+
Herpetoviridae	Epstein-Barr virus [7]	DNA	$\approx 90 \times 10^6$	Burkitt's lymphoma; nasopharyngeal carcinoma	B-cell lymphoma	+	−	+
Poxviridae	Shope fibroma [8]	DNA	$\approx 120 \times 10^6$	Fibroma	Fibroma, myxoma	+	−	+
Retroviridae: oncovirinae	Vertebrate leukemia-sarcoma viruses [9]	RNA	$10\text{–}12 \times 10^6$	Leukemia, sarcoma	Leukemia sarcoma	+	+	+
	Mouse mammary tumor virus [10]	RNA	$10\text{–}12 \times 10^6$	Mammary carcinoma	Mammary carci-noma	+	+	+

nism for establishment of latency, in which virus genome is harbored in surviving cells, in one of three modes noted in FIGURE 2. From those viruses able to accomplish the latter could have come the tumor viruses capable of effecting integration of their genome into the genetic complement of somatic cells and/or of germ cells.

At the cellular level, optimal evolutionary survival would be expected to take the course of selection of virus that produces a latently infected cell with a complete viral integrate, inherited as a stable conserved component of the cellular genome (FIGURE 2, Mode III). At the level of the whole organism, the most successful chromosomal parasite might be expected to produce genetic transformation of the germ cells; a somewhat less effective tumor virus being one which modifies the genome of only a select population of somatic cells.

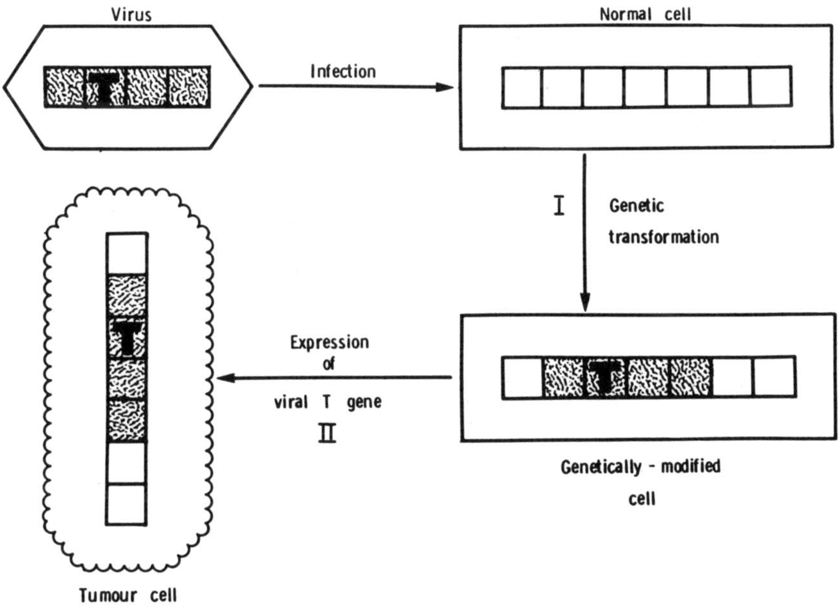

FIGURE 1. Neoplastic transformation by a virus: a model.

Maximum evolutionary pressure on such chromosomal parasites should give rise to an integrated tumor virus genome, whose capacity for autonomous replication and expression is totally under host cell control. Such a virus–cell genome interaction would clearly be compatible with the selection of integrated tumor virus genomes which are mutated in, or deleted for, the genes of virus replication and/or transformation. Equally favored would be the evolution of host cell genes that prevent virus multiplication [12] and transmission.[13]

Over long periods, evolutionary pressures towards maximum silencing of chromosomal parasites would result in the loss of effective tumor–virus genetic information, the evolution of an apparent cellular gene of unknown coding potential, and ultimately the loss of the viral species. The latter possibility has been circumvented in nature by the clearly demonstrated requirement for a stage of horizontal transmission in the life cycle of all tumor viruses.

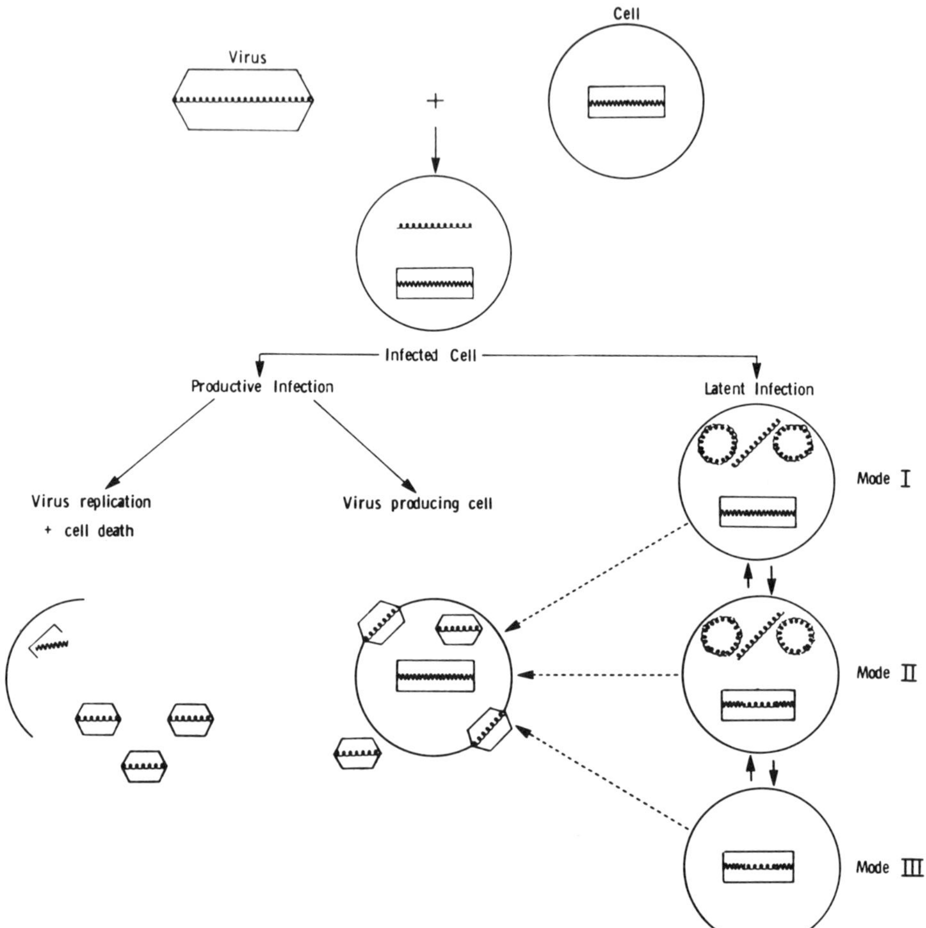

FIGURE 2. Modes of virus infection of eukaryotic cells. Virus and cell genome are respectively designated as and ; episome-like forms and integrated viral DNA as and . The hatched arrows indicate possible interconversion.

DETERMINANTS OF VIRAL ONCOGENESIS

It has become clear over many years that our understanding of the mechanism of action and evolution of oncogenic viruses depends in large part on knowing the structure and function of that genome segment designated as T in FIGURE 1. Investigations with all tumor virus classes, summarized below, reveal that two genetic components may function as codeterminants of neoplastic conversion; one originating from the virus, the virogene; [15] and the other from the cell, the oncogene.[16]

The virogene component, essential for the chromosomal parasitism of the tumor viruses would comprise at a minimum, the origin of replication of the viral genome and part or all of that genome domain which encodes information for a protein which acts in initiation of viral chromosome replication, derepression of the machinery for replication of the host cell chromatin, and for initiation and maintenance of the genetically-transformed state.[2, 6, 9, 17–20]

In some instances it would appear that the virogene component is adequate to determine neoplastic expression of the chromosomal parasite. In others this requires, or is enhanced by, expression of an oncogene; e.g. the *src, erb, myb,* and *mac* genes of some of the oncogenic retroviruses [9, 21, 22] (see FIGURE 3). At a minimum this would necessitate the development of tumor viruses with clear transducing potential,[23] carrying virogene information for genome replication coupled directly or indirectly to cellular DNA sequences.

The mechanisms by which the multitude of tumor virus isolates have evolved, to effect their particular form of chromosomal parasitism and neoplastic conversion are not yet entirely clear. It is anticipated that the precise details will emerge from our understanding of the molecular mechanisms for each virus or virus class, with respect to virus genome replication during the reproductive cycle; integration of virogene into the cell genome; virus genome replication from the integrated state directly, through a replicative intermediate or via excision of virogene from its chromosomal housing.

In this context it is of interest to examine the interrelatedness of the machinery of replication of virus and host cell genomes (TABLE 2). With the exception of the poxviruses, all other classes of tumor viruses show major dependence of virus on host cell genetic information. This dependence is manifested as physical association of virus genome with that of the cell in the nucleus, and sequestration of the enzymes and proteins of cell chromatin replication and modelling. Indeed there is a remarkable similarity between the nucleosomal organization of the viral minichromosome and that of the eukaryotic chromatin.[34–37] In the case of papovavirus group, the analogy has been shown to be marked, in that histones H3, H4, H2A, and H2B are the major determinants of the chromatin substructure [34] (TABLE 3). Such histones have not been detected in the subunit structures of adenoviruses,[35] herpesviruses,[36] and poxviruses.[37] Their function in the adenovirus chromosome appears to be served by other highly basic [38] and/or phosphorylated proteins,[39] one of which is covalently bonded to the termini of the linear, double-stranded DNA genome.[40] The significance of these observations for the evolution of the oncogenic viruses remains to be established. However they may have major import for the mechanism of chromosomal parasitism of each.

A number of possible mechanisms for integration of tumor virus genome into host cell chromatin have long been entertained. These include site-specific integration analogous to that effected by the transducing bacteriophage

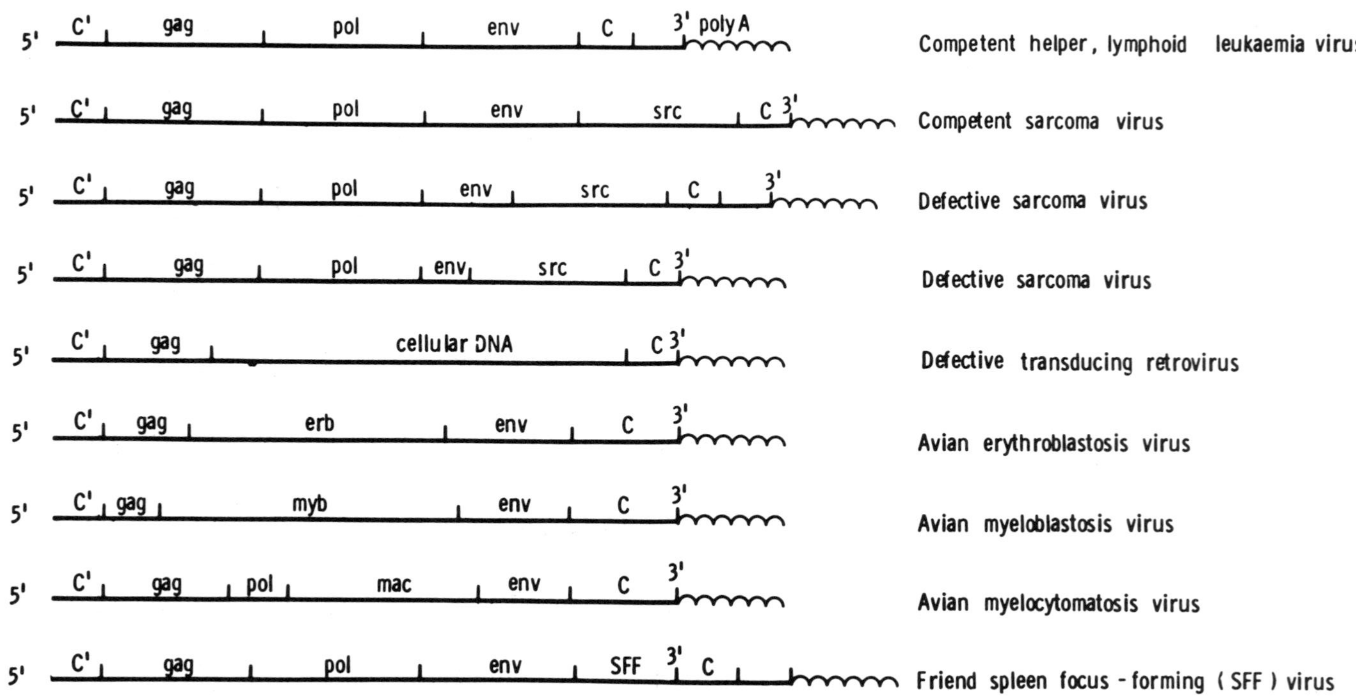

FIGURE 3. Models of genomes of avian and murine retroviruses. Virogene components C, C′, *gag*, *pol*, *env*, respectively encoding reiterated DNA sequences (which probably function in genome replication), virion core proteins (which determine group-specific antigens), the RNA-directed DNA polymerase (reverse transcriptase), and the glycoproteins of the virion envelope. Oncogene components; cellular DNA sequences, e.g. *src*, *erb*, *myb*, and *mac* associated with specificity for fibroblasts, erythroblasts, myeloblasts, macrophages, and spleen cells, respectively. Poly A, cell-derived polyadenylated sequence conferring translation capacity on the RNA molecule. (See section on Retroviruses.)

TABLE 2

SOME PROPERTIES OF TUMOR VIRUS GENOMES

		Genome Replication		Chromosomal Parasitism of		Viral DNA in Transformed Cell		
Virus [14]	Genome Form	Site	Host Genome Interaction	Somatic Cells	Germ Line	Integrate	Copy Number	Episome
Hepatitis B [184–186]	ds, circular	Nucleus	Major dependency	+	−	+	1–few	?
Papovavirus [2, 3, 17, 24]	ds, CC*-DNA	Nucleus	Major dependency	+	−	+	1–few	Rare [25, 26]
Papillomavirus [5, 19]	ds, CC-DNA	Nucleus	Major dependency	+	−	+	10– >1000	+
Adenovirus [6, 27]	ds, linear DNA	Nucleus	Major dependency	+	−	+	1–few	?
Herpesvirus [7, 20, 28–30]	ds, linear DNA	Nucleus	Major autonomy	+	−	+	1–many	+
Poxvirus [8, 31, 32]	ds, linear DNA	Cytoplasm	Autonomy	+	−	+	Unknown	
Retrovirus [9, 21, 33]	RNA dimer	Nucleus	Major dependency	+	+	+	1– >100	±

* ds, double-stranded; CC, covalently-closed.

TABLE 3

THE PAPOVAVIRIDAE: GENUS B, AND THEIR RELATEDNESS

Virus	Designation	Animal of Origin	Relatedness Revealed at Level of Genome	T-antigen	Core	Virion Proteins Histone	Virion Proteins Capsid
Papova-type virus [46]	PTV	American leopard frog					
Polyoma [2]	Py	Mouse	SV40 [2, 47, 48]	SV40 [90]	SV40 [50]	+	—
K-virus [51]	KV	Mouse					
Hamster papovavirus [52]	Ha PV	Syrian hamster					
Simian virus 40 [47]	SV40	African Green monkey	Py [2, 47] BK [53] SA12, STMV [54]	HD, Py [49] BK[40]	Py [2, 50] BK [56] HD, SA12, STMV [54]	+	—
Lymphotropic primate papovavirus [57]		African Green monkey	—				—
HD Papovavirus [58]	HD	Vero monkey	SA12, STMV [54]				—
Primate papovavirus [54]	SA12	Chacma baboon	HD, SA12, SV40 [54]		SV40 [54]		—
Stumptail papovavirus [54]	STMV	Stumptail macaque	HD, SA12, SV40 [54]		SV40 [54]		—
Human Papovavirus	BK [59]	Human beings	JC, SV40 [47, 53]	SV40 [55, 60]	+ [50]	+ [58]	+ [59]
	JC [59]		BK, SV40 [47, 53]		+	+	+
	DAR[61]						
	GS [62]		BK [62]				

lambda;[23] the non–site-specific lysogenization characteristic of the general transducing bacteriophages;[41] the integration associated with gene transposing elements like bacteriophage Mμ,[42] bacterial plasmids,[43] transposons and insertion sequence elements of bacteria;[42, 43] and the analogous DNA segments in eukaryotic microorganisms[44] and corn.[45] Specific evidence obtained with individual tumor viruses is compatible with one or more of these models.

ONCOGENIC DNA-CONTAINING VIRUSES

Considerable bodies of knowledge relevant to the present discussion have been accumulated for the DNA-containing papovaviruses (TABLE 3), the papillomaviruses (TABLE 4), the human adenoviruses,[6, 27] and the herpesviruses (TABLE 5). Analogous data are now emerging rapidly for the hepatitis B family of viruses. By comparison, little is as yet known about the oncogenicity of poxviruses, perhaps because the cytoplasmic site of autonomous viral genome replication (TABLE 2) has made chromosomal parasitism an extremely rare evolutionary event for the *poxviridae.*

The key questions with respect to these viruses are perhaps the following: (i) What is the information encoded in the T locus shown in FIGURE 1? (ii) Does it carry viral information, cellular information or both?

Extensive studies with Py, SV40, BK virus (TABLE 3) and the human adenoviruses[6, 27] suggest that the oncogenic papovaviruses, adenoviruses, and, by implication, papillomaviruses[19] (TABLE 4) must carry into a target cell DNA of the origin of virogene replication and of the domain encoding information for the virus-specific T (for tumor) antigens and/or other related proteins. These are families of phosphorylated, DNA-binding proteins (DBPs) of molecular weight 20–120 kilodaltons,[2, 6, 47, 81–84] with affinity for the origin of virogene replication,[85–87] other sites on the viral genome,[85–90] and for cell DNA sequences.[88–90] It is known that these early viral proteins are required for initiation of viral genome replication,[91–93] for derepression or activation of synthesis of cellular chromatin,[17, 18, 91–95] and for initiation and maintenance of the transformed cell state;[25, 26, 91, 95–97] i.e., for establishing and maintaining chromosomal parasitism in the form of integrated viral DNA. Thus cells transformed by papovaviruses and adenoviruses continuously synthesize large amounts of T antigen homologues and/or other early DBPs, readily detectable in the nucleus,[98, 99] and smaller amounts associated with the plasma membrane.[100, 101] The mechanism of action of these DBPs proteins is not understood, although two enzyme functions have been detected in association with the T antigen; an ATPase[102] and a protein kinase,[103] analogous to that encoded in the *src* region of the sarcomagenic retroviruses, discussed below.

The importance of cellular DNA as part of the oncogenic papova- and adenovirus genomes is not entirely clear. Fully competent and defective virus particles carrying cointegrates of viral–cell DNA have been isolated and their transforming potential has been demonstrated.[104–110] Py[111] and SV40[106] pseudovirions, with only cell DNA in a virus-like capsid shell have also been obtained; but their role in genetic transformation remains unknown. However it seems likely that in the absence of viral genome origin sequences, and information for at least part of the protein for initiation of genome replication, the resulting virus particle will be defective for integration of virogene into host genome, and therefore for chromosomal parastism. Where viruses have evolved to the stage of loss of these genome regions they may re-appear in the form of a recombinant with the analogous segment from a different virus strain [106, 112] or species.[113]

Analogous conclusions are beginning to derive from studies of the oncogenic herpesviruses (TABLE 5). Thus cells transformed by Epstein-Barr virus (EBV) and the closely related analogues of Old World primates [78, 114] carry full or partial integrates of virus DNA [7, 20, 30] and continuously synthesize a chromatin-associated,[115, 116] virus-specified protein termed the Epstein-Barr nuclear antigen (EBNA). Other oncogenic herpesviruses appear to function in a similar manner, giving rise to genetically transformed cells carrying viral DNA

TABLE 4

THE PAPILLOMAVIRUSES AND THEIR RELATEDNESS

Papillomavirus	Designation	Relatedness Revealed at Level of Genome	Relatedness Revealed at Level of Viron Proteins
Mouse [63]	Mn PV		
Chaffinch [64]			
Cottontail rabbit [8, 65]	CRPV		
Domestic rabbit [8, 65]	DRPV		
Canine (oral) [187]	COPV		
Equine [67]	EPV		
Deer [68]			
Sheep [68]	OPV		
Bovine [4, 69, 70]	BPV-1	BPV-2 [70, 71]	
	BPV-2	BPV-1 [70, 71]	
	BPV-3		
Bovine alimentary [4]	BAPV		
Hamster [68]	HaPV		
Chimpanzee [68]			
Human [72, 73]	HPV-1		CRPV [66]
	HPV-2		
	HPV-3		
	HPV-4		
	HPV-5		
	HPV-6		

integrates and continuously synthesizing analogous virus-specified nuclear DNA-binding antigens; e.g., the HUPNA specified by Herpesvirus *papio*,[117] the acid-fixed nuclear binding (AFNB) antigens of HVS- and HVA-transformed cells,[118] and perhaps HSV-I and HSV-II–specified proteins in transformed cells.[119, 120]

Reports of recombinant papillomaviruses,[121] EBV-strains,[114] and HSV-I and -II[122] have been made. However, little is yet known about the significance of such recombination for the generation and/or evolution of oncogenic viruses. Evidence for the presence of oncogenes in the genomes of these viruses has as yet not been obtained.

TABLE 5

HERPESVIRUSES WITH ONCOGENIC POTENTIAL

Virus	Designation	Animal of Origin	Natural Target Cells for: Replication	Natural Target Cells for: Latent Infection	Natural Target Cells for: Transformation	*In vitro* Transformation
Luckés virus [46, 74]		*Rana pipiens* *R. sylvatica*	Epithelial cells of kidney	Epithelial cells of kidney	Epithelial cells of kidney	
Marek's disease virus [75]	MDV	Chicken	Feather follicle cell [76]	T-lymphocyte	T-lymphocyte	T-lymphocyte
Herpesvirus of turkeys [75]	HVT	Turkey		T-lymphocyte	T-lymphocyte	—
Herpesvirus Sylvilagus [77]	—	Cottontail-rabbit	—	Lymphocyte	Lymphocyte	—
Herpesvirus Saimiri [78]	HVS	Squirrel monkey	Oropharyngeal epithelium	T-lymphocyte	T-lymphocyte	—
Herpesvirus Ateles [78]	HVA	Spider monkey	Oropharyngeal epithelium	T-lymphocyte	T-lymphocyte	T-lymphocyte
Epstein-Barr [7]	EBV	Human	Epithelial cells of oropharynx	B-lymphocyte, epithelial cell of nasopharynx	B-lymphocyte, epithelial cell of nasopharynx	B-lymphocyte of human and other primates
Herpes simplex virus, type I [28]	HSV-I	Human	Epithelium of upper respiratory tract	Trigeminal ganglion [79]	—	+
Herpes simplex virus, type II [29]	HSV-II	Human	Epithelium of genitalia	Sacral ganglion [79]	Cervical epithelial cell [80]	+

Genetic Relatedness Amongst DNA-Containing Viruses

It has long been recognized that the tumor viruses, at least within classes, are related. Yet their uniqueness as a major group of oncogenic agents within, and outside of the families of viruses has been equally compelling. Clarification of these interrelationships has come from studies of the relatedness between viral genes and the sequence homology between viral genome and the integrated viral DNA carried as chromosomal parasite. In a majority of instances such experiments have rested on the technique of molecular hybridization as applied to whole genomes, or to fragments derived by cleavage of the viral genome with a spectrum of site-specific restriction endonucleases. Equally compelling are studies that rest on amino acid and/or peptide homology of gene products and those that assess relatedness by immunological cross-reactivity amongst proteins.

Study of the structure and interrelatedness between the genomes of Py, SV40, and the human papovaviruses (Table 3) has revealed marked sequence homology in that region carrying the origin for genome replication, somewhat less in the domain determining the primary structure of the T-antigen, and perhaps less in the genes for major virion protein. Little cross-reactivity has been detected amongst the capsid proteins except in the case of SV40 and some human papovaviruses isolates.[50, 59]

The genetic relatedness amongst the so-called SV40-PY papovaviruses subgroup[50] may reflect a common evolutionary ancestor, or it may be an example of quite recent horizontal transmission between species. Of relevance to these considerations will be the rapidly emerging studies on DNA sequence homology amongst the papovaviruses of vertebrate families and species; e.g. amongst the primate papovaviruses (Table 3).

Particularly intriguing in this context is the apparent total unrelatedness between the genomes of viruses of Genus A and Genus B *Papovaviridae.* And yet the papillomaviruses are structurally similar to the papovaviruses and both appear to replicate in like manner.[2, 19] Multiple papillomavirus isolates have been made from individual host species. These are particularly interesting from the evolutionary viewpoint because so many tend to exhibit genetic uniqueness (Table 4). On the other hand, molecular hybridization data could argue for relatedness between BPV-1 and -2 and an equine papillomavirus that may play a role in the etiology of equine sarcoid fibromas.[123]

Amongst the adenoviruses, the most extensively studied are the >35 human isolates.[6, 124, 125] These have been grouped according to their oncogenic potential as assessed by sensitivity of hamster neonates to inoculation. Type A, B, and C are respectively highly-, weakly-, and non-oncogenic. The genetic relatedness within each group is stronger than between groups, with type A virus genomes showing marked DNA sequence homology. Little is known about the evolution of the human adenoviruses; however one might expect recombination between species to have been important.

Structural analyses reveal extensive likeness in specific DNA sequences in the genomes of many isolates of HSV-I and and -II;[122] a close relationship amongst the growing number of EBV-isolates and EBV-like herpesviruses derived from the higher primates;[7, 20, 78, 114] and nucleotide sequence homology between MDV and HVT.[75, 126, 127] In this context it is of interest that herpesviruses entered the genetic pool of *Homo sapiens* recently on the paleontological time scale,[128] and evolved into agents that can establish latency in select somatic cells in highly protected sites within the host.[79] This aspect of the natural history of herpesvirus infection may be conducive to the generation of many, and varied viral recombinants with oncogenic potential.

The Oncovirinae

The oncogenic retroviruses have perhaps yielded the most salient information on tumor viruses as chromosomal parasites, as mobilizers of cellular genetic material, and as possible determinants of evolution. This large and growing family of microorganisms (Table 6) is unique amongst the oncogenic viruses in that it has an RNA genome [9] that is obligatorily replicated through a DNA provirus [9, 148] integrated into the cellular chromatin (Figure 4). Provirus formation is mediated by a virus-encoded, reverse transcriptase [149] whereas replication of progeny virus RNA from the integrated protovirus or virogene is entirely mediated by the cell-determined, DNA-directed RNA polymerase II.[9]

The major evolutionary importance of the oncogenic retroviruses is that they are able to effect chromosomal parasitism of germ cells. Their entry into the vertebrate germ line inheritance appears to have occurred early [150] (Figure 5); perhaps as an unique event or, more likely, in multiple stages. The present retrovirus spectrum indicates marked divergence within *oncovirinae* subgroups not necessarily along directions of host speciation.[21, 136, 141, 143, 150, 152] Such divergence may have occurred by mutation within the chromosomal parasite and/or by re-arrangement of virogene within the host genome.[153] What appears to be an even likelier route for retrovirus evolution is recombination during genome replication; between endogenous virogenes, between virogenes and cellular genes, between endogenous viral genome and that of a super-infecting exogenous virus, and between exogenous viruses.[21, 136, 143, 152, 154] Such recombinational events, which are observed to occur very rapidly in nature, in laboratory animals, and in *in vitro* cultures, may explain the enormous diversity in form and in function of the genomes of oncogenic retroviruses (Figures 3 & 6). The possible mechanisms for such recombination are several.[21, 153–156] They involve both DNA replication and transcription, and provide for major movement of viral and cellular genes within a cell, between cells, and between organisms.

Perhaps the most exciting aspect of retrovirus study derives from the recently emerging facts confirming the original models of tumor viruses as transducing agents.[157] The data for the ALSVs is already extensive,[21, 22, 148, 156, 158, 159] that for the MuLSVs is catching up rapidly,[160, 161] that for the FeLV-FeSVs [141, 162] and the primate analogoues [163] is building, and that for the human retroviruses is ready for the plucking.[164]

The vertebrate sarcoma viruses, which show specificity for *in vivo* and *in vitro* transformation of fibroblasts (often across species barriers) are now known to carry a genome segment designated as *src* (Figure 3), which bears nucleotide sequence homology with a *sarc* cellular gene.[9, 21, 158–171] These genetic regions have been shown to code for similar phosphoproteins that exhibit protein kinase activity [166–169] with specificity for the hydroxyl group on tyrosine residues [172] of a very small group of cellular proteins.[172, 173] The implication is that the sarcomagenic retroviruses evolved through recombination between a prototype retrovirus and a specific cellular DNA sequence that confers upon the new virus, cell tropism with respect to target for genetic modification.

TABLE 6

THE ONCOGENIC RETROVIRUSES

Virus or Complex	Designation	Type	Animal of Origin	Vertebrate Order
C-type virus [129]	—	C	Boney fish	Pisces
Frog Oncovirus [46, 130]		C	Frog	Amphibia
C-type virus [131]	VV-1, etc.	C	Vipers, snakes	Reptilia
Avian leukosis sarcoma viruses [21]	ALSVs	C	Chicken	Aves
Reticuloendothelial virus [132]	REV	C	Chicken	Aves
Spleen necrosis virus [133]	SNV	C	Chicken	Aves
Bovine leukosis leukemia viruses [134]	BLV	C	Cattle	Ungulata
Murine leukemia sarcoma viruses [135]	MuLV, MSV	C	Mouse	Rodentia
Mouse mammary tumour virus [10, 136]	MMTV	B	Mouse	Rodentia
Rat sarcoma virus [137]		C	Rat	Rodentia
Hamster sarcoma virus [138]	HaSV	C	Hamster	Rodentia
Guinae Pig C-type virus [139]		C	Guinea Pig	Rodentia
Agouti C-type virus [140]	DPC-1	C	Agouti	Dasyprocta
Pig sarcoma virus [139]		C	Pigs	Artiodactyla
Feline leukemia sarcoma virus [141]	FeLV, FeSV	C	Cats	Carnivores
Canine leukemia sarcoma viruses [142]	CaLV, CaSV	C	Dogs	Carnivores
Primate leukemia sarcoma complex [143]		C	Chimpanzee, gorilla, etc.	Primates
Simian sarcoma virus	SiSV	C	Woolly monkey	Primates
Baboon endogenous virus	BaEV, M28, PP-1	C	Baboon	Primates
Gibbon ape leukemia virus	GaLV	C	Gibbon ape	Primates
M. arctoides type C	MAC-1	C	Stumptial monkey	Primates
Mason-Pfizer virus	MP-MV	D	Rhesus monkey	Primates
Langur virus	PO-1-Lu	D	Langur	Primates
Human C-type retrovirus [144]	HL-23V,[145] SH-DHL-1,[146] HEL-12 [147]	C	Human being	Hominids

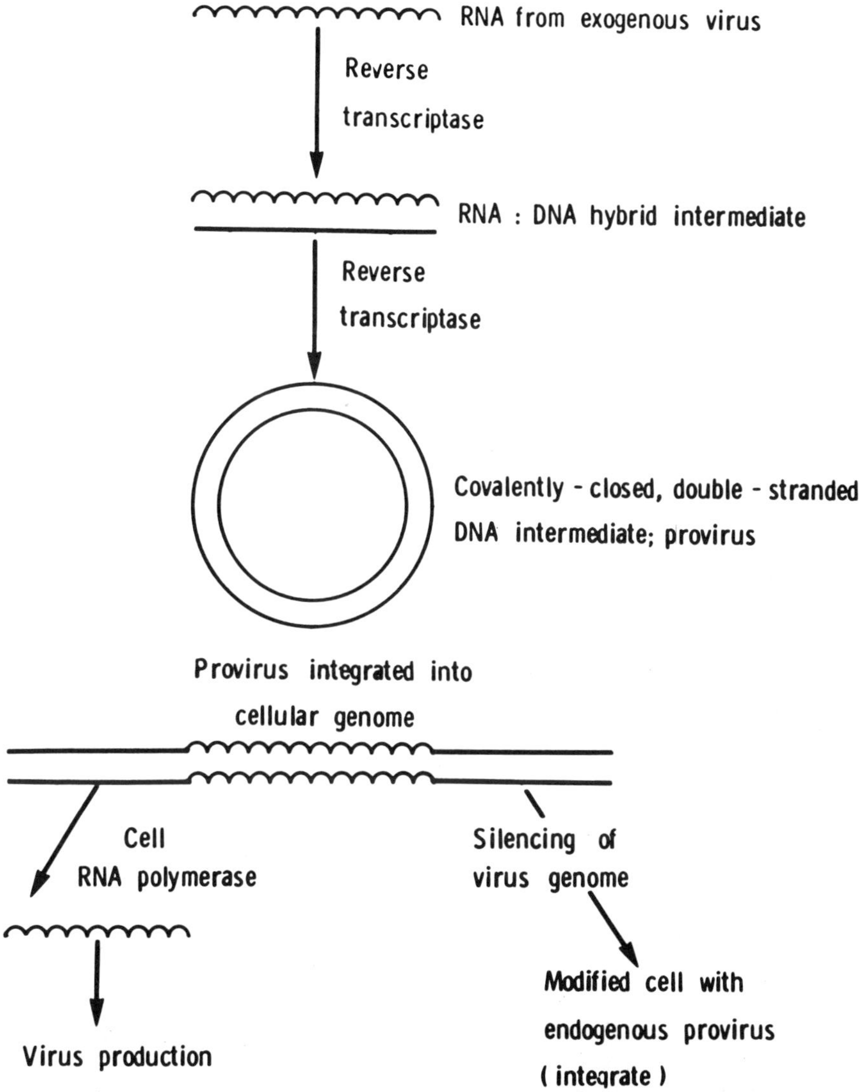

FIGURE 4. Model for replication of retrovirus genomes and establishment of endogenous provirus.

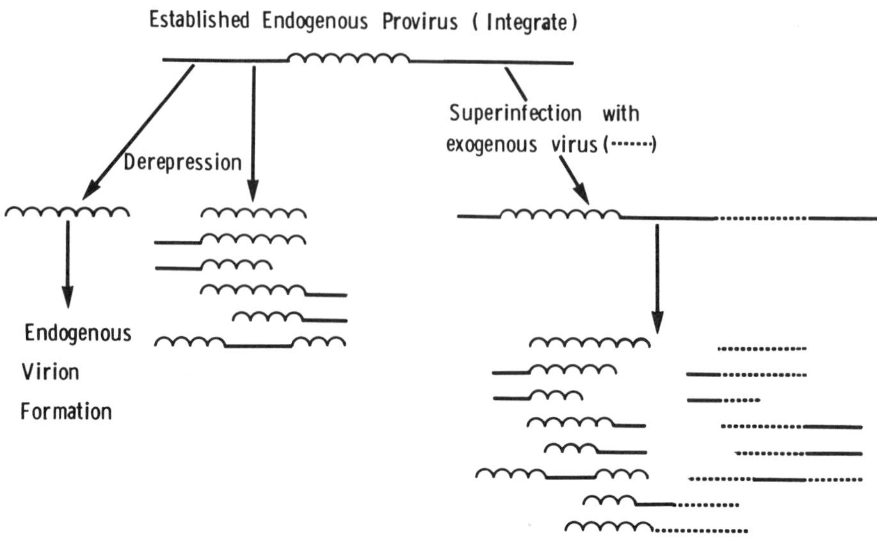

FIGURE 5. Evolutionary trees indicating retrovirus hosts. Model modified from Figures 7–24 of Reference 151. Underlinings indicate detected retrovirus.

By analogy, similar recombination events may have resulted in the appearance of the avian erythroblastosis, myeloblastosis, and myelocytomatosis viruses which have been shown to carry specific, and distinct, cell-derived DNA sequences designated as *erb, myb,* and *mac,* respectively.[21, 22, 174] These terms indicate that genetic transformation and induction of lymphoid leukemia viruses are targeted on erythroblasts, myeloblasts, and macrophages. A similar recombinational event may underlie the relatedness between host genome and the murine sarcoma viruses,[160, 168, 169] the Abelson murine lymphoma virus,[165, 175] the Friend leukemia virus, strains A and P,[176] and the Friend spleen focus-forming virus.[177] Indeed it may provide for a general mechanism for conferring tissue cell specificity on the retroviruses.

Genetic Relatedness Amongst Retroviruses

Strong genetic ties have been established amongst members of major retrovirus complexes of separate vertebrate orders and families. Individual virus strains induce *in vitro* genetic transformation in specific target cells and/or elicit neoplastic disease in analogous cells within different host species. This has made it possible to test for the presence of any given virus or complex, as genetically inherited endogenous protovirus or virogene.

From such investigations have come a number of important observations, some totally unexpected. The B-type retroviruses, determinants of mammary carcinoma are genetically distinct from the leukemogenic-sarcomagenic C-type viruses and their related D-type endogenous viruses.[9, 10, 136, 178] Retrovirus genomes, although largely evolving with the genome of the vertebrate hosts, have undergone major divergence, some of which clearly derives from horizontal transmission between species, between genuses, and between orders.[21, 136, 143, 152, 179]

Several years ago evidence began to accumulate pointing to the major relatedness between endogenous type-C retroviruses of gibbon apes and orangutans, whose origin is traced to sites in Asia.[143, 180] Such DNA sequences showed no homology with the C-type protoviruses of the African true ape, gorilla, and chimpanzee. Because the probes derived from the endogenous viruses of the higher primates of Asia, but not of their African relatives, appear to detect sequences in human tissues, it was suggested that hominids may have originated in Asia. This conclusion has received some support from the recent fossil find of an anthropoid primate in Burma.[181]

Studies on endogenous retrovirus sequences and their germ-line inheritance in mammals generally, in primates and particularly amongst the higher apes have major relevance for our understanding of the retroviruses of *Homo*

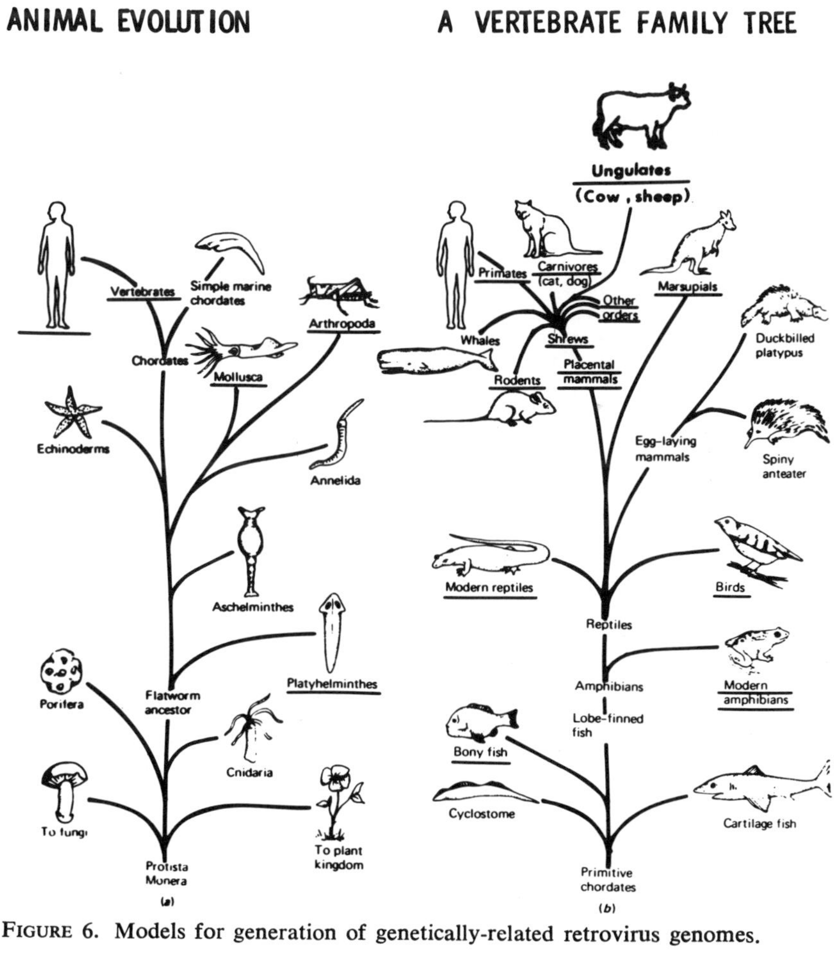

FIGURE 6. Models for generation of genetically-related retrovirus genomes.

sapiens and our present search for them. Clearly the evidence is overwhelming that all mammalian species carry such chromosomal parasites, making human beings an unlikely exception. Equally important is the finding that there is major genome homology amongst the leukemia–sarcoma retrovirus complex of the hominid-related higher apes and at least one human retrovirus isolate.[164] These findings will re-open the study of oncogenic human retroviruses with new insights, new tools, and renewed expectations.

Summary, Synthesis, and Perspectives

A number of important unifying principles are emerging from the study of the various classes of tumor viruses. First and foremost is the fact that all are lysogenizing viruses, able to effect integration of viral DNA into the genome of a susceptible host cell. In the case of the DNA-containing viruses such genetic modification is confined to somatic cells; whereas supreme chromosomal parasitism of germ cells is the prerogative of the oncogenic retroviruses.

The molecular mechanisms for integration of viral into cellular DNA remains to be established for every tumor virus class. By analogy with the lysogenizing bacteriophages, plasmids, and transposable elements, attention is however drawn to the possible obligate role of covalently closed double-stranded DNA genome forms, already identified amongst the papovaviruses, the papillomaviruses, the herpesviruses, and the retroviruses (Table 2). Sporadic reports of analogous intermediates in adenovirus replication [182] become more interesting in this context. However more than one general mode of integration may well be uncovered.

There appears to be a spectrum of tumor virus isolates that shows some correlation with the neoplastic disease evoked. At one extreme are the papovaviruses which establish chromosomal parasitism as a very rare outcome of virus–cell interaction *in vivo* or *in vitro*. At the other are the extremely efficient *oncovirinae*.

In the case of the papovaviruses and the adenoviruses, the virogene is integrated in one, or very few sites (Table 2), which appear to be randomly distributed over the chromosomal complement; but may exhibit nucleotide sequence specificity not yet appreciated. Little evidence exists for natural recombination between cell and Py virus genomes, or between viral genomes. On the other hand, recombinant viruses have been obtained by super-infection of SV40-transformed cells by human adenovirus,[113] perhaps between primate papovaviruses after SV40 was inadvertently introduced into the human population [183] and by induction of SV40 rescue from transformed cells.[106] In contrast, the capacity of many of the oncogenic retroviruses for integration into multiple chromosomal sites (Table 2), for recombination between virus genomes, and for recombination between virogenes and cellular DNA segments is quite remarkable;[21, 22, 136, 148, 152–156, 158–162, 165–167, 174–179] perhaps explaining their efficiency as tumor viruses and as chromosomal parasites.

It is interesting that no neoplastic disease has, as yet, been identified as an outcome of natural infection by the papovaviruses or the human adenoviruses (Table 1). This contrasts markedly with the demonstration that the papilloma viruses, the oncogenic herpesviruses, the Shope fibroma, and the *Molluscum contagiosum* viruses elicit benign and malignant disease in very specific target cells of their hosts of origin. The nature of the determinants for such host-cell selectivity are not yet understood. They may or may not result from transduction of specific cellular DNA sequences that appear to play a role in cellular tropism characteristic of the *oncovirinae*. Most interesting and promising from the latter point of view are the oncogenic herpesviruses with B- or T-lymphocyte tropism (TABLE 4), and the tissue-specific human and bovine papilloma viruses (TABLE 3). On the other hand target specificity may be determined through the generation of recombinant viruses; e.g. thc human papilloma virus HPV-3+ HPV-5A isolate derived apparently in the course of neoplastic conversion of dysplastic epidermal cells.[121]

A totally unexpected fallout from the study of tumor viruses has been the realization that these can be used as tools for the study of vertebrate evolution. This is particularly so for the retroviruses that are transmitted vertically. The evidence obtained bears not only on the relatedness of vertebrate species, but on their paleontological origins.[143, 180]

REFERENCES

1. MELNICK, J. L. 1976. Progr. Med. Virol. **22:**211–221.
2. FAREED, G. C. & D. DAVOLI. 1977. Ann. Rev. Biochem. **46:**471–522.
3. CONSIGLI, R. A. & M. S. CENTER. 1978. Crit. Rev. Microbiol. **6:**263–299.
4. JARRETT, W. F. H., P. E. MCNEIL, W. T. R. GRIMSHAW, I. E. SELMAN & W. T. H. MCINTYRE. 1978. Nature **274:**215–217.
5. LANCASTER, W. D. & C. OLSON. 1980. *In* Viruses in Naturally-Occurring Cancers. G. J. Todaro, M. Essex & H. zur Hausen, Eds. :223–232. Cold Spring Harbor Laboratory. Cold Spring Harbor, New York.
6. WOLD, W. S. M., M. GREEN & W. BÜTTNER. 1978. *In* Adenoviruses. D. P. Nayak, Ed.: 673–678. Marcel Dekker. New York, N.Y.
7. EPSTEIN, M. A. & B. G. ACHONG, Eds. 1979. The Epstein-Barr Virus. Springer Verlag. New York, N.Y.
8. SHOPE, R. E. 1966. Perspect. Biol. Med. **9:**258–274.
9. BISHOP, J. M. 1978. Ann. Rev. Biochem. **47:**35–88.
10. MOORE, D. H., C. S. LONG & A. B. VAIDYA. 1979. Adv. Cancer. Res. **29:** 347–418.
11. SHEININ, R. 1975. Laryngoscope **85:**468–486.
12. JOLICOEUR, P. 1979. Curr. Top. Microbiol. Immunol. **86:**67–122.
13. NOTKINS, A. L. & D. L. LODMELL. 1975. Perspect. Virol. **9:**115–134.
14. GROSS, L. 1970. The Oncogenic Viruses. Pergamon Press. Oxord.
15. HUEBNER, R. J. & R. V. GILDEN. 1971. *In* RNA Viruses and Host Genomes in Oncogenesis. P. Emmelot & P. Bentvelzen, Eds.: 197–219. North-Holland Publishing Co.
16. HUEBNER, R. J. & G. J. TODARO. 1969. Proc. Natl. Acad. Sci. USA **64:** 1087–1094.
17. SHEININ, R. 1967. *In*: The Molecular Biology of Viruses. J. S. Colter & W. Paranchych, Eds.: 627–643. Academic Press, Inc. New York, N.Y.
18. SHEININ, R., J. HUMBERT & R. E. PEARLMAN. 1978. Ann. Rev. Biochem. **47:** 331–380.
19. FAVRE, M., F. BREITBURD, O. CROISSANT & G. ORTH. 1977. J. Virol. **21:** 1205–1209.
20. KAPLAN, A. S., ED. 1979. The Herpesviruses. Academic Press, Inc. New York, N.Y.
21. ROBINSON, H. L. 1978. Curr. Top. Microbiol. Immunol. **83:**1–36.
22. ROUSSEL, M., S. SAULE, C. LAGROU, C. ROMMENS, H. BEUG, T. GRAF & D. STEHELIN. 1979. Nature **281:**452–455.

23. LANDY, A. & W. ROSS. 1977. Science **197:**1147–1160.
24. BOTCHAN, M., W. TOPP & J. SAMBROOK. 1978. Cold Spring Harbor Symp. Quant. Biol. **43:**709–719.
25. TAKEMOTO, K. K., H. LINKE, T. MIYAMURA & G. C. FAREED. 1979. J. Virol. **29:**1177–1185.
26. BIRG, F., R. DULBECCO, M. FRIED & R. KAMEN. 1979. J. Virol. **29:**633–648.
27. KELLY, T. J. JR. & R. L. LECHNER. 1978. Cold Spring Harbor Symp. Quant. Biol. **43:**721–728.
28. RAPP, F. & E. J. SHILLITOE. 1978. *In* Oncogenesis and Herpesviruses III. G. DeThé, W. Henle & F. Rapp, Eds. IARC Sci. Publ. 22 (Pt. 2): 431–450. Lyon, France.
29. SABIN, A. B. 1976. Adv. Pathobiol. **5:**136–150.
30. ADAMS, S. 1979. *In* The Epstein-Barr Virus. M. A. Epstein & B. G. Achong, Eds.: 155–183, Springer Verlag. Heidelberg.
31. TAYLOR, J. L. & H. ROUHANDEH. 1977. Biochim. Biophys. Acta **478:**59–67.
32. MCFADDEN, G., W. E. PACE, J. PURRES & S. DALES. 1979. Virology **94:** 297–313.
33. COHEN, J. C. & H. E. VARMUS. 1979. Nature **278:**418–423.
34. CREMISI, C. 1979. Microbiol. Rev. **43:**297–319.
35. SERGEANT, A., M. A. TIGGES & H. J. RASKAS. 1979. J. Virol. **29:**888–898.
36. SHAW, J. E., L. F. LEVINGER & C. W. CARTER, JR. 1979. J. Virol. **29:**657–665.
37. DALES, S. & B. T. POGO. 1980.
38. SUNG, M. T., M. A. LISCHWE, J. C. RICHARDS & K. HOSOKAWA. 1977. J. Biol. Chem. **252:**4981–4987.
39. JENG, Y-H., W. S. M. WOLD, K. SUGAWARA, Z. GILEAD & M. GREEN. 1977. J. Virol. **22:**402–411.
40. COOMBS, D. H., A. J. ROBINSON, J. W. BODNAR, C. J. JONES & G. D. PEARSON. 1978. Cold Spring Harbour Symp. Quant. Biol. **43:**741–753.
41. SUSSKIND, M. M. & BOTSTEIN, D. 1978. Microbiol. Rev. **42:**385–413.
42. SHAPIRO, J. A. 1979. Proc. Natl. Acad. Sci. USA **76:**1933–1937.
43. REANNEY, D. R. 1976. Bacteriol. Rev. **40:**552–590.
44. STINCHCOMB, D. T., K. STRUHL & R. W. DAVIS. 1979. Nature **282:**39–43.
45. FINCHAM, J. R. S. & G. R. K. SASTRY. 1974. Ann. Rev. Genet. **8:**15–50.
46. GRANOFF, A. 1969. Curr. Top. Microbiol. Immunol. **50:**107–137.
47. KELLY, T. J. JR. & D. NATHANS. 1977. Adv. Virus Res. **21:**85–173.
48. SOEDA, E., G. KIMURA & K.-I. MIURA. 1978. Proc. Natl. Acad. Sci. USA **75:**162–166.
49. FRIEDMAN, T., R. F. DOOLITTLE, & G. WALTER. 1978. Nature **274:**291–293.
50. SHAH, K. V., H. L. OZER, H. N. GHAZEY & T. J. KELLY, JR. 1977. J. Virol. **21:**179–186.
51. LAW, M-F., K. K. TAKEMOTO & P. M. HOWLEY. 1979. J. Virol. **30:**90–97.
52. SCHERNECK, S., M. BÖTTGER & J. FEUNTEUN. 1979. Virology **96:**100–107.
53. YANG, R. C. A. & R. WU. 1979. Nucleic Acid Res. **7:**651–668.
54. NEWELL, N., K. V. SHAH & T. J. KELLY, JR. 1979. J. Virol. **30:**624–636.
55. LAI, C-J., N. D. GOLDMAN & G. KHOURY. 1979. J. Virol. **30:**141–147.
56. LAW, M-F., J. D., MARTIN, K. K. TAKEMOTO & P. M. HOWLEY. 1979. Virology **96:**576–587.
57. ZUR HAUSEN, H., L. GISSMANN, A. MINCHEVA & J. F. BÖCKER. 1980. *In* Viruses in Naturally-Occurring Cancers. G. J. Todaro, M. Essex & H. zur Hausen, Eds. :365–372. Cold Spring Harbor Laboratory. Cold Spring Harbor, N.Y.
58. BOSSLET, K. & G. SAUER. 1978. J. Virol. **25:**596–607.
59. TAKEMOTO, K. K. 1978. Intl. Rev. Exptl. Pathol. **18:**281–301.
60. SIMMONS, D. T. & M. A. MARTIN. 1978. Proc. Natl. Acad. Sci. USA **75:** 1131–1135.
61. DHAR, R., C.-J. LAI & G. KHOURY. 1978. Cell **13:**345–358.
62. PATER, M. M., A. PATER & G. DIMAYORCA. 1979. J. Virol. **32:**220–225.
63. MÜLLER, H. & L. GISSMANN. 1978. J. Gen. Virol. **41:**315–323.

64. Osterhaus, A. D. M. E., D. J. Ellens & M. C. Horzinek. 1977. Intervirology **8**:351–359.
65. Stevens, J. G. & F. O. Wettstein. 1979. J. Virol. **30**:891–898.
66. Orth, G., F. Breithburd & M. Favre. 1978. Virology **91**:243–255.
67. Fulton, R. E., F. W. Doane & L. W. MacPherson. 1970. J. Ultrstr. Res. **30**:328–343.
68. zur Hausen, H., L. Gissmann, W. Steiner, W. Dippold & J. Dreger. 1975. Bibl. Haemat. **43**:569–571.
69. Lancaster, W. D. & C. Olson. 1978. Virology **89**:372–379.
70. Pfister, H., U. Linz, L. Gissmann, B. Hutchhausen, D. Hoffmann & H. zur Hausen. 1979. Virology **96**:1–8.
71. Law, M-F., W. D. Lancaster & P. M. Howley. 1979. J. Virol. **32**:199–207.
72. Orth, G. S., S. Jablonska, F. Breitburd, M. Favre & D. Croissant. 1978. Bull. Cancer **65**:151–164.
73. zur Hausen, H. 1977. Curr. Top. Microbiol. Immunol. **78**:1–30.
74. Granoff, A. & R. F. Naegeley. 1978. *In* Persistent Viruses. J. G. Stevens, Ed.: 15–25. Academic Press, New York, N.Y.
75. Nazerian, K. 1979. Biochim. Biophys. Acta **560**:375–395.
76. Calnek, B. W., H. K. Adldinger & D. E. Kahn. 1970. Avian Disease **14**: 219–233.
77. Hinze, H. C. 1971. Int. J. Cancer **8**:514–522.
78. Fleckenstein, B. 1979. Biochim. Biophys. Acta **560**:301–342.
79. Openshaw, H., S. Puga & A. L. Nathans. 1979. Fed. Proc. **38**:2660–2664.
80. Jones, K. W., C. M. Fenoglio, M. Shevchuk-Chaban, N. J. Maitland & J. K. McDougall. 1978. *In* Oncogenesis and Herpesviruses III. G. DeThé., W. Henle & F. Rapp, Eds. IARC Sci. Publ. 24 (Pt. 2): 917–925. Lyon, France.
81. Simmons, D. T., C. Chang & M. A. Martin. 1979. J. Virol. **29**:881–887.
82. Smith, A. E., R. Smith & E. Paucha. 1979. Cell **18**:335–346.
83. Lassam, N. J., S. T. Bayley & F. L. Graham. 1979. Cell **18**:781–793.
84. Farrell, M. P., R. A. Mäntyjärvi & J. S. Pagano. 1978. J. Virol. **25**:871–877.
85. Livingston, D. M., D. G. Tenen, D. Jessell, L. L. Haines, V. Woodard, A. P. Modest, A. Maxam & J. Hudson. 1977. *In* Genetic Manipulation as It Affects the Cancer Problem. J. Schultz & Z. Brada, Eds.: 103–109. Academic Press, Inc. New York, N.Y.
86. Tjian, R. 1978. Cold Spring Harbor Symp. Quant. Biol. **43**:655–662.
87. Soeda, E., J. R. Arrand & B. E. Griffin. 1979. Nucleic Acid Res. **7**:839–858.
88. Ito, Y., N. Spurr & R. Dulbecco. 1977. Proc. Natl. Acad. Sci. USA **74**: 1259–1263.
89. Roberts, T. M., I. Bikel, R. R. Yocum, D. M. Livingston & M. Ptashne. 1979. Proc. Natl. Acad. Sci. USA **76**:5596–5600.
90. Klein, H., W. Maltzmann & A. J. Levine. 1979. J. Biol. Chem. **254**:11051–11060.
91. Eckhart, W. 1977. Comp. Virol. **9**:1–26.
92. Hutchinson, M. A., T. Hunter & W. Eckhart. 1978. Cell **15**:65–77.
93. Tegtmeyer, P., M. Schwartz, J. K. Collins & K. Rundell. 1975. J. Virol. **16**:168–178.
94. Hiscott, J. B. & V. Defendi. 1979. J. Virol. **30**:590–599.
95. Seif, R. & F. Cuzin. 1977. J. Virol. **24**:721–728.
96. Prasad, I., D. Zouzias, & C. Basilico. 1978. Virology **85**:328–331.
97. Basilico, C., S. Gattoni, D. Zouzias & G. Della Valle. 1979. Cell **17**: 645–659.
98. Williams, M. G. & R. Sheinin. 1961. Virology **13**:368–370.
99. Tevethia, S. S. & M. J. Tevethia. 1978. *In* Cancer: A Comprehensive Treatise. F. F. Becker, Ed.: 185–207. Plenum Press. New York, N.Y.
100. Schaffhausen, B. S., J. E. Silver & T. L. Benjamin. 1978. Proc. Natl. Acad. Sci. USA **75**:79–84.

101. Soule, H. R. & J. S. Butel. 1979. J. Virol. **30:**523–532.
102. Giacherio, D. & L. P. Hager. 1979. J. Biol. Chem. **254:**8113–8116.
103. Lassam, N. J., S. T. Bayley, F. L. Graham, S. Mak & P. E. Branton. 1980. J. Virol. **31**
104. Bendig, M. M. & W. R. Folk. 1979. J. Virol. **32:**530–535.
105. Winocour, E., N. Frenkel, S. Lavi, M. Osenholts & S. Rozenblatt. 1974. Cold Spring Harbor Symp. Quant. Biol. **39:**101–108.
106. Winocour, E., M. Oren, S. Lavi, T. Vogel & Y. Gluzman. 1977. *In* Genetic Manipulation as It Affects the Cancer Problem. J. Schultz & Z. Brada, Eds.: 181–196. Academic Press, Inc. New York, N.Y.
107. Tjia, S., E. Fanning, J. Schick & W. Doerfler. 1977. Virology **76:**365–379.
108. Hammarskjöld, M-L., G. Winberg, E. Norrby & G. Wadell. 1977. Virology **82:**449–461.
109. Werner, G. & H. zur Hausen. 1978. Virology **86:**66–77.
110. Doerffler, W. 1977. Comp. Virology. **10:**279–399.
111. Türler, H. 1975. J. Virol. **15:**1158–1167.
112. Brockman, W. W. 1977. Prog. Med. Virol. **23:**69–95.
113. Patch, C. T., A. S. Levine & A. M. Lewis. 1979. Comp. Virol. **13:**459–542.
114. zur Hausen, H. & K-O Fresen. 1979. Biochim. Biophys. Acta **560:**343–353.
115. Luka, J., W. Siegert & G. Klein. 1977. J. Virol. **22:**1–8.
116. Yamamoto, K., F. Mizuno, T. Matsuo, A. Tanaka, M. Nonoyama & T. Osato. 1978. Proc. Natl. Acad. Sci. USA **75:**5155–5159.
117. Ohno, S., J. Luka, L. A. Falk & G. Klein. 1978. Europ. J. Cancer **14:**955–960.
118. Ohno, S., J. Luka, G. Klein & M. D. Daniel. 1979. Proc. Natl. Acad. Sci. USA **76:**2042–2046.
119. Kit, S., M. Kurchak, W. Wray, D. R. Dubbs. 1978. Proc. Natl. Acad. Sci. USA **75:**3288–3291.
120. Donner, L., D. L. Dubbs & S. Kit. 1977. Int. J. Cancer **20:**256–267.
121. Orth, G., S. Jablonska, M. Jarzabek-Chorzelka, S. Obalek, G. Rzesa, M. Favre & O. Croissant. 1979. Cancer Res. **39:**1074–1082.
122. Roizman, B. 1979. Ann. Rev. Genet. **13:**28–57.
123. Lancaster, W. D., G. H. Theilen & C. Olson. 1979. Intervirology **11:** 227–233.
124. Tibbetts, C. 1977. J. Virol. **24:**564–579.
125. Green, M., K. Brackmann, W. S. M. Wold, M. Cartas, H. Thornton & J. H. Elder. 1979. Proc. Natl. Acad. Sci. USA **76:**4380–4384.
126. Sugden, B., C. R. Kintner & W. Mark. 1979. Adv. Cancer Res. **30:**239–268.
127. Hirai, K., K. Ikuta & S. Kato. 1979. J. Gen. Virol. **45:**119–131.
128. Nahmias, A. J. 1974. *In* Viruses Evolution and Cancer. E. Kurstak & K. Maramorosch, Eds.: 605–624. Academic Press, Inc., New York, N.Y.
129. Sonstegard, R. A. 1979. Bibliothec. Haematol. (In press.)
130. Ashne, W. 1977. Tieraerztl Prax. **5:**529–540.
131. Lunger, P. D. & H. F. Clark. 1978. Adv. Virus Res. **23:**159–204.
132. Battula, N. & H. M. Temin. 1978. Cell **13:**387–398.
133. Battula, N. & H. M. Temin. 1977. Proc. Natl. Acad. Sci. USA **74:**281–285.
134. Burny, A., F. Bex, H. Chantrenne, Y. Cleuter, D. Dekegel, J. Ghysdael, R. Kettmann, M. Leclercq, J. Leunen, M. Mammerickx & D. Portetelle. 1978. Adv. Cancer Res. **28:**251–311.
135. Levy, J. A. 1978. Curr. Top. Microbiol. Immunol. **79:**111–213.
136. Schlom, J., W. Drohan, Y. S. Teramoto, J. Young & P. H. Hand. 1980. *In* Viruses in Naturally Occurring Cancers. G. J. Todaro, M. Essex & H. zur Hausen, Eds.: Cold Spring Harbor Laboratory. Cold Spring Harbor, New York.
137. Rasheed, S., M. B. Gardner & R. J. Huebner. 1978. Proc. Natl. Acad. Sci. USA **75:**2972–2976.
138. Gregerson, D. S., P. Russell & T. W. Reid. 1979. J. Gen. Virol. **43:**327–337.

139. Aaronson, S. A. & J. R. Stephenson. 1976. Biochim. Biophys. Acta **48:** 323–354.
140. Sherwin, S. A., T. I. Bonner, U. Heine & G. J. Todaro. 1979. Virology **94:**409–416.
141. Essex, M., S. M. Cotter, J. R. Stephenson, S. A. Aaronson & W. D. Hardy, Jr. 1977. *In* Origins of Human Cancer. H. H. Haitt, J. D. Watson & J. S. Winston, Eds.: 1197–1214. Cold Spring Harbor Laboratory. Cold Spring Harbor, New York.
142. Shifrine, M., M. S. Bulgin, N. E. Dollarhide, H. G. Wolf, N. J. Taylor & F. D. Wilson. 1971. Nature **232:**405–406.
143. Benveniste, R. E. & G. J. Todaro. 1978. Arthritis Rheumatism **21:**S2–S16.
144. Hehlmann, R. 1976. Curr. Top. Microbiol. Immunol. **73:**142–215.
145. Reitz, M. S., N. R. Miller, F. Wong-Staal, R. E. Gallagher, R. C. Gallo & D. H. Gillespie. 1976. Proc. Natl. Acad. Sci. USA **73:**2113–2117.
146. Goodenow, R. S. & H. S. Kaplan. 1979. Proc. Natl. Acad. Sci. USA **76:** 4971–4975.
147. Panem, S. & J. T. Reynolds. 1979. Fred. Proc. **38:**2674–2678.
148. Temin, H. M. 1974. Ann. Rev. Genet. **8:**155–177.
149. Gilboa, E., S. W. Mitra, S. Goff & D. Baltimore. 1979. Cell **18:**93–100.
150. Beneviste, R. E. & G. J. Todaro. 1974. Nature **252:**170–173.
151. Davis, P. W. & E. P. Solomon. 1979. The World of Biology, 2nd Edit. McGraw Hill, Inc. New York, N.Y.
152. Frisby, D., R. MacCormick, S. Pennie, M. Roussel, D. Stehelin & R. Weiss. 1979. FEBS Proceedings. pp. 509–518.
153. Coffin, J. M. 1979. J. Gen. Virol. **42:**1–26.
154. Voga, P. K. 1977. Comp. Virol. **9:**341–455.
155. Hunter, E. 1978. Curr. Top. Microbiol. Immunol. **79:**295–309.
156. Steffen, D., P. Anderson, M. Goldfarb & R. A. Weinberg. 1978. Cold Spring Harbor Symp. Quant. Biol. **43:**875–879.
157. Luria, S. E. 1964. U.S. Cancer Inst. Monogr. **14:**93–106.
158. Hughes, S. H., F. Payvar, D. Spector, R. T. Schimke, H. L. Robinson, G. S. Payne, J. M. Bishop & H. E. Varmus. 1979. Cell **18:**347–359.
159. Bister, K. & P. H. Duesberg. 1979. Proc. Natl. Acad. Sci. USA **76:**5023–5027.
160. Canaani, E., K. C. Robbins & S. A. Aaronson. 1979. Nature **282:**378–383.
161. Stephenson, J. R., A. S. Khan, J. M. van der Van & F. H. Reynolds, Jr. 1979. J. Natl. Cancer Inst. **63:**1111–1119.
162. Koshy, R., F. Wong-Staal, R. C. Gallo, W. Hardy & M. Essex. 1979. Virology **99:**135–144.
163. Sefton, B. M., T. Hunter & K. Beemon. 1980. Proc. Natl. Acad. Sci. USA **77:**220–229.
164. Gallo, R. C. & F. L. Meyskens, Jr. 1978. Prog. Exp. Tumour Res. **22:**216–249.
165. Witte, O. N., N. E. Rosenberg & D. Baltimore. 1979. Nature **281:**396–398.
166. Collett, M. S. & R. L. Erikson. 1978. Proc. Natl. Acad. Sci. USA **75:**2021–2024.
167. Brugge, J. S., M. S. Collett, A. Siddiqui, B. Marczynska, F. Deinhardt & R. L. Erikson. 1979. J. Virol. **29:**1196–1203.
168. Sen, A., G. J. Todaro, D. G. Blair & W. G. Robey. 1979. Proc. Natl. Acad. Sci. USA **76:**3617–3621.
169. Porzig, K. J., K. C. Robbins & S. A. Aaronson. 1979. Cell **16:**875–884.
170. Rohrschneider, L. R., R. N. Eisenman & C. R. Leitch. 1979. Proc. Natl. Acad. Sci. USA **76:**4479–4483.
171. Anderson, G. R., K. R. Marotti & P. A. Whitaker-Dowling. 1979. Virology **99:**31–48.
172. Hunter, T. & B. M. Sefton. 1980. Proc. Natl. Acad. Sci. USA **77:**1311–1315.
173. Radke, K. & G. S. Martin. 1979. Proc. Natl. Acad. Sci. USA **76:**5212–5216.

174. Duesberg, P. H., K. Bister & C. Moscovici. 1979. Virology **99:**121–134.
175. Van de Ven, Wim J. M., F. H. Reynolds, J., R. P. Nalewaik & J. R. Stephenson. 1979. J. Virol. **32:**1041–1045.
176. Friend, C. 1976/1977. The Harvey Lecture **72:**253–281.
177. Mak, T. W., A. A. Axelrad & A. Bernstein. 1979. Proc. Natl. Acad. Sci. USA **76:**5809–5812.
178. Cohen, J. C., J. E. Majors & H. E. Varmus. 1979. J. Virol. **32:**483–496.
179. Bonner, T. I. & G. J. Todaro. 1976. Virology **94:**224–227.
180. Beneviste, R. E. & G. J. Todaro. 1976. Nature **261:**101–108.
181. Maw, B. A., G. J. Todaro & D. E. Savage. 1979. Nature **282:**65–67.
182. Bourgaux, P. 1974. *In* Viruses, Evolution and Cancer. E. Kurstak & K. Maramorosch, Eds.: 119–149. Academic Press, Inc. New York, N.Y.
183. Shah, K. & N. Nathanson. 1976. Amer. J. Epidemiol. **103:**1–12.
184. Marion, P. L. & W. S. Robinson. 1980. *In* Viruses in Naturally Occurring Cancers. M. Essex, G. Todaro & H. zur Hausen, Eds.: 423–434. Cold Spring Harbor Laboratory. Cold Spring Harbor, N.Y.
185. Snyder, R. L. & J. Summers. 1980. *In* Viruses in Naturally Occurring Cancers. M. Essex, G. Todaro & H. zur Hausen, Eds.: 447–458. Cold Spring Harbor Laboratory. Cold Spring Harbor, N.Y.
186. Marion, P. L., F. H. Salazar, J. J. Alexander & W. S. Robinson. 1980. J. Virol. **32:**796–802.
187. Pfister, H. & J. Meszaros. 1980. Virology **108:**243–246.

Discussion of the Paper

Dr. K.-S. Chiang: You talk about the oncogenes that are carried by some of the viruses, how big is that piece, or can that be used as a probe to study the evolution?

Dr. R. Sheinin: These DNA pieces vary in size. They can go from one thousand base-nucleotide residues to half of a retrovirus genome, which is about 1.5×10^6. So this evolutionary tree that I showed you is in fact done by DNA probe studies. Those hybridizations are relatively easy to do once you have the probes. You can do both, you can follow the oncogene development and you can follow the virus gene development, which is what makes this a really interesting tool.

Dr. M. V. Simpson: Do you want to say anything more about the episome as a possible mechanism or model for a mechanism for evolution and the origin of organelle DNA?

Dr. Sheinin: I really can't. The reason being that no one has really looked at this problem from this point of view.

Now only the Herpes viruses as a group establish that kind of latency. Not only is the viral DNA integrated into the genome but the virus is there as an episome. I put that up as a very tentative scheme as the kind of steps you might expect for the evolution of what I call the supreme chromosomal parasite, but no one has really looked at that. If you are going to have a chromosomal parasite or a tumor virus, the replication of the viral genome must be under the control of the host cell. When that breaks down you have a manifestation that is quite different—viruses produced in the cells die—so that the episomes when present, are under the hosts control.

Dr. S. S. Cohen: When it became clear that the endosymbiotic hypothesis had to be able to answer this strange phenomena in which a number of the genetic elements that control the synthesis of the so-called endosymbiont were actually in the nuclear chromosomes, the answer was it's going to be relatively easy to move a genetic element from symbiont to the nucleus. Of course this fitted in very nicely with what we knew about a number of viral tumor developments, but it was a very strange thing that those people who were engaged at the same time in the discussions of the possible dangers of recombinant DNA never referred to all of the written material on this. They never referred to the position that we were put in with respect to the endosymbiontists and rarely referred directly to the known phenomena whereby a variety of DNAs, episomes, etc., moved into this DNA. I wonder whether you'd care to comment on this type of thing Dr. Sheinin?

DR. SHEININ: I have used the term "chromosomal parasitism" because this is really what it is; these viruses are essentially carried by their animals of origin. In some situations, the cellular target that carries these parasites is known. As I said, in the Herpes viruses, it's known. We don't know where polyoma sits. In the case of Merricks disease of turkeys, it's quite a fascinating thing: the virus sits for the lifetime of the bird in the feather follicle epithelial cell whereas, it transforms a T-lymphocyte. In looking for the harbor cell for these viruses one really has to use one's imagination. These viruses are always there and what are they doing there? Why are they there? What selective advantage do they confer upon an organism, a cell, a what-have-you. I don't think too much thought has been given to this. But if you look at the retrovirus families—and that's the only way you can think about them as whole families of viruses—they have the capacity to move within a cell from one place to another and move along with them cellular DNA. They are able to carry this cellular DNA to another cell. Not only that, they can carry this same cellular DNA to another species and to other orders by the horizontal mechanism of transmission.

The point that you made, Dr. Cohen, about the biohazards is quite right. I don't think anyone thinks about tumor viruses in this way at all.

DR. M. V. SIMPSON: Just one point that you made in terms of selective advantage to the cell. What you say is true, but what isn't thought of very often is that it has a selective advantage for the virus.

DR. SHEININ: In terms of the chromosomal parasitism of the virus, Nature has been very, very shrewd, because to be a supreme chromosomal parasite the virus has to be completely silenced, and under those conditions you would favor the selection of mutants in the genes, or in the origin of replication you would favor deletions of all kinds. Eventually you would favor the destruction of that particular species of virus. Nature has guarded against this by providing for recombination between viruses and cellular genome, and it has the ultimate guarantee from the point of view of the virus in that there is an obligatory horizontal transmission phase in the natural history of these viruses. All of them are transmitted by horizontal transmission so that they don't rely for their survival on the genetic pathway.

DR. R. A. LEWIN: A few years ago I read a paper in which somebody indicated that certain sequences in human nucleic acid matched exactly those in three species of cat, the domestic cat, Egyptian, and Abyssinian cat. The suggestion is then that this is what you call horizontal transmission presumably virus mediated, is this right?

DR. SHEININ: Yes. I always make a plea to consider human tumor viruses from simply the biological point of view. I think that once you see a vertebrate evolutionary scheme of that kind where the c-type retroviruses have been present for a very long time and have evolved within every vertebrate species then you have to consider the human. The analogous chromosomal parasites in the human system show sequence homology with the retroviruses of the specific class of higher apes that live out in the Far East. The cat virus is an interesting example of cross-species horizontal transmission and then recombination, and this is very common, particularly where you have families of vertebrates living together.

DR. L. MARGULIS: I may have missed something but when you say that these obligatory chromosomal parasites (retroviruses) can go from one site in the cell, to another, is there direct evidence that they get into mitochondria?

DR. SHEININ: No, the evidence is against this; they do not get into mitochondria at all.

DR. MARGULIS: Cytoplasm?

DR. SHEININ: No, I'm sorry I'm referring exclusively to the nuclear organelle. There really is no evidence that these viruses integrate into the mitochondria.

DR. SIMPSON: I guess there are about a dozen varied reports of seeing viruses in mitochondria, but no reports of any kind of integration. I don't know any of these reports of the presence of viruses in mitochondria that really have been confirmed.

DR. S. S. COHEN: I might point out that there is one example of a virus that is considered to multiply in very close association with chloroplasts and I've been working with this for the last few years.

DR. A. CRONQUIST (*New York Botanical Garden, Bronx, N.Y.*): I gather from what has been said that our species acquired certain DNA sequences from cats by the medium of viruses. If so, does it mean that this has been spread throughout all of our species now, so you'll find it in China and Africa and the United States and South America?

DR. SHEININ: No. The data from DNA sequence homology would suggest that some of the cat virus sequences became mixed with sequences of a baboon quite a long time ago and the human analog has sequence homology with the baboon. There is no evidence that there are cat sequences in the human genome, but the evidence is that one part of the cat sequence somehow got into the baboon. Who bit whom is not entirely certain and what kind of cat did the biting or scratching is not certain.

COMPARISON OF THE *EUGLENA* CHLOROPLAST PROTEIN SYNTHESIZING SYSTEM WITH THOSE OF PROKARYOTES AND EUKARYOTES

Linda L. Spremulli, Caroline A. Breitenberger, Mary C. Graves, Carol M. Beck, and Deborah A. Steege *

Department of Chemistry
University of North Carolina
Chapel Hill, North Carolina 27514
and
** Department of Biochemistry*
Duke University Medical School
Durham, North Carolina 27710

Introduction

Chloroplasts are known to contain a protein-synthesizing system distinct from that of the cell cytoplasm.[1] In many respects the chloroplast protein-synthesizing machinery more closely resembles that of prokaryotes than that of the eukaryotic cell cytoplasm. In general, chloroplast ribosomes are smaller than those of the eukaryotic cell cytoplasm and sediment very close to the 70S sedimentation value observed for bacterial ribosomes.[2] The antibiotic sensitivity of chloroplast ribosomes is similar to that of prokaryotic ribosomes and quite distinct from that observed with eukaryotic ribosomes. Furthermore, elongation factors from *E. coli* are active on chloroplast ribosomes while those from the eukaryotic cell cytoplasm are not.[3-6] These observations on the similarity between chloroplast and prokaryotic protein-synthesizing systems have been cited as evidence for an endosymbiotic origin for chloroplasts.

We are interested in carrying out a detailed study of organellar protein synthesis and have chosen to work with the chloroplast translation system of *Euglena gracilis*. The chloroplast protein synthesizing system of this organism is induced by exposure of the cells to light, permitting many types of studies that are more difficult in other organisms. We summarize here our detailed observations on a number of features of *Euglena* chloroplast ribosomes. Several of these features are characteristic of prokaryotic ribosomes, others of eukaryotic ribosomes, and still others are not observed with either of these. In addition, we report studies comparing a number of properties of *Euglena* chloroplast elongation factor G ($EF\text{-}G_{chl}$) with those of prokaryotic, eukaryotic, and mitochondrial translocases. These data are then discussed in terms of theories on the origin and evolution of chloroplasts.

Materials and Methods

Materials

[^{14}C]Phenylalanine and *E. coli* tRNA were purchased from Schwarz-Mann. [^{14}C]Phenylalanyl-tRNA (150 cpm/pm) was prepared as described by Ravel

0077-8923/81/0361-0461 $01.75/0

and Shorey,[7] with synthetases prepared by the procedure of Muench and Berg.[8] Cultures of *Euglena gracilis* Klebs var. *bacillaris* Cori (*Euglena* B) and its aplastidic mutant W_3BUL were generously provided by Dr. J. Schiff of Brandeis University. Yeast (S288 cα) was from the collection of Dr. R. K. Mortimer and *Chlamydomonas reinhardtii* was from American Type Culture Collection (ATCC # 18798). The sodium salt of fusidic acid was generously provided by Dr. W. O. Godtfredsen of Leo Pharmaceuticals (Denmark). Protein concentrations were determined by the method of Lowry *et al.*[9]

Preparation of Cell Extracts

Euglena was grown and light-induced basically as described previously.[10, 11] Whole cell extracts were prepared by the procedure of Breitenberger *et al.*,[11] except that the Sephadex G-25 chromatographic step was replaced by an overnight dialysis. Cell extracts from *Chlamydomonas,* yeast, and spinach were prepared in a similar manner.

Preparation of E. coli and Wheat Germ Ribosomes and Elongation Factors

E. coli EF-G and a complex containing EF-Tu and EF-Ts (EF-Tu/Ts) were prepared from *E. coli* MRE 600.[12] *E. coli* ribosomes were prepared from strain W and high-salt–washed as described previously,[13] except that the first high-salt wash contained 1 M NH_4Cl. They were freed of any residual EF-G contamination when necessary by a further 0.6 M NH_4Cl wash and collected through a 20% (w/vol) sucrose cushion. Wheat germ extracts were prepared as described previously[14] and a 30–70% ammonium sulfate fraction of the postribosomal supernatant was used as a source of EF-1 and EF-2 which were then separated by chromatography on Sephadex G-200.[15] Wheat germ high-salt–washed ribosomes were prepared as described previously,[14] except that an additional low-salt wash was performed.

Preparation of Euglena Chloroplast Ribosomes and EF-G

The preparation of *Euglena* chloroplast ribosomes free of endogenous elongation factors will be described in detail elsewhere.[16] *Euglena* chloroplast EF-G was purified by the fusidic acid affinity-binding procedure previously described[17] and further purified by chromotography on DEAE-cellulose. On SDS-polyacrylamide gel electrophoresis this preparation has one major band migrating with a molecular weight of 85,000 and several minor bands, and is estimated to be 70% to 80% pure. Antibodies were raised in rabbits against chloroplast EF-G by standard immunological procedures.[18] *Euglena* mitochondrial EF-G was prepared from dark-grown *Euglena* B or its aplastidic mutant W_3BUL by a procedure to be described elsewhere.[19]

Assays

E. coli Ribosomes

The activities of the translocases from *Euglena* chloroplasts and mitochondria, yeast mitochondria, *E. coli, Chlamydomonas,* spinach, and the eukaryotic cell cytoplasm were tested in the poly(U)-directed polymerization of phenylalanine on *E. coli* ribosomes in the presence of a saturating amount of *E. coli* EF-Tu/Ts. The activities of *E. coli* EF-Tu/Ts and wheat germ EF-1 were tested on *E. coli* ribosomes in the presence of a saturating amount of *E. coli* EF-G. The assay conditions described previously[11] were used except that the concentration of Mg^{2+} was 6 mM and 15 μg of *E. coli* W ribosomes were used unless otherwise indicated.

Wheat Germ Ribosomes

The activities of various translocases on wheat germ 80S ribosomes were determined by the poly(U)-directed polymerization of phenylalanine in the presence of a saturating amount of the high molecular weight form of EF-1 from wheat germ as described previously.[11] *E. coli* EF-Tu/Ts and wheat germ EF-1 were assayed for activity in polymerization on wheat germ 80S ribosomes in the presence of saturating amounts of wheat germ EF-2 under polymerization conditions optimal for wheat germ ribosomes.[11]

Chloroplast Ribosomes

The activities of chloroplast, bacterial, and cytoplasmic elongation factors on *Euglena* chloroplast ribosomes were tested in the presence of saturating amounts of the complementary factor so that the activity of only one elongation factor was tested at any one time. The complementary factors used in each case were fully active on the chloroplast ribosomes. Conditions optimal for polymerization on chloroplast ribosomes were used.[16]

RNA Sequencing

Chloroplast 16.5S rRNA was prepared by phenol extraction[20] of whole cells and purified on denaturing 2.5% to 5% polyacrylamide gels.[21] Sequencing procedures were primarily those described by Barrell.[22]

RESULTS

Recently we have developed procedures which permit the isolation of chloroplast ribosomes free of endogenous elongation factors from whole cell extracts of *Euglena*.[16] It became apparent during the preparation of these ribosomes that very high concentrations of Mg^{2+} ions are required to maintain the chloroplast ribosome as the 68S monosome. At 5 mM Mg^{2+} the majority of the chloroplast ribosomes dissociate into 30S and 50S subunits. Under the same conditions, both *E. coli* 70S Type A ribosomes and wheat germ 80S ribosomes remain fully associated. This observation suggests that the interaction of the chloroplast ribosomal subunits requires exceptionally high concentrations of Mg^{2+}. The role of Mg^{2+} in facilitating subunit association is not fully understood. However, one major theory suggests that Mg^{2+} ions neutralize the negative charges on the phosphodiester backbone of the rRNA and thereby reduce the electrostatic repulsion between the subunits.[23] Since rather high concentrations of Mg^{2+} are required to maintain the *Euglena* chloroplast ribosome as the intact monosome, there may be considerable stretches of exposed rRNA that must be neutralized at the subunit interface in the chloroplast ribosome. High Mg^{2+} ion concentrations are required not only to prevent the dissociation of the *Euglena* chloroplast ribosome during preparation but also to prevent the conversion of the 68S ribosomes to a 53S defective particle.[16, 24]

TABLE 1 summarizes the approximate monovalent and divalent cation optima for poly(U)-directed polymerization of phenylalanine on chloroplast, *E. coli,* and wheat germ cytoplasmic ribosomes. The high divalent cation concentration apparently required to permit the association of the chloroplast ribosomal subunits is reflected in a high requirement for Mg^{2+} in the elongation steps of protein synthesis. Both wheat germ and *E. coli* ribosomes are most active at Mg^{2+} ion concentrations below 10 mM, while *Euglena* chloroplast ribosomes exhibit almost no activity at these concentrations and are most active at 18 to 22 mM Mg^{2+}. In contrast to *E. coli* and wheat germ ribosomes, the *Euglena* chloroplast ribosomes display an unusually low monovalent cation

TABLE 1

CATION REQUIREMENTS FOR POLY(U)-DIRECTED POLYMERIZATION ON CHLOROPLAST, PROKARYOTIC AND EUKARYOTIC RIBOSOMES

Ribosomes *	$[Mg^{2+}]$ optimum †	$[M^+]$ optimum
Chloroplast	18–22 mM	40–80 mM NH_4^+
E. coli	6–9 mM	100–160 mM NH_4^+ or K^+
Wheat Germ	4–6 mM	100–160 mM K^+

* *E. coli* and *Euglena* chloroplast ribosomes were tested in the presence of saturating amounts of *E. coli* EF-Tu/Ts and *E. coli* EF-G. Wheat germ ribosomes were assayed in the presence of saturating amounts of wheat germ EF-1 and EF-2.

† Done in the presence of 0.1 mM spermine.

optimum for polymerization and are more active in the presence of NH_4^+ than in the presence of K^+. Monovalent ions, unlike divalent ions, tend to cause dissociation of ribosomal subunits and K^+ is much more effective in this than is NH_4^+.[25] Thus, the low monovalent cation optimum for polymerization is probably related to the ease with which chloroplast ribosomes are dissociated and to the requirement for subunit association during the elongation steps of protein synthesis.

Although *Euglena* chloroplast ribosomes have ionic optima for polymerization quite different from those of *E. coli* ribosomes, elongation factors from *E. coli* are fully active on them. In contrast, eukaryotic elongation factors display little activity on the chloroplast ribosomes. TABLE 2 compares the activity in polymerization of the bacterial EF-Tu/Ts and wheat germ cytoplasmic EF-1 on chloroplast, *E. coli,* and wheat germ ribosomes. The bacterial factor has little activity on wheat germ 80S ribosomes but is as active on chloroplast ribosomes as it is on the homologous ribosomes. Similar results are observed when the *Euglena* chloroplast ribosomes are tested with prokaryotic and eukaryotic translocases (TABLE 3). *E. coli* EF-G has similar activity on

TABLE 2

SPECIFIC ACTIVITY OF EF-Tu/Ts AND EF-1 ON CHLOROPLAST, E. COLI, AND WHEAT GERM RIBOSOMES

Source of Ribosomes *	Specific Activity † *E. coli* EF-Tu/Ts	Wheat Germ EF-1_H
E. coli	15	<0.2
Wheat Germ	<0.2	8
Euglena Chloroplast	18	<0.2

* Assays were performed under ionic conditions optimal for the ribosomes being tested and in the presence of saturating amounts of the appropriate translocase.

† Units per μg of protein.

both *E. coli* and chloroplast ribosomes but is inactive on eukaryotic ribosomes. Wheat germ EF-2 is specific for eukaryotic ribosomes and has no activity on either *E. coli* or chloroplast ribosomes.

The data on the elongation factor specificity of the *Euglena* chloroplast ribosomes are another indication that these organellar ribosomes are prokaryotic in nature. One of the many differences between prokaryotic and eukaryotic ribosomes is found in the nucleotide sequence near the 3′ end of the ribosomal RNA from the small subunit. A polypyrimidine tract is present very close to the 3′ end of prokaryotic 16S rRNAs.[26] This polypyrimidine tract has been shown to base-pair with a sequence of nucleotides in the mRNA on the 5′ side of the AUG initiation codon, thus stabilizing the interaction of the ribosome with the initiation signal on the mRNA.[27] Eukaryotic 18S rRNAs, on the other hand, do not contain this type of sequence and there is little evidence that base-pairing between the mRNA and rRNA is important in the formation of a stable initiation complex.[29] This difference between the rRNAs suggests that there is a fundamental difference in the mechanism by which prokaryotic and eukaryotic ribosomes interact with the initiation signal on the mRNA. We have determined the sequence of the T_1 RNase 3′ terminal oligonucleotide

TABLE 3

SPECIFIC ACTIVITY OF PROKARYOTIC AND EUKARYOTIC TRANSLOCASES ON CHLOROPLAST, E. COLI, AND WHEAT GERM RIBOSOMES

Source of Ribosomes *	*E. coli* EF-G	Specific Activity † Wheat Germ EF-2	*Euglena* EF-G_{chl}
E. coli	50	<0.1	40
Wheat Germ	<0.1	4.5	<0.1
Euglena Chloroplast	70	<0.1	60

* Assays were performed under optimal ionic conditions for the ribosomes being tested and in the presence of saturating amounts of *E. coli* EF-Tu/Ts (*E. coli* and chloroplast ribosomes) or wheat germ EF-1 (wheat germ ribosomes).

† Units per μg of protein.

from *Euglena* chloroplast 16.5S rRNA and compared this sequence to the corresponding sequences from prokaryotic and eukaryotic ribosomal RNAs. The results are summarized in TABLE 4. The 3′ terminal oligonucleotide from the *Euglena* chloroplast 16.5S rRNA does not contain a polypyrimidine tract such as that observed in prokaryotic 16S rRNAs. Rather, the chloroplast rRNA has an oligonucleotide sequence in this position that is virtually identical to the sequence found in the 18S rRNA of all eukaryotes for which sequences have been determined to date.[28] These data indicate that the 3′ terminal oligonucleotide sequence of this chloroplast rRNA is essentially eukaryotic in nature and suggests that the mechanism used to stabilize the interaction between the *Euglena* chloroplast ribosome and the initiation signal on the mRNA may be similar to that used by eukaryotes and may be quite different from the mechanism found in prokaryotes.

A second surprising observation made during the sequence analysis mentioned above was the presence of high amounts of the minor base pseudouracil in the chloroplast rRNA. Pseudouracil, in significant quantities, has been found only in eukaryotic rRNAs and is present in barely detectable levels in prokaryotic rRNAs.[34] This is an additional indication that the *Euglena* chloroplast rRNA has some features more characteristic of eukaryotic than of prokaryotic rRNAs.

The *Euglena* chloroplast translocase ($EF\text{-}G_{chl}$) recently has been purified extensively in this laboratory.[17, 19] This factor is specific for prokaryotic or organellar ribosomes and displays essentially the same activity on ribosomes from *E. coli* as on homologous ribosomes (TABLE 3). We have compared a number of the properties of *Euglena* $EF\text{-}G_{chl}$ to those of translocases from other sources. These comparisons include sensitivity to inhibition by fusidic acid and immunological cross-reactivity.

The steroidal antibiotic fusidic acid generally inhibits the translocation step of protein synthesis by the formation of a [fusidic acid·ribosome·translocase·GDP] complex. We have determined the sensitivity of the *Euglena* $EF\text{-}G_{chl}$ to inhibition by fusidic acid and compared this sensitivity to that observed with prokaryotic, eukaryotic, mitochandrial, and other chloroplast translocases. The

TABLE 4

COMPARISON OF THE NUCLEOTIDE SEQUENCE FROM THE 3′ TERMINAL OLIGONUCLEOTIDE OF EUGLENA CHLOROPLAST 16.5S RRNA WITH CORRESPONDING PROKARYOTIC AND EUKARYOTIC SEQUENCES

Ribosomal RNA	T_1 RNase Oligonucleotide Sequence from the 3′ end	Reference
E. coli 16S	(G) $AUCACCUCCUUA_{OH}$	26
B. stearothermophilus 16S	$AUCACCUCCUUUCUA_{OH}$	30
Rabbit Reticulocyte 18S	$AUCAUUA_{OH}$	31
Yeast 18S	$AUCAUUA_{OH}$	32
Wheat Germ 18S	$AUCAUUG_{OH}$	28
Euglena Chloroplast 16.5S	$AUCAU\Psi(X)^{*}_{OH}$	33

* Preliminary evidence indicates the 3′ nucleotide is G_{OH}.

TABLE 5

FUSIDIC ACID INHIBITION OF TRANSLOCASES FROM VARIOUS SOURCES

Source	Translocase	Concentration of Fusidic Acid Required for 50% Inhibition
Euglena	Chloroplast EF-G	3 μM
Chlamydomonas	Chloroplast EF-G *	3 μM
Euglena	Mitochondrial EF-G	25 μM
Yeast	Mitochondrial EF-G	1,000 μM
E. coli	EF-G	35 μM
Euglena	Cytoplasmic EF-2	550 μM
Yeast	Cytoplasmic EF-2	70 μM
Wheat Germ	Cytoplasmic EF-2	9 μM

* This preparation consists largely of chloroplast EF-G but some mitochondrial EF-G is probably also present.

results, summarized in TABLE 5, indicate that *Euglena* EF-G_{chl} is extremely sensitive to inhibition by fusidic acid. A similar sensitivity is seen with the EF-G activity in whole cell extracts of *Chlamydomonas.* This EF-G activity probably consists primarily of the chloroplast enzyme but may have some mitochondrial EF-G as well. The high sensitivity of the *Euglena* and *Chlamydomonas* EF-G_{chl} to inhibition by fusidic acid is also observed with spinach EF-G (data not shown). Compared to the *Euglena* EF-G_{chl}, about 8-, 300-, and 10-fold higher concentrations of fusidic acid are required to inhibit the translocases from *Euglena* mitochondria, yeast mitochondria, and *E. coli,* respectively. The cytoplasmic translocases from a variety of organisms display a wide range of sensitivities but all those we have tested are less sensitive to inhibition than *Euglena* EF-G_{chl}. Although all translocases that have been examined are inhibited by fusidic acid, their greatly differing sensitivities suggest that they must differ considerably in structure near the fusidic acid- or ribosome-binding sites.

As an additional probe for structural similarities or differences between *Euglena* EF-G_{chl} and other translocases, we have prepared antibodies against this factor and determined their ability to cross-react with translocases from other systems. When tested by immunodiffusion, this antiserum formed a well-defined precipitin band with *Euglena* EF-G_{chl} but did not form precipitin bands with *Chlamydomonas* chloroplast EF-G, *Euglena* or yeast mitochondrial EF-G, or *E. coli* EF-G. The lack of structural homology is somewhat surprising since all of these enzymes catalyze translocation efficiently on *E. coli* ribosomes. The antiserum also failed to show any cross-reaction with cytoplasmic EF-2 from *Euglena,* yeast, or wheat germ. This observation is not unexpected since EF-2 is specific for eukaryotic ribosomes whereas the chloroplast EF-G is active only on prokaryotic or organellar ribosomes.

If the antigenic sites recognized by the antibodies are required for the activity of chloroplast EF-G, then the antiserum should inhibit the activity of the factor in poly(U)-directed polymerization of phenylalanine. This method of testing for immunological cross-reactivity should be much more sensitive than the immunodiffusion plates for determining structural similarities at sites required for catalytic activity since slight changes in activity are easily measured. The antiserum against *Euglena* EF-G_{chl} was tested in this manner against a variety of other translocases and the results are summarized in TABLE 6. The *Euglena* chloroplast EF-G was completely inhibited by a low level of the antiserum. Neither the mitochondrial nor cytoplasmic translocases from *Euglena* were inhibited by the antiserum. The EF-G activity of a whole cell extract of *Chlamydomonas* was slightly inhibited by high levels of the antiserum against *Euglena* EF-G_{chl}. The EF-G activity of this extract probably consists primarily of EF-G_{chl} and these results suggest a small amount of homology between the chloroplast factors of these two organisms. The EF-G activity

TABLE 6

EFFECT OF ANTISERUM AGAINST EUGLENA CHLOROPLAST EF-G ON THE ACTIVITY OF TRANSLOCASES FROM VARIOUS SOURCES

Translocase	Amount of Antiserum (μl)	% Inhibition of Activity
Euglena chloroplast EF-G	0.5	>95%
Euglena mitochondrial EF-G	1.0	< 2%
Euglena cytoplasmic EF-2	1.0	0%
Chlamydomonas EF-G *	0.5	10%
	2.0	30%
Spinach EF-G *	0.5	0%
	2.0	5%
Yeast mitochondrial EF-G	2.0	0%
E. coli EF-G	0.5	2%
	2.0	15%

* This preparation is believed to contain largely chloroplast EF-G but also probably has some mitochondrial EF-G.

from spinach leaves was only very slightly affected by the antiserum. The yeast mitochondrial EF-G did not cross-react with the antibodies, but *E. coli* EF-G was inhibited slightly. The degree of homology between the *Euglena* EF-G_{chl} and *E. coli* EF-G detected by this procedure was very slight since a level of the antiserum that completely inhibited the chloroplast factor produced only a 2% inhibition of *E. coli* EF-G. These results suggest that the *Euglena* chloroplast EF-G has little if any overall structural homology with mitochondrial or cytoplasmic translocases but does have some homology with the *Chlamydomonas* chloroplast factor and perhaps some with *E. coli* EF-G.

DISCUSSION

The results presented here indicate that the *Euglena gracilis* chloroplast protein-synthesizing system has some features in common with prokaryotes, other features that seem characteristic of eukaryotes, and yet others that are unlike either of these systems. Any theory concerned with the origin of the *Euglena* chloroplast protein-synthesizing system must somehow resolve two apparently contradictory pieces of evidence—the "prokaryotic-like" elongation-factor specificity displayed by these ribosomes and the "eukaryotic-like" sequence found near the 3′ end of the *Euglena* chloroplast 16.5S rRNA. The prokaryotic elongation-factor specificity of the *Euglena* chloroplast ribosomes might be construed as evidence for an endosymbiotic origin of chloroplasts. This theory argues that certain sites on both bacterial and organellar ribosomes have been highly conserved over the past billion years or so of evolution. This model could then account for the eukaryotic nature of the 3′ terminal oligonucleotide of the 16.5S rRNA by postulating genetic recombination between an extrachromosomal genetic element carrying the genes for the cytoplasmic 18S rRNA and the chloroplast DNA. Such an idea may receive some support from the observation that small closed-circular DNA molecules carrying ribosomal RNA genes occur in certain organisms.[35]

On the other hand, it could be argued that the most primitive cells from which both bacteria and higher cells evolved had a protein synthetic system more like that seen in present-day eukaryotes. In some respects the "prokaryotic-like" protein biosynthetic system seems more efficient than the one found in the eukaryotic cell cytoplasm. It certainly accomplishes the same reactions with over a million daltons less protein and RNA. One might, therefore, argue that the eukaryotic system is the less efficient, more primitive system and that both prokaryotic and organellar systems evolved from it or its predecessor. A selective pressure for efficient protein synthesis in bacteria and subcellular organelles may have led to the natural evolution of certain "prokaryotic-like" characteristics in these translational systems.

References

1. Boulter, D., R. Ellis & A. Yarwood. 1972. Biol. Rev. Cambridge Phil. Soc. **47:**113–175.
2. Kroon, A. M., E. Agsteribbe & H. Devries. 1972. *In* The Mechanism of Protein Synthesis and Its Regulation. L. Bosch, Ed.: 539–582. North-Holland. Amsterdam.
3. Grivell, L. A. & G. S. P. Groot. 1972. FEBS Lett. **25:**21–24.
4. Sala, F., S. Sensi & B. Parisi. 1970. FEBS Lett. **10:**89–91.
5. Tiboni, O., G. Di Pasquale & O. Ciferri. 1976. Plant Sci. Lett. **6:**419–429.
6. Tiboni, O., G. Di Pasquale & O. Ciferri. 1978. Eur. J. Biochem. **92:**471–477.
7. Ravel, J. M. & R. Shorey. 1971. *In* Methods in Enzymology. L. Grossman & K. Moldave, Eds. Vol. 20, Part C: 306–316. Academic Press, Inc. New York, New York.
8. Muench, K. & P. Berg. 1966. *In* Procedures in Nucleic Acid Research. G. Cantoni & D. Davies, Eds.: 375. Harper & Row, Inc. New York, N.Y.
9. Lowry, O., N. Rosebrough, A. Farr & R. Randall. 1951. J. Biol. Chem. **193:**265–275.
10. Holowinsky, A. & J. Schiff. 1970. Plant Physiol. **45:**339–347.
11. Breitenberger, C. A., M. C. Graves & L. L. Spremulli. 1979. Arch. Biochem. Biophys. **194:**265–270.
12. Ravel, J. M., R. Shorey, S. Froehner & W. Shive. 1968. Arch. Biochem. Biophys. **125:**514–526.
13. Remold-O'Donnell, E. & R. Thach. 1970. J. Biol. Chem. **245:**5737–5742.
14. Spremulli, L. L., B. J. Walthall, S. R. Lax & J. M. Ravel. 1977. Arch. Biochem. Biophys. **178:**565–575.

15. GOLINSKA, B. & A. LEGOCKI. 1973. Biochim. Biophys. Acta **324:**156–170.
16. GRAVES, M. C., C. A. BREITENBERGER & L. L. SPREMULLI. 1980. (Manuscript in preparation.)
17. BREITENBERGER, C. A., M. N. MOORE, D. W. RUSSELL & L. L. SPREMULLI. 1979. Anal. Biochem. **99:**434–440.
18. CAMPBELL, D. H., J. S. GARVEY, N. E. CREMER & D. H. SUSSDORF. 1964. Methods in Immunology. W. A. Benjamin, Inc. New York, N.Y.
19. BREITENBERGER, C. A. & L. L. SPREMULLI. 1980. (Manuscript in preparation.)
20. ZELDIN, M. H. & J. A. SCHIFF. 1967. Plant Physiol. **42:**922–932.
21. SPRADLING, A., M. L. PARDUE & S. PENMAN. 1977. J. Mol. Biol. **109:**559–587.
22. BARRELL, B. G. 1971. *In* Procedures in Nucleic Acid Research. G. L. Cantoni & D. R. Davies, Eds. Vol 2: 751–779. Harper & Row, Inc. New York, N.Y.
23. WISHNIA, A., A. BOUSSERT, M. GRAFFE, PH. DESSEN & M. GRUNBERG-MANAGO. 1975. J. Mol. Biol. **93:**499–515.
24. SCHWARTZBACH, S. D., G. FREYSSINET & J. SCHIFF. 1974. Plant Physiol. **53:** 533–542.
25. SPERRAZZA, J., D. W. RUSSELL & L. L. SPREMULLI. 1980. Biochemistry. (In press.)
26. SHINE, J. & L. DALGARNO. 1974. Proc. Natl. Acad. Sci. USA **71:**1342–1346.
27. STEITZ, J. & K. JAKES. 1975. Proc. Natl. Acad. Sci. USA **72:**4734–4738.
28. HAGENBÜCHLE, O., M. SANTER, J. A. STEITZ & R. J. MANS. 1978. Cell **13:**551–563.
29. STEITZ, J. A. 1979. *In* Biological Regulation and Development. R. F. Goldberger, Ed. Vol 1: 349–399.
30. SPRAGUE, K. U., J. A. STEITZ, R. M. GRENLEY & C. E. STOCKING. 1977. Nature **267:**462–465.
31. HUNT, J. A. 1970. Biochem. J. **120:**353–363.
32. DALGARNO, L. & J. SHINE. 1973. Nature New Biol. **245:**261–263.
33. GRAVES, M. C., L. L. SPREMULLI & D. STEEGE. 1980. (Manuscript in preparation.)
34. JEANTEUR, P., F. AMALDI & G. ATTARDI. 1968. J. Mol. Biol. **33:**757–775.
35. MEYERINK, J., J. KLOOTWIJK, R. PLANTA, A. VAN DER ENDE & E. VAN BRUGGEN. 1979. Nuc. Acid Res. **7:**69–76.

DISCUSSION OF THE PAPER

DR. L. NEWELL (*Harvard University, Cambridge, Mass.*): Is there a color sensitivity of the 16S RNA?

DR. L. L. SPREMULLI: We're trying to find out.

DR. L. NEWELL: It's a very high magnesium concentration for the poly(U). Do you have any data on the misreading of poly(U) by the ribosomes? Because when you have very high magnesium concentrations you tend to override a lot of controls.

DR. SPREMULLI: With leucine we don't appear to have misreading, but we've only checked it with leucine.

DR. N. WEEDEN (*University of California, Davis, Calif.*): The chloroplast has been shown to have a high concentration of magnesium under light. I was wondering, have you tested blue-green algae, which also show this high concentration of magnesium? Do their ribosomes also require high magnesium?

DR. SPREMULLI: We haven't tested it but that's a very interesting point. That's true of all the blue-green species of algae?

DR. N. WEEDEN: I don't know if it's true of all of them.

DR. K.-P. KOLLER: We have been trying to translate messenger RNA from blue-green algae in the *E. coli* system and we can say that it requires really high amounts of magnesium to get nice translation.

DR. H. LYMAN (*State University of New York, Stony Brook, N.Y.*): That light-induced synthesis of translocase, is that chloramphenicol-sensitive?

DR. SPREMULLI: It's not chloramphenicol-sensitive, it is cyclohexamide-sensitive and it's also light-inducible in a plastid mutant of *Euglena*. So we believe it's a nuclear gene.

DR. H. LYMAN: That's sort of interesting that the chloroplast enzyme is in the nucleus.

DR. SPREMULLI: It's been suggested that the *Chlorella* translocase is a chloroplast gene, and that the spinach translocase is a chloroplast gene. So that would suggest it's quite different in the *Euglena* system.

DR. S. AARONSON: I wonder how this was determined, that there was no genome for chloroplasts in the mutant.

DR. SPREMULLI: I can refer you to the work of Dr. Schiff who I believe derived that mutant and supplied it to us.

DR. AARONSON: If you used Dr. Schiff's mutant I can't argue but I do know that there are apoplastid or at least apochlorotic mutants that are available that I've examined under the electron microscope that have a plastid which is colorless, a remnant of a plastid, so I wonder if we can actually always say that they are missing their genome?

DR. J. A. SCHIFF: I take it this was the W3 mutant? First of all in the mutants that lack chloroplast DNA or at least in which chloroplast DNA is undetectable, there is a plastid remnant but it's composed of nuclear-coded materials. All of the mutants which have no chloroplast DNA by our list of criteria, also had plastid remnants, and we assume that all of the evidence indicates that they're nuclear-coded constituents that are still being made. The evidence by cesium chloride gradients shows no chloroplast DNA satellites. On Bernardi columns there is no chloroplast DNA. Hybridization with very high specific activity iodinated homologous ribosomal RNA and tRNA shows no hybridization.

EVIDENCE FOR AFFINITIES AMONG MAJOR TAXA FROM THE ANALYSIS OF AMINO ACID FREQUENCIES IN GLYCOPROTEINS FROM CELL SURFACE STRUCTURES

Robert E. Calhoon and Sheldon Aaronson

Biology Department, Queens College
City University of New York
Flushing, New York 11367

INTRODUCTION

As this conference deals with the evolution of intracellular organelles, we thought we might not be too far amiss if we considered the evolution of the glycoproteins found in or on the cell surface structures of higher plants and eukaryotic microorganisms. Further, we hoped that we might find information that would help us to bridge relationships among the major taxa where gaps presently exist. The glycoproteins (glycopeptides) found in cell walls, scales, or other surfaces of eukaryotic organisms have their polypeptides assembled on membrane-bound ribosomes; sugars are then added either in the rough endoplasmic reticulum followed by further processing in the Golgi apparatus or the sugars are added in the Golgi apparatus.[1] The glycopeptides destined for the cell surface are usually packaged for export in membrane-bound vesicles in the Golgi from which they are moved to the plasma membrane and ultimately they are incorporated into the surface structure of eukaryotic microorganisms and metaphytes.[1, 2] Prokaryotes (bacteria) do not require synthesis and packaging of surface polypeptides in organelles such as the ribosomes and Golgi but as they represent the putative ancestors of eukaryotes, we thought it would be appropriate to include some of the surface polypeptides for comparison. Recently evidence has been presented that polypeptides for export to the outer membrane of *Escherichia coli* may be synthesized on plasma membrane-bound polysomes.[3, 4] A similar site may function for other secreted proteins in bacteria.[5]

MATERIALS AND METHODS

The data on the surface glycoproteins are in a very rudimentary state and relatively few species have been carefully analyzed. For example, much of the data is still in the form of the amino acid composition of the cell wall, rather than the amino acid composition of isolated cell-wall proteins. Nevertheless, as glycoproteins have been found in every eukaryotic surface structure (wall, scale, coat) examined, we analyzed the existing amino acid composition data to see if useful classification functions could be derived. Earlier, Calhoon and Aaronson [6] demonstrated the usefulness of analyzing amino acid frequencies in collagen and collagen-like proteins. We were able to show that information contained in the amino acid frequencies is sufficient to differentiate among major animal taxa and that the amino acid frequencies of an individual species

0077–8923/81/0361–0472 $01.75/0 © 1981, NYAS

could be used successfully to identify samples with their appropriate taxa. The methodology is described in Calhoon and Aaronson.[6]

The data analyzed in this study were derived from published analyses of amino acid composition of cell walls, other surface structures, and surface glycoproteins (TABLE 1). The species for which data exist do not represent a random sample of all possible species; an attempt was made, however, to locate data on as many species from as many taxa as possible. A total of five major taxa were analyzed (metaphyta, algae, fungi, protozoa, and bacteria). The preparation of purified samples and the subsequent hydrolysis of polypeptides were not consistent from sample to sample, and there is reason to suspect that certain forms of acid hydrolysis at high temperature may have destroyed some of the amino acids. Amino acid frequencies were reported in several different units; therefore samples usually could not be compared directly. Furthermore, amino acid frequency expressed in any of these units does not have a multivariate normal distribution. In order to approximate the appropriate distribution, the data were first converted to percent of total amino acid residues and then transformed to *arc sine*.

In this study, discriminant analysis was employed as described by Anderson.[7] The mathematical derivations and adaptations to amino acid frequency data were described by Calhoon and Aaronson.[6] All calculations were performed at the City University of New York University Computer Center, using software developed by the Health Science Computing Facility, University of California, Los Angeles (BMDP).[8]

RESULTS AND DISCUSSION

The mean frequencies and standard deviations of 19 amino acids in five plant and microbial taxa are given in TABLE 2. In most sources aspartic acid and asparagine frequencies were not listed separately, nor were glutamic acid and glutamine; therefore the pooled frequencies are given here as Asx and Glx respectively. Tryptophan was not detected in most analyses and is not listed here. Hydroxyproline and hydroxylysine were present in several samples.

The results of the discriminant analysis are given in TABLE 3 for nineteen measurements on 102 species in five major taxa. All four of the discriminant axes generated by this analysis are statistically significant ($P \leq 0.01$) and combined they account for 100% of the total variance. Hence, the null hypothesis that taxa centroids on this set of discriminant functions are equal is rejected. The four discriminant axes are orthogonal to each other, and differences among taxa centroids on each axis are statistically significant. This means that taxa differ in the absolute frequencies of their amino acids, and also in the relative frequencies of these variables.

FIGURE 1 is a plot of the taxa centroids on the first and second discriminant axes. The D_1 (horizontal) axis represents the line that maximally distinguishes among the five taxa, and accounts for 55.92% of the total variance. The D_2 axis represents the line orthogonal to D_1 that maximally distinguishes among groups and accounts for 23.49% of the total variance. The six vectors radiating from the grand mean represent some of the original coordinates associated with the 19 amino acids used in the analysis. The length of each vector is proportional to one standard deviation of the measurement and the orientation of the vector indicates the direction in which an increase in the

TABLE 1

MAJOR TAXON, SPECIES AND SOURCE OF DATA USED IN THE DISCRIMINANT ANALYSIS

Major Taxon	Species	Source	Date
Bacteria	Acinetobacter sp.	10	1978
	Bacillus polymyxa	10	1978
	Bacillus sphaericus	10	1978
	Clostridium thermohydrosulfuricum	10	1978
	Clostridium thermosaccharolyticum	10	1978
	Escherichia coli	11	1978
	Halobacterium salinarium	12	1976
	Halobacterium sp.	13	1969
	Spirillum serpens	10	1978
	Sporosarcina ureae	14	1979
Algae	Bryopsis maxima	15	1972
	Bryopsis plumosa	15	1972
	Caulerpa okamurai	15	1972
	Caulerpa prolifera	16	1973
	Chaetomorpha sp.	17	1967
	Chlamydomonas globosa	18	1969
	Chlamydomonas gymnogama	19	1972
	Chlamydomonas reinhardii	20	1976
	Chlorococcum oleofaciens	21	1978
	Cladophora sp.	17	1967
	Codium adharens	15	1972
	Codium fragile	15	1972
	Codium latum	15	1972
	Codium sp.	17	1967
	Cyclotella cryptica	22	1973
	Cyclotella stelligira	22	1973
	Eudorina californica	23	1978
	Halimeda discodea	15	1972
	Hymenomonas sp.	24	1969
	Melosira granulata	22	1973
	Melosira nummuloides	22	1973
	Navicula pelliculosa	22	1973
	Navicula sp.	25	1971
	Nitella sp.	17	1967
	Nitzschia breverostris	22	1973
	Pleurochrysis scherffelii	16	1973
	Polyphysa cliftonii	67	1974
	Porphyridium cruentum	27	1977
	Volvox carteri	28	1974
	Volvox sp.	19	1972
Metaphyta	Acer pseudoplatanus	29	1965
	Armoracia lapathifolia	30	1974
	Arena sativa	31	1965
	Centaurea cyanus	29	1965
	Cumumis melo	32	1973
	Daucus carota	33	1972
	Gingko biloba	29	1965

Major Taxon	Species	Source	Date
	Gossypium sp.	16	1973
	Helianthus tuberosus	31	1965
	Lilium longiflorum	16	1973
	Lycopersicon escalentum	29	1965
	Lycopersicon sp.	30	1974
	Nicotiana tabacum	29	1965
	Phaseolus vulgaris	29	1965
	Pisum sativum	29	1965
	Pisum sp.	30	1974
	Rosa sp.	29	1965
	Santalum album	34	1974
	Solanum tuberosum	29	1965
	Solanum sp.	35	1973
	Vicia faba	36	1969
	Zea mays	37	1967
Fungi	Atkinsiella dubia	38	1969
	Blastomyces dermatitidis	39	1972
	Candida albicans	40	1968
	Cordyceps militaris	41	1969
	Epidermophyton floccosum	42	1975
	Fusarium sulphureum	43	1977
	Fuscrium sp.	26	1975
	Fusicoccum amygdali	44	1971
	Helminthosporium sativum	45	1969
	Histoplasma capsulatum	46	1967
	Leptosphaeria albopunctata	47	1971
	Leptosphaeria allorgei	47	1971
	Leptosphaeria discors	47	1971
	Leptosphaeria nitschkei	47	1971
	Leptosphaeria orae-maris	47	1971
	Leptosphaeria robusta	47	1971
	Nadsonia elongata	48	1964
	Paracoccidioides brasiliensis	49	1977
	Pencillium notatum	50	1967
	Pencillium roquefortii	51	1968
	Phytophthora magasperma	52	1974
	Piricularia oryzae	53	1970
	Saprolegnia diclina	54	1975
	Sapromyces elongatas	55	1970
	Schizochytrium aggregatum	56	1973
	Sporothrix schenkii	57	1979
	Tremella mesenterica	54	1975
	Trichophyton mentagrophytes	58	1971
	Trichophyton sp.	59	1977
Protozoa	Acanthamoeba sp.	60	1969
	Eimeria tenella	61	1978
	Physarum polycephalum	62	1979
	Physarum sp.	63	1970
	Trypanosoma brucei	64	1979

TABLE 2

MEAN FREQUENCIES IN RESIDUES PER THOUSAND *

Amino Acids	Bacteria	Algae	Metaphyta	Fungi	Protozoa
Ser	83.5 ± 42.6	103.3 ± 64.9	98.1 ± 32.3	81.8 ± 27.6	59.6 ± 15.3
Thr	112.1 ± 34.9	61.0 ± 15.3	57.5 ± 13.6	90.0 ± 31.6	92.1 ± 42.3
Ala	115.3 ± 30.0	96.2 ± 25.1	76.0 ± 26.3	97.5 ± 23.6	91.0 ± 28.1
Gly	107.1 ± 41.5	123.6 ± 41.2	109.4 ± 48.2	92.1 ± 31.6	76.8 ± 33.4
Pro	27.4 ± 17.3	39.4 ± 27.7	56.6 ± 30.0	78.1 ± 46.8	62.2 ± 30.2
Arg	18.5 ± 11.3	23.6 ± 18.1	22.6 ± 13.7	39.2 ± 30.5	37.2 ± 6.7
Asx	154.1 ± 34.6	108.8 ± 25.4	86.1 ± 26.8	102.1 ± 30.3	98.1 ± 38.1
Cys	0.4 ± 1.0	9.6 ± 19.6	8.1 ± 20.7	7.7 ± 13.2	29.2 ± 30.5
Glx	91.9 ± 31.7	106.0 ± 38.4	81.2 ± 27.9	104.4 ± 38.8	126.5 ± 26.7
His	4.8 ± 2.4	7.6 ± 9.0	19.0 ± 12.0	19.7 ± 18.4	17.5 ± 9.7
Ile	39.7 ± 16.7	33.9 ± 12.9	35.7 ± 12.4	35.0 ± 12.9	26.2 ± 14.9
Leu	62.7 ± 14.2	63.6 ± 19.1	61.6 ± 28.7	61.3 ± 23.4	57.3 ± 19.1
Lys	55.9 ± 25.7	34.5 ± 24.2	69.2 ± 26.3	49.3 ± 18.6	74.7 ± 28.1
Met	6.7 ± 6.7	6.4 ± 6.0	6.7 ± 6.1	13.3 ± 27.0	6.8 ± 8.8
Phe	28.5 ± 12.8	30.3 ± 14.7	25.5 ± 12.2	31.7 ± 13.3	65.1 ± 87.0
Trp	13.5 ± 11.9	21.3 ± 10.3	21.1 ± 11.3	30.7 ± 12.9	29.6 ± 21.3
Val	77.8 ± 31.0	69.3 ± 23.8	61.4 ± 25.3	55.6 ± 20.6	49.9 ± 22.0
Hypro	0.0 ± 0.0	60.2 ± 77.5	104.3 ± 66.6	10.6 ± 37.2	0.0 ± 0.0
Hylys	0.0 ± 0.0	1.2 ± 7.0	0.0 ± 0.0	0.0 ± 0.0	0.0 ± 0.0

* For 19 amino acids in surface glycoproteins of plant and microbial taxa.

TABLE 3

DISCRIMINATORY ANALYSIS *

Removed	Eigenvalue	% of Trace	Approximate F	Degrees of Freedom	Probability
0	2.5402	55.92	4.85	68, 390	<0.0001
1	1.0669	23.49	3.39	48, 250	<0.0001
2	0.5649	12.44	2.66	30, 170	<0.0001
3	0.3703	8.15	2.33	14, 85	<0.01

* Analysis of 19 measurements on 102 species in five taxa. All roots not shown are not statistically significant, $\alpha = 0.01$.

TABLE 4

PAIR-WISE COMPARISON OF GROUP CENTROIDS FOR FIVE TAXA *

	Algae	Fungi	Protozoa	Metaphyta
Fungi	9.38			
Protozoa	3.88	*1.77*		
Metaphyta	3.27	7.66	3.92	
Bacteria	5.26	4.38	2.99	6.31

* Each F ratio has 17 d.f. in the numerator and 80 d.f. in the denominator. All values except Fungi × Protozoa (*italic*) exceed the 0.01 critical value (2.20).

variable would displace a group centroid. That is, the vectors indicate the direction and the distance a group centroid would move if a given amino acid were increased in frequency by one standard deviation. To interpret the discriminant graph, one must keep in mind that the location of any group on the graph is a function of all 19 variables and therefore attempting to relate differences in any one amino acid is often confusing.

The purpose of a discriminant analysis is to accurately differentiate among groups. To assess how successfully this derived set of discriminant functions

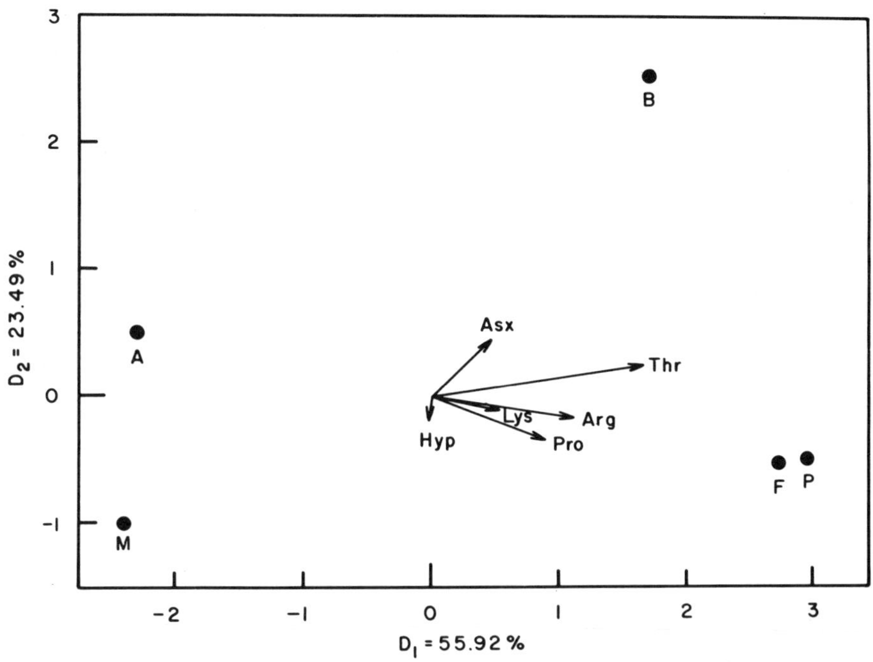

FIGURE 1. Discriminant analysis of the amino acid composition of cell surface structures in five major taxa. M = Metaphyta, B = Bacteria, F = Fungi, P = Protozoa. The centroid of each taxon is plotted on the first (D_1) and second (D_2) discriminant axes. Six of the original nineteen coordinates are represented by vectors emanating from the grand mean. The length of each vector is proportional to one standard deviation of the measurement.

distinguishes among taxa the group centroids on the four variate space were compared in a pair-wise fashion using D_2 as described by Rao.[9] Of the ten possible comparisons of group centroids, only one pair (protozoa × fungi) failed to show highly significant differences (TABLE 4). We interpret this to mean that possible polypeptide homologies exist in these two groups and that further primary structure analysis of the proteins of these two taxa might support this hypothesis. Further evidence supporting possible relationships between these two taxa may be adduced from some of the biochemical and ultrastructural characteristics listed in TABLE 5.

TABLE 5

COMPARISON OF BIOCHEMICAL CHARACTERISTICS IN SEVERAL MICROBIAL TAXA AND METAPHYTA

Trait	Bacteria	Alga	Metaphyta	Fungi	Protozoa
DNA(GC mole%)	37–64	34–70	30–49	26–70	19–66
Histones	—	+	+	+ or —	+ or —
Carotenoid biosynthetic pathway	A,B,H,K, M,N,O	C,D,E,F, J,N,O,P, Q,R,S,T, U	C,D,N,O, R	A,J,L,N, O,V	J,L,N
Unsaturated fatty acid biosynthesis	Desat. I Anaerobic	Desat. I Desat. II	— Desat. II	Desat. I —	Desat. I —
Polyunsaturated fatty acid biosynthesis	— —	ω3 ω6	ω3 —	— ω6	— ω6
Type of glycopeptide synthesized	o	IIIB	IA, IC IIIB	IA, IC IIC	IIA
G-3-P dehydrogenase	NAD NADP	NAD NADP	NAD NADP	NAD —	NAD —
Lysine biosynthesis					
DAP	+	+	+	+	+
AAA	—	+	—	+	—

REFERENCES

1. MORRÉ, D. J. & W. J. VAN DER WOUDE. 1974. *In* Macromolecules Regulating Growth and Development. E. D. Day, T. J. King & J. Papaconstantinou, Eds.: 81–111. Academic Press, Inc. New York, N.Y.
2. EVANS, L. V. 1974. Cytoplasmic Organelles. *In* Algal Physiology and Biochemistry. W. D. P. Stewart, Ed.: 86–123. University of California Press. Berkeley, Calif.
3. RANDALL, L. L. & S. J. S. HARDY. 1977. Eur. J. Biochem. **75:**43–53.
4. RANDALL, L. L., S. J. S. HARDY & L. G. JOSEFSSON. 1978. Proc. Natl. Acad. Sci. USA **75:**1209–1212.
5. BECKWITH, J., T. SILHAVY, H. INOUYE, H. SHUMAN, S. EMR, P. BASSFORD & E. BRICKMAN. 1978. *In* Transport of Macromolecules in Cellular Systems. S. C. Silverstein, Ed.: 299–314. Dahlem Konferenzen. Berlin.
6. CALHOON, R. E. & S. AARONSON. 1979. J. Theor. Biol. **78:**225–239.
7. ANDERSON, T. W. 1958. An Introduction to Multivariate Analysis. p. 158. Oliver & Boyd. London.
8. DIXON, W. T., Ed. 1971. Biomedical Computer Programs. p. 196. University of California Press. Los Angeles, Calif.
9. RAO, C. R. 1965. Linear Statistical Inference and its Applications. Sect. 8e. Wiley & Sons, Inc. New York, N.Y.
10. SLEYTR, U. B. 1978. Int. Rev. Cytol. **53:**1–64.
11. ESHDAT, Y., I. OFEK, Y. YASHOUV-GAN, N. SHARON & D. MIRELMAN. 1978. Biochem. Biophys. Res. Commun. **85:**1551–1559.
12. MESCHER, M. F. & J. L. STROMINGER. 1976. J. Biol. Chem. **251:**2005–2014.
13. STEENSLAND, H. & H. LARSEN. 1969. J. Gen. Microbiol. **55:**325–336.

14. Beveridge, T. V. 1979. J. Bacteriol. **139:**1039–1048.
15. Maeda, M. & Nisizawa. 1972. Contributions to the Systematics of Benthic Marine Algae of the North Pacific. I. Abbott & M. Kuragi, Eds.: 33–44. Japanese Society of Phycology. Kobe, Japan.
16. Brown, R. M., Jr., W. Herth, W. W. Franke & D. Romanovicz. 1973. *In* Biogenesis of Plant Cell Wall Polysaccharides. F. Loewus, Ed.: 207–257. Academic Press, Inc. New York, N.Y.
17. Thompson, E. W. & R. D. Preston. 1967. Nature **213:**684–685.
18. Pakhomova, M. V. & G. N. Zaitseva. 1969. Microbiology **38:**575–578.
19. Miller, D. H., D. T. A. Lamport & M. Miller. 1972. Science **176:**918–920.
20. Catt, J. W., G. J. Hills & K. Roberts. 1976. Planta **131:**165–171.
21. Miller, D. H. 1978. J. Phycol. **14:**189–194.
22. Hecky, R. E., K. Mopper, P. Kilham & E. T. Degens. 1973. Mar. Biol. **19:** 323–331.
23. Tautvydas, K. J. 1978. Planta **140:**213–220.
24. Isenberg, H. D., S. D. Douglas, L. S. Lavine, S. S. Spicer & H. Weissfellner. 1966. Ann. N. Y. Acad. Sci. **736:**155–190.
25. Nakajima, T. & B. E. Volcani. 1969. Science **164:**1400–1402.
26. Barran, L. R., E. F. Schneitler, P. T. Wood, C. Malhosingh & R. W. Miller. 1975. Biochim. Biophys. Acta **393:**148–158.
27. Heaney-Kieras, J., L. Roden & D. J. Chapman. 1977. Biochem. J. **165:**1–9.
28. Starr, R. C. & L. Jaenicke. 1974. Proc. Natl. Acad. Sci. USA **71:**1050–1054.
29. Lamport, D. T. A. 1965. The Protein Component of Primary Cell Walls. Adv. Bot. Res. **2:**151–218.
30. Lamport, D. T. A. 1974. *In* Macromolecules Regulating Growth and Development. Soc. Develop. Biol. **30:**113–130.
31. King, N. J. & S. T. Bayley. 1965. J. Exp. Bot. **16:**294–303.
32. Esquerré-Tugaye, M.-T. 1973. C-R. Acad. Sc. Paris **276:**525–528.
33. Brysk, M. M. & M. J. Chrispeels. 1972. Biochim. Biophys. Acta **257:**421–432.
34. Mani, U. V. & A. N. Radhakrishnan. 1974. Biochem. J. **141:**147–153.
35. Allen, A. K. & A. Neuberger. 1973. Biochem. J. **135:**307–314.
36. Pusztai, A. & W. B. Watt. 1969. Eur. J. Biochem. **10:**523–532.
37. Boundy, J. A., J. S. Wall, J. E. Turner, J. H. Woychek & R. J. Dimmler. J. Biol. Chem. **242:**2410–2415.
38. Aronson, J. M. & M. S. Fuller. 1969. Arch. Mikrobiol. **68:**295–305.
39. Roy, I. & J. W. Landau. 1972. Can. J. Microbiol. **18:**473–478.
40. Chattaway, F. W., M. R. Holmes & A. J. E. Barlow. 1968. J. Gen. Microbiol. **51:**367–376.
41. Marks, D. B., B. J. Keller & A. J. Guarino. 1969. Biochim. Biophys. Acta **183:**58–64.
42. Kitajima, Y. & Y. Nozawa. 1975. Biochim. Biophys. Acta **394:**558–568.
43. Schneider, E. F., L. R. Barran, P. J. Wood & I. R. Siddiqui. 1977. Can. J. Microbiol. **23:**763–769.
44. Buck, K. W. & M. A. Obaidah. 1971. Biochem. J. **125:**461–471.
45. Applegarth, D. A. & G. Bozoian. 1968. Can. J. Microbiol. **14:**489–490.
46. Kobayashi, G. S. & P. L. Guiliacci. 1967. Sabouraudia **5:**180–188.
47. Szaniszlo, P. J. & R. Mitchell. 1971. J. Bacteriol. **106:**640–645.
48. Dyke, K. G. 1964. Biochim. Biophys. Acta **82:**374–384.
49. San-Blas, G., F. San-Blas, E. Ormaechae & L. E. Serrano. 1977. Sabouraudia **15:**297–303.
50. Applegarth, D. A. 1967. Arch. Biochem. Biophys. **120:**471–478.
51. Applegarth, D. A. & G. Bozoian. 1969. Arch. Biochem. Biophys. **134:**285–289.
52. Lippman, E., D. C. Erwin & S. Bartnicki-Garcia. 1974. J. Gen. Microbiol. **80:** 131–141.
53. Nakajima, T., K. Tamari, K. Matsuda, H. Tanaka & N. Ogasawara. 1970. Agric. Biol. Chem. **34:**553–560.
54. Cameron, D. S. & I. E. P. Taylor. 1976. Biochim. Biophys. Acta **444:**212–222.

55. PAO, V. M. & J. M. ARONSON. 1970. Mycologia **62:**531–547.
56. DARLEY, W. M., D. PORTER & M. S. FULLER. 1973. Arch. Mikrobiol. **90:**89–106.
57. PREVIATO, J O., P. A. J. GORIN & L. R. TRAVASSOS. 1979. Exptl. Mycol. **3:**83–91.
58. NOGUCHI, T., Y. KITAZIMA, Y. NOZAWA & Y. ITO. 1971. Arch. Biochem. Biophys. **146:**506–512.
59. WU-YUAN, C. D. & T. HASHIMOTO. 1977. J. Bacteriol. **129:**1584–1592.
60. NEFF, R. J. & R. H. NEFF. 1969. Symp. Soc. Exp. Biol. **23:**51–81.
61. STOTISH, R. L., C. C. WANG & M. MEYENHOFER. 1978. J. Parasitol. **64:**1074–1081.
62. ZAAR, K., P. BEYER & H. KLEINIG. 1979. Biochim. Biophys. Acta **582:**21–32.
63. MCCORMICK, J. J., J. C. BLOMQUIST & H. P. RUSCH. 1970. J. Bacteriol. **104:** 1119–1125.
64. JOHNSON, J. G. & G. A. M. CROSS. 1979. Biochem. J. **178:**689–697.
65. RAGAN, M. A. & D. J. CHAPMAN. 1978. A Biochemical Phylogeny of the Protists. Academic Press, Inc. New York, N.Y.
66. SINOHARA, H. 1972. Tohoku J. Exp. Med. **106:**93–107.
67. PARADISE, H. H., L. GOKE & G. WURZ. 1977. *In* Progress in Acetabularia Research. C. L. G. Woodcock, Ed.: 305–315. Academic Press, Inc. New York, N.Y.

DISCUSSION OF THE PAPER

DR. J. L. LEHMAN (*Public Health Research Institute, New York, N.Y.*): I have a couple of questions on this data. The first is that I always get very uncomfortable when I see standard deviations larger than averages and I notice that that was true with the hydroxyproline data.

DR. R. E. CALHOON: That's right.

DR. LEHMAN: Was that true with any of the other amino acids that you didn't show?

DR. CALHOON: Yes.

DR. LEHMAN: How do you account for that?

DR. CALHOON: Well in our paper we discussed that the frequency data is not normally distributed, it is definitely skewed. The large standard deviations were a clue to the fact that we were dealing with highly skewed data.

DR. LEHMAN: In your analysis, did you take into account the fact that some amino acids are more closely related to others in terms of the codons that specify them? In other words it might mean less if you are switching between two where there is only one nucleotide different as opposed to ones that are different by all three.

DR. CALHOON: We thought about that and although some codons may be more similar than others, the chance of fixation by selection pretty well would confound that so that you don't really have any way of knowing whether that's a significant consideration or not. So we didn't consider it in the analysis.

TWO DATA SETS: ALTERNATIVE EXPLANATIONS AND INTERPRETATIONS *

Thomas Uzzell and Christina Spolsky

Academy of Natural Sciences
19th and the Parkway
Philadelphia, Pennsylvania 19103

For those biologists interested in the evolutionary origin of the higher taxa of organisms, the origin of eukaryotic cells and their organelles is one of the most puzzling events in the history of the living world. For those biologists studying the eukaryotic organelles themselves, an understanding of the origin of the organelles is an essential complement to a description of their structure and function. Although two competing hypotheses concerning the origin of eukaryotic organelles (at least of the chloroplasts and mitochondria) have been available since 1890,[2, 19] it is only in the last 30 years that basic data bearing on these hypotheses have begun to accumulate. One of these basic facts is our discernment of the fundamental structural differences between eukaryotes and prokaryotes.

A traditional view of the origin of cellular organelles is that they arose autogenously, by small evolutionary steps, within some prokaryotic lineage. Although this view is traditional, that is not evidence for or against its correctness. The traditional view may, however, involve a rather unthinking acceptance of the idea that there has been a single origin of life, followed by subsequent divergence. Such assumptions should never remain unchallenged.

A worthy challenge to this traditional view is posed by the hypothesis that eukaryotic organelles arose by endosymbiosis, a hypothesis picked up, dusted off, shaken into conformity with a lot of modern data, and stated vigorously by Lynn Margulis in her book, *Origin of Eukaryotic Cells*.[17] This hypothesis has been a fruitful scaffolding on which to hang much of the data on structure and function of cellular organelles and their constituents, and is consonant with much new data on the relations of prokaryotic groups. One need only read widely in several journals, or examine many of the presentations at this symposium, to see the effectiveness of this hypothesis and to realize the wide acceptance it has won.

And yet history repeats itself. Rather than being an alternative hypothesis for challenging the tradition, endosymbiosis has become a new tradition in itself. It is sometimes taken as proven, at least for chloroplasts,[32] and often appears to be almost unthinkingly accepted. We find, however, that many of the data so far adduced in favor of endosymbiosis are equally supportive of the earlier traditional view.

PRINCIPLES

Because so much of the early evidence for the endosymbiotic origin of organelles involved morphological data, it seems worthwhile, especially at this

* Supported by National Science Foundation Grant DEB 77–13472 to T.U.

0077–8923/81/0361–0481 $01.75/0

symposium, to repeat certain principles important in the analysis of morphological data in trying to determine phylogeny.

1. Primitive character states may be retained independently in a variety of lineages in a phylogeny (FIGURE 1). By character we mean a feature that varies among organisms. The different manifestations are called states [25] and may be classified as primitive (present in the common ancestor of a set of taxa) or derived (present in descendent taxa as a result of the fixation of mutations during the course of evolution).

Seemingly identical derived states may evolve in parallel, but this is less likely than the parallel retention of primitive states. Derived states have shorter phylogenetic histories and thus carry more phyletic information than primitive states.[13, 26] They should therefore be emphasized in reconstructing phylogenies;

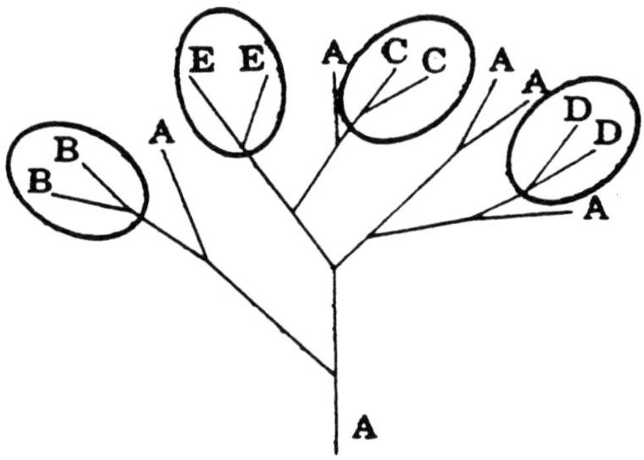

FIGURE 1. Phylogeny and taxonomic groupings of an array of organisms show the distribution of primitive and derived character states for a single character. The primitive character state A is present in the common ancestor of the array. More recent manifestations of this character—B, C, D, and E—are derived states. Organisms sharing state B—or C, D, or E—are more closely related to each other (that is, they share a more recent common ancestor) than they are to other organisms. Organisms having primitive character state A cannot, however, be grouped into a meaningful taxon, because state A is independently retained by several organisms with varying degrees of relatedness to each other. (From *American Scientist* **62:**335.)

conversely, primitive states should be deemphasized. For example, the animal kingdom is divided into vertebrates and invertebrates. The backbones of vertebrates are clearly a derived state, while lack of them is primitive; vertebrates thus form a coherent grouping, a single branch on a phylogenetic tree. Invertebrates, however, have no such coherence; they include numerous independent lineages, but especially, they include both the phyla in the arthropod–mollusk–annelid lineage as well as backboneless members of the very dissimilar echinoderm–chordate lineage to which vertebrates belong.

Criteria for determining which states of characters are primitive, which derived, fall into four categories: [18, 26, 30] (1) states that appear only late in the fossil record are derived; (2) states widely spread in groups outside the group in question are likely to be primitive for that group; (3) states that in their distribution in a group of taxa show high correlation with derived states are also likely to be derived states; and (4) states that by logical necessity must have preceded other states in a phylogeny are primitive. The assessment of states as primitive, using these criteria, varies from the logically compelling through various degrees of probability to the uncertain.

2. The larger the number of interrelated states (for example, details of a morphological structure or steps in a biosynthetic pathway) shared by a group of organisms, the more likely it is that the states are shared as a result of common inheritance rather than of parallel evolution from distinct ancestors.

Proposition 2 must be interpreted in light of proposition 1: A group of organisms that share a set of derived character states probably have a more recent common ancestor than a group that share an equal number of primitive states.

3. A state that is derived for the common ancestor of a group of taxa may be primitive for a descendent group.

The effect of these three propositions is to minimize the number of evolutionary steps invoked in reconstructing a phylogeny. We have no reason to assume that evolution is a parsimonious process,[27] but when biologists appeal to multiple parallelisms in reconstructing a phylogeny without independent supporting evidence, we should be skeptical of these reconstructions.

There seems little reason to doubt that the features by which eukaryotic organelles resemble prokaryotes are primitive features for eukaryotes. We discussed this earlier,[29] and it is, we believe, accepted by proponents of both the endosymbiosis hypothesis and the autogenous origin hypothesis. Statements that the prokaryotic nature of the chloroplast or of the mitochondrion has been established require this acceptance. Regardless of how the eukaryotic cell got its organelles, their features are primitive; it is the nucleus and what it controls that is full of derived character states (FIGURE 2).

We repeat these principles to emphasize, once again, that establishing the "prokaryotic nature" of cellular organelles does not establish the endosymbiosis hypothesis; the prokaryotic nature of the organelles is consistent with either hypothesis.

CYTOCHROMES *c*

Recently, in trying to unravel the phyletic history of organisms, more and more emphasis has been placed on informational macromolecules. Such methods work quite well in the group we know best, the vertebrates. Whether they will work as well in prokaryotes is not known. Certainly the sequence data do not correlate well with the morphological data.[3,4] We cannot share the defeatism voiced by Ambler, and rather agree with Woese and with Drouet[10] that the morphological characteristics available to classify prokaryotes are limited and probably subjected both to much parallel and convergent evolution and to much ecophenotypic variation. Although it is possible that there is genetic transfer among prokaryotes, if it is assumed to be frequent we cannot determine their phylogeny; we prefer to assume that such genetic transfer is rare.

The greatest difficulty in constructing phylogenies using amino acid sequence data, aside from not having enough data, is in making the distinction between orthologous and paralogous polypeptides.[12] Orthologous proteins derive from a single cistron differentiating in various organisms, while paralogous polypeptides result from duplication of a given cistron. After cistronic duplication, each cistron has its own evolutionary history and forms its own evolutionary lineage. The two lineages may or may not have similar rates of evolution. After duplication, either cistronic lineage may disappear in time, becoming a nonfunctional unit or being removed by deletion. Since the phyletic branching sequence is determined by the phylogeny of the organisms, it is the same for each cistron. Thus by looking at amino acid sequence divergence in orthologous proteins (i.e., proteins coded by a single evolutionarily continuous cistron), the

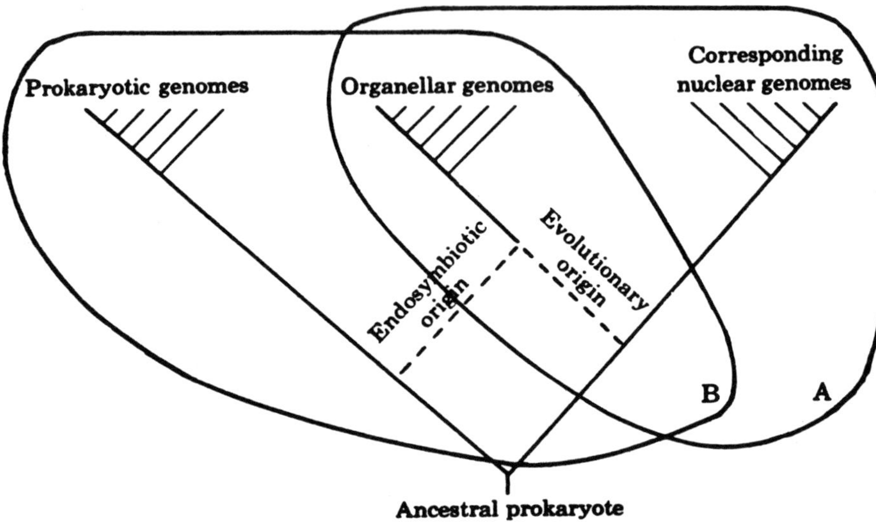

FIGURE 2. There are two alternative origins of eukaryotic organelles based on their genomes. If mitochondrial or plastid (organellar) genomes are more similar to nuclear genomes (indicated by line A), the organellar and nuclear genomes probably share a more recent common ancestor with each other than either does with prokaryotic genomes; this similarity would support the autogenous hypothesis only. In contrast, if organellar genomes are no more similar to nuclear genomes than to prokaryotic genomes (indicated by line B), the similarity of organellar genomes to prokaryotic genomes could be due to the retention of primitive states by organellar genomes and would be consistent with both the autogenous and endosymbiotic origins hypothesis. (From *American Scientist* **62:**336.)

phyletic branching sequence can be determined. One must be careful, however, to compare orthologous and not paralogous proteins within a phyletic grouping.

The phylogenetic tree drawn by Schwartz and Dayhoff [23] for c-type cytochromes is a case in point. Much of what we have to say about this tree echoes comments by Demoulin.[9] Schwartz and Dayhoff, for sound methodological reasons, constructed independent trees for classes of c-type cytochromes such as c_2 and c_6. They then connected these independent trees using a matrix of data for all of the cytochromes, excluding differences due to multiple residue insertions and deletions. The composite tree places the mitochondrial cytochrome c of eukaryotes with the cytochromes c_2 and cytochrome c_{550} of

Paracoccus denitrificans. The cytochromes c_6 of *Euglena, Porphyra, Alaria,* and *Monochrysis* fall with the c_6s of the blue-green bacteria. If all of these were orthologous proteins, this would be strong evidence in favor of an endosymbiotic origin, either of chloroplasts or of mitochondria.

Schwartz and Dayhoff [24] correctly pointed out that "None of the available sequence data contradict our supposition that the cytochrome c_2 and c_6 genes are descended from the same gene present at the time of the divergence of the chloroplast–blue-green algal line from the pseudomonad–Rhodospirillaceae–mitochondrial line." Their statement does not, however, establish that cytochromes c_2 and c_6 are orthologous rather than paralogous, and there is at least some evidence that they are not. In any case, it is certain that duplications involving *c*-type cytochrome cistrans have occurred several times during the evolution of prokaryotes.

Within the genus *Pseudomonas,* two *c*-type cytochromes occur: c_5 in *P. mendocina* and c_{551} in *P. stutzeri.* We cannot be certain that *Pseudomonas* is not taxonomically a composite genus, nor can we be certain that the c_5 is not derived from c_{551} (or the reverse) by small evolutionary steps within the genus. The possibility of a transfer of genetic material between taxa, introducing c_5 into *P. mendocina,* is impossible to rule out on present evidence. We are inclined, however, to accept Schwartz and Dayhoff's suggestion that the *c*-type cytochrome diversity in *Pseudomonas* resulted from a gene duplication, with subsequent loss of one or the other form in various lineages.

The situation in the Rhodospirillaceae is even more striking. According to Ambler *et al.*,[3] two of the 12 species of this family (*Rhodospirillum tenue* and *Rhodopseudomonas gelatinosa*) have a *c*-type cytochrome resembling the small c_{551} of denitrifying pseudomonads. Five other species have a small c_2 (95–107 amino acid residues, in the size range of vertebrate cytochrome *c*), while the remaining five have a larger c_2 (112–124 residues, in and above the size range of plant cytochrome *c*). No member of the Rhodospirillaceae is known to have both the c_{551}-type and the c_2 cytochromes. There are of course several possible interpretations of these data: that the Rhodospirillaceae are taxonomically composite; that c_2 is derived, within the Rhodospirillaceae, from c_{551} by small evolutionary steps; that there has been transfer of genetic material into certain species of Rhodospirillaceae, specifically of c_{551} cistrons (the hypothesis favored by Ambler *et al.*); or that some ancestor of extant Rhodospirillaceae had both c_2 and c_{551} cytochrome cistrons. We prefer the latter hypothesis.

There is, moreover, explicit evidence of multiple *c*-type cytochrome cistrons within the Rhodospirillaceae. Both *Rhodospirillum molischianum* and *R. fulvum* each produce two distinct cytochromes c_2. The two isocytochromes (1 and 2) within each species differ from each other at numerous amino acid positions, although the iso-1-cytochromes of the two species differ from each other by only 12 residues and the iso-2-cytochromes of the two species also differ by only 12 residues. Clearly the cistronic duplication that gave rise to the two isocytochromes must have been present in the common ancestor of the two taxa.

There is additional evidence suggesting cistronic duplication in the Rhodospirillaceae, although not so clearly. The iso-2-cytochromes c_2 of both *Rhodospirillum molischianum* and *R. fulvum* appear to contain a mixture of phenylalanine and tyrosine at the position that corresponds to the highly conserved tyrosine-67 of mitochondrial cytochromes *c*. If this does in fact represent another gene duplication, it would appear to have occurred independently in the two species.

Among the blue-green bacteria, available amino acid sequence data appear to be for molecules that are orthologous to each other, the photosynthetically important c_6 cytochromes. In *Anacystis nidulans,* however, Holton and Myers[14, 15] observed three *c*-type cytochromes. There is no possibility that this is due to taxonomic heterogeneity, and we assume that there was no contamination of the *A. nidulans* cultures by other organisms. Either three cistrons for *c*-type cytochromes exist here, representing at least two cistronic duplications, or there have been at least two transfers of genetic material into *A. nidulans* from some other taxon. We believe that these data are best interpreted as showing duplication of *c*-type cytochrome cistrons.

The spectral and physiological properties of the three *c*-type cytochromes are significant. Two of them, cytochromes c_{549} and c_{552}, are like mammalian cytochrome *c* in α absorbance. Two, cytochromes c_{549} and c_{554}, are like cytochrome c_6 in ratio of reduced Soret absorbance to α absorbance. Two, cytochromes c_{549} and c_{554}, have acidic isoelectric points, while c_{552} resembles mammalian cytochrome *c* in having a basic isoelectric point. As pointed out by Demoulin,[9] these data suggest that cytochrome c_{552} is orthologous to cytochromes *c* and c_2, and that the c_6-type cytochromes, c_{549} and c_{554}, are not. It is c_{554} that has been sequenced, and that was used by Schwartz and Dayhoff.[23]

Photosynthetic eukaryotes have both *c* and c_6 cytochromes. While it is possible that these represent two separate transfers of genetic material into the protoeukaryotic host cell, it is much more likely that they represent a cistronic duplication present in the common ancestor of eukaryotes, purple nonsulfur bacteria, and blue-green bacteria. We present in FIGURE 3 a diagrammatic phylogeny of these three groups, showing both cistronic duplications and losses. No sequence data available contradict this interpretation, but even if it could be demonstrated that cytochromes *c* and c_6 are paralogous, this would not exclude an endosymbiotic origin of eukaryotic organelles.

In our diagram, we have followed Almassy and Dickerson[1] by starting with a small cytochrome of the c_{551} type; we do not mean to imply anything about the spectral characteristics of this ancestral form in calling it c_{551}. Cytochrome c_6 arose from this ancestral form by cistronic duplication. Later, cytochrome c_2 arose by another cistronic duplication. We are thus postulating an ancestral organism with three *c*-type cytochromes. In the lineage leading to blue-green bacteria, the original c_{551} type is lost unless it is represented by one of the three present in *Anacystis*. Blue-greens thus inherit both cytochrome c_2 and c_6 cistrons from this common ancestor. In the lineage leading to purple nonsulfur bacteria, in contrast, the ancestral c_{551} locus is retained as well as the c_2 cistron. In some extant purple nonsulfur bacteria, c_{551} is lost, while in others, c_2 is lost. Among those in which c_{551} is lost, c_2 differentiates into a larger form and a smaller form. In the lineage leading to eukaryotes, cistrons for cytochromes c_2 and c_6 are present.

There are certain advantages for this scheme that we would like to emphasize: 1. If one considers the common ancestor of extant photosynthetic and nonphotosynthetic aerobic bacteria other than the Flexibacteria, it is clear that they had a very elaborate metabolism including glycolysis, a complete Krebs cycle, and an electron transport chain with a *c*-type cytochrome and a cytochrome oxidase. Furthermore, it had an elaborate photosynthetic system, although without the ability to photolyse water. On the other hand, the common ancestor of extant blue-greens and aerobic bacteria obviously had even photolysis of water. If *Beggiatoa* is really a colorless blue-green and if it indeed has a complete Krebs cycle as suggested by Almassy and Dickerson's[1] FIGURE 5, it is almost certain that a complete Krebs cycle was present in the common ancestor of extant blue-greens and of aerobic photosynthetic and nonphotosynthetic bacteria, as we suggested earlier.[29] It follows rather clearly then that somewhere in prokaryote phylogeny, before the separation leading to extant blue-greens and aerobic bacteria, organisms existed that could carry out

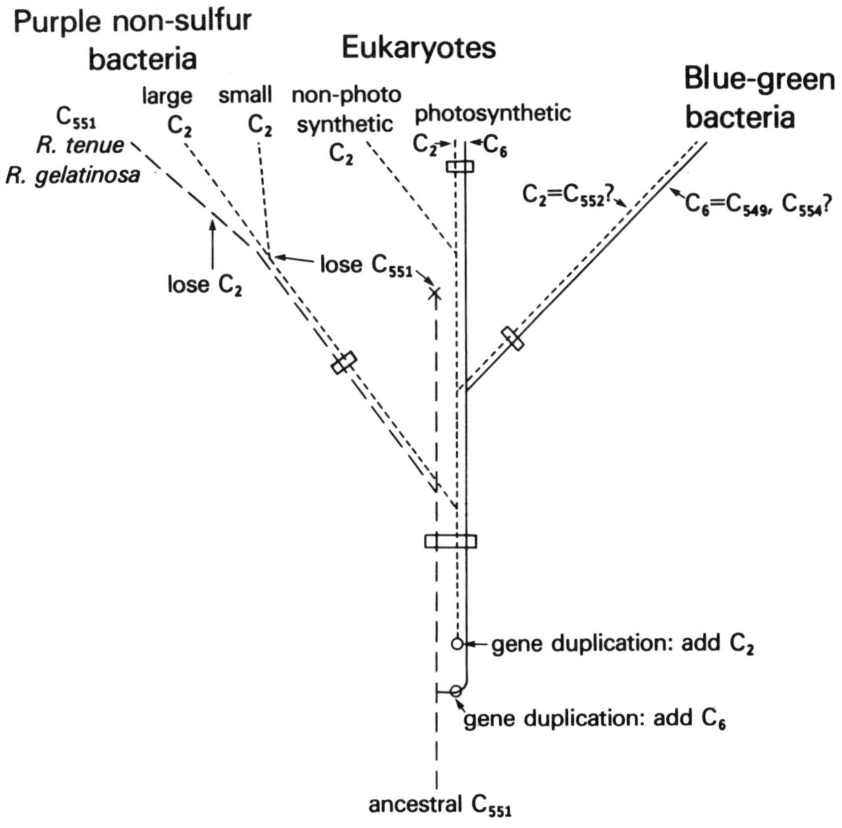

FIGURE 3. A diagrammatic phylogeny of the eukaryotes and the blue-green and purple nonsulfur bacteria, indicating cistronic duplications and deletions for c-type cytochromes. C_{551} present in the common ancestor is supposed only to be a small cytochrome, resembling neither the photosynthetic cytochromes c_6 nor the larger c and c_2 cytochromes that have a large insertion below the heme cleft; no assignment of spectral characteristics is intended.

in themselves the functions of both the chloroplast and the mitochondrion. These are ideal ancestors of eukaryotic organisms.

We believe that this provides an optimal resolution to the problem raised by Schopf,[22] who pointed out that it is difficult to imagine the development of oxygen releasing photosynthesis prior to the development of a rudimentary mechanism for detoxifying oxygen. We believe that these two developments occurred nearly simultaneously, although probably over a protracted period of time, in a single lineage. The multiplicity of *c*-type cytochromes probably represents part of this process.

2. One need not, this way, postulate large transfers of genetic material from endosymbiont genomes to nuclear genomes. What is required instead is cistronic duplications and selective deletions within the prokaryotes. That such duplications have occurred among the prokaryotes is clear in both the blue-green and purple nonsulfur bacteria. In order to evolve the eukaryotic organelles, genomic duplications must occur, followed by sequestration of duplicate genomes in organelles with special functions (FIGURE 4). Subsequently, many of the duplicated cistrons of organellar genomes are lost by deletion.

SMALL SUBUNIT (16S–18S) RIBOSOMAL RNA

Although *c*-type cytochromes are found in all of the groups of interest for determining the origin of mitochondria and chloroplasts, the cistrons for the mitochondrial form at least are nuclear. Ribosomal RNAs, however, occur both in the nuclear and organellar genomes of eukaryotes, as well as in all the prokaryotes of interest. They therefore appear to be particularly useful as possible sources of resolution to the question of organellar origin.

If cellular organelles arose autogenously by slow evolutionary steps, the small subunit rRNA cistrons of mitochondrion, chloroplast, and nucleus are paralogous within the eukaryotic cell. For constructing a phylogeny of prokaryotes and eukaryotes, any one of these eukaryotic rRNAs should do as well as any other in determining branching sequences among the eukaryotes, although the branch lengths among the eukaryotes might not be comparable to the branch lengths among the prokaryotes, because the particular eukaryotic rRNA cistron chosen might have had a different evolutionary rate from that used in the prokaryotes. Using a mixture of paralogues among the eukaryotes could, however, easily yield a faulty branching sequence for them, especially if the rates of evolution among the paralogues are diverse. In contrast, if cellular organelles arose through endosymbiosis, we may assume that the small subunit rRNA cistrons of prokaryotes, organelles, and nuclei are all orthologues; comparisons among them should yield correct phylogenies for these independently derived genomes.

We were first attracted to the flurry of work on 16S–18S rRNA and its bearing on the origin of cellular organelles by the work of L. Bonen and W. F. Doolittle. We were particularly struck by the results of their study of the mitochondrial 18S rRNA of *Triticum vulgare*.[6] By their evidence, the mitochondrial rRNA of *Triticum* shares no more oligonucleotides (pentamers or larger) with the cytosolic 18S rRNA of *Triticum* than can be expected by chance. It does, however, share more than expected by chance with 16S rRNA of *Rhodopseudomonas spheroides, Gloeocapsa aipicola, Escherichia coli, Bacillus subtilis,* and the rather large 16S rRNA of chloroplasts from *Porphyridium cruentum*. There seemed to us some possibility that there was a greater similarity between the rRNA oligomer catalogues of *Triticum* mitochondrial rRNA and *Porphyridium* chloroplast rRNA than between mitochondrial rRNA and the rRNAs of the prokaryotes that were compared with it.

These data suggested to us a test of the autogenous origin of cellular organelles. While the *Triticum* mitochondrial 18S rRNA has only randomly

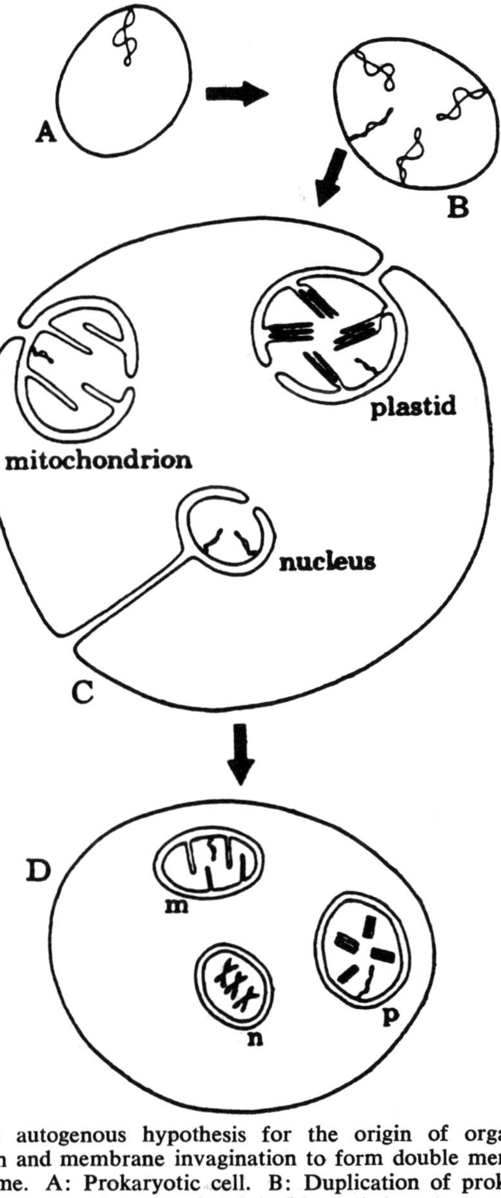

FIGURE 4. The autogenous hypothesis for the origin of organelles postulates genome duplication and membrane invagination to form double membrane structures around each genome. A: Prokaryotic cell. B: Duplication of prokaryotic genomes. C: Invagination of cell membrane to form double-walled nucleus, mitochondrion and plastid. D: Eukaryotic cell: the nuclear genome enlarges and differentiates; the organellar genomes specialize by losing many duplicated cistrons. (From *American Scientist* **62:**338.)

shared oligonucleotides with the cytosolic 18S rRNA of *Triticum,* the greater similarity to prokaryotic 16S rRNAs may be due to functional constraints that operate on prokaryotes and organelles, as pointed out by Bonen *et al.*[6] Given an autogenous origin, chloroplast and mitochondrial 16S and 18S rRNAs might have greater similarity to each other than to prokaryotic 16S rRNAs. They might appear not to, however, if they were taken from disparate eukaryotes or if the rates of evolution in the two paralogous cistrons were quite dissimilar.

The endosymbiosis hypothesis makes quite different predictions. It is clear that the mitochondrial cytochromes *c* are most similar in sequence and structure to the cytochromes c_2 of purple nonsulfur bacteria and of *Paracoccus.*[1, 3] The mitochondrial rRNAs should therefore have their greatest shared number of oligonucleotides with this group of bacteria. It is equally clear that chloroplasts and blue-green bacteria share numerous features, and therefore, according to the endosymbiosis hypothesis, chloroplast 16S rRNA should have its strongest homologies with this group.

We have undertaken such a comparison, but are minimally satisfied with the results. We offer them here, not as an answer, but as a challenge to people more familiar with these data, and in hopes of stimulating further work in the area.

We have taken our data from published sources. To simplify data handling, we have restricted our attention to ten taxa: *Aphanocapsa* 6308 and *Anacystis nidulans,* blue-green bacteria; *Porphyridium cruentum* chloroplast, red alga; *Euglena gracilis* chloroplast, euglenoid;[7] *Rhodopseudomonas spheroides,* purple nonsulfur bacterium;[33] *Bacillus stearothermophilus,* bacillus;[31] *Escherichia coli,* enteric-group bacterium;[28] *Triticum vulgare* mitochondrion, angiosperm;[6] *Methanobacterium arbophilicum* and *Methanospirillum hungatii,* methanogenic bacteria.[4] For both blue-green and the methanogenic bacteria we chose a distantly related pair of species as judged by published estimates of 16S rRNA homology.

A joint catalogue for these ten taxa was prepared by combining the catalogues of all of the taxa listed in the above papers. We restricted our compilation to the pentamers, hexamers, and heptamers. As pointed out by Bonen and Doolittle,[5] pentamers are more likely to be shared by chance than hexamers and larger oligonucleotides. Their work on *Triticum* mitochondrial 18S rRNA[6] shows, however, that there is nevertheless considerable information in addition to noise in the pentamers. We have excluded the octamers and larger nucleotides, not by preference, but because the sequence of many of these larger oligomers was uncertain in the *Triticum* mitochondrial 18S rRNA, and it was with this rRNA that we particularly wanted to make comparisons. A summary catalogue for the ten taxa chosen is given in TABLE 1.

In order to determine the similarity of the oligonucleotides, we have determined the number of matches for pentamers, hexamers, and heptamers between each pair of taxa. We have also determined the number of nucleotides present in the published pentamer, hexamer, and heptamer catalogues for each of the ten taxa considered. From these numbers we have calculated an S_{ab}; for each pair, S_{ab} equals twice the number of nucleotides in the shared oligonucleotides, divided by the total number of nucleotides in pentamers, hexamers, and heptamers in the two individual taxa. All of these values are indicated in TABLE 1.

A phenogram clustering the ten rRNA catalogues was generated using the unweighted pair-group method with arithmetic averaging option of NT-SYS.[21]

TABLE 1

OLIGONUCLEOTIDE CATALOGUES OF PENTAMERS, HEXAMERS, AND HEPTAMERS OF SMALL SUBUNIT RIBOSOMAL RNAs OF SEVEN PROKARYOTES AND THREE EUKARYOTIC ORGANELLES *

Oligonucleotide Sequence	Present in Organism	Oligonucleotide Sequence	Present in Organism
Pentamers			
CCCAG	1–3, 5–8	CUCUG	3, 5, 6, 8, 10
CCACG	1, 7	CCUUG	2–9; 6
CACCG	7	CUUAG	3, 8
ACCCG	1, 2	UCUAG	5
CCAAG	8	UUCAG	1–3, 10
CAACG	1–7, 9, 10; 1–3, 5, 7, 10	CUAUG	3
ACACG	1–3, 5–8, 10	UAUCG	1, 10
AACCG	2, 7–10	UACUG	3, 4, 8, 10
ACCAG	7	ACUUG	4, 6–9
AACAG	3, 4, 8	UUAAG	1–7, 9, 10; 1–4, 7, 9, 10; 4, 7
AAACG	1–4, 7, 10	UAAUG	3, 5, 7, 9
AAAAG	1, 3, 4, 8, 9	UAUAG	1, 8; 1
UCCCG	1–7	AAUUG	1–10; 5, 9
CUCCG	2, 8	AUUAG	1–10; 1–7, 9
CCUAG	4	UUCUG	3, 8
CUCAG	1–10	UCUUG	1, 4, 6–8, 10; 1, 7
UCCAG	7, 10	AUUUG	4, 6; 4
CCAUG	2, 5, 7–10	CCGCG	5, 7
CUACG	8	CCCCG	1–10; 5, 8–10; 5, 9
UCACG	1, 9	CAAAG	5
CACUG	4	ACCUG	5, 6
CAUCG	1, 3–5, 8	AUAAG	5, 6, 10
UACCG	1, 2, 6, 9, 10	ACAAG	6, 9; 9
CUAAG	6	CCCUG	8
UCAAG	1, 2, 4, 5, 7, 8; 4	ACUCG	8
CAAUG	1–4, 6, 8, 9	AUAAG	10
UAACG	2, 5–8, 10	CUUCG	10
ACAUG	5, 8	UACUG	10
AACUG	3, 5, 7, 8	AUCUG	7, 10
ACUAG	1–3, 10	UUUCG	8
AAUCG	1–8; 1	CUUUG	6–10
AUACG	2	UUUUG	8, 10
AUCAG	1–8	CUCUG	6
UAAAG	1–8; 1, 3, 5, 7, 8	CUCAG	7
AAAUG	1–4, 6, 7; 3, 4, 7; 3, 4	AACUG	7
AAUAG	3, 5, 8	UCUCG	7
UUCCG	2, 9, 10	UUACG	8
UCCUG	9		
Hexamers			
CCCCAG	1	AAACCG	6, 10
CCAACG	2	ACAAAG	1
AACCCG	2, 6	AAACAG	3
CAAACG	1	CCUCAG	3, 8
CACAAG	1–8	CCUAAG	3, 4, 8, 9
ACACAG	4	CUAACG	3, 6–8

Oligonucleotide Sequence	Present in Organism	Oligonucleotide Sequence	Present in Organism
CAACUG	6	UCUCUG	5
ACACUG	1, 3, 10	AUCUUG	5
AACUCG	1	CCCAAG	8
AAUCCG	1, 8	CAACCG	10
UCAAAG	1, 2, 4	ACACCG	10
CAAUAG	4	CCCACG	6
UAAACG	1–10	CCCUCG	10
AAACUG	1, 2, 5–7, 9, 10; 1, 10	UCCCAG	10
AAUACG	2, 4–7, 9	ACCUCG	7
UAAAAG	1	CAUACG	9
AAAAUG	4, 8	AACCUG	5, 7–9; 7, 9
AAUAAG	2–4, 9	CUAAAG	10
UUCCCG	1–3, 5–7	AACAUG	8
CCUUCG	6–8; 7	AAAUCG	8, 9
CCUCUG	2, 10	CCAUCG	7
CCUUAG	1, 5, 10	CCCUUG	7, 8
CUCUAG	2, 9	UCCCUG	9, 10
UCCUAG	9	CCUAUG	8
CUACUG	2, 9	UCACUG	10
CAUCUG	4, 7	CUAUCG	10
UACCUG	2, 3	ACUCUG	10
AUCCUG	1–3, 5, 6, 8–10	UCUAAG	10
CUAAUG	4	UUACAG	10
CAUUAG	3, 6	ACAUUG	9
UAAUCG	1–10	UCUCUG	9
AACUUG	3	CUAUUG	9
AAUCUG	1–4, 8, 10	AUUCUG	10
ACUUAG	5	UAUAUG	10
AAUUCG	3	CUUUUG	6, 8, 9; 9
AUCUAG	4, 5	UAAACG	8
UUAAAG	3, 4	UUUUCG	8
AUAAUG	5	UCAUUG	6
UUCUCG	4	CAAAAG	7
CUCUUG	1	AUCCAG	8
CCUUUG	4–6, 8, 9	CCUCCG	7
UCUUAG	1, 3	AUCAUG	7, 8
UUUACG	1	CCACCG	8
UUUUAG	3	ACCAAG	8
UAUUUG	4	AAAAAG	8
UUUUUG	2	CCUCCG	8
ACCACG	5	UACCCG	8
UCCACG	5–7	UACCAG	8
ACUACG	5	CCAAUG	8
UAACAG	5, 7	CUAAAG	8
AUAUUG	3	AUCAAG	8
ACAAUG	5, 8–10	CUCUCG	8
UUCCAG	5, 9	UCUACG	8
UCUCAG	5	UCCAUG	8
AUACUG	5, 7, 8, 10	AAAUUG	8
AUUAAG	5, 9, 10	UCUUCG	3, 8

Oligonucleo- tide Sequence	Present in Organism	Oligonucleo- tide Sequence	Present in Organism
Heptamers			
CCCAAAG	2	CCCUAAG	10
AACACCG	7	CUACACG	10
CAAACAG	4–6	AUACCCG	9, 10
AACAAAG	3, 8	AACCUCG	10
CCUACAG	2	AAUCCAG	10
CAACUCG	2, 5, 8	AAAUCCG	8–10
CCUAAAG	1, 9	UAAAAAG	9
UAACACG	3, 6, 9, 10	CCCUUAG	9
AAACUCG	1, 3, 4	CAUCCUG	10
AAACAUG	3	ACCUUCG	7
AAUACAG	3	UUACCAG	9
UAUCCCG	1, 3, 4	UAACUCG	9, 10
CCUAUAG	3	CUUAAAG	10
CUAUACG	4	AAUCUAG	9, 10
CAUACUG	1, 2	CUCCUUG	9, 10
ACUCUAG	1	UACUUCG	10
AAUCCUG	1	UCUCUUG	10
CAUUAAG	1	UUUAUCG	9
UAAUACG	1–7	UAUUUCG	9
AUACUAG	1, 3, 9	UUUUUUG	9
CUCUUCG	2	CACUCCG	6
CCUUUAG	1, 3	CAACCUG	6
UUAUCCG	1–4	CAACUCG	6
AUCCUUG	3, 4	UACAAAG	6
UAUCUAG	4	CUCUCAG	6
AAUCUUG	8	CCUCUAG	6
AAUUUCG	4, 8	CUCUCUG	6
AUAUCUG	2	UCACUUG	6
UAAUAUG	4	CAUAACG	7
UUUAAUG	4	CACAAUG	7
UUUUCUG	3	AAUACCG	7
UUUUUAG	2	UUACCCG	7
CACACAG	5	CCUCUUG	7
UAACAAG	1–5, 7	CUCAUUG	7
CAUUCCG	5	UUAAUCG	7
CUCACUG	5	AUUCAUG	7
CAUUCAG	5, 6	AAUAUUG	7
AAUUCCG	5	UUUUCAG	7
AAAUUCG	5	CCAAAAG	8
CCCUUUG	5	ACCAAUG	8
CUCUUAG	5	CUCCUAG	8
AUAUUCG	5	AUCUACG	8
UUAUUCG	5	CCUUUCG	8
ACCCACG	8–10	CUUUCAG	8
CCAACAG	10	AUUCCUG	8
CAACACG	9	ACUUAUG	8
CAAACCG	10	ACUAUUG	8
CCCUACG	9, 10	AUAUAUG	8

* The taxa and numbers assigned are: (1) *Aphanocapsa* 6308, (2) *Anacystis nidulans*, (3) *Porphyridium cruentum* chloroplast, (4) *Euglena gracilis* chloroplast, (5) *Rhodopseudomonas spheroides*, (6) *Bacillus stearothermophilus*, (7) *Escherichia coli*, (8) *Triticum vulgare* mitochondrion, (9) *Methanobacterium arbophilicum*, (10) *Methanospirillum hungatii*. Numbers following semicolons represent additional copies.

The phenogram is illustrated in FIGURE 5. The S_{ab} values were also used to calculate a network using Fitch and Margoliash's[11] procedure. In order to do this, each S_{ab} value was subtracted from unity. Only one network was constructed. The two methods produced identical clustering patterns, although with differing branch lengths.

The clustering pattern obtained is not particularly supportive of either hypothesis. The two blue-green bacteria, not surprisingly, share the greatest number of oligonucleotides. The two chloroplast 16S rRNAs also cluster with each other. This may suggest that, even if endosymbiotic, these two organelles had a single origin. Further, since they cluster with each other, this is at least compatible with an autogenous origin. On the other hand, this clustering partly reflects the procedures used. *Porphyridium* chloroplast 16S rRNA is as similar to *Aphanocapsa* 16S rRNA as it is to the 16S rRNA of *Euglena* chloroplasts, and slightly more similar to *Anacystis* 16S rRNA than the 16S rRNA of *Euglena* chloroplast is to either blue-green rRNA (TABLE 2). Red algae are a peculiar group in any case. They alone of eukaryotic photosynthesizers have phycobilins as well as chlorophyll *a*, and their cytosolic ribosomes, while certainly eukaryotic in size, are the smallest eukaryotic ribosomes known.[16]

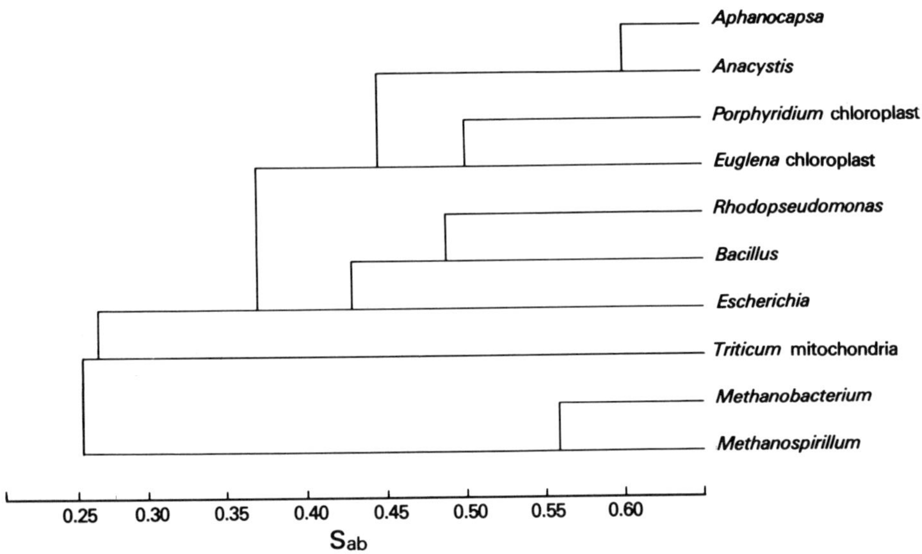

FIGURE 5. An unweighted pair-group method phenogram interconnecting small subunit rRNA catalogues of seven prokaryotes and three eukaryotic organelles. The taxa are *Methanospirillum hungatii* and *Methanobacterium arbophilicum* (methanogenic bacteria); *Triticum vulgare* mitochondrion (wheat); *Escherichia coli* and *Bacillus stearothermophilus* (aerobic bacteria); *Rhodopseudomonas spheroides* (purple non-sulfur bacterium); *Euglena gracilis* chloroplast (euglenoid); *Porphyridium cruentum* chloroplast (red alga), and *Anacystis nidulans* and *Aphanocapsa* 6308 (blue-green bacteria). S_{ab} values for each pair were calculated using twice the number of nucleotides shared in pentamers, hexamers and heptamers, divided by the sum of the nucleotides in the published pentamer, hexamer and heptamer catalogues. A network calculated using Fitch and Margoliash's (1967) method has an identical clustering sequence.

TABLE 2

MEASURES OF SIMILARITY BETWEEN OLIGONUCLEOTIDE CATALOGUES FOR SMALL SUBUNIT RIBOSOMAL RNAS OF SEVEN PROKARYOTES AND THREE EUKARYOTIC ORGANELLES *

	Ap	An	P	Eu	R	B	Es	T	Mb	Ms
1. *Aphanocapso* (Ap)	388	191	212	154	141	128	154	116	98	123
2. *Anacystis* (An)	.60	348	174	142	153	145	160	117	118	121
3. *Porphyridium* chl. (P)	.51	.44	444	200	154	146	154	166	104	105
4. *Euglena* chl. (Eu)	.42	.40	.50	353	116	118	133	112	85	57
5. *Rhodopseudomonas* (R)	.40	.46	.41	.35	312	165	184	144	120	113
6. *Bacillus* (B)	.34	.41	.36	.33	.49	356	156	112	108	97
7. *Escherichia* (Es)	.37	.40	.34	.30	.48	.38	454	144	106	104
8. *Triticum* mito. (T)	.24	.25	.33	.24	.32	.24	.28	573	121	110
9. *Methanobacterium* (Mb)	.25	.32	.25	.23	.34	.29	.25	.25	389	151
10. *Methanospirillum* (Ms)	.29	.30	.23	.14	.30	.24	.23	.21	.56	448

* The taxa are listed in the caption for TABLE 1. The values in the upper right are the number of nucleotides shared between pairs for pentamers, hexamers, and heptamers. The numbers on the diagonal are the numbers of nucleotides in pentamers, hexamers, and heptamers in the published catalogues of the individual taxa. The numbers in the lower left are values of S_{ab}, twice the number of nucleotides in pentamers, hexamers, and heptamers shared between each pair of taxa divided by the sum of the number of nucleotides in the published catalogues of pentamers, hexamers, and heptamers for each member of the pair.

It is thus not surprising to find, in these primitive eukaryotes, additional similarity to prokaryotic photosynthesizers.

The two chloroplast 16S rRNAs cluster with the blue-green bacteria (FIGURE 5), as predicted by the endosymbiosis hypothesis. We have suggested earlier, however,[29] particularly on the basis of DNA hybridization data,[20] that the eukaryotes probably evolved from the prokaryotic ancestors of blue-green bacteria soon after their separation from the aerobic bacteria (FIGURES 3 & 6).

Rhodopseudomonas, also not surprisingly, falls with the other aerobic bacteria, closer to *Bacillus* than to *Escherichia.* This is as predicted by the endosymbiosis hypothesis, but is also what we would expect if eukaryotic organelles arose autogenously. On the other hand, the 16S rRNA of *Rhodopseudomonas* is as similar to the 16S rRNAs of the two blue-greens as is the 16S rRNA of *Euglena* chloroplasts, which might not be expected if chloroplasts are endosymbiotic blue-greens.

The clustering of the mitochondrial 18S rRNA of *Triticum* is not particularly supportive of either hypothesis (FIGURE 5). It does not fall with *Rhodopseudomonas,* as predicted by the endosymbiosis hypothesis, nor does it fall with the chloroplasts as we speculated it might in proposing our test. Rather, the aerobic bacterial group, including *Rhodopseudomonas,* joins to the chloroplast–blue-green bacterial group, and then the mitochondrial 18S rRNA joins this large cluster. Only the methanogenic bacteria, which group with each other, are added to the main cluster after the mitochondrial 18S rRNA.

Mitochondrial 18S rRNA thus appears to be very different from the 16S rRNA of each of the relevant groups of prokaryotes. Its greatest similarities, while quite low (TABLE 2), are, however, with *Porphyridium* chloroplast 16S rRNA and with *Rhodospirillum* 16S rRNA. It appears to us that the best interpretation of these data is that the mitochondrial 18S rRNA has diverged extensively from its prokaryotic ancestor; this is consistent with either an autogenous or an endosymbiotic origin of mitochondria. Some of this divergence is by increase in size, which would add shared oligonucleotides only at random.

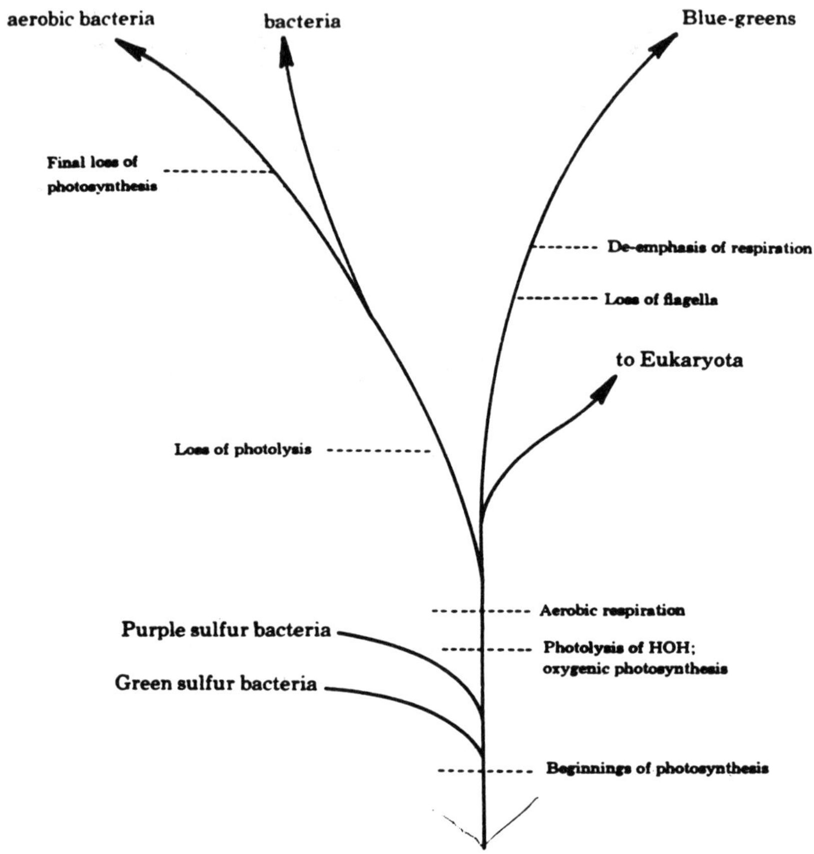

FIGURE 6. A plausible phylogeny of certain prokaryotic and eukaryotic taxa indicates a branching sequence with adaptive changes at various stages in the phylogeny. (From *American Scientist* **62**:341.)

This extreme divergence is consistent with either the endosymbiotic or the autogenous hypotheses. As pointed out by Bonen *et al.*,[6] the oligonucleotides shared by *Triticum* mitochondrial 18S rRNA and 16S rRNA of *Escherichia coli* are in highly conserved parts of the molecule. These probably are retained because of functional requirements. The remainder may have undergone rapid nucleotide substitution paralleling the rapid nucleotide substitution observed in mitochondrial DNAs;[8] mitochondrial rRNA may thus no longer share many oligonucleotides with its ancestral rRNAs.

Bonen *et al.*[6] discovered no more than random sharing of oligonucleotides between the mitochondrial and cytosolic 18S rRNAs of *Triticum;* this is also true of *Porphyridium* chloroplast 16S and *Porphyridium* cytosolic 18S rRNAs.[5] Certainly these rRNAs have diverged extensively from the homologous prokaryotic rRNAs in size. The cytosolic ribosomes of eukaryotes, even of the simpler forms, carry out functions far different from those of prokaryotic and organellar ribosomes. The more complex the eukaryote, the more this is true. It is hardly surprising, therefore, to find little sequence homology between the organellar and nuclear cistrons. As noted earlier, it is the nuclear genomes of eukaryotes that are highly divergent, not their organellar ones, and this relation holds regardless of whether the organelles arose autogenously or by endosymbiosis (FIGURE 2).

OVERVIEW

We find that the morphological features invoked to support the endosymbiosis hypothesis are largely if not exclusively primitive states, and are therefore compatible with either an autogenous or an endosymbiotic origins hypothesis. The cytochrome *c* phylogeny proposed by Schwartz and Dayhoff[23] at first glance appears to offer strong support for the endosymbiosis hypothesis, but is readily reinterpreted in terms of paralogous polypeptides, and the evidence of duplications of cistrons for *c*-type cytochromes in both the blue-green and the purple nonsulfur bacteria makes this possibility too great to ignore. We suspect that the same phenomenon exists in ferridoxins, given the great differences between the 2Fe-2S and 4Fe-4S ferridoxins.[24]

Analysis of some small subunit rRNA catalogues yields ambiguous answers but the data are at least as compatible with an autogenous as with an endosymbiotic origin of eukaryotic organelles.

On the other hand, an autogenous origin of eukaryotic organelles is especially appealing given the kinds of features that must have been present in common ancestors of extant aerobic photosynthetic and nonphotosynthetic bacteria, and in the common ancestor of these and extant blue-green bacteria. Furthermore, this avoids appealing to massive transfers of genetic material from endosymbionts into the genome of what must have been an essentially characterless, featureless protoeukaryotic host.

With the evolution of both aerobic respiration and oxygenic photosynthesis, protoeukaryotes would have had significant adaptive advantages over other prokaryotes. Elaboration of membranes permitted them to sequester these functions in different organelles. Subsequently, prokaryotes have specialized as much by losing capacities as by changing their metabolisms.

We expect that future acquisition of nucleic acid sequence data of a wide variety of prokaryotes and of many eukaryotic organelles will provide an answer to this question. In the meanwhile, acceptance of either hypothesis as demonstrated is a matter of belief; both are at present viable alternatives.

REFERENCES

1. ALMASSY, R. J. & R. E. DICKERSON. 1978. *Pseudomonas* cytochrome c_{551} at 2.0Å resolution: enlargement of the cytochrome *c* family. Proc. Natl. Acad. Sci. USA **75:**2674–2678.

2. ALTMANN, R. 1890. Die Elementarorganismen und ihre Beziehungen zu den Zellen. Veit und Comp. Leipzig. 145 pp.
3. AMBLER, R. P., M. DANIEL, J. HERMOSO, T. E. MEYER, R. G. BARTSCH & M. D. KAMEN. 1979. Cytochrome c_2 sequence variation among the recognized species of purple non-sulfur photosynthetic bacteria. Nature **278:**659–660.
4. BALCH, W. E., G. E. FOX, L. J. MAGRUM, C. R. WOESE & R. S. WOLFE. 1979. Methanogens: reevaluation of a unique biological group. Microbiol. Rev. **43:** 260–296.
5. BONEN, L. & W. F. DOOLITTLE. 1975. On the prokaryotic nature of red algal chloroplasts. Proc. Natl. Acad. Sci. USA **72:**2310–2314.
6. BONEN, L., R. S. CUNNINGHAM, M. W. GRAY & W. F. DOOLITTLE. 1977. Wheat embryo mitochondrial 18S ribosomal RNA: evidence for its prokaryotic nature. Nucleic Acids Res. **4:**663–672.
7. BONEN, L., W. F. DOOLITTLE & G. E. FOX. 1979. Cyanobacterial evolution: results of 16S ribosomal ribonucleic acid sequence analyses. Can. J. Biochem. **57:**879–888.
8. BROWN, W. M., M. GEORGE, JR. & A. C. WILSON. 1979. Rapid evolution of animal mitochondrial DNA. Proc. Natl. Acad. Sci. USA **76:**1967–1971.
9. DEMOULIN, V. 1979. Protein and nucleic acid sequence data and phylogeny. Science **205:**1036–1038.
10. DROUET, F. 1973. Revision of the Nostocaceae with cylindrical trichomes. Hafner Press. New York, N.Y.
11. FITCH, W. M. & E. MARGOLIASH. 1967. Construction of phylogenetic trees. Science **155:**279–284.
12. FITCH, W. M. 1970. The usefulness of amino acid and nucleotide sequences in evolutionary studies. Evol. Biol. **4:**67–109.
13. HENNIG, W. 1966. Phylogenetic Systematics. D. D. Davis & R. Zangerl, translator. University of Illinois Press. Chicago, Ill.
14. HOLTON, R. W. & J. MYERS. 1967. Water-soluble cytochromes from a blue-green alga. I. Extraction, purification, and spectral properties of cytochromes *c* (549, 552, and 554, *Anacystis nidulans*). Biochem. Biophys. Acta **131:**362–374.
15. HOLTON, R. W. & J. MYERS. 1967. Water-soluble cytochromes from a blue-green alga. II. Physiological properties and quantitative relationships of cytochromes *c* (549, 552, and 554, *Anacystis nidulans*). Biochem. Biophys. Acta **131:**375–384.
16. HOWLAND, G. P. & J. RAMUS. 1971. Analysis of blue-green and red algal ribosomal RNAs by gel electrophoresis. Arch. Mikrob. **76:**292–298.
17. MARGULIS, L. S. 1970. Origin of eukaryotic cells. Yale University Press. New Haven, Conn.
18. MARX, H. & G. B. RABB. 1970. Character analysis: an empirical approach applied to advanced snakes. J. Zool., London **161:**525–548.
19. MERESCHKOWSKY, C. 1905. Über Natur und Ursprung der Chromatophoren im Pflazenreiche. Biol. Zentralbl. **25:**593–604.
20. PIGGOTT, G. H. & N. G. CARR. 1972. Homology between nucleic acids of blue-green algae and chloroplasts of *Euglena gracilis*. Science **175:**1259–1261.
21. ROHLF, F. J., J. KISHPAUGH & D. KIRK. 1974. NT-SYS. Numerical taxonomy system of multivariate statistical programs. State University of New York. Stony Brook, N.Y.
22. SCHOPF, W. J. 1975. Precambrian paleobiology: problems and perspectives. Ann. Rev. Earth Planet Sci. **3:**213–249.
23. SCHWARTZ, R. M. & M. O. DAYHOFF. 1978. Origins of prokaryotes, eukaryotes, mitochondria, and chloroplasts. Science **199:**395–403.
24. SCHWARTZ, R. M. & M. O. DAYHOFF. 1979. Protein and nucleic acid sequence data and phylogeny. Science **205:**1038–1039.

25. SOKAL, R. R. & P. H. A. SNEATH. 1963. Principles of Numerical Taxonomy. W. H. Freeman & Co. San Francisco.
26. THROCKMORTON, L. H. 1968. Concordance and discordance of taxonomic characters in *Drosophila* classification. Syst. Zool. **17**:355–387.
27. THROCKMORTON, L. H. 1978. Molecular phylogenetics. *In* Beltsville Symposia in Agricultural Research. 2. Biosystematics in Agriculture. J. A. Romberger *et al.*, eds.: 221–239. John Wiley & Sons, Inc. New York, N.Y.
28. UCHIDA, T., L. BONEN, H. W. SCHAUP, B. J. LEWIS, L. ZABLEN & C. WOESE. 1974. The use of ribonuclease U_2 in RNA sequence determination. J. Molec. Evol. **3**:63–77.
29. UZZELL, T. & C. SPOLSKY. 1974. Mitochondria and plastids as endosymbionts: a revival of special creation? Am. Sci. **62**:334–343.
30. WAGNER, W. H., JR. 1969. The construction of a classification. *In* Systematic Biology. Charles G. Sibley, ed.: 67–90. National Academy of Sciences. Washington.
31. WOESE, C., M. SOGIN, D. STAHL, B. J. LEWIS & L. BONEN. 1976. A comparison of the 16S ribosomal RNAs from mesophilic and thermophilic bacilli: some modifications in the Sanger method for RNA sequencing. J. Molec. Evol. **7**: 197–213.
32. WOESE, C. R. 1977. Endosymbionts and mitochondrial origins. J. Molec. Evol. **10**:93–96.
33. ZABLEN, L. & C. R. WOESE. 1975. Prokaryote phylogeny IV: concerning the phylogenetic status of a photosynthetic bacterium. J. Molec. Evol. **5**:23–34.

DISCUSSION PAPER

Arthur Cronquist

The New York Botanical Garden
Bronx, New York 10458

I am in basic agreement with the position of Drs. Uzzell and Spolsky that the presently available data do not require acceptance of the endosymbiotic hypothesis as a general explanation for the origin of the eukaryotic condition, and that said hypothesis has some serious unresolved problems. I reach this position by a somewhat different route, however, for which I would like to present a road map.

I should start out by saying that I am a consumer, not a producer of data regarding the evolutionary origin of the eukaryotic condition. Whatever status I may have to comment on the subject comes from my interest and publications in phylogeny and in the general system of classification of organisms. Most of my work has been on the angiosperms, the flowering plants, and I have only marginal competence in the detailed taxonomy of algae, flagellates, and prokaryotes. Yet I think there is room, and perhaps need, for an approach to the matter by a broad-scale synthesizer, and with that apologia I will go ahead.

Occam's Razor is a useful tool in evaluating evolutionary theory. Explanations should not be unnecessarily complicated. Occam's Razor does not always provide the right answer, of course. Maybe the explanations one is choosing among are all incorrect. In that case the simplest one that is compatible with the known facts is likely to require less change to become correct, when more data are available, than any of the more complicated known alternatives. Maybe a complication that now seems unnecessary will turn out to be necessary when we know more, but in any particular case the odds are against it. So I try to take the simplest explanation that will provide for the known facts. Another way of putting it is that I will not trade a familiar interpretation for a newer and more complicated one unless the newer one gets me out of more trouble than it gets me into.

As Dr. Uzzell and others have pointed out, an endosymbiotic interpretation of both chloroplasts and mitochondria has been in the literature since about the turn of the century. Not many people took it seriously, however, until about two decades ago. It probably had less scientific approval, and certainly enjoyed less public attention, than the more recent cosmological miasmas of Velikovsky.

It was the increase in our knowledge of DNA that led to the great interest in the endosymbiotic hypothesis. The recognition during the 1940s that DNA is the essential nuclear material, followed by the dramatic discovery of the essential structure of DNA in 1953, led biologists to equate DNA with nucleus. Then when DNA was discovered in chloroplasts and mitochondria in the early 1960s, it was easy to suppose that these organelles must somehow be nuclear in nature, and that if they are nuclear they must be endosymbionts. Then when the DNA of chloroplasts and mitochondria turned out to resemble that of prokaryotes rather than eukaryotes in a number of respects, proponents of the endosymbiotic hypothesis seemed to sweep the field. It mattered not that some of us cautioned that the DNA data were equally compatible with the classical

0077–8923/81/0361–0500 $01.75/0 © 1981, NYAS

hypothesis of the origin of intracellular organelles. By 1970 Peter Raven could write that virtually everyone who was writing about the subject accepted the endosymbiotic hypothesis, and that therefore it must be correct. I am sure that we are all acquainted with Margulis' hypothesis that not only chloroplasts and mitochondria, but also flagella are of endosymbiotic origin.

Still, the apparent victory for the endosymbiotic hypothesis was not total and final. It was discovered that although the DNA of chloroplasts and mitochondria resembles that of prokaryotes in some important respects, it differs in others, and that the genetic system of these organelles is incomplete, requiring also the participation of nuclear genes. This made it necessary for proponents of the endosymbiotic hypothesis to suppose that part of the genetic function of the endosymbiont had been ceded to the nucleus of the host. This complication, although a bit worrisome, is not theoretically impossible or even improbable. As Dr. Schiff has pointed out during this symposium, such a change would be useful and perhaps even necessary for the evolutionary integration of the endosymbiont into its host.

An attempt to understand the presently available data on organelle genetic systems has led a number of people, among them Dr. Bogorad at this symposium, to the conclusion that the data are in fact compatible with both the classical and the endosymbiotic interpretation. Under either hypothesis, the separation of organellar DNA from nuclear DNA occurred a very long time ago, when both types were at the prokaryotic level. Since that time nuclear and organellar DNA have been subject to different evolutionary pressures, and have evolved in partial independence and partial interconnection. So the data that were responsible for the rising acceptance of the endosymbiotic hypothesis in fact prove nothing.

So we look for other kinds of data. Dr. Olson has a very interesting hypothesis, presented at this symposium, that ties the origin of modern photosynthesis into endosymbiosis. It would be possible to accept large parts of his hypothesis, however, without accepting endosymbiosis. He must surely be right that photosystem II evolved gradually, a step at a time, rather than in one great leap forward. Many interpretations that we now accept as scientific truths originated as creative, imaginative syntheses, comparable in that respect to Olson's hypothesis on the evolution of photosynthesis, but many other such creative, imaginative syntheses have failed and been forgotten. We do not yet know which category Olson's hypothesis falls into, or how much of it can be salvaged if it turns out not to be correct in all respects.

The sequence of amino acids in vital proteins such as cytochrome *c* is another possible source of data. Computer-generated phylogenetic trees for a limited number of vertebrates, based on the structure of cytochrome *c*, are tolerably close to what we believe on other grounds, with the notable exception of the position of rattlesnakes, in which the reported sequence is more similar to that of man and the chimpanzee (the latter two are identical) than to anything else that has been sequenced. Aside from the fanciful hypothesis of the lonesome cowboy, the most likely explanation here is that the reported sequence for rattlesnakes is simply wrong. I understand that it is now being done again, and we shall soon know.

The apparent success of the cytochrome *c* method for vertebrates led to efforts to elucidate the phylogeny of angiosperms by the same method. The efforts have been notably unsuccessful to date, giving results that are not harmonious with anything else we know or believe. Part of the problem is that cytochrome *c* of angiosperms is more labile, in an evolutionary sense, than that of vertebrates. If changes occurred in the angiosperms at the same rate as in vertebrates, dicotyledons with triaperturate pollen would have to take their origin somewhere around the Devonian period, instead of in the Cretaceous as indicated by the fossil record.

Several years ago I did look in some detail into the efforts to use cytochrome *c* in angiosperm phylogeny. The number of possible sources of error is so large, and their consequences so difficult to foresee, that I do not believe any *a priori* confidence can be placed in the method. One of the problems is that the phylogenetic trees are so far removed from the data. Different methods of handling the data are evaluated essentially on how close they come to giving us the result we want. In at least one of the methods there is a fudge-factor that can be adjusted at one extreme to peel off units one at a time from the remaining mass, and at the other extreme to break the group into two essentially equal halves at each dichotomy. For a more complete exposition of the problems, anyone who is interested may look up my paper in *Brittonia* (**28:** 1–27. 1976).

I must admit that I have not looked into Dr. Dayhoff's data and data-handling to anywhere nearly the depth that I did for Boulter's work on angiosperms. I cannot escape the feeling, however, that similar problems may exist. In comparing the evolutionary history of chloroplasts, mitochondria, and nuclei we may be doing something like comparing cats, catbriars, and catfishes, under the assumption that they must all behave in basically the same way because they are all cats. Certainly the reports at this symposium of rapid evolution of mitochondrial DNA in some groups should be grounds for caution. Given the creative possibilities available to the user of amino acid sequence data, I believe I could take Dayhoff's data, start with some different assumptions, use a different but theoretically defensible program, and come up with a very different tree, which might not be any more nearly correct than I think hers is.

It may well be, as one of our speakers has suggested, that neither of the two alternative hypothesis on the origin of the eukaryotic cell is ultimately falsifiable, no matter what new data may be produced. Given the creative ingenuity that has already been shown by evolutionary speculators, either hypothesis may be able to absorb any data that might be produced without fatal damage. Yet one or the other, or some combination of them, presumably must be correct.

In such a situation, we should start thinking in terms of probabilities, and be willing to use Occam's Razor. Certainly the classical theory is simpler than the endosymbiotic hypothesis. Is there any way in which the endosymbiotic explanation, or some part of it, might be helpful? I have no *a priori* objection to admitting the existence of endosymbionts in particular cases. It has been clearly demonstrated that they do indeed exist, and the present symposium includes a fascinating paper on endosymbiotic bacteria.

For such a consideration I would like to emulate Caesar and divide the endosymbiotic concept of cell organelles into three parts; chloroplasts, mitochondria, and flagella. I think a better case can be made for chloroplasts than for either of the other two. As Dr. Whatley has so eloquently reminded us, there are modern endosymbiotic algae, so considered by all, that serve as functional chloroplasts for their nonalgal hosts. Might not a further mutual adaptation of the endosymbiont and the host be possible? There is no theoretical reason why it should not sometimes happen. Very likely it did happen in *Euglena,* which Dr. Schiff and I and some others agree presents the best case for an endosymbiotic interpretation of undoubted chloroplasts. I was interested to hear him say that the chloroplast of *Euglena* is not so fully integrated into the physiology of the cell as are chloroplasts of most other plants. More important, from my standpoint, is that the well developed but apparently nonfunctional gullet of *Euglena* is highly compatible with the idea that the chloroplast is an endosymbiont, and difficult to explain by the classical hypothesis. So I will buy the idea, or at least make a down payment on it, for *Euglena*. [Dr. Gibbs' more detailed presentation of the case that the chloroplasts of *Euglena* are endosymbionts came only on the following day of the symposium.]

Does that logically commit me to the position that all chloroplasts are endosymbionts? I think not. When Thor Heyerdahl took his raft, the Kon-Tiki, from the coast of South America to the Polynesian islands, he showed that such a trip is possible, but anthropologists are still very skeptical of the idea that such journeys were a regular and routine thing that can be used to explain whatever cultural similarities there may be between the two regions. Demonstration of any single event suggests that other such events may have occurred, but it does not show that they were routine.

Next we have the question of mitochondria. There is enough similarity in structure and enzymology between chloroplasts and mitochondria so that acceptance of chloroplasts as endosymbionts, if you do so accept them, suggests a similar explanation for mitochondria. Yet it is harder to visualize what the free-living precursor of a mitochondrial endosymbiont might have been like. If there are any modern candidates for the job, I don't know what they might be. My gut reaction is that these two organelles require a common evolutionary explanation, and my difficulty in seeing mitochondria as endosymbionts contributes to my difficulty in seeing chloroplasts as such.

Finally we come to flagella. I find the endosymbiotic interpretation of these organelles the most difficult of all. Clearly the components of the flagellum are homologous with the microtubules that are free in the cytoplasm, and the basal body is homologous with the centriole. Yet if I understand correctly, both microtubules and centrioles (the latter of a perhaps primitive type) are known in the red algae, which give no evidence at all of ever having a flagellated ancestry. (Here I use the term flagellated to refer to eukaryotic flagella, not bacterial ones). Again, if I understand correctly, early reports of the possible presence of DNA in centrioles have not withstood re-examination. Of course, I don't think that the presence of DNA in centrioles would prove very much even if it were found, but the proponents of endosymbiosis have stressed this item with regard to chloroplasts and mitochondria.

One of the things that has attracted much comment from proponents of the endosymbiotic hypothesis is the origin of the double membrane, which some people evidently think represents the plasma membrane of the endosymbiont, plus the plasma membrane of the host. One problem with that interpretation is that the nuclear membrane and the endoplasmic reticulum are also double membranes.

I would like to propose another explanation. It may already be in the literature somewhere, but if so I am not aware of it. The thylakoid is the prototype of the double membrane. Among existing prokaryotes we can see many or all of the stages in the evolution of a thylakoid from dispersed quantasomes. In the early evolution of eukaryotes, we wrap one thylakoid around some others, to produce a protochloroplast, which also contains respiratory enzymes and some DNA. Then we could have an evolutionary differentiation of mitochondria, specialized for respiration, from true chloroplasts, specialized for photosynthesis. The nuclear membrane *cum* endoplasmic reticulum represents a sort of evolutionary afterthought, turning existing double membranes to another function. Thus all double membranes have a common evolutionary origin, without the need for endosymbionts. This is my own leap of creative imagination, which will of course be subject to the same sort of skeptical critical analysis as other such leaps.

PATTERNS OF ENZYME DIVERSITY AND ORGANELLE EVOLUTION *

P. D. J. Weitzman

Department of Biochemistry
University of Bath
Bath BA2 7AY, England

The citric acid (Krebs) cycle occurs almost universally in living cells of quite diverse types. The cycle lies at the very heart of cellular metabolism and fulfills a multifunctional role in the provision of energy, reducing power, and several starting materials for a range of biosynthetic pathways. The chemistry of the intermediates of the cycle and their conversions, one to another, remains constant throughout Nature. However, diverse organisms may make different uses of the cycle and may place different emphasis on its multiple metabolic functions. These differences may, in turn, be reflected in the fine detail of enzyme structure and function. Thus structural, catalytic, and regulatory features of cycle enzymes might be similar within one group of organisms but distinctively different in another. Enzyme 'patterns' might thereby be discernible and these might aid the establishment of natural relationships between different organisms.

FIGURE 1 serves as a reminder of the essential steps in the citric acid cycle and FIGURE 2 is a much simplified representation of the carbon flow. The cycle is initiated by condensation of a 2-carbon fragment with a 4-carbon compound to form a 6-carbon compound. This condensation is catalyzed by the enzyme citrate synthase, according to the equation:

$$\text{acetyl-CoA} + \text{oxaloacetate} + H_2O \rightarrow \text{citrate} + \text{CoA-SH}.$$

After isomerization of citrate to isocitrate there follow two successive decarboxylation reactions resulting in the formation of the 4-carbon compound succinyl-CoA (designated C_4* in FIGURE 2). The remaining arc of the cycle consists of a sequence of reactions that serves to rearrange C_4* back into the original starting C_4 compound (oxaloacetate) so that a new round of metabolic conversions can be undertaken. This 'rearrangement' phase of the cycle is initiated by the action of the enzyme succinate thiokinase, which catalyzes the reaction:

$$\text{succinyl-CoA} + \text{NDP} + \text{phosphate} \rightarrow \text{succinate} + \text{NTP} + \text{CoA-SH}.$$

NDP and NTP represent nucleoside di- and tri-phosphates respectively, and may be derivatives of either adenosine or guanosine.

Citrate synthase and succinate thiokinase may thus be considered as catalyzing the starting reactions of the 'outward' and 'return' phases of the cycle. They also have in common that they both utilize an acyl-CoA substrate. The following discussion will concentrate on these two enzymes, will describe the structural and functional diversity that they display between different organisms

* Supported by the Science Research Council, U.K.

0077–8923/81/0361–0505 $01.75/0

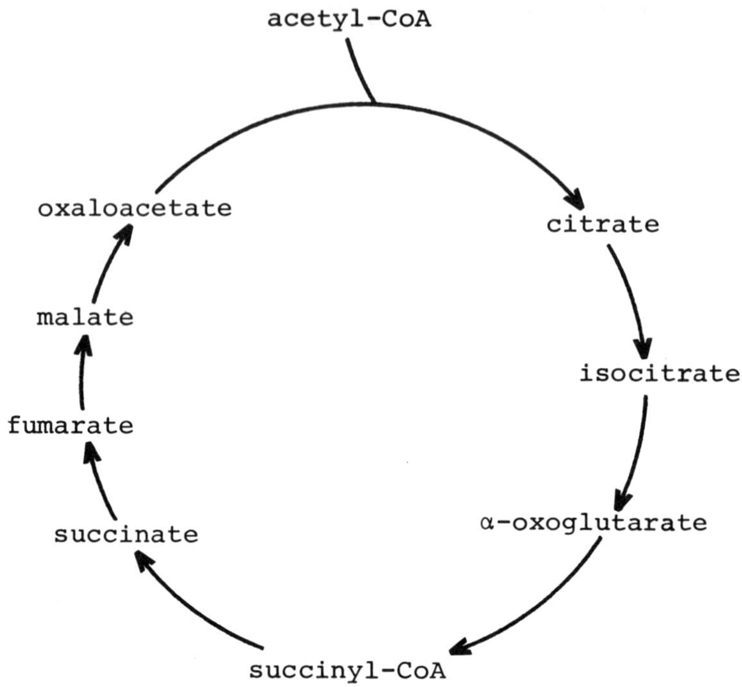

FIGURE 1. The citric acid cycle.

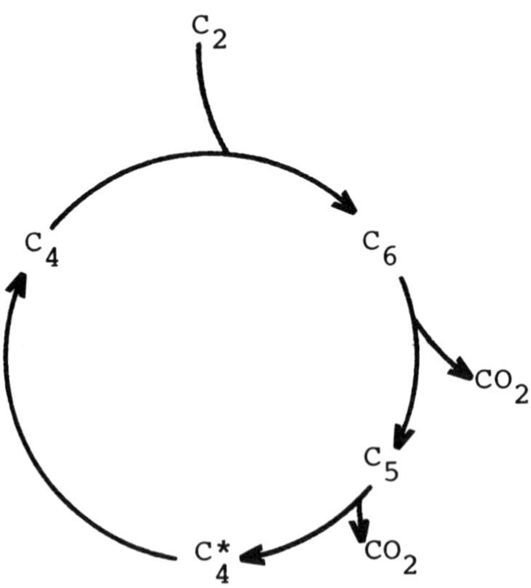

FIGURE 2. Simplified citric acid cycle scheme.

and organelles, and will indicate possible evolutionary connections suggested by the enzymological data.

The oxidative reactions of the cycle result in the formation of reduced pyridine nucleotide, NADH, whose reoxidation to NAD^+ is coupled to ATP production via the electron transport chain. NADH is thus, in a sense, the 'end-product' of the cycle and would be expected to exert a feedback inhibitory effect on the initial enzyme—citrate synthase.

Studies on citrate synthases from a wide range of prokaryotes and eukaryotes have shown that NADH is indeed a powerful inhibitor of the enzyme, but only from gram-negative bacteria.[1, 2] Citrate synthases from gram-positive bacteria and eukaryotes are insensitive to inhibition by NADH. Other regulatory patterns have also been observed. Within the gram-negative bacteria, the NADH inhibition of citrate synthase is overcome by AMP in the strict aerobes, but not in the facultative species.[1, 2] The latter group, on the other hand, are distinguished by showing feedback inhibition of their citrate synthases by α-oxoglutarate.[2, 3] Clearly, the citrate synthases of gram-negative bacteria are functionally more complex than are those from gram-positive bacteria and eukaryotes, suggesting that they may also be of more complex molecular structure. This prompted an investigation of the molecular sizes of various citrate synthases by simple examination of their gel filtration behavior on columns of Sephadex G-200 with appropriate marker proteins.[4] What emerged from these studies, and was extended and confirmed by other investigations,[2] was that all citrate synthases fall into one of two groups—'large' or 'small'. The 'large' enzymes have a molecular weight in the region of 250,000 and are eluted from the Sephadex column well ahead of lactate dehydrogenase, whereas the 'small' enzymes have a molecular weight around 100,000 and are eluted after lactate dehydrogenase. Typical gel filtration elution profiles are shown in FIGURE 3. The striking fact is that the 'large' citrate synthases occur only in gram-negative bacteria, while the enzymes from gram-positive bacteria and eukaryotes are, without exception, of the 'small' type. It thus appears that the more complex regulatory properties exhibited by gram-negative bacterial citrate synthases are necessarily associated with a more complex molecular structure. It is likely that their more extensive quaternary structure confers on the 'large' enzymes the ability to undergo subtle conformation and subunit rearrangements in response to their interaction with the various activity modifiers. Such structural changes probably underlie the inhibition and activation effects.[5]

Shifting attention now to succinate thiokinase, this is an enzyme which has been investigated in some detail,[6] in particular from *Escherichia coli* and pig heart. The succinate thiokinases from these two sources differ strikingly in their molecular weights. That of the *E. coli* enzyme is in the region of 140,000–150,000 whereas that of the pig enzyme is approximately half this size, with a molecular weight of 70,000–75,000. The *E. coli* enzyme appears to be a tetramer and the pig enzyme a dimer.

Our own experience with the pattern of molecular sizes of citrate synthases sensitized us to the possibility that succinate thiokinases might also display a definite size distribution and prompted the examination, by gel filtration, of a range of bacterial and eukaryotic succinate thiokinases.[7] In agreement with the molecular weights quoted above, the *E. coli* succinate thiokinase was eluted from the Sephadex column slightly ahead of lactate dehydrogenase, whereas the pig enzyme was eluted well after lactate dehydrogenase (see FIGURE 3). Once again, it was found that all the succinate thiokinases examined fall into one of these two types, i.e., 'large' or 'small' and, moreover, the incidence of these two types parallels precisely that of the two types of citrate synthase. Thus the 'large' succinate thiokinase occurs only in gram-negative bacteria while the 'small' enzyme is a feature exclusive to gram-positive bacteria and eukaryotes.

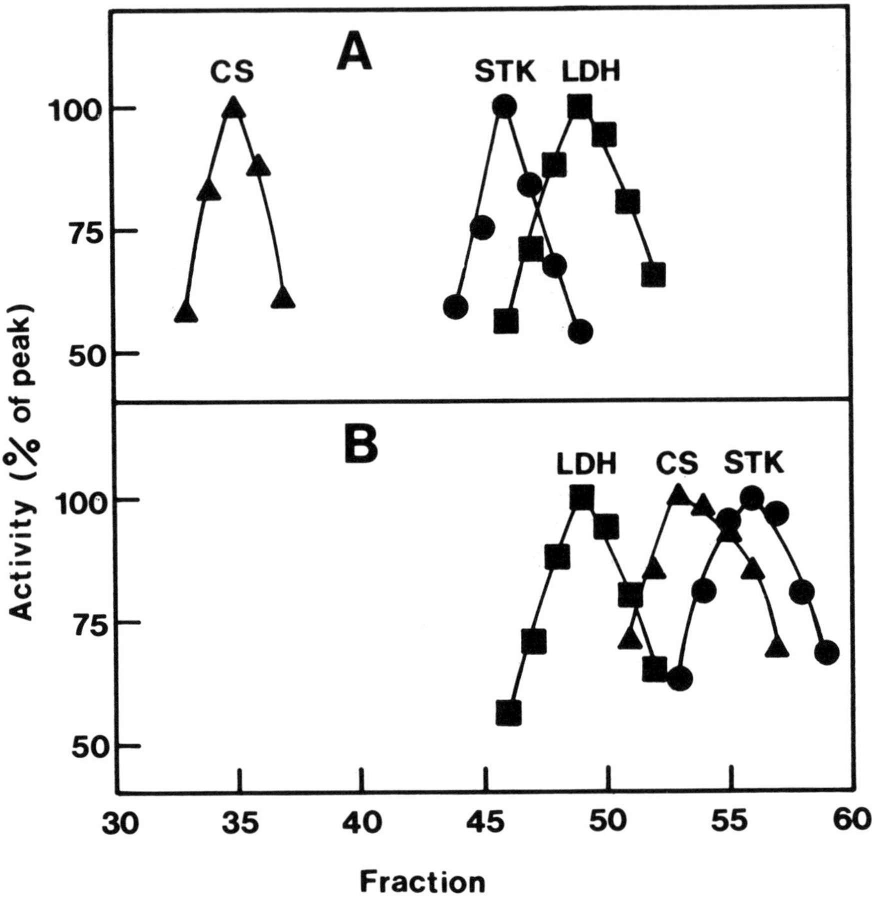

FIGURE 3. Gel filtration of citrate synthase and succinate thiokinase. Cell extracts were run on a column (2.5×35 cm) of Sephadex G-200, equilibrated with 0.1 M phosphate pH 7, and 2 ml fractions were collected. Lactate dehydrogenase served as a marker. (A) Extract of *E. coli*; (B) extract of pig heart. (▲) citrate synthase (CS); (●) succinate thiokinase (STK); (■) lactate dehydrogenase (LDH). From Weitzman & Kinghorn.[7] By permission of *Federation of European Biochemical Societies Letters.*

These results indicate a remarkable similarity between the natural occurrence of structural variants of citrate synthase and succinate thiokinase and also show a clear correlation between the molecular sizes of these two enzymes and the taxonomic status of the source organism. This correlation has been helpful as a tool for bacterial taxonomy.[8,9] Examination of these enzymes may therefore furnish information having some bearing on the relatedness of different organisms and organelles and hence pertinent to the subject matter of this conference.

The classification of cyanobacteria as gram-negative prokaryotes [10] makes it desirable to examine their citrate synthase and succinate thiokinase. Consistent with this classification, we found all the citrate synthases of a number of cyanobacteria to be of the 'large' type.[11, 12] Although succinate thiokinase has been reported to be absent from cyanobacteria,[13] we have detected its presence in all of several species of cyanobacteria tested and have attributed this difference in results to our use of a very sensitive assay method.[14] The molecular size of the cyanobacterial succinate thiokinase was in all cases of the 'large' type, again confirming the relationship between cyanobacteria and other gram-negative bacteria.

In eukaryotes, both citrate synthase and succinate thiokinase are strictly mitochondrial enzymes and are not found in the cytosol. The enzyme similarities described above indicate a resemblance between mitochondria and gram-positive bacteria and show that these stand clearly distinct from the gram-negative prokaryotes. On the basis of a prokaryotic origin for eukaryotic organelles one might therefore anticipate that an evolutionary connection is to be sought between mitochondria and gram-positive bacteria.

It has, however, been proposed that the bacterium *Paracoccus denitrificans* more closely resembles mitochondria than do other bacteria, such resemblance being based on comparisons between their apparatus for electron transport and oxidative phosphorylation,[15-18] and it was suggested that an ancestral bacterium similar to *P. denitrificans* may have evolved into the mitochondrion. As *P. denitrificans* is a gram-negative bacterium, this proposal runs counter to the evolutionary connections suggested by our comparative enzyme studies. We therefore examined the citrate synthase and succinate thiokinase of *P. denitrificans* to see whether their properties were those expected of a gram-negative organism or whether there was any anomaly. Both *P. denitrificans* enzymes were found to be of the 'large' type and, furthermore, the regulatory properties of the citrate synthase were precisely those associated with gram-negative aerobes, i.e., inhibition by NADH and de-inhibition by AMP. Professor Whatley, in his paper,[18] has also included consideration of *Rhodopseudomonas* within the context of the evolutionary origins of the mitochondrion. It is pertinent to note that the citrate synthase and succinate thiokinase of this organism are also both 'large' and therefore typically gram-negative prokaryotic.[2, 7]

Of course, these enzyme comparisons cannot establish an evolutionary link between mitochondria and gram-positive bacteria, nor do they detract from the marked and persuasive resemblances identified between mitochondria and *P. denitrificans*. Nevertheless, the pattern of citrate synthase and succinate thiokinase molecular sizes is a striking one and should perhaps be borne in mind when considering prokaryotic–mitochondrial similarities.

The examination of enzyme features as a possible clue to the evolutionary thread may also be extended to the chloroplast. Unfortunately (for our purposes!) chloroplasts do not appear to contain citrate synthase. There has been some controversy over whether they contain succinate thiokinase. Employing the very sensitive polarographic enzyme assay which we have developed [7, 19] it has been possible to detect succinate thiokinase activity in chloroplasts prepared from leaves of lettuce, spinach, pea, and cauliflower. Furthermore, gel filtration of these preparations showed that all the chloroplast succinate thiokinases are of the 'large' type.[20] On the other hand, the succinate thiokinases of mitochondria from these leaves are all of the 'small' type, as are the plant mitochondrial citrate synthases. Thus higher plant chloroplasts appear to contain a typically gram-negative prokaryotic type of succinate thiokinase and it is irresistibly tempting to associate this feature with an evolutionary link between the chloroplasts and the cyanobacteria.

The results discussed here are summarized in TABLE 1. These enzyme patterns are based only on gross molecular structural features. Further information on the similarities between the enzymes from different organisms and organelles, and on the evolutionary significance of the molecular size patterns, will have to await the results of comparative fine structural investigations.

TABLE 1

COMPARISON OF ENZYME MOLECULAR SIZES

Source	Citrate Synthase	Succinate Thiokinase
Gram-negative bacteria and cyanobacteria	large	large
Gram-positive bacteria and mitochondria	small	small
P. denitrificans	large	large
Chloroplasts	absent	large

REFERENCES

1. WEITZMAN, P. D. J. & D. JONES. 1968. Nature **219**:270–272.
2. WEITZMAN, P. D. J. & M. J. DANSON. 1976. Curr. Top. Cell. Regul. **10**:161–204.
3. WEITZMAN, P. D. J. & P. DUNMORE. 1969. FEBS Lett. **3**:265–267.
4. WEITZMAN, P. D. J. & P. DUNMORE. 1969. Biochim. Biophys. Acta **171**:198–200.
5. ROWE, A. J. & P. D. J. WEITZMAN. 1969. J. Mol. Biol. **43**:345–349.
6. BRIDGER, W. A. 1974. *In* The Enzymes. 3rd Edit. P. D. Boyer, Ed. Vol. **10**: 581–606.
7. WEITZMAN, P. D. J. & H. A. KINGHORN. 1978. FEBS Lett. **88**:255–258.
8. JONES, D. & P. D. J. WEITZMAN. 1974. Int. J. Syst. Bacteriol. **24**:113–117.
9. WEITZMAN, P. D. J. & D. JONES. 1975. J. Gen. Microbiol. **89**:187–190.
10. STANIER, R. Y. & G. COHEN-BAZIRE. 1977. Ann. Rev. Microbiol. **31**:225–274.
11. LUCAS, C. & P. D. J. WEITZMAN. 1975. Biochem. Soc. Trans. **3**:379–381.
12. LUCAS, C. & P. D. J. WEITZMAN. 1977. Arch. Microbiol. **114**:55–60.
13. PEARCE, J., C. K. LEACH & N. G. CARR. 1969. J. Gen. Microbiol. **55**:371–378.
14. WEITZMAN, P. D. J. & H. A. KINGHORN. 1980. FEBS Lett. **114**:225–227.
15. JOHN, P. & F. R. WHATLEY. 1975. Nature **254**:495–498.
16. JOHN, P. & F. R. WHATLEY. 1977. Biochim. Biophys. Acta **463**:129–153.
17. JOHN, P. & F. R. WHATLEY. 1977. Advan. Bot. Res. **4**:51–115.
18. WHATLEY, F. R. 1980. This volume.
19. WEITZMAN, P. D. J. 1976. Biochem. Soc. Trans. **4**:724–726.
20. WEITZMAN, P. D. J. 1979. Abst. XIth Int. Cong. Biochem. Toronto, p. 219.

Index of Contributors

(Italicized page numbers refer to comments in Discussions)